Introduction to Insect Biology and Diversity

SECOND EDITION

HOWELL V. DALY
Professor Emeritus of Entomology
University of California, Berkeley

JOHN T. DOYEN
Professor Emeritus of Entomology
University of California, Berkeley

ALEXANDER H. PURCELL, III
Professor of Entomology
University of California, Berkeley

Illustrations by **Barbara Boole Daly**

Featuring photographs by **Edward S. Ross**
Curator Emeritus of Entomology
California Academy of Sciences
San Francisco

Oxford   New York
OXFORD UNIVERSITY PRESS
1998

OXFORD UNIVERSITY PRESS

Oxford   New York
Athens   Auckland   Bangkok   Bogota   Bombay
Buenos Aires   Calcutta   Cape Town   Dar es Salaam
Delhi   Florence   Hong Kong   Istanbul   Karachi
Kuala Lumpur   Madras   Madrid   Melbourne
Mexico City   Nairobi   Paris   Singapore
Taipei   Tokyo   Toronto   Warsaw

and associated companies in

Berlin   Ibadan

Published by Oxford University Press, Inc.
198 Madison Avenue, New York, New York 10016

Library of Congress Cataloging-in-Publication Data
Daly, Howell V.
Introduction to insect biology and diversity / Howell V. Daly,
John T. Doyen, Alexander H. Purcell III ; illustrations by Barbara
Boole Daly ; featuring photographs by Edward S. Ross. — 2nd ed.
p.   cm.
Includes bibliographical references (p.   ) and index.
ISBN 0-19-510033-6 (cloth)
1. Entomology.   I. Doyen, John T.   II. Purcell, Alexander H.
III. Daly, Barbara Boole.   IV. Title.
QL463.D34   1998
595.7—dc21   97-27059

2 4 6 8 9 7 5 3 1
Printed in the United States of America
on acid-free paper

To the memory of our friend
*R*OBERT *L. U*SINGER

CONTENTS

Color insert between pages 152 and 153

PREFACE

This text is an introduction to the study of insects for students who have completed at least a basic course in biology. It is designed for modern courses emphasizing the major features of insects as living systems. Lectures in such courses are commonly devoted to insect structure and function, behavior, ecology, evolution, and the applied aspects, and the laboratory work is focused on the recognition of different kinds of insect and their biology and morphology. Parts One and Two of the text are thus appropriate as readings for the lectures. Part Three supplies information on the insect groups and keys for their identification. Here we have added biological information on the orders and a final chapter on collecting methods. Many new references to reviews and the primary literature are included for readers who wish further information. We have tried to limit the terminology and discussions to topics required for a broad knowledge, but not at the expense of oversimplifying complicated subjects. Nor have we avoided controversial topics. The study of the world's most varied animals is complex. With the application of new physical, molecular, and mathematical techniques for the analysis of insect biology, this complexity is increasing.

In Part One, the treatment of insect biology begins with structure and function and progresses through behavior to the social insects. Part Two describes the different relationships of insects to their environment. We have added a chapter on pest management and distilled the sections on population biology from the first edition. In the usual sense, this is not a text on applied entomology. Part Two mentions in numerous places those aspects of basic insect biology that affect human welfare or the environment. The principal pest control methods are summarized in the chapter on pest management. Examples of combining these methods for integrated pest management are given. Future entomologists will require detailed knowledge of insect biology to understand, implement, and improve the increasingly sophisticated practices now being developed for the management of beneficial and harmful insects. In this special sense, we hope our text provides a foundation for professional training.

To a large extent, the principles of modern ecology and evolution are based on studies of insects. Mimicry, competition, and coevolution are only three out of many important concepts that have been developed largely through an understanding of the biology of various groups of insects. We have attempted to integrate this information into Part Three, so that the various taxa may be viewed in a broad biological context. Throughout, we have cross-referenced discussions of biological phenomena to more detailed treatments in other parts of the book. Also, for the convenience of the reader, illustrations are extensively cross-referenced, a practice that occasionally disrupts the normal numerical sequence of figures in the text.

Even though we agree that a cladistic approach to classification is ultimately the method of choice, we have continued to present the higher classification of the class Insecta in a traditional, evolutionary form. Many issues of phylogeny are unsettled, and traditional names for higher taxa that are now considered by cladists to be paraphyletic, like Exopterygota, are still frequently used in the literature to designate insects that look alike to beginning students. The keys include nearly all the families of insects known to occur in North America. However, they are not designed to accommodate exceptional species that are normally encountered only by specialists. Likewise, extremely difficult taxa that require special preparation or collection techniques are mentioned, but not keyed. Discussions of the large endopterygote orders are organized by suborder and superfamily. These are important taxonomic categories whose members usually share many biological features.

We wish to acknowledge the generous assistance of many colleagues in preparing the first and second editions of the text. We regret that Professor Paul Ehrlich was unable to participate in this edition. The second edition features the excellent photographs of Edward S. Ross, who added useful ideas and comments as well to the text. All photographs in the text not otherwise acknowledged are those of Dr. Ross. Other photographs in the second edition were provided by Kim Hoelmer, George O. Poinar, Frank E. Skinner, L. L. Strand, Ronald E. Stecker, Gerard M. Thomas, Clyde Willson, and the late Howard E. Hinton. Barbara Daly returned to the drawing table to prepare additional line drawings. Christina Jordan was responsible for the computer graphics. We deeply appreciate their contribution to the illustrations. Special thanks go to Lowell Etzel for information on biological control and his patient advice on computer-based bibliographic methods.

For the first or second editions, or both, the following persons gave us valuable information, key references, or specimens: Paul H. Arnaud, Herbert G. Baker, Cheryl Barr, James Baxter, Leo Caltagirone, John Casida, Byron Chaniotis, Frank R. Cole, David Durbin, Wyatt Durham, Mark Eberle, Deane P. Furman, David Furth, Kenneth S. Hagen, Michael Haverty, Steve Hendrix, Marjorie A. Hoy, Carlton S. Koehler, Jarmila Kukalová-Peck, Robert Lane, Andrew Liebhold, E. Gorton Linsley, Vernard Lewis, Werner J. Loher, Richard Merritt, P. S. Messenger, Nicholas Mills, Robert Ornduff, William Peters, George O. Poinar, Jerry A. Powell, Vincent H. Resh, Helmut Riedl, Edward Rogers, Ray F. Smith, John Sorensen, Felix Sperling, Keith Waddington, William E. Waters, Ward B. Watt, Steve Welter, and David L. Wood. Brenda Griffing, as copyeditor, gave the text clarity and consistency of style. We also thank Joseph D. Culin, Gordon Pritchard, and Daniel K. Young, who read many chapters of the first or second edition and offered constructive comments.

Finally, we express our appreciation to our wives Barbara, Lisa, and Rita, for their patience and encouragement during the preparation of the text.

Berkeley, CA   H. V. D.
June 1997   J. T. D.
                                                 A. H. P., III

PART ONE

Insects as Organisms

1
INTRODUCTION

The acorn weevil, *Curculio uniformis*. Snout beetles or weevils of the family Curculionidae (Coleoptera) surely must be considered the most successful of insects, for the family includes over 60,000 species, making it the largest family of animals. Nearly all are phytophagous. The adults use the tiny mandibles at the tip of the snout to bite holes into seeds, fruits, and other parts of plants. Their eggs are often inserted in the holes.

In this first chapter we briefly view the biology of insects from a variety of aspects: the relative numbers of species, the smallest and largest insects, and the limits to the geographic distribution of insects; the kinds of place in which they live and the food they eat; elementary facts about their anatomy; major features of their evolution and classification; and factors in their success as terrestrial organisms. We conclude with a discussion of insects in relation to human welfare. This will provide some sense of the scope of insect life and introduce you to the first of the many terms, facts, and concepts that comprise a working knowledge of entomology.

A GLIMPSE OF THE WORLD OF INSECTS

Insects are the earth's most varied organisms. About half of the described species of living things and almost three-quarters of all animals are insects (Fig. 1.1). Of all the animal phyla, only the arthropods and the chordates have succeeded extensively in adapting to life in dry air. Insects now inhabit virtually all land surfaces of the globe except the extreme polar regions and the highest mountain peaks. Wherever they occur they tend to dominate the small fauna, being rivaled only by another group of arthropods, the mites, in some habitats.

In body size, most insects are 1 to 10 mm in length. Some tiny insects are smaller than some Protozoa, and some giant insects are larger than the smallest mammals. Competitors for the status of "smallest insect" are two species of wasps. *Megaphragma mymaripenne* (Trichogrammatidae, Hymenoptera) measures 0.17 or 0.18 mm in body length and lays its eggs in the eggs of other tiny insects, the thrips. The other wasp, *Alaptus* sp. (Mymaridae, Hymenoptera), measures 0.2 mm in length but is much more slender-bodied. *Alaptus* also lays its eggs in the eggs of other small insects,

the psocids. Adult female wasps of both species are smaller than the period at the end of this sentence. They are smaller than even a large one-celled, ciliate animal. Yet, the miniaturized wasps are fully functional insects with multicellular organs and complex behavior like their larger relatives (Fig. 1.2a). Other minute insects, measuring 0.25 mm, are beetles of the family Ptiliidae (Coleoptera).

At the other extreme, several giant insects compete in two classes for titles. In terms of wingspread, the champions of all time are the extinct dragonflies *Meganeura* and *Meganeuropsis* (Meganeuridae, Protodonata), both insects with wings measuring 700 mm or more from tip to tip. Among living insects, two very large moths are *Erebus agrippina* (Noctuidae, Lepidoptera) from Brazil, which measures 280 mm, and the Asian *Attacus atlas* (Saturniidae, Lepidoptera), which spans 240 mm. The tropical American grasshopper, *Tropidacris latreillei* (Acrididae, Orthoptera), has

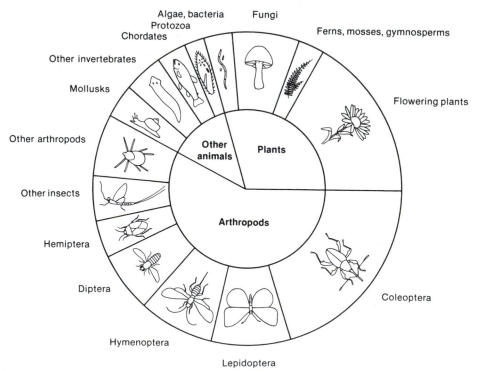

Figure 1.1   Relative numbers of species of arthropods, other animals, and plants.

a wingspread of 240 mm. In terms of body length, among the scarab beetles are the African *Goliathus* spp. (up to 110 mm; Fig. 1.2b), the Central American *Megasoma elephas* (130 mm), and the South American *Dynastes hercules* (160 mm). Another huge insect is the Australian phasmid, *Acrophylla titan* (Phasmatidae, Phasmatodea), with a body length of up to 250 mm (a little longer than this page). Compare these to some shrews, mice, and bats that measure less than 50 mm.

The limits to the geographic distribution of insects are still incompletely known. In Antarctica, for example, insects are represented by two species of flies, a species of bird flea, and several louse species that parasitize birds and seals. In the Arctic, at least 300 insect species, mostly flies, live in the Canadian islands north of the 75th parallel. At an altitude of 6000 m (19,685 ft) in the Himalaya Mountains, surprising numbers of insects have been found as permanent residents. Insects have also been found living in deep caves, hot springs, salt lakes, and pools of petroleum. About 3 percent of all species of insects live in freshwater, and perhaps 0.1 percent are found in the marine intertidal zone. One genus of water striders, *Halobates* (Gerridae, Hemiptera), lives permanently on the surface of the open ocean, and many other gerrids inhabit the surfaces of tropical seas closer to land. Only a few kinds of insects are found in the offshore waters of deep lakes, and they are missing entirely in the deeper waters of the ocean.

The vast majority of insects, therefore, are terrestrial. They occur in an enormous variety of habitats, including arid deserts. Of course, they must have access to water for drinking or in food to restore that lost by evaporation from their body surfaces. As in other small terrestrial organisms, the surface area of an insect's body is relatively large in proportion to its volume, so the risk of desiccation is always great in warm, dry atmospheres. This loss is reduced in insects by a thin wax coating on the body surface and by certain water-conserving mechanisms associated with excretion and respiration. Insects avoid unfavorably hot or dry environments by seeking cooler, shady places or moist crevices, and many species satisfy

Figure 1.2 Examples of minute and giant insects: **a**, Among the smallest of insects are parasitoids of the eggs of other insects. Here a female wasp, *Trichogramma* sp. (Trichogrammatidae, Hymenoptera), attacks the egg of the corn earworm, *Helicoverpa zea* (Noctuidae, Lepidoptera). Excluding the wings, the insect is about 0.3 mm in body length. These wasps are widely used for the biological control of pests. **b**, African goliath beetles, *Goliathus goliathus* (Scarabaeidae, Coleoptera), are among the world's largest insects. The female is on the left, the male on the right. Their larvae live in rotten wood. Adults feed on tree sap and fruits. (a, courtesy of and with permission of University of California at Davis, IPM Education and Publications.)

their water requirements by metabolizing carbohydrates.

Some small invertebrates and algae survive drought by suspending metabolic activity and becoming dehydrated. Insects usually die if their water content falls below about 20 percent. One exceptional insect, however, is able to tolerate almost total dehydration. The larvae of immature stages of the African midge *Polypedilum vander-*

plankei (Chironomidae, Diptera) live in pools that periodically become dry. The larvae can survive drought for several years by becoming dehydrated. In this state they are resistant to extremes in the physical environment. For example, Hinton (1977) has immersed them in liquid gases at $-270°C$ and heated them to $102°C$ for a minute without injury.

The critical upper limit in temperature for many insects is in the narrow range of 40 to 45°C. The record is held by the ant *Ocymyrmex barbiger* (Formicidae, Hymenoptera), which forages in the sun at temperatures up to 67°C on the surface of the Sahara desert. The lower limits for activity and survival of insects are much broader. Some tropical species become inactive and do not live for long at 5°C; yet some temperate species are normally active on snow at temperatures barely above 0°C. Insects of the temperate and colder regions survive winter by finding suitable shelter and by physiological adjustments that prevent desiccation and freezing. Some species can tolerate the formation of ice crystals in their tissues and survive -35 to $-40°C$. These are the most complex animals to tolerate freezing.

HABITATS AND FOOD HABITS

Insects can be found in forests, grasslands, deserts, cultivated lands, urban areas, bodies of fresh and salt water, and flying in the air. In other words, they occupy all major habitats within the limits of geography and physical environment just discussed. Specifically, they are found in or on soil, in caves, in surface or underground water, immersed in decaying plant or animal matter, and on or inside the bodies of fungi, living plants, animals, and other insects. Many insects are able to occupy more than one habitat and feed on different foods during different stages of their lives.

In soil, insects are especially rich in variety in the litter of leaves and dead plant matter. Many of the insects are **scavengers** (Fig. 1.3a) that feed on dead plants or animals or on the living microbes that flourish in decaying matter. Some insects pen-etrate the upper layers of soil by following natural crevices or entering mammal burrows, while others are adapted to digging burrows themselves. Rocks and other "cover objects" provide shelter for ground-dwelling forms. Decaying logs are occupied by insects of many types, especially beetles and termites. The "subcortical" habitat beneath the bark of dead and, less frequently, living trees is favored by beetles. Animal waste in the form of dung or carcasses on the ground is highly attractive to certain scavenging beetles and flies, and to the insects that prey on the scavengers.

Insects that feed on green plants are termed **phytophagous** (Fig. 1.3b; Color photos 5a–5f). All parts of green plants are attacked: roots, trunks, stems, twigs, leaves, flowers, seeds, fruits, and sap in the vascular system. Insects either feed externally by chewing tissues or by sucking sap or cell contents, or they feed internally by boring into the plant's tissues. The sucking insects, especially Hemiptera, are the only animals that are able to extract sap in quantity from the vascular systems of plants. Likewise, except for some tropical bats, some mammals, and some birds such as hummingbirds, insects are the main consumers of pollen and nectar in flowers. Insects also live among and feed on fungi and primitive plants such as algae, lichens, mosses, and ferns.

Insects feed on most other kinds of terrestrial animals. Insects that kill other insects are termed **entomophagous** (Color photos 6a–f). Of these, the **predators** (Fig. 1.3c) kill their prey more or less immediately, while **parasitoids** (Fig. 1.3d) feed externally or internally in their host for some period, before finally killing it. Insect predators and parasitoids also devour other small invertebrates such as snails, millipedes, spiders, and earthworms. Mammals, birds, reptiles, and amphibians are **parasitized** by bloodsucking insects but usually not killed. Some parasites live on the host (biting and sucking lice, adult fleas), burrow in its flesh, or inhabit the alimentary or respiratory tract (fly maggots). Other parasites (e.g., mosquitoes) visit the host only to suck blood.

In the aquatic environment there are scavenging, phytophagous, and predatory insects, but few

Figure 1.3 Examples of insects with different food habits: **a**, Scavenger: the German cockroach, *Blattella germanica* (Blattellidae, Blattodea), is a common household pest. **b**, Phytophagous: the adults and larvae of the cucumber beetle, *Diabrotica soror* (Chrysomelidae, Coleoptera), are phytophagous. Several species of *Diabrotica* (e.g., the corn rootworms) are serious pests on cucurbits and corn. **c**, Predator: the larvae of most species of ladybird beetles (Coccinellidae, Coleoptera), like the adults, are voracious predators. Here a larva feeds on aphids in Peru. **d**, Parasitoid: a wasp, *Apanteles* sp. (Braconidae, Hymenoptera), emerging from one of the cocoons on a moribund host, which is a moth caterpillar.

parasitoids and true parasites. Some species are able to skate about supported by the surface film, while others float in the film or swim for brief periods below the surface and return for air. Certain groups are adapted to remain beneath the surface indefinitely, obtaining their oxygen from water by means of gills. Of these, some are free-swimming, some crawl about on vegetation and the bottom, and some burrow deeply into the bottom gravels, sand, or silt. Some insects are the only truly aquatic animals capable of prolonged

flight—a distinct advantage for inhabitants of temporary bodies of water.

Insect architects are capable of building complex structures out of their secretions and bits of natural objects (Color photos 2e, 4a, 4b, 4d, 4f). Some construct about their bodies shelters of rolled leaves, plant debris, or carefully selected grains of sand. Many that have a resting pupal stage in their life history construct a special cocoon of silk or a chamber in soil or decayed wood. Shelters such as these provide protection against desiccation, freezing, and reduce detection by enemies. Social ants, termites, and bees may construct elaborate nests that permit some degree of control over temperature, humidity, and light, as well as providing a defensible fortress against enemies.

By their feeding activities and being fed upon, insects have evolved complex **trophic** relationships among themselves and with other animals, plants, and microbes. The origin and diversification of flowering plants, terrestrial vertebrates, and disease agents of plants and animals are due in part to the interactions of these organisms with insects. The blossoms of flowering plants that we enjoy were evolved to attract pollinators, primarily insects. Spices and flavorings originated in plants as chemical defenses against plant-feeding insects. The abundance of insects as food on land probably lured the ancestors of the reptiles to forage far from water, thus leading to the adaptations that made the reptiles the first fully terrestrial chordates. The first birds, mammals, and primates were also primarily insectivores. It was not until insects began feeding on the blood of animals or the sap of plants that certain disease agents, especially viruses, could be transferred directly from the vascular system of one host to another.

ELEMENTS OF INSECT ANATOMY

A typical adult insect has a segmented body with an external skeleton, or **exoskeleton,** composed of chitin (a nitrogenous polysaccharide) and protein. The exoskeleton provides not only strong support and protection for the body but also a large internal area for muscle attachments. It is subdivided into plates, or **sclerites,** that are separated by joints or lines of flexibility and are movable by muscles. In proportion to weight, the tubular construction of the body segments and appendages gives relatively great resistance to bending because of the mechanical strength of a hollow cylinder.

The adult body is divided into three main regions: head, thorax, and abdomen (Fig. 1.4a). The **head** bears a pair of large **compound eyes** and as many as three simple eyes, or **ocelli,** a pair

of sensory **antennae,** and the feeding appendages or mouthparts surrounding the mouth. The **thorax** is composed of three segments, each bearing a pair of **legs.** Each of the last two segments may also bear a pair of **wings.** The **abdomen** is composed of no more than 10 or 11 visible segments and lacks appendages except for a pair of cerci and the reproductive **external genitalia,** which may be present near the tip.

Internally the body cavity, called the **hemocoel,** is filled with muscles, viscera, and the blood or **hemolymph** (Fig. 1.4b). The hemolymph is circulated by a **dorsal vessel** that is differentiated into a chambered **heart** in the abdomen and a tubular **aorta** in the thorax and head. Transverse membranes, the **dorsal** and **ventral diaphragms,** subdivide the abdominal hemocoel. The tubular **alimentary canal,** or **gut,** extends from the mouth to the anus at the extreme tip of the abdomen. The gut is divided into three regions: **foregut, midgut,** and **hindgut.** At the junction of the last two regions are the filamentous **Malpighian tubules,** which serve the excretory function. The paired **salivary glands** open beneath the mouth in a single **salivary duct.** The **central nervous system** consists of a large **brain** in the head in front of the mouth, and a **ventral nerve cord** of segmental **ganglia** extending through the body with nerves to each segment. Large clusters of fat cells commonly create a **fat body** near the gut or the body wall. Here food reserves are stored, symbiotic microorganisms may be accommodated, and a variety of biosynthetic processes take place. Paired openings, or **spiracles,** are situated on either side of the last two thoracic segments and the first eight or fewer abdominal segments. These openings lead to a system of tubular **tracheae** that convey oxygen in air to the internal tissues.

The male and female reproductive organs are located in the abdomen. In both sexes the paired gonads usually open to the exterior by a single, tubular duct in association with the external genitalia. The duct of the male opens beneath, on the ninth abdominal segment, and is surrounded by an intromittent organ, or **aedeagus,** that transfers the sperm directly to the female. The duct of the

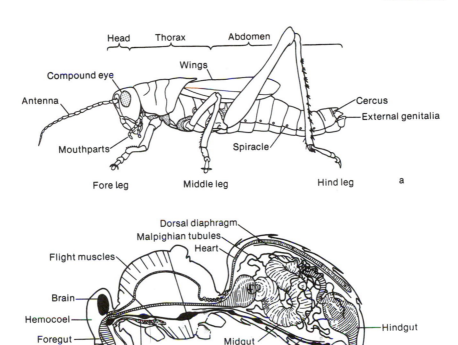

Figure 1.4 Elements of insect anatomy: **a**, external anatomy of the grasshopper, *Romalea microptera* (Acrididae, Orthoptera); **b**, internal organs of the honey bee, *Apis mellifera* (Apidae, Hymenoptera). (a, redrawn from original by M. Lynn Morris, by artist's permission; b, redrawn from *The Hive and the Honey Bee,* Dadant and Sons, 1975, by permission of Dadant and Sons, Inc., Hamilton, IL.

female usually opens beneath, on the eighth abdominal segment. On the eighth and ninth abdominal segments, the female may have specialized appendages that form an **ovipositor** for placing the eggs in soil, plant tissues, or other media.

ELEMENTS OF INSECT CLASSIFICATION AND EVOLUTION

The hardest task in beginning a study of entomology is learning the scientific names of the different kinds of insects. Recall that the hierarchic system for classifying the animals includes six primary categories: Phylum, Class, Order, Family, Genus, and Species. You will be mainly concerned with learning the "higher classification" of insects, that is, the names of the classes, orders, and common families. To assist in this task, we list the family for each kind of insect discussed in the text and, in the first seven chapters, we also list the order. The identification of genera and species requires special training and is often difficult. To represent accurately the evolutionary lineages of insects, authorities augmented the higher classification with intermediate categories, such as superclass or subclass. Unfortunately, authorities are not in universal agreement on the choice of names for these categories or on the best arrangement. In this text we treat 32 orders and generally follow conservative classifications of insects. Our classification is outlined in Table 1.1. It will be helpful

TABLE 1.1
Approximate Size and Diversity of Classes and Orders of Hexapodous Arthropods

Category	Taxon	No. of families[1]		No. of species	
		World	NA	NA[2]	World[3]
Class	Parainsecta				
Order	Protura	3	3	20	500
	Collembola (springtails)	4	3	700	6,000
Class	Entognatha				
Order	Diplura	5	5	64	800
Class	Insecta				
Subclass	Apterygota				
Order	Archeognatha (bristletails)	2	2	35	350
	Thysanura (silverfish)	5	3	30	370
Subclass	Pterygota				
Infraclass	Paleoptera				
Order	Ephemeroptera (mayflies)	17	17	614	2,500
	Odonata (dragonflies, damselflies)	24	11	415	4,900
Infraclass	Neoptera				
Division	Exopterygota				
Superorder	Orthopteroidea				
Order	Blattodea (roaches)	5	5	66	4,000
	Mantodea (mantids)	8	1	20	2,000
	Isoptera (termites)	6	4	41	2,300
	Grylloblattodea	1	1	13	25
	Dermaptera (earwigs)	7	6	23	1,800
	Plecoptera (stoneflies)	8	9	578	2000
	Embiidina (webspinners)	8	3	13	~2000
	Orthoptera (grasshoppers, katydids)	13	12	1,800	>20,000
	Phasmatodea (walkingsticks)	2	1	31	>2,500
Superorder	Hemipteroidea				
Order	Zoraptera	1	1	3	30
	Psocoptera (booklice, barklice)	17	26	257	>3,000
	Phthiraptera (biting lice, sucking lice)	16	9	776	>3,000
	Hemiptera (true bugs)	104	75	3,834	~55,000
	Thysanoptera (thrips)	5	5	700	4,500
Division	Endopterygota				
Superorder	Neuropteroidea				
Order	Megaloptera (alderflies, dobsonflies)	2	2	43	300
	Raphidioptera (snakeflies)	2	2	21	175
	Neuroptera (lacewings, antlions)	17	11	312	5,000
	Coleoptera (beetles)	135	117	23,640	>300,000
	Strepsiptera	5	4	109	525
Superorder	Mecopteroidea				
Order	Mecoptera (scorpionflies)	7	5	75	500
	Diptera (flies)	117	104	19,562	~150,000
	Siphonaptera (fleas)	17	7	258	2,400
	Lepidoptera (moths, butterflies)	102	80	11,300	~150,000
	Trichoptera (caddisflies)	24	22	1,340	>7,000

TABLE 1.1 (*continued*)

Category	Taxon	No. of families[1]		No. of species	
		World	NA	NA[2]	World[3]
Superorder	Hymenopteroidea				
Order	Hymenoptera (ants, bees, wasps)	73	71	17,429	~150,000
		762	627	84,122	>883,475

[1]Number of families estimated from various sources. Numbers are approximate because different classifications vary in the number of families recognized.

[2]Number of species in North America according to Kosztarab and Schaefer (1990). These numbers are based on described species; some groups have many undescribed species that are not included in the estimate.

[3]Number of described and undescribed species in the world estimated from various sources, especially CSIRO (1991); symbol > indicates that the true number probably is higher than the one given; approximate numbers are preceded by symbol ~.

to remember that the category names used here have the following endings:

> Subclass (-ota)
>> Infraclass (-ptera)
>>> Division (-ota)
>>>> Superorder (-pteroidea)
>>>>> Order (-ptera, -ura, or -odea, except Collembola, Archeognatha, Odonata, Embiidina)
>>>>>> Superfamily (-oidea)
>>>>>>> Family (-idae)
>>>>>>>> Subfamily (-inae)
>>>>>>>>> Tribe (-ini)

The names of genera and species are italicized, and, in this text, the family is indicated in parentheses. Family names are sometimes abbreviated. Thus members of the Acrididae, or grasshopper family, become "acridids."

To assist you in obtaining an overall view of insect diversity, let us briefly trace the key steps in their evolution and match these steps with the higher classification, as shown in Table 1.1. In the broad sense, an insect is a terrestrial arthropod that has three pairs of walking legs. This distinguishes insects from the arachnids (spiders, ticks, mites, and scorpions), which have four pairs of walking legs, and from the myriapods (millipedes and centipedes), which have more than four pairs. Insects presumably evolved from arthropods such as the

millipedes and their allies by a reduction in the number of legs. The six-legged condition, however, probably evolved more than once from the same ancestral line. For this reason, the six-legged arthropods are placed in the **Superclass Hexapoda** or "hexapods" (meaning six-legged organisms) and three classes are recognized: Parainsecta, Entognatha, and Insecta. **Class Parainsecta** designates small, wingless, soil-dwelling arthropods of the Orders Protura and Collembola. The latter, also called springtails, are distributed worldwide and can be extremely abundant. **Class Entognatha** comprises another group of small, wingless, soil-dwelling arthropods, of the Order Diplura, that have their mouthparts retracted in a cavity in the head (hence the name Entognatha). **Class Insecta** is used for the insects in the strict sense. Insects do not have their mouthparts retracted in the head.

The word "insect" is derived from the Latin *insecare,* meaning "to cut into," referring to the bodies of some insects that are almost cut in half by constriction at the neck or waist. Although "insect" is properly applied to the Class Insecta, all the hexapods are commonly called insects and we will do so frequently in this text.

The first members of the Insecta were wingless, and some wingless representatives are common today. These are placed in the **Subclass Apterygota** (meaning "without wings"; Fig. 1.5a). The silverfish is a familiar example.

The winged insects are placed in the **Subclass**

Figure 1.5   Representatives of major kinds of insects: **a**, Apterygota: silverfish (Lepismatidae, Thysanura). (Photo by H. V. Daly.) **b**, Paleoptera: dragonfly, *Libellula luctuosa* (Libellulidae, Odonata), **c**, Orthopteroidea: grasshopper, *Trimerotropsis* sp. (Acrididae, Orthoptera). **d**, Hemipteroidea: milkweed bugs, *Oncopeltis fasciatus* (Lygaeidae, Hemiptera). **e**, Neuropteroidea: lacewing, *Eremochrysa* sp. (Chrysopidae, Neuroptera). **f**, Mecopteroidea: buckeye butterfly, *Junonia* sp. (Nymphalidae. Lepidoptera), from South America, attracted to sweat on shirt.

Pterygota (meaning "with wings"). The first winged insects were unable to fold their wings flat over their backs. Some survivors from this stage in evolution are also alive today. These are the dragonflies and mayflies, and they are grouped together in the **Infraclass Paleoptera** (meaning "with an ancient type of wing"; Fig. 1.5b). Insects that can fold their wings flat over their backs are classified as the **Infraclass Neoptera** (meaning "with a modern type of wing"). Within this infraclass, two groups can be distinguished on the basis of the manner in which the wings develop. In the **Division Exopterygota,** the wings develop externally and are visible on the young insect as small wing pads—hence the prefix *exo-*. Insects in the **Division Endopterygota** have wing rudiments that develop internally during the early life of the insect (the prefix *endo-* means inside). The life history is divided into stages that are strikingly different in form and habits: **larva,** resting **pupa,** and winged adult. This type of life history is called complete or **holometabolous metamorphosis** (Color photos 1a–f).

The orders of the Exopterygota may be divided into two groups: **Superorder Orthopteroidea** (meaning "orders resembling the Order Orthoptera"; Fig. 1.5c) and **Superorder Hemipteroidea** ("orders resembling the Order Hemiptera"; Fig. 1.5d). The Orthoptera are grasshoppers, crickets, and katydids, and they rank as the sixth largest order in number of species. The Hemiptera are the true bugs, such as stink bugs and aphids, and rank as the fifth largest order.

The orders of the Endopterygota are divided into three groups: **Superorder Neuropteroidea** (Fig. 1.5e), **Superorder Mecopteroidea** (Fig. 1.5f), and **Superorder Hymenopteroidea** (Fig. 1.6). The largest order of the Neuropteroidea is not Neuroptera (lacewings, antlions, and their allies) but the Coleoptera, or beetles. The superorder is so named because at one time the Neuroptera were considered ancestral to the neuropteroid group of orders. The Coleoptera, which comprise the largest of all insect orders, include the largest family of insects, the Curculionidae, or weevils, with over 60,000 species. The Mecopteroidea are named after the Mecoptera (scorpionflies) because the order probably resembles the ancestral stock that gave origin to the Lepidoptera (butterflies and moths) and Diptera (true flies). These orders are the second and fourth largest orders, respectively. The last superorder, Hymenopteroidea, contains only the Order Hymenoptera (ants, bees, wasps, and their allies). The Hymenoptera are the third largest order.

FACTORS IN THE SUCCESS OF INSECTS

The factors that are probably the most significant in the success of insects are the following: (1) a highly adaptable exoskeleton, (2) colonization of the terrestrial environment before the chordates, (3) small body size, (4) high birthrate and short generation time, (5) highly efficient flight, and (6) life history with complete metamorphosis. To a

Figure 1.6 Representative of major kinds of insect: Hymenopteroidea: a worker (sterile female) yellowjacket wasp, *Vespula* sp. (Vespidae), hangs upside down as she adds plant fibers to the wall of her nest.

large extent these factors are interdependent and must be considered in combination.

The arthropods were the first animal phylum to overcome the problems of locomotion, respiration, and water conservation in a terrestrial environment. Their success can be traced to their adaptable exoskeleton: the jointed, paired walking legs were suited for locomotion on land as well as in water. Insects and some arachnids solved the problem of respiration without losing water by evaporation. In these species, a wax covering was added to the exoskeleton, and the respiratory surfaces were restricted to tubular invaginations of the exoskeleton, the tracheae. The tracheae open to the exterior by spiracles that admit air but reduce the loss of moisture. Thus equipped, insects had ample opportunity to occupy various habitats on land long before any other animal phylum offered serious competition.

In comparison to the calcareous endoskeleton of chordates, the chitin-protein exoskeleton is a distinct asset to small terrestrial animals such as insects. The tubular design and chemical composition combine strength with light weight. Small size itself has several physiological and ecological advantages. Muscles can be used more effectively because their power increases as the square of their cross-sectional area, whereas body volume (and weight) increases as the cube of body dimensions. Because of their size, insects are able to occupy an enormous variety of small places that are not accessible to larger animals. A single piece of shelf fungus on a dead tree, for example, provides food and shelter for several generations of fungus beetles, numbering hundreds of individuals.

Small size also permits short generation time, because less time is required to grow to maturity. When combined with a high birthrate per generation, an increased number of generations per year increases the potential for genetic changes in populations. Such populations may adapt to changes in the environment more quickly. If reproductively isolated from one another, they may differentiate into separate species more frequently than animals with longer generation times and lower reproductive rates.

The appendages of the exoskeleton are greatly varied in structure and function in different groups of insects. The mouthparts may be adapted for taking solid or liquid food; the legs may be adapted for walking, jumping, clinging, grasping, swimming, or digging; and the ovipositor may be modified according to the media into which the eggs are deposited. Wings are also parts of the exoskeleton. Insects are the only invertebrates and historically the first of any animal group to possess wings. The power of flight has allowed insects to escape unfavorable habitats and to colonize new habitats, sometimes at great distances. Flying insects are at an advantage in escaping enemies and in finding mates, food, and places to lay eggs. Indeed, except for the Coleoptera, the many species in the large endopterygote orders are strongly dependent on flight for most of the requisites of life. The social ants and termites represent an interesting exception. The reproductive castes fly during mating, dispersal, and establishment of new nests; worker castes are completely flightless, and yet their numbers and ecological impact are enormous. In terms of weight, these insects were found to be about one-third of the biomass of all animals in a rain forest in the Amazon.

The majority of the insect species are in the division Endopterygota (Table 1.1). In addition to all the features discussed above, the endopterygotes have a complete metamorphosis. The larva is specialized for feeding and in many instances can reach food that is inaccessible to the adult. For example, larvae of certain species are able to bore into stems or leaves or into the bodies of other animals that the winged adult is unable to enter. This specialization also reduces the competition between the larva and the adult for the same resources.

INSECTS AND HUMAN WELFARE

The relationships of early humans to insects were similar to those between insects and other primates and mammals. Insect parasites fed on humans, annoyed them, and transmitted diseases among them. In common with other mammals, humans have acquired host-specific parasites: for example, the head and body louse, *Pediculus humanus,* and the crab louse, *Pthirus pubis* (both in the family Pediculidae, Phthiraptera). Insects also destroyed crops in the field, as well as stored food and shelters, and articles made of wood, plant fibers, and animal hides. In turn, insects were eaten by people, sometimes as a regular part of the diet. In some cultures today, insects continue to provide an important source of nutrition. Honey was widely sought in widely varying regions. Honeybees are native to the Old World, and stingless bees produce honey in the tropical regions of the New World.

With development of agriculture and cities, humans came into cooperation and conflict with insects. Despite the devastating effects of some insects as destroyers of crops and wooden structures, and as carriers of diseases, it is generally agreed that the majority of insects are directly or indirectly beneficial to human society. In economic terms, insects contribute more of value to the agricultural economy than they subtract in damage. Certain species are of direct benefit because they provide useful products, pollinate crops, or can be used as natural enemies to reduce the populations of insect pests or weedy plants. Insects serve as food for many game birds, mammals, and fish, especially trout, on which our recreational hunting

and fishing industry depends. The indirect benefits of the work of insects in recycling nutrients in streams and soils, in pollinating wild flowers, and in regulating the populations of plants and animals in our wildlands are impossible to evaluate.

Beneficial Insects

Crane (1990) lists 177 crops in the world that are dependent on or are benefited by pollination by honey bees. In 1983 the total value of 129 such crops in the United States was estimated to be $18.9 billion (U.S. billion = thousand million) if all the products that depend on pollinated plants were included (Levin, 1983). The value of meat and dairy products was included because domestic animals eat plants grown from seed that required pollination. Honey bees (Fig. 1.7) are the main pollinators in modern agriculture, but some other kinds of bees and other insects are more effective pollinators on certain crops. Honey bees also produce honey (annual world production estimated at 1 million tons in 1984) and beeswax (47,600 tons in 1982).

Other valuable products of insects are silk and shellac. Silk is secreted from the salivary glands of the larva of the silkworm moth, *Bombyx mori* (Bombycidae, Lepidoptera), when it spins its co-

Figure 1.7 Honey bees are among the most valuable insects. In addition to the production of honey and wax, bees are widely used in the pollination of crops. Beekeepers maintain bees in wooden boxes called "supers." The bees construct their combs on removable wooden frames. A veil keeps bees from stinging the beekeeper's face. The device on the right, called a "smoker," generates puffs of smoke into the hive thus stimulating the bees to engorge with honey and become less defensive.

coon. Although the silkworm has been in culture for over 6000 years, beginning in China, it no longer exists as a wild insect in the natural state, outside human care. The annual value of raw silk in the world has been estimated to be about $1.3 billion (Metcalf and Metcalf, 1993). Shellac is used for finishing wood and as a component of many industrial products. It is a secretion of the lac insect, *Kerria lacca* (Coccoidea, Hemiptera), which feeds on trees in Myanmar (formerly Burma) and India. The resinous material is collected, refined, and dissolved in denatured alcohol to make the familiar clear paint. Annual sales of shellac in the United States were estimated to be $20 to $30 million (Metcalf and Metcalf, 1993).

The control of plant and animal pests by natural enemies is termed **biological control.** Populations of organisms are kept in check by a combination of physical and biotic factors, a phenomenon sometimes called the "balance of nature." In the natural terrestrial environment, insects play a significant role in this regard by feeding on other insects, mammals, and vascular plants, and by transmitting diseases. If insect and weed pests of agriculture were not checked naturally by entomophagous and phytophagous insects, it would be difficult to grow crops at all. In biological control, such natural enemies are manipulated to maximize their regulatory effects and reduce pest populations.

Most insect and weed pests of agriculture are exotic; that is, they originated in other countries. To find their natural enemies, it is necessary to locate the native home of the pest and then unravel the details of its biology and enemies. Colonies of enemies are brought to special laboratories under quarantine to test for harmful attributes and to eliminate any diseases or enemies of the desired species. The insects are then bred in large quantities for release. In some instances the release is inoculative. That is, the intent is to establish self-sustaining populations that will continue to provide control indefinitely. In other cases, the release is inundative: intended to bring about immediate control without necessarily establishing resident populations.

The cost/benefit ratio of biological control is difficult to estimate because some of the benefits are not readily assessable in dollars. Biological control agents normally cause no environmental pollution or harm to other organisms. They are usually chosen to attack a specific pest. Once the natural enemies are established in the pest-infested region, the control is essentially permanent and entails no further cost. Only rarely, if ever, does the pest evolve the ability to avoid or survive attack by natural enemies, thus requiring a renewed effort. In terms of savings over previous losses (total losses in products and control costs), Huffaker and Messenger (1976) cite savings totaling $274,942,000 in the biological control of seven agricultural pests in California for the years 1928 to 1973. It is estimated that for each dollar spent in biological control research by the University of California, about $30 was saved in the state.

Theoretically, biological control could be applied to any agricultural pest. This management strategy remains underdeveloped, however, because of the relatively slower rate of control and the lower market value of products with minor defects. The rising cost for the development of chemical pesticides (currently estimated at $50 to $100 million each), the cost of repeated applications (machinery, labor, fuel), the evolution of resistance by pests, and the undesirable effects of pesticides on beneficial natural enemies and other organisms, however, are making biological control more attractive.

Injurious Insects

About 3000 species of insects are directly destructive or harmful to human welfare. These include insects that (1) damage crops directly or by transmitting diseases, thus lowering their yield and the value of the product, (2) destroy agricultural products in storage, (3) damage wooden structures and household goods, (4) attack, parasitize, annoy, and transmit diseases to domestic animals and reduce their value, and (5) attack, parasitize, annoy, and transmit diseases to humans. The total annual loss of agricultural products and household goods damaged by insects has been estimated at $14.4 billion (Metcalf and Metcalf, 1993). Here we limit discussion to a few examples.

The most destructive agricultural pests are those that feed on stored grains. Damage to grains in storage is especially costly because energy and other resources have already been invested in producing, harvesting, and storing the crop. Two beetles, the rice weevil, *Sitophilus oryzae,* and the granary weevil, *S. granarius* (Curculionidae, Coleoptera), are the most harmful of the grain insects. They attack rice, wheat, corn, oats, barley, sorghum, and other grains. With over two-thirds of the world's population dependent on rice as a staple food, these weevils must be considered to be the most destructive insects in the world.

Various species of the moth genera *Heliothis* and *Helicoverpa* (Noctuidae, Lepidoptera) are of worldwide importance as pests of crops in the field. In the United States, *Helicoverpa zea* is known by several common names according to the crop attacked by the larvae: corn earworm, tomato fruitworm, or cotton bollworm. This species also attacks tobacco, beans, alfalfa, and various garden plants. In 1975 over $66.9 million in agricultural losses was attributed to *H. zea* in California, making it the single most destructive species in the state. *Heliothis virescens,* the tobacco budworm, is similar in habits and appetite.

Grasshoppers and locusts (Acrididae, Orthoptera) are of great importance in the more arid regions of the world. They devour nearly all kinds of plants, both cultivated and wild. The migratory locusts, such as *Schistocerca gregaria* and *Locusta migratoria* of Europe, Asia, and Africa, can totally devastate plant landscapes when they land in swarms and feed. In the United States, *Camnula pellucida* and four species of *Melanoplus* do 90 percent of the grasshopper damage of rangelands and cultivated plants.

In timberlands insects attack trees at every stage of growth, from seeds to mature trees, as well as the freshly cut logs and lumber in storage. Graham and Knight (1965) state that "more wood has been destroyed by insects, fungi, and fire than has ever been cut and used." They estimate the total yearly losses due to insects for the United States and coastal Alaska to be 8.6 billion board feet (a board foot equals the volume of a board measuring $12 \times 12 \times 1$ in.).

The most important forest insect pest in the eastern United States is the gypsy moth, *Lymantria dispar* (Lymantriidae, Lepidoptera), which was introduced from Europe during the last century. Although the moth was well known as a forest pest, a naturalist who was experimenting with silkworms brought eggs of the moth from Europe to his home in Medford, Massachusetts. In 1868 or 1869 eggs or some young larvae apparently escaped his care and started an infestation near his house. The destructive caterpillars went largely unnoticed, except by the neighbors, until 1889, when the first severe outbreak occurred. Fruit and shade trees in Medford were completely defoliated, and the citizens subjected to hordes of migrating caterpillars. Later research revealed that the caterpillars would eat over 450 different species of plants, many of which were of economic value. The pest now occurs over most of the northeastern states, Canada, the upper Midwest, and recently the Pacific Northwest. The young larvae are carried considerable distances on wind. The egg masses of 100 to 1000 eggs, normally laid on trees, may be laid in vehicles as well, thus increasing the chance for dispersal over long distances. The favored hosts of the caterpillars are oaks, gray birch, and poplar. Conifers growing in mixed stands with hardwoods also are attacked. Over $500 million has been spent by government agencies in the United States to combat the gypsy moth by means of research, control, and preventive measures.

The most important pests of living coniferous trees are bark beetles (Scolytidae, Coleoptera). The beetles attack in large numbers and overcome the resistance of mature trees. The loss of mature trees is especially detrimental to forestry management because the optimal schedule for long-term harvesting is disrupted.

In the urban environment insects attack not only living ornamental and garden plants but also the framework of houses and their contents. Virtually any article of plant or animal origin can be damaged or destroyed by insects, including wooden structures and furniture, clothing and furs, food, leather goods, books and paper products, and drugs and tobacco. The most common home-infesting insects are termites, cockroaches, ants, fleas, wasps, flies, silverfish, powderpost beetles, and various pantry pests. For wood-destroying insects alone, the estimated cost of control and repair of damage is nearly $5 billion per year in the United States. Furthermore, people are usually annoyed by the sight of live insects in their homes, even when the insects are harmless intruders from outdoors.

Insects that transmit disease agents are called **vectors.** The most important vectors of human diseases are the mosquitoes (Culicidae, Diptera). Indeed, the mosquitoes are the most important family of all insects in relation to human welfare. They transmit the microbes that cause malaria, yellow fever, dengue, encephalitis, and filariasis. These diseases not only kill but also cause sicknesses that result in lost working days and reduced working efficiency. Malaria is considered the most important human disease. Despite a decline in the United States, malaria continues to cause increasing mortality and debilitation elsewhere in the world. Some other important vectors and the diseases they carry are the human body louse (epidemic relapsing fever, typhus); several flea species (plague, murine typhus); tsetse flies, or *Glossina* spp. (Glossinidae, Diptera), of Africa (trypanosomiasis or "sleeping sickness"); and kissing bugs, or *Triatoma* spp. (Reduviidae, Hemiptera), of the New World (trypanosomiasis or Chagas' disease).

Pest Management

Extensive plantings of single crops (i.e., monocultures) favor the rapid population growth of phytophagous insects and mites and of plant diseases. Prior to World War II, agricultural pest control was a combination of several methods because no single approach was usually satisfactory. Cultural methods such as early or late planting avoided the seasonal appearance of certain pests. Destruction of crop residues and plowing exposed overwintering pests to adverse physical factors. Varieties of plants that were known to be resistant to pest attack were often selected for cultivation in certain areas. The role of natural enemies was widely acknowledged. In 1888 the Australian lady beetle or vedalia, *Rodolia cardinalis* (Coccinellidae, Coleoptera), was deliberately brought to California to control the cottony-cushion scale, *Icerya purchasi* (Coccidae, Hemiptera). The success of this measure proved that such enemies could be manipulated by man. Chemical pesticides were limited in variety and effectiveness, and some were toxic to humans and domestic animals. In 1914 scientific evidence was presented that pests could develop resistance to insecticides. Yet for annual crops, where prompt relief from pests meant increased production and profits, pesticides were generally expected to be the most economical approach in the future.

In 1942 the chlorinated hydrocarbon DDT (**d**ichloro**d**iphenyl**t**richloroethane) was demonstrated to have major medical importance in the military effort of World War II. It was effective against body lice, mosquitoes, and flies, and its acute toxicity for humans and mammals was low. Furthermore, the residue had a persistent action against insects.

After the war, DDT and other newly developed synthetic organic insecticides were shown to be highly effective on a wide range of agricultural pests. For a period of about 10 years DDT was produced and used at a rate of some 100,000 tons per year. Pesticides were frequently applied according to a fixed schedule of calendar dates during the growing season and without evaluation of the actual damage taking place. Nonchemical methods of pest control were largely neglected.

The success of insecticides, however, began to be offset by several factors. Pests evolved resistance to specific insecticides. In other words, with each application of an insecticide at a given dosage, some individuals of the pest were not killed.

If the survivors possessed a genetically based trait that allowed them to avoid the insecticide or to detoxify it, and they lived to reproduce, their offspring carried the resistant trait. Over time, with repeated applications of the insecticide, the number of resistant individuals in the population increased. Thus, stronger and more frequent doses or new synthetic compounds were required to control pests. Species that had not been pests became economically important because their natural enemies were eliminated by pesticides. In addition, evidence began to accumulate that DDT and its breakdown products not only were widely distributed in soil and water but had entered the food chains of other animals, including humans. Food chains may act as "biological amplifiers," concentrating a toxic substance in higher amounts at each trophic level. The chlorinated hydrocarbon insecticides were especially likely to be concentrated because of their high solubility in fatty substances and their low solubility in water. In 1962 Rachel Carson argued the case against reliance on pesticides in her book *Silent Spring.* Ten years later, the use of DDT was restricted in the United States to applications specially authorized by the Environmental Protection Agency.

The reported instances of resistance to pesticides continue to mount. Whereas in 1948 only 14 species of insects and mites exhibited resistance to one or more pesticides in some populations, the total in 1989 was 481 pest species of insects, mites, and ticks. Included are nearly all the major agricultural and medically important pests. Resistance is known to all types of insecticides: DDT and other organochlorines such as the cyclodienes (e.g., chlordane), the organophosphates (e.g., diazinon), the carbamates (e.g., carbaryl), the pyrethroids (e.g., permethrin), and fumigants (e.g., methyl bromide). Many species of mosquitoes exhibit resistance to several pesticides. Some species that are among the most important vectors of human diseases are resistant to all types of organochlorines, organophosphates, carbamates, and pyrethroids (Georghiou and Lagunes-Tejeda, 1991; Metcalf and Metcalf, 1993).

Endotoxins from the soil bacterium *Bacillus thuringiensis* have been used as a relatively safe insecticide for more than 20 years without the appearance of substantial resistance in pests. Resistance has now been found in some pest populations, and the future of this highly effective insecticide is threatened.

Resistance to one insecticide may confer resistance to chemically related compounds, and resistance to several different classes of chemicals may show up in the same population. This means that resistance mechanisms that evolved in response to current insecticides may predispose pests to survive insecticides developed in the future as well. In a few species of pests, no insecticide now available is sufficient to give economical control.

The multidisciplinary approach called **integrated pest management** (IPM) is based on ecological principles. It integrates into a single program control methods of all kinds against pests of all types (insects, mites, nematodes, plant pathogens, weeds). Thus entomologists, nematologists, plant pathologists, and weed scientists as well as agricultural economists are being encouraged to collaborate on pest problems. The intent is to provide an effective and economical approach that optimizes farm profits while protecting both public health and the quality of the environment. This diversified but coordinated approach requires constant innovation and adjustment to cope with newly established pest species, new insecticides and crop varieties, changes in agricultural practices, and new information on environmental impacts of pesticides and farming practices.

Most, if not all, new approaches to pest management have the same limitation that was encountered during the pesticide era: insects placed under strong selection pressure are capable of rapidly evolving resistance to substances that threaten their survival. Thus, natural evolutionary processes tend to undermine pest control efforts. Only by a thorough understanding of these processes can we expect to develop long-term programs in pest management.

SELECTED REFERENCES

General Texts

Arnett, 1985; Borror, et al., 1989; CSIRO, 1991; Elzinga, 1987; Evans, 1984; Gillott, 1995; Gullen and Cranston, 1994; Richards and Davies, 1977; Romoser and Stoffolano, 1994.

History of Entomology

CSIRO, 1991 (in Australia); Essig, 1930; Knight, 1989 (in United States); Mallis, 1971 (in North America); R. F. Smith et al., 1973.

Applied Entomology

Caltagirone, 1981 (biological control); Caltagirone and Doutt, 1989 (biological control); Croft, 1990 (biological control); DeBach and Rosen, 1991 (biological control); Denholm and Rowland, 1992 (pesticide resistance); Georghiou and Lagunes-Tejeda, 1991 (list of insects resistant to pesticides); Liebhold et al., 1989 (Gypsy moth); Metcalf and Metcalf, 1993 (text); Sosa, 1989 (yellow fever); Tabashnik, 1994 (insecticide resistance).

Literature (a journal of General interest, Books, and Articles)

American Entomologist (published by the Entomological Society of America); Adams, 1992; Cherry, 1987 (silk); DeFoliart, 1989 (insects as food); Evans, 1985; Hubbell, 1993; Stokes, 1983.

2
THE INSECT BODY

A lubber grasshopper from arizona, *Taeniopoda eques* (Acrididae, Orthoptera), exhibits the major features of insect anatomy. The striking aposematic colors and slow movements of this species are characteristic of insects that are distasteful.

This chapter introduces the general body plan and external anatomy of insects. The description of each part of the anatomy begins with the features seen on common adult insects, such as grasshoppers, and concludes with a brief discussion of some distinctive modifications of other insects. The functional relationships of the parts, the evolution of certain structures, and the anatomy of each order of insects are reserved for later chapters.

HOMOLOGY

The application of anatomical terms to different insects requires the determination of **homology,** or the identification of anatomical parts that have structural likenesses as a result of a common evolutionary origin. In other words, a structure possessed by two or more organisms is said to be homologous because the genes responsible were inherited from a common ancestor. The same homologous structure may differ in function from one organism to another. For example, the mandibles of all insects are homologous; yet in different species the mandibles may differ in appearance, and they are used for various purposes (gnawing, digging, courtship, defense, etc.).

Organisms without a common ancestor may independently evolve similar structures by convergent evolution, the process in which structures become similar because they have similar functions. The result of this process is called resemblance by **analogy.** The wings of insects and of birds are analogues.

Homologies are established in insects by matching the structures with regard to segmental position, location on the segment, time of embryonic development, size, and shape, and the association of other structures whose homologies have been determined. The structures named in this chapter are easily recognized, yet some insects possess unique parts whose homology is unknown.

Another kind of homology, called **serial homology,** is applied to the same structures on different segments of an individual. For example, each body segment of an arthropod usually has one pair of serially homologous, jointed appendages and one large ganglion of the ventral nervous system. It is generally agreed that in insects these appendages are modified into mouthparts, legs, external organs for reproduction, and cerci. Several ganglia may fuse together to appear as one. Thus, the serial homology of structures on different segments is sometimes not immediately apparent; careful study of embryology and of ancient fossil insects is then required. Considerable argument, for example, has concerned the nature of antennae and wings. Are the antennae special sense organs found only on the head or, as argued currently, serially homologous to legs? Are wings unique expansions from the sides of the thoracic skeleton or, rather, highly modified structures from the bases of legs?

Terms for Anatomical Position

Before proceeding further, you may wish to review the following terms that denote anatomical position (the suffix -ad means toward):

anterior, cephalad, or cephalic: pertaining to the front end or head.
posterior, caudad, or caudal: pertaining to the hind end or tail.
dorsal or dorsad: pertaining to the upper surface or back.
ventral or ventrad: pertaining to the lower surface or belly.
lateral or laterad: pertaining to the side or outer part.
median, medial, mesad, or mesal: pertaining to the middle or inner part.
proximal, basad, basal, or base: pertaining to the part nearest the body or base of a part.
distal, distad, apical, or apex: pertaining to the part farthest from the body or tip of a part.
longitudinal: oriented parallel to the length or anteroposterior axis of the body or, in the case of an appendage, the direction parallel to a line extending from base to tip.
transverse: oriented perpendicular to the longitudinal axis of the body or appendage.

PROPERTIES OF THE EXOSKELETON

The body wall of an insect is called the **integument.** The outer layer is a noncellular **cuticle,** chemically composed of chitin and proteins that are secreted by the inner layer of **epidermal cells** (Fig. 2.1). In some areas the cuticle is chemically tanned, or sclerotized, by the epidermis to form rigid, platelike **sclerites.** Areas of the cuticle between sclerites remain tough and flexible, creating membranous joints or articulations. The sclerites and membranes together provide an external support, or **exoskeleton,** for the insect body.

Single cells of the epidermis may be modified into hairs, or **setae.** The seta-forming cell is the **trichogen,** and it is accompanied by a socket-forming cell, or **tormogen.** Setae vary greatly in

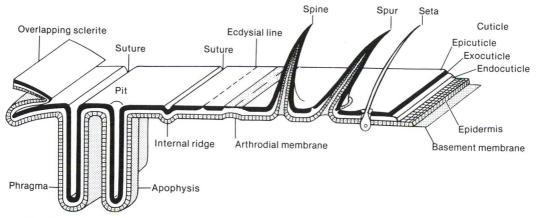

Figure 2.1 Diagram of integumental features. (Modified from Metcalf et al., 1962. Reproduced with permission of The McGraw-Hill Companies.)

both form and function. The common types are hairlike or **simple** setae, featherlike or **plumose** setae, and platelike **scales. Poison setae** are hollow and filled with toxic fluids for defense. Setae of diverse shapes are innervated by the peripheral nervous system and function as sensory organs called **sensilla** (Figs. 3.1a, 6.6a–d).

Many cells are involved in local outgrowths or ingrowths of the integument. Rigid outgrowths are commonly called **spines.** Some outgrowths, especially on the legs, are movable and called **spurs.** Rigid fingerlike ingrowths of the integument associated with muscle attachments are commonly called **apodemes** or, if larger and armlike, **apophyses.** In entomology, the latter term is also used for external protuberances. Large internal apophyses in the head and thorax buttress the exoskeleton and provide areas for muscle attachment. These apophyses function as the **endoskeleton.** Platelike apophyseal invaginations associated with the dorsal flight muscles are **phragmata.**

The integument may also be infolded linearly to produce internal ridges for strengthening the body wall or for muscle attachment. The internal ridge is usually traced externally by a line or groove. Virtually any line seen on the surface of an insect is loosely called a **suture,** regardless of whether it is a line of flexibility (e.g., an articulation), a line marking an internal ridge, or a deep phragma. Sutures are used as convenient boundaries to delimit areas of the exoskeleton. In the strict sense, a suture is a seam along which two plates have fused. The lines between plates of the vertebrate skull are properly called sutures. The cuticle of the insect integument, however, is a continuous layer; adjacent sclerites usually become unified in the course of evolution without leaving a trace. A "suture" transversing a sclerite, therefore, should not be interpreted as an indication that the adjacent areas were once separate.

EVOLUTION OF THE BODY REGIONS

The organization of an insect's body is more readily understood if we recall that insects and other arthropods evolved from the segmented worms, or Annelida. Snodgrass (1935) proposed a sequence of steps leading to the insects that is instructive in this context. In the earliest stage (Fig. 2.2a), the wormlike ancestor has a cylindrical body, with the tubular intestine running nearly the full length. Between the mouth and anus the body is divided transversely into segments. In front

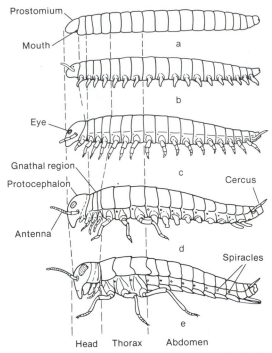

Prostomium

Mouth

a

b

Eye

Gnathal region

Protocephalon

c

Cercus

Antenna

d

Spiracles

e

Head Thorax Abdomen

Figure 2.2 Diagrams of the steps in the hypothetical origin of the insect body from a wormlike ancestor. See text for explanation. (Redrawn from R. E. Snodgass, *The Principles of Insect Morphology.* Copyright © by Ellen Burden and Ruth Roach. Used by permission of the publisher, Cornell University Press.)

of the mouth is an unsegmented part called the **prostomium.** The last part of the body encircles the anus and is the **periproct.** This is the endpiece and also is not a true segment. In a second hypothetical stage (Fig. 2.2b), the prostomium acquires a pair of light-sensitive receptors, or eyes, and a pair of sensory antennae. Each true body segment develops a pair of movable, ventrolateral appendages. In the third stage (Fig. 2.2c), the ancestor is an arthropod with jointed legs and enlarged sensory apparatus. Evidently it has become terrestrial, because most of the segments have a pair of openings, called **spiracles,** leading to the internal air tubes, or **tracheae,** of the respiratory system. The appendages of the last true segment are called **cerci.**

The features of the insects are foreshadowed in the fourth stage (Fig. 2.2d) as regions of the body become specialized. The modification of segments into functional units is called **tagmosis** (see also Chap. 15). At this stage, four such units, or **tagmata,** can be recognized: **protocephalon** (prostomium plus one to three segments), **gnathal region** (three segments), **thorax** (three segments), and **abdomen** (eleven segments plus periproct).

The exact number of segments that have become fused to form the insect head is a matter of much debate. Snodgrass defended the simple theory that the prostomium housed the primitive brain. This was a mass of nervous tissue in front of the mouth. As the sense organs on the prostomium enlarged, the part of the brain innervating the eyes became the **protocerebrum** and the part innervating the antennae became the **deutocerebrum.** Each segment behind the mouth had a pair of primitive **ganglia,** or masses of nerve cells. Each ganglion innervated, respectively, the appendage on its side of the body. The ganglia communicated transversely within each segment by nervous **commissures,** and longitudinally from segment to segment by nervous **connectives** (Fig. 2.3a).

Snodgrass speculated that the appendages of the first segment behind the mouth became sensory in function and shifted forward. The ganglia also moved forward to either side of the mouth and became more closely associated with the brain (Fig. 2.3b). In the insects, the appendages atrophied and were lost in the course of evolution, yet the ganglia of the first segment still persist in insects as the third part of the brain, or **tritocerebrum.** The tritocerebral commissure is evidence of its former segmental position behind the mouth (Fig. 2.3c). In Crustacea the appendages of this segment still persist as the second pair of antennae.

The appendages of the next three segments are specialized for feeding. This is the gnathal region. The three segments following are modified for walking and became the thorax. The last of the four tagmata, the abdomen, contains no more than 11 true segments plus the periproct. Most of the visceral organs are located here. With the walking appendages concentrated in the thorax, the appendages of the abdomen became modified for other functions or atrophied.

In the final stage (Fig. 2.2e), the body regions have been reduced to three by the tagmosis of the protocephalon and gnathal region to form the **head** tagma. The head combines the sense organs and brain, with the gnathal appendages, to form a center for sensing the environment and ingesting food. The thorax is the center for locomotion and houses the large muscles that attach to the leg bases, as well as others that act on the thoracic skeleton to move the wings. The abdomen is the center for the reproductive organs and intestines. By pumping movements, the abdomen also aids airflow in the tracheal system.

After study of fossils and living insects, Kukalova-Peck (1991) argued that the insect head has no unsegmented prostomium as described by Snodgrass. She proposed that the head has six segments, each originally with a pair of appendages that have become highly modified. These are named the labral, antennal, postantennal, mandibular, maxillary, and labial segments. In this scheme, the combined clypeus and labrum are derived from the appendages of the labral segment, the antennae are the appendages of the antennal segment, and the appendages of the postantennal segment are probably incorporated in the hypopharynx.

Secondary Segmentation

Among arthropods a distinction is made between **primary segments,** which correspond to true em-bryonic segments, and **secondary segments,** which are functional subdivisions. The trunk segments of annelid worms and soft-bodied insect larvae are clearly defined by transverse constrictions on which the principal longitudinal muscles insert. These are primary segments, and the transverse folds to which the muscles attach are true intersegmental boundaries. The contraction of the muscles serves both to deform the soft segmental walls and to produce wriggling movements (Fig. 2.4a).

With the evolution of a rigid exoskeleton, a mechanical problem becomes evident: If the soft body wall becomes sclerotized, how can the muscles continue to move the body segments? The evolutionary solution was to (1) keep the muscles attached to true intersegmental lines, (2) develop the rigid sclerites to include the true intersegmental lines so that the sclerites can be moved by the muscles, and (3) place lines of flexibility just in advance of the true intersegmental lines. The insect still has a segmented appearance, but the lines of flexibility between the "segments" no longer trace the true intersegmental lines. Accordingly, they are called secondary segments (Fig. 2.4b).

The dorsal plate of a secondary segment is the **tergum,** and the ventral plate is the **sternum.** The transverse, muscle-bearing ridge that marks the primary intersegmental fold is the **antecosta.** Externally, the ridge is traced by the **antecostal suture.** The margin of the tergum anterior to the antecostal suture is the **acrotergite,** and the corresponding sternal margin is the **acrosternite.**

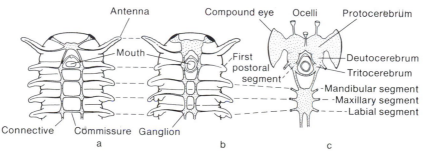

Figure 2.3 Diagrams of the steps in the hypothetical origin of the insect brain and head segmentation. See text for explanation. (Redrawn from R. E. Snodgrass, *The Principles of Insect Morphology.* Copyright © by Ellen Burden and Ruth Roach. Used by permission of the publisher, Cornell University Press.)

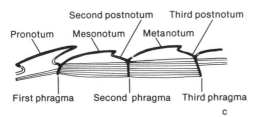

Figure 2.4 Types of body segmentation: **a**, primary segmentation; **b**, secondary segmentation; **c**, secondary segmentation and formation of dorsal phragmata in pterothorax. (Redrawn from R. E. Snodgrass, *The Principles of Insect Morphology*. Copyright © by Ellen Burden and Ruth Roach. Used by permission of the publisher, Cornell University Press.)

THE HEAD

The head contains the feeding appendages, the most important sense organs, and the brain. As the first, or most anterior, body division, the head leads the way when the insect moves forward. It is in a position to detect the changing physical and chemical properties of the environment. Sensations of images in color, moisture, touch, sounds, odors, and flavors travel but a brief distance to the brain, where the insect's behavioral reactions are controlled. The feeding appendages by which the body is fueled are appropriately located in the midst of this battery of sensing devices.

The head articulates with the body at the membranous neck, or **cervix.** Small **cervical plates** are sometimes present at each side in the cervical membrane. These function in the extension and retraction of the head. The rear of the head opens into the cervix and the body beyond by a large hole, the **foramen magnum.** Through this opening pass the ventral nerve cord, salivary ducts, foregut, aorta, tracheae, various neck muscles, and the hemolymph.

Among the various insects, three positions of the head can be distinguished. The original position relative to the body is with the mouthparts directed downward, immediately ventral to the head capsule. This is the **hypognathous** position of grasshoppers (Fig. 2.5d), roaches, some Heteroptera (Hemiptera), and many other insects. In the **prognathous** position, seen in some beetles (Fig. 10.2b), the mouthparts are directed forward and project anterior to the eyes. The last is the **opisthognathous** position possessed by some Heteroptera and all Homoptera (Fig. 34.1b). Here the sucking beak is directed toward the rear, beneath the thorax.

Cranium

The sclerotized head capsule, minus the appendages, is the **cranium.** In the course of evolution the distinction between the head segments has been lost almost entirely. Most of the lines seen now on the cranium represent secondary modifications unrelated to segmentation. Some lines trace internal ridges that strengthen the exoskeleton, while others are actually lines of weakness that allow the cuticle to split during ecdysis. In most insects only the postoccipital suture, described below, is believed to be a segmental boundary. Here a strong internal ridge has been retained between the maxillary and labial segments, serving as a place for the attachment of neck muscles that move the head. Fossils of the order Monura (Fig. 15.3) and the living fossil *Tricholepidion* (Thysanura; Fig. 19.4) exhibit intersegmental lines on the cranium.

The **compound** eyes are easily recognized as the large paired organs located dorsolaterally on the cranium (Fig. 2.5a). The surface of the eye is covered with minute facets, each of which represents the lens of an individual eye unit, or **ommatidium.** Situated between the compound eyes on the face are as many as three simple eyes, or **ocelli.** Each has a single lens. One is usually present on the midline, and the others are placed above and at each side.

On some insects a pale line, shaped like an inverted Y, can be seen medially on the top of the head. This is the **ecdysial line,** along which the old cuticle first splits when it is shed by the immature insect. One or two ventral ecdysial lines occur on the underside of the head of many endopterygote larvae. The lines serve no function in the adult.

The cranium has an internal, sclerotized structure called the **tentorium** (Fig. 2.5b). This endo-

Figure 2.5 Insect head: **a**, cephalic view of head of *Romalea microptera* (Acrididae, Orthoptera); **b**, diagram of tentorium of a grasshopper with most of the cranium removed; **c**, caudal view of head of *Romalea*; **d**, lateral view of same. (b, redrawn from R. E. Snodgrass, *The Principles of Insect Morphology*. Copyright © by Ellen Burden and Ruth Roach. Used by permission of the publisher, Cornell University Press.)

skeleton is created by paired invaginations of the integument, at both the front and rear of the head, that fuse together inside the head. Externally the invaginations are indicated by pits or narrow slits. The **anterior tentorial pits** mark the entry of anterior invaginations that become the internal **anterior tentorial arms.** Similarly, the **posterior tentorial pits** mark the invaginations of the **posterior tentorial arms.** The arms meet medially to form the **tentorial bridge.** In some insects, each anterior arm has a **dorsal arm.** The tentorium is variously shaped in different insects. It strengthens the cranium and provides attachment for muscles to the antennae and the gnathal appendages.

Names have been given to different areas of the cranium. In some instances sutures serve as convenient boundaries between areas, but elsewhere the limits are undefined. This should not cause confusion because the names refer merely to topographic areas, not to discrete sclerites.

Beginning at the top of the head and descending in front, the names of the facial areas are as follows: the summit of the head, between and behind the compound eyes, is the **vertex.** The next area, the **frons,** is between the antennae and the compound eyes. No definite boundary exists between the frons and the vertex. Below the frons is the **clypeus.** The clypeus may be defined as the area of the cranium to which the labrum is attached. The **labroclypeal suture** is the line of articulation between the labrum and clypeus. The frons and clypeus are sometimes separated by a suture extending between the anterior tentorial pits. This is the epistomal or **frontoclypeal suture.**

Returning to the top of the head and descending on one side (Fig. 2.5d), the **gena** is the area extending from below the vertex and behind the compound eyes, to the ventral edge of the cranium. The gena is not delimited above from the vertex, but along other boundaries sutures may exist. The gena may be separated anteriorly from the frons by a **frontogenal suture** or a subocular suture in *Romalea* (Acrididae, Orthoptera), and from the clypeus by a **clypeogenal suture.** Ventrally, a **subgenal suture** may extend between the anterior and posterior tentorial pits, separating off a marginal

area, the **subgena.** Posteriorly, the gena may be delimited by the **occipital suture.**

Turning now to the back of the head, the areas can be seen best if the head is removed at the neck, or cervix (Fig. 2.5c). At each side near the base of the labium are the slitlike posterior tentorial pits. Extending from the posterior pits and running dorsally around the foramen magnum is the **postoccipital suture.** The area anterior to the suture is the **occiput.** The narrow sclerotized rim of the cranium, posterior to the suture, is the **postocciput.** The labium is suspended at each side from the lower edges of the postocciput, indicating that this is the tergum of the labial segment. The postocciput also bears the **occipital condyles** on which the cervical plates articulate. Returning to the occiput, an **occipital suture,** or at least an incomplete ridge, may separate the occiput dorsally from the vertex and laterally from the genae. The lower part of the occiput, near the posterior tentorial pits, is called the **postgena.**

Modifications of the Cranium

The sutures and areas of the cranium vary considerably in extent and position largely as a result of modifications of the mouthparts and associated musculature. It is entirely possible for distinctive, new areas of the cranium to develop in certain groups of insects. A special terminology is often applied to these areas.

The faces of Hemiptera (suborder Homoptera) and Psocoptera (Fig. 32.2a) are thus altered by the development of a sucking pump. The prominent bulge between the eyes that houses the muscles for this pump is called the **postclypeus,** and the smaller area below, the **anteclypeus.** In reality this is a secondary modification of the entire frontoclypeal region where the original frontoclypeal boundary has been obscured.

Insects such as Orthoptera lack sclerotization of the integument beneath the foramen magnum and posterior to the labium. In other words, the throat behind the mouthparts is membranous. However, in hypognathous insects such as some Coleoptera this area may be secondarily sclerotized to form either a **hypostomal** or a **postgenal** bridge (called the **gula** in Coleoptera). Recall that the subgenal suture runs near the edge of the cranium and terminates in the posterior tentorial pit. That part of the suture posterior to the mandible is also known as the **hypostomal suture,** and the subgenal area thus delimited is the **hypostoma.** Recall also that the adjacent area of the occiput is the postgena. In caterpillars of Lepidoptera, sclerotic lobes of the hypostoma extend mesally onto the throat, but do not meet. In adult Diptera and some Hymenoptera the hypostomal lobes unite to form a hypostomal bridge beneath the foramen magnum (Fig. 2.6b). In other adult Hymenoptera and Hemiptera (suborder

Heteroptera), the postgenae also unite, forming a postgenal bridge (Fig. 2.6a). In the formation of bridges of both types, the posterior tentorial pits remain unaltered in position.

In prognathous insects such as some Coleoptera and Neuroptera, the underside of the head is substantially modified by the forward position of the mouthparts. The postgenae are usually enlarged, and the posterior tentorial pits are shifted forward. The posterior pits, however, retain their connections at each side with the postoccipital suture. The midventral area of the head between the pits corresponds to the membranous throat of Orthoptera. In Coleoptera and Neuroptera, this area is secondarily sclerotized to form the gula

Figure 2.6 Modifications of the posteroventral region of the head: **a**, postgenal bridge of a wasp, *Vespa* sp. (Vespidae, Hymenoptera); **b**, hypostomal bridge of the honey bee, *Apis mellifera* (Apidae, Hymenoptera); **c**, narrow gula of beetle, *Staphylinus* sp. (Staphylinidae, Coleoptera); **d**, broad gula of the dobsonfly, *Corydalus* sp. (Corydalidae, Megaloptera). (Redrawn from R. E. Snodgrass, *The Principles of Insect Morphology.* Copyright © by Ellen Burden and Ruth Roach. Used by permission of the publisher, Cornell University Press.)

(Fig. 2.6c,d). The gula is continuous with the postocciput and sometimes also the labial base. The parts of the postoccipital suture that delimit the gula laterally are correspondingly renamed the **gular sutures.** The sutures are external indications of the internal tentorium.

Antennae

The antennae are paired, segmented appendages that articulate with the cranium between or below the compound eyes. Three parts can usually be recognized: the basal, or first, segment is the **scape**; the second segment is the **pedicel**; and the remaining segments constitute the **flagellum** (Fig. 2.7). Each segment of the flagellum is called a flagellomere. The whole antenna is moved by muscles from the head that insert on the scape. In the **antennal socket** the antenna may pivot on an articular process, or **antennifer.** The flagellum is moved by muscles from the scape that insert on the pedicel but not on the segments beyond. In accordance with its sensory function, the surface of the flagellum is supplied with many sensory receptors that are innervated by the deutocerebrum of the brain. In addition, within the pedicel is a mass of sense cells that detects movements of the flagellum. This is called **Johnston's organ** (Fig. 6.6f). Exceptions to the foregoing account are

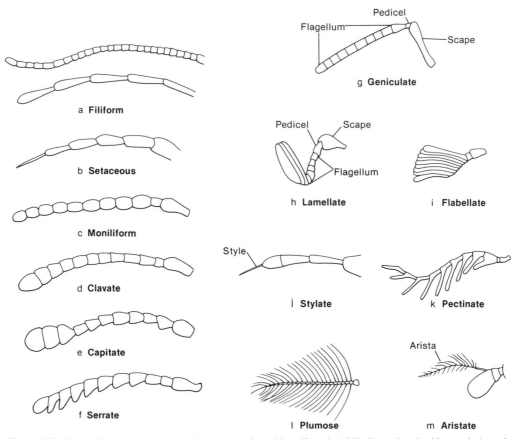

Figure 2.7 Types of antenna. (a–g, j, and l redrawn from Metcalf et al., 1962. Reproduced with permission of The McGraw-Hill Companies.)

found in the other six-legged arthropods: in Collembola (class Parainsecta) and Diplura (class Entognatha) the flagellar segments have intrinsic muscles and Johnston's organ is absent. In Protura (class Parainsecta) antennae are missing entirely.

Modifications of Antennae

Special terms for different shapes of antennae are indicated in Figure 2.7 (see also Chap. 39, Fig. 39.6).

Mouthparts

The mandibulate, or chewing, type of mouthpart is the basic type from which the specialized types have been derived. The mandibulate type consists of an upper lip (or labrum), paired mandibles, paired maxillae, a lower lip (or labium), and a median tonguelike hypopharynx (Fig. 2.8). The mouthparts enclose the preoral cavity, within which are the true mouth and the opening of the salivary glands.

The **labrum** is a flap that closes the preoral cavity in front (Figs. 2.5a, 2.8b). It articulates with the cranium by the membranous **labroclypeal suture.** The inner surface forms the roof of the preoral cavity and may have a median lobe called the **epipharynx** (Fig. 2.8a). Muscles from the head capsule insert near the base of the appendage and provide a variety of motions. The labrum is innervated by the tritocerebrum of the brain.

The **mandibles** are heavily sclerotized jaws that represent limbs that have lost their distal segments (Fig. 2.8c). The inner edge is modified for biting and may be hardened by deposits of zinc, manganese, or iron in some insects. Toward the tip the edge has cutting teeth, or incisors; at the base is a grinding surface, variously called the **mola, molar,** or **molar lobe.** Each mandible has two basal points, or **condyles,** where it articulates with the cranium. The posterior condyle is larger and fits against the gena or postgena. This was the original articulation point of the limb base. The anterior condyle is a secondary modification and

fits against the base of the clypeus, thus forming a hinge line that permits the mandibles to swing laterally. Powerful adductor muscles from the cranium insert at the base of the mandibles on the inner side of the hinge line and by their action bring the opposing mandibles together in a pinching motion. Smaller abductor muscles, attached to the outer margin of the mandibles outside the hinge line, open the mandibles.

The **maxillae,** a second pair of jaws, are much less massive and retain the segmentation of an appendage (Fig. 2.8d). Each maxilla has a limb base that articulates with the cranium by a single condyle. The base is divided into a proximal sclerite, or **cardo,** that bears the condyle, and a distal sclerite, the **stipes.** At the apex of the stipes are two lobes: an inner **lacinia,** which may be toothed, and an outer **galea.** Lateral to the galea is the **maxillary palp,** which represents the distal part of the limb and is sensory in function at the apex. The area on the stipes where the palp articulates may be differentiated into a segmentlike structure called the **palpifer.** Muscles from the cranium insert on the cardo and stipes, producing various motions of the maxilla about the single articulation. The lacinia, galea, palp, and palpal segments are also individually movable by muscles.

The **labium** closes the preoral cavity to the rear. The structure of the labium resembles that of the maxillae except that the labial appendages are fused medially with each other and basally with the segmental sternum (Fig. 2.8e). The lobed apex of the labium, called the **ligula,** consists of an inner pair of lobes, the **glossae,** and a lateral pair, the **paraglossae.** The glossae and paraglossae are serial homologues, respectively, of the maxillary lacineae and galeae. Similarly, the **labial palp** corresponds to the maxillary palp and is likewise sensory in function. The palps of the maxilla and labium function to taste and feel food. The labial palp may be borne on a segmentlike **palpiger.** The sclerite to which the ligula and palps attach is the **prementum.** The prementum articulates basally with the postmentum by the **labial suture** and contains all the muscles of the ligula. The postmentum in turn may be subdivided into a distal sclerite,

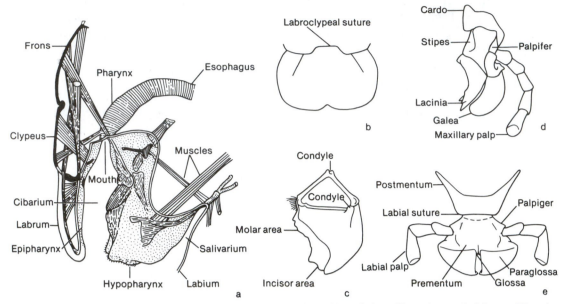

Figure 2.8 Mouthparts and preoral cavity: **a**, diagram of preoral cavity and lateral view of hypopharynx; **b**, labrum of *Romalea microptera* (Acrididae); **c**, mandible of same; **d**, maxilla of same; **e**, labium of same. (a, redrawn from R. E. Snodgrass,*The Principles of Insect Morphology*. Copyright © by Ellen Burden and Ruth Roach. Used by permission of the publisher, Cornell University Press.)

or **mentum,** and a proximal sclerite, the **submentum.** The labium is complexly musculated such that the ligular lobes, the palps, and their segments, and the prementum are individually movable.

The **hypopharynx** is a large median lobe situated between the mouth and the labium (Fig. 2.8a). The **cibarium** is a special chamber in front of the mouth that is formed by the anterior surface of the hypopharynx. On the posterior side, the **salivary duct** opens in a cavity, or **salivarium,** between the hypopharynx and the labium. The hypopharynx is movable by muscles from the tentorium that insert on sclerites in the wall of the hypopharynx. Nerves to the mandibles, maxillae, labium, and hypopharynx are supplied by the subesophageal ganglion.

Modifications of Mouthparts

Insect mouthparts vary in structure and function according to the nature of the food eaten. As a consequence, they not only provide important characters to be used in taxonomic classification, but also indicate the feeding relationships of the insect in its ecological community. The latter is of special importance to human welfare. The successful transmission of a disease, for example, may be determined by the minute structure of an insect's beak. Or an insecticide residue on a leaf surface may or may not be ingested by a pest, depending on the creature's exact manner of feeding.

Some insects do not feed in the adult stage, and the mouthparts are atrophied and functionless. Examples are Ephemeroptera and at least some members of the following orders: Plecoptera, Trichoptera, Megaloptera, Lepidoptera, Strepsiptera, Coleoptera, Diptera, and Hemiptera (male scales).

Various schemes for classifying functional mouthparts have been proposed. Here we have adopted distinctive categories (indicated in bold print) and listed examples and exceptions, with brief descriptions of the mechanisms. Additional information can be found elsewhere in this text under the treatments of the different taxonomic groups.

A. **Entognathous mouthparts.** The mandibles and maxillae are recessed and largely hidden from view by lateral folds of the head. Examples: Protura (Fig. 17.1c,d), Diplura (Fig. 18.1b), and Collembola.

B. **Ectognathous mouthparts.** The mandibles and maxillae are visible or secondarily recessed, but lateral folds of the head are absent. Examples: all members of the class Insecta.

1. **Mandibulate mouthparts.** The feeding appendages are more or less complete (Figs. 2.8, 2.9), freely movable, and not united in a beak. The mandibles are usually conspicuous and suited for seizing objects or chewing solid food. Those of predators are fitted with sharp apical teeth for catching prey, while the mandibles of plant feeders have more obtusely angled cutting teeth together with basal grinding molars. Mandibles are also used in defense, courtship, and the construction of nests or shelters. Most adult and immature insects have mandibulate mouthparts or some modification of this type. The exceptions are listed under the suctorial and derived types below. Unusual modifications of the mandibulate type deserve mention:

   a. Odonata. The labium of the nymph is greatly enlarged (Fig. 21.5), hinged at the labial suture, and held retracted beneath the head. By means of quick forward thrusts, prey are seized in the pincerlike labial palps and brought to the mandibles near the mouth.

   b. Hymenoptera. The adult mouthparts are basically mandibulate, but the maxillae and labium are united in a mobile, extendible structure (Fig. 46.2b). The glossae are fused medi-

ally, forming a lapping organ, so that liquids can be taken readily in addition to solid food. An elongation of the maxillolabial structure forms the sucking tongue of certain families of bees [see below: item (2) of 2, b].

   c. Hymenoptera, Lepidoptera, Trichoptera. In the larvae the maxillae, labium, and hypopharynx are united into a lower lip with the opening of the salivary duct at its tip. In these orders the salivary glands produce

Figure 2.9 Examples of insects with mandibulate mouthparts: **a**, The bulldog ants, *Myrmecia* sp. (Formicidae, Hymenoptera), of Australia use their large mandibles to gently pick up their eggs and, in combination with powerful stinging, to defend their nest. **b**, The hornworm caterpillar, *Manduca sexta* (Sphingidae, Lepidoptera), has mandibulate mouthparts suited for eating leaves. The notch on the labrum may serve as a guide while the insect eats along the edge of a leaf. The massive head capsule contains the large mandibular muscles.

the silk used in making cocoons or larval shelters. The palps and other distal parts are reduced or wanting in some groups, but the mandibles are usually retained (Fig. 2.9b).

d. Isoptera (Fig. 24.2b), Megaloptera (Corydalidae, Fig. 36.1b), Coleoptera (Lucanidae and several other families), and Hymenoptera (Formicidae, Fig. 2.9a). Adults of certain insects in these taxa have greatly enlarged mandibles that are not used for feeding. The jaws of male Lucanidae actually bite less severely than the small jaws of the female. The mandibles of some male beetles are used in jousting (W. G. Eberhard, 1979, 1982). The other taxa listed use the mandibles primarily for defense.

2. **Suctorial mouthparts.** In this type, one or more feeding appendages are modified into a tubular organ for taking liquid food. Insects in different, unrelated orders have independently evolved suctorial, or haustellate, mouthparts, so the exact mechanisms vary. Parts are commonly reduced or missing. The cibarium and pharynx, either alone or in combination, provide a sucking pump in most insects of this type. Note that suctorial mouthparts are present in both the immature and adult stages of exopterygote orders listed below, whereas among the endopterygotes only one stage of the life history has developed suctorial mechanisms.

a. **Piercing-sucking mouthparts** include one or more appendages that are sharp at the apex and suited for piercing the surface of plant or animal bodies. Saliva is usually injected while feeding. Insects that attack vertebrates often have anticoagulants in the saliva to aid the flow of blood. We list examples of insects with piercing-sucking mouthparts in eight categories.

(1) Thysanoptera. Sharp stylets are used to rupture leaf cells and the exuding sap is sucked up by a ciberial pump (Fig. 35.1d). The mouthparts form a short conical beak and are uniquely asymmetrical. Inside the beak the right mandible and inner processes of the two maxillae are slender and sharp, forming piercing stylets. Thrips feed mostly on leaves, seeds, fungi, or pollen, but some are predatory (Chap. 35).

(2) Hemiptera. An elongate proboscis is formed by the mandibular and maxillary stylets and the labium (Fig. 34.2e,g; Color photo 5d). The labium ensheaths the stylets but does not penetrate the host. The maxillary stylets interlock and create two canals by matching their opposing grooves. The anterior canal, which leads to the cibarial pump and mouth, is appropriately named the "food canal." The posterior, or salivary, canal carries saliva under pressure from a syringelike organ within the hypopharynx. At each side of the united maxillae are the separate mandibles. By alternate thrusts, the mandibular blades pierce the host's tissues and provide an entry for the maxillae. The vascular systems of both plants and vertebrates are penetrated by the mouthparts of Hemiptera.

(3) Phthiraptera (suborder Anoplura). The mouthparts of sucking lice are concealed in a

long pouch beneath the foregut in the head (Fig. 33.2a–c). Three stylets, possibly representing the united maxillae, the hypopharynx, and the labium, make up the piercing organ. Suction is applied to the feeding wound by a muscular region of the foregut just behind the mouth and including the pharynx.

(4) Siphonaptera. The adult mouthparts of fleas lack mandibles. The piercing organ is formed by the bladelike maxillary laciniae and stylet-shaped epipharynx (Fig. 43.1b). Both the maxillary and the labial palps are present. The latter surround the laciniae but do not penetrate the wound. Once the skin has been penetrated, blood is sucked up by a cibarial and pharyngeal pump.

(5) Neuroptera and Coleoptera (Dytiscidae). The larvae of these insects are predatory, and each larval mandible is elongate, sickle-shaped, and grooved on the inner surface. In Dytiscidae the mandibles function alone, but in Neuroptera both maxillae are similarly elongated and fit against the mandibular groove to form a closed canal (Figs. 38.1c, 38.2c). The body of the insect victim is pierced by the opposing mandibles, enzymes are injected, and the predator sucks the digested body contents as fluid from the victim. The cibarium of Dytiscidae and the pharynx of Neuroptera provide the sucking pumps.

(6) Diptera (Culicidae, Tabanidae, Asilidae). The piercing organ of adult female "biting flies" that take blood meals (Culicidae, Tabanidae) consists of six stylets: labrum-epipharynx (enclosing the food canal), hypopharynx (containing the salivary duct), paired mandibles, and paired maxillae (Fig. 42.3b). The labium surrounds and supports the piercing stylets but does not penetrate the feeding puncture. A cibarial pump provides suction. Mosquitoes (Culicidae; Fig. 2.10b) force the slender piercing organ into the skin, leaving the blunt, flexible labium at the surface. Blood is taken directly into the food canal. Horse flies (Tabanidae) cut the skin surface with their broad, bladelike stylets. The apex of the labium is greatly enlarged in two lobes called the **labellum.** The undersurfaces of the lobes are traversed by fine grooves, called **pseudotracheae.** These lead mesally to a cleft within which the labral food canal opens. The pseudotracheae direct the flow of blood to the tip of the labrum, and the blood is sucked into the canal. The males of Culicidae and Tabanidae do not suck blood and feed instead on nectar, plant juices, or other liquids with their reduced mouthparts. The mouthparts of the predaceous robber flies (Asilidae, Fig. 2.10c) are also of the piercing-sucking type, but differ from those of the bloodsucking flies. Both sexes of robber flies have the bladelike hypopharynx adapted for stabbing their prey and injecting proteolytic en-

zymes. Mandibles are absent. The digested contents of the prey are sucked up through a food canal formed by the elongate labrum, maxillae, and hypopharynx, all of which are partly enclosed in a heavily sclerotized labium.

(7) Diptera (*Stomoxys* and *Haematobia* in Muscidae; *Glossina* in Glossinidae). The piercing organ of the stable flies, horn flies, and tsetse flies (Fig. 2.10d) is a

secondary modification of the sponging type described below for adult Diptera. Both sexes of these flies bite mammals. In contrast to the flies described in item (6), the labium of these flies penetrates the host. The labium is slender and stiff, with the labellum reduced in size and bearing small rasping teeth at the tip. The labrum and hypopharynx are partly enclosed by the labium, so that the combined

Figure 2.10   Diversity of mouthparts in Diptera: **a**, Mouthparts of a flesh fly, *Sarcophaga* sp. (Sarcophagidae), is representative of flies that "sponge" liquid food such as sweat. **b**, The piercing-sucking mouthparts of a female mosquito, *Aedes* sp. (Culicidae), deeply penetrate human skin, as the labium is bent backward. **c**, Piercing-sucking mouthparts of a predaceous robber fly (Asilidae) with its grasshopper prey. **d**, Both sexes (shown mating) of the African tsetse fly, *Glossina* sp. (Glossinidae), have piercing-sucking mouthparts and feed on blood. Blood-feeding mouthparts of highly evolved flies were derived from the sponging type.

appendages form the piercing organ.

(8) Lepidoptera (certain Noctuidae, Pyralidae, Geometridae). The piercing organ of the fruit-piercing moths of Asia is a secondary modification of the nonpiercing-sucking proboscis of adult Lepidoptera described below. The proboscis is formed by the elongation of the maxillary galeae (as in Fig. 44.2c). These are interlocked to form a tubular food canal. The apex of the piercing proboscis is sharp and has hundreds of movable barbs that aid in penetration. By alternate movements of the galeae, the tip is forced into fruit. Some species also visit the eyes of mammals for eye discharges and take blood from wounds. The unique *Calpe eustrigata* (Noctuidae) pierces the skin of mammals and sucks blood.

b. **Non-piercing-sucking mouthparts.** The following items give examples of insects that take water or nutrients dissolved in water, such as the nectar of flowers (Fig. 10.1) or the honeydew (plant sap) excreted by aphids and leafhoppers.

(1) Lepidoptera. Adult butterflies and moths take water, nectar, or honeydew with a long proboscis that is coiled beneath the head when not in use. The proboscis represents the galeae, greatly elongated and tightly interlocked to form a tube (Fig. 44.2c; Color photo 5a). Suction is created by the pumping action of muscles inserted in front of and behind the mouth. The other mouthparts are reduced or want-

ing, except the labial palps and the smaller maxillary palps.

(2) Hymenoptera (Megachilidae, Apidae, Anthophoridae). The bees in these families have the maxillolabial structure elongated and are appropriately named the "long-tongued bees" (Fig. 46.2b; Color photo 5b). The long galeae and labial palps cover the glossal tongue to form a tubular proboscis. The pharynx is modified into a sucking pump, allowing the bee to withdraw nectar from deep inside flowers.

(3) Diptera. Most adult flies that do not suck blood have a flexible proboscis consisting of the labrum (enclosing the food canal) and slender hypopharynx (containing the salivary duct) (Figs. 2.10a, 42.3c). Together these fit in an anterior groove on the labium. The labium is enlarged apically into the bilobed labellum, which functions as a sponging organ. The structure of the labellum is described above for the Tabanidae. Liquid food taken up by the labellum is directed into the food canal and sucked up by a cibarial pump. Small particles of food, such as pollen, can also be eaten by some flies. Maxillary palps are present, but mandibles and labial palps are missing. Flies of the genus *Melanderia* (Dolichopodidae), however, have unique horny processes on the labellum that function like mandibles to grasp prey.

3. **Mouth hooks of maggots.** Larvae of Diptera (division Cyclorrhapha) have the head highly reduced and in-

vaginated into the thorax, leaving only a neck fold to form a conical snout or functional head (Fig. 42.4c). Within the snout is a preoral cavity, or atrium, from which projects a pair of mouth hooks. The hooks move vertically and arise from the lips of the atrium. No trace of the normal feeding appendages remains. The hooks are secondary modifications of the atrium and are unique to fly maggots. Within the prothorax are sclerotized pharyngeal plates that, together with the hooks, make up the cephalopharyngeal skeleton. Maggots feed on a variety of soft or semi-liquid foods, including decaying matter as well as living tissues of plants and animals.

THE THORAX

The second body division, or thorax, is modified for locomotion. The three segments and their ventrolateral appendages are designated from front to rear as follows: **prothorax,** bearing the **forelegs: mesothorax,** bearing the **middle legs**; and **metathorax,** bearing the **hind legs.** Internally each segment has a segmental ganglion of the ventral nerve cord. A pair of lateral spiracles is present at the anterior edge of each of the last two segments. In some insects the mesothoracic pair actually opens on the prothorax, but this is a secondary condition.

In the adult pterygote, a pair of wings may be located dorsolaterally on the mesothorax and metathorax. The two wing-bearing segments are called the **pterothorax.** Wings are absent in the prothorax of insects living today, but certain extinct insects had winglike appendages on the prothorax.

Thoracic Nota

The terga of thoracic segments are called **nota** to distinguish them from abdominal terga. Accordingly, the nota of the respective segments are desig-nated **pronotum, mesonotum,** and **metanotum.**

The pronotum is usually a simple plate (Fig. 34.13a), but in some insects like the cockroach (Fig. 22.1) it is greatly enlarged to cover the head or, as in the treehoppers (Color photo 7e), it covers other parts of the body. In the grasshopper the pronotum extends back over the pterothorax and has several transverse sutures (Fig. 2.11a). The sutures are related to the musculature (which is peculiar to the prothorax) and are not homologous with sutures on the pterothoracic nota.

Each pterothoracic notum is divided into a wing-bearing sclerite, the **alinotum,** and a phragma-bearing sclerite, the **postnotum** (Fig. 2.12a). Each alinotum is usually traversed by one or more sutures. These trace the position of internal supporting ridges or lines of flexibility. The sutures vary greatly in extent and position, according to the nature of the flight mechanism. Commonly the alinotum is divided anteriorly by a transverse **prescutal suture** that delimits an anterior **prescutum.** Posteriorly a V-shaped **scutoscutellar suture** separates an anterior **scutum** and a smaller, posterior **scutellum.** The lateral edges of the alinotum are modified for the articulation of the wings. At each wing base are **anterior** and **posterior notal wing processes** (Figs. 2.12a, 2.14a).

To understand the origin of the postnota and phragmata, it is necessary to recall the terminology of secondary segmentation (Fig. 2.4c). The postnotum corresponds to the acrotergite, and the phragma corresponds to the antecosta. The acrotergite and the antecosta anterior to the pronotum were apparently lost during the evolution of the cervix, leaving the pronotum as a simple plate. The narrow acrotergite and the **first phragma** anterior to the mesonotum are continuous with the alinotum to the rear, just as expected in a secondary segment. In the metathorax, however, the **second postnotum** and its **second phragma** are detached from the metathoracic alinotum and associated anteriorly with the mesothorax. Likewise the **third postnotum** and **third phragma** are detached from the first abdominal tergum and associated anteriorly with the metathorax. As a result of this reassociation, the first abdominal tergum is left as a simple plate.

Thoracic Sterna

The ventral plate of a thoracic segment is the **eusternum.** In the intersegmental regions between the euserna of the prothorax and mesothorax, and the mesothorax and the metathorax, there may be small, separate sclerites. These are named **spinasterna** because each carries a median, internal apodeme called a **spina.** One or both of the spinasterna may fuse anteriorly with the euserna of the preceding segments (Fig. 2.11b). The euserna plus spinasterna are designated segmentally as **prosternum, mesosternum,** and **metasternum,** respectively. Each eusternum may have a pair of large invaginations, the **sternal apophyses.** The apophyses may be well separated and their positions externally marked by a pair of furcal or **apophy-** seal pits (Figs. 2.11b, 2.12b) or fused medially to form a **sternal furca** (Fig. 2.12c). The longitudinal, midventral line that traces the invagination of the furca is the **discrimen.** The lateral boundaries of the eusternum are usually fused partly or entirely with the pleuron. The exact limits are often indeterminate, but sometimes a **pleurosternal suture** can be identified.

Thoracic Pleura

The lateral wall of a thoracic segment between the notum and sternum (Fig. 2.12a) is the **pleuron.** Sclerites of the pleura are **pleurites.** When the pleurites are unified in a solid plate to provide support and points of articulation for the legs

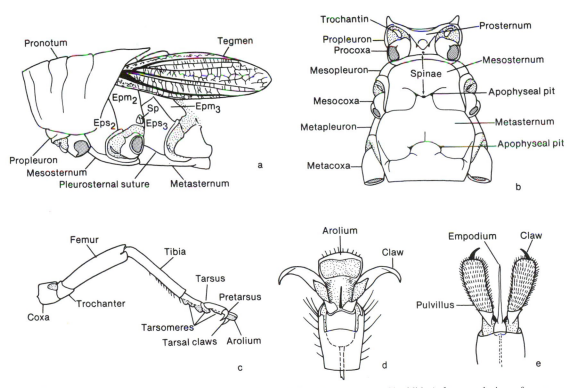

Figure 2.11 Insect thorax and legs: **a**, lateral view of thorax of *Romalea microptera* (Acrididae); **b**, ventral view of same; **c**, foreleg of same; **d**, pretarsus of a roach, *Periplaneta* sp. (Blattidae, Blattodea); **e**, pretarsus of an asilid fly (Asilidae, Diptera), ventral view. Epm, epimeron; Eps, episternum; Sp, spiracle. (d and e, redrawn from R. E. Snodgrass, *The Principles of Insect Morphology*. Copyright © by Ellen Burden and Ruth Roach. Used by permission of the publisher, Cornell University Press.)

Figure 2.12 Diagrams of the pterothorax: **a**, lateral view of a mesothoracic segment; **b**, cross section of a pterothoracic segment, showing separate sternal apophyses; **c**, same, showing apophyses fused medially to form furca, Scl₂, mesoscutellum; Sct₂, mesoscutum. (Redrawn from R. E. Snodgrass, *The Principles of Insect Morphology.* Copyright © by Ellen Burden and Ruth Roach. Used by permission of the publisher, Cornell University Press.)

and wings, the plate is called the **pleuron.** The segmental designations are **propleuron, mesopleuron,** and **metapleuron.**

The pleuron of a pterothoracic segment has a dorsal **pleural wing process,** on which the wing articulates, and a ventral **pleural coxal process,** which provides an articulation for the coxa, or leg base. Extending between the two pleural processes is a strong internal **pleural ridge.** This provides mechanical support for the pleuron. The ridge may be invaginated at one point to form an internal arm, or **pleural apophysis.** The arm is connected by muscles or elastic cuticle to the sternal apophysis, and together the apophyses from the pleuron and sternum form a thoracic endoskeleton. Externally the position of the ridge is traced by the **pleural suture.**

The pleural suture divides the pleuron into two areas, an anterior **episternum** and posterior **epimeron.** Small sclerites beneath the wing base are the **epipleurites.** The epipleurite above the episternum is the **basalare,** and that above the epimeron is the **subalare.** The small crescent-shaped sclerite anterior to the coxa is the **trochantin.**

The propleuron is smaller and simpler in structure than a pterothoracic pleuron. The wing process and epipleurites, of course, are lacking, but the pleural suture may be present.

Modifications of the Thorax

The pterothoracic segments are similar in structure and nearly equal in size in insects that provide muscular power about equally to both pairs of wings. Odonata and Orthoptera, for example, fly with the wings unattached to each other. The pterothoracic segments of insects that fly with the wings coupled together tend to have an enlarged mesothorax and a reduced metathorax to accommodate the dominant role assumed by the forewings in flight. In the Diptera, where the hind wings no longer function in flight, the metathorax is almost absent (Fig. 42.2). An opposite trend is seen among insects that fly mainly with hind wings. The metathorax is thus enlarged in Dermaptera, Phasmatodea, Strepsiptera, and Coleoptera (Fig. 39.1b).

Legs

The paired appendages of the thoracic segments are the legs. Each pair has the same general structure, but modified for different functions in different insect groups. Beginning at the base, the six segments of the leg are **coxa, trochanter, femur, tibia, tarsus,** and **pretarsus** (Fig. 2.11c). Each segment is individually movable by muscles (Fig. 5.11). The tarsus is divided into no more than five subsegments, or **tarsomeres.** The tarsomeres are commonly called "segments," but with the exception of the muscles attaching to the basal tarsomere, they do not have individual muscles. Hence the entire tarsus is considered to be the morphological equivalent of the other leg segments. The pretarsus is the smallest tarsal segment. It is movable by muscles that originate in the tibia and

femur and pass to the pretarsus by a slender apodeme. In the grasshopper the pretarsus is represented by a pair of **ungues,** or **tarsal claws,** and a median **arolium** (as in Fig. 2.11d). In contrast to the arolium, which functions as an adhesive pad on smooth surfaces, the claws grip rough surfaces.

Modifications of Legs

The various named modifications of legs for locomotion include the following: **ambulatory,** or walking, legs (Fig. 39.1b; forelegs and middle legs in Fig. 1.4a); **cursorial,** or running, legs (Fig. 2.13a; Color photo 6a); **fossorial,** or digging, legs (Figs. 2.13e, 9.1); **saltatorial,** or jumping, legs (hind legs in Fig. 1.4a; Figs. 2.13c, 29.3); and **natatorial,** or swimming legs (Figs. 2.13d, 34.10a, 39.5d). Legs modified for grasping are **prehensile** (Figs. 2.13f, 33.4), and for seizing prey, **raptorial** (forelegs in Figs. 2.13b, 23.1a; Color photos 8a–f). The legs of female bees are used to collect and transport pollen from flowers. Worker honey bees and bumble bees have a pollen basket, or **corbiculum,** of long setae on the hind legs (Fig. 46.18c; Color photo 5b). Many other species of bees have a brush, or **scopa,** of plumose setae on the hind legs (Fig. 46.18d). Male bees do not collect pollen and are not so modified. Larvae of advanced Diptera (Fig. 42.5c,d) and Hymenoptera (Fig. 46.4b) lack thoracic legs and are called **apodous.**

The coxa normally articulates dorsally with the pleural coxal process. In many endopterygotes the coxa also has a ventral articulation with the sternum. In Neuroptera, Mecoptera, Trichoptera, Lepidoptera, and Diptera, the coxa has a conspicuous posterior subdivision called the meron. In an extreme development the meron is actually incorporated into the pleural wall of Diptera where it is called the hypopleuron (Fig. 42.18d).

The trochanter is usually a single segment, but in Odonata it is divided into two subsegments. In some parasitic Hymenoptera, the base of the femur is divided into a short subsegment that resembles a second trochanter (Fig. 46.10b). Wasps with the arrangement are commonly said

Figure 2.13 Modifications of insect legs: **a**, cursorial legs of Cicindelidae (Coleptera); **b**, raptorial forelegs of Mantidae (Mantodea); **c**, saltatorial hind legs of Gryllacrididae (Orthoptera); **d**, natatorial legs of Corixidae (Hemiptera); **e**, fossorial legs of Gryllotalpinae (Gryllidae, Orthoptera); **f**, prehensile or cheliform legs of Anoplura (Phthiraptera).

to have two trochanters. Trochanters are missing entirely on one or more pairs of legs in some insects. In the saltatorial hind leg of grasshoppers, for example, the trochanter is fused with the femur. The femur of insects contains major muscles for locomotion and is usually large and thick. In contrast, the tibia is often slender and armed with large movable **tibial spurs** near the apex (Fig. 29.3c).

The number of tarsomeres varies from none to five. The basal tarsomere, often longer than the others, is named the **basitarsus.** The undersurface of the tarsus may be modified to grip smooth surfaces. The tarsi of Orthoptera and some Coleoptera have pads, or **tarsal pulvilli.** Male beetles of the families Gyrinidae and some Dytiscidae even have suckers on the fore tarsi for clinging to the smooth bodies of their mates. The entire tarsus and pretarsus of larval Lepidoptera and Coleoptera (Polyphaga) is represented by a single claw.

The pretarsus of Diptera has a pair of lateral lobes, or pulvilli, under the claws. The arolium is usually absent, but a spinelike median structure, the **empodium,** is present (Fig. 2.11e). The tarsal claws may be reduced to a single claw; for example, in Anoplura or nymphal Ephemeroptera. In the mammal-infesting lice, the single-clawed legs are modified for gripping the hairs of their host (Fig. 33.4c,d). Both claws are reduced in Thysanoptera, but the arolium is much enlarged into an adhesive organ.

Wings

The evolutionary origin of wings is not definitely known; theories are discussed in Chapter 15. During ontogenetic development, the wings grow as saclike expansions of the lateral body walls. The upper and lower layers of the sac partly fuse, leaving a system of narrow, blood-filled channels that are continuous with the body cavity. Tracheae and nerves grow into the channels. At maturity, most of the channels are transformed into wing veins with rigid sclerotized walls that support the fused or membranous portions.

At the base of the wing are small articulatory sclerites (Fig. 2.14a). These include the three **axillary sclerites,** the **humeral plate,** and the **median plate** (mp). The main, or **longitudinal,** wing veins are named as follows, beginning at the anterior or leading edge of the wing: **costa** (C), **subcosta** (Sc), **radius** (R), **media** (M), **cubitus** (Cu), **anal** (A), and **jugal** (J).

The scheme shown in Fig. 2.14b was devel-

oped by J. H. Comstock and J. G. Needham, and is called the Comstock system (Comstock, 1918). Although there is not universal agreement among entomologists on the homologies of the veins from order to order, or even on the basic number of veins (see Hamilton, 1972; Kukalová-Peck, 1977), the Comstock system remains the most widely used scheme. We have adopted it for general use in this text.

The longitudinal veins commonly fork, and the branches are indicated by subscripts. The radius often has a main anterior branch, R_1 and a main posterior branch, the **radial sector** or R_S,

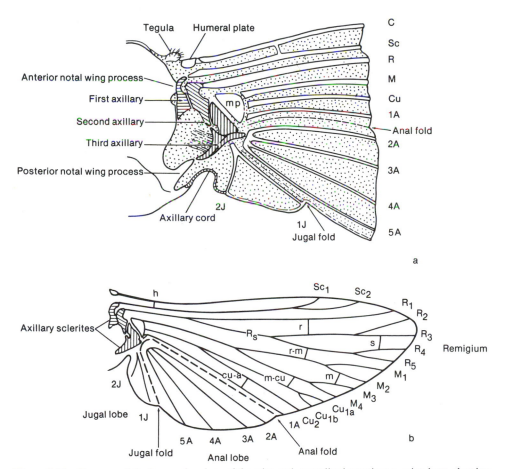

Figure 2.14 Diagrams of the base and regions of the wing and generalized venation: **a,** wing base, showing articulation and bases of major veins; **b,** diagram of venation. See text for explanation of abbreviations. (a, modified from R. E. Snodgrass, *The Principles of Insect Morphology.* Copyright © by Ellen Burden and Ruth Roach. Used by permission of the publisher, Cornell University Press.)

that is further subdivided into R_2, R_3, R_4, and R_5. When two or more veins are thought to have fused together to form a single vein, the names of the veins are joined by a plus sign (see, e.g., Sc + R_1 + R_S in Fig. 44.12b). Some longitudinal veins are joined by crossveins that are usually designated according to the longitudinal veins involved (Fig. 2.14b): **humeral** (h), **radial** (r), **sectorial** (s), **radiomedial** (r-m), **medial** (m), **mediocubital** (m-cu), and **cubitoanal** (cu-a).

The membranous areas enclosed by veins are called cells and are formally named according to the longitudinal vein just anterior to the cell. For example, the veins M_2 and M_3 define a cell, called cell M_2. This may be divided by a median crossvein to form a proximal cell, first M_2, and a distal cell, second M_2. For taxonomic purposes the cells may be informally referred to as **basal, marginal,** or **costal,** and so on (Fig. 46.7d), or be given special names such as **discal** (Fig. 44.10c), depending on the taxonomic group. A **closed** cell is completely surrounded by veins, whereas an **open** cell has one side bordering the wing margin.

The area of the wing anterior to the anal vein and its branches is called the **remigium** (Fig. 2.14b). This is the area, stiffened by veins, that delivers the power of the wing stroke against the air. The anterior edge close to the base of the wing is the **humeral** region. In the area of the anal veins, the wing is more flexible and may be expanded into a fanlike vannal or **anal lobe.** An **anal fold** may separate the anal lobe from the remigium. A smaller lobe at the posterior base of the wing is sometimes present; it is called the **jugal lobe** and is separated from the anal lobe by the jugal fold.

Modifications of Wings

Wings that are transparent are called **membranous** (Fig. 2.15a). Wings may be modified to serve protective functions. Thickened, leathery wings such as the forewings of grasshoppers are called **tegmina** (Figs. 2.11a, 2.15b). The hard forewings of beetles are the **elytra** (Fig. 2.15d). Hemiptera (suborder Heteroptera) have only the bases of the forewings thickened. These are called **hemelytra** (Fig. 2.15c).

Wings may be reduced partly or entirely in size or discarded after use. Wings that are functional but are shed after flight are said to be **deciduous** (compare alates and dealates, Fig. 7.2). Adult insects with short wings are **brachypterous** (Figs. 1.4a, 22.2); those with vestigial wings are **micropterous,** and those that never develop wings are **apterous** (Figs. 25.1, 33.4). The small structures representing the reduced hind wings of flies (Figs. 2.15f, 42.2a,c) and male coccids are **halteres** (Fig. 34.16b). The same term is sometimes used for the reduced forewings of Strepsiptera (Figs. 2.15g, 40.1a).

An evolutionary tendency exists for the two pairs of wings to be coupled together and function as a single unit. The simplest coupling is made by the overlap of the adjacent margins. For example, the jugal lobe of the fore wing overlies the humeral region of the hind wing in Megaloptera (Corydalidae) and many Neuroptera. In the **amplexiform** coupling of butterflies the overlap is much broader (Figs. 2.15e). Among some other Lepidoptera, two kinds of modification are found: in **jugate** coupling, the jugal lobe is fingerlike in certain moths and clasps the hind wing (Fig. 44.3b); in **frenate** coupling of other moths, the humeral region of the hind wing has one to several large bristles, called the **frenula** (Fig. 44.3a). These are engaged by a hooklike flap or a cluster of special bristles called the **retinaculum,** located near the anterior base of the forewing. The hind wings of Hymenoptera (Fig. 2.15i) and some Trichoptera have a row of hook-shaped setae, or **hamuli,** along the leading edge of the hind wing. In the so-called **hamate** coupling, the hamuli engage the folded posterior edge of the forewing. Small hooks also are found in Psocoptera (hooks on the forewing engage the hind wing) and some Homoptera; that is, alate aphids (hooks on the hind wing catch the forewing). Other Homoptera have various lobes and folds along the adjacent wing margins to couple the wings. Heteroptera have a groove under the clavus of the forewing that grips the leading edge of the hind wing.

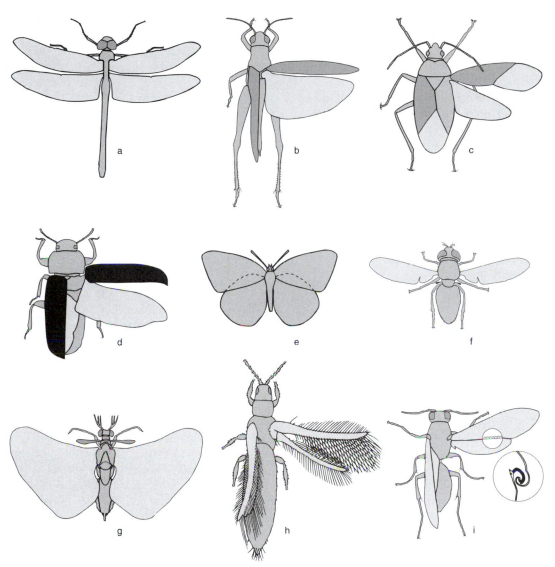

Figure 2.15 Modifications of wings: **a**, the membranous wings are of equal size in Odonata; **b**, the forewing is a tegmen and the hind wing is larger and membranous in Acrididae (Orthoptera); **c**, the forewing is a hemelytron and the hind wing is membranous in Coreidae (Hemiptera); **d**, the forewing is an elytron and the hind wing is larger and membranous in Tenebrionidae (Coleoptera); **e**, both wings are membranous, covered with scales, and overlap (amplexiform coupling) in Pieridae (Lepidoptera); **f**, the forewings are membranous and the hind wings are reduced to halteres (shown in black, between mid and hind legs) in Syrphidae (Diptera); **g**, the forewings are halteres and the hind wings are large and membranous in Stylopidae (Strepsiptera); **h**, both wings are fringed with long setae in minute insects such as Thripidae (Thysanoptera); **i**, both wings are membranous and the larger forewing is coupled to the smaller hind wing by hamuli in Vespidae (Hymenoptera); this is called hamate coupling (in the magnified detail, black hooks on the leading edge of hind wing engage a fold on the trailing edge of the forewing).

The number of veins and crossveins in wings tends to become reduced in the more advanced orders and especially among the more derived members. The homologies of the veins with the basic pattern are sometimes controversial. A standardized scheme has usually been established within an order, but differences in terminology exist between orders.

The drag forces caused by the viscosity of air exceed aerodynamic lift forces on wings of minute insects. Such insects have long hairs along the wings that increase drag-induced friction and allow minute flying insects to literally swim through the air. Examples of such wings are found on thrips (Fig. 2.15h), on certain parasitic wasps of the superfamily Chalcidoidea, and on beetles of the family Ptiliidae.

THE ABDOMEN

The last body division, or abdomen, contains the viscera (Fig. 2.16b). Included are most of the alimentary canal and dorsal circulatory vessel, as well as the Malpighian tubules, fat body, and reproductive organs. The total number of segments in the abdomen remains controversial: arguments have been made that 12 are theoretically possible; yet no more than 11 segments can be counted in the abdomen of adult insects, and frequently only 10 or 9 are clearly evident on close examination. Each of the first 8 segments may have a pair of lateral spiracles. Internally, each of the first 7 segments may each have a ganglion of the ventral nerve cord, but the ganglion of the eighth and the ganglia of the remaining segments are fused together in a single large, compound ganglion (Fig. 6.3).

Paired appendages, homologous to legs, are absent on the first through seventh, or **pregenital,** segments of adult pterygotes. On the eighth and ninth segments of the female and the ninth of the male, the appendages are modified as external organs of reproduction, or **genitalia.** These segments are known as the **genital** segments. The tenth and remaining segments are the **postgenital**

segments. The eleventh may have a pair of lateral appendages called **cerci.**

Except where appendages are present, each abdominal segment typically has the **tergum** and **sternum** as single plates. The plates overlap posteriorly and often exhibit the features of secondary segments. Externally near the anterior margin is the antecostal suture that delimits the anterior acrotergite and traces the internal ridge or antecosta. Muscles attach to the antecosta and to lateral apodemal lobes of the sterna. On the first abdominal segment of the grasshopper is a pair of lateral acoustical organs, or tympana (Figs. 2.16b, 29.2b).

Ovipositor

The paired appendages of the eighth and ninth segments of females may be modified into an appendicular **ovipositor** for depositing eggs (Fig. 2.16c). On segment 8 the sternum may be present, but it is absent in segment 9. The appendicular ovipositor consists of basal **valvifers,** on which muscles insert, and elongate **valvulae.** The eighth segment bears the **first valvifers** and **first valvulae,** and the ninth segment bears the **second valvifers** and the **second** and **third valvulae.** The valvifers correspond to the paired coxae or gonocoxites on each segment. The first and second valvulae correspond to median lobes or gonapophyses of the respective coxae and together may form an interlocking, tubular shaft. The third valvulae correspond to apical lobes of the coxae on the ninth segment and often form a protective sheath for the first and second valvulae. Eggs issue from the genital opening on the eighth segment, pass down a channel in the shaft formed by the first and second valvulae, and are deposited in soil, plant tissues or, in the case of insect parasitoids, into the bodies of insect hosts. The tips of the valvulae have sensory chemoreceptors to aid the female in placing the eggs. In grasshoppers and their allies the ovipositor is differently constructed for digging in soil (Figs. 2.16a, 2.17d). Here the organ is formed by the first and third valvulae. The first valvulae are ventral, the third valvulae are dorsal,

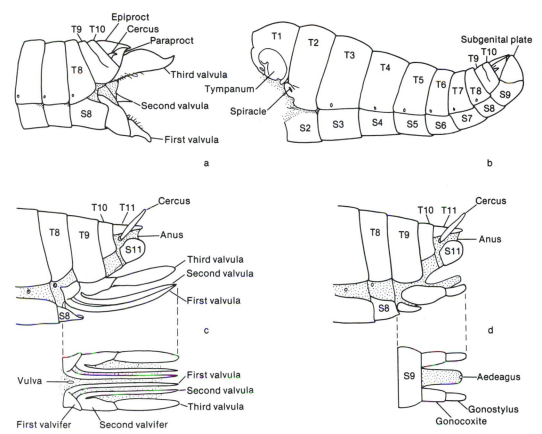

Figure 2.16 Abdomen and external genitalia: **a**, lateral view of female genitalia of *Romalea microptera* (Acrididae); **b**, same of male abdomen and genitalia; **c**, diagram showing general structure of female genitalia; **d**, same of male genitalia, S, sternum; T, tergum. (c, redrawn from R. E. Snodgrass, *The Principles of Insect Morphology*. Copyright © by Ellen Burden and Ruth Roach. Used by permission of the publisher, Cornell University Press.)

and the second valvulae are small, median in position, and concealed by the other valvulae.

In the great majority of insects, however, the appendicular ovipositor is replaced by reduced structures or by special modifications of the tip of the abdomen (Chap. 4).

Male Genitalia

The modified appendages of the ninth segment of males form a copulatory organ (Fig. 2.16d). The basic structure of many insects consists of a median intromittent penis, or **aedeagus,** and lateral

clasping **parameres.** The ninth sternum may be present as the **subgenital plate.** The parameres are commonly two-segmented and are thought by most morphologists to be modified legs. The proximal segment is the **gonocoxite,** and the distal segment, if any, is the **gonostylus.** These sclerotized structures engage the genital opening of the female and maintain the connection during intromission.

The aedeagus is formed by a median fusion of median coxal lobes and is often partly sclerotized. The ejaculatory duct opens apically. During copulation the membranous portion of the aedeagus is inflated by blood pressure, everts into the vagina of the female, and discharges the sperm

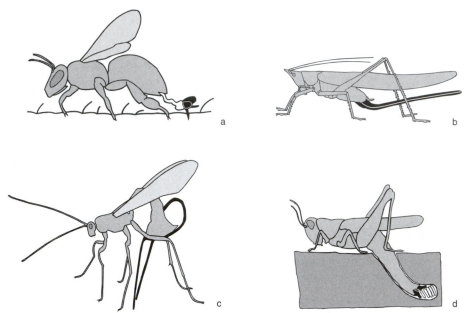

Figure 2.17 Modifications of ovipositors (in black): **a**, the sting of a female honey bee (Apidae, Hymenoptera) is a modified ovipositor with barbs on the valvulae (the barbs stick in the skin of the victim and as the bee struggles, the sting is pulled out of her body); **b**, the long ovipositor of Tettigoniidae (Orthoptera) is used to place eggs into a substrate; **c**, the first and second valvulae of the ovipositor of Ichneumonidae (Hymenoptera) are driven into wood and, if a host larva is detected, the wasp's eggs pass down the hollow shaft to the host (the third valvulae, also long and slender, are pushed to the side and do not enter the wood); **d**, the ovipositor of Acrididae (Orthoptera) is used to dig in soil, where the eggs are laid in rows.

and seminal secretions from the ejaculatory duct. Further discussion of reproduction is found in Chapter 4. The specialized male genitalia of grasshoppers, seen externally, consist of a ventral, cup-shaped subgenital plate (Fig. 2.16b) and a dorsal, membranous pallium.

Postgenital Segments

The segmentation beyond the genitalia may be difficult to determine. The eleventh segment, when present, surrounds the anus and may bear paired appendages or cerci. Sclerotized plates, if any, on the eleventh segment are divided into a dorsal **epiproct** and lateral **paraprocts** below the cerci. In the grasshopper both sexes have the ninth and tenth terga fused and the small cerci situated between the epiproct and paraproct (Fig. 2.16a,b).

Modifications of the Abdomen

The number of visible segments is often reduced in the endopterygotes by the telescoping of the genital and postgenital segments into the apex of the abdomen. In some cases the first segment is reduced or fused with the second, or in advanced Hymenoptera, the first segment is incorporated into the posterior wall of the thorax and known as the **propodeum** (Fig. 46.1). Also in the Hymenoptera, the second and sometimes the third segments are constricted into a **petiole,** leaving the remaining segments as the **gaster** (Fig. 46.10c).

Paired appendages are present on the pregenital segments of Parainsecta, Entognatha, Apterygota, and the immature stages of certain pterygotes. The appendages are clearly serially homologous with legs in the orders Protura, Collembola, Diplura, Archeognatha, and Thysanura.

Those of Archeognatha, for example, have paired, flattened coxae, each within a stylus that represents the distal part of the limb (Fig. 19.1a). The abdominal appendages of pterygotes are much more highly modified and less readily identified as limbs. Examples include the abdominal tracheal gills of nymphal Ephemeroptera (Fig. 20.4b through d), larval Megaloptera (Fig. 36.2), and larval Sisyridae (Neuroptera), and the fleshy abdominal **prolegs** of larval Lepidoptera (Fig. 44.5a,b) and Hymenoptera (Fig. 46.4a). The larvae of Diptera, especially Nematocera, may have a variety of locomotory (prolegs, suction disks, spiny ridges called creeping welts), respiratory (anal gills, tracheal gills), or osmoregulatory (anal papillae) structures on the abdomen.

The appendicular ovipositor is well developed in certain Odonata, Orthoptera, Hemiptera (Suborder Homoptera), Thysanoptera, and Hymenoptera. The appendicular ovipositor is absent in many other insects (e.g., Coleoptera, Diptera, or Lepidoptera) but is replaced in these orders by special modifications of the terminal abdominal segments.

The size and shape of the ovipositor are closely correlated with the nature of the medium into which the eggs are inserted. Insects that deposit their eggs in living plant tissues usually have stout valvulae with sawlike ridges. Parasitic wasps may use long, slender valvulae to insert eggs into insect hosts that are in burrows deep in solid wood (Figs. 2.17c, 46.3a, 46.10a).

The ovipositors of predatory wasps, ants, and bees are modified into an organ for defense, the **sting** (Figs. 2.17a, 46.2d). Thus, only females can sting. The second valvulae are fused to form a stiff piercing organ, with the first valvulae locked beneath in sliding grooves. Glands associated with the ovipositor release toxic venom into the wound created by the valvulae. In the stinging Hymenoptera, the ovipositor no longer functions to deposit eggs; instead, eggs are released from the vaginal opening on the eighth segment.

Worker honey bees have tiny hooks at the tip of the second valvulae (Fig. 2.17a). Once the sting has been plunged into the skin of a person, the barbs prevent the sting from being withdrawn. The bee struggles to pull away, but it loses its sting, plus the poison glands and attached viscera. The

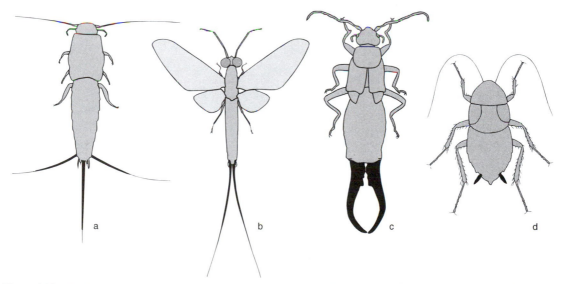

Figure 2.18 Modifications of cerci (in black): **a**, Thysanura have paired, filamentous cerci plus a single median caudal filament; **b**, Ephemeroptera also have filamentous cerci and some species (not shown) have a median caudal filament; **c**, the cerci of Dermaptera are heavy pincers; **d**, the small cerci of Blattodea have sense organs that are able to detect the movement of air near the insect.

bee flies off, only to die shortly thereafter. In the person who has been stung, the muscles of the sting continue to contract, and venom continues to be injected. The best first aid is to scrape the sting off with a finger nail rather than pinch the stinger, which will only force more venom into the wound.

The male genitalia of insects are so greatly varied that they are used by insect taxonomists to classify species. Special terminologies have been applied in some orders in which the structures are complex and the homologies uncertain. Male Odonata have the sterna of segments 2 and 3 modified as a secondary genitalia for the transmission of semen to the female (Fig. 21.2b,c). Male Ephemeroptera have paired intromittent organs on the seventh segment These correspond to the paired openings of the lateral oviducts on the seventh abdominal segment of the female. Some Dermaptera have paired ejaculatory ducts instead of a single duct. The genitalia are especially complicated in Diptera, where in some groups the terminal abdominal segments and genitalia actually undergo a clockwise rotation of 90°, 180°, or even 360° during pupal development.

The cerci are appendages of the eleventh abdominal segment. They may be long, slender, and subsegmented as in Thysanura (Fig. 2.18a), Arch-

eognatha, and Ephemeroptera (Fig. 2.18b); shorter and with fewer subsegments as in Blattodea (Fig. 2.18d); or one-segmented as in Orthoptera (Fig. 29.3). The cerci of Dermaptera (Fig. 2.18c) and Diplura (Japygidae) (Fig. 18.1e) are stout and forcepslike. Cerci are absent in the Hemipteroidea. The epiproct is greatly elongated into a cercuslike, **median caudal filament** in Thysanura (Fig. 2.18a), Archeognatha (Fig. 19.1), and Ephemeroptera (Fig. 20.1a). In nymphal Odonata (suborder Zygoptera) the epiproct and paraprocts are greatly enlarged into tracheal gill plates (Fig. 21.3b).

SELECTED REFERENCES

General References on Insect Anatomy and Morphology

Chapman, 1982; CSIRO, 1991; Hamilton, 1971, 1972 (wings); Kukalová-Peck, 1991 (anatomy of fossils); Matsuda, 1965 (head), 1970 (thorax); Scudder, 1971 (genitalia); Snodgrass, 1935, 1952, 1958, 1963; Wootton, 1992 (wings).

Glossaries of Terms

Nichols, 1989 (general); Tuxen, 1970a (genitalia).

3
THE INTEGUMENT

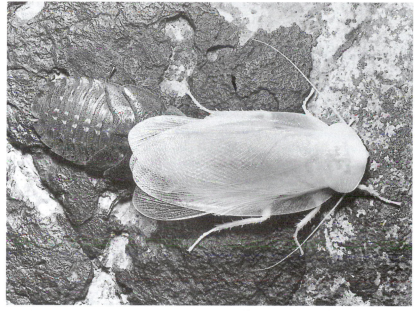

This cockroach has just completed the final ecdysis (cast skin of the nymph at the left). It will soon darken as its cuticle sclerotizes and melanin pigments are deposited.

The organ most suitable as a starting point for a microscopic study of the insect body is the integument. It is physically the largest of all organ systems and exhibits the greatest diversity in structure and in function. The integument has three basic components: a tough, outer, noncellular **cuticle**; a single layer of cells, or **epidermis**; and an inner sheet of connective tissue, or **basement membrane** (Figs. 2.1, 3.1a).

THE EXOSKELETON

The most easily seen portion of the integument is the hardened and jointed cuticle that covers the outside of the body and forms an external skeleton, or **exoskeleton,** as described in detail in the preceding chapter. Our classification of insects is based mainly on the appearance of the exoskeleton. The special arrangements of the hairs and

51

Figure 3.1 Structure of the integument and the molting cycle. See text for explanation.

other features provide nearly all the diagnostic characters for the classification of insects down to the level of species. This means that easily a million or more distinctive exoskeleton designs exist. We are indeed fortunate that the exoskeleton is so simply preserved and that so much information concerning insect habits can be gained from it.

The exoskeleton is extremely resistant to decomposition. When protected from pests, dried specimens of insects may outlast the pins and cabinets of the museum, usually outlasting the collectors and curators as well. When preserved as fossils, pieces of the exoskeleton may remain intact. The chemical component chitin, for example, has been detected in the fossil of an extinct eurypterid arthropod from the Silurian Period, over 409 million years ago.

In addition to functioning in the support of internal organs and as attachments for muscles, the exoskeleton is involved in such important activities as feeding, locomotion, and reproduction. The intricate shapes of the mouthparts permit the exploitation of a wide variety of foods. Similarly, walking and flying depend in large degree on the strength and flexibility of the exoskeleton. In reproduction, parts of the exoskeleton provide the means by which the male may transmit sperm to the female and the female, in turn, may deposit her eggs in special places.

The exoskeleton also serves many of the same

functions as the skin of other animals. It is the main barrier to the loss of water, a problem that is severe in small terrestrial organisms with a relatively large evaporative surface area in proportion to the volume of their bodies. In addition, the exoskeleton may prevent the entrance of disease organisms and the penetration of chemicals such as insecticides. Assaults by predators are not only resisted by the tough outer covering, but also avoided by camouflage or warning coloration that is created by pigments in the integument. The sense organs of the nervous system are intimately connected with specialized areas of the exoskeleton because all the stimuli from the environment must be transmitted through the exoskeleton to the receptors beneath.

The exoskeleton determines the form and maximum size of the insect. At intervals during growth, parts of the cuticle are digested and a new, larger cuticle is deposited. This process is termed the **molt,** and the actual shedding of the old cuticle is called the **ecdysis** (Color photo 2a). At times of molting, a large proportion of the chemical components of the old exoskeleton is digested and used again. For this reason, the exoskeleton is actually a kind of food reserve.

The integument includes not only the conspicuous exoskeleton but also thin-walled, tubular structures hidden from view inside the body. These structures have the same basic components: cuticle, epidermis, and basement membrane. The tracheal system, functioning in respiration, is created by invaginations of the integument. The ultimate subdivisions of these tubular ducts number literally in the millions and supply oxygen to virtually every cell of the body. The elimination of carbon dioxide is accomplished through the tracheae and also through the general body surface. In the alimentary canal, the foregut and hindgut are formed by invaginations of the integument during development. The integument also lines the ducts leading from the internal reproductive organs, providing not only passageways and storage vessels for the gametes, but also glandular secretions that function in their protection and transmission. Finally, a rich variety of other glands, including the salivary glands, is derived from cells of the epidermis. Their products function in nearly every phase of the insect's life history by supplying enzymes for digestion, materials for shelter and support, and other chemical compounds for attracting the sexes, communication, defense, and offense. Among these integumental secretions, human beings have found some of great value (e.g., silk, shellac, and beeswax).

The Origin of the Integument

Of the integument's three major components, only the epidermis is cellular. In the embryo, the epidermis is derived from the ectoderm, as described in Chapter 4. The cuticle is periodically secreted by the epidermis, both daily as additional thin layers and at longer intervals when the cuticle is replaced throughout nearly all the body during molting. Certain parts of the cuticle are thought to be secreted by oenocytes (cells of ectodermal origin in the blood). The basement membrane on the inner surface of the epidermis is believed to be secreted by certain hemocytes (blood cells) in the hemolymph and possibly by other cells of the body.

THE CUTICLE

The discussion in this section focuses on the cuticle as it exists in the exoskeleton. The cuticle may equal up to half the dry weight of the insect's body, and it always exists external to the epidermis. When thin sections are made and viewed microscopically, it is seen to be made up of layers. These can be shown to have special chemical and physical properties and to be secreted by the various cells of the epidermis. Two major divisions of the cuticle are the thin outer **epicuticle,** no more than about 4 μm thick and composed of proteins and lipids, but with chitin rarely detected, and the inner **procuticle** (combined **exocuticle** and **endocuticle**), which may measure a millimeter or more in thickness and contains protein and chitin. Each division may have sublayers, depending on the species of insect and the location of the cuticle on the insect's body.

Epicuticle

The epicuticle is created by secretions from several different sources that are deposited outside the procuticle. Four layers are currently recognized: the outermost **cement layer,** the **wax or lipid layer,** the **outer layer of epicuticle,** and the **inner layer of epicuticle.** The cement layer is derived from certain modified epidermal cells, called **dermal glands,** which empty their secretions on the outer surface of the epicuticle by means of gland ducts through the procuticle. The cement protects the critical wax layer from being scratched or lost by absorption onto foreign objects. The copious secretions of the lac insect, *Kerria lacca* (Coccoidea, Hemiptera), yield commercial shellac, a high-quality, clear wood finish. The secretions are probably quite similar in origin to the cement produced normally and in much smaller quantities on the body surfaces of other insects.

The lipids for the wax layer are produced by epidermal cells and must pass through the cuticle to reach the outer surface. The most common route is thought to be by means of the pore canals (see below, "Epidermis") of the procuticle and the wax canals of the epicuticle. Thin **wax filaments,** about 6 nm in diameter, extend through the canals from the epidermal cells to the surface. The filaments, possibly of tanned protein, are believed to function in the transport of the lipids. The wax secretions vary in chemical composition but are mainly mixtures of hydrocarbons, alcohols, and fatty acids. Depending on the composition and temperature, the physical properties of the wax vary: the greasy surface of the cockroach and the firm honeycombs of beeswax in the beehive are both products of the wax-producing cells of the epidermis.

The wax layer is of crucial importance in preventing the loss of water from an insect in a dry atmosphere. Even thin layers of lipids are sufficient to provide protection. Much interest now centers on the molecular organization of these thin films in relation to their impermeability. Wax may mix with the cement layer or appear on the outside surface as a powdery wax bloom or a greasy film. In such cases a water-repellent or hydrophobic surface is created which has important consequences for the insect in contact with water. For example, a small insect can be supported by the surface tension of water rather than trapped by it. The wax layer also is important in the resistance to invasion by pathogens such as fungi and is probably the main barrier to the penetration of some insecticides. Evidence is accumulating that the various mixtures of compounds that make up the wax layer can serve other functions, namely, as pheromones in sexual behavior and as recognition odors among social insects.

The names of the remaining layers of the epicuticle continue to change as our technology allows more detailed studies. The "outer layer" of the epicuticle is beneath the wax layer and is always present in insect cuticle. This layer was formerly called the cuticulin layer or protein epicuticle. The outer layer is composed of tanned lipoproteins (lipoproteins stabilized by molecular cross-links). It is the first layer of a new cuticle to be formed by the epidermis during the molting cycle. The layer is only slightly thicker than a cell membrane but may function as a critical barrier between the new and old cuticles. In addition to the wax canals that pass through it, the outer layer has small pores that may filter the digestion products of the old cuticle. Once in place, the

outer layer may determine the maximum expansion of the cuticle, that is possible before another molt.

The "inner layer" of the epicuticle, formerly included with the outer or cuticulin layer, now is recognized as distinct in some insects. Much thicker than the outer layer, it is penetrated by wax canals and also probably composed of tanned lipoproteins.

Procuticle

"Procuticle" (meaning early cuticle) specifically refers to the soft, chitin-rich cuticle secreted by the epidermis early in the molting cycle. The term also is used generally to refer to the second major division of cuticle in contrast to the epicuticle. Later in the molting cycle the procuticle becomes differentiated into two layers, an outer hardened and often darker **exocuticle** and an inner, flexible **endocuticle,** which is usually lighter in color. These layers are of much the same composition, namely, long molecules of proteins and chitin.

Various proteins exist in the procuticle (the water-soluble proteins are collectively called **arthropodin**) and may contribute more than half the dry weight. The proteins may differ from instar to instar and from place to place on the insect. For example, one protein may be associated with rigid cuticle and another with flexible cuticle. Among the latter, a rubberlike protein called **resilin** (from the Latin *resilire,* to jump back) may occur in special places in almost pure form. Resilin confers an exceptional elasticity to certain wing articulations and muscle insertions.

The second major component of the procuticle, chitin, is thought to be widely distributed among organisms, being reported in fungi, diatoms, sponges, hydroids, bryozoans, brachipods, mollusks, annelids, and arthropods. Chitin may be up to 60% of the total dry weight of an insect's cuticle. Biosynthesis begins with the sugar trehalose and ends with chitin, a nitrogenous polysaccharide of *N*-acetylglucosamine plus some glucosamine. The long molecules of chitin resemble cellulose, a polysaccharide widespread in the supportive structures of plants. Like cellulose, chitin is one of the most abundant organic materials in the world and new commercial uses are being developed. Pure chitin has been described as a colorless, amorphous solid, insoluble in water, dilute acids, dilute alkalies, alcohol, and all organic solvents. It may be dissolved in concentrated mineral acids, but it is rapidly degraded.

The process of hardening, called **sclerotization,** involves the tanning or cross-linking of the proteins by the action of sclerotizing agents (one or both of the compounds *N*-acetyldopamine and *N*-β-alanyldopamine) into a stable, three-dimensional molecular structure. The sclerotizing agents are derived from tyrosine or phenylalanine, amino acids that are essential to the diet of insects. Proteins modified by the agents are called **sclerotins.** The hardened zone becomes the exocuticle and the general area where the exocuticle occurs in the integument becomes a discrete, rigid sclerite in the body wall. The unsclerotized inner part of the procuticle remains flexible and is the endocuticle. At joints, or **arthrodial membranes** (Fig. 2.1), the exocuticle is absent or broken into small blocks, and the endocuticle predominates. This permits bending between sclerites. Elastic properties may be conferred to hinges by the presence of resilin.

The exocuticle is extremely resistant to digestion, both by the insect itself during molting and by other organisms, like pathogens, equipped with hydrolytic enzymes. The chitin chains are also cross-linked together and bound to the tanned proteins, thus becoming part of the three-dimensional network. However, the tanned proteins rather than the chitin are responsible for the special hardness of the exocuticle. For this reason a hard piece of the exoskeleton is correctly termed "sclerotized" rather than "chitinous" because soft cuticle also has chitin. Hard exocuticles are usually dark, partly from sclerotin and partly from pigments like melanin that are deposited in the cuticle. Hard exocuticle is not necessarily dark: albino mutants of insects may have hard cuticles lacking dark pigments.

Within an insect's cuticle, the chitin molecules are laid down in parallel **microfibers,** which, in turn, are arranged in thin layers or **lamellae.** The orientation of the microfibers shifts during the process and produces a banded pattern when the cuticle is viewed in cross section. Some lamellae are deposited at times of molting, and additional lamellae may be added each day as distinct **growth layers.** The characteristics of the growth layers are influenced by temperature and by the relative lengths of day and night.

Sclerotization occurs elsewhere in insect life histories where strong, light-weight, flexible structures are produced: eggshells, oothecae, puparia, cocoons, and silk. The same biochemical process is widespread among the lower animals.

Although calcium is present in the cuticles of crabs and other crustaceans, it is rarely incorporated in the cuticle of insects. When this element is present in insects, calcification does not modify the hardness of the sclerotized parts. By geological standards, some sclerotized cuticles can scratch the mineral calcite. On a scale of 10 ranked by increasing hardness, these cuticles would be placed third; that is, in hardness they are between a thumbnail and a copper penny. In California, the beetle *Scobicia declivis* (Bostrichidae, Coleoptera) easily chews through the lead sheathing of aerial telephone cables, earning it the name "lead-cable borer." Other insects can penetrate thin sheets of tin, copper, aluminum, zinc, or silver. Sclerotized cuticle is, however, not indestructible, and its surface together with the epicuticle may show extensive abrasion and wear, especially in burrowing insects. Wear is also found on the cutting and grinding surfaces of the mandibles of chewing insects. Renewal of the cuticle at each molt may serve not only to promote increases in body size, but also to restore the valuable properties of the cuticle.

Because the production and health of the integument is essential for the survival of an insect, opportunities exist for the development of pesticides that disrupt normal processes. Abrasive and absorptive materials may be used to destroy the wax layer. Such desiccant dusts may be essentially inert and nontoxic, yet able to provide long-term pest control for insects like roaches. The biosynthesis of chitin in insects and fungi is inhibited by benzoylphenyl ureas such as diflubenzuron, known also as the pesticide dimilin. The biosynthetic pathways leading to sclerotization also can be disrupted by metabolic inhibitors, or the process can be abnormally accelerated with a toxic effect. Cryomazine is an insecticide whose mode of action is still unknown, but it acts on feeding larvae to alter the cuticle.

THE EPIDERMIS

Epidermal cells are organized in a single layer with an apical-basal polarity. Thus the cells have an apical surface, next to the cuticle, bearing microvilli for the secretion of the cuticle. Between the cells are intercellular spaces and at the base the cells are next to the basement membrane. Once secretion of the new cuticle is under way, the physical routes by which the epidermal cells pass their secretions to the procuticle and the epicuticle are the **pore canals** (Fig. 3.1). These are minute channels, 1.0 to 0.15 μm in diameter, extending from the surface of the epidermal cell to the interface between the epicuticle and the procuticle. They are not lined with the plasma membrane of the cell, but wax filaments pass from the cell through the pore canal. The density of canals may be quite high, ranging from 15,000 to 1,200,000 per square millimeter. The canals are not straight but seem either to be coiled about an axis perpendicular to the cuticular surface or to curve with the microfibers in each cuticular lamella.

THE BASEMENT MEMBRANE

A thin layer of connective tissue covers all the internal organs and the internal surfaces of the integument. So situated, the membrane separates the various organs from the hemolymph and must regulate the flow of molecules to and from the organs. Consequently, the idea that the cells of the insect body are directly bathed by the blood in an "open circulatory system" is incorrect. When they lie beneath the epidermis, the connective tissues are called the basement membrane and consist of an amorphous granular layer up to 1 μm in thickness. In some cases a fibrous structure is apparent in insect connective tissues. In certain chemical and physical properties, these fibers resemble collagen of the vertebrate connective tissues.

THE PROCESS OF MOLTING

The integument of insects undergoes cyclic activities throughout the life of the insect. The events that trigger the molting cycle and the hormonal control of molting are treated in Chapter 4. The most conspicuous changes are associated with the periodic shedding of the old cuticle. The growing insect is restricted in size by the maximum area to which the cuticle will expand. The new cuticle that is laid down beneath the old is highly wrinkled, thus providing potentially greater area for expansion after the molt. The major growth in the size of the exoskeleton is therefore limited to times of molting and is discontinuous. Changes in the form of the exoskeleton at metamorphosis and in patterns of setae, wound healing, and the regeneration of lost parts are likewise limited to the molt. In addition to these changes in morphologic characteristics, the valuable physiological properties of the cuticle that may become weakened by age and wear must be renewed, and the molt serves this purpose as well. Between molts, the epidermis continues to be active in secreting additional layers of endocuticle and in wax production.

Apolysis

To begin our discussion of the molting cycle, our attention is directed to the interface between the epidermis and cuticle. This is an area of exceptional physiological activity. Molting is initiated by ecdysteroid hormones in the hemolymph. The onset is signaled by **apolysis,** or the detachment of the epidermis from the old cuticle. Just prior to the detachment, several preliminary events occur. The epidermis changes from a thin layer of polygonal cells to a thick columnar layer. Mitotic figures, indicating cell divisions, may also be seen at this time. The detachment may possibly be created by a retraction of the cells or by a change in the consistency of the cuticle immediately next to the epidermis. Muscle attachments and connections of the nervous system to sensory receptors in the cuticle remain in place so that their functions continue. The space created between the old cuticle and epidermis is termed the **exuvial space.** Into this space is deposited the **molting fluid** from the epidermal cells. The fluid may exist in this space for some time before becoming active (Fig. 3.1c).

New Cuticle

The secretion of the new cuticle begins with the outer layer of the epicuticle. This first layer is believed to be critical in determining the surface features and area of the new cuticle. Then the inner layer of epicuticle is deposited, followed by the procuticle. Once the protein epicuticle is in place, the molting fluid becomes active. The main function of the fluid is the digestion of the endocuticle. Proteases and chitinases dissolve up to 90 percent, leaving only the resistant epicuticle and exocuticle. In some insects a resistant part of the old endocuticle survives as an ecdysial membrane. Amino acids and sugars, the products of digestion, are probably absorbed through small pores in the outer layer of epicuticle. The products are recycled into the new procuticle by the epidermis. As the old endocuticle is digested, the new procuticle increases in thickness. The proteins destined for the procuticle are produced by the epidermal cells or produced elsewhere in the body and transported to the epidermis by the hemolymph. The lamellae, characteristic of the procuticle, probably are the result of cyclic secretory activities of the epidermal cells. The pore canals are initiated with the first layer and become progressively longer with increasing thickness of the procuticle. By means of

these canals, the cells remain in communication with the epicuticle/procuticle interface.

Sclerotizing of the outer zone of the procuticle to form the exocuticle may begin now in certain areas of the insect's body; in other areas, it may be delayed until after ecdysis. Tanning and hardening of the cuticle take place from the outside inward, by means of sclerotizing agents or their precursors, which are produced in the fat body or hemocytes and transported by the hemolymph to the epidermal cells. The epidermis also is capable of producing the sclerotizing agents from precursors. The sclerotizing agents then diffuse outward from the epidermis through the pore canals to the surface of the cuticle. The sclerotizing agents become activated by enzymes that catalyze oxidation and diffuse inward, transforming the protein and chitin into a rigid polymerized structure, and creating the exocuticle. Just before the old cuticle (now called the **exuviae**) is shed, a layer of wax is deposited on the surface of the new epicuticle. The molecular mechanisms by which a hydrophobic substance such as wax may pass through the hydrophilic procuticle are not known. The pore canals probably provide the exit passageway, but wax is secreted on some cuticles lacking these canals. Finally, a cement layer is deposited over the wax layer by the dermal glands.

Ecdysis

The insect is now ready to shed the exuviae by the process of **ecdysis.** This behavioral feat requires coordinated movements of the body in a definite sequence. The behavioral acts are programmed into the central nervous system and initiated by the eclosion and Mas-ETH hormones in the hemolymph (Chap. 4). By taking in air or water in the gut, the insect increases its body volume, and this, together with muscular contractions, splits the old cuticle along preformed lines of weakness. Certain areas of the cuticle, even within the boundaries of sclerites, have the exocuticle greatly reduced. When the endocuticle has been dissolved by the molting fluid, only the epicuticle remains along these lines. Such **ecdysial lines** (Fig. 2.1) provide the initial lines for breakage of the old cuticle. A lubricant may be demonstrated on the inner surface of the exuviae and probably aids in the escape of the emerging insect. It is mucilaginous and may be derived from the molting fluid.

Once freed of the exuviae, the cuticle is expanded to the limits predetermined by the epicuticle. Blood pressure is increased by the swallowing of air or water and by contractions of various parts of the body. Certain regions of the cuticle are plastic and expand under this pressure. By this means a dramatic change in body size and shape takes place over a relatively brief time. This is especially marked at the last molt, when the insect's wings are inflated to full size. After expansion, the cuticle rapidly hardens and darkens. Sclerotization is under control of the hormone bursicon (Chap. 4). When the molting cycle is over, additional layers are added to the endocuticle by the epidermis, and wax may continue to be produced.

THE PRODUCTION OF COLOR

Beautiful colors, or even the general darkening of the cuticle, are created by processes that may be quite apart from the events leading to hardening (Color photos 3a–f). In some wild bee species, for example, certain areas of the cuticle may be perfectly transparent and reveal the pale colors of the tissues beneath. Yet they are as rigid as the adjacent black cuticle. Except for special mutant forms or some subterranean species, all insects exhibit colors. The functions of these colors in survival are treated in Chapter 12 under protective coloration. Here color is discussed from three points of view: (1) the anatomical location of the color-producing structure, (2) the manner in which the color is generated, and (3) changes in color during the life of the insect. With regard to the first topic, most colors are produced in the cuticle. When the cuticle is translucent or transparent, the underlying epidermal cells, hemolymph, or other internal tissues may produce color. Both the cuticle

and the deeper tissues may contribute to the production of a single color pattern.

White light is composed of various wavelengths of light, commonly identified as red, orange, yellow, green, blue, indigo, and violet. When an object or insect absorbs all but the red portion of the spectrum, we perceive a red color, and so on. The selective reflection of certain colors may be due to the physical structure of the object, producing **structural colors**; or it may be the result of selective absorption by certain chemical compounds, producing **pigmentary** colors; or a color may be produced by a combination of structure and pigments together. When the cuticle contains the pigment or physical structure for color, the colors often remain lifelike in museum specimens. On the other hand, when the color resides in the cells or hemolymph beneath the cuticle, one may expect fading and discoloration as the internal organs decompose after death.

Structural Colors

To produce a structural color, white light is altered in one of three ways: by the interference, by diffraction, or by the scattering of light waves. Interference colors are familiar as the iridescence of a thin film of oil on water or a soap bubble. Both the upper and lower surfaces of the film reflect light. The thickness of the film determines the distance these reflecting surfaces are separated. Since light reflected from the lower surface must travel a minute distance farther to reach the eye of the observer than the light from the upper surface, the former is said to be retarded. When waves are retarded by an odd number of half-wavelengths, they will be out of phase with the waves of the same length reflected from the upper surface. In this case the two reflections cancel each other, and the only visible part of the spectrum will be the one with the waves in phase. Colors vary because of the varying thickness of the film. Changes in the angle of observation will also change the thickness traversed by the retarded waves and likewise produce iridescence.

Interference colors are widespread among the insects. Least conspicuous, but extremely common, is the play of iridescent colors on the transparent wings of insects. The membranes of the wings have the same effect as thin films. More spectacular are the butterflies and moths whose individual scales are intricately modified to produce striking colors. The blue of the *Morpho* (Nymphalidae, Lepidoptera) arises from a series of vertical vanes on each scale. Each vane has a series of delicate ribs spaced at the correct distance for canceling or transmitting without reflection all wavelengths but the blue which is reflected. A series of superimposed laminae on each scale of the diurnal moth *Urania* (Uraniidae, Geometroidea, Lepidoptera; Color photo 3f) has a similar effect. Lamellate scales or lamellae within the cuticle itself may produce interference phenomena in certain metallic beetles (e.g., tiger beetles). When underlain by dark pigments, the interference colors are especially sharp, since stray reflections from internal organs do not detract from the purity of the reflected colors.

Color produced by diffraction is rare in insects. The beetle *Serica sericea* (Scarabaeidae, Coleoptera), produces color by a parallel series of minute grooves, or striae, that are evenly spaced, about 0.8 μm apart. When viewed in direct sunlight, these produce an iridescence on the elytra, which function as a diffraction grating, operating by reflection. An artificial impression of the close-spaced striae, made of plastic and peeled from the elytron, will exhibit the same iridescence.

The last type of structural color, that produced by scattering, is not iridescent and does not change with the angle of observation. Minute particles, less than 0.6 nm in diameter and reflecting more of the short waves of light than long waves, produce a blue color. The sky is blue, for example, because dust particles and atmospheric gases reflect more blue and violet than other colors. Without the impurities in the atmosphere to reflect these colors, the sky would appear black. Such a color is called **Tyndall blue** after an early student of sky color. Among the damselflies, granules in the epidermal cells produce a blue color (Color photo 3e). Granules in cells in combination with yellow pigments will give green. Waxy secretions on the surface of the cuticle in other species of dragonflies also give the Tyndall effect. Larger granules reflect all wavelengths of light, producing a structural white. White human hair is white because of minute air bubbles in the hair that scatter light. Depending on the arrangement of the reflecting surfaces, a dull, pearly, or silvery white can be produced in insects.

Pigmentary Colors

The second major category, pigmentary colors, includes a much wider range of colors and a diverse assortment of chemical compounds serving as pigments. These colors are not metallic or iridescent, but otherwise vary enormously. They are often excretory products that are no longer involved in biochemical pathways but now serve an important function in the ecology of the insect. White and yellow colors of butterflies like the pierids are of this type; the pterine pigments responsible for the colors may be derived from uric acid (Color photo 3c). Pigments may also be absorbed from the food of the insect.

One of the most common of animal pigments is melanin. The colors produced vary from black and brown to light brown and yellow (Color photo 3a). The synthesis of each color pigment is incompletely known. Nearly all colors in this range were formerly attributed to melanin. Some are now known to be the result of other chemical compounds, such as sclerotin or ommochromes (Color photo 3b). Melanogenesis of the black pigment involves the oxidation of tyrosine by the enzymatic action of tyrosinase, plus several additional steps, one of which requires air by way of the tracheal system. Once formed, melanin is insoluble in ordinary solvents, but it can be bleached if a transparent exoskeleton is desired for study.

Tyrosine and tyrosinase are present in insect blood but somehow kept apart. When an insect is injured, the blood that is exposed to air often blackens. Such melanization in experimental insects is usually fatal but may be prevented by excluding oxygen from the surgery or by blocking the action of tyrosinase. The hereditary lack of tyrosinase produces the familiar albino mutant.

Melanogenesis may be partly influenced by the environment. For example, insects reared in the laboratory at lower temperatures often show increased melanization. In nature, an insect with two or more generations per year often displays distinctive colors associated with the seasons during which pigments were developed. The interaction of the environment and melanogenesis is not easily unraveled. The presence of darker populations of an insect in a geographic region of lower temperatures cannot be simply explained as entirely an environmental effect. Such individuals may be genetically predisposed to greater melanin deposition, and the darker colors may have survival advantages peculiar to the geographic region, irrespective of the lower temperatures. High humidity and darker colors have also been associated in the study of geographic variation of insects, but oddly there is little evidence for a direct physiological influence. Some environmental factors have indirect effects through the nervous or endocrine systems: certain wavelengths of light received by the eyes of an insect in the pupal stage may promote melanin deposition or, as in the migratory phase locusts, the darker gregarious phase is associated with crowding.

Yellow, orange, and red colors are produced by carotenoid pigments derived from α- and, more commonly, β-carotene, which occurs in plants. Insects are believed to be unable to synthesize these pigments and therefore must depend on plants for their supply. Some predators acquire their pigment via plant-feeding victims (Color photo 3d). The pigments are absorbed and deposited in the epidermis or fat cells or in secretions such as silk or wax.

Green colors that closely resemble the chlorophyll of leaves are produced by insects of many kinds (Color photo 8d). Thus far no experiments have shown that chlorophyll is absorbed and used

directly for the color. In several different insects, however, the green is produced by a combination of a yellow carotenoid and a blue or blue-green pigment belonging to a group known as a bilin or bile pigment. The origin of the latter is not clear, but it is probably synthesized by the insect. Curiously, a diet of fresh green leaves seems to be important for the synthesis, but it is thought that some component other than chlorophyll is the needed constituent.

The color patterns of insects are usually fixed for the life of the instar. An exception is the walking stick, *Carausius morosus* (Phasmatodea), in which the color changes daily under neuroendocrine control: dark at night and pale in day. This physiological color change is the result of dark granules in the epidermis. When the granules are clustered together at the base of the cell, the insect is pale; when the granules migrate apically to form a sheet, the insect is dark. Darkening can also be stimulated by low temperature or high humidity.

Finally, pigments produced by some insects are of commercial value. The red pigment carmine or cochineal, used in lipstick and rouge, is extracted from female *Dactylopius coccus* (Coccoidea, Hemiptera), a scale insect that feeds on cactus. Kermes, a crimson color and also from a scale insect, *Kermes* sp. (Coccoidea), was used as a dye in ancient Greece and Rome.

SELECTED REFERENCES

Integument

Chapman, 1982; numerous authors in the volumes edited by Binnington and Retnakaran, 1991; Kerkut and Gilbert, 1985.

Color

Fox and Vevers 1960 (general); Cromartie, 1959 (pigments); Fuzeau-Braesch, 1972 (pigments); Rowell, 1971 (color changes).

4
CONTINUITY OF THE GENERATIONS: DEVELOPMENT AND REPRODUCTION

During part of their annual cycle, aphid "stem mothers" (Aphididae) may reproduce by viviparous parthenogenesis. Here an apterous female aphid on a rosebud gives birth to a female nymph. The other young aphids are also her progeny.

This chapter traces the major events in the life history of insects. We begin with the egg as it leaves the female and follow development through the larval stages to sexual maturity. Then we return to the reproduction system, describe the structure and function of the male and female organs and mating (Fig. 4.1), discuss different types of reproduction, and conclude with the polymorphic life histories of cynipid wasps, aphids, and locusts.

EGGS AND EMBRYONIC DEVELOPMENT

The majority of insects are diploid animals that reproduce sexually and are **oviparous** (i.e., they lay eggs). Haploid eggs and sperm are produced by meiosis. The eggs of the class Insecta and the order Diplura are **centrolecithal,** or rich in yolk. The early cleavage, or division, of the nuclei in the fertilized egg does not involve division of the

Figure 4.1 The male (above) and female (below) of mantids, *Stagomantis californica* (Mantidae), during copulation. Direct transfer and storage of sperm are important adaptations that permit insects to reproduce in terrestrial environments, including deserts.

large yolk mass into separate cells. This is known as **meroblastic** or **superficial** cleavage. The eggs of Collembola, like those of the myriapods, are **microlecithal** (have less yolk), and the early cleavage is **holoblastic**; that is, the egg is completely divided. The embryology of eggs of Protura has not been studied.

Egg Structure

The eggs of insects are usually elongate and ovoid, with rounded ends (Fig. 4.2a; Color photos 1b, 1c), but other shapes also occur. Eggs of exopterygotes tend to be slightly larger than those of endopterygotes, but some of the largest eggs are those of large carpenter bees, *Xylocopa* (Anthophoridae, Hymenoptera), which measure as much as 16.5×3.0 mm. From the outset, eggs are bilaterally symmetrical and have a dorsal and a ventral surface.

The outer shell, or **chorion,** is secreted by the follicle cells of the ovary. Passages are provided for the sperm to enter, for respiration and, in some, for water absorption. At the anterior end is the **micropyle** of one or more holes through the chorion that admit sperm. **Aeropyles** are pores of one to a few micrometers that permit the entry of oxygen and the exit of carbon dioxide, with rather little loss of water. In eggs that are laid in water

or moist places such as decomposing organic matter or in areas that are temporarily submerged, the chorion is often modified into a **plastron** (Chap. 5). This is a spongy meshwork that retains a film of gas and resists wetting. The gas film operates as a physical gill, permitting oxygen dissolved in water to pass into the gas and then into the embryo (Fig. 5.6a,b). Special areas called **hydropyles** are sometimes present on terrestrial eggs to absorb water.

Beneath the chorion is the **vitelline membrane,** which in most insects is secreted by the **oocyte,** or egg cell. Most of the substance of the egg is **yolk,** consisting of lipid, protein, and carbohydrate. A yolk-free island of cytoplasm surrounds the female nucleus, which is usually located peripherally in an anterodorsal position. A thin yolk-free **periplasm** often surrounds the egg beneath the vitelline membrane.

Fertilization

Apparently in all hexapods sperm are stored by the female and some are released to fertilize the eggs as they pass out the genital tract. Compared to the sperm wastage of other animals, this means of ensuring fertilization is highly efficient and allows fertilization to continue, as in honey bees and some ants, even years after mating. When the

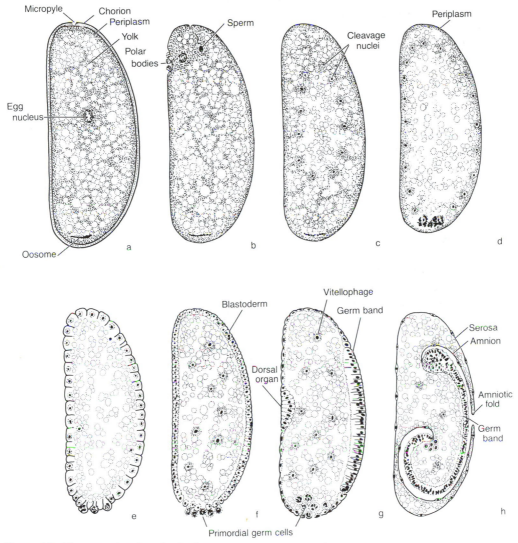

Figure 4.2 Diagrams of embryonic development: **a**, egg structure before fertilization; **b**, entrance of sperm; **c**, cleavage of zygote; **d**, migration of cleavage nuclei; **e**, beginning of blastoderm formation; **f**, blastoderm complete; **g**, formation of germ band; **h**, elongation of germ band. (Redrawn from Johannsen and Butt, 1941.)

egg is laid by the female, it is usually in the metaphase of the first maturation division of meiosis. During the exit of the egg from the female's reproductive tract, sperm are released near the micropyles, and one to several enter the egg (Fig. 4.2b). The penetration of more than one sperm, known as **polyspermy,** is unusual among animals but common among insects. The fruit fly *Drosoph-*

ila (Drosophilidae, Diptera) is an exception; only one sperm enters the egg.

After penetration of the egg by the sperm, the egg nucleus completes meiosis, resulting in two or three polar bodies and the haploid female pronucleus. The latter migrates toward the interior of the egg, and the polar bodies remain grouped at the surface of the oocyte. The female pronucleus unites with one sperm pronucleus to produce the diploid zygote. The other sperm degenerate.

Formation of the Blastoderm

The early cleavages, or mitotic divisions, of the zygote result in nuclei, each of which is surrounded by an island of cytoplasm but without a cell membrane (Fig. 4.2c). The cleavage nuclei spread through the yolk, emerging into the periplasm and covering the yolk surface (Fig. 4.2d) as a **syncytial blastoderm.** In *Drosophila* at this stage, the nuclei are totipotent (each capable of developing into a whole organism). Development up to now has been under control of gene products that originated in the female parent (i.e., were maternally encoded and stored in the egg). The basic spatial organization of the embryo (anterior-posterior and dorsal-ventral axes), DNA replication, nuclear assembly, and nuclear migration up to this point have all been supported by maternal products.

After cell membranes have surrounded each nucleus, the resulting cells form the **cellular blastoderm,** one cell thick (Fig. 4.2e,f). In *Drosophila,* once the cells have formed, zygotic genes begin transcription and dominate control of development. Depending on their position in the embryo, embryonic cells now begin to express different sets of genes to develop into different tissues. The effects of maternal products can still be detected: for example, in the development of the nervous system. Some cells remain in the yolk or return to it as **vitellophages,** which digest yolk for the embryo.

Formation of the Germ Band and Germ Layers

Along the midventral line, the blastoderm thickens into a single layer of columnar cells producing the **germ anlage,** while the remainder of the blastoderm becomes a thin membrane enclosing the yolk (Fig. 4.2g). Also at this time, a cluster of cells migrates through a special region at the posterior pole of the egg, called the **oosome,** becoming the **primordial germ cells** (the sex cells). These cells can usually be traced through development and will give rise to eggs and sperm in the adult insects.

The membranous portion of the blastoderm will develop into the serosa and amnion that will cover the yolk and fold over to enclose the developing germ band (Fig. 4.2h). The germ anlage next develops a longitudinal gastral groove, corresponding roughly to gastrulation in other animals, and becomes the **germ band.** The band differentiates into two **germ layers,** the outer **ectoderm** and inner **mesoderm.**

At this point it is useful to distinguish three major categories of insect embryos based on the relative length and development of the germ band. **Short germ band** insects have a band a third or less of the length of the egg. At first the band consists of only the protocephalon (preantennal and antennal segments) plus a small region that will gradually elongate by cell proliferation into the rest of the body. **Long germ band** insects have a band that extends over the length of the egg. All the segments of the body appear almost at once as the germ band develops by subdivision of the anlage. Short germ band insects tend to develop more slowly and are usually Exopterygota with panoistic ovaries (see "Types of ovary," below). Such ovaries lack nurse cells, which probably means that the oocytes synthesize essential components. Long germ band insects tend to develop more quickly and are usually Endopterygota with meroistic ovaries. In this case nurse cells supply the oocyte with large amounts of maternal products that may serve to allow rapid development. *Drosophila* is a long germ band insect. Most insects, however, belong to the third or **intermediate germ band** category in which the band is from one-third to half the length of the egg. Otherwise difficult to characterize, this category includes a mixture of species in the Pterygota.

When part of the ectoderm becomes the epidermis, transverse divisions into segments appear. Research on *Drosophila,* a highly specialized insect, has revealed an unexpected and confusing aspect of segmentation. The primary segments of insects that have been identified by comparative morphology (see Figs. 2.2, 2.4) are not the visible "segments" of the fly's embryonic epidermis. One would have expected that the primary segments could be traced back to the embryo. However, in the embryo of the fly, the developmental unit defined by transverse grooves is called a **parasegment** and consists of the posterior part of one morphological segment and the anterior part of the following segment. The embryo initially has 14 parasegments, corresponding approximately to 3 for the mouthparts, 3 for the thorax, and 8 for the abdomen. For example, parasegment 6 will develop into the posterior part of the metathorax and the anterior part of the first abdominal segment.

Returning now to a generalized description of embryonic development, we note that the ectoderm differentiates into the epidermis of the integument, tracheal system, endocrine and exocrine glands, sense organs, and the nervous system. Paired appendage rudiments become visible on the epidermis of the anterior segments for insects in which the larvae have well-developed mouthparts and legs (*Drosophila* larvae have highly reduced mouthparts and no legs). At this stage, the insect embryo consists of a strip of tissues representing the ventral part of the body and is open to the yolk above. As the appendages lengthen, the ectoderm grows dorsally, finally meeting along the middorsal line and closing the embryo with the yolk now inside the body. Each segment (parasegment of *Drosophila*) initially has a bilateral cavity in the mesoderm that corresponds to the **coelom** of other animals. The coelom later breaks down as the mesoderm differentiates into muscles, heart and dorsal diaphragm, gonads (including the primordial germ cells), fat body, and blood cells. Invaginations appear at the anterior and posterior ends of the band of ectoderm. These blind sacs elongate to become the foregut and hindgut, respectively. Near each invagination is a mass of cells that probably corresponds to the **endoderm** of other animals. The cells will grow inward from each of the gut invaginations, forming a tubular sheet

around the yolk. This tube becomes the midgut. The Malpighian tubules develop from the blind end of the invaginated hindgut. In this position it would appear that the tubules arise from ectoderm, but the true embryonic origin has not been determined. Near the time of hatching from the egg, the blind ends of the foregut and hindgut break down to become continuous with the tubular midgut, thus forming the complete alimentary canal. In the tobacco hornworm moth, *Manduca sexta* (Sphingidae, Lepidoptera), the embryonic epidermis undergoes two molting cycles before the larva hatches from the egg.

POSTEMBRYONIC DEVELOPMENT

Embryonic development ends and **postembryonic** development begins when the insect hatches from the eggshell. All hexapods except Protura emerge from the egg with a body that has the adult number of segments. This is called **epimorphic** development, in contrast to the **anamorphic** development of Protura, in which segments are added after hatching. Although Collembola are epimorphic, they have a reduced number of body segments in comparison with other hexapods. As explained in Chapter 3, the integument undergoes repeated molting cycles during the life of an insect. The number of molts may be as low as 4 or 5 or may exceed 30. Only at times of molting can the exoskeleton be increased in size or altered structurally. Other organ systems are less clearly synchronized with molting, but changes in their size and organization are under the same hormonal controls. Therefore, the postembryonic development of an insect is conveniently marked by the ecdysis at the end of each molt. The form of the insect between ecdyses is called an **instar** and is designated by numbers. For example, the first instar is the insect after hatching from the egg and before the first ecdysis.

Arthropods, amphibians, and certain other animals may have a distinctive, actively feeding, sexually immature stage that is generally called a **larva**. The change in appearance from the larva to the sexually mature adult, or **imago,** is broadly termed **metamorphosis.** When the insect reaches sexual maturity, molting ceases in the Pterygota, but continues in certain of the Apterygota, Para-insecta, and Entognatha.

We note here several words that may cause confusion. The term **eclosion** has been used for the act of hatching from the egg, for the emergence of the adult from the pupal cuticle at ecdysis, and for the emergence of any instar from the old cuticle at ecdysis. Among entomologists, the designation "immature insects," or "immatures," refers to the larval instars of insects, not to sexually immature adult insects. Further, some entomologists use the word "larva" for all immature instars that are active, while others restrict the term to the immature instars of holometabolous insects. The larvae of terrestrial, hemimetabolous exopterygotes are commonly called **nymphs.** Finally, the distinctive aquatic larvae of Odonata, Ephemeroptera, and Plecoptera are often called **naiads.**

Types of Metamorphosis

Among the insects, three general types of metamorphic development can be recognized, based on the degree to which the larva differs from the imaginal form. In the most extreme metamorphosis, the tissues of virtually the entire larva are destroyed and replaced by imaginal tissues.

Ametabolous development literally means "without change." The life history typically has the egg, several larval instars, and several imaginal instars (Fig. 4.3a–d). The living hexapods exhibiting this type of growth are all primitively apterous: Collembola, Diplura, Thysanura, and Archeognatha. The larvae change little except in size and proportions as they grow from instar to instar and mature sexually. Upon reaching adulthood, insects in these orders continue to molt an indeterminate number of times until death.

Hemimetabolous, or incomplete, metamorphosis involves a partial change in appearance from the larva to imago. The life history typically has the egg, several larval instars, and the final winged and sexually mature imago (Fig. 4.3e–i; Color photos 2a, 2b). Hemimetabolous metamorphosis is characteristic of Paleoptera and nearly all Exopterygota (including female Coccidae).

In hemimetabolous insects, the most conspicuous change in the imago is the development of functional wings and external genitalia. The larvae usually possess well-developed exoskeletons with legs, mouthparts, antennae, compound eyes, and, except in Hemipteroidea, ocelli and cerci. These structures are carried over without much change into the imago. The transition in form is gradual up to the last molt; each successive larval instar increasingly resembles the form of the imago. Wings develop as external pads on the pterothoracic nota and then expand enormously at the last molt (Fig. 4.3e–i).

Most Exopterygota have larvae and adults that live in the same places and eat the same food (Color photo 2b). Except for the wings and genitalia, the alterations at maturity are usually slight. The larvae of Odonata (Fig. 21.3), Ephemeroptera (Fig. 20.4; Color photo 2f), and Plecoptera (Fig. 27.1b), however, are aquatic, and the adults are aerial. Consequently the change is marked, involving the loss of gills and modification of the mouthparts as well as the acquisition of wings and new body proportions.

All living pterygotes differ from the living ametabolous orders in that when sexual maturity has been reached, molting ceases. Furthermore, functional wings are restricted to the imago stage of all living pterygotes except the Ephemeroptera. In this order the first winged instar, or **subimago,** is sexually immature except in a few species, where it is sexually mature. After a brief aerial existence, the subimago molts and the second winged instar is the sexually active imago.

Holometabolous, or complete, metamorphosis is characterized by larvae that undergo histological reorganization before becoming sexually mature adults. The life history typically has an egg, several larval instars, one pupal instar, and the final winged and sexually mature imago (Fig. 4.4; Color photos 1a–f, 2c–e). The **pupa** is an inactive, nonfeeding instar during which metamorphosis

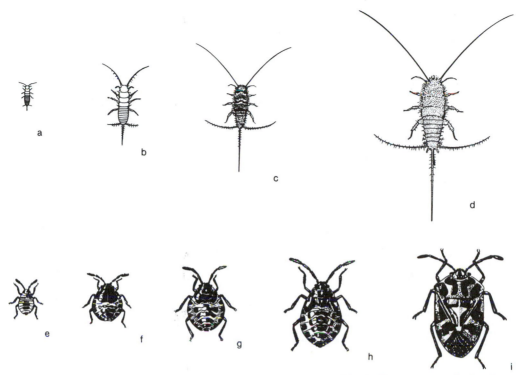

Figure 4.3   Types of life history: **a–d**, ametabolous development, illustrated by the firebrat, *Lepismodes inquilinus* (Lepismatidae) (**a–c**, nymphs; **d**, imago); **e–i**, hemimetabolous development, illustrated by the harlequin bug, *Murgantia histrionica* (Pentatomidae) (**e–h**, nymphs; **i**, imago). (a–d, redrawn from Metcalf et al., 1962. Reproduced with permission of The McGraw-Hill Companies. e–i, redrawn from U.S. Department of Agriculture, Farmers Bulletin 1061.)

transforms the larva into the adult form. In contrast to the hemimetabolous exopterygotes, the holometabolous endopterygotes delay the development of wings until the last one to three larval instars. At this time the wings develop externally while the body undergoes reorganization. In most holometabolous insects, the wings do not become visible until the pupa. All Endopterygota and certain exceptional Exopterygota (Thysanoptera, Fig. 35.2; Aleyrodidae, Fig. 34.14c-e; male Coccidae, Fig. 34.16b) have holometabolous development.

The last larval instar of a holometabolous insect usually seeks a pupation site and often secretes a silken cocoon or constructs a special cell of earth or wood particles. The cocoon or cell probably functions to reduce loss of moisture, provide protection from ice crystals in winter, and make possible the avoidance of predators. Certain

specialized endopterygotes do not construct pupal shelters. For example, cyclorrhaphous Diptera pupate within the sclerotized last larval cuticle, called a puparium.

The molt into the pupal instar is called **pupation.** During this time, after apolysis and before ecdysis, the insect is called a **pharate pupa.** Metamorphosis begins and the larval tissues are partly or entirely destroyed by histolysis and replaced by growth of imaginal tissues. The latter are derived from clusters of cells, called **imaginal disks** or anlagen, that are distributed in the larval body near where the adult organs develop. The imaginal disks become visible early in the embryo before hatching from the egg, but remain inactive during the larval instars.

In larval *Drosophila*, the cells of the imaginal disks remain diploid while larval cells become

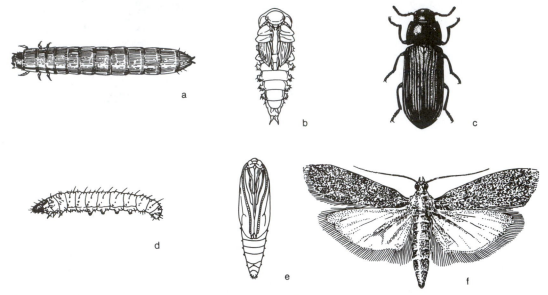

Figure 4.4 Holometabolous life histories: **a–c**, the mealworm, *Tenebrio molitor* (Tenebrionidae) (**a**, eruciform larva; **b**, adecticous exarate pupa; **c**, imago); **d–f**, the Mediterranean flour moth, *Ephestia kuhniella* (Pyralidae) (**d**, eruciform larva; **e**, obtect pupa; **f**, imago). (Redrawn from California Agricultural Experimental Station Bulletin 676.)

polyploid and generally do not divide, although the larva grows larger. The fly larva has pairs of imaginal disks that will develop into adult mouthparts, antennae and compound eyes, thorax, legs, wings, and halteres. A single disk develops into the genitalia. Other clusters of imaginal cells are scattered in the brain, along the gut, along the salivary glands, and in the abdominal segments. At metamorphosis the disks and gonads are activated and begin to divide and differentiate into adult organs.

After ecdysis, the pupa emerges. For most insects one can see that the features of the larva have been replaced by imaginal structures such as long antennae, compound eyes, ocelli, mouthparts, legs, and miniature wings. Pregenital abdominal appendages of the larva, if any, are now missing and rudimentary genitalia are present. The pupa has a special pupal cuticle, which differs from that of the larva and the adult.

With apolysis and the onset of the final molt, the insect is called a **pharate adult.** The wings grow enormously in size within the wing sacs of the detached pupal cuticle. Internally, the alimentary canal and nervous system are reorganized, the growth of the endoskeleton apophyses is completed, and the musculature assumes its final distribution and attachment. Setae, wing scales, and pigmentation characteristic of the imago are completed. Following the final ecdysis, the body usually contracts in volume as the wings are inflated by hemolymph and the cuticle is sclerotized. At this time a fluid substance, called the **meconium,** is sometimes excreted from the alimentary canal.

HORMONAL CONTROL OF MOLTING, ECDYSIS, METAMORPHOSIS, AND REPRODUCTION

Postembryonic development, sexual maturation, and some aspects of behavior and physiological functions are regulated by **hormones.** These are substances secreted by one part of the body that have a profound effect on another part of the body.

Endocrine cells or organs produce hormones that may be stored and released by specialized **neurohemal** structures. Our understanding of insect endocrinology is constantly being revised as new details are revealed by research taking advantage of modern biochemical, immunocytochemical, and molecular techniques. Earlier models involved single hormones from single glands and with single targets. We now recognize that groups of related hormones, some from more than one source, may have different targets depending on the stage of the insect's life history. The hormones specifically involved in development are listed below (boldface type), together with other hormones from the same organs (Fig. 6.2). Abbreviations are provided for organs and hormones commonly mentioned in the scientific literature.

1. Several paired clusters of **neurosecretory cells** (NSC) in the brain produce at least two groups of peptide hormones: Prothoracicotropic hormones and eclosion hormones. **Prothoracicotropic hormones** (PTTH) exist in different molecular weights and may prove to be produced in different clusters of NSC. PTTH was known first simply as the "brain" hormone that stimulates the prothoracic glands to synthesize ecdysteroids, hence the name. PTTH also probably has other functions in the brain and even in embryonic development. In Lepidoptera, nerves pass from the NSC to the **retrocerebral complex,** consisting of the corpora cardiaca and corpora allata. PTTH is released in the hemolymph by neurohemal endings of the secretory cells located in the corpora allata. In some other insects, the corpora cardiaca are reported to release the hormone. **Eclosion hormones** (EH) are produced by NSC in the brain and the proctodeal region of the ventral nervous system. These hormones exist in different molecular forms. EH is released just before ecdysis by neurohemal sites of the proctodeal nerves. At this time ecdysteroids are at a low level. The eclosion hormones belong to a series of hormones that ultimately trigger the central nervous system to begin the programmed sequence of muscular movements that comprise ecdysis behavior.

2. **Neurosecretory cells** in the brain or ventral ganglia secrete the peptide hormone **bursicon,** which initiates sclerotization following ecdysis. Bursicon is released by neurohemal organs in the abdomen. NSC in the brain or ventral ganglia also secrete peptide diuretic and antidiuretic hormones that control water balance.

3. Paired **corpora cardiaca** (CC) are located behind the brain and connected to NSC in the brain. The CC are closely associated with the corpora allata in the retrocerebral complex. As neurohemal organs, the CC are reported to store and release PTTH to the hemolymph in some insects. Hormones are also produced that regulate blood sugar (trehalose) concentration, lipid mobilization in blood (adipokinetic hormone), heartbeat, and Malpighian tubule movement.

4. Paired **prothoracic glands** (PG), or ecdysial glands, are located in the thorax or head. Upon stimulation by PTTH in the hemolymph, the PG secrete the hormone **ecdysone** (E), which is synthesized from cholesterol or a related steroid obtained in the diet. In moths, E is converted to the more active hormone **20-hydroxyecdysone** (ecdysterone or 20-E) by the fat body, gut, and Malpighian tubules. Collectively, these hormones, plus other forms that have been identified, are called **ecdysteroids.** The ecdysteroids initiate molting in the epidermis and stimulate differentiation of imaginal tissues.

5. Paired **corpora allata** (CA) are located behind the brain in the retrocerebral complex. CA are innervated by nerves from the brain that pass through the CC and also by nerves from the subesophageal ganglion. The CA secrete sesquiterpenoid compounds called **juvenile hormones** (JH). Several structures are known and designated JH 0 to JH III, plus others. Hemiptera have JH I; Diptera have a *bis*oxide of JH III, Lepidoptera have up to five structural forms in a single species; the other orders appear to have only JH III. The hormones serve to maintain metabolism for growth, preserve larval characteristics during the molting cycle, and inhibit metamorphosis. In the adult the hormones

stimulate vitellogenesis in the female and accessory glands in the male.

6. **Epitracheal glands** are associated with each spiracle of *Manduca sexta*. The gland is a cluster of cells, including a large cell that produces a peptide hormone ("*Manduca sexta* ecdysis-triggering hormone" or **Mas-ETH**). When released in the hemolymph, Mas-ETH directly acts on the ventral nervous system to initiate the programmed ecdysial behavior. Analogues of this hormone have not yet been reported in other insects.

Hormonal Control of Molting

The endocrine organs are ultimately under the control of the central nervous system. Depending on the kind of insect involved, certain events serve to stimulate molting. External stimuli such as a pattern of changing temperature or light, or internal cues such as the stretching of the abdomen after feeding, are received by the brain, which then causes PTTH to be released. PTTH stimulates secretory activity in the prothoracic glands, and ecdysteroids increase in the body. This in turn stimulates the epidermis to begin a molting cycle, starting with apolysis. The targets of the ecdysteroids are genes in epidermal cells. Minutes after ecdysteroids have been injected into the midge, *Chironomus* (Chironomidae, Diptera), chromosomes in the salivary glands exhibit puffs indicating RNA synthesis.

Hormonal Control of Ecdysis

The final step of the molting cycle, ecdysis, is accompanied by a stereotyped pattern of muscular movements programmed into the ventral nerve cord. The behavior is currently thought to be controlled by a sequence of hormonal interactions. In *Manduca sexta,* the eclosion hormone is released into the hemolymph when 20-hydroxyecdysone declines. In turn, the eclosion hormone stimulates the epitracheal glands to release the peptide Mas-

ETH. The last hormone acts directly on the nervous system to initiate the ecdysial behavior. Sclerotization or tanning of the new cuticle is stimulated by release of bursicon from neurohemal organs.

Hormonal Control of Imaginal Differentiation

Molting and differentiation of imaginal tissues are stimulated by ecdysteroids. Juvenile hormone released from the corpora allata inhibits differentiation as long as the hormone is present in the hemolymph of the larva. The exact means of control of the corpora allata is not understood. Nervous connections or neurosecretions (called allatotropins) probably serve to regulate the secretion of juvenile hormones. When molting is stimulated by ecdysteroids in a young larva, sufficient juvenile hormone is normally present to cause the cells to retain larval characteristics. The larva increases in body size in the next instar without metamorphosis.

In hemimetabolous insects, juvenile hormone disappears in the last larval instar, and the insect differentiates into the adult at the next or final molt. Holometabolous insects are similar, but the process of metamorphosis is more complex. In the last larval instar, juvenile hormone drops to an undetectable level followed by secretion of ecdysteroid, which commits the insect to metamorphosis into the pupal instar.

Metamorphosis of Holometabolous Insects

The steps in the hormonal control of holometabolous metamorphosis have been determined in detail in *Manduca sexta* (Fig. 4.5; Color photos 2c, 2d). After ecdysis to the fifth larval instar, juvenile hormone drops to undetectable levels. Then two small pulses of ecdysteroids, called the "commitment pulse," are released, initiating metamorphosis and triggering stereotyped behavior. The larva stops feeding, wanders about, becomes positively geotactic, burrows in soil, and constructs a chamber in

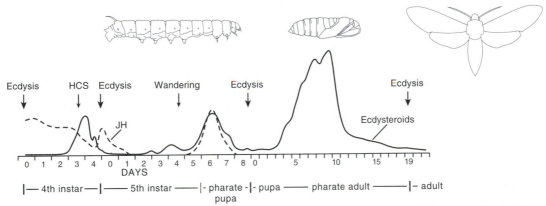

Figure 4.5 Titers (concentrations of substances determined by titration) of ecdysteroid (solid line) and juvenile hormone (dashed line) during late larval life and metamorphosis of *Manduca sexta* (Sphingidae). Above the graph are illustrations of a fifth instar larva, a pupa, and an adult. The times of ecdysis are marked with large arrows. Head capsule slippage (HCS) is the separation of the old cuticle of the head from the epidermis. This event, which can be seen in larvae reared in the laboratory, marks the beginning of apolysis in the fourth-instar larva. The fourth-instar larva then molts and after ecdysis becomes the fifth-instar larva. The molt of the fifth-instar larva occurs after a period of crawling about or wandering in preparation for pupation. Following apolysis, the pharate pupa develops inside the cuticle of the fifth-instar larva. At ecdysis, the insect becomes the pupa. Likewise, after apolysis in the pupa, the pharate adult develops inside the pupal cuticle. The adult emerges from the pupal cuticle after the final ecdysis. (Modified from Goldsmith and Wilkins, 1995. Reprinted with the permission of Cambridge University Press.)

which to pupate. The commitment pulse is followed by much larger pulse of ecdysteroid, called the "prepupal peak." The second release of ecdysteroids initiates apolysis and the molt. A pulse of juvenile hormone is also released at the same time and presumably regulates the differentiation of imaginal disks. The pharate pupa, while still hidden in the last larval cuticle, has begun metamorphosis. The epidermis secretes the distinctive pupal cuticle. Ecdysis occurs after the ecdysteroids and juvenile hormones have again declined. The dark brown pupa exhibits some of the external features of the adult, although the appendages are glued to the body (see discussion of obtect type of pupa in "Characteristics and Types of Endopterygote Pupa," below). Ecdysteroids again are secreted, now accumulating to high titers in the hemolymph as the differentiation of imaginal tissues continues. Juvenile hormone is absent. The final apolysis detaches the pupal cuticle, and the epidermis begins to secrete the adult cuticle. The pharate adult, concealed in the dark pupal cuticle, completes development as the ecdysteroids decline to low levels.

The final ecdysis, as in the preceding ecdyses, is controlled by the eclosion hormone followed by the Mas-ETH hormone. After burrowing out of the soil, the moth perches and begins to pump hemolymph into the wings. The hormone bursicon is now released from neurohemal organs in the abdomen, causing the cuticle of the wings to become plastic during their expansion and later to harden by sclerotization.

Hormones and Reproduction

In adult pterygote insects, the prothoracic glands degenerate and, although ecdysteroids are produced elsewhere, the integumental epidermis no longer responds by molting. After regulating the growth and molting of the immature instars, juvenile hormones and ecdysteroids assume new roles in the maturation and function of the adult reproductive organs. The corpora allata renew secretion of juvenile hormone and ovarian tissues secrete 20-hydroxyecdysone. These hormones now stimulate organs involved in reproduction; that is, they become **gonadotropic.**

In female insects, both hormones have been associated with the synthesis of yolk proteins in the fat body and deposition of yolk in developing eggs. Both hormones may act in vitellogenesis in the same species of insect, or only one of the hormones may be present. Juvenile hormone, alone or in combination, is the most common. Juvenile hormone is also responsible for development of the ovarian follicles and preparation of the oocyte to receive yolk. Ecdysteroids are present in embyos.

In males, the production of fluids by the genital accessory glands is dependent on stimulation by juvenile hormone. The males of the moth *Hyalophora cecropia* (Saturniidae, Lepidoptera) have an extraordinary amount of juvenile hormone stored in the abdomen, but the function remains unknown. There is some evidence that neurosecretory peptide hormones from the brain also may be involved in the function of adult female and male organs.

Ecdysteroids in male pupae stimulate the development of the accessory glands and testes and the differentiation of the germ cells. For example, in certain male Lepidoptera spermatogenesis begins in the pupa. Here ecdysteroids stimulate the testes, control spermatogenesis, and inhibit the release of sperm.

In the ametabolous Thysanura, sexual maturation occurs gradually. Once adulthood has been reached, the neurosecretory cells of the brain, ecdysial glands, and corpora allata remain active. Molting cycles, initiated by the brain, occur regularly. If females are mated within a few days after ecdysis, the corpora allata become active and yolk is deposited in the eggs. If the females are not mated, the eggs are resorbed at the onset of the next molting cycle. Thus molting and reproduction alternate in their demands on the body reserves.

CHARACTERISTICS OF ENDOPTERYGOTE LARVAE

In addition to having concealed wing buds, endopterygote larvae differ in other respects both from hemimetabolous larvae and from their own imaginal instars. Megaloptera (Fig. 36.2) and Raphidi-

optera (Fig. 37.1b) exhibit the least difference between larvae and imagoes, while in Diptera the headless, legless maggots are totally unlike adult flies (Fig. 42.5). In general, larvae differ from imagoes in being soft-bodied and in having simplified sensory, feeding, and locomotory structures. The larval antennae are reduced to but a few segments in all endopterygotes except beetles of the family Scirtidae (= Helodidae). Compound eyes are present in some larval Mecoptera (Fig. 41.1c), but in other larvae the light receptors consist of lateral clusters of single-faceted stemmata (Fig. 16.1b). The curious berrylike eyes of Strepsiptera are sometimes called compound eyes, but are stemmata. Ocelli are absent in larvae, except possibly in certain Mecoptera (Fig. 41.1). The mouthparts and legs are often reduced in structural complexity. Legs are absent in larval Diptera (Fig. 42.5c) and Hymenoptera (suborder Apocrita; Figs. 4.6, 46.4b). Pregenital appendages are present on the abdomen as gills in Megaloptera (Fig. 36.2) and Sisyridae (Neuroptera), or as fleshy prolegs in Lepidoptera (Fig. 44.5a,b) and Hymenoptera (suborder Symphyta; Fig. 46.4a).

Types of Endopterygote Larva

Three general types of larva have been recognized: campodeiform, eruciform, and vermiform. **Campodeiform** larvae have prognathous mouthparts, elongate flattened bodies, long legs, and usually caudal appendages of some kind. Larvae with these characteristics resemble the Campodeidae (Fig. 18.1a) of Diplura. They are usually active predators. Examples are found in Neuroptera (Fig. 38.1b), Megaloptera (Fig. 36.2), Raphidioptera (Fig. 37.1b), Trichoptera (Rhyacophilidae, Fig. 45.2a), and many Coleoptera (Dytiscidae, Fig. 39.5b; Carabidae; Staphylinidae).

Eruciform larvae have hypognathous mouthparts, cylindrical bodies, short thoracic legs and abdominal prolegs, and reduced or no caudal appendages (Fig. 4.4a,d). Examples are caterpillars of Lepidoptera (Fig. 44.5a,b; Color photos 2c,7d) and larvae of Mecoptera (Fig. 41.1c) and Hyme-

Figure 4.6 Queen cells at the bottom edge of comb in a honey bee hive (Apidae). The pupa (left) and fully grown larva (right) are both females that have been fed a diet of royal jelly. Both will develop into queens. The pupa is of the exarate type and the larva is vermiform, without appendages.

noptera (suborder Symphyta, Fig. 46.4a). Larvae of this type are rather inactive, and most feed on plant materials.

Vermiform larvae are wormlike and cylindrical, and they lack locomotory appendages. Most larvae or maggots of Diptera (Fig. 42.5; Color photo 6c) are of this type, as are the larvae of Siphonaptera (Fig. 43.1d), certain wood-boring beetles (Fig. 39.11c), and aculeate Hymenoptera (Figs. 4.6, 46.4b). Larvae of other types are described under the various orders in Part Three.

Hypermetamorphosis

In some insects the larvae pass through two or more instars that differ markedly in appearance. The first instar is often campodeiform and actively seeks food or a host. The subsequent instars are almost always relatively inactive and grublike. Examples are commonly parasites or parasitoids and are found in Neuroptera (Mantispidae), Coleoptera (certain species of Carabidae and Staphylinidae; Meloidae and Rhipiphoridae), Strepsiptera, Diptera (Bombyliidae, Acroceridae, Nemestrinidae, and some Tachinidae and Calliphoridae), Hymenoptera (certain Chalcidoidea), and certain ectoparasitic Lepidoptera (Cyclotornidae and Epipyropidae). When the first instar has legs, as in Coleoptera, Mantispidae, or Strepsiptera, it

is called a **triungulin** (Fig. 40.1 b,c). The legless **planidium** larvae of Diptera and Hymenoptera use long "walking" setae or jumping movements for locomotion (Fig. 46.5).

CHARACTERISTICS AND TYPES OF ENDOPTERYGOTE PUPAE

Pupae with movable mandibles are called **decticous.** Such pupae are found in Megaloptera, Neuroptera, Mecoptera, Raphidioptera, Trichoptera, and certain Lepidoptera (suborders Zeugloptera and Dacnonypha). The pupal mandibles are used by the pharate adult to cut through the cocoon or gnaw out of the pupal cell. The other appendages are free from the body in a condition called **exarate** (Figs. 4.4b, 4.6; Color photo 2e). In some of the taxa just named, the pharate adult uses the legs to walk or swim before the pupal cuticle is shed.

Pupae without movable mandibles are **adecticous.** The pupae of this type may be exarate, or the appendages may be cemented to the body in a condition described as **obtect** (Fig. 4.4e; Color photos 1f, 2d). Exarate adecticous pupae are found in Hymenoptera and Coleoptera. To escape the cocoon or cell, the insect first sheds the pupal cuticle and then uses the adult mandibles and legs to exit. In Strepsiptera and certain Diptera (Cyclorrhapha) the pupa is exarate, but it is enclosed in a **puparium,** or sclerotized cuticle of the last larval instar. Obtect pupae are found in certain Lepidoptera (suborders Ditrysia and Monotrysia, Fig. 44.5e) and Diptera (Nematocera, Orthorrhapha). Backwardly directed spines on the cuticle of some obtect pupae help to force the emerging insect out of the cocoon or cell.

REPRODUCTIVE ORGANS AND THEIR FUNCTION

Male Organs

The functions of the male organs of insects are to produce sperm and to transfer the sperm to the female of the same species (Fig. 4.7a,b). The mesodermal gonads are the **testes,** which are paired but may be secondarily united in a single body. **Spermatogenesis,** or the production of sperm, takes place in the **sperm tubes.** In insects with a short adult life, spermatogenesis may be completed as early as the larval or pupal instars, while in long-lived adults sperm may not be produced before adulthood. The sperm tubes may be single as in some Apterygota, Coleoptera, and Diptera, or multiple as in other insects. The sperm tubes of each testis are usually held together by a **peritoneal sheath.**

Mature sperm pass from the sperm tube through a short **vas efferens** to the **vas deferens.** The vas deferens is usually mesodermal in origin, but in Diptera it is ectodermal. It may be dilated into a **seminal vesicle,** where sperm are stored prior to their exit via the ectodermal **ejaculatory duct. Accessory glands** of mesodermal or ectodermal origin may be associated with the vas deferens or ejaculatory duct. These are especially well developed in insects that produce a **spermatophore,** or jellylike capsule in which the sperm are transferred to the female.

Sperm Tube Structure and Spermatogenesis

The sperm tube has a cellular **epithelial sheath** of one or two layers. Inside the tube, the germ cells, or **spermatogonia,** are located at the apex in the **germarium.** Here also is a large cell, the **apical cell.** The spermatogonia multiply mitotically. The cells from consecutive divisions remain clustered and enclosed in a cellular capsule or **sperm cyst.** Within the cyst the spermatogonia produce **spermatocytes,** which undergo meiotic division to form the haploid spermatids. The spermatids then differentiate into flagellated **spermatozoa** or sperm. Insect sperm are quite slender and long. Their narrow diameter is probably correlated with the diameter of micropyles in the egg chorion, an opening that must be kept small to reduce water loss. The several steps in spermatogenesis can be seen in successive zones arranged lengthwise in the sperm tube. Near the base, where the vas efferens is located, the cyst disintegrates and the sperm are released, still grouped in bundles. The sperm pass into the vas deferens and are stored in the seminal vesicles until ejaculation.

Transfer of Sperm

Among the Pterygota, the mates find each other by sight, sounds, or chemical sex pheromones. Following courtship, sperm are transferred by the male to the female directly and internally by a copulatory organ (Fig. 4.1). Odonata are unique in having the male transfer sperm to a secondary organ on his third sternum (genital fossa, Fig. 21.2b,c) from which the female receives the sperm.

The Apterygota and Entognatha have an indirect or external transfer of sperm, not involving copulation. The sperm are contained in a spermatophore and deposited on the substrate or on threads to be picked up by the female. This method of transfer requires humid conditions to prevent the drying of the spermatophore. In Collembola and Diplura the sperm are contained in a stalked spermatophore that is deposited by the male on the substrate. In some Collembola the female picks up the spermatophore with her vulva in the absence of the male; in other species the spermatophore is picked up after courtship and in the male's presence. Diplura males deposit the spermatophore only in the presence of the female. In Thysanura

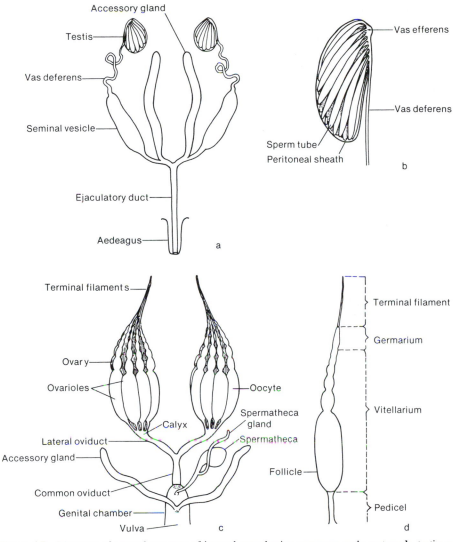

Figure 4.7 Diagrams of general structure of internal reproductive organs: **a**, male system; **b**, testis; **c**, female system; **d**, ovariole. (Redrawn from R. E. Snodgrass, *The Principles of Insect Morphology*. Copyright © by Ellen Burden and Ruth Roach. Used by permission of the publisher, Cornell University Press.)

and Archeognatha, the male spins silk threads during courtship. In Thysanura the spermatophores are placed on the substrate under the threads and the female is guided under. In the Archeognatha the spermatophores are deposited on a thread held by the male (Fig. 19.2), and the female is guided to pick up the spermatophores.

In the Pterygota, sperm are transferred directly to the female's genital tract during copulation. Internal fertilization results in less wastage of sperm and is much less dependent on humid conditions. The spermatophore is retained in many orders: Orthoptera, Blattodea, Mantodea, Dermaptera, Psocoptera, Neuroptera, Lepidoptera, and some Diptera (suborder Nematocera). Mecoptera and most Diptera transfer sperm freely to the female without a spermatophore. Depending on the species, spermatophores may be present or absent in Hemiptera, Trichoptera, Coleoptera, Hymenoptera, and Neuroptera.

Once the spermatophore has been deposited in the genital opening, the sperm are freed by the partial rupture or digestion of the spermatophore. The sperm then migrate or are moved to the spermatheca. The empty spermatophore is dissolved in the vagina or eaten by the female.

Traumatic or **hemocoelic insemination** occurs in the Strepsiptera; in the bedbugs (Cimicidae) and some of their allies, the Anthocoridae and Polyctenidae; and in some Nabidae (Hemiptera). The male genitalia are used to pierce the integument of the female, and the sperm are injected into the hemocoel. In some bedbugs a **spermalege,** or specialized organ of the integument, is developed where the puncture normally is made.

Reproductive females of social insects (Chap. 7) may mate multiple times during a single mating bout (honey bees) or repeatedly over a lifetime (termites). Female insects of most species, however, become less receptive or unreceptive to further mating and increase production of their eggs after they have been inseminated. Changes in receptivity can be traced to substances contained in the seminal fluids or spermatophores. The spermatophore of male crickets, *Teleogryllus commodus* (Gryllidae, Orthoptera), contains a prostaglandin-synthesizing complex. When transferred to the spermatheca of the female, this enzyme complex synthesizes prostaglandin from the precursor arachidonic acid. As a result, the female initiates massive oviposition. In another example, the experimental injection of seminal fluid into the hemocoel of female leafhoppers makes recipient females reject mating advances by males.

Female Organs

The functions of the female organs of insects are to produce eggs and to ensure their fertilization and placement in the environment (Fig. 4.7c,d). The mesodermal gonads are the paired **ovaries,** each of which may be single or divided into **ovarioles. Oogenesis,** or the production of eggs, takes place in the ovarioles, beginning in some insects in the pupa or last nymphal instar. The number of ovarioles is commonly four to eight on a side, depending on the species, and is correlated with the number of eggs produced. In Lepidoptera the number is usually four on each side. The highly fertile queen ants or termites, for example, may have more than a thousand on each side.

At **ovulation** the eggs pass from the ovariole and follow ducts toward the exterior. The paired **lateral oviducts,** each with a **calyx,** lead to the single **common oviduct,** which empties into the **genital chamber.** Attached to the genital chamber are **accessory glands** and the **spermatheca** with its **spermathecal gland.** Fertilization takes place in the genital chamber before the eggs are oviposited. The external opening of the reproductive tract is the **vulva.** Except for the lateral oviducts, which may be partly mesodermal, the other ducts and glands are ectodermal in origin and are lined with cuticle. The arrangement of the female tracts varies in the Lepidoptera (see Fig. 44.4), in which separate openings for copulation and oviposition may be present.

Ovariole Structure and Oogenesis

The ovariole is essentially a tube, attached anteriorly in the body by a **terminal filament.** The combined filaments of the ovarioles of each ovary may form a **suspensory ligament.** The wall of the tube is the **tunica propria,** which is lined internally by a **follicular epithelium.** The ovariole is divided along its length into the **germarium, vitellarium,** and **pedicel.** In the germarium are the **oogonia,** which are derived from primordial germ cells that can be traced back to the embryo of the female insect. The oocytes, produced by mitotic divisions of the oogonia, pass into the vitellarium. Each oocyte becomes surrounded by follicular cells, forming a cystlike **follicle.** Within the follicle, yolk is deposited in the oocyte during **vitellogenesis,** and the chorion is added. As the follicles swell, arranged one behind the other, the vitellarium elongates and becomes beaded in appearance.

The first or most advanced follicle is separated from the opening of the pedicel by a **follicular plug.**

Studies on *Drosophila* indicate that a large fraction of the genome is required during oogenesis. The oocytes rapidly increase in size as protein, carbohydrate, and lipid yolk bodies are formed by vitellogenesis. Recall that this process is under endocrine control by juvenile and ecdysteroid hormones. The process of vitellogenesis is complex and not fully understood. The nutrients that make up the yolk are derived from food eaten by the female or, if the adult does not feed, from food eaten and stored in the fat body during the immature instars. The female insect deposits in the egg not only nutrients for the future embryo, but also products of maternal genes (proteins, ribosomal RNA, messenger RNA) on which embryonic development will depend up to the stage of the cellular blastoderm. Some of the protein components of yolk are synthesized in the fat body and enter the oocyte from the hemolymph by way of the follicle cells. Protein synthesis inside the oocyte may involve additional nucleic acids contributed by the trophocytes or follicle cells. Lipids and the glycogen found in the yolk of many insects are transported to the oocyte by the trophocytes and follicle cells, especially the latter.

When yolk deposition is complete, the chorion is secreted around the oocyte by the follicle cells. The chorion is layered like a cuticle with respect to the follicle cells: next to the follicle cells is protein, and the layer next to the vitelline membrane of the oocyte is lipoprotein. The sculpturing and various passages through the chorion are created by the follicle cells. The vitelline membrane is usually considered to be the cell membrane of the oocyte, but in some Diptera it also is secreted by the follicle cells.

Types of Ovary

Two major structural types of ovariole are recognized: **panoistic** and **meroistic.** The former is the simple ovariole just described and found in Entognatha, Apterygota, Paleoptera, most Orthopteroidea, Thysanoptera, and in one endopterygote order, Siphonaptera. The meroistic type has associated with each oocyte additional cells, called **nurse cells** or trophocytes. These are usually derived from the oogonia and function in vitellogenesis.

Meroistic ovarioles are further divided into **polytrophic** and **telotrophic** types. In the former the trophocytes accompany each oocyte into the vitellarium and are included in the follicle. Polytrophic ovarioles are found in the exopterygotes Phthiraptera, Dermaptera, and Psocoptera, and in most endopterygotes (except Siphonaptera), and Coleoptera (Adephaga). In the telotrophic ovarioles, the trophocytes remain in the germarium and are connected with the oocytes by **nutritive cords.** Ovarioles of this type are found in Hemiptera and Coleoptera (suborder Polyphaga).

Ovulation and Fertilization

At ovulation the follicular covering breaks down and the egg pushes past the follicular plug into the pedicel and to the calyx beyond. The degenerate follicular covering forms a reddish or yellowish mass, which persists for some time in the ovariole. The presence of the colored mass can be used to discriminate between **parous** females that have ovulated and **nulliparous** females that have not. In insects that lay their eggs in groups, such as many Orthopteroidea, the eggs may be accumulated in the pedicels or oviducts before fertilization and oviposition.

As the eggs pass into the genital chamber from the common oviduct, they are usually oriented in a manner to receive sperm directly from the opening of the spermathecal duct. The mechanism of sperm release involves the special muscles that are associated with the spermatheca. In most insects the release is probably stimulated by ovulation or the passage of eggs in the oviduct. In Hymenoptera, however, the female can control the release so that some eggs are fertilized and others are not. The latter become males.

Oviposition

The segmental position of the genital opening varies among the orders. All insects have a common oviduct except the Ephemeroptera, in which the lateral oviducts open separately just behind the seventh sternum. In Dermaptera the short common oviduct opens behind the seventh sternum. Those insects with an ovipositor derived from the appendages of the eighth and ninth segments usually have the genital opening on the eighth segment. The great majority of insects, however, have the appendicular ovipositor reduced or replaced by special modifications of the abdominal apex: most Odonata, Plecoptera, Phthiraptera, Thysanoptera (suborder Tubulifera), Coleoptera, most Neuroptera, Mecoptera, Trichoptera, Lepidoptera, and Diptera. In these insects the opening is in the eighth or ninth segment. Many have the caudal abdominal segments tubular and telescopic, with the genital opening at the apex. This permits the eggs to be placed in crevices or cemented to leaves or other surfaces. Before the eggs are released, they may receive adhesive substances or a protective coating from the accessory glands. Groups of eggs of Mantodea (Fig. 23.1b,c; Color photo 8c) and Blattodea (Fig. 22.1) are enclosed in a protective case, or **ootheca,** derived from the accessory (also called colleterial) glands.

Oviposition behavior may be quite simple, as in Phasmatodea, which merely drop the eggs at random. More complex behavior involves the selection of appropriate sites through the use of various sense organs, including those on the ovipositor or abdominal tip. The sites are usually near the food required by the first instar of the offspring. Depending on the food habits of the species, the favored site may be in or on water, soil, decaying organic matter, certain live plant or animal species, or other insects. Oviposition may occur only at certain times during the 24 hours. All the eggs may be laid at once, as in some Ephemeroptera, or singly or in small groups over a considerable period. Social relationships, involving the special care of eggs and offspring, are discussed in Chapter 7.

SEX DETERMINATION

Sex in insects, as in other organisms, is commonly determined by **heterogamy,** or the production of gametes of two types. The types differ in genic sex determiners and may also differ in the presence and number of sex chromosomes. The exact nature of the sex-determining mechanism is not yet clear, but the ratio of males to females in insects is usually 1:1. Sex chromosomes have been identified in many orders. Sex chromosomes that occur in pairs in one sex are designated X; those that occur as single chromosomes are Y; and the missing chromosome of a pair is O. Males are heterozygous XY and XO and females homozygous XX in most Diptera, Neuroptera, Mecoptera, Hemiptera, Odonata, Coleoptera, and Orthoptera. The reverse can occur within the same order. Females are the heterozygous XY or XO and males homozygous XX in Lepidoptera and Trichoptera.

In Hymenoptera, fertilized eggs develop into females and unfertilized eggs into males. The females are therefore diploid and the males haploid. The sex ratio may deviate considerably from 1:1. This mechanism of sex determination, called **haplodiploidy,** is also known in some Thysanoptera and *Micromalthus* (Micromalthidae, Coleoptera). Male Coccidae (Hemiptera, suborder Homoptera) are also haploid, but apparently this is a secondary condition, since the male zygote is initially diploid.

TYPES OF REPRODUCTION

Thus far we have discussed the most common type of reproduction, oviparity, in which eggs are deposited by the adult female shortly after fertilization. A variety of other reproductive methods, however, exist among the insects; these are listed below.

Ovoviparity

In ovoviviparous reproduction, the eggs are retained in the female's genital tract until embryonic development is complete. The female then deposits a larva or nymph instead of an egg. The embryo is nourished solely by the yolk initially deposited in the egg. Examples are found in some Thysanoptera, Blattodea, Coleoptera, Diptera (some Tachinidae and Muscidae), and other taxa.

Viviparity

In contrast to the process just described, the embryo of viviparous insects is fed by the female after development has begun. Three types of viviparity have been recognized: adenotrophic, pseudoplacental, and hemocoelous.

Adenotrophic viviparity is found in the tsetse fly, *Glossina* (Glossinidae, Diptera), and in several other ectoparasitic flies, Hippoboscidae, Nycteribiidae, and Streblidae. The female carries one embryo at a time and deposits it as a larva, which soon pupates. After the egg yolk has been consumed, the larva ecloses from the thin chorion and is fed orally by special glands of the genital chambers.

Pseudoplacental viviparity involves eggs that have little yolk; the embryo is nourished by the wall of the genital tract, but not orally. A placenta-like organ for the transfer of nutrients may be developed by maternal or embryonic tissues or both. Examples are found in Hemiptera (Polyctenidae, Aphidoidea), Dermaptera (suborders Arixeniina and Hemimerina), and *Diploptera* (Blaberidae, Blattodea) and *Archipsocus* (Archipsocidae, Psocoptera).

Hemocoelous viviparity occurs in all Strepsiptera and in the paedogenetic larvae of certain Diptera (Cecidomyiidae), discussed below. In these insects the ovaries disintegrate and the embryos in their egg membranes take nourishment directly from the maternal tissues. The female is consumed as a result.

Paedogenesis

Reproduction by larval insects occurs in the beetle *Micromalthus* (Micromalthidae) and the flies *Miastor, Mycophila,* and certain other Cecidomyiidae (Diptera). In the flies, the life history may be that of normal, sexual adults when the food supply is average. When the food supply changes, larvae may give birth to more larvae for several generations without the appearance of the adult instar. The ovaries within the larva become functional, the eggs develop parthenogenetically, and the young larvae devour their mother (hemocoelous viviparity). The life cycles of these species are quite complex.

Parthenogenesis

Development without fertilization occurs in at least some species in all orders that have been investigated except the Odonata and Hemiptera (suborder Heteroptera). The types of parthenogenesis are classified by the sex produced, by the cytological mechanism, and by the frequency of occurrence in the species. According to the sex produced, three types are recognized. **Arrhenotoky** is the parthenogenetic production of males. This is the sex-determining mechanism in all Hymenoptera, some Thysanoptera, and *Micromalthus* (Coleoptera). **Thelytoky,** the parthenogenetic production of females, is the most common type. **Amphitoky,** the production of both sexes, is known in certain aphids and cynipid wasps.

The cytological mechanisms fall into two main types: apomictic and automictic. In **apomictic** parthenogenesis the egg fails wholly or in part to undergo meiosis. As a result, there is no reduction in chromosome number, and no opportunity for new gene combinations. Except for mutations, the offspring retain the genes of the mother. This is the most common type of mechanism in insects. In **automictic** parthenogenesis, the egg undergoes meiosis and the diploid condition is restored in a variety of ways: by fusion of two cleavage nuclei, by fusion of two polar bodies, or by fusion of a polar body with the egg pronucleus. The last is probably the most frequent method. A given lineage will become increasingly heterozygous or homozygous depending on which nuclei fuse.

Parthenogenesis may be **facultative,** that is, the eggs if fertilized develop normally and if not fertilized develop parthenogenetically. Thus the eggs of Hymenoptera are facultatively arrhenotokous. **Obligatory** parthenogenesis involves only thelytoky; males are rare or altogether lacking in the population. Such populations are often morphologically closely similar to a normal bisexual population and apparently arose as a parthenogenetic race. Without males or the need for mating, a thelytokous race can rapidly reproduce, but at the expense of genetic variability. Examples are found in some Phasmatodea, Lepidoptera (Psychidae), and Coleoptera (Curculionidae). A number of species of minute wasps of the genus *Trichogramma* (Trichogrammatidae, Chalcidoidea, Hymenoptera) have both sexual and thelytokous populations. Thelytoky in these wasps may arise from either hybridization between species or, more commonly, by microbial infection. In the latter, wasps become thelytokous when infected by *Wolbachia* bacteria, but treatment by temperature or antibiotics can render the wasps permanently sexual.

Polyembryony

In certain Strepsiptera and Hymenoptera (Chalcidoidea, Braconidae, Dryinidae), a single egg results in two or more individuals. This unusual phenomenon probably is made possible by the totipotent property of the early cleavage nuclei in the insect embryo. The insects are endoparasites of other insects. Polyembryony permits a large number of offspring, sometimes several thousand, to emerge from oviposition on one host (Fig. 11.2b). The female of *Copidosoma floridanum* (Encyrtidae, Hymenoptera) lays one or two eggs in the egg of a host caterpillar. If one egg is laid, one sex emerges; if two eggs are laid, both males and females emerge. Eggs of both sexes develop into numerous larvae, but of two types. First to differentiate are the "precocious" larvae or defender morphs, which eliminate the larvae of other species of competing parasitoids inside the host. The precocious larvae fail to pupate and die, thus making an altruistic sacrifice of their own chance to reproduce (see discussion of altruism in Chap. 7). The second type, or "reproductive" larvae, complete their development, emerge from the host body, mate, and disperse. Female precocious larvae may kill male reproductive larvae, thus altering the sex ratio of emerging wasps. In *Copidosomopsis* (Encyrtidae), the defender morphs have heavily sclerotized mouthparts and have been observed to attack other parasitoids (Cruz, 1981).

Functional Hermaphroditism

Individual insects are sometimes found or experimentally produced that have both male and female characteristics. **Gynandromorphs** are genetic mosaics of male and female tissues derived from a zygote plus other, genetically different, nuclei. All tissues of **intersexes** have the same genetic composition, but unstable development results in the differentiation of male and female features. Gynandromorphs and intersexes are not normal in any insect species. However, in the cottony-cushion scale *Icerya purchasi* (Coccoidea, Hemiptera) and related species, a **functional hermaphroditism** does occur normally. Both male and female gonads develop in the female scale, and the eggs are self-fertilized. Haploid male scales are rarely produced, and no pure females are known.

GENERATION TIME AND LONGEVITY OF INSECTS

The number of generations per year in insects is described in terms of **voltinism. Univoltine** insects have one generation per year, and this is probably the most common type of life history. **Bivoltine** insects have two generations per year, and **multivoltine** insects have more than two per year. Some insects require one or more years to complete their life cycle. For example, Thysanura and Archeognatha live for 2 to 3 years. The long-lived 13- and 17-year periodical cicadas (Cicadidae) spend most of their life in soil, sucking the juice of roots.

The number of generations per unit time tends to be inversely related to body size. In other words, smaller insects have more generations than larger insects during the same period. This relationship is only approximate because other factors such as moisture and nutrition influence the rate of growth. The number of generations per year for some species depends on suitable temperatures during development. This may vary markedly with location and to a lesser extent from year to year. The European corn borer, *Ostrinia nubialis* (Pyralidae, Lepidoptera), for example, has three or more generations in southern North America but only one at the northern extent of its range.

Insects that live in dry environments and eat nutritionally deficient food require longer to mature. Wood-boring beetles (Cerambycidae, Bostrichidae, Buprestidae), wasps (Siricidae), and moths (Cossidae) take 1 to several years. Records of 23 years for a cerambycid and 22 years for a bostrichid were obtained when the beetles emerged from wood that had been dried and made into furniture.

The adult life of insects is devoted to mating and reproduction. Adults may last only a few hours, as in male Strepsiptera (Chap. 40), some species of Ephemeroptera, or some species of the *Clunio* midges (Chironomidae, Diptera). On the other hand, univoltine insects that diapause in the adult stage may have adults that live for almost a year. Adult *Cryptocercus* (Cryptocercidae, Blattodea) roaches live 3 to 6 years.

Among the longest lived adults are the reproductive castes of social insects. Queen honey bees normally live 2 to 3 years, and lifetimes of up to 5 years or more have been recorded. The maximum age of an ant queen is 18 years; for a termite queen, it is 12 years.

Several families of beetles are long-lived as adults: Elmidae (up to 9 years), Scarabaeidae (up to 2 years), and Dytiscidae (up to 2.5 years). Marked individuals of tenebrionid beetles have been recovered in the field after 3 years, and some may live for up to 10 years. The tenebrionid illustrated in Figure 5.10, after serving as the artist's model for the first edition of this book, lived for 15.5 years on a diet of wheat bran, lettuce, and green peas.

INSECT SEASONALITY AND ARRESTED DEVELOPMENT

The life cycles of insects are synchronized on a long-term or seasonal basis. They survive annual periods of winter cold, drought, summer heat, or food shortage by entering a state of dormancy. When activity is suspended at the onset of the unfavorable period and resumed immediately afterward, the temporary arrest is called **quiescence.** During quiescence the hormonal stimulation of growth or reproduction remains unaffected. A few chilly days in spring, for example, may lower metabolism, but growth continues with the return of sunny weather.

Development or reproduction may be arrested on a long-term basis in winter or summer, however, by the failure of the endocrine organs to secrete appropriate hormones. A prolonged dormancy regulated by hormones is called **diapause.** Depending on the species of the insect, diapause may occur in any stage of the life. In eggs, larvae, and pupae the diapause is an interruption of growth and development; in adults the diapause is a period of no reproduction. The stage at which diapause occurs is genetically determined, but whether it will occur may be influenced by environmental conditions. In moth eggs, diapause is controlled by a maternal diapause hormone stored in the eggs. In moth larvae, juvenile hormone is the main regulatory agent for the beginning and duration of diapause. In moth pupae, PTTH is not released by neurohemal organs in the retrocerebral complex. Consequently ecdysteroids are not produced, and molting and the associated developmental changes do not occur. In diapausing adults the corpora allata are inactive and reproduction is not stimulated by juvenile hormone.

An **obligate** diapause is one that takes place regardless of the environment. Univoltine insects (e.g., many Lepidoptera of the Temperate Zone) commonly have an obligate diapause in an immature stage. The onset, or **induction,** of a **facultative** diapause is dependent on certain environmental cues. A winter diapause, for example, can be triggered by the shorter days and longer nights of autumn. Bivoltine and multivoltine insects usually have a facultative diapause. During spring and summer, development and reproduction continue without interruption.

The onset of diapause usually begins at a time when environmental conditions are still favorable for continued growth. The brain senses the photoperiodic cues, possibly directly and not through the eyes, and the neurosecretory cells fail to stimulate the appropriate endocrine organs. Diapause is recognized by a variety of symptoms. Active insects become relatively inactive, but not necessarily immobile or unresponsive to prodding. Oxygen consumption falls as the rate of metabolism is lowered and no longer is correlated with temperature. Water loss decreases. Thus in diapause a resistant state is attained before the beginning of unfavorable weather or food shortage.

Diapause is induced in most insects by a specific proportion of day length to night length, called the **critical day length.** This is, strictly speaking, insect **photoperiodism,** in contrast to the brief cues that entrain circadian rhythms (Chap. 6). It is not yet clear whether the underlying mechanism of photoperiodism resembles a circadian rhythm or is better explained as a physiological hourglass, "tipped over" by light each day, and with an additional mechanism for accumulating the products of daily cycles. Some insects respond to long days/short nights, others to short days/long nights, and some require more complex schedules in which long days are followed by short days, or vice versa. Near the equator only slight differences in day length may be sufficient. The stage of the insect that experiences the critical day length is not necessarily the stage that will diapause. For example, it is the female parent that determines whether eggs will diapause.

The **intensity** of diapause and its probable duration may be judged by the extent to which metabolism is reduced. The length of time that an insect remains in diapause is the result of a complex interaction involving seasonal changes in photoperiod and temperature, the insect's changing responses to these physical factors, and the rate of **diapause development.** The term "diapause development" is intended to indicate that a physiological process, the nature of which is unknown, must be completed before diapause is terminated. Ordinarily diapause development takes place only at temperatures below the minimum temperature for growth. A period of chilling, therefore, is often required for diapause development. Once diapause has been broken, the brain resumes neurosecretory activity and development resumes or, in adults, the reproductive functions are restored.

Diapause in the egg stage may take place late enough to be controlled by the young larva's neuroendocrine system. In the embryo, however, the nervous system is not developed. In the silkworm, *Bombyx mori* (Bombycidae, Lepidoptera), a **diapause hormone** is secreted by the subesophageal ganglion. Some genetic races of silkworms have an obligate diapause in the eggs of each generation; others have a facultative diapause. The eggs begin development in the female pupa. In the facultative races, the environment experienced by the embryo and larva of the parent female influences the type of eggs she will lay. Females reared under spring conditions (shorter day lengths, lower temperatures) lay eggs that will hatch and produce a second generation. Females reared under summer conditions (longer day lengths, higher temperatures), on reaching the pupal instar, will add the diapause hormone to the eggs. The species survives the winter in the egg stage and will not hatch until after chilling at 0°C for several months.

In insects with more than one generation each year, those that develop in different seasons may exhibit different characteristics. Seasonal forms are known in Lepidoptera, Gryllidae, and Cicadellidae. Early taxonomists frequently described the different forms as separate species. For example, the European butterfly *Araschnia levana* (Nymphalidae, Lepidoptera) has two forms. Caterpillars reared under short-day photoperiods diapause and emerge as the spring, or levana, form (Fig. 4.8a). When reared under 16 or more hours of light, the caterpillars do not diapause, and the summer, or prorsa, form develops (Fig. 4.8b).

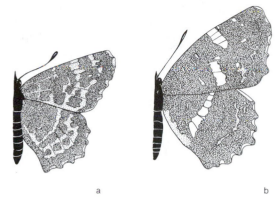

a   b

Figure 4.8 Seasonal forms of the European nymphalid butterfly *Araschnia levana:* **a**, spring, or levana, form; **b**, summer, or prorsa, form.

POLYMORPHISM

The term **polymorphism** in the broad sense means the presence of two or more forms of an organism. The differences are not restricted to physical appearance, but may be physiological, behavioral, or genetic. The forms may be different individuals in a population or different stages in the life history of a single individual. For example, the holometabolous life history is polymorphic. Here we are concerned with different forms other than the usual two sexes (sexual dimorphism) among adults of the same species. Adult forms may be genetically different, as in mimetic butterflies, or individuals may be genetically similar and develop differently according to external influences. The several forms of aphids or cynipid wasps are correlated with the season of the year. As mentioned in the discussion of seasonality, the spring and summer generations of bivoltine butterflies may differ in appearance. Quality and quantity of food or pheromones emitted by other members of the colony influence the development of morphologically and behaviorally different forms, called castes, in social insects (Chap. 7). In most of these examples, developmental hormones play a central role in directing the development of different forms of the same species.

Cyclical parthenogenesis or heterogony is a kind of polymorphism involving the alternation of parthenogenetic and bisexual generations. Examples are found in the Hymenoptera (some Cynipidae), Hemiptera (Aphidoidea), and Diptera (paedogenetic Cecidomyiidae, mentioned above). Here the advantages of sexual and asexual reproduction are combined in an annual cycle, and the life histories may be complex.

Seasonal Forms of Cynipid Wasp

Female cynipids place their eggs in the meristematic tissues of plants, mainly in the oak and rose families. Larval feeding induces the plant host to develop a characteristic gall. The adults later escape from the gall by chewing an exit hole. The life histories of cynipids vary greatly: some species are bisexual and univoltine; others are mainly thelytokous and without males; still others are cyclic, alternating between a sexual generation and a parthenogenetic generation (called the agamic generation). The females of each generation in the cyclic type and their galls on plants may differ in appearance. In some instances, the females were originally assigned to different species. The annual cycle is as follows. In the spring, males and females of the sexual generation develop quickly and emerge from their galls and mate. The sexual females lay fertilized eggs that become agamic females. These develop slowly in their galls, overwinter, and emerge the following spring to produce unfertilized eggs that become either males or females of the sexual generation. In some species of aphids and cynipids, one type of female produces both males and females by amphitoky, while in other species the females are of two types: producers of males and producers of females.

Seasonal Forms of Aphid

In aphids usually one or more parthenogenetic generations take place in spring, followed by a bisexual generation that produces overwintering eggs. Most aphid species live throughout the year on one species of plant. The typical cycle begins in the spring when the eggs hatch to produce the first generation, or fundatrices (singular, fundatrix). These become alate females that reproduce by parthenogenetic viviparity. Several successive generations are produced by this means until autumn. As colonies become crowded or plants become senescent or dry, individuals may fly to other plants of the same species and establish new colonies. In autumn, alate males and apterous oviparous females are produced parthenogenetically. These individuals mate, and the females lay eggs that overwinter. Some species of aphids, however, live on woody perennial plants (primary hosts) in the autumn, winter, and spring, and fly to herbaceous annual plants (secondary hosts) in the summer.

For example, the annual cycle of the apple-grain aphid, *Rhopalosiphum padi* (Aphididae, Hemiptera), is shown in Figure 4.9. The fundatrices (a) are found in the spring on various trees, including apple, pear, and plum. The fundatrices and most of the second generation (b) are apterous, but by the third generation, alates are produced (c). These migrate to various grasses such as the grains, wheat, oats, and corn, or weedy grasses. New colonies of apterous aphids (d) are established by parthenogenetic viviparity. When the colonies become crowded, alates (e) are again produced, and they fly to new grasses. In autumn, alate males (g) and females (f) are produced that fly to the primary host. The females produce apterous females (h) that mate with the alate males. The apterous females are oviparous and lay eggs (i) that overwinter in diapause.

Aphids maximize their reproductive rate during favorable circumstances by channeling morphological and metabolic resources that could be

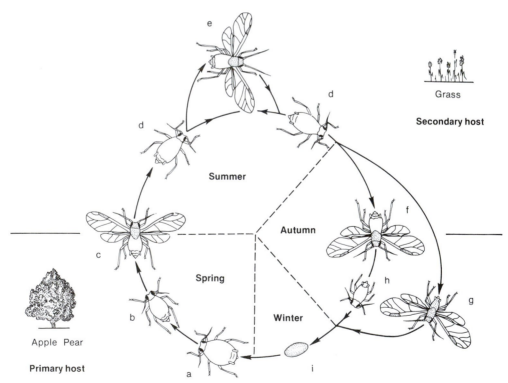

Figure 4.9   Annual cycle of the apple-grain aphid, *Rhopalosiphum padi*. See text for explanation. (Redrawn from Dixon, 1973, by permission of Edward Arnold/Hodder & Stoughton Educational Publishers, Ltd., London.)

used for flight into producing offspring. Alates produce fewer total offspring than do wingless adults. Yet aphids retain the critical ability to produce alates in response to environmental signals (e.g., crowding) that predict deteriorating resources. Parthenogenesis further maximizes production of progeny by eliminating the time, effort, and males needed for mating. The annual sexual cycle preserves sexual recombination. Viviparity minimizes the time for development to sexual maturity, but the egg stage is retained to survive the winter's cold and absence of suitable plants for food.

Phase Polymorphism

We conclude this chapter with a discussion of **phase polymorphism** in the species of Acrididae (Orthoptera) known as locusts: *Anacridium aegyptium*, *Dociostaurus maroccanus* (Moroccan locust), *Locusta migratoria* (African migratory locust), *Nomadacris septemfasciata* (red locust), and *Schistocerca gregaria* (desert locust). Theirs is one of the most extraordinary transformations in insects. Under certain ecological circumstances, a harmless, even rare, insect becomes a threat to human survival equal to a military invasion.

In locusts, individuals of each species may develop into one of a spectrum of adult forms or **phases,** depending on the circumstance under which the young nymph or hopper grows to maturity. The extreme forms are called the **solitary** phase and the **gregarious** phase (Fig. 4.10). The visible differences are in coloration, proportions of body parts, and behavior, but these are only reflections of underlying physiological differences.

Hoppers of the solitary phase of *Schistocerca*, for example, are uniformly colored green, whereas the gregarious hoppers have a black pattern on a yellow or orange background. In the adults, the coloration of the phases is more or less reversed: the solitary phase has a more pronounced dark pattern and the gregarious phase a less pronounced one. Furthermore, male adults of the gregarious phase continue to change color, but the solitary

adults do not. Structurally, the head of the solitary locust is narrow, the pronotum is convex or arched in profile, and the hind femur is relatively long. In the gregarious phase the head is wider, the pronotum is depressed in profile, and the hind femur is relatively shorter.

In terms of behavior, the solitary hoppers spend relatively less time walking and walk at a slower rate, whereas the gregarious hoppers form groups that march together for long periods and at a faster rate. The most dramatic difference is in the adult behavior. The solitary adults remain in the vicinity of their birthplace, but the gregarious adults aggregate in immense swarms and take flight. Like great clouds of smoke that darken the sun, swarms of *Schistocerca* may have up to 10,000 million or more individuals. Over a season, a swarm may migrate for distances totaling up to 2000 miles (3,226 km). In October 1988, desert locusts from western Africa were found in the West Indies of North America after an exceptional flight over the Atlantic Ocean estimated to be at least 3100 miles (5000 km) and 6 days in duration. The locusts probably were able to glide some of this distance by taking advantage of upcurrents in the wind. Normally, the direction of flight over land is with the prevailing winds that carry the insects to areas of recent rainfall and new plant growth. To sustain this flight, locusts consume green plants voraciously. Large swarms, which may weigh up to 100,000 tons, eat their weight in green vegetation each day. When a large swarm settles on the ground, vegetation and crops are completely defoliated.

Even today, locust plagues are among the most feared of natural disasters. Locusts are figured in Egyptian tombs as early as the sixth dynasty (2354–2181 BC), and the Old Testament contains vivid references to them (e.g., Book of Joel). Famine and disease epidemics with high human mortality often followed the devastation of crops. It is only through international cooperation and research that we are beginning to understand the biology and ecology of locusts.

Until 1921, entomologists considered the migratory locusts to be several species that were

Figure 4.10 An outbreak of plague locusts (Acrididae) in Argentina. The large nymphs have developed into the migratory form and after their final molt will form a migratory swarm in the air.

distinct from solitary acridids. About this time and almost simultaneously, J. C. Faure in South Africa, V. Plotnikov in Central Asia, and Boris Uvarov in the Russian Caucasus came to the same independent conclusion: each species of locust may exist in two phases, of which one is the destructive migratory form and the other is a relatively harmless, nonmigratory or solitary form. Largely through the research of Uvarov and his associates at the Anti-Locust Centre in London, it is known that locusts are influenced in their development and sexual maturation by the presence of other locusts (see Uvarov, 1966). Hoppers that are reared in isolation develop into the solitary phase; but when as few as two hoppers are reared together, the development is shifted partly toward the characteristics of the gregarious phase. Up to a certain density, increases in the number of hoppers reared together increases the shift toward characteristics of the gregarious phase. Beyond a certain density, no further change is noticed. As a result, a continu-

ous series of adult forms can be reared experimentally.

A pheromone called the gregarization pheromone has been identified that stimulates the young instars to exhibit gregarious behavior. Changes also take place in the adult stage when locusts are placed together. If sexually immature gregarious females and males are placed with sexually mature males, their rate of maturation is increased, and the young males also change color from pink to yellow. Loher (1960) found that the epidermal cells of mature males secrete a volatile pheromone that is stimulating to other locusts, even at a distance. The pheromone, now called the maturation pheromone, stimulates the yellowing of males and the maturation of both sexes. Crowding of hoppers and adults, therefore, increases the sensory stimuli (probably including tactile, visual, and auditory stimuli as well as olfactory stimuli) that, acting through the corpora allata of the endocrine system, promote the development of the gregarious phase.

Migratory swarms regularly emerge from certain geographic areas, known as "outbreak areas." These are usually mixtures of habitats favorable to oviposition and offering food and shelter. In such areas, the locusts are sparsely distributed in the solitary phase and cause no economic damage. At intervals several years apart, however, conditions such as rainfall become more favorable for increased breeding. As a result of population buildup, the locusts are concentrated and the gregarious phase is produced.

Similar phase changes in response to crowding are known in about two dozen other acridid species, but the gregarious phase is less regularly produced. Up until the late nineteenth century, *Melanoplus sanguinipes* (Rocky Mountain locust) and *Melanoplus spretus* developed migratory forms and moved in vast swarms from areas east of the Rocky Mountains to the Mississippi Valley and Texas. For hundreds of years before European colonization, some of these migratory grasshoppers landed on glaciers high in the mountains of Wyoming, and their frozen bodies have been found entombed. Since the alteration of the natural environment by agriculture, conditions have not again favored the formation of the migratory phase. *Melanoplus spretus* is now considered extinct, but *Melanoplus sanguinipes* continues to be an agricultural pest over wide areas of the semiarid West.

SELECTED REFERENCES

General Embryology

Johannsen and Butt, 1941; Hagan, 1951 (viviparous insects).

Development of *Drosophila*

Campos-Ortega and Hartenstein, 1985; Hoy, 1994; P. A. Lawrence, 1992; Poulson, 1950; Wilkins, 1993.

Development of Lepidoptera

Goldsmith and Wilkins, 1995.

Hormonal Control of Metamorphosis and Reproduction

Gupta, 1990; Happ, 1992 (male reproductive system); Holman et al. 1990 (neuropeptides); Nijhout, 1994; Ohnishi and Ishizaki, 1990; Raikhel and Dhadialla, 1992 (yolk); Riddiford, 1995 (hormones and gene expression); Truman, 1990 (eclosion hormone), 1995 (hormone action on nervous system), 1996 (comment on epitracheal gland); Westbrook and Bollenbacher, 1990 (PTTH); Zitnan et al., 1996 (epitracheal gland and Mas-ETH).

Physiology and Anatomy of Reproductive Organs

Chapman, 1982; Kerkut and Gilbert, 1985 (volumes 1, 2, 7, 8); Loher et al., 1981 (prostaglandins); Snodgrass, 1935.

Parthenogenesis and Polyembryony

Cruz, 1981; Grbic et al., 1992; Stouthamer et al., 1990 *(Wolbachia* and *Trichogramma).*]

Cynipid Biology

Askew, 1984.

Aphid Biology

Dixon, 1985; Minks and Harrewijn, 1987; Moran, 1992.

Locust Biology

Baron, 1972 (history of recent plagues); J. A. Byers, 1991; Lockwood et al., 1994 (locusts in glaciers); Ritchie and Pedgley, 1989 (locusts cross Atlantic).

5
MAINTENANCE AND MOVEMENT

Transparent larva of a skipper butterfly (Hesperiidae) hidden in a leaf fold. Concealed from visual predators like birds, such larvae have no integumental coloration, bristles, or hairs for protection. The heart is visible as a dark, middorsal line running the length of the larva. Also visible are the trachea in each segment. Although safe from birds, this larva was attacked by a female parasitoid, probably a tachinid fly, that was guided by chemical clues. The parasitoid's egg was inserted at the place indicated by a black spot behind the head (at left). The white area in the middle of the host is the internal larva of the parasitoid.

This chapter is devoted to the nutrition of insects and the various organs involved in the general flow of nutrients and oxygen into the insect's body, the utilization of energy to produce movements, and the elimination of waste products (Fig. 5.1). Other aspects of feeding behavior are treated in chapters 6, 10, 11, and 12.

NUTRITION

Considering the infinite number of possible dietary combinations, the remarkable similarity in general nutritional requirements throughout the animal kingdom is evidence of the ubiquity of the same biosynthetic systems. In the study of insect nutri-

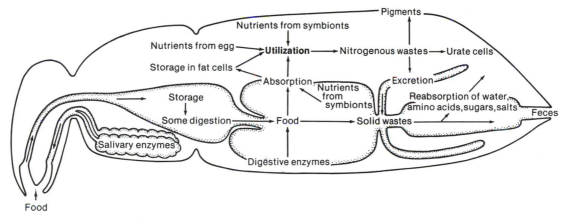

Figure 5.1 Diagram of routes of nutrients and wastes in an insect.

tion, enough insects of different kinds have been surveyed to indicate a common need for many of the same chemical nutrients as other animals (see Table 5.1 for a general list of essential nutrients). Yet the specific dietary requirements for a given species in terms of origin, molecular structure of nutrients, and relative amounts remain difficult to define. This is not only because of technical problems in experimentation, but also because of the dynamic nature of internal metabolism and the ecological relationships involved in feeding and nutritional symbiosis.

Nutritional Ecology

Some insect species have a close association with microorganisms that provide essential nutrients. In this relationship, called **symbiosis** (meaning "living together"), the microorganisms, or **symbionts,** are passed to succeeding insect generations by special devices. The arrangement is prevalent especially among insects that feed on diets lacking essential nutrients and among insects that cannot digest the nutrients that are available. Some symbiotic relationships are external to the insect, as in the fungus gardens cultivated by ambrosia beetles, some gall midges, fungus termites, wood wasps, and fungus ants. In other insects, the microorgan-

isms are present in the intestinal tracts of insects, in their hemolymph, or even in special organs called **mycetomes** (see also Chap. 13). Dependence on symbionts is associated most often with feeding on wood, plant sap, vertebrate blood, or other food sources that may be deficient in essential amino acids, sterols, or vitamins.

For example, *Glossina* (Glossinidae, Diptera), *Rhodnius* (Reduviidae, Hemiptera), *Cimex* (Cimicidae, Hemiptera), and *Pediculus* (Pediculidae, Phthiraptera) feed exclusively on vertebrate blood in all stages of life. Blood is deficient in several B vitamins. Microbes are believed to supply these and other missing nutrients. In contrast, fleas and mosquitoes, which feed on blood only during the adult stage, lack obligate symbionts. Special experimental cultures of insects devoid of other organisms may be prepared for nutritional studies. Such insects, deprived of their normal microbes, fail to grow or may die, but they will complete development if the diet is supplemented with vitamins or other nutrients.

Plant-feeding Hemiptera that suck plant juices exist on a diet deficient in nitrogen and vitamins. Microorganisms probably supply vitamins and may aid in recycling nitrogen from waste products inside the insect. Some wood-feeding termites depend on flagellate protozoa in the hindgut to digest cellulose and make oxidizable energy sources

TABLE 5.1
Nutritional Requirements of Insects

Water

Energy sources

Amino acids (L-isomers)
  Arginine
  Lysine
  Leucine
  Isoleucine
  Tryptophan
  Histidine
  Phenylalanine
  Methionine
  Valine
  Threonine

Lipid growth factors

Sterols
  Cholesterol, 7-dehydrocholesterol,
  β-sitosterol, or ergosterol
Polyunsaturated fatty acids
  Linoleic or linolenic acids
  Palmitoleic acid (some Diptera)
Fat-soluble vitamins
  Vitamin A (retinol)
  Vitamin E (α-tocopherol)

Water soluble growth factors

B Vitamins
  Thiamine (B_1)
  Riboflavin (B_2)
  Pyridoxine (B_6)
  Nicotinic acid (niacin)
  Pantothenic acid
  Biotin
  Folic acid
Ascorbic acid (vitamin C)
Lipogenic growth factors
  Choline and myoinositol
Ribonucleic acid (Diptera)

Minerals

Potassium
Magnesium
Phosphorus
and trace amounts of
  Sodium
  Calcium
  Zinc
  Copper
  Sulfur
  Aluminum
  Manganese
  Iron
  Cobalt
  Chlorine

available, commonly in the form of acetic acid. Most termites depend on bacteria as symbionts, although termites are probably able to produce their own cellulases.

Insects may store nutrients in their fat body, including special proteins called storage hexamers, to be utilized in a subsequent stage of the life history. Female mosquitoes, for example, suck blood only as adults and use essential nutrients that were obtained while they were aquatic larvae. The use of stored food is a major evolutionary strategy in many Lepidoptera and Ephemeroptera, and in some Diptera (Oestridae) of which the adults do not feed. Adults of other Lepidoptera, Diptera, and Hymenoptera also depend to a large extent on reserves built up by the larvae but need additional nutrients for egg production and longevity of females, as well as carbohydrates for energy.

In other insects, eggs taken from well-fed parents may contain enough reserves of certain essentials to permit the resulting offspring to survive on an experimentally deficient diet. Effects of the maternal nutrition on offspring are known in ants, locusts, and cockroaches. Some nutrients can replace others, such as many sterols for cholesterol, the requirement for which can be satisfied by a part of the molecule. Substitutes or combinations may also replace certain functions of essential nutrients or have a "sparing action" when a structural similarity exists. Utilization of a diet depends not only on the chemical constituents, but also on the quantity of each with respect to others. Excesses of a single constituent may inhibit growth; deficiencies in one can decrease the use of others.

From an ecological and evolutionary viewpoint, dietary requirements are subject to change over time. For example, natural selection will favor genetic strains of a phytophagous insect (plus its symbionts) that are able to thrive and reproduce on the unique combination of nutrients supplied by a specific host plant. When the success of the insect becomes detrimental to the survival of the plant population, the plant host may evolve defensive responses such as protective structures, lowered dietary value to the insects, antifeeding compounds, or toxins. This may provoke corresponding changes in the insect that allow survival on the altered plant host (see Chap. 10).

Essential Nutrients

Water

Depending on the insect's behavior and physiological equipment, water can be obtained directly from drinking or from moisture in food, oxidative metabolism, or absorption of water vapor. The small amounts of water required by an insect usually are generally available from such sources as ponds, temporary rain puddles, and dewdrops.

The water content of food varies greatly. For phytophagous insects, water is available in plant leaves, sap, and nectar; but seeds, nuts, grains, and pollen are low in water content. When large amounts of water are taken in by feeding directly on the plant's vascular system, the insect's excretory system must regulate the excess and conserve needed substances in the hemolymph (see below, "Excretion"). The honeydew excreted by Homoptera makes water available to other insects. Living fungi chosen as food usually have a high water content. Predators and parasitoids find water in fresh insect prey and their hemolymph, while insect parasites of mammals and birds suck blood.

For insects otherwise associated with mammals, wounds and secretions like sweat or mucus around the eyes, mouth, or nose are attractive sources of moisture. Scavengers on dry animal matter or wood feeders must deal with food low in water content.

Oxidative metabolism is an important source of water that can be conserved by the excretory system. Substantial water is lost during vigorous flight, while the spiracles remain open to supply oxygen to the flight muscles; yet metabolic water from the same muscular effort helps to offset the loss. By conserving metabolic water, insect pests in dry, stored grains have a higher water content than the food they eat.

In a moist atmosphere, water can be absorbed passively by insects, but usually not in significant quantities because of the waxy epicuticle. On the other hand, special structures capable of active absorption by expending energy allow some insects to take water from unsaturated atmospheres. Larvae of *Tenebrio molitor* (Tenebrionidae, Coleoptera), the mealworm, are able to absorb water vapor in the rectum as a function of their cryptonephric excretory system. Bladders in the mouth of the desert cockroach, *Arenivaga investigata* (Polyphagidae, Blattodea), absorb water vapor. Likewise, specialized mouthparts are involved in water vapor absorption by Psocoptera and Phthiraptera.

Energy

Energy is available from the oxidation of carbohydrates, fats, organic acids, and suitable amino acids. Carbohydrates are readily available in a variety of sources, including plant tissue, nectar from flowers, and honeydew (plant sap) excreted by aphids. Glucose, fructose, and sucrose commonly are utilized among the monosaccharides and disaccharides, but some sugars, such as pentose sugars, are not used and may be toxic. Energy from sugars is often stored in the fat body as glycogen. Larvae of Diptera require little sugar during growth, but adult Coleoptera, Diptera, and Hymenoptera require sugar to fuel the flight muscles. Insect flight muscles of the fibrillar type convert more energy per unit weight than any other animal tissue. For this reason many beetles, flies, wasps, and bees frequent flowers for sucrose-rich nectar. Long-distance fliers [many Lepidoptera, locusts, Coleoptera, and the bug, *Lethocerus* (Belostomatidae, Hemiptera)] usually consume their stored sugars and then utilize fat. The latter is suited as a reserve energy source because it yields twice as many calories as would the same weight of carbohydrate. Fat is frequently stored in the fat body of larval insects for later use during pupation, energy for nonfeeding adults, or development of eggs in females. Trehalose, the major sugar in insect blood, also is used in the synthesis of chitin in the cuticle.

Amino Acids

Enzymes and structural components of cells are proteins. Hence, proteins or amino acids are required in the diet for the growth and maintenance of a living insect. Insects require at least the same essential 10 amino acids required by many animals such as the white rat (Table 5.1). The total nitrogen in food, measured as a percentage of dry weight, is a useful guide to the amount of nitrogen-containing compounds in the food. Predatory, parasitoid, and scavenger insects that feed on the bodies of other insects or animals have rich sources of nitrogen. Phytophagous insects that feed on leaves, seeds, pollen, or fungi are supplied with less nitrogen. The nitrogen content of leaves generally declines with age, making young leaves the most nutritious. Insects that feed on decay-free dead wood (i.e., without fungi as an additional source of food) or leaf litter receive still less nitrogen. Nitrogen and other compounds in phloem fluid, and especially in xylem fluid, are so dilute that a sap-feeding insect must take in large amounts of liquid to extract adequate nutrients.

Methionine in the diet is important because insects cannot synthesize a sulfur-containing amino acid. Phenylalanine or its derivative, tyrosine, is required because tyrosine is a precursor of acetyldopamine, the tanning agent in sclerotization of the cuticle. Some species have been found also to require asparagine, aspartic acid or glutamic acid, and proline. Bacterial symbionts of aphids synthesize both methionine and tyrosine.

Lipids

Lipid requirements include fatty acids and sterols. Reserve energy sources in the form of fat, mentioned above, are derived from either sugars or dietary saturated and monosaturated fatty acids. Fatty acids for energy are not deemed essential because alternate sources of energy are available. Polyunsaturated fatty acids, on the other hand, are dietary essentials for most insects because insects are unable to synthesize them. These substances are involved in the formation of phospholipids of cell membranes and in the production of hormonelike prostaglandins. Dietary linoleic and linolenic acids usually satisfy this need, which tends to appear late in the insect's life (at pupation in many Lepidoptera; in reduced fecundity of adult boll weevils) or, in cockroaches, in the next generation, because of a deficiency due to inadequate provisioning of the eggs. Diptera have a unique lipid metabolism not dependent on polyunsaturates. Their major fatty acid is palmitoleic, and their phospholipids are of the ethanolamine phosphoglyceride type.

In contrast to mammals, insects are unable to synthesize sterols. Besides other functions, a sterol is required for the production of the growth hormone ecdysone. Foods vary in the predominant kind of sterol supplied to insects: animal foods have cholesterol, plants have β-sitosterol, and fungi have ergosterol. Regardless of the origin, the presumed biogenetic pathway to ecdysone involves cholesterol to 7-dehydrocholesterol. Cholesterol is ordinarily adequate in an artificial diet, but some insects with highly specific diets have apparently lost the ability to convert cholesterol: the fly *Drosophila pachea* (Drosophilidae, Diptera) requires the cactus sterol, schottenol, from its host *Lophacereus schotti*.

Vitamins and Growth Factors

Vitamins are substances essential for living processes but are not used for energy or basic structures. Until relatively recently, insects were thought to need only water-soluble vitamins. Among these, seven B vitamins are required for most insects (Table 5.1) and some species also require cyanocobalamin, B_{12} and carnitine, B_T. The physiological role of ascorbic acid (vitamin C) is controversial, but this substance evidently is required by some insects.

Among the fat-soluble vitamins are D (calciferol), involved in calcium metabolism; E (α-tocopherol), an antisterility factor for mammals; A (retinol) or precursors, involved in carotenoid visual pigments; and K (phylloquinone, etc.), essential for clotting mammalian blood. Vitamin D is not required because insects do not have calcareous skeletons and are apparently unable to use it even as a sterol. The clotting of insect blood differs from that of mammals and does not involve vitamin K, but some experimental evidence exists that vitamin K stimulates growth in insects. Vitamin E deficiencies in insects have been difficult to demonstrate consistently, but both egg and sperm production, as well as larval growth, may be affected. A need for vitamin A has long been suspected because the visual pigments of insects are rhodopsins. Morphological and visual defects have been found in several insect species reared on vitamin A deficient diets.

The growth factors choline and myoinositol are required as constituents of phospholipid molecules. Ribonucleic acid has been found to be required by some Diptera.

Minerals

The mineral requirements of insects are approximately the same as those of vertebrates, because minerals, of course, cannot be synthesized. In comparison to mammals, insects require much less calcium (not needed for the skeleton), less iron (few insects have hemoglobin, but otherwise iron is essential for many cellular processes), and much less sodium and chloride (many insects do not use these elements in homeostatic functions of blood). Although required in only trace amounts, zinc is apparently so critical in the moth *Heliothis virescens* (Noctuidae, Lepidoptera), that it is con-

served by transferring it in the male semen to the female during copulation.

ANATOMY AND FUNCTION OF THE ALIMENTARY CANAL

The alimentary canal of an insect is a tube passing through the body that provides the special internal environment for food to be mechanically and chemically disintegrated and brought to the vicinity of absorbent cells. Despite the diversity in physical consistency and composition of food, the gut exhibits certain general anatomical features throughout the insects (Figs. 5.2, 5.3).

Anatomy of the Alimentary Canal

The gut is attached at each end to the body wall but otherwise is held in place in the hemocoel largely by the pressure of adjacent organs and flexible tracheae. The three major divisions are the **stomodaeum** or **foregut** (Fig. 5.2a), derived from embryonic ectoderm; the **mesenteron,** or **midgut,** derived from endoderm, and the **proctodaeum,** or **hindgut,** derived from ectoderm. Attached at the juncture of the midgut and hindgut are the filamentous excretory organs called **Malpighian tubules.** In the embryo, the Malpighian tubules arise from the anterior end of the developing hindgut.

The wall of the alimentary canal is composed of a single layer of cells separated from the hemolymph by a basement membrane. Both the foregut and the hindgut are lined with a thin cuticle next to the food, called the **intima,** that is continuous with the integumental cuticle. These portions of the gut are properly classed as organs of the integument, and the cell layer is homologous with the epidermis. The intermediate section, the midgut, is differentiated by its endodermal origin and by the lack of an intima. The cells of the midgut are called **epithelial cells** or the **epithelium.** In place of an intima, however, most insects secrete a loose film, the **peritrophic membrane,** that, in addition to other functions, protects the delicate cells.

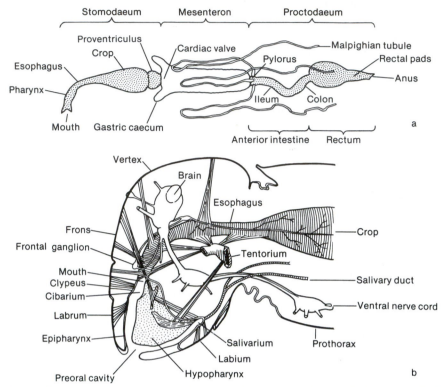

Figure 5.2 Alimentary canal: **a**, general structure; **b**, mouth and foregut. (Redrawn from R. E. Snodgrass, *The Principles of Insect Morphology.* Copyright © by Ellen Burden and Ruth Roach. Used by permission of the publisher, Cornell University Press.)

Muscular connections between the gut and the exoskeleton are limited to near the mouth and pharynx. Layers of circular and longitudinal muscles in the gut wall push the food along by peristaltic contractions. These are controlled by sensory and motor innervations from the stomodaeal nervous system (Chap. 6) to the foregut and, in some insects, the midgut. The hindgut receives nerves from the posterior abdominal ganglia of the ventral nerve cord. The scarcity of sensory nerves probably means that insects rarely suffer stomachaches. Tracheae from visceral trunks supply oxygen to the alimentary canal according to the metabolic needs of the different regions.

The true mouth is hidden from view and opens into the **preoral cavity** above the hypopharynx and below the epipharynx (Fig. 5.2b). The preoral cavity directly in front of the mouth is the **cibarium,** where food is passed to the mouth. The preoral cavity also contains another chamber beneath the hypopharynx and above the labium. This is the **salivarium,** where the saliva is emptied into the cavity from the labial or salivary glands.

The first distinctive region of the foregut is the **pharynx.** This heavily muscled organ is situated behind the mouth. The **esophagus** is merely a tubular portion of the foregut leading to the saclike **crop** (Fig. 5.3a). The release of food from the crop is controlled by an intricate organ, the **proventriculus.** Like the pharynx, the proventriculus may have a heavy muscular coat. Internally the cuticle is often shaped into heavy teeth, plates, or spines, but in fluid-feeding insects and many larvae, it is a simple sphincter. The teeth act as a filter, regulating the passage of food and permitting digestive enzymes from the midgut to enter the crop. The teeth also serve to grind the food further into smaller particles. The last part of the foregut is invaginated in the midgut and is called the **cardiac valve** (Fig. 5.3b). Special secretory cells between the valve and the midgut wall are partly responsible for the formation of the peritrophic membrane.

The midgut may be a simple tube or variously modified. The epithelium may be differentiated into distinctive histological regions. The columnar cells are especially active, since the secretory processes that yield the digestive enzymes often involve the eventual disintegration and rejuvenation

of the cellular lining. Next to the lumen, the cells characteristically are densely covered with short, hairlike filaments, or microvilli. Secretory cells lost during enzyme production or by molting are replaced by regenerative cells. Substances may be accumulated in epithelial cells for storage.

In insects such as grasshoppers, true bugs, and larval scarab beetles, hollow, fingerlike pouches may be present at the anterior or posterior ends of the midgut. These are the gastric caeca and are known to shelter symbiotic bacteria. At the junction of the midgut and hindgut are the Malpighian tubules. These are closely associated with the anterior end of the hindgut and serve as a convenient

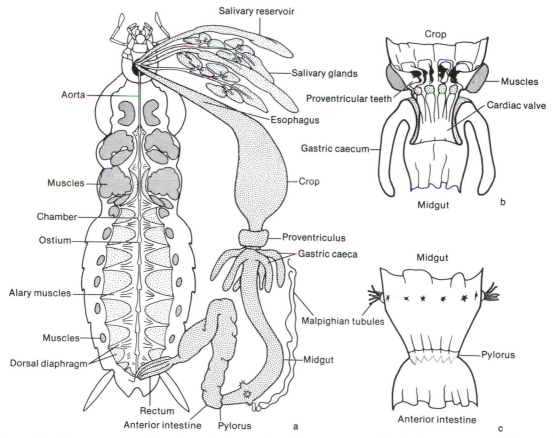

Figure 5.3 Alimentary canal and dorsal vessel of the roach *Periplaneta americana* (Blattidae): **a**, ventral view of dissected roach with fat body, reproductive organs, and most of Malpighian tubules removed; **b**, interior of proventriculus and cardiac valve; **c**, interior of pylorus.

external marker to distinguish the posterior end of the midgut. Also in this area is the pylorus valve (Fig. 5.3c). The valvelike function of this opening is best developed in caterpillars and beetles.

The hindgut may have several recognizable sections: (1) an **anterior intestine,** composed of the **ileum,** which is larger in diameter, and a narrow posterior portion, the **colon**; and (2) the **posterior intestine,** or **rectum,** which is the final section before the anus. The hindgut is lined with cuticle, which is usually impermeable. The **rectal pads** (see below: Fig. 5.8), however, have a cuticular lining that is permeable to water, salts, sugars, and amino acid molecules. Together with the Malpighian tubules, the ileum and rectal pads form parts of the excretory system. In the rectum, the waste is dehydrated and often compressed by muscles into a pellet that bears the shape of the rectal walls.

Modifications of the Alimentary Canal

Primitive insects such as Collembola, as well as many larvae of Endopterygota, have alimentary canals that are simple, direct tubes with but slight modifications even in the regions of the cardiac and pyloric valves. More advanced chewing insects, including Orthoptera, Odonata, Hymenoptera, and many Coleoptera, have an enlarged crop and proventricular specializations to handle particulate meals, plus frequent caecal pouches and enlarged rectal pads. The ingestion and passage of a fluid diet poses other engineering problems. Some fluid feeders, such as some Diptera and Lepidoptera, receive liquids by means of lateral, bladderlike diverticula protruding from the crop. The midgut is enlarged in Siphonaptera and many Hemiptera (suborder Heteroptera). The most elaborate arrangement is the filter chamber of many Hemiptera (suborder Homoptera), which is discussed further on.

During the larval instars of Neuroptera and some Hymenoptera, the midgut is a blind sac, not connected to the hindgut until pupation. As a result, waste products are retained in the larval gut and excreted on emergence as adults. For insects such as larval bees that develop in a confined space, this accommodation has an obvious sanitary function.

The ileum is modified to contain symbiotic microbes in termites and larval scarab beetles. In the aquatic, immature stages of dragonflies, tracheated gills project into the rectum and serve in the acquisition of oxygen from water circulated in the enlarged rectum. Forceful contractions of the rectum also eject the water and propel the nymphs forward.

Salivary Glands

The enlarged integumental glands of the head function in a great variety of life processes but are discussed here because they contribute generally to ingestion, movement of food, and digestion. The largest and most common glands that secrete saliva are the paired **labial (salivary) glands** (Fig. 5.3a). The single, median salivary duct opens in the salivarium between the labium and hypopharynx. The glandular cells like clusters of grapes, extend back into the thorax, or even far into the abdomen in some caterpillars, and may be differentiated into regions. In addition to saliva, the modified labial glands are responsible for the silk secreted by larvae of Lepidoptera, Trichoptera, and Hymenoptera. These fibrous proteins are used in construction of cocoons, webs, and nets. Other glands of the head found in various insects are **mandibular glands,** which are the main salivary glands of larval Lepidoptera and are also known to secrete the pheromone called **queen substance** in the queen honey bee (Chap. 7); **maxillary glands,** which are small and probably provide lubricants for the mouthparts; and **pharyngeal glands,** which are best developed in worker honey bees for production of the food royal jelly, fed to larvae (Chap. 7).

Functionally, the saliva of the labial glands moistens and lubricates the mouthparts, dissolves food, and transports flavors to gustatory (taste) sensilla; it also contains enzymes, which act on the food before and after ingestion (Fig. 5.1), and serves as an excretory organ. Aphids, mirids, a leafhopper, and a lygaeid bug are aided in the penetration of their piercing-sucking mouthparts into plants by a pectinase in the saliva. This enzyme hydrolyzes the pectin in the cell walls of plant tissues. Hemiptera (suborder Homoptera) and the stinkbugs (Pentatomidae) and their allies secrete a salivary sheath around their feeding punctures. The sheath presumably functions to retain the turgor pressure of the plant's vascular system, thus enhancing the insect's sucking mechanism.

In bloodsucking insects, such as the biting flies and parasitic Hemiptera, ingestion is facilitated by the inclusion in the saliva of anticoagulins and antiagglutinins. Blood clots would otherwise plug the fine canals of the insect's mouthparts. The anticoagulant factors may find uses in human medicine. Digestive enzymes are notably absent in the saliva of these insects, and some salivas contain a local anesthetic, perhaps to avoid irritating the host while the insect feeds.

The injection of saliva by insects into plant or animal hosts provides a natural entry for pathogens. Complex cycles of transmission have been evolved by viruses and microbes that take advantage of the insect vector's feeding behavior (Chap. 13).

Peritrophic Membrane

The peritrophic membrane (meaning membrane around the food) is a tubular film about 1 μm thick that surrounds the food and forms a loose lining inside the midgut and hindgut. The membrane is often multilayered or laminar. Structurally it is a fibrous network, or an amorphous sheet of polysaccharide, including chitin, glycoprotein, and protein. In most insects that have been examined, the membrane is produced from microvillate columnar epithelial cells that occur in either or both of two places: (1) secretory cells located in a ring or rings between the invaginated cardiac valve and the anterior end of the midgut, forming the "cardiac press," and (2) secretory cells distributed

locally or generally in the midgut epithelium. In certain Diptera, for example, the secretory cells in the cardiac press extrude a tubular membrane of several layers around the incoming food bolus as it passes through the cardiac valve. In some other Diptera, Coleoptera, and Lepidoptera, membranes are produced from the epithelial surface. In Dermaptera, both methods have been reported; in mosquitoes, the membrane is produced by the cardiac press in larvae and by the epithelium in adults. Regardless of the method, the transparent membrane continues to move posteriorly along with the food and is ultimately discarded, often still in place around the fecal pellets. The activation of the secretory cells is related to the hormonal controls of metamorphosis and in response to a feeding stimulus or gut extension.

The membrane has not been reported in some insects, but not all life stages have been examined. The fluid-feeding Hemiptera lack a peritrophic membrane, yet they have a unique multilayered matrix on the apical surfaces of the midgut epithelial cells that may function like the peritrophic membrane of other insects.

The membrane presumably acts to protect the exposed midgut epithelium from abrasion by food particles, and to lubricate and facilitate movement of food. In this, the membrane functions like mucus, a substance not found in insects. The membrane also resembles mucus in chemical composition and in the nature of the secretory cells that produce it.

The peritrophic membrane covers the food and separates it from direct contact with the epithelium. It has minute pores that appear to be selectively permeable. The pores may act as a filter controlling the movement of enzymes into the ingested food and the movement of the products of digestion into the midgut cells.

Disease organisms that are ingested commonly invade the bodies of insects by penetrating the midgut and entering the hemocoel. These include useful microbial agents for biological control of pests as well as agents of harmful human diseases carried by insect vectors. The peritrophic membrane is a critical barrier to the invasion and may effectively exclude certain microbes. The pore sizes are usually smaller than bacteria, yet some bacteria, viruses, trypanosomes, and filarial worms are able to penetrate the membrane either mechanically or by action of their enzymes.

Digestion

Digestion of food may take place outside the insect (i.e., extraorally) or in the crop or midgut. The salivary labial glands and midgut epithelium are the primary sources of digestive juices. For solid food, mechanical breakdown begins with the mouthparts and may be continued with the cuticular proventriculus. The reduced size of food particles means that larger surface areas are exposed to the action of digestive enzymes. The enzymes are hydrolases, commonly including proteases, lipases, carbohydrases, and nucleases. Thus, complex organic constituents in the food are reduced stepwise to a molecular structure that can be absorbed by the epithelial cells of the midgut.

As a tubular membrane enclosing the food within the tubular midgut, the peritrophic membrane is believed to have important functions in digestion. Enzymes initially break down the polymers in the food while it is covered by the membrane. Intermediate oligomers pass through the membrane to be finally digested into monomers next to the apical surfaces of the epithelial cells and immediately absorbed. It is possible that enzymes are conserved and recirculated in the following manner. In the anterior midgut, the enzymes are secreted by epithelial cells, pass through the peritrophic membrane, and begin hydrolyzing the food. The enzymes move posteriorly with the food and, on reaching the posterior midgut, pass back through the membrane. Fluids secreted by the epithelium in the posterior midgut move anteriorly between the membrane and the epithelium. This countercurrent flow of fluids outside the membrane carries forward the enzymes and products of digestion. Once in the anterior midgut, the enzymes

pass again through the membrane to the food and repeat the cycle.

Extraoral digestion is widespread among the animal phyla and is common among predatory insects. Enzymes, especially proteases, in saliva or regurgitated from the midgut are injected into prey. The body of the victim becomes a container in which the tissues are liquefied and sucked out. The enzymes also are recovered, thus conserving protein, and they continue their action after ingestion. Paralysis and extraoral digestion of prey are accelerated when assassin bugs (Reduviidae) inject hyaluronidase into prey. This spreading agent probably dissociates the cells of the prey and hastens the action of digestive enzymes, which are concurrently injected. It is because of the enzymes that the bites of predatory Hemiptera are painful to people. The venomous saliva can also be defensively ejected at vertebrate enemies. The housefly ejects saliva on solid food, and the partly digested liquid is eaten. An enzyme such as amylase (which converts starch to maltose) or invertase (which converts sucrose to glucose and fructose) is commonly present in saliva.

Digestion in Orthoptera, Blattodea, and some Coleoptera is mainly in the crop, with enzymes produced by the salivary glands or regurgitated from the midgut. In Diptera, Lepidoptera, and Trichoptera, digestion apparently is limited to the midgut.

The honey bee adds salivary invertase to sucrose-rich nectar that is carried back to the hive in the crop. The digested mixture of glucose and fructose is regurgitated in the hive, dehydrated, and stored in the comb as honey.

Digestion of Unusual Foods

Insects are exceptional in being able to digest some abundant, yet resistant, compounds. Structural proteins in vertebrate tissues, such as collagen of connective tissues or keratin in wool, hair, and feathers, are highly resistant to degradation by common proteases. This is evident when a dead bird or mammal decays and only the resistant tissues remain. The feces of blow fly larvae (Calliphoridae) feeding on such corpses contain a collagenase that attacks the collagen and elastin of muscle tissues, resulting in extraoral digestion. Bird lice (Mallophaga), the larvae of clothes moths (Tineidae), house moths (Oecophoridae), and dermestid beetles possess a keratinolytic enzyme that attacks keratin. The moths and beetles are household pests, infesting woolen clothing and carpets. A strong reducing agent is able to break the sulfur bonds of keratin because poor tracheation in the insect's midgut provides a low oxidation-reduction potential. Once the peptide chains are free, the keratinase is effective.

Beeswax is ordinarily resistant to digestion, but the larval wax moth is able to utilize some wax components. At an optimal pH of 9.3 to 9.6 in the gut, the wax is hydrolyzed by enzymes originating from the larva's cells, from intestinal bacteria, or from both.

Woody plant tissue is regularly exploited as a food source by larvae of some wood-boring beetles (some Anobiidae, some Scarabaeoidea, Buprestidae, Cerambycidae, Lyctidae, Bostrichidae, some Tenebrionidae, and others), termites, wood-feeding roaches, and silverfish. Cellulase, hemicellulase, lignocellulase, and lichenase have been demonstrated in insects, some or all of which have been identified in members of the foregoing list. Microbes in the intestine also doubtless play a role in producing these enzymes. Although termites probably are capable of producing cellulases, the symbiotic flagellate protozoa or bacteria housed in the gut are critical in digesting cellulose. Bacteria in the midgut caeca of Scarabaeoidea beetles are thought to digest cellulose. Some of these enzymes, however, are directly secreted by insects (e.g., termites and the larvae of some cerambycid beetles). Among wood-boring insects, the ability to digest cellulose to varying degrees permits partitioning of the food resource: some thrive only on the more easily digested, but scarce, starch and sugar in the wood (Lyctidae, Bostrichidae); others use also the hemicelluloses (Scolytidae); and a third group consumes the whole cell wall, minus the lignin (Anobiidae, Cerambycidae).

Absorption

The midgut epithelium and the hindgut are the principal absorptive regions. Although digestion may take place in the crop, absorption is largely prevented by the cuticular intima. In the midgut, monosaccharide sugars, commonly glucose, rapidly pass through the permeable epithelial membrane without an active transport mechanism as long as the concentration inside the cells is lower than outside. A steep osmotic gradient is maintained by converting the simple sugar to the disaccharide trehalose. In the migratory locust this conversion takes place in the fat body, which is near the midgut. In the silkworm, oligosaccharides may be absorbed and further hydrolyzed in the epithelium.

Amino acids are similarly absorbed by passive diffusion processes related to the uptake of water into the hemolymph. In *Rhodnius* (Reduviidae, Hemiptera), even proteins that are not fully degraded to amino acids may be absorbed and then digested further by the midgut cells.

The absorption of lipids is more difficult to analyze because these molecules may be able to pass through the cuticular intima of the crop or between epithelial cells of the midgut, in addition to the usual route via the epithelial cells. Lipoidal material has been seen as droplets in cells of both the midgut and crop. Although cholesterol has been recently demonstrated to be absorbed by the crop of a cockroach, the midgut is known to be the major absorptive region. The fats need not be fully degraded before absorption. No evidence exists that an active transport mechanism is involved.

The rectal pads of the hindgut are specialized for absorption of water from the feces before defecation. Active transport of ions across cellular membranes attracts water osmotically into the region of higher salt concentration.

TRACHEAL SYSTEM AND GAS EXCHANGE

Oxygen for cellular respiration in insects is physically transported by an internal system of air-filled tubes, the **tracheae,** to within a few cell diameters of each cell (Fig. 5.4a). Whereas oxygen transport in vertebrates depends on the pigment hemoglobin as an oxygen carrier and the rapid circulation of blood throughout the body, respiratory pigments and blood are not involved in gas exchange in most insects. The respiratory system of terrestrial vertebrates has clearly evolved from that of aquatic ancestors, but the aerial systems of insects and certain other arthropods are thought to be pecu-liarly terrestrial adaptations that may have independently evolved in several arthropod stocks. Tracheae are ramifying and interconnecting tubular invaginations of the integument that arise from paired, lateral openings, the **spiracles,** and ultimately terminate in fine fluid-filled tubules, the **tracheoles** (Fig. 5.4f).

Small insects, such as Protura and most Collembola that live in damp places, or some aquatic larvae of Chironomidae (Diptera), lack a developed tracheal system and exchange gases through the cuticle. In larger, free-living, terrestrial insects, the demand for oxygen cannot be met by the body surface, partly because the area is inadequate for sufficient transfer and because the thicker cuticle

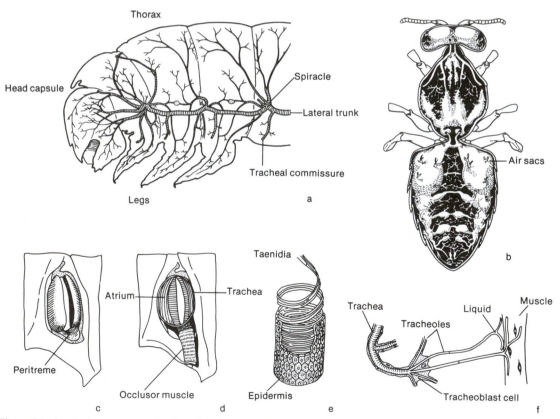

Figure 5.4 Tracheal system: **a**, tracheation of the head and thorax of a noctuid caterpillar; **b**, air sacs in a honey bee; **c**, external view of closing mechanism of a thoracic spiracle; **d**, internal view of same; **e**, structure of a trachea; **f**, structure of a tracheole. (a, c–f, redrawn from R. E. Snodgrass, *The Principles of Insect Morphology.* Copyright © by Ellen Burden and Ruth Roach. Used by permission of the publisher, Cornell University Press; b, redrawn from Snodgrass, 1956.)

permits only a negligible diffusion of oxygen. More importantly, since a molecule of water is smaller than a molecule of oxygen gas, a body wall permeable for oxygen diffusion would also be permeable to water loss. The tracheal system solves this problem by creating an enormous internal surface of permeable cuticle where loss of water can be controlled. In the fifth-instar larva of the silkworm, the number of tracheoles has been estimated to be 1.5×10^6.

In trachea diffusion plays the dominant role in oxygen transport and is a millionfold more rapid in the gas phase than in tissue. Calculations of oxygen consumption, diffusion rates, and measurements of tracheae indicate that diffusion alone could supply small insects (< 3 mm long) even with active flight muscle. Ventilation (i.e., moving drafts of air through the system) can greatly reduce the distance that oxygen must diffuse over a gradient from atmospheric concentration to respiring tissue. However, the tracheal system imposes maximum limits to the size that insects can attain. Even forced movements of air cannot overcome the size limitations of diffusion as the primary means of gas exchange. Diffusion is like sound communication. For conversation within a room, sound (330 meters per second) is as effective as electronic communication (300,000 kilometers per second), but it is not adequate for distances greater than hundreds of meters. Similarly, diffusion aided by ventilation is adequate over distances measured in millimeters but not more than a few centimeters.

The passage of air into the tracheal system is regulated by the spiracles. In some apterygotes, such as the springtail *Sminthurus* (Sminthuridae, Collembola), the opening is simply a hole in the body wall. The rate of gas diffusion through this pore, as through the stomata of plant leaves, is a function of the perimeter rather than the cross-sectional area. Elaborate modifications of the spiracle, therefore, are possible without necessarily reducing the intake of air. In most insects the tracheal orifice is recessed in the integument, creating a small chamber, or **atrium,** which communicates with the exterior. The atrium may be situated on a separate sclerite, the **peritreme,** and provided

internally with filtering hairs or projecting, valvelike lips, closable by muscles (Fig. 5.4c). The latter are characteristic of thoracic spiracles and are not found on the abdomen. The atria of abdominal spiracles remain open, with a closing device located inside at the tracheal orifice. The opening is pinched shut by the action of an occlusor muscle on one or two sclerotic bars. When the muscle relaxes, the elastic bars spring open, or they are pulled apart by a second muscle, the dilator. Spiracular muscles are innervated from their segmental ganglion or the ganglion immediately anterior. Tracheae leading from the spiracles may be connected longitudinally by **lateral trunks** and transversely by **tracheal commissures.**

The histological parts of tracheae are homologous to corresponding parts of the integument: an inner, cuticular intima secreted by a surrounding layer of epidermal cells that are bounded outside by a basement membrane. The intima is thin, permeable to gases, and marvelously flexible, because of fine, spiral thickenings called **taenidia** (Fig. 5.4e). The ultrastructure of the intima is still being debated. The epicuticle contains at least an outer layer and probably a hydrofugic (water-repellent) wax layer next to the lumen. Corresponding to the external procuticle is a chitin and protein layer that is largely (if not entirely in smaller tracheae) involved in constructing the taenidia. At each molt, portions of the intima are routinely shed along with the exuviae. During metamorphosis of flying insects the entire system is reorganized to accommodate the increased demand for oxygen by the flight muscles. New tracheal branches or fusions are developed, and old tracheal pathways are not replaced; yet the design of major trunks and anastomoses is consistent in taxonomically related forms and may be traced from the immature stages to the adult.

The tubular network of tracheae may be interrupted at intervals by enlarged dilatations, or **air sacs** (Fig. 5.4b). Taenidia may be discernible on the walls of some sacs as if the trachea is merely inflated, but other cavities are reinforced by reticulate thickenings. The sacs are collapsible, a feature of importance in ventilation and one that also permits internal organs to expand in the body cavity, as when the ovaries are enlarged by eggs or the gut is engorged with food.

Associated with the tracheae in Diptera, Lepidoptera, and Hymenoptera are **peritracheal glands,** which consist of several cells each and are distributed pairwise in thoracic and abdominal segments. The cells are thought to secret exuvial fluids during molting and possibly hydrofugic lipids that coat the intima. Of special note are the large, polytene chromosomes in the cells that undergo "puffing" or enlargement at certain times.

Tracheoles

Through repeated branching, the tracheae are reduced in diameter and terminate in tiny subdivisions, the **tracheoles** (Fig. 5.4f). These are tubules up to 350 μm length and 1 μm in diameter, tapering to 0.2 to 0.1 μm. Whereas the tracheal tube is enclosed in a multicellular coat, the epidermis, the tracheole is intracellular in a **tracheoblast cell.** The intima of the tracheole exhibits taenidia and an outer layer, and it may be exempt from renewal at each molt. When tissues are experimentally deprived of tracheae or organs are implanted in a new host, tracheoles migrate to the affected tissues and establish an oxygen supply. The same processes are presumably involved in normal development and injury repair. Although the general surface of the tracheal intima probably is diffusible by gases, the tracheole is believed to be the major site of oxygen transfer. Organs of high metabolic activity are richly tracheated. For example, the flight muscles of fast-flying insects are enveloped in tracheae and air sacs. The profuse tracheoles indent the plasma membranes of muscles to bring oxygen to the mitochondria deep in the fibers and only 3 μm part.

Tracheoles are often filled with a lymphlike liquid. During the depletion of oxygen in surrounding cells, this liquid is withdrawn against a capillary force by the tracheoblast cell, bringing gas to the tracheolar tip. The physical mechanism of withdrawal is correlated with changes in osmotic pressure of surrounding tissues, although the details are not clear. A porous structure observed in the tracheolar outer layer may facilitate this process. Oxygen moves the final distance from the tracheoles to respiring cells by diffusing through tissue. The carbon dioxide produced by respiration exits by the reverse course, hastened by a greater solubility in water and a permeability through tissue 36 times faster than that of oxygen but diffusing slightly slower than oxygen in gaseous phase. The greater mobility of carbon dioxide in tissue favors some loss of this gas directly through the cuticle, especially at less sclerotized intersegmental membranes.

Ventilation

The distance gases must diffuse is shortened by ventilation of the tracheal system. Motions of the body or churning of internal organs of active insects incidentally move air through the interconnected major trunks. Air movement through the system is greatly accelerated by pumping movements of the telescoping abdominal segments, creating inspiratory and expiratory flows. When the spiracles remain open, air may enter and exit tidally through the same orifice. The closing valves attached to spiracles in most insects permit coordinated opening and closing of the tracheal orifices, resulting in unidirectional flow. Ordinarily air is taken in through the anterior spiracles and expelled via the posterior spiracles. Air sacs increase the volume of air pumped and, since they are partly collapsible, may have other mechanical functions in ventilation. During flight, abdominal pumping is inadequate to supply the flight muscles. Movements of the thoracic segments, in conjunction with open spiracles, ventilate the air sacs and tracheae surrounding the muscles. Large, quiescent insects like diapausing moth larvae and pupae as well as temporarily inactive insects may have discontinuous ventilation that may span up to several hours. This cyclic process begins by closing the spiracles while a slight vacuum develops as oxygen is consumed, then partly opening and closing the spiracles to allow air to be sucked into the system, and finally fully opening the spiracles only briefly to release a burst of carbon dioxide.

Ventilatory movement of air in and out of the insect's body entails a decided risk of water loss. Air in the tracheal system is near saturation, while outside the humidity may be quite low. The structure of the atrium and closing mechanisms reduce or prevent the diffusion of water vapor from the system. It is therefore advantageous for the spiracles to remain closed, with the occlusor muscle contracted, until breathing is demanded. Spiracles may be controlled separately or in concert by the central nervous system. Cellular respiration leads to lowered levels of oxygen and higher levels of carbon dioxide. Changes in both gases will reduce impulses from the central nervous system to the occlusor muscle (thus relaxing the muscle and opening the spiracle by its elastic hinge) or, where dilator muscles are involved, there will be an increase in the frequency of impulses to the dilator to stimulate opening. Occlusor muscles are also influenced directly by carbon dioxide and will relax when the local concentration is high. As a consequence of this reaction, care must be exercised when carbon dioxide is used as an anesthesia to immobilize insects: the open spiracles permit water loss that may cause death.

Hemoglobin

Hemoglobin functions in the gas exchange of only three insects studied thus far: the bloodworm, *Chironomus* (Chironomidae, Diptera); the back-swimmer, *Anisops pellucens* (Notonectidae, Hemiptera); and the horse bot fly larva, *Gasterophilus intestinalis* (Gasterophilinae, Oestridae, Diptera). The hemoglobin of insects contains two heme groups, half as many as the vertebrate pigment. In each case, hemoglobin provides a temporary respite from oxygen deprivation. The bloodworm lives in stagnant or polluted water that is deficient in oxygen. Movements of the larva in its burrow refresh the cutaneous supply and charge the hemoglobin with oxygen, which is not released unless

the tension in tissues is low. The bloodworm's hemoglobin has a high affinity for oxygen, and the uptake of oxygen is more rapid than diffusion in the hemolymph; hence recovery from hypoxia is aided and aerobic respiration is sustained under poorly oxygenated conditions. The horse bot fly maggot is similarly situated in a liquid, oxygen-poor medium: a horse's stomach. Hemoglobin is in richly tracheated fat cells, permitting rapid uptake and storage of oxygen derived from passing gas bubbles. The bug *Anisops* utilizes hemoglobin not only to extend its period under water, but also to sustain a favorable specific gravity, which allows submerged floating.

Types of Tracheal System

The structure, number, and position of spiracles vary taxonomically: the apterygotes exhibit the most diverse arrangements and provide some insight to the evolution of the system. As already mentioned, most Collembola and Protura altogether lack tracheae. Some springtails (i.e., *Sminthurus* spp.) have a single pair of simple openings on the cervix, between the head and thorax, leading to branched tracheae at each side. The proturan *Eosentomon* (Eosentomidae) has two thoracic pairs, likewise leading to independent tracheal tubes. *Campodea* (Campodeidae) of the Diplura is similar, with three thoracic pairs of spiracles; but the system of *Japyx* (Japygidae) has four thoracic (the mesothorax and metathorax each have two pairs) and seven abdominal pairs, making a total of eleven, the maximum for any insect. Moreover, the spiracles are connected at each side by a longitudinal trunk, thus permitting movement of air lengthwise, but not transversely.

The Machilidae are relatively less specialized, with two thoracic and seven abdominal spiracles individually leading to mostly independent tracheae. The Lepismatidae, on the other hand, exhibit features that also characterize many of the winged insects: a pair of spiracles on each of the mesothorax, the metathorax, and the first eight abdominal segments, plus longitudinal trunks and transverse segmental commissures. Ventilating drafts of air through this interconnected system are forced by movements of the body and internal organs.

Three major types of respiratory system are recognized. The **holopneustic** respiratory system (ten pairs of functional spiracles) described for the Lepismatidae is found in the immature stages and adults of many terrestrial insects, including Orthopteroidea, Hemipteroidea, and some Hymenoptera. The **hemipneustic** system is derived by the loss of one or more functional spiracles and is characteristic of larval Neuropteroidea. The hemipneustic system is subdivided into the peripneustic arrangement, prevalent among terrestrial larvae (metathoracic spiracles nonfunctional, as in larvae of Cecidomyiidae), and the oligopneustic type of many Diptera (only one or two pairs of functional spiracles), which is specialized for life in water or liquid media. The oligopneustic type is further subdivided into two types: the amphipneustic (mesothoracic pair plus a pair of posterior abdominal spiracles functional, as in larvae of Psychodidae) and the metapneustic (only the last pair of abdominal spiracles functional, as in larvae of Culicidae). Finally, the **apneustic** system of submerged aquatic insects and most endoparasitic larvae has no functional spiracles but does have a closed tracheal system. Oxygen enters the body by cutaneous diffusion, either over the general body surface or in special, integumental gills. The apneustic system is found in immature stages of Ephemeroptera, Odonata, Plecoptera, Trichoptera, Sialidae (Megaloptera), some Hymenoptera, and many larval Diptera.

Gas Exchange in Aquatic Insects

Nearly all aquatic insects (Table 9.1) have a gas-filled tracheal system. The few exceptions are the young fly larvae of *Chironomus* (Chironomidae, Diptera) and *Simulium* (Simuliidae, Diptera), and the young caterpillars of *Ancentropus* (Pyralidae, Lepidoptera). In some groups the system is **open** and air can be taken inside the body. In other groups, however, the system is **closed** (apneustic) to the entry of air and oxygen must be absorbed from water. In either case, many aquatic insects have the body surface well supplied with tracheae and are able to obtain at least part of their oxygen by **cutaneous gas exchange.**

In contrast with air, the supply of oxygen from water is sharply limited by a series of physical constraints. Oxygen dissolves in water in an amount inversely proportional to temperature (warm water has less oxygen than cold water). The concentration even when saturated is many times less than in air, and movement by diffusion in solution is exceedingly slow. Immediately next to a submerged insect's body is a thin layer of still water called the boundary layer in which oxygen is renewed only by the slow process of diffusion. Finally, the diffusion of oxygen through cuticle is many times slower than in water.

With the exception of the stonefly, *Utacapnia* sp. (Capniidae, Plecoptera), closed tracheal systems are found only in the immature stages. The adults as well as immature stages of the unusual stonefly live deep in Lake Tahoe, California, and have a closed system throughout their life. Some insects, such as many larval Diptera, depend mainly on cutaneous gas exchange, but others have special **tracheal gills** that augment gas exchange. These are thin-walled outgrowths of the integument that are richly tracheated. Oxygen in water diffuses through the cuticle and into the gas-filled tracheoles. Gills may also be important in the elimination of carbon dioxide. Platelike gills are found in Ephemeroptera (Fig. 9.6a,b; Color photo 2f) and Odonata (suborder Zygoptera, Fig. 21.3b). Filamentous gills are also found in Ephemeroptera (Fig. 20.4b) and are characteristic of Plecoptera (Fig. 27.1b), Megaloptera (Fig. 36.2), Neuroptera (Sisyridae, not visible in Fig. 38.2b), Trichoptera (Fig. 45.2; Color photo 2e), and certain Coleoptera (e.g., Gyrinidae, Scirtidae, Psephenidae), Diptera (*Phalacrocerca,* Tipulidae), and Lepidoptera (Nymphulinae, Pyralidae). The rectum of Odonata is thin-walled and tracheated as a rectal gill. In general, gills occur most commonly on the sides of the abdomen, less commonly on the thorax, and rarely on the head.

Adult and immature insects with an open tracheal system obtain their oxygen by frequent trips to the surface for air, by tapping air spaces in submerged plants, or by carrying a bubble or film of gas with them while they are submerged. The metapneustic system of larval Diptera is suited to obtaining air by returning to the surface (Fig. 5.5b). The spiracular opening is usually surrounded by hydrofuge areas or hairs that prevent entry of water into the spiracle.

Air spaces in plants are punctured by the specialized respiratory tubes of various unrelated insects: certain larval beetles (Donaciinae, Chrysomelidae; *Noterus,* Noteridae; *Lissorhoptrus,* Curculionidae) and flies (*Chrysogaster,* Syrphidae; *Notiphila,* Ephydridae; *Mansonia* and *Taeniorhynchus,* Culicidae).

Bubbles of air are carried beneath the surface by many aquatic bugs and beetles (Fig. 5.5a). Such temporary air stores are held in place by special hairs or are carried in cavities beneath the wings or elytra. The spiracles are situated to open into the air bubble. All aquatic Hemiptera in North America that swim beneath the water carry air stores. The manner of obtaining air varies: through a caudal siphon (Nepidae, Fig. 34.10b), caudal flaps (Belostomatidae), or abdominal tip (Naucoridae, Notonectidae), or through the pronotum (Corixidae). Aquatic beetles of the suborder Adephaga carry air under the elytra and renew the store by breaking the surface film with the tips of their elytra and abdomen. Aquatic members of the suborder Polyphaga have elytral air stores and usually also a coating of hydrofuge hairs on the ventral surface that holds a film of gas. Hydrophilid beetles break the surface with an antenna.

The bubble that is taken below serves not only as a store of atmospheric oxygen and as a hydrostatic organ, but also as a **physical gill,** which functions as follows. The gas in the bubble initially has the composition of air (viz., about 20

Figure 5.5 Examples of aquatic respiration: **a,** Aquatic beetle, *Tropisternus* sp. (Hydrophilidae), submerged with a shiny film of air that functions as a physical gill. **b,** Larvae and a pupa of a mosquito, *Culiseta* sp. (Culicidae), "hang" from the surface tension of water and obtain atmospheric air. The larvae have the metapneustic type of respiratory system with only the terminal spiracles open. The pupa has respiratory "horns" on the thorax.

percent oxygen, 79 percent nitrogen, and less than 1 percent carbon dioxide). The carbon dioxide produced by the insect rapidly diffuses into the water and is of negligible importance as a gas in the bubble. As oxygen is used by the insect, the content in the bubble falls below that in the surrounding water. Oxygen will then diffuse into the bubble from the water. The bubble of gas persists for a time because nitrogen diffuses out of the bubble relatively slowly. As a result, the insect is able to obtain considerably more oxygen than was originally contained in the bubble.

A remarkable modification that utilizes the principle of the physical gill is the **plastron.** In this device, a thin film of gas is held on the body surface indefinitely by dense hairs or a fine cuticular meshwork (Fig. 5.6). The European bug *Aphelocheirus* (Naucoridae, Hemiptera) has fine, short hairs over much of the body surface. The density of hairs is 200 to 250 million per square centimeter. The hairs are bent at the tips in an arrangement that supports the gas/water interface and resists compression of

Figure 5.6 Examples of plastrons: **a**, plastron crater of eggshell of fly, *Musca sorbens* (Muscidae); **b**, respiratory horn of egg of fruit fly, *Drosophila melanogaster* (Drosophilidae); **c**, metasternum of aquatic beetle, *Tylctelmis mila* (Elmidae); **d**, elytron of aquatic beetle, *Portelmis nevermanni* (Elmidae). (Scanning electron microscope photos by Howard E. Hinton.)

the gas film. The bug and similarly equipped beetles of the families Elmidae and Dryopidae can remain submerged for months.

The aquatic pupae of certain beetles (*Psephenoides,* Psephenidae) and flies (some Tipulidae, Blephariceridae, Deuterophlebiidae, Simuliidae, Empididae) have long cuticular processes developed from the peritreme and atrial region of one or more spiracles. These processes are called **spiracular gills.** The gill is associated with a plastron that is supported by a cuticular meshwork rather than by dense hairs. Likewise, the chorion of the eggs of some insects has a plastron that allows the embryo to respire when the egg is flooded or covered by a film of moisture.

The hair or cuticular plastron functions equally well when the insect or egg is out of the water, because the spiracles or embryonic respiratory organs are then in direct contact with air. In water that is deficient in oxygen, however, the physical gill fails, because oxygen no longer diffuses into the bubble. For this reason, insects with plastron gas exchange typically are slow-moving or are found in swiftly flowing streams or other well-oxygenated waters.

Various adaptations for gas exchange similar to the foregoing can be observed among insects in semiliquid media: scavengers in decaying plant or animal bodies or animal waste, phytophagous insects that burrow in juicy fruits or other plant tissues, and endophagous parasites and parasitoids immersed in hosts' blood. For example, small Hymenoptera larvae living as internal parasitoids of other insects may depend on cutaneous gas exchange but still retain tracheae inside their bodies.

ANATOMY AND FUNCTION OF THE CIRCULATORY SYSTEM

The circulatory system has multiple functions, including transportation of nutrients, waste products, and hormones; storage of nutrients and waste; hydrostatic pressure for molting and locomotion; and resistance to disease, wounds, and freezing.

The fluid, or **hemolymph,** that circulates in the bodies of insects differs in many respects from that of vertebrates. It is a single fluid with cells, not divisible into blood and lymph systems, that moves freely around the internal organs without a pressurized, closed network of blood vessels. The cells, or **hemocytes,** are highly variable in size, shape, number, and functions. The plasma is biochemically unique in the regulation of osmotic pressure by amino acids and organic acids instead of inorganic ions. Neither hemocytes nor plasma is significantly concerned with the transport of oxygen and carbon dioxide in most species. Consequently, insect blood is rarely red (see discussion of hemoglobin above) but may be colored yellow, blue, or green by various pigments, including α-carotene, riboflavin, and chromoproteins, but probably not chlorophyll.

The blood is limited to the major cavity of the body, the hemocoel. Insect hemolymph, however, does not directly bathe the cells of the body, as is often claimed. The inner surface of the integument and all the organs are covered in a basement membrane of connective tissue. This membrane probably is interposed between the hemolymph and all other tissues. Thus situated, a "blood-cell barrier" is created that regulates the exchange of materials between the body cells and the hemolymph.

Dorsal Vessel

The circulation of the hemolymph is not haphazard but follows a pattern of flow. Motion is imparted to the fluid by movements of the body segments and of the gut, and by the peristaltic contractions of the **dorsal vessel** (Fig. 5.7a). This is a tube developed middorsally from the embryonic mesoderm and extending almost the full length of the body. In the abdominal region the vessel is closed at the posterior end but perforated by paired segmental openings, or **ostia.** Insects may have as many as 12 pairs, bearing valvelike flaps that

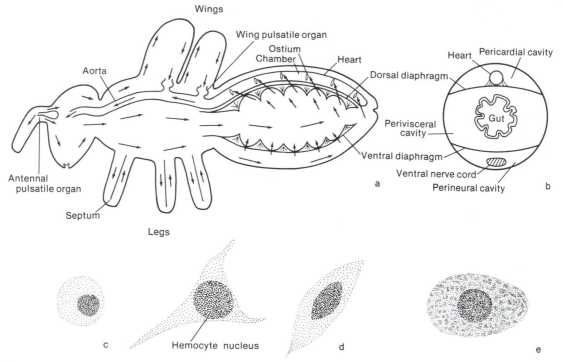

Figure 5.7 Circulatory system and hemocytes: **a**, diagrammatic longitudinal section of general structure and direction of circulation; **b**, cross section of abdomen; **c**, prohemocyte; **d**, plasmatocyte; **e**, granulocyte. (a modified from Weber, 1933.)

project into the lumen of the vessel. The portion of the dorsal vessel with the ostia is called the heart. It is slightly inflated, especially between the pairs of ostia. These dilations create the chambers of the heart. The heart is suspended middorsally in the abdomen by a transverse membrane of connective tissue and thin muscles. This **dorsal diaphragm,** with its **alary muscles** (meaning "winglike" in shape), is not complete but is interrupted segmentally by openings. By this means blood can pass from the vicinity of the gut or **perivisceral** cavity to the space above the heart, the **pericardial** cavity. Also in the abdomen is a **ventral diaphragm,** formed by an incomplete sheet of connective tissue. This membrane extends transversely, usually above the ventral nerve cord, and partly separates the perivisceral cavity from the ventral **perineural cavity.**

Movement of Hemolymph

Muscles embedded in the fibrous tissues of the heart wall provide the sequential contractions that impel the blood forward. As the wall of a chamber begins to contract (the systolic phase), the ostia close to block the rear chamber. Thus, blood moves forward to the next chamber. When the wall relaxes (the diastolic phase), the ostia open and the flaps admit blood from the outside and from the chamber to the rear. The flow of blood is normally toward the head, but reversals can be observed. The rate of heartbeat varies widely with age, species, state of nervous excitement, and environmental factors such as temperature. The initiation of contraction may come from an independent but diffuse "pacemaker" in the heart, or it may be partly or entirely controlled by the central nervous system.

The **aorta,** a continuation of the dorsal vessel, conveys blood to the head. This part of the vessel lacks ostia and is generally smaller in diameter. It is not supported by the dorsal diaphragm, but it may be suspended by a vertical sheet, or **dorsal septum,** in the thorax. The aorta opens beneath the brain and above the pharynx. Blood released here moves posteroventrad, entering the appendages of the head and the thorax. Flow into and out of these cylindrical structures is made possible by a septum that divides the appendage lengthwise into two channels. Blood also enters the wing veins, flows in an orderly but variable pattern throughout the wing, and returns. **Accessory pulsating membranes** aid local movement of the blood. After passing around the leg muscles and flight muscles in the thorax, the blood returns to the abdomen. The movement here is upward from the perineural and perivisceral cavities through the dorsal diaphragm and back to the heart. Although there is almost no blood pressure, complete mixing time in an insect has been estimated at from 8 to 10 or up to 30 minutes. In humans the time is 2 to 4 minutes.

Hemocytes

Unlike the blood cells of humans, the cells or hemocytes of insects vary greatly in size, cytological details, and the numbers contained in a sample of hemolymph. Common types have been named prohemocytes, plasmatocytes, and granulocytes (Fig. 5.7c–e). Hemocytes function in resisting disease and injury, storing and distributing nutrients, and probably in many other aspects of internal physiology.

The cuticle of the body, gut, and tracheal system is the main barrier to entry of disease organisms. After entry of such pathogens insects are able to resist the progress of disease by **humoral reactions** and by hemocytic **phagocytosis** and **encapsulation.** Plasmatocytes and granulocytes protect against infections by phagocytizing viruses, bacteria, fungi, and protozoa. Immune reactions of the hemolymph also appear to regulate numbers of symbiotic microorganisms within insects. In the human louse, *Pediculus humanus* (Pediculidae, Phthiraptera), for example, symbionts move from protected intracellular locations within the mycetome to quickly invade the base of the ovarioles and ultimately the developing eggs (Fig. 13.2). Symbionts that do not succeed in penetrating the ovarioles are soon engulfed by hemocytes (Eberle and McLean, 1983).

Foreign objects, microbes, and endoparasites too large to be engulfed by a single cell are immobilized or killed by hemocytes of the same kind in a process called encapsulation. A capsule of connective tissue is secreted as the inner layer of cells flatten over the object's surface and new cells are added outside. The capsule may remain clear or become darkened by melanization.

Insects do not have the equivalent of mammalian lymphocytes or their serum proteins, the immunoglobins. However, insects do have substances in the hemolymph or hemocytes that function in a humoral response to infection by acting on foreign antigens. Lectins have been identified in insect hemolymph, are inducible by injury, and are known to agglutinate microbes. Also inducible by bacterial infection are defensive proteins such as the peptides called cecropins, which disrupt the cell membranes of bacteria and other pathogens. Not inducible, but probably damaging to microbes, are phenoloxidase and lysozyme in the hemolymph.

The risk of bleeding is greatly reduced among insects because of the exoskeleton and the lack of blood pressure. However, mechanisms exist in the hemolymph of many, but not all, insects for forming clots and repairing wounds. Prohemocytes gather at a wound, causing local coagulation of the hemolymph. Later, epidermal cells and possibly hemocytes seal the wound and repair the cuticle.

Composition of the Plasma

The cell-free fluid, or plasma, of the hemolymph contains water, inorganic ions, and organic molecules, including metabolic wastes. Some of these molecules may be associated with hemocytes. The volume of hemolymph and water content varies widely, up to 94 percent of the body weight. In vertebrate blood, inorganic ions determine osmotic pressure; but as the evolution of insects is traced, inorganic ions are replaced in the more advanced orders by amino acids and other small organic molecules. Thus, in Lepidoptera, Hymenoptera, and many Coleoptera, amino acids are the primary osmotic effectors. The high concentration of amino acids is the most distinctive feature of insect blood.

Organic molecules in the hemolymph include carbohydrates, amino acids, peptides, proteins, lipids, organic acids, organic phosphates, and glycerol. These substances are derived ultimately from nutrients in the diet and from symbionts. The main carbohydrate is the sugar trehalose, although glucose and some other sugars have been found in quantity in the honey bee and in certain flies. Glycogen and lipids are present, but are largely stored in the fat body. Depending on the stage of the life history, large amounts of amino acids are in the hemolymph, both from the diet and from synthesis, and are utilized for growth and metamorphosis. Insect hemolymph has a remarkable variety of peptides, polypeptides, and proteins. These function in a wide array of physiological processes, including immunity, regulation of growth, development, reproduction, metabolism, and homeostasis. Hemolymph components with hormonal or neurotransmitter functions are in low concentration.

Organic acids, especially those associated with the tricarboxylic acid cycle, are most prevalent in larval endopterygotes. Acid-soluble organic phosphates are synthesized and found in high concentrations in certain moths. Levels of waste end products of protein metabolism in plasma—including uric acid, allantoin, allantoic acid, urea, and, in some immature aquatic insects, ammonia—are sometimes high. Glycerol is accumulated in the plasma of certain insects as an antifreezing agent (see further on).

ANATOMY AND FUNCTION OF THE EXCRETORY SYSTEM

Excretory processes expel or render harmless metabolic wastes and other substances that are harmful, and at the same time they maintain water balance (Fig. 5.8a). Excretion thereby plays the key role in physiological **homeostasis** (i.e., the maintenance of a constant internal environment) by regulating body chemistry. In contrast to defecation, the physical elimination of food not digested or absorbed, excretion involves the movement of waste molecules across one or more plasma membranes. The principal wastes are nitrogenous compounds resulting from protein metabolism and the breakdown of nucleic acids. Also included as waste are excessive amounts of other substances, such as water or salts. Most of the excretory products in insects are regulated by the Malpighian tubules and the rectum acting in concert.

Nitrogenous Waste

Nitrogenous end products may be discussed in three groups: (1) ammonia, (2) urea, and (3) uric acid, allantoin, and allantoic acid. Ammonia is a direct product of the oxidative deamination of amino acids. Containing 82 percent nitrogen, it is an efficient waste product, but it is also highly soluble in water and highly toxic. Terrestrial animals must conserve water; hence ammonia is usually transformed to the less toxic urea (e.g., as in mammals, amphibians, some reptiles) or uric acid (e.g., as in birds). Among insects, ammonia is the primary waste of species with access to water, such as the xylem sap feeders and the aquatic immature stages of some dragonflies (Aeshnidae), alderflies (Sialidae), and blow flies (Calliphoridae). The rapid flow of water through the gut in such insects rapidly dilutes and eliminates excreted ammonia to avert its toxicity.

Urea, containing 46 percent nitrogen, is the primary waste of carpet beetles (Dermestidae).

Most insects excrete uric acid, one of its breakdown products (allantoin and allantoic acid), or a combination of these substances. These are the least soluble, as well as the least toxic, end products, and they contain less nitrogen (32 to 35 percent). Some uric acid is derived from the breakdown of nucleic acids, but the bulk of these wastes must be synthesized by energy-consuming pathways.

Waste Storage and Reuse

Certain insects expel only some of the nitrogenous waste, storing the remainder in a harmless state or using it elsewhere in the body. Uric acid is deposited in special **urate cells** or **urocytes** of the fat body in the roach *Periplaneta americana* (Blattidae, Blattodea), in adult *Culex* (Culicidae, Diptera) mosquitoes, in the silkworm *Bombyx mori* (Bombycidae, Lepidoptera), and in the wasp *Habrobracon juglandis* (Braconidae, Hymenoptera). The rate of accumulation of uric acid thus stored provides an index of aging. Adult males of *Blattella germanica* (Blattellidae, Blattodea) store uric acid in accessory sex glands in amounts up to 5 percent of body weight. During mating, this accumulation is discharged as a white covering on the spermatophore. The white color of crystalline uric acid is used as a pigment in the epidermis of some insects or conspicuously in the wing scales of butterflies in the family Pieridae.

Waste Elimination

Most of the nitrogenous waste is eliminated from the insect's body as part of regulatory processes involving other substances of small molecular size. The organs responsible are the **Malpighian tubules** (named for the seventeenth-century insect anatomist M. Malpighi) and the hindgut, especially the **rectum.** The filamentous, blind tubules insert and empty at the anterior end of the hindgut.

The walls of the Malpighian tubules and rectum are both one cell thick. Thus an internal and an external plasma membrane are situated between the hemolymph and ducts leading to the exterior. The cells of the Malpighian tubules are richly tracheolated and bounded by a basement membrane. The plasma membrane next to the hemolymph is deeply and complexly infolded; the lumi-

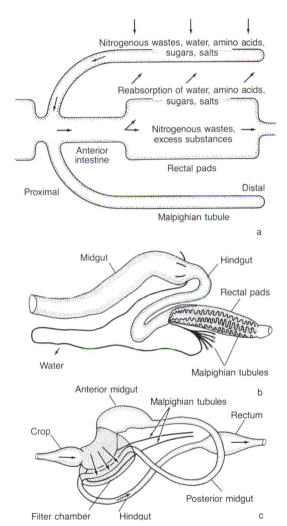

Figure 5.8 Excretory systems: **a**, generalized diagram of flow of nitrogenous wastes in the insect excretory system; **b**, cryptonephric system of *Tenebrio molitor* larva (Tenebrionidae) with five of the six tubules partly removed; **c**, diagram of filter chamber of Homoptera. (c modified from Weber, 1933.)

nal membrane is microvillate and associated with long mitochondria. The tubules are usually free in the hemocoel and may move by means of slender muscles placed lengthwise in a spiral fashion. Malpighian tubules are lacking in Collembola, *Japyx* (Japygidae, Diplura), and most aphids; are rudimentary in certain Protura, Diplura, and Strepsiptera; and are only two in number in Coccoidea and some larval parasitoid Hymenoptera. Other insects usually have multiples of 2, commonly for a total of six. Fifty or more tubules are formed in some Ephemeroptera, Plecoptera, Odonata, Orthoptera, and aculeate Hymenoptera.

The composition of the hemolymph, is regulated in the following manner (Fig. 5.8a). Water, salts, sugars, amino acids, and nitrogenous wastes pass into the lumen of the Malpighian tubules, beginning at the tip or distal end. The filtrate then moves through the tubule to the hindgut and rectum. Along this route useful components are selectively reabsorbed back into the hemolymph while the waste and surplus are voided. In contrast to vertebrates in which the initial plasma filtration in the glomerulus uses hydrostatic pressure, the process in insects involves energy-consuming active transport of ions. Potassium or sodium ions or both are transported across a steep electrochemical gradient from hemolymph to lumen. The movement of these ions by a cation pump and the transport of chloride ions is linked to the secretion of urine in the lumen. Water, sugars, amino acids (except proline, which is transported), and nitrogenous wastes are thought to be passively carried along with this flow, while some other substances are actively transported. Glucose and trehalose may be returned to the hemolymph at once from the tubule. The fluid is nearly iso-osmotic with the hemolymph when it is initially secreted.

Selective reabsorption may begin in the tubule and the anterior parts of the hindgut, but in most insects the rectal pads comprise the major site of activity. Active transport is again involved in the movement of potassium, sodium, and chloride ions against an osmotic gradient. Here the molecules must move from the lumen of the rectum, through the rectal cuticle, through or between the rectal

cells, and finally to the hemolymph. The recovery of useful amounts of water, amino acids, salts, and sugars is linked to this process, but the molecular pathways remain unclear.

In larvae and adults of Coleoptera and larval Lepidoptera, the distal ends of the Malpighian tubules may be held in contact with the rectum. This **cryptonephric** arrangement of the tubules has been carefully studied in the mealworm, *Tenebrio molitor* (Tenebrionidae, Coleoptera). The adults and larvae live in dry, stored grains, where conservation of water is required. The rectal complex is specially developed to dehydrate feces before elimination. Water is even extracted from air around the feces if the humidity is greater than 88 percent. The six tubules are held in contact with the rectum by a multilayered perinephric membrane and an outer membrane (Fig. 5.8b). This arrangement allows the potassium pumps in the tubules to generate an osmotic gradient strong enough to draw water from the rectum across the rectal cells and into the tubule. The fluid in the tubule is moved proximally into the free portion of the tubule, where it is passed to the hemolymph. The waste is eliminated as a dry powder.

Insects that feed directly on the sap in the vascular tissues of plants have the opposite problem: excessive fluid and dilute nutrients. In the Cicadoidea (Hemiptera, suborder Homoptera), the basal portions of the Malpighian tubules, the posterior end of the midgut, and the beginning of the hindgut are folded into and surrounded by the anterior midgut (Fig. 5.8c). The membrane-enclosed area of contact is called the **filter chamber.** This structure concentrates ingested plant sap by shunting water from the anterior midgut to the hindgut. As a result, the density of nutrients in the absorptive regions of the midgut is increased. Cations from the hindgut are actively transported from the hindgut contents into the distal ends of the Malpighian tubules and then down into the proximal parts of the tubules. Water moves osmotically from the proventriculus into proximal parts of the Malpighian tubules because the proximal parts contain increased salt concentrations. From the lumen of the tubules, the fluid is expelled into the hindgut. Thus, the bulk of the liquid is excreted, bypassing the rest of the midgut where nutrients are concentrated, digested, and absorbed.

MUSCLES AND LOCOMOTION

Insects move from place to place in virtually every natural medium on the earth's surface except solid rock, ice, or rarely, salt water. On the surface of the ground they wiggle, hop, walk, or run, often with surprising agility. By their feeding and burrowing activities, they move through the flesh of animal hosts, the tissues of plants, including solid wood, and the leaf litter, soil, or subsoil of the earth. In the aquatic environment some skate on the surface, propelled by their legs or by chemicals that alter the surface tension. Under the surface they swim with the legs or wings, with body movements, or by means of jets of water expelled from the anus. The best speeds, the longest distances, and probably the safest trips are achieved in the air. Yet, compared to locomotion on the ground or in water, flight consumes much more energy and requires a lightweight body equipped with wings and muscles for flight, together with keen senses and a nervous system for coordination.

Insects are the only invertebrates to fly actively and to control the duration and direction of their flights. The only other airborne invertebrates are the ballooning spiders and some mites, which, however, are passive drifters. As shown by the fossil record, insects had wings and presumably were flying in the Pennsylvanian Period (323 to 290 million years ago; see Table 15.3), over 150 million years before the first flying reptiles (Pterosauria) appeared in the Jurassic (208 to 146 million years ago). Much later the birds and mammals (bats) each independently evolved flying mechanisms.

Insect Muscle

The power for locomotory and other movements is supplied by the muscular system. Internal organs, such as the gut, heart, or Malpighian tubules, also have sets of muscles. All muscles of insects are **striated**; that is, each fiber has a crossbanded appearance when viewed under the microscope (Fig. 5.9a). In contrast, vertebrates and many other invertebrates have both the striated type and a second type, smooth muscle, which is composed of single spindle-shaped cells.

The muscular system of insects differs in several important respects from that of vertebrates. The most conspicuous difference is the relation to the skeleton: the muscles of insects are inside the skeleton, while those of the vertebrates are outside. As a consequence of the tubular construction, which resists bending, and the general absence of calcium salts, the insect skeleton is relatively lighter and stronger than the vertebrate skeleton and provides greater space and attachment areas for the muscles. The attachment of muscles to the skeleton is not by connective tissues, as in the vertebrates, but by **tonofibrillae.** Tonofibrillae are special secretions of the epidermal cells. The attachments may extend through the cells into the cuticle as far as the epicuticle. If so, the muscles remain attached to the old cuticle even while molting is in progress, thus allowing movements during this critical period. At the point of attachment, the cuticle may be invaginated into a tendon-like apodeme. Such apodemes lack elasticity when composed of normal cuticle; when the special cuticular protein resilin is present, however, a highly elastic tendon is formed.

Although insects perform feats of unusual strength, the power of their muscles is closely similar to that of vertebrates. Power is measured by the cross-sectional area of the muscle in relation to the greatest load it can lift. The explanation for the relative strength of insects lies in the effective use of lever systems and the small body size. As the body of an organism decreases in size, the volume or mass decreases as the cube root, whereas the cross-sectional area decreases as the square root. If the absolute power of a muscle per unit of area remains the same, an exactly scaled reduction in the size of an organism will result in relatively lighter loads for the smaller muscles.

Miniaturization of organisms is rarely linear, but the general relationship remains valid. Reductions in the sizes of insect muscles involve mainly reductions in the numbers of fibers rather than reductions in the sizes of the fibers. A species of small insect may have only a single fiber functioning in the position occupied by hundreds of fibers in a species of large insect.

Structure of Muscles

Each muscle is the product of many cells that share a common plasma membrane and an outer sheath, the **sarcolemma.** This outer covering may be invaginated into the muscle in certain places, forming a system of intracytoplasmic **intermediary tubules** (called the IT or T, for transverse, system). Tracheoles entering these tubules may reach the interior of the muscle without penetrating the sarcolemma. When the muscle is stimulated by an excitatory nerve, the T system allows the depolarization of the cell membrane to pass from the surface to the deeper regions of the muscle. Within the muscle are the contractile **myofibrils** traversing the length of the muscle. These striated units may exist as flat sheets (lamellae) or as threadlike fibers. Under great magnification with an electron microscope, each fibril is seen to be composed of protein filaments of two sizes. The thick filaments are myosin, and the thin filaments are actin. In cross section each actin filament of an insect muscle is usually situated between two myosin filaments. In vertebrate muscle the actin filament is between three myosin filaments. When the muscle contracts, the two sets of filaments are believed to slide past each other. The distribution of the two kinds of filament also provides an explanation for part of the banded, or striated, appearance. The I bands represent only actin filaments, the H band only myosin filaments, and the A band represents both together. The Z line has a disklike shape. Dispersed among the myofibrils are numerous nuclei and mitochondria and various amounts of a sarcoplasmic (endoplasmic) reticulum.

Individual muscles are grouped together into a "muscle unit," which has a separate tracheal supply. A single muscle unit or several units may be separately innervated by branches of one, two, or three motor axons (Chap. 6), making a motor unit. Finally, a single motor unit or several motor units may be grouped closely together into a functional "muscle," which is anatomically discrete from other such units, has definite areas of attachment, and exerts a pulling force on the attachments in a definite direction. Unlike vertebrate muscles, insect muscles lack connective tissues among the fibers. Ordinarily, one of the attachments of a muscle, called the origin, is located on an immovable part of the skeleton, and its insertion is found on a movable part such as a segment of an appendage or an articulating sclerite. Some muscles are bifunctional: that is, under certain circumstances the structures at either end may be moved. For example, some thoracic muscles function during walking by moving the leg bases and in flying by moving sclerites at the opposite ends of the muscles. The large indirect flight muscles move large portions of the thoracic skeleton simultaneously. The total number of muscles and the arrangements vary with the different kinds of insect. A caterpillar may have more than three times the 529 muscles present in humans.

Histological Types of Muscle

Although all insect muscle is striated, evolutionary changes in structure and function have led to several distinctive kinds of muscle. Differences become pronounced even in a single individual of a holometabolous insect, in which larval musculature differs histologically from that of the adult and the flight mechanism may include muscles with unique histological and physiological properties. Intermediate conditions in the anatomy of muscles may be discovered as the ontogeny and evolution of a taxonomic group are traced. For the purposes of discussion, five basic types can be recognized: larval, visceral, tubular, microfibrillar, and fibrillar.

1. **Larval** muscles, known from immature bees and flies, are distinguished by minute myofibrils with the nuclei located externally beneath the sarcolemma (Fig. 5.9a).

2. **Visceral** muscles are composed of parallel or anastomosing fibers similar to larval muscles in construction. They have larger myofibrils, however, and are associated with the internal organs of adult insects.

3. **Tubular** fibers are so named because the myofibrils are arranged in lamellae around a central canal filled with nuclei (Fig. 5.9b): these are the muscles in the appendages and between sclerites in many insects. The flight muscles of some Orthoptera are tubular and resemble the tubular muscles elsewhere in the body. The flight muscles of Odonata, however, represent an extreme modification of the tubular design; the lamellae are greatly thickened, and the enormous mitochondria measure up to 7 μm.

4. **Microfibrillar** muscles have small myofibrils measuring 1 to 1.5 μm in diameter and are found in the flight musculature of Ephemeroptera, Dermaptera, and some Hemipteroidea

(some Homoptera), and in the endopterygote orders Hymenoptera (restricted to some Symphyta), Lepidoptera, Mecoptera, Neuroptera, Raphidioptera, and Trichoptera. Microfibrillar and fibrillar muscles possess threadlike myofibrils with the nuclei and mitochondria scattered among the fibrils.

5. **Fibrillar** muscles have large myofibrils measuring 1.5 to 5.4 μm in diameter (Fig. 5.9c). These coarsely fibered muscles are pinkish when freshly dissected. They are the main source of power for the flight mechanism of the orders Hymenoptera (most species), Diptera, and Coleoptera, as well as certain Hemipteroidea (some Psocoptera, Thysanoptera, Heteroptera, some Homoptera). The muscles that vibrate the tymbals of cicadas are also fibrillar.

Tubular, microfibrillar, and fibrillar muscles are involved with movements of the exoskeleton. Visceral, larval, tubular, and microfibrillar muscles respond to a nerve impulse by contracting **synchronously** with the impulses. Fibrillar muscles contract many times for each nerve impulse and are said to be **asynchronous** with the impulses. The asynchronous type has evolved independently from the synchronous type in 9 or 10 different lineages of insects.

Walking

Hexapods are distinct among the arthropods for having three pairs of locomotory appendages. In terrestrial locomotion, each leg must function in two ways: to support the body's weight above the substrate, and to thrust against the substrate for propulsion. The best explanation for the evolutionary choice of six legs lies in the fact that a minimum of three legs can provide support in a stable tripod arrangement while the other three legs are stepping forward (Fig. 5.10). The insect walks by shifting from alternate tripods of leg contact with the substrate. The sequence or rhythm of leg movements changes as the insect increases

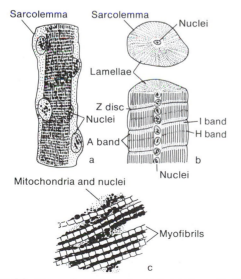

Figure 5.9 Types of muscles; **a**, larval, **b**, tubular, **c**, fibrillar. (Redrawn from Snodgrass, 1956)

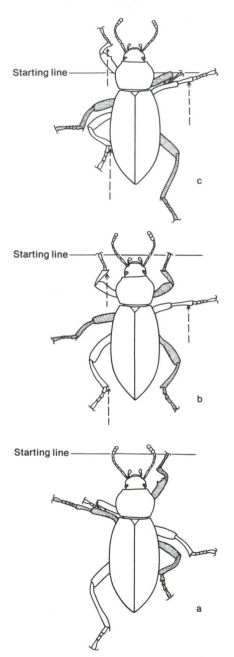

Figure 5.10 Diagram of walking movements of the beetle *Eleodes dentipes* (Tenebrionidae). Shaded legs are supporting the beetle as the unshaded legs step forward in the sequence **a** to **c**. At **c** the unshaded legs assume support of the body and the shaded legs begin stepping forward.

speed. Depending on the gait, more than three or fewer than three legs may be on the substrate at once. At high speeds, cockroaches run by alternate steps of only the hind legs. The rhythmic movements of insect legs originate in the central nervous system and are modified by feedback from sensory receptors in the legs. The segmental ganglia control their respective legs but are coordinated by intersegmental nervous coupling.

The forelegs, in front of the center of gravity, exert their force in a pulling direction, while the middle and hind legs, which are near or behind the center of gravity, exert a pushing thrust. The propulsive thrust of the hind legs is greatly developed as a jumping device in Orthoptera (Figs. 29.1, 29.3), many Homoptera (Fig. 34.13a,b), fleas (Figs. 43.1a,2), and flea beetles (Alticinae, Chrysomelidae). The middle legs of Encyrtidae (Hymenoptera) are modified for jumping. Jumping is also accomplished in other ways: click beetles snap their bodies at the joint between the prothorax and mesothorax, and springtails utilize an abdominal appendage (Figs. 17.2, 17.3a).

During walking, a stepping motion involves movements of the whole leg as well as movements of individual leg segments. The whole leg is moved by a complex musculature that originates inside the thorax and exerts pulling forces on the basal rim of the coxa. By the positions and insertions of the muscles, the whole leg can be moved forward, backward, away from the body, or toward the body. Movements of the individual segments of the leg are made by muscles operating in pairs that arise in one segment and exert a pulling force on a more distal segment. One member of the pair operates as a levator, raising the distal segment, and the other member operates as a depressor, lowering the distal segment. As shown by the skeletomuscular mechanism of a grasshopper's hind leg (Fig. 5.11), the femur is moved by a levator muscle from the coxa (1) and two depressor muscles, one from the coxa (2) and the other from the thorax (3). The tibia is moved by a levator muscle (4) and depressor muscle (5), both originating in the femur. The tarsus is moved by a levator

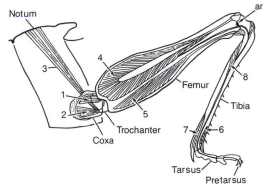

dinal indirect flight muscles extend between skeletal phragmata and, by their contraction, raise the center of the flexible notum. The **dorsoventral indirect flight muscles** operate as antagonists and depress the notum by their contraction. Mechanically, the wing base rests on the pleural wing process in the same manner as a lever on a fulcrum (Fig. 5.12a). Thus, contraction of the longitudinal indirect muscles causes an upward movement of the thoracic notum, which lifts the wing base and forces the rest of the wing downward (Fig. 5.12c). Likewise, contraction of the dorsoventral indirect muscles causes depression of the notum which lowers the wing base and forces the wing upward (Fig. 5.12b).

A simple up-and-down movement of flat wings would not, however, generate the forward thrust needed for flight. For propulsion, the wings must twist at the base so that the surface of the wing is at an angle to the body. On the downstroke, the leading (anterior) edge of the wing is lowermost and the wing is said to be pronated. On the upstroke, the leading edge is uppermost and the wing is supinated. At the anterior edge of the wing base is located a small sclerite, the basalare, and at the posterior base of the wing is the subalare (Fig. 2.12a). Each sclerite has a direct muscle from the pleuron. Contraction of the **direct basalare muscle** depresses the leading edge of the wing to produce **pronation** (Fig. 5.12a,c). Similarly, contraction of the **direct subalare muscle** depresses the trailing edge, producing **supination.**

On meeting the resistance of the air, and depending on the structure of the wing veins and flexibility, the wings may bend and fold in a regular pattern of deformation during each stroke. In the grasshopper, the hind wing traverses a slightly greater amplitude and leads the forewing, thus avoiding its turbulence. The elastic properties of the thorax and the antagonistic muscles are greatly enhanced by special deposits of the elastic protein resilin on the pleural wing process.

The general design of the flight mechanism of the grasshopper is found throughout the Orthopteroidea and in some Neuropteroidea. The muscles of

Figure 5.11 Skeleton and muscles of the hind leg of a grasshopper. Note that on the hind leg the trochanter is fused to the base of the femur. Numbers 1 to 8 are muscles described in the text; ar, articulation of femur and tibia. (Redrawn from R. E. Snodgrass, *The Principles of Insect Morphology.* Copyright © by Ellen Burden and Ruth Roach. Used by permission of the publisher, Cornell University Press.)

muscle (6) and a depressor muscle (7) that originate in the tibia and insert on the base of the tarsus. The individual tarsomeres do not have muscles. The pretarsus has only a depressor muscle (8) that originates in the femur and tibia and inserts by means of a long tendon on the base of the pretarsus.

Flight

The historical origin and evolution of wings are discussed in Chapter 15. Here we are concerned with the skeletomuscular mechanism. Flight in an insect such as the grasshopper is powered by muscles acting within the mesothoracic and metathoracic segments. Each of the thoracic segments is like a box that is partly rigid and partly elastic. The large flight muscles are divided into **indirect** muscles, attached broadly within a thoracic segment and affecting the wing indirectly by deforming the thoracic box; and **direct** muscles, inserting directly on the basalare and subalare sclerites near the wing base.

Lengthwise in the thorax, the **dorsal longitu-**

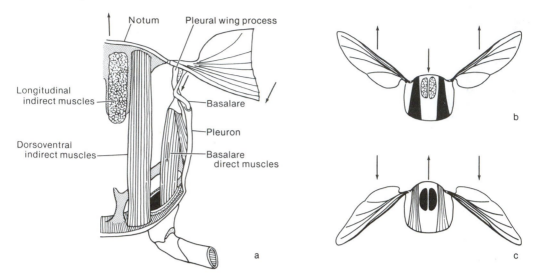

Figure 5.12   Flight mechanism: **a**, cross section of a mesothoracic segment during the downstroke and pronation of the wing (view is of left side of insect as seen from in front); **b**, wing upstroke and supination (contraction of dorsoventral indirect muscles indicated in black); **c**, wing downstroke and pronation (contraction of longitudinal indirect muscles indicated in black). (Redrawn from R. E. Snodgrass, *The Principles of Insect Morphology.* Copyright © by Ellen Burden and Ruth Roach. Used by permission of the publisher, Cornell University Press.)

Odonata and Blattodea are of the tubular type, but other Orthoptera exhibit transitional states between tubular and microfibrillar types. Microfibrillar muscle is found in the neuropteroid orders Neuroptera, Raphidioptera, Lepidoptera, and Trichoptera. Exceptions are the Odonata, Blattodea, and Mantodea, where the downstroke is powered by direct muscles rather than the longitudinal indirect muscles.

In summary, with the exceptions just noted, the orders discussed above have two sets of synchronous muscles, the indirect and direct muscles, which provide the power and the twisting of the wings. Both the mesothoracic and metathoracic segments are usually so equipped. The wing beat frequency and, consequently, the speed of flight in all these orders are limited to the rate at which the nervous system can send impulses to each of the participating muscles in a coordinated fashion (i.e., about 50 per second).

A second major type of flight mechanism is found in the Diptera, Hymenoptera, Coleoptera, and some Hemipteroidea. These orders have a high wing beat frequency that depends on fibrillar flight muscles contracting asynchronously with many contractions per nerve impulse. The wing beat frequencies of bees and flies is on the order of 100 to 300 beats per second. With wings shortened experimentally and at a high temperature, the midge *Forcipomyia* (Ceratopogonidae, Diptera) was rated at a frequency of 2218 strokes per second.

In the second group of orders, the main source of power for the wings resides in one thoracic segment and is often limited to one set of longitudinal and one set of dorsoventral muscles operating antagonistically. In flies, elastic parts of the thoracic box are stretched during one part of the wing stroke, storing energy, then quickly released by a "click" mechanism to assist the next movement. The direct muscles to the basalare and subalare may be fibrillar in some insects, contributing their usual roles; but in the most advanced fliers, these muscles are tubular and no longer are responsible for the pronation and supination of the wing. The complex path of the wing is now determined mechanically by the shapes of the articulatory sclerites at the base of the wing and the adjoining thoracic processes. You can demonstrate this for yourself in a freshly killed bee or fly. Depression of the mesonotum will automatically supinate the wings as they are forced upward. Or, raise the mesonotum with a pin and watch the wings pronate as they are forced downward. The direct muscles of these insects serve to modify the wing stroke over many beats. This allows bees and flies to hover and maneuver in flight.

The Venturi effect of lowered static pressure created by the higher velocity of air over a wing's upper surface creates lift for flight. Many insects, especially those with large wings, use their momentum to glide between wing strokes. Diptera have the most accurate control of hovering flight. Their greatly reduced hind wings, the halteres, do not provide lift. Instead, they vibrate to act as gyroscopic sensors to detect any change in body position relative to the earth. Reductions in the amplitude of the wing beat on one side reduce lift and cause a rotation toward that side. Varying the angle of pronation of the wings, in conjunction with wing beat frequency, changes lift to climb or dive.

Minute insects, like Thysanoptera and some Coleoptera (Ptiliidae) and Hymenoptera (Mymaridae, Trichogrammatidae), are so small that the viscosity of air relative to their size acts more like a fluid than a gas. Their tiny paddle-shaped wings are fringed with fine bristles that utilize drag forces on the wings rather than lift, so that their flight is analogous to swimming through the air.

THERMOREGULATION

Insects are known to be active within definite ranges of environmental temperature. The snow scorpionflies *Boreus* (Boreidae, Mecoptera; Fig. 9.5), snow fly *Chionea* (Tipulidae, Diptera), snow "fleas" *Achorutes* (Poduridae, Collembola), and winter stoneflies *Allocapnia* (Nemouridae, Plecoptera) are active at low temperatures on snow. *Boreus* dies if held for long in the human hand. At the other extreme, the firebrat *Lepismodes inquilinus* (= *Thermobia domestica;* Lepismatidae,

Thysanura) prefers 12 to 50°C and dies at 51.3°C. At unfavorably high ambient temperatures, an insect is severely limited in the extent to which it can cool itself. Cooling by the evaporation of water quickly leads to desiccation unless water is freely accessible. Heat is readily regained because of the surface/volume relationship. The tsetse fly *Glossina morsitans* (Glossinidae, Diptera) feeds while exposed to sun on the hot skin of African mammals. The opening of the spiracles at temperatures above 39°C permits evaporative cooling within the tracheal system to lower the body temperature about 2°C. This is possible because the meal contains ample liquids to restore lost water. Honey bees transport water to their hives, deposit it in the combs, and vigorously fan with their wings so that evaporative cooling takes place. Most insects, however, escape heat by seeking shade or cooler places.

Ectothermy

Although insects usually cannot lower their temperature below that of ambient, many can actively regulate their temperature above that of the environment and benefit from a higher metabolism rate, the distinct disadvantage of small size notwithstanding. The high surface/volume ratio results in rapid transfer of body heat to the environment. At rest, the body temperature of an insect soon becomes nearly the same as the surrounding medium.

Insects are able to increase their body temperature well above ambient by absorbing heat from the sun. Organisms that derive heat almost entirely from their environment are called **ectothermic.** Butterflies assume distinctive postures while basking in the sun. "Dorsal baskers" spread their wings and point their heads away from the sun. The skippers (Hesperiidae) commonly spread only the hind wings, leaving the forewings vertical. "Lateral baskers" keep both wings vertical over the body and orient themselves to present one side to the sun's rays.

Endothermy

Energy metabolism is a second source of heat. The flight muscles of insects are metabolically the most active tissues known. A substantial part of the energy that undergoes conversion is degraded to heat. Insects with insulating hairs or scales, or with free access to high-energy food, or both, are able to raise and maintain their bodies above ambient temperature by metabolic heat for varying periods. Such insects are called **endothermic.** Essentially all the heat comes from muscular activity.

The flight muscles of larger insects require higher temperatures, in some species near or above 40°C, to generate wing beat frequencies sufficient to lift the insect. Various species, but not all, in Odonata, Hemiptera, Orthoptera, Lepidoptera, Coleoptera, Hymenoptera, and Diptera have been observed warming up before flight by contracting the flight muscles. Those with indirect flight muscles of the synchronous type visibly quiver or "shiver" the wings as the muscles contract against each other. Flight mechanisms with asynchronous muscles may be uncoupled from the wings so that no movement is seen. Basking can be combined with shivering to bring the thorax up to the required temperature.

Once the insect is in flight, the muscles continue to produce heat. Thoracic temperatures of flying butterflies and locusts are 5 to 10 C° above ambient, while well-insulated moths and bumble bees fly at 20 to 30 C° above ambient. A thoracic temperature of 36°C was recorded in a large queen bumble bee while it was flying in air at 3°C.

Thoracic temperatures of flying endothermic insects are regulated within narrow limits even though a wide range of ambient temperatures may be experienced. Temperatures in other parts of the body vary. A study of thermoregulation in the sphinx moth *Manduca sexta* (Sphingidae, Lepidoptera) reveals how this is accomplished. Overheating is avoided in the thorax by transferring heat to the abdomen. The abdomen is thermally isolated from the thorax by a large air sac and is less well insulated by scales. Hemolymph flows posteriorly from the hot thorax to the abdomen, where the heat is dissipated by convection. The cooler hemolymph is pumped forward by the heart to the thorax and travels a sinuous course in the aorta between the flight muscles, thus cooling them. The rate of heartbeat is apparently controlled by the nervous system in the thorax or head. During warm-up, the flow of hemolymph is restricted between the thorax and abdomen, allowing heat to accumulate. When overheated, the heart pumps strongly.

INSECT COLD HARDINESS

Insects suffer varying consequences when temperatures fall below their growth and activity thresholds. They may become immobilized at temperatures above freezing and, if kept in this state, will die from desiccation because they are unable to restore lost water by drinking. Insects living in temperate or colder regions, such as the Arctic or high mountains, risk freezing when air temperatures drop below 0°C. Yet some insects do survive subzero weather, and others live normally after being frozen.

Some insects avoid low air temperatures by virtue of their habitat or behavior. Aquatic insects and those in deep, moist soil escape because freezing usually occurs only at the water or soil surface. In winter, honey bees cluster together in a mass in the combs of their hive. Even when air temperatures outside are well below 0°C, they maintain 20 to 30°C inside the cluster. Heat is derived from the metabolism of stored honey. Winged adult insects that migrate, such as the monarch butterfly, fly south in autumn to warmer regions and return northward in spring.

Immature stages, however, are often on the soil surface, on plants or other objects, and must endure exposure to cold weather. Diapause is a physiological state that is associated with overwintering (Chap. 4). We now describe other processes that are usually combined with diapause to ensure survival even in the Arctic. These processes confer **cold hardiness.**

Susceptibility and Tolerance to Freezing

When water is cooled to 0°C and below, freezing does not necessarily take place even though 0°C is ordinarily considered the freezing point. Before ice can be formed, some molecules of water must be arranged in a certain pattern to provide a nucleus for growth of the crystal. Until by chance the pattern occurs, the liquid can be further cooled. In careful experiments with small droplets of pure water, −40°C has been reached before crystallization. Water that remains liquid below its freezing point is said to be **supercooled.** The temperature at which a crystal begins to form is the supercooling point. Usually crystallization is instantaneous and only one crystal is formed. During the process of crystallization heat is released, raising the temperature toward the freezing point. When the crystal is complete, the temperature can be further lowered.

Mineral particles, especially dust, supply surfaces on which ice nuclei readily form. In other words, the particles serve as "nucleators." The presence of dust nucleators in the gut of an insect or of other nucleator substances in the hemolymph raises the supercooling point by increasing the likelihood of freezing. On the other hand, the addition of dissolved substances in water lowers the points of freezing and supercooling. Even without special antifreeze substances, the body fluids of overwintering insects can often be supercooled to −20°C. Thus many insects survive subzero weather by evacuating their guts in autumn and supercooling without harm, to temperatures above the supercooling point. When the supercooling point of the insect is reached, an ice crystal will form internally around a nucleator.

Insects that die when frozen are called **freezing-susceptible.** Those that can survive freezing during winter are called **freezing-tolerant.** Freezing generally begins in the hemolymph and may or may not include the cells. Extracellular freezing is not as severe as intracellular freezing, yet the cells may become dehydrated. The nature of the fatal damage to susceptible insects is not clear. Possibly crystals mechanically rupture cells. Alternatively, the removal of water by ice formation may cause an imbalance in the concentration of chemicals in the hemolymph and cell fluids.

Glycerol and Freezing Tolerance

Glucose, trehalose, low molecular weight lipids, and sorbitol in the hemolymph give additional protection against freezing. No natural chemical, however, equals glycerol as an antifreeze. Glycerol has been found in most overwintering, freezing-tolerant insects, especially larvae and pupae. The action of glycerol is to extend the temperature range of supercooling without freezing, to retard the rate of freezing, and to reduce the size of crystals. When the insect is indeed frozen, glycerol presumably reduces deleterious osmotic effects and prevents the intracellular freezing that would be fatal.

Studies of overwintering adults of the Arctic ground beetle *Pterostichus brevicornis* (Carabidae, Coleoptera) provide an example of the ecology, behavior, and physiology of an insect in a cold environment. The beetle is a scavenger that lives in forest litter in circumpolar regions. Near Fairbanks, Alaska, summer is spent in the upper 2 to 3 cm of surface litter that is warmed to 10 to 14°C by air and intermittent sunshine. Only a few centimeters below the surface, the temperature of litter drops to 0 to 1°C because of the underlying permafrost. After the first frost in autumn and independently of photoperiod, the adult beetles migrate by the thousands into the old galleries of wood-boring beetles in decaying tree stumps. The beetles continue to move toward the interior of the stump, favoring regions of −1 to 2°C, until immobilized and finally frozen by the approaching winter.

In winter the stumps may be exposed to −40°C for several weeks and may reach −60°C for shorter periods. Temperatures in the interior of the stumps lag behind the outside temperature. Stumps also are often covered by an insulating blanket of snow. Thus the hibernating beetles are buffered against rapid temperature changes. When beetles are kept at 5°C, no glycerol is detectable in the hemolymph. Without glycerol, the freezing

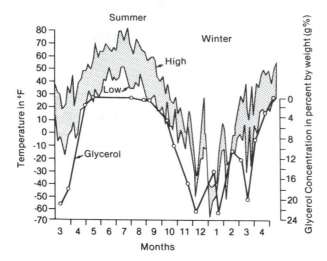

Figure 5.13 Seasonal variations in the average glycerol concentration in hemolymph of the Arctic carabid beetle, *Pterostichus brevicomis,* and high/low ambient temperatures near Fairbanks, Alaska. Note that the glycerol scale is inverted. (Data from Baust and Miller, 1970.)

point of hemolymph is −2°C and the supercooling point is −7.3°C. Synthesis of glycerol is stimulated by exposure to 0°C for more than 24 hours. Glycerol then continues to accumulate at temperatures below freezing (Fig. 5.13). A sudden drop in temperature to −15°C or more is fatal. Beetles inside stumps, however, are protected against such rapid changes. If given time to synthesize glycerol, the beetles become freezing-tolerant.

Other adult insects are generally not known to be freezing-tolerant. In this respect the carabid is unique, but the annual cycle of glycerol production and loss resembles that known in less detail for overwintering larvae and pupae of other insects. It is also interesting to note that the carabid is able to digest food, develop eggs, and exhibit other signs of physiological activity while in a cold-hardened and frozen state. Although only the adults have thus far been analyzed, the beetles also overwinter as larvae or pupae. Some individuals overwinter twice and live up to 36 months.

Beetles start hibernation with a water content at about 62 percent of the body weight, abundant fat, and the guts filled with the summer diet of small insects and other arthropods. During the first few months of hibernation, digestion and excretion continue until the gut is emptied. Body weight declines as the gut is cleared, fat is consumed, and

water content drops about 10 percent. Then in January and February whenever temperatures exceed −3 or −4°C, the beetles began to crawl about and feed. The diet is now rotten wood containing bacteria and fungi. The ovaries become active and eggs start to develop. Fat droplets in fat cells fluctuate in abundance as a result of the renewed intake of food (increasing fat) and the growth of eggs (decreasing fat). Internal water content begins to rise as water in the stump thaws and becomes available. In the spring, the glycerol concentration declines steadily. After one day at 21°C, the beetles have completely lost their tolerance to freezing; all die if returned to below-freezing temperatures.

SELECTED REFERENCES

General Structure and Function of Internal Organs

Chapman, 1982, 1991; Kerkut and Gilbert, 1985; Snodgrass, 1935; Wigglesworth, 1972.

Nutrition

Scriber and Slansky, 1981.

Alimentary Canal, Digestion, Absorption

Billingsley, 1990; Chapman and de Boer, 1995 (regulation of feeding); Cohen, 1995; Retnakaran and Binnington, 1991 (peritrophic membrane); Ribeiro, 1987; Shapiro et al., 1988; Terra, 1990.

Tracheal System and Respiration

Hinton, 1969a,b; Nicolas and Sillans, 1989.

Circulatory System

Gupta, 1985 (hemocytes); Haunerland and Shirk, 1995 (fat body); Telfer and Kunkel, 1991.

Excretory System

Bradley, 1987; Cochran, 1985; Hadley, 1994; Pannabecker, 1995

Muscles and Locomotion

Brackenbury, 1992; Brodsky, 1994; Delcomyn, 1985; Wootton, 1992

Thermoregulation, Endothermy

Heinrich, 1973, 1974, 1993; Heinrich and Bartholomew, 1972; cold hardiness of *Pterostichus:* Baust, 1972a,b, 1976; Baust and Miller, 1970; Kaufmann, 1971; L. K. Miller, 1969.

6
RECEPTION OF STIMULI AND INTEGRATION OF ACTIVITIES

Male western cecropia moth, *Hyalophora euryalis* (Saturniidae) with large plumose antennae that detect minute quantities of airborne sex attractants from the female moth. American physiologist C. M. Williams used *Hyalophora cecropia* for experiments on metamorphosis. The adult males were unexpectedly found to have large amounts of juvenile hormone in the abdomen.

In this chapter we describe the last of the organ systems, the nervous system, and then shift our emphasis to the varied relationships of insects to their environment. Beginning with the structure and function of nerve cells, we progress through the major subsystems, including the sense organs by which insects detect changes in their external world. The responses of insects to internal and external stimuli are discussed in the context of the physiological state of the individual, the specific anatomical characteristics, the capacity of the nervous system to exhibit complex behavior, and the adaptive significance of the behavioral acts.

After an introduction to the kinds of behavior exhibited by insects, we conclude the chapter with a discussion of communication and circadian rhythm.

ORGANIZATION OF THE NERVOUS SYSTEM

Neurons and Neuronal Connections

The nervous system has cells of two kinds: **neurons,** which are excitable by chemical or electric signals and transmit electric impulses rapidly from one part of the insect to another, and **glial cells,** which support, protect, and nourish the neurons. The nucleated cell body of a neuron is the **perikaryon** (Fig. 6.1a). Extending from the perikaryon are one or more slender processes, the fibers or **axons,** by which the neuron connects with other neurons or muscles, or conveys neurosecretions to other organs or the hemolymph.

The usual point of contact, or **synapse,** between connecting axons or between axons and muscles is at the axonal tips. Unlike vertebrate nerve cells, insects do not have synaptic terminals on the perikaryon. In insects, the tips of the axons are finely divided into **terminal arborizations** and separated by a gap. For communication between neurons, impulses commonly are transmitted from the **presynaptic** axon across the gap to the **postsynaptic** axon by a chemical substance called a

neurotransmitter. The postsynaptic or receptive axon is also called the **dendrite.** Except at synaptic junctions, individual neurons are surrounded by layers of glial cells.

The chemical composition of the neurotransmitter may vary in different parts of the nervous system. Compounds that have been identified include acetylcholine, catecholamines (dopamine and noradrenaline), and others associated with neuromuscular synapses. In cholinergic transmission, the impulse crosses the gap by means of quickly released acetylcholine. Normally, the transmitter is then destroyed by the enzyme acetylcholinesterase and the impulse is stopped. Certain insecticides, such as the organophosphates, are anticholinesterases. They derive their lethal effect by acting as inhibitors of the enzyme, thus prolonging the action of acetylcholine at each impulse and disrupting nervous control of the insect's body. Organophosphate insecticides such as parathion have a similar effect on the human nervous system and are highly dangerous. However, some other organophosphates, such as malathion, have a different mode of action and can be used with much less risk to humans.

Neurons that have one axon issuing from the perikaryon are called **monopolar.** The single axon is at least forked so that one branch serves as a dendrite. Some monopolar neurons, however, are highly branched and interconnected with other neurons in complex ways. Neurons having two separate axons extending from the perikaryon are called **bipolar.** Again, one serves as the dendrite and the other leads to the next neuron. Multipolar neurons with more than two axons are rare in insects (stretch receptors; see below, "Mechanoreceptors").

Nerve impulses commonly begin in a sense organ, pass to the central nervous system, and then out to an appropriate **effector** (i.e., a muscle or gland that provides a response when stimulated). In the stable or resting state, the nerve cell is slightly deficient in sodium (Na^+) ions. This difference in resting potential between the interior and exterior of the cell membrane is about -70 mV. A nerve impulse is the wavelike change in

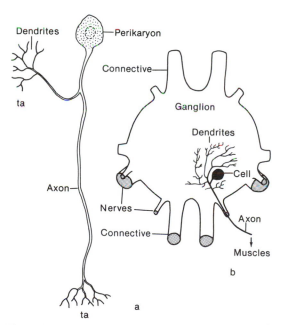

Figure 6.1 Structure of neurons and ganglia: **a**, diagram of a monopolar neuron; **b**, diagram of a monopolar motor neuron inside a thoracic ganglion. ta, terminal arborizations. (a, modified from R. E. Snodgrass, *The Principles of Insect Morphology.* Copyright © by Ellen Burden and Ruth Roach. Used by permission of the publisher, Cornell University Press; b, simplified and modified from Altman and Tyrer, 1974.)

permeability of the cell membrane to Na^+ ions that depolarizes the voltage difference to drop to just above zero millivolts for a few milliseconds, then recover.

In the simplest pathway, the impulse begins in the dendrite of a bipolar **sensory** neuron closely associated with a sense organ. The sensory perikaryon is usually located just beneath the sense organ, and it communicates with the segmental ganglion by a long sensory fiber. The neuron initiates impulses only in response to stimuli of certain kinds, depending on the structure of the sense organ and the special sensitivity of the dendrites. Thus only certain mechanical disturbances stimulate some neurons. For others, light waves, temperature, or chemicals trigger a nerve impulse in the dendrite.

Within the ganglion the sensory axon transmits impulses either directly by one synapse to a monopolar **motor** neuron or indirectly by two or more synapses involving other monopolar neurons called the internuncial neurons or **interneurons.** The speed of transmission of an impulse is delayed by each synapse even though the synapses are much shorter than the neurons. This is because neurotransmitter substances diffuse much more slowly than the depolarization impulse along the neuronal cell membrane. The motor neuron then completes the arc by sending the impulse over a long motor axon to stimulate the effector.

The nervous stimulation of muscle fibers in insects differs from that of mammals. In the latter the axon terminates in a single motor end plate, and the transmitter substance to the muscle is acetylcholine. Motor axons of insects are branched, providing many endings on a muscle fiber, and the transmitter is not acetylcholine. A muscle may receive stimulation from two or three different kinds of motor axon. Stimulation by a **fast** axon produces a "fast" response in the form of a brief powerful contraction. Each stimulation by a **slow** axon results in a small contraction, but when the stimulation is increased in frequency, a "slow" contraction of increasing force is produced. Slow axons thus give greater control over movements. The transmitter substance for these neuromuscular junctions is probably L-glutamate. The

third type of axon, if present, may inhibit the contraction of muscle and the transmitter is probably γ-aminobutyric acid.

Most nervous pathways are probably much more complex than the simple reflex arc. The various segmental responses to environmental stimuli are regulated by the brain and by the interactions of other segmental ganglia. Synapses may involve more than two neurons, and a given neuron may have more than two synapses.

Communication between ganglia is by internuncial axons. Some of these, called **giant interneurons,** are larger in diameter and span many segments without intervening synapses, thus bringing distant segments of the body into direct communication for rapid responses. For example, the cockroach *Periplaneta americana* (Blattidae, Blattodea) has giant fibers connecting mechanical receptor hairs on the cerci with ganglia as far anterior as the prothorax. Sense organs attached to the hairs are sensitive to movements of the hairs caused by air currents. When an approaching predator creates air currents, impulses from the receptors travel directly forward via the giant fibers and connect with motor neurons, resulting in a quick escape reaction.

Nerves and Ganglia

On a larger scale, the bundles of sensory and/or motor axons that issue from the segmental ganglia are called **nerves.** A **ganglion** is a mass of nervous tissue (Fig. 6.1b) often including thousands of neurons. The word is used loosely to apply not only to a single large mass but also to smaller masses that may be present inside. Thus each segmental ganglion is formed by the median fusion of a pair of embryonic ganglia. The central portion is the **neuropile,** consisting entirely of axons and processes of their associated glial cells. Included are sensory, motor, and internuncial axons and their various synaptic junctions. The cell bodies of the neurons and glial cells are located only around the outside of the neuropile. The paired longitudinal connectives between ganglia are bundles of internuncial axons plus sensory and motor axons that communicate with other segments.

The ganglia, nerves, and connectives are surrounded by a supportive sheath composed of an outer, noncellular **neural lamella** of connective tissue and an inner **perineurium** of glial cells. The sheath is a "blood-brain barrier," isolating the neurons from the chemical components of the hemolymph and maintaining around the axons a chemical environment that is essential for impulse transmission. The glial cells also pass nutrients to the neurons and probably prevent contact between neurons except at synapses.

The nervous system is divided into four subsystems: **sense organs,** including the light, chemical, temperature, and mechanical receptors; **peripheral nervous system,** including the sensory and motor neurons that communicate with the integument and muscles (this system is not discussed further); **central nervous system,** including the brain and the segmental ventral ganglionic chain that generally regulate the bodily activity; and **stomatogastric nervous system,** including several small ganglia that regulate the foregut, the midgut, and several endocrine glands.

CENTRAL NERVOUS SYSTEM

The central nervous system of insects includes the **brain,** situated above the mouth, and the **ventral nerve cord,** beginning behind the mouth. The evolution of the central nervous system was discussed earlier (Fig. 2.3). Recall that the basic structure of the system consists of paired embryonic ganglia connected transversely by commissures and longitudinally by connectives.

Brain

The brain has three ganglionic masses (Fig. 6.2): the protocerebrum, the deutocerebrum, and the tritocerebrum. The **protocerebrum** is the largest and most complex region. Laterally it receives nerves from the large optic lobes of the compound eyes; dorsally, it receives nerves from the ocelli. Internally, the protocerebrum has a pair of mushroom-shaped neuropiles, the **corpora pedunculata,** which are important centers for the integration of behavior. In most insects the main sensory input is from the antennal lobes. In social Hymenoptera the corpora pedunculata are especially large, suggesting that they are important in the complex behavior of these insects. Medially lies the **pars intercerebralis,** a complex cellular mass. Clusters of neurosecretory cells here and elsewhere in the brain communicate by long nerves with the neuroendocrine corpora cardiaca and corpora alata. Three other neuropiles are present in the protocerebrum: the median **protocerebral bridge** and **central body,** and the ventral, paired **accessory lobes,** which are connected by a commissure.

The **deutocerebrum** has motor and sensory axons to the antennae. The paired neuropiles, or **antennal lobes,** are connected by a commissure. Recall that both the protocerebrum and deutocerebrum were primitively in front of the mouth.

The **tritocerebrum** has sensory and motor axons to the labrum and to the frontal ganglion of the stomatogastric system. The paired neuropiles, or **tritocerebral centers,** are connected by the **tritocerebral commissure,** which is visible behind the esophagus. The tritocerebrum was primitively the first ventral ganglion behind the mouth and secondarily became incorporated into the brain.

Estimates of the number of neurons in the brain are about 850,000 for a worker honey bee, 1.2 million for a drone honey bee, 330,000 for a house fly, and 360,000 for a locust.

Ventral Nerve Cord

The ventral nerve cord is a chain of segmental ganglia connected anteriorly with the tritocerebrum by the paired **circumesophageal connectives.** The ganglia of the three gnathal segments are fused into a compound mass named the **subesophageal ganglion.** Motor and sensory nerves extend to the mandibles, maxillae, labium, salivary glands, and cervical muscles. The three ganglia of

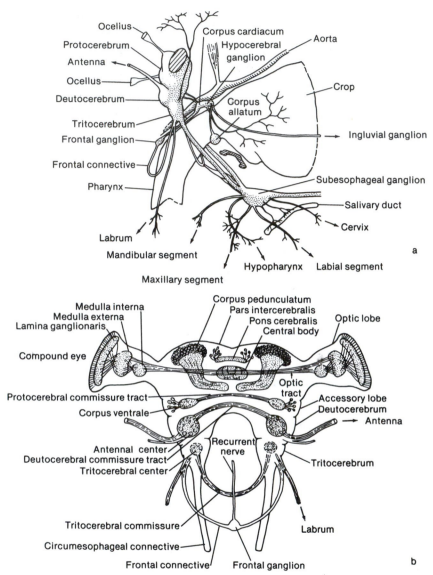

Figure 6.2 Insect brain: **a**, lateral view of nervous system in head of a grasshopper; **b**, diagrammatic cross section of brain, showing ganglia and commissures. (Redrawn from R. E. Snodgrass, *The Principles of Insect Morphology*. Copyright © by Ellen Burden and Ruth Roach. Used by permission of the publisher, Cornell University Press.)

the thoracic segments serve the legs and flight mechanism and are usually the largest ganglia of the ventral nerve cord. Sensory and motor nerves extend to the various sense organs and muscles. In the abdomen are no more than eight ganglia. The last is a large, compound ganglion serving the eighth and following segments, including the genitalia and cerci.

An evolutionary tendency exists for segmental ganglia to fuse anteriorly into compound ganglia (Fig. 6.3). In the grasshopper, for example, the first three abdominal ganglia are fused with the metathoracic ganglion. Coleoptera and Diptera have variable amounts of fusion among the ganglia. In nearly all the Hemiptera, all the ganglia of the thorax and abdomen are fused into a single compound ganglion, with nerves extending to the respective segments. During the life history of an insect, some separate ganglia in the larva also tend to be fused in the adult. This arrangement presumably permits more rapid communication among the ganglia by reducing the lengths of internuncial axons.

STOMATOGASTRIC NERVOUS SYSTEM

The stomatogastric system includes two median ganglia connected to the brain and a single ganglion or paired ganglia on the surface of the foregut. The **frontal ganglion** is situated medially in front of the pharynx, usually just behind the frontoclypeal suture. It receives the paired **frontal connectives** from the tritocerebrum and the single **recurrent nerve** from the hypocerebral ganglion. Some small nerves from the frontal ganglion innervate the foregut and mouth cavity. The frontal ganglion controls the emptying of the crop. If the ganglion is removed experimentally, undigested food accumulates in the crop. The ganglion probably also contributes neurosecretions to the endocrine system.

The **hypocerebral ganglion** is situated behind the brain, between the aorta and the foregut. It usually has three major connections: the single recurrent nerve running from the frontal ganglion; one or two esophageal nerves, running posteriorly to the ingluvial ganglia on the surface of the foregut and connections dorsally to the brain by way of the neuroendocrine corpora cardiaca. The hypocerebral ganglion also sends nerves directly to the wall of the esophagus.

The single or paired **ingluvial ganglia,** located posteriorly on the foregut, control gut movements through nerves to the foregut wall, proventriculus, and midgut. The neuroendocrine corpora cardiaca and the corpora allata make up the retrocerebral complex and are closely associated with the stomatogastric system (Chap. 4).

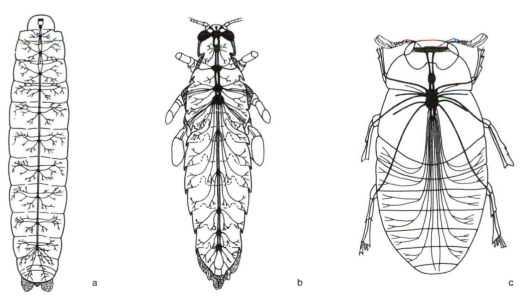

a   b   c

Figure 6.3 Modifications of the ventral nerve chain: **a,** system of a caterpillar, *Malacosoma americana* (Lasiocampidae); **b,** system of a grasshopper, *Dissoteira carolina* (Acrididae); **c,** system of a beetle, *Lachnosterna fusca* (Scarabaeidae). (a and b, redrawn from R. E. Snodgrass, *The Principles of Insect Morphology.* Copyright © by Ellen Burden and Ruth Roach. Used by permission of the publisher, Cornell University Press; c, redrawn from Packard, 1898.)

Often discussed along with the stomatogastric system are two groups of small nerves closely associated with the ventral nerve cord and called the **visceral nervous system:** (1) the unpaired ventral visceral nerves that originate from each ventral ganglion and run caudally to innervate the tracheal system, spiracles, and some somatic muscles; and (2) the caudal visceral nerves that originate from the last abdominal ganglion and run to the reproductive organs and hindgut.

SENSE ORGANS

Insect Eyes

Light is detected by specialized **photoreceptor** cells that are grouped into visual organs on the head (Fig. 6.4) or may exist as small, light-sensitive sense organs elsewhere on the body. On the head, cells are typically clustered in a **retina** beneath a transparent lens-shaped area of the cuti-

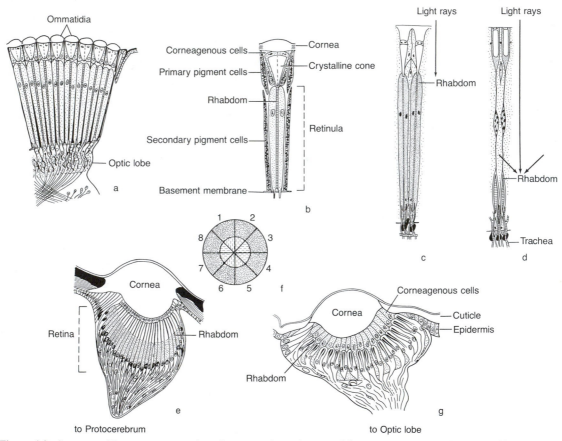

Figure 6.4 Structure of insect eyes: **a**, section of compound eye; **b**, ommatidium; **c**, ommatidium of the apposition type; **d**, dark-adapted ommatidium of the superposition type; **e**, dorsal ocellus; **f**, diagrammatic cross section of a retinula with eight cells and the central rhabdom [Each cell of the retinula forms a rhabdomere of microtubules. In the honey bee, as shown here, the cells are fused pairwise (1-2, 3-4, 5-6, 7-8). Microtubules in each pair are parallel, as are the microtubules of opposite pairs); **g**, stemma or lateral ocellus. In c light rays that pass through the cornea of an ommatidium strike only the rhabdom beneath, but in d the light rays pass into adjacent ommatidia and strike other rhabdoms. (a and b, Redrawn from R. E. Snodgrass, *The Principles of Insect Morphology.* Copyright © by Ellen Burden and Ruth Roach. Used by permission of the publisher, Cornell University Press; c and d, redrawn after Nowikoff, 1931; e, redrawn after Caesar, 1913; g, redrawn after Corneli, 1924.)

cle called the **cornea.** Two general kinds of visual organ are recognized: compound and simple. **Compound eyes** are located dorsolaterally on the head, and each is composed of one to many thousands of closely packed photoreceptive units, the **ommatidia** (Fig. 6.4b). Each ommatidium has its own corneal lens, which usually forms a hexagonal facet on the eye surface. Compound eyes are directly connected to the optic lobes of the brain.

Simple eyes each have a single round corneal lens with several to many light-sensitive cells beneath. Simple eyes are of two types. **Dorsal ocelli** (Fig. 6.4e) are located on the top of the head in a triangle and number no more than three. They are innervated by the median part of the protocerebrum and commonly occur together with compound eyes in both adult and larval stages (ocelli absent in endopterygote larvae). **Stemmata,** or lateral ocelli, are located laterally and number usually less than a dozen at each side. They are innervated by the optic lobes and commonly occur in endopterygote larvae in the place of compound eyes (Fig. 6.4g). Insects do not have devices like eyelids to shield their light-sensitive receptors from direct sunlight.

Protura (Fig. 17.1b) and Diplura (Fig. 18.1) have no discrete eyes. Compound eyes are found on at least some adults of all other orders except Siphonaptera and Strepsiptera. The lateral eyes of Siphonaptera, if present, are simple (Fig. 43.1a), and the berrylike eyes of adult Strepsiptera are stemmata. Compound eyes are also found on immature stages of Exopterygota, but they are missing in the larval Endopterygota. The lateral eyes of endopterygote larvae are usually stemmata (Fig. 16.1b,d). Larval Mecoptera are the exception: they may have stemmata or unique compound eyes.

The number of ommatidia in compound eyes ranges from 8 or fewer per eye in Collembola to 28,000 in adult Odonata (Fig. 6.5). Ocelli commonly occur together with compound eyes in adult insects and in immatures of Exopterygota. However, compound eyes and ocelli may be secondarily lost in various groups of insects, especially in apterous insects such as the parasitic Phthiraptera, female scale insects, cave insects (Fig. 9.3b), or worker castes of some termites and ants.

Compound Eyes and Vision

The general structure of an ommatidium (Fig. 6.4b) includes two lenses, the **corneal lens** and **crystal-**

Figure 6.5 Dragonfly head with huge compound eyes. Such aerial predators have a large field of vision for seeing prey.

line cone, and a set of six to eight sensory, or **retinula,** cells. Surrounding the crystalline cone are **primary pigment cells,** and around the retinula are **secondary pigment cells.** The biconvex cornea is secreted by two modified epidermal cells, the **corneagenous cells.** The cone is composed jointly of four cells whose internal structure serves to focus light from the cornea on the retinula. The elongate cells of the retinula are usually packed together and modified around the central axis into a **rhabdom.** Each cell contributes a **rhabdomere** of tightly packed microtubules oriented at right angles to the central axis. Microtubules within a rhabdomere are parallel to one another and usually also parallel to those on the opposite side of the axis. In the honey bee, for example, the eight retinula cells are fused pairwise into four compound units. The opposite units have parallel microtubules such that adjacent units have microtubules at right angles (Fig. 6.4f).

Rhodopsins, the visual pigments, are located on the membranes of the microvilli in the rhabdomeres. A nerve impulse is initiated when the rhodopsins react to light waves between 300 and 700 nm in length. In a single insect, different rhodopsins may be sensitive to different portions of the spectrum, thus providing a mechanism for color vision. Through behavioral and electrophysiological studies it is known that many insects are sensitive to 3 or 4 spectral classes: ultraviolet (340 to 360 nm), violet to blue (420 to 460 nm), blue-green (490 to 550 nm), and, in addition in butterflies, red (585 to 610 nm). Although the color vision of insects can be described as a trichromatic system like our own, their visual world is different because the spectrum is shifted toward the shorter wavelengths to include ultraviolet. Except for butterflies, most insects lose the ability to distinguish differences in wavelengths between 550 and 650 nm, even though their vision extends into this region.

Some insects are able to discriminate the plane of polarized light coming from the sky and to use the information for navigation during walking or flight. Polarized light is light vibrating in a single plane, whereas direct sunlight has planes of vibration in all directions around the line of travel. Depending on atmospheric conditions, the proportion of light that is polarized increases in certain regions of the sky up to about 70 percent. You can see the differences in polarized light in the sky with Polaroid sunglasses. The regular orientation of the microtubules plus the configuration of the rhodopsin molecule provides an inherent mechanism for a sensitivity to polarized light. Certain ommatidia along the dorsal margin of the eye appear to be specialized for this function.

Image Formation in Compound Eyes

Insects evidently detect the form, color, and movement of objects in their environment, but the mechanism of image formation is not fully understood. Each ommatidium receives light from a small spot in the total visual field. According to the **mosaic theory** of insect vision, each ommatidium responds only to the wavelengths and intensity of light from its own limited view. The function of the lens is to gather light and make possible a much smaller eye than otherwise would be required. Although ommatidia have separate retinula cells, the signals are reduced on the way from the ommatidium to the protocerebrum. For this reason, the details of the tiny image, if any, formed inside each ommatidium by the lens system are lost. The overall image of the visual field is constructed as a spotted mosaic of the different responses of the ommatidia to the different light intensities.

The anatomical types of ommatidium and the physiological types of image formation vary in different taxonomic groups of insects. In most insects, although the retinula cells remain as distinct cells, their rhabdomeres appear to be fused together to form a single, central rhabdom. Two physiological types of images have been identified: the **apposition image** is formed when each ommatidium is stimulated only by light passing through its own lens system, and the **superposition image** is formed when ommatidia also receive light passed by the lenses of adjacent ommatidia.

In the anatomical **apposition type** of eye, the rhabdom is placed immediately below the crystalline cone. Structurally the ommatidia of these insects are shielded from stray light inside the eye by a complete jacket of pigment cells so that only light from the lens reaches the rhabdom immediately beneath (Fig. 6.4c). These eyes form a physiological apposition image and are found among insects that are active during the strong light of day. The mosaic image formed by apposition gives the highest resolution of detail. In terms of visual acuity, such insects can distinguish two objects that are separated by an angle of about 1°, which corresponds to the smallest angular separation of ommatidia. Human eyes by comparison can distinguish objects separated by as little as one-ninetieth of 1°. Not all ommatidia are identical. Day-active predators such as dragonflies and wasps have clusters of anterior ommatidia that subtend smaller angles and have more overlap with adjacent ommatidia, permitting the resolution of greater detail in the forward field of vision.

Although the visual acuity of insects would seem to be adequate for seeing objects in some detail, behavioral experiments with honey bees demonstrate that discrimination of form is poorly developed in these insects. The compound eye, however, is well adapted for the detection of movement because successive ommatidia are stimulated as the object sweeps across the visual field. Depth perception is possible when the visual fields of the paired eyes overlap, triangulating the position of the object seen by both eyes. The wide separation of the eyes in predators such as mantids and dragonfly naiads probably improves distance perception.

In a second anatomical type of eye, called the **clear zone type,** the exact structure of the ommatidia varies from group to group. In general, the fused rhabdoms are relatively short and separated from the lens system by a gap, or clear zone. The migration of the secondary pigment cells permits the eye to form images over a wide range of light intensities. During day the pigment and to some extent the individual pigment cells move proximally into the clear zone and limit the passage of light from the lenses to single rhabdoms. This is the **light-adapted** eye, and the image is formed by apposition. At dusk, the pigments and secondary pigment cells contract distally, per-

mitting light from adjacent ommatidia to stimulate the rhabdoms. This is the **dark-adapted** eye and the image is formed by superposition (Fig. 6.4d). A superposition image has much less resolution of detail. It is not even certain whether an image, if formed, is sensed in certain types of dark-adapted superposition eye. The detection of overall light intensity, however, would seem to be increased because stray light rays within the eye are not blocked by pigment. In some moths, tracheal tubes are arranged between and parallel to the ommatidia, forming a **tapetum** that may aid in reflecting light onto the rhabdoms. Most insects of this type are Lepidoptera, Trichoptera, Coleoptera, and Neuroptera that are active at twilight or night.

Last, a third anatomical type of eye called the **open rhabdom type** is found in the higher Diptera and some Hemiptera. Here the rhabdomeres within an ommatidium are separate and not fused into a single rhabdom. The image formation is probably by apposition, but the sensory unit is formed by rhabdomeres in adjacent ommatidia that share the same view, rather those in a single ommatidium.

Ocelli and Stemmata

Simple eyes have a single, biconvex, cuticular lens that focuses light on the rhabdoms of sense cells beneath (Fig. 6.4e,g). Ocelli do not detect images because the optical image is formed below the level of the sense cells. Ocelli are quite sensitive to changes in the intensity of light, especially in the ultraviolet and green part of the spectrum. In flying insects, the ocelli, situated on the top of the head, may monitor the position of the horizon and function quickly through connections to the motor centers to maintain a stable orientation during flight.

On the other hand, the stemmata of larvae are clearly visual organs like the eyes of spiders and also may discriminate the plane of polarized light. The lens of a stemma forms an image on the rhabdoms. When several stemmata are clustered at each side of the head, the mosaic image may be equivalent in resolution to that of a compound eye, yet the sensory unit is much smaller. This poses a mystery as to why compound eyes, rather than stemmata, are the main visual organs of adult insects.

Mechanoreceptors

Mechanical distortion of the body is detected by a variety of simple sense organs, or **sensilla.** In the usual arrangement, the dendrite of a single bipolar neuron is attached to a movable part of the body, often by a minute cuticular sheath called a **scolopale.** Movements of the body part in response to mechanical stress, touch, wind, or vibrations in air, water, or solid substrates initiate nerve impulses that are transmitted to the central nervous system. Some sensilla respond to disturbances from the environment as **exteroceptors,** while others respond to the position of one part of the body relative to another, thus functioning as **proprioceptors.** These responses are important in maintaining posture and in orienting to gravity.

Four types of mechanoreceptor are recognized. (1) **Trichoid** sensilla are hairlike setae with a single dendrite attached at the base. Some project well above the cuticle surface and serve as **tactile,** or touch, organs (Fig. 6.6a). (2) **Campaniform**

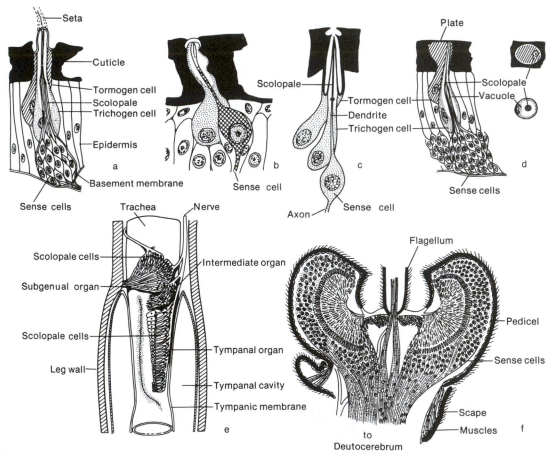

Figure 6.6   Sensilla: **a,** trichoid chemoreceptor; **b,** campaniform mechanoreceptor; **c,** coeloconic chemoreceptor; **d,** plate organ; **e,** chordotonal acoustical organ in tibia of right leg of a tettigoniid; **f,** Johnston's organ in antenna of the fly *Chaoborus.* (a, b, d, and e, redrawn from R. E. Snodgrass, *The Principles of Insect Morphology.* Copyright © by Ellen Burden and Ruth Roach. Used by permission of the publisher, Cornell University Press; c, redrawn from Snodgrass, 1925; f, redrawn after Child, 1894.)

sensilla are oval, domelike areas of cuticle that raise or lower as the adjacent exoskeleton is mechanically strained (Fig. 6.6b). The dendrites detect movements of the dome. (3) **Chordotonal** organs are completely internal and formed by units or **scolopidia** consisting of three cells: a bipolar neuron, a scolopale cell, and an attachment cell. The dendrite sometimes ends inside the scolopale in a process that resembles a cilium. One or more scolopidia cluster into a chordotonal organ and are variously attached to the epidermis. Some are suspended between two points in membranous areas or between leg joints, thus permitting detection of movements between the points of attachment. **Subgenual organs** of the tibia, however, have a single attachment and do not span a joint. They are especially sensitive to vibrations in the substratum. **Johnston's organ** is a chordotonal organ in the antennal pedicel (Fig. 6.6f): by means of its attachment to the base of the flagellum, movements of this appendage are detected. **Tympanal** organs are specially designed for the reception of sound (Fig. 6.6e). A chordotonal organ is attached to a thin cuticular membrane or **tympanum** that freely vibrates when struck by sound waves. (4) **Stretch** receptors are attached to connective tissue or muscles, registering the tension of such soft tissues, especially in the abdomen. These receptors differ from other sensilla in having multipolar rather than bipolar neurons.

All four types of sensillum may function as proprioceptors. Groups of trichoid sensilla located at each side behind the head on the cervical sclerites register the position of the head with respect to the body. Similar groups at leg joints register the positions of the limbs. Campaniform sensilla are widely scattered over the body surface at points of stress and bending. The halteres of Diptera and insect wings in general are well supplied. Strategically placed chordotonal organs and stretch receptors permit detection of movements of the limbs, segments, and viscera. The positions of the antennae are detected by Johnston's organ.

Vibrations are detected mainly by trichoid sensilla and chordotonal organs. Hairs in the cerci of cockroaches, for example, respond to low-frequency sound waves as well as to wind currents. Subgenual organs are found in many insects but are lacking in Archeognatha, Coleoptera, and Diptera. In male Culicidae and Chironomidae (Diptera), the antennae have fine, long hairs that vibrate in response to the flight tone of the females. This stimulates a mating response. The hair vibrations move the flagellum, which in turn stimulates Johnston's organ in the pedicel. The same behavioral response can be evoked with a tuning fork that vibrates at 100 to 800 Hz.

The special acoustical receptors of insects are the tympanal organs found in various unrelated taxa and in different places on the body. The organs are on the abdomen of Acrididae (Orthoptera, Fig. 29.2b), Cicadidae (Hemiptera, Fig. 6.8a–c), and Pyraloidea and Geometroidea (Lepidoptera, Fig. 44.6f); on the tibia of the forelegs in Gryllidae and Tettigoniidae (Orthoptera, Fig. 29.2a); and on the mesothorax of the bugs *Corixa* (Corixidae, Hemiptera) and *Plea* (Pleidae, Hemiptera), and on the metathorax of Noctuoidea (Lepidoptera, Fig. 44.6e).

The Lepidoptera listed above are night-flying moths. Their tympanal organs are used to detect and avoid approaching bats, the sounds of which can be sensed at about 30 m. In the Noctuoidea the organ has only two scolopidia at each side, and these respond to sounds well over 100 kHz. The tympanal organs of Orthoptera and Hemiptera are used in courtship behavior. Sounds produced usually by the male, or by both sexes, are detected and serve to bring mates together. In *Plea* the organ has but one scolopidium per organ, but in others the organ may be extremely complex, involving up to 1500 scolopidia in Cicadidae. These organs respond to a wide range of sound frequencies, generally less than 20 kHz. The average upper limit for humans is about 14 kHz. Insects discriminate poorly, if at all, among the different frequencies, but respond mainly to the intensity and duration of sound and the intervals between sounds. Insects can detect an interval as brief as 0.01 second, whereas the human ear fails to detect intervals less than 0.1 second.

Chemoreceptors

The sensilla that respond to chemicals in air or liquids are modified setae. Typically the dendrites of several to many bipolar neurons pass through one scolopale, divide into fine branches, and terminate at one or more minute pores through the cuticle of the hair. Molecules embedded in the dendrites' membranes react with specific molecules in the surrounding air or fluid to stimulate a nerve impulse. Each neuron of the cluster responds to a different, but possibly overlapping, array of chemical compounds that are important in the insect's life. In addition to specific sensitivities, the neurons may also respond to general irritant substances, such as ammonia fumes.

Setae of four types are commonly associated with chemoreceptor neurons: trichoid, basiconic, and coeloconic sensilla, and plate organs. **Trichoid** sensilla may serve both as mechanoreceptors and chemoreceptors (Figs. 6.6a, 6.7g). Such hairs on the labella and legs of blow flies (Calliphoridae) have a mechanoreceptor attached to the base and the dendrites of a few chemoreceptor neurons situated at a single pore at the tip. Trichoid sensilla are commonly gustatory (taste) receptors. **Basiconic** sensilla are peglike and thin-walled, have many pores, and project above the cuticle. They are the most common type of olfactory (smell) receptors, being found on the antennae of most insects, as well as on the mouthparts and other places on the body (Fig. 6.7f). **Coeloconic** sensilla have the sensory peg sunk in a cuticular pit (Figs. 6.6c, 6.7h). These are found on the antennae and mouthparts of some insects. **Plate organs** are found on the antennae of aphids and honey bees (Figs. 6.6d, 6.7g). Each is a round, flat cuticular plate perforated by pores. **Sensory pits** contain many sensilla in a subcuticular pit. Such pits have been identified on the third antennal segment of Diptera (suborder Cyclorrhapha) and labial palpi of Lepidoptera and Neuroptera. Sensilla that sense odors usually have multiple pores, often opening into a chamber with a network of channels that lead to multibranched dendrites. Taste sensilla often have a single pore that leads to an interior space where a viscous liquid surrounds the dendrite. This arrangement presumably aids the processes by which the dendrites react to specific molecules.

Chemoreceptors are found on the antennae, mouthparts, legs, and ovipositor. They play crucial roles in many activities. The detection of odors, or **olfaction,** is involved in locating food and sites for oviposition, and in finding mates through chemical sex attractants. Social insects recognize colony members by odor. The ability of insects to detect minimal concentrations of chemicals varies with the physiological state of the insect and, of course, with the specific chemical compound or mixture of compounds. Tests with honey bees involving many odorous compounds have led to the general conclusion that the threshold concentration detected is on the average about equal to that of humans. Bees do sense certain substances important to them at lower concentrations, such as the odors of beeswax and certain glandular secretions.

Certain key compounds that are crucial in the life of an insect may be sensed at unbelievably low concentrations. Calculations for the male silk moth, *Bombyx mori* (Bombycidae, Lepidoptera), indicate that single molecules of the sex attractant bombykol produce impulses in the appropriate antennal chemoreceptors. The moth responds behaviorally to as few as 200 bombykol-induced impulses per second. This is barely above the theoretical minimum needed to transmit information in a system that already has 1600 impulses per second of spontaneous nerve activity or "noise."

A distinction is sometimes drawn between olfaction of odors and the taste or **contact chemoreception** of substances in liquid. However, the sensilla and physiological mechanisms are the same. By behavioral and electrophysiological analyses, insects have been shown to discriminate among sweet, salt, acid, and bitter compounds in solution and to exhibit a sense for water. The minimum taste thresholds are generally lower in insects than in humans. A convenient behavioral response has aided in surveying chemicals in various concentrations. When sugar solutions are ap-

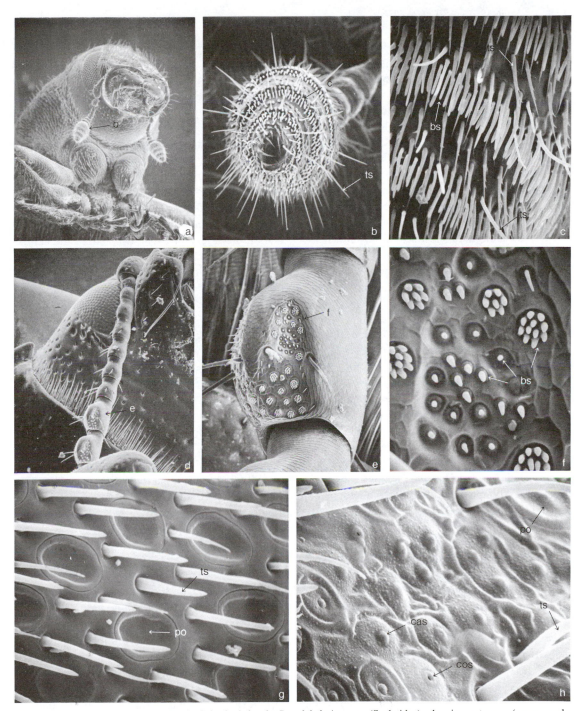

Figure 6.7 Antennal sensilla: **a**, head of the bark beetle *Pseudohylesinus* sp. (Scolytidae), showing antennae (arrow marks position of photo b; 37×); **b** antennal club of same showing sensory bands (arrow marks position of photo c; 304×); **c**, sensilla of club of same, blunt basiconic sensilla form dense bands (743×); **d**, head of a male beetle *Temnochila virescens* (Ostomidae), showing antenna (arrow marks position of photo e; 32×); **e**, ninth antennal segment enlarged (arrow marks position of photo f; 169×); **f**, basiconic sensilla of several types whose functions are unknown (878×); **g**, eleventh antennal segment of the worker honey bee, *Apis mellifera* (Apoidea; 1439×); **h**, apical (twelfth) antennal segment of same (2483×). bs, basiconic sensilla; cas, campaniform sensilla; cos, coeloconic sensilla; po, plate organ; ts, trichoid sensilla. (Photos a–f courtesy of and with permission of Clyde Willson; photos g and h courtesy of and with permission of Kim Hoelmer.)

plied to the tarsi of flies, butterflies, and honey bees, the insects respond by extending the proboscis. The proboscis is not extended for water or solutions too weak to detect. As a consequence, responses to sugars have been extensively studied.

Honey bees respond positively to only seven, and marginally to two, of 30 substances that taste sweet to humans. The seven are sucrose, glucose, fructose, melezitose, trehalose, maltose, and α-methylglucoside. Of these, five are naturally found in the bee's diet: nectar contains the first three and honeydew includes the first five. The last two sugars, as well as the 23 tasteless or nearly tasteless sugars, are not in the bee's natural food. Sucrose solutions ranging in concentration from 0.06 to 0.12 M can be discriminated from water by starved bees with their mouthparts. The antennae are much more sensitive. With behavioral training, bees respond to sugar solutions applied to their antennae in concentrations as low as 0.0001 M.

The trichoid sensilla on the tarsi and labella of blow flies (Calliphoridae) can be stimulated individually with solutions. They accept a much greater variety of carbohydrates than do honey bees. When a sucrose solution is applied to the tarsi, the acceptance threshold is 0.01 M in *Phormia regina,* but as low as 0.0006 M in *Calliphora erythrocephala* (Calliphoridae, Diptera). *Pyrameis* butterflies (Nymphalidae, Lepidoptera) respond by tarsal stimulation to 0.00008 M sucrose when starved. By comparison, the threshold for humans at 0.02 M is 250 times poorer.

BEHAVIOR

What an insect does in response to external and internal stimuli depends not only on the nature of the stimulus, but also on the insect's physiological state. The response to food may depend not only on the taste of food and hunger, but also on the hormonal milieu within which the nervous system functions. At the onset of molting, for example, feeding may cease altogether. Thus an insect's behavior is controlled by two interrelated systems: nervous and endocrine. Both function to sequence

behavior and prevent conflicting behavior, with the result that the survival of the individual and the propagation of the species are favored. Both communicate directly to various tissues; the nervous system communicates by physical contact, and the endocrine system by hormones in the hemolymph. The small size of insects is probably an advantage in this internal communication. The systems differ in that the nervous system functions on a moment-to-moment basis, integrating specific sensory and motor impulses, while the endocrine system reacts more slowly and has long-term effects in a variety of tissues, including the central nervous system itself.

The behavior of insects under natural conditions is a complex mixture of learning and stereotyped responses to stimuli that are controlled and coordinated by the central nervous system. Segmental ganglia may be largely independent in controlling the responses to certain stimuli. Stepping movements or reflexes of the legs, for example, are controlled directly by the respective thoracic ganglia but are coordinated by the inhibition and stimulation of the brain and other ganglia. Stereotyped responses are programmed into the nervous system by heredity, but the "program" is not as "fixed" as once thought. Considerable flexibility may exist in the timing, coordination, and sequencing of acts, depending on sensory input.

Taxes

A complex but stereotyped movement that orients the whole body to an environmental stimulus is called a **taxis.** Commonly this is an orientation to gravity (**geotaxis**), to light (**phototaxis**), to wind currents (**anemotaxis**), or to sound (**phonotaxis**). An orientation may be either **positive** (toward the stimulus) or **negative** (away from the stimulus). In a positive phototaxis, for example, the body is oriented so that both eyes are equally stimulated in front. The orientation to light may also involve positioning the body so that the light is dorsal or ventral. The **dorsal light reaction** is common among aquatic insects, providing a means to orient to the earth's surface when the effect of gravity on proprioceptors is reduced by the buoyancy of water. The backswimmer *Notonecta* (Notonectidae, Hemiptera), so named because it swims upside down (Fig. 9.6d), appropriately has a **ventral light reaction.** Orientation at a fixed angle to sunlight is called the **light compass reaction.** When it is used as a means for navigation to and from a distant point, compensation must be made for the sun's movement. Honey bees, some ants, and a few other insects have such a compensatory mechanism, probably involving their sense of time or circadian rhythm (discussed further on in this chapter). Orientation to a celestial body combined with compensation for its movement is called an **astrotaxis.**

Although a taxis is a stereotyped behavior, a given taxis is not an invariant response to a given stimulus. Taxes are under the control of the nervous system and can be modified by learning. Depending on the circumstances, a taxis can be quickly switched on or off, the orientation switched from positive to negative (or vice versa), and receptivity to one stimulus replaced by another. Positive geotaxis and negative phototaxis, and each with opposite signs, are widely coupled among insects. In honey bees, gravity and the sun are interchanged as stimuli for orientation.

Using this terminology, an exploratory foraging trip by a worker ant might be described as follows. The ant is negatively geotactic and positively phototactic as she makes her way to the surface from her underground colony. As she walks over the ground she is oriented at an angle to the sun by astrotaxis. On her return, the astrotaxis is switched 180°, permitting her to navigate back to the nest, and on entering the nest, she becomes positively geotactic and negatively phototactic.

Instincts

Many of the activities of insects are series of stereotyped behaviors of greater or lesser complexity. Such sequences are loosely called instincts or fixed action patterns. Examples include feeding, migration, mating, oviposition, nest building, cocoon spinning, and movements during ecdysis. Here the physiological and developmental state of the insect, especially the hormonal milieu, is important in determining whether a stimulus will trigger, or release, a given program of behavior. The program may be complex and flexible, including alternate sequences of behavior depending on the response of the mate, the suitability of food presented, or other circumstances.

Learning

A change in behavior as a result of experience is **learning.** Learning can be shown to occur even in the segmental ganglia of insects. Headless cockroaches and grasshoppers can be trained to lift a leg to avoid an electric shock. Such elementary forms of learning may prove to be widespread. The most important centers for learning are the brain's mushroom bodies, or corpora pedunculata. In insects that exhibit complex behavior and learning, the bodies are larger in volume and have more cells and more synapses than in insects with simpler and clearly stereotyped behavior.

Communication

The behavior interactions of insects among themselves and with other organisms usually involve some sort of communication. **Communication** occurs when an organism provides a visual, tactile, auditory, and/or chemical signal that influences the physiology or behavior of another organism. The distance over which the communication takes place may be such that only one sense will effectively receive the signal. This is **long-range** communication, in contrast to **short-range** communicatio, which may involve more than one sense.

Instances of communication may be broadly divided into those that occur between members of the same species and those that occur between members of different species. The former, or **intra-**

specific, cases of communication are for the purpose of attracting or recognizing mates, aggression or defending territories, assembling or dispersing aggregations, or, in social species, for various cooperative activities and to control caste development and behavior.

Communication that is **interspecific** (i.e., between different species) may be divided further into those instances that benefit the sender, those that benefit the receiver, and those that result in mutual benefit. In the first type of relationship the sender is usually under threat of attack by a predator and produces one or a combination of warning signals to discourage attack (see Chap. 12 for additional details of this strategy). Other instances in which the sender benefits involve signals that confuse the communication of a predator or competitor. High-frequency clicks produced by the tymbals of night-flying Arctiidae moths (see below) probably confuse the echolocation system of bats. In the second type of relationship, the sender may become a victim when some aspect of its activities signals its presence to a predator or parasitoid. Or a potential victim may escape by sensing the enemy's presence. The detection of the sounds of bats by the tympanal organs of moths is an example. In either case the receiver of the signal is the beneficiary. The insect pollination of flowers supplies an example of the third type of relationship. The flower provides visual, tactile, and chemical signals that attract insects. The insects obtain food, and the flower achieves cross-pollination (see Chap. 10).

As we shall see, communication may become quite complex: a given signal may be sensed both by members of the same species and by members of different species; different species of organisms may produce identical signals; and the same signal may be used for different purposes by sender and receiver.

Communication by Sight
Insects that are active by day commonly depend on visual signals. For example, the color patterns, size, and motion of female butterflies are known to attract males of the same species. Visual signals

also carry the risk of attracting predators and parasitoids.

Visual communication is not restricted to diurnal insects. Bioluminescent organs are found in some Collembola (Onychiuridae, Poduridae), Diptera (Mycetophilidae), and Coleoptera (Elateridae, Homalisidae, Lampyridae, Phengodidae, Telegeusidae). The organs commonly function for communication between the sexes while mating, but they also may serve other purposes such as illumination, defensive warning, or attraction of prey.

Lightning bugs or fireflies (Lampyridae, Coleoptera) have received the most study. Both sexes use flashes of light to find each other at night. Males emit flashes while flying that have a species specific duration and frequency. The females, which are less active or even wingless and larviform, respond by flashes. Depending on the species, the female's flashes may be emitted after a brief pause of precise duration after the male signal. The luminescent organs consist of large, well-tracheated cells in the abdomen. Light is generated by oxidation of luciferin in the presence of the enzyme luciferase. The reaction is highly efficient; 98 percent of the energy is released as light. A product of the reaction inhibits luciferase and stops further reaction. Nerve impulses stimulate light production, probably by removing the inhibition of luciferase.

The flash signals of fireflies also may function interspecifically in what is called aggressive mimicry. The females of *Photuris versicolor* and related species are known to flash-respond correctly to mating signals of males of other species and to eat the males that are attracted. *P. versicolor* preys on 11 other firefly species.

Communication by Sound
Of all the animals, only arthropods and vertebrates communicate by producing vibrations in air, water, or solid substrates. In insects the vibrations may be created as a by-product of some other activity, especially flying. We have already mentioned how the flight tone of female mosquitoes attracts males, thus serving an intraspecific function. Loud buzz-

ing by bees and wasps serves interspecifically as an effective warning signal that is even imitated by other insects. Vibrations in wood or soil are created by striking the body against the substratum. The deathwatch beetle *Xestobium* (Anobiidae, Coleoptera) bores in furniture and wooden houses. At sexual maturity it strikes its head against the burrow walls, producing a rapping sound that at night influences the behavior of humans as well. Many species of termites drum their heads in response to disturbances of their colonies.

Special devices for creating sound are prevalent in numerous orders, but are especially frequent in species of Orthoptera and Hemiptera (Fig. 6.8). Vibrations are generated mostly by one of two mechanisms: movement of roughened surfaces together or vibration of a membrane. In the first type of mechanism, one surface may be modified into a scraper and the other into a file (Fig. 6.8d–f). The movement of the scraper on the file is called **stridulation.** Some adult as well as immature insects stridulate to produce arrhythmic, rasping warning sounds. These are heard for only a short distance or sensed by contact. The devices are simple and may be formed wherever two parts of the body normally rub together.

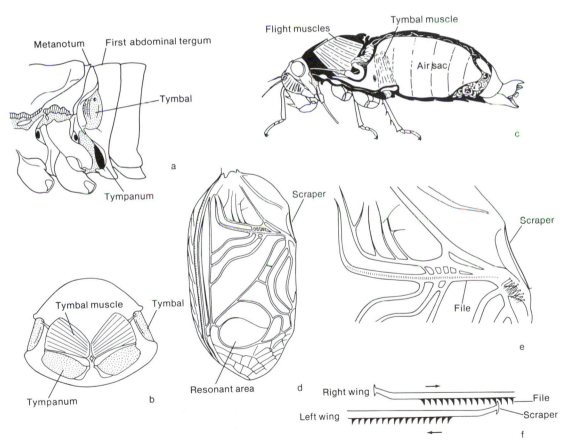

Figure 6.8   Sound-producing organs: **a**, lateral view of anterior abdomen of male cicada, *Okanagana vanduzeei,* showing sound-producing organ (tymbal) and sound receptor (tympanum); **b**, cross section of same; **c**, longitudinal section of *Magicicada septendecim* (Cicadidae) showing large resonant air sac; **d**, ventral view of right wing of cricket *Acheta assimilis* (Gryllidae); **e**, enlargement of stridulatory organ of same; **f**, diagram of relationship of right wing and left wing during stridulation. (c, redrawn from R. E. Snodgrass, *The Principles of Insect Morphology.* Copyright © by Ellen Burden and Ruth Roach. Used by permission of the publisher, Cornell University Press.)

In Orthoptera stridulation by adults can be heard at long range and serves to bring males and females of the same species together for mating. The devices for sound production and reception are complex, and the sounds of each species have a characteristic frequency, rhythmic spacing, and duration. In Gryllidae and Tettigoniidae, one or both forewings have a resonant area of cuticle that vibrates as the scraper of one wing strikes the teeth of the file on the opposite wing. Tympanal organs of hearing are located in the fore tibiae (Fig. 29.2a). Receptive females orient to the calling songs of the males and are attracted to them. Acrididae rub a row of pegs on the inner side of the hind femur against raised veins of the forewings, causing the wings to vibrate. The tympanal organs are on the first abdominal segment (Fig. 29.2b). Both males and females produce long-range calling songs and approach each other.

Sound-producing membranes called **tymbals** are situated in the metathorax of Arctiidae (Lepidoptera) and in the abdomen of the following Hemiptera: male Auchenorrhyncha; both sexes of Cercopidae and some Cicadellidae and Cicadidae; and some Pentatomidae. Tymbals are best developed in male Cicadidae (Fig. 6.8a–c), in which they produce long-range calling songs. In these insects the tymbal is a thin resilient area of cuticle situated dorsolaterally on the first abdominal segment. Contraction of an asynchronous fibrillar muscle attached to the center causes the tymbal to click inward. When the muscle relaxes, the elastic cuticle then clicks outward. Large air sacs in the abdomen resonate with the tymbal vibrations and increase the intensity of sound. The tympanal organs are located posterolaterally in the same segment. Orthoptera and Cicadidae are unique among insects in producing rhythmic "songs" and in the synchronous singing of many individuals to form a "chorus."

The tymbals of the much smaller leafhoppers (Cicadellidae) produce weaker vibrations that are transmitted for longer distances along plants than through the air as sound waves. This communication requires that the insects be on the same or touching plants. Plant hoppers (Fulgoroidea) communicate with vibrations produced by rapidly drumming their abdomen against a plant surface.

Communication by Chemicals

Chemical messengers are used for communication throughout the animal kingdom, especially among insects. The term **semiochemicals** is applied to all chemicals produced by one organism that incite responses in another organism (Matthews and Matthews, 1978). Most semiochemicals of insects are produced by **exocrine glands** of the integument. The chemicals, either single compounds or mixtures, are usually active in minute quantities and are usually species-specific. Nonvolatile chemicals are limited to contact chemoreception. Volatile chemicals function in either short or long range communication, but the direction and distance may be limited by wind.

Semiochemicals that function intraspecifically are called **pheromones.** These are often blends of several compounds and, at least in social insects, may have more than one effect on the behavior and physiology of the recepient. Pheromones may be grouped broadly into **primer** pheromones, which have a long-term effect on the physiology and development of the receiver that is mediated by hormones but may not have an immediate effect on behavior, and **releaser** pheromones, which have immediate, reversible effects on the receiver's behavior. Among insects, primer pheromones are well known in social species, where they are involved in caste determination and the inhibition of reproduction by workers (see Chap. 7). In other insects, especially those that aggregate, primer pheromones function to bring adults into reproductive synchrony. For example, the presence of mature males of the locust *Schistocerca gregaria* (Acrididae, Orthoptera) increases the rate of sexual maturation of immature males. Experimental evidence indicates that the pheromone responsible for promoting maturation is transmitted by air and by contact to the young adults and affects their corpora allata (Chap. 4).

Releaser pheromones are greatly varied in function and widespread taxonomically. The best known are the sex attractants, or **sex pheromones,** which bring the sexes together. In the most common situation, the female emits odors at a certain time of day or night as a long-range signal to advertise her receptivity to mating. The odors are carried away from her by wind currents, forming an odor trail that becomes wider and less concentrated with increasing distance. A male downwind that perceives the odor with chemoreceptors on his antennae becomes active and flies against the direction of the wind, oriented by a positive anemotaxis. The flight is a zigzag pattern toward the female, as the male turns back and forth to remain in the odor trail. In the vicinity of the female where the odor concentration is high, the male arrests his forward flight and searches for the female by other senses, especially sight. Courtship and copulation are then stimulated by close-range communication involving visual, tactile, or auditory signals, alone or together with pheromones. The long-range sex attractant released by the female may continue to excite the male, and the male, on close contact, may produce his own releaser pheromone or aphrodisiac that makes the female more receptive.

The foregoing sequence of sex attraction is by no means universal. In some insects males produce long-range sex attractants, in some the females produce aphrodisiacs, and in some, one sex produces pheromones or **aggregating scents** that attract both sexes to a common place (see behavior of bark beetles, below).

The sensory nerves from the antennae of moths and other insects can be monitored with electrical devices that record nerve impulses triggered by specific molecules in a stream of air. Such records, called electroantennagrams, can be used to monitor a detached antenna or an antenna on a living insect. To identify chemicals that stimulate various responses, the output from a chemical mixture analyzed by gas chromatography can be blown across an electroantennagram. The correspondence of peak stimulation from the monitored antennae with the detection of particular compounds by the gas chromatograph identifies specific compounds that induce the firing of olfactory sensilla in the antenna. This simple approach has aided the identification of many pheromones.

The active compounds in the sex pheromones of a number of insects have been isolated, identified, and synthesized. The common names for the active synthetic compounds include grandlure for the boll weevil, *Anthonomus grandis* (Curculionidae, Coleoptera); disparlure for the gypsy moth, *Lymantria dispar* (Lymantriidae, Lepidoptera); and riblure for the red-banded leafroller moth, *Argyrotaenia velutinana* (Tortricidae, Lepidoptera); hexalure for pink bollworm moth, *Pectinophora gossypiella* (Gelechiidae), Lepidoptera; bombykol for the silkworm moth, *Bombyx mori* (Bombycidae, Lepidoptera); and trimedlure for the Mediterranean fruit fly, *Ceratitis capitata* (Tephritidae, Diptera). Sex pheromones are extremely useful in pest management (Chap. 14).

Intraspecific pheromones also have nonsexual functions. In some insects, the aggregations of immatures or of females at oviposition sites are probably assembled and maintained by chemical releasers. Rapid dispersal from an assemblage may likewise be signaled by chemicals, in this case by **alarm** pheromones. When disturbed, the aphid *Myzus persicae* (Aphididae, Hemiptera) discharges a secretion that disperses nearby aphids.

Social insects, especially ants, employ an array of chemical signals for aggregation, alarm, and other communications among colony members. Odor trails that recruit and guide foragers to food are known in certain ants, termites, and stingless bees (*Trigona*, Apidae, Hymenoptera). On returning to the nest after discovering a food source, a forager ant lays down a minute trace of a volatile, species-specific chemical, or trail pheromone. Other ants are stimulated to follow the odor trail from the nest to the food, and on their return add their own trail pheromone. A rich food source is soon visited by an increasing number of foragers. When the food becomes covered by ants or diminishes in size or quality, some ants return without contributing to the trail. Unless renewed by successful foragers, the trail declines in attractiveness because the odor is rapidly dissipated to a level below the ant's threshold of perception. Thus the foragers readily shift in appropriate numbers to exploit new food sources.

Semiochemicals involved in interspecific communication are called **allelochemicals** and are divided according to the effect on the sender or ;receiver as follows: **allomones** are substances released by the sender that evoke a response in the receiver that is beneficial to the sender; **kairomones** are substances that, when transmitted, benefit the receiver; and **synomones** are substances that are mutually beneficial to the sender and receiver. Theoretically the same allelochemic compound can be an allomone, kairomone, or synomone depending on the context in which it is transmitted.

Allomones include the venoms injected by the sting of female aculeate Hymenoptera; defensive secretions of nasute termites and various beetles and true bugs; and toxic chemicals in the body, such as cardiac glycosides in the monarch butterfly, that discourage ingestion by vertebrate predators (Chap. 12). Plants also produce allomones, such as the secondary metabolites, juvenile hormone mimics, and phytoecdysteroids that protect themselves against plant-feeding insects.

Kairomones include odors or tastes that attract and stimulate attack, be it an insect victim or a plant host. Sex pheromones become kairomones when predators or parasitoids are attracted to feed on the mating insects. Secondary plant compounds can serve as kairomones for insects that tolerate the toxic compounds and attack the plant (Chap. 10).

Odors of flowers that attract pollinating insects are synomones; both benefit because the plant is cross-pollinated and the insect obtains food (Chap. 10). Here again, the same odors may be kairomones by attracting flower feeding insects that destroy the flowers and seeds.

Aggregation Behavior of Bark Beetles

A complex system of semiochemicals coordinates the mass attacks on living trees by the bark beetles *Ips, Dendroctonus,* and *Scolytus* (Scolytidae, Coleoptera) (Raffa et al., 1993; Wood, 1982). Coniferous trees resist attack by expelling or drowning boring insects in sap (Fig. 6.9). This resistance is overcome by bark beetles when they attack simultaneously in large numbers and introduce a fungus that interrupts the water-conducting system of the tree. In the usual sequence of events, beetles of both sexes emerge from the old host or overwintering site and fly for some period before orienting to a new host. Visual and possibly odor cues guide the beetles toward trees. "Pioneer" beetles (males of *Ips;* females of *Dendroctonus*) initiate attacks

Figure 6.9   A female bark beetle, *Dendroctonus brevicomis* (Scolytidae), crawling in a pitch tube on a pine tree, *Pinus ponderosa.* Female beetles initiate the attack on the host tree by boring through the bark. The production of the sticky pitch by the tree is a defensive response caused by the mechanical injury to the resin ducts by the tunneling beetles. Other beetles are attracted by the mixture of pheromones: the female's *exo*-brevicomin, the male's frontalin, and myrcene from the host pine. (Photo courtesy of and with permission of Clyde Willson.)

to feed. Galleries are excavated in the bark of acceptable hosts, probably as a direct result of chemical feeding stimuli in the bark and/or phloem. During this initial boring activity, an aggregation pheromone or a species-specific mixture of such pheromones is produced both by de novo biosynthesis (Seybold et al., 1995) and by conversion of host terpenes (Hendry et al., 1980) at some unknown site inside the insect and liberated with the feces. The compounds ipsenol, ipsdienol, *cis*-verbenol and *trans*-verbenol, alone or in combination, are pheromones of various species of *Ips*; *exo*-brevicomin, *trans*-verbenol, verbenone, frontalin, and others are found in *Dendroctonus* species; multistriatin has been identified in *Scolytus multistriatus*. These compounds may be derived from bacterial, fungal, and/or insectan biosynthetic pathways. Mycangial fungi are closely associated with the beetles as symbionts in the gut or gallery (Chap. 13). The waste, or frass, from the borings is made attractive at a long range to other beetles. Thus the pioneers attract more beetles of both sexes. Their boring activity serves to increase the concentration of aggregation pheromone. In some cases certain volatile compounds from the host tree itself also contribute to the overall attractiveness of the tree (e.g., myrcene for *D. brevicomis* and α-cubebene for *S. multistriatus*). The total aggregation pheromone now consists of products from the beetle's gut and from the tree.

Mates are located and recognized by both sound and odor. *Ips* females are attracted to frass marked with a male pheromone, and they stridulate before being permitted entry into the gallery occupied by a male. Males of *Dendroctonus pseudotsugae* are similarly attracted to frass marked with female-produced pheromones, and they also stridulate at the entrance to a female's gallery before entering. The factors that terminate a mass attack are not well understood. Possibly the aggregation pheromone declines in production or one or more compounds from either or both sexes increase in concentration and interrupt the chemoreception of the pheromone.

Birch and Wood (1975) discovered that *Ips pini* and *I. paraconfusus* both attack *Pinus jeffreyi*

and *P. ponderosa* at the same season in the Sierra Nevada of California, but not the same individual trees. The volatile compounds produced by male *I. paraconfusus* inhibit the response of *I. pini* to its own pheromone and vice versa. Thus the first males to arrive at a tree effectively prevent colonization by the other species. The pheromone components simultaneously act as intraspecific attractants and as an allomone against a competitor.

The aggregations of bark beetles attract a large number of other insects that feed on the beetles, on other insects, or on the weakened host tree. Certain of the predators and parasitoids of bark beetles are specifically attracted by one or more of the constituents of the aggregation pheromones. In this situation, the pheromone components function simultaneously as kairomones (Borden, 1982; Raffa et al., 1993; Wood, 1982). Recent studies indicate that fungi inoculated into trees during host colonization may attract parasitoids that generally arrive on the tree when the new bark beetle broods have developed to late-stage larvae (Stephen et al., 1993).

Learning and Communication in Honey Bees

The Greek philosopher Aristotle noticed that a bee forages on one species of flower on each foraging trip and is attended by other bees on her return to the hive. These simple observations describe aspects of what has proved to be a remarkable ability for learning and communication in the honey bee, *Apis mellifera* (Apidae, Hymenoptera). Much of our knowledge has resulted from the investigations of the late Karl von Frisch of the University of Munich. For his efforts, he shared the 1973 Nobel Prize in physiology or medicine with N. Tinbergen and K. Lorenz (see von Frisch, 1974).

After about 3 weeks spent as an adult in the hive (Fig. 1.7), a worker honey bee begins the last phase of its life foraging for pollen and nectar in flowers (Chap. 7). In an extensive series of experiments that began prior to 1914, von Frisch demonstrated that bees could be trained to seek a sugar solution and to associate specific stimuli with this food source. The following experiment for color discrimination is typical. A table was placed near a hive. A blue square of cardboard was placed flat on the table and surrounded by cards of different shades of gray. On each square was placed a glass dish. The dish on the blue card was then filled with sugar solution. After a few hours during which the bees foraged on the food, the cards and dishes were removed and replaced by clean cards and empty dishes arranged in a different pattern. The bees returned and landed on the blue card, demonstrating that they discriminated the blue color from all shades of gray. This result indicates that bees have true color vision, and that they had learned to associate the color with food.

In similar experiments, von Frisch trained bees to one or a combination of odors, flavors, geometric figures, and colors within their range of discrimination (1971). He also trained bees to visit a feeding station at a specific time. Some bees were even trained to seek food at three to five separate times of day, thus demonstrating a time sense in these insects.

The most famous of von Frisch's discoveries was the complex signaling system that communicates the distance and direction to food sources up to about 14 km from the hive. On returning to the hive from a rich food source at 100 m or more away, the worker inside the dark hive moves excitedly in a **waggle dance** on the vertical comb (Fig. 6.10c). Afterward she may move to a different area on the comb and repeat the performance. Although it is dark, some bees follow the dance with their antennae, and they detect the flower odor on the waxy surface of the forager's body. The forager may also offer some of the nectar she has collected and carries in her crop. A rich concentration of sugar is highly attractive and stimulating to other bees. In experiments with feeding stations located at various distances and directions from the hive, von Frisch demonstrated that bees recruited by the dancer search in a certain direction and at a certain distance for food of the same odor and nectar flavor as that collected by the dancer. After successful foraging trips, the newly recruited bees begin to dance too. More bees are recruited, and the number of foragers rapidly increases. When the food supply diminishes, the frequency of dancing declines.

During part of the waggle dance the worker bee makes a straight run, vigorously wiggling her abdomen and emitting about 32 sound pulses per second. This is the called the waggle run of the waggle dance (Fig. 6.10d–g). Then she circles back to the starting point and repeats a waggle run of the same duration. The time spent during each run and the total number of pulses are directly proportional to the distance traveled to the food source. For a flight of 100 m, the time of a single waggle run is about 0.25 second, and for a 1200 m flight, 2 seconds. Thus, with increasing distance the waggle runs take longer, and fewer runs are completed within a given period.

The direction of movement during the waggle run on the vertical comb has a consistent orientation with respect to gravity (Fig. 6.10d–g). This is related to the direction from the hive to the food with respect to the compass direction of the sun, when it is more than about 2° from the zenith position. Recall that positive phototaxis and negative geotaxis are commonly linked. For example, if the forager flew toward the compass direction of the sun en route to the food, her waggle run is oriented vertically up the comb, and her head points up during the run. If she flew at an angle to the right of the sun, the run is oriented at the same angle to the right of vertical.

As the food site is moved closer to the hive at distances of about 100 m or less, the dance shifts from the waggle dance to a transition form (Fig. 6.10b) and then to the **round dance** (Fig. 6.10a). The waggle runs become less conspicuous, but the forager continues to emit sounds as she moves on the comb in a direction appropriate for the direction of the food Kirchner et al., 1988). It is likely that bees following a round dance obtain information about the direction of food quite close to the hive. However, the duration of the sound is too variable to indicate different distances to food less than 100 m from the hive.

Wenner and associates (1971, 1990) criticized the design of earlier experiments and argued that their own experiments indicated odor (specific

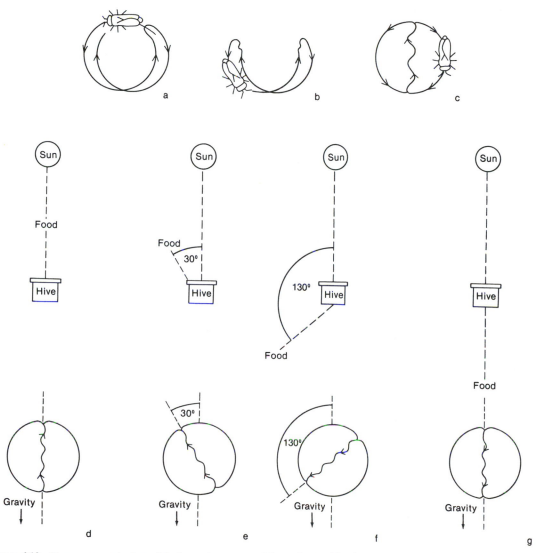

Figure 6.10   Dance communication of the honey bee; **a**, round dance; **b**, transition between round and waggle dance; **c**, waggle dance; **d–g**, diagrams of waggle run (on vertical surface of comb in dark hive) in relation to the direction of food from the hive and the compass direction of the sun.

food odor, Nassanoff gland scent, and overall odor of the food area) constituted the main stimulus, both for recruitment and for the finding of food at some distance from the hive. These criticisms stimulated others to look more closely at the behavior of the dancers and recruits.

By means of ingenious experiments, Gould (1975) demonstrated that distance and directional information is indeed passed from the dancer to other bees. He was able to have a forager send recruits to a food site the forager had never visited and consequently could not have indicated by carrying its odor on her body. Here are the facts on which Gould based his design. When bees dance on the comb in a light beam of sufficient brightness, they will orient to the light instead of gravity. The bees that follow the dancer also orient to the light and correctly interpret the direction to the food source. If, however, the ocelli of the dancer are covered by paint, she becomes much less sensitive to light and a much brighter light is required to stimulate her to do a reorientation. In Gould's experiments, the light was adjusted to not affect the ocelli-covered dancer, but to cause reorientation of the recruits to the light. Suppose the waggle run of the recruiter's dance is vertically upward, correctly indicating a food station directly toward the sun. If the light beam is directed from the left at 90° from the vertical (i.e., horizontally), the recruits with normal ocelli will orient to the light and interpret the direction of the food station to be 90° to the right of the sun. If the recruits rely on odor alone, Gould reasoned, they should appear at the dancer's station, but if they rely on their interpretation of her dance, they should be "misdirected" to other stations situated at 90° to the right of the sun. Gould found that normal recruits were "misdirected" according to the dance information of an ocelli-covered dancer. The light was placed at various angles and the recruits appeared at the predicted stations with an error of about 11.9°, or 31 m at 150 m, and 4.2°, or 29 m at 400 m. It is not known why the more distant stations are indicated with less error.

Kirchner, Michelsen, and their associates (Kirchner and Towne, 1994) carried the analysis further with the construction of a computer-controlled mechanical model bee that danced with waggle runs, vibrated, and dispensed sucrose to recruits. The researchers were able to show that both the waggle run and the pulsed vibrations of the dancer's wings were required to successfully direct a recruit to the food site. Honey bees were thought to be deaf to airborne sounds, but they demonstrated that Johnston's organ in the antennal pedicel was responsive to the movement of air particles in the vicinity of the dancer's wings.

The dance is important in initially recruiting new foragers to a new, distant food source. By following the dance, the recruits learn the distance and direction, discovering the flower odor on the dancer's body and nectar flavor from samples given by the dancer; but they do not know the color or shape of the flower or the landmarks and obstacles en route. Once they have made the trip, foragers are able to return to the food by their memory of landmarks, obstacles, and various odors encountered en route, distance and the compass direction with reference to the sun's position, and color pattern and shape of the flower. For orientation to the sun, they utilize an astrotaxis and compensate for the sun's movement. Cells connected to the nervous system in the abdomen contain magnetite and are believed to function in magnetic orientation.

On returning to the hive, foragers unload the liquid carried in their crop to nest bees. The rate at which the nest bees accept the liquid regulates how quickly each forager can depart for another trip. When the hive is overheated and water is used for evaporative cooling, the nest bees readily take water or dilute nectar rather than concentrated nectar from the returning foragers. This change in behavior stimulates the foragers to collect water or dilute nectar on their next trip rather than concentrated nectar. When the hive temperature is normal, nest bees favor foragers with concentrated nectar over those with dilute nectar. At times, the amount of incoming nectar may exceed the capacity of the nest bees. When delayed in unloading concentrated nectar, a forager performs a **tremble dance,** a distinctive shaking motion, plus a vibration against the comb called a stop signal. This signals to nest bees to increase the reception of nectar and inhibits other foragers from dancing (Nieh, 1993; Seeley, 1992).

Rhythmic Activity and Sunlight

Sunlight plays a key role in the orientation of insects in their environment and in the timing of events in their life cycles. Momentary responses to light, perception of images, color vision, and behavioral reaction to light and polarized light were discussed earlier in this chapter. Of interest here is the influence of the natural alternation of light and darkness in the rhythmic activity of insects.

As a consequence of the earth's inclined rotation, physical factors important to the life of terrestrial organisms undergo regular daily and annual fluctuations. Recall that the earth's axis is tilted 23.5° from a plane vertical to the plane of the earth's annual orbit around the sun. Any given area of the earth's surface receives sunlight for only a part of the 24-hour rotation. With the appearance of the sun's heat each day, temperature usually rises and relative humidity falls. The tilted axis produces an annual variation in the quantity of heat reaching the nonequatorial regions, resulting in the seasons. Furthermore, the earth's rotation around the tilted axis precisely changes the proportion of light and darkness each 24 hours. This results in long summer days and short winter days in the Temperate and Polar Zones.

The daily activities of insects usually take place during specific periods of the 24-hour cycle. Such synchronization, for example, permits the sexes to find each other, limits the search for food to periods of peak abundance, and allows insects to avoid enemies or to take advantage of a favorably high humidity for ecdysis or oviposition. Insects usually spend a part of each day inactive. For example, at night some bees and wasps (Fig. 6.11) clasp leaves or the stems of plants with their mandibles or crawl into flowers and seem to "sleep." During their regular rest period at night, honey bees exhibit behavioral and physiological changes comparable to sleep in birds and mammals.

An activity during the day is termed **diurnal**; at dawn or dusk, **crepuscular**; at night, **nocturnal**; and at dawn only, **matinal.** These terms should be used in reference to specific activities or events rather than loosely to overall activity: the emergence from pupae of adult *Drosophila pseudoobscura* (Drosophilidae, Diptera) is matinal; attraction to food baits, mating, and oviposition are crepuscular; and pupation is mainly crepuscular or nocturnal.

One might interpret a daily rhythm in activity as simply a response to a certain level of light or other physical factor that occurs regularly each day. In other words, the insect's activity is a response to **exogenous** factors. This is true for oviposition behavior in the walkingstick *Carausius morosus* (Phasmatidae, Phasmatodea). Most eggs are laid at or near dusk, or at the time the lights are switched off in laboratory experiments. After *C. morosus* has been kept only a day in constant light or darkness, the laying of eggs no longer occurs at 24-hour intervals but is distributed over

Figure 6.11 Gripping with their mandibles, these male eucerine bees (Anthoporidae) in Peru cling to the underside of a leaf. They may remain together during midday heat or all night. Male wasps and bees often assemble at dusk in such sleeping aggregations.

the 24-hour period. A direct response to an environmental signal may also be delayed a specific interval as if timed by a "sand hourglass."

Circadian Rhythms

In most insects activity rhythms have a physiological, or **endogenous,** basis that is partly independent of environmental signals. In a light/dark cycle of 12:12 hours, mature male crickets, *Teleogryllus commodus* (Gryllidae, Orthoptera), begin to stridulate their mating song 1 to 2 hours before the onset of darkness. Singing continues during darkness and is terminated 2 to 3 hours before the lights switch on. When these crickets are placed in constant darkness, singing begins at intervals of 23.5 hours and lasts the normal period. In constant light, singing occurs at intervals of about 25.3 hours (Fig. 6.12, days 1 to 12). A return to the 12:12 hour cycle of light/dark results in a return to the 24-hour rhythm, beginning 2 hours before darkness (Fig. 6.12, days 13 to 31).

When a given activity or event can be demonstrated experimentally to recur at about 24-hour intervals after all environmental signals have been excluded, it is said to occur in a **circadian rhythm** (meaning "approximately daily"). Under constant conditions the time interval is nearly always slightly more or less than 24 hours, resulting in a progressive drift each day in the time of initiation (Fig. 6.12, days 1 to 12). The presence of the drift is evidence that other possible signals with 24-hour periodicities are not influencing the insect. Cosmic radiation and geomagnetic forces, for example, fluctuate on an exact 24-hour cycle with the earth's rotation.

The physiological basis for circadian rhythm is largely a mystery. Individual cells exhibit rhythmic changes in nuclear volume and protein synthesis. Specific organs function cyclically; endocrine glands, for example, synthesize hormones cyclically. Susceptibility to insecticide poisoning is rhythmic, as is general oxygen consumption. Clearly, rhythms are evident in a variety of physiological processes. The rhythm itself, however, is not determined by the metabolic rate of the process. Unlike nearly all physiological processes, circadian rhythms are largely independent of temperature within normal limits of tolerance.

The underlying mechanism is conceptualized as a "biological clock" or an "oscillator." The clocks of circadian rhythms are reset daily or entrained by brief environmental signals. The most

Figure 6.12 Circadian rhythm of stridulation in a male cricket, *Teleogryllus commodus* (Gryllidae). Marks in rows are tic marks from automatic recorder that is actuated by sound of stridulation. Each row records stridulation during one day. When the record began, the cricket had been in constant light and the stridulation rhythm was free-running. On day 13 the lights were switched off at 1200 and remained off for 12 hours. Thereafter, the cricket was exposed to 12 hours of light and 12 hours of dark. Note that stridulation is temporarily interrupted when the lights are switched off at 1200 each day. (Data courtesy of Werner Loher.)

common cues are dawn or dusk, when light intensity varies rapidly. The location of the photoreceptors varies with the species. Compound eyes or photosensitive tissues in the brain have been shown to be essential for entrainment, but experimental evidence points to the brain, not sensilla, as the site of the oscillator. Even insects inside fruits or leaves respond to light changes if the surrounding media are translucent.

An entrained circadian rhythm is called a **diel** rhythm. It is possible that more than one clock controls the different rhythms of cells, development, general physiology, and behavior. The clock or clocks serve to coordinate life processes within the insect while at the same time scheduling the insect's activity in a changing environment. Locomotor activity, feeding, mating, and the insect's varying responsiveness to stimuli are programmed to take place at opportune times in the daily cycle.

A different type of diel rhythm involves populations and "once-in-a-lifetime" events. If *Drosophila melanogaster* (Drosophilidae, Diptera) is reared in constant darkness, the adults emerge from pupae at any time. When the pupae are exposed to light for a period as brief as 1 minute, emergence becomes synchronized to occur during one period each 24 hours. The period of emergence has been compared to a developmental "gate" that opens or closes. If an individual misses the open gate, its emergence is delayed 24 hours. Gated phenomena are known for egg hatching in the pink bollworm *Pectinophora gossypiella* (Gelechiidae, Lepidoptera), pupation of the mosquito *Aedes taeniorhynchus* (Culicidae, Diptera), and emergence of the intertidal midge *Clunio marinus* (Chironomidae, Diptera) in Europe.

The genetic basis of circadian rhythms is illustrated by the mating behavior of moths. Most moths have specific periods for mating activity. Certain closely related species have identical sex pheromones. Consequently, the males can be attracted to females of different species. Some of these species are allopatric (i.e., do not occur in the same place) or mate at different seasons. In species that are sympatric (i.e., occurring in the same place), mating is at different times of day

under circadian control, thus preventing cross-attraction.

SELECTED REFERENCES

General Structure and Function of the Nervous System

Chapman, 1982, 1991; Homberg et al., 1989 (deutocerebrum); Kerkut and Gilbert, 1985 (vols. 5 and 6).

Sense Organs

Gilbert, 1994 (stemmata); Payne et al., 1989 (olfaction); Roelofs, 1984 (electroantennagrams); Zacharuk and Shields, 1991 (sensilla).

Behavior

Atkins, 1980 (general text); Kerkut and Gilbert, 1985 (vol. 9); Matthews and Matthews, 1978 (general text); Papaj and Lewis, 1993 (learning); Wehner, 1984 (astronavigation).

Communication

Aldrich, 1988 (Heteroptera semiochemicals); Blum, 1987 (exocrine compounds); W. L. Brown et al., 1970 (allomones and kairomones); Byers, 1991 (locust maturation); Claridge, 1985a,b (acoustic signals of Homoptera); David and Birch, 1989 (pheromones and behavior); Lewis, 1984 (general); Lloyd, 1971, 1983, 1984 (bioluminescence and Lampyridae); Pickett et al., 1992 (aphid semiochemicals); Schneider, 1992 (pheromones of Lepidoptera); Stowe et al., 1995 (kairomones used by predators and parasitoids).

Host Selection by Bark Beetles

Borden, 1984; Schowalter and Filip, 1993; Wood, 1982.

Sensory Physiology, Behavior, and Communication of Honeybees

Frisch, 1967 (classic reference), 1971, 1974; Gould, 1975; Hsu and Li, 1994; Kaiser and Steiner-Kaiser, 1987 (sleep); Kirchner, 1993; Kirchner and Towne, 1994; Kirchner et al., 1988 (round dance); Nieh,

1993; Seeley, 1992; Wenner, 1971, 1990 (role of dance questioned).

Circadian Rhythms

Lavialle and Dumortier, 1975 (*Carausius* behavior); Loher, 1972 (*Teleogryllus* behavior).

See Also:

Discussions of behavior elsewhere: reproductive and ecdysis behavior (Chap. 4); cold hardiness (Chap. 5); social behavior (Chap. 7); feeding behavior by phytophagous insects (Chap. 10); behavior of predators and parasitoids (Chap. 11); cryptic and aposematic behavior (Chap. 12).

Color photos 1.   Life history of a holometabolous insect, the checkerspot butterfly, *Euphydryas chalcedona* **(Nymphalidae, Lepidoptera). a.** Male (left) and female (right) mating in the spring. **b.** Female attaching eggs with an adhesive on underside of a leaf of the favorite larval food plant, monkey flower, *Mimulus aurantiacus* (family Scrophulariaceae). **c.** Eggs and first-instar larvae just hatching. **d.** Later in the summer the food plant becomes less palatable and the fourth-instar larvae become inactive in leaf litter until early spring when they complete development on a diet of fresh leaves. **e.** When fully grown, the larvae hang head down and attach themselves by the tip of the abdomen to a twig in preparation for the metamorphic transformation to the pupal instar. **f.** The pupa, also called a chrysalis, is an intermediate stage between the wormlike larva and the adult butterfly. Adult features are visible such as the wing pads, eyes, and banded antennae. The cast skin of the larva is at the top.

Color photos 2. Immature insects. a. Ecdysis of the cicada, *Platypedia* sp. (Cicadidae, Hemiptera). The adult (left) has just emerged from the nymphal skin (right). The nymph has spent many years underground sucking sap of roots. **b.** Parent female and nymphs (note wing pads) of the stink bug, *Edessa rufimarginatus* (Pentatomidae, Hemiptera). Females of some insects may stay with their young and protect them in a subsocial relationship. **c.** Larva of the tomato hornworm, *Manduca sexta* (Sphingidae, Lepidoptera) feeding on its favored host, the tomato. Much of our knowledge of the hormonal control of metamorphosis is based on experimental studies of this insect. **d.** Pupa of *Manduca sexta.* The larva excavates a chamber in soil in which it pupates. The old larval skin is at the right. The pupa is lying on its back, head to the left, with its long adult proboscis encased in the "jug handle." A pupa with the appendages cemented to the body is called an obtect pupa. **e.** Life history of an aquatic insect, the caddisfly, *Limnephilus* sp. (Limnephilidae, Trichoptera). From left to right are: the case of small stones constructed by the larva with its silk; the larva that lives inside the case (note abdominal gills for respiration); a pale pupa that has recently transformed from the larval stage; and a darker, older pupa with advanced development of its cuticle (note the highly folded wing within the wing case). A pupa with the appendages free from the body is called an exarate pupa. **f.** The nymph of an aquatic insect, the mayfly, *Rhithrogena* sp. (Heptageniidae, Ephemeroptera), is highly flattened for life in swift water. Note that beneath the filamentous abdominal gills are additional membranous gills that help the nymph to adhere to rocks. (e and f by H.V. Daly.)

Color photos 3.   Insect pigments and structural colors. a. The beetle, *Moneilema* sp. (Cerambycidae, Coleoptera), has the dark pigment melanin in the cuticle. Melanin may protect insects from ultraviolet light and help absorb heat. The beetle is a Batesian mimic of black tenebrionid beetles, which have a repugnatorial secretion. A black color, in contrast to the pale desert sands, may function as aposematic coloration during both day and night. **b.** Yellow, red, and brown pigments of nymphalid butterflies are ommochromes deposited in the scales. One such butterfly is the monarch, *Danaus plexippus* (Danaidae, Lepidoptera), seen here basking on a sunny winter day. **c.** White and yellow colors of the wings of pierid butterflies, as shown in a pair of mating *Pieris rapae* (Pieridae, Lepidoptera), are pterine pigments. **d.** Red colors of the ladybird beetles, *Hippodamia convergens* (Coccinellidae, Coleoptera), are carotenoid pigments derived from plants. In this case, because the beetles are predators, the pigments are derived from their prey, which in turn are plant feeders. Red colors commonly function as aposematic coloration, warning that the insects are distasteful. The gregarious behavior of ladybird beetles helps to advertise their presence. **e.** Blue colors of the damselfly (Coenagrionidae, Odonata) are created by microscopic granules in the cuticle that reflect blue light in a phenomenon called Tyndall scattering, a type of structural color. **f.** Green, blue, and purple iridescent colors of the day-flying moth, *Urania* sp. (Uraniidae, Lepidoptera), are interference colors. The colors are created by the reflection of different light waves from a series of 5 to 10 very thin, transparent layers on the wing scales. Like other interference colors, the colors change as one views the moth from different angles.

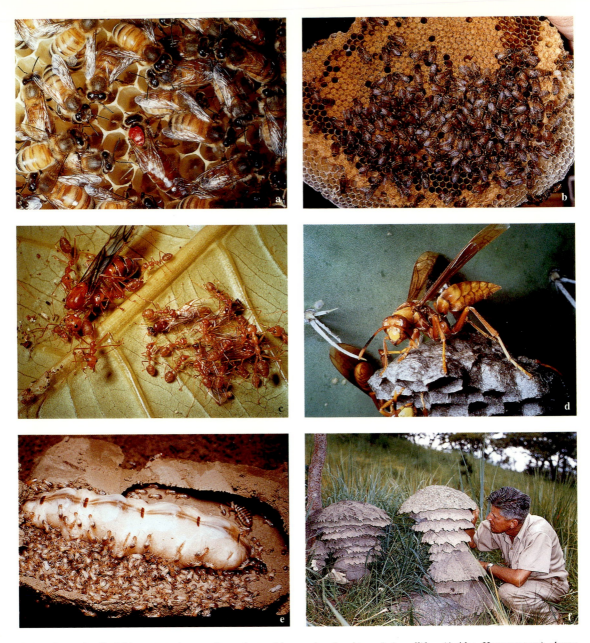

Color photos 4. Social insects. a. A queen honey bee and her worker daughters, *Apis mellifera* (Apidae, Hymenoptera), shown on the comb of hexagonal wax cells. The queen is marked with red paint for easy recognition, and her left wing clipped to prevent escape from the hive. The abdomen of the queen is longer and darker than that of the worker. **b.** A cluster of workers on a comb with capped brood. When a larva is fully grown and ready to transform into the pupa, the worker bees seal the cell with a wax cap. Cells are also used to store pollen and honey. **c.** Some members of a colony of African weaver ants, *Oecophylla smaragdina* (Formicidae, Hymenoptera), including a virgin queen (large winged ant at left), males (small winged ants at right), and female workers of two sizes (wingless ants). After mating in the air, the queen will break off her wings and establish a new nest. **d.** Female paper wasps (Vespidae, subfamily Polistinae, Hymenoptera) build their nests of masticated plant fibers. The queen is usually larger than but otherwise similar in appearance to the female workers. **e.** The huge queen of the African fungus-growing termite, *Macrotermes* sp., (Termitidae, Isoptera), is located in a "royal chamber" along with the smaller male or king (partly visible at upper right) and some of the many workers of the colony. Worker and soldier termites may be either male or female. The spots along the queen's back are the abdominal tergites stretched apart by the enormous ovaries in her abdomen. **f.** The mushroom-shaped nests of the African termite, *Cubitermes* sp. (Termitidae, Isoptera) are designed to shed rain water. Nests are shown here with the photographer, Dr. Edward S. Ross.

Color photos 5.   Insect/plant relationships. Many insects visit flowers for protein-rich pollen or for nectar, a source of sugar for energizing adult activities. Some floral visitors transport pollen from flower to flower, an act that benefits the plant by providing cross-fertilization. **a.** The alfalfa butterfly, *Colias eurytheme* (Pieridae, Lepidoptera), takes nectar from a daisy with a specialized sucking proboscis formed from the maxillary galea. **b.** The sucking mouthparts of the honey bee, *Apis mellifera* (Apidae, Hymenoptera), are modified from the maxilla and a tonguelike labium. Female bees are equipped to carry pollen back to their nests as food for their larvae. In this species, pollen is packed in a "basket" of hairs on the hind leg. **c.** Some insects damage flowers, as do these soft-winged flower beetles (Malachiidae, Coleoptera), shown here eating pollen. **d.** Many insects destructive to plants, such the Hemiptera, have evolved piercing-sucking mouthparts (thin stylets of the mandibles and maxillae) that tap directly into the vascular system. Here a coreid bug (Coreidae) is inserting the stylets, which are supported by the thicker, jointed labium. **e.** Nearly all species of Lepidoptera are phytophagous in the larval stages. A group of young woolly bears (Arctiidae) are shown skeletonizing a pear leaf. **f.** Some tiny moth larvae are specialized to feed inside a leaf between the upper and lower layers. The yellowish mature larva of the leafblotch miner (Gracillariidae) is visible at the leaf edge on the left. Here it will spin a cocoon and pupate. The young larva began the mine on the right side of the leaf and was limited by the larger veins until it was large enough to chew through veins. The dark line tracing the mine is the frass or fecal matter.

Color photos 6.   Entomophagous insects. a. Many mandibulate insects such as this large beetle, *Calosoma scrutator* (Carabidae, Coleoptera), are aggressive predators of other insects. The prey here is a hornworm caterpillar, *Celerio lineata* (Sphingidae, Lepidoptera). **b.** Some Hemiptera with piercing-sucking mouthparts are predators. Powerful digestive enzymes injected into the prey, liquiefy internal tissues, allowing them to be sucked out. Here a reduviid bug (Reduviidae) has attacked a stingless bee (Apidae, Hymenoptera). **c.** The larva of the flower fly, *Syrphus* sp. (Syrphidae, Diptera), has reduced mouthparts and no legs, but nevertheless is an important predator of aphids. **d.** Antlions are larvae of the neuropteran family Myrmeleontidae. Using tonglike mandibles, they are formidible ambush predators. Some make pits in sand to trap prey. This African antlion, *Palpares* sp., does not make a pit, but hunts on the soil surface, or just beneath. **e.** Social insects that are predatory have the advantage of cooperation in subduing a large prey. Weaver ants, *Oecophylla* sp. (Formicidae, Hymenoptera), here dismember a coreid bug (Hemiptera, Coreidae). **f.** Some insects capture prey not for themselves, but to feed their offspring. This thread-waisted wasp, *Ammophila wrighti* (Sphecidae, Hymenoptera), has captured and paralyzed a cutworm caterpillar (Noctuidae, Lepidoptera). The immobilized but living caterpillar will be placed in a burrow as food for the larvae.

Color photos 7. **Protective coloration**. Insects have evolved diverse ways to avoid being eaten by their enemies. **a.** The black body, red wings, and distinctive odor of the tarantula hawk, *Pepsis* sp. (Pompilidae, Hymenoptera), are aposematic or warning signals. After an experience with the sting of the female wasp, an enemy with the ability to learn will avoid this insect. The wasp in the photo is a male. Male wasps have no sting, but the warning coloration stills provides some protection because the predator cannot distinguish between males and females (automimicry). **b.** This mydas fly (Mydidae, Diptera) frequents the same habitats as *Pepsis* and exhibits the same coloration, but apparently is edible (Batesian mimicry). **c.** The African grasshopper, *Phymateus morbillosus* (Acrididae, Orthoptera) feeds on poisonous *Euphorbia* plants and thus acquires the toxins. The taste and a froth emitted by the insect discourage predators. **d.** If disturbed, the larva of the swallowtail butterfly, *Papilio zelicaon* (Papilionidae, Lepidoptera), everts scent glands behind the head and releases foul-smelling isobutyric acids. This chemical defense is probably effective against birds, parasitic flies, wasps, and possibly ants. **e.** Because the pronotum of the treehopper, *Spongophorus* sp. (Membracidae, Hemiptera), resembles a dried twig, the species is ignored by predators (protective resemblance). **f.** The Io moth, *Automerus* sp. (Saturniidae, Lepidoptera), normally rests with its wings together so that its cryptic coloration blends with the surroundings. If discovered, the moth suddenly exposes the large eyespots on the hind wings. The eyespots may startle the predator and allow the moth to escape. Alternatively, the eyespots may be nonvital targets for the pecks of birds, giving time for the moth to escape.

Color photos 8. Diversity of coloration in mantids (Mantidae, Mantodea). Mantids are "sit-and-wait" predators. Theoretically, their coloration may allow them to catch their prey more easily or protect the mantids themselves from attack by insectivorous predators or both. Here are examples of some of the many patterns exhibited by mantids throughout the world. **a.** The flower mantis, *Hymenopus coronatus,* closely resembles a common orchid in the forests of Malaya, yet the pink and green colors are not of plant origin. Note the petal-like femora. On the hind legs, the tarsi, which might distract from the orchid shape, are pale and inconspicuous. **b.** Immature instars of this African mantid in Zaire resemble lichens. **c.** Another African mantid is shown with her egg case or ootheca. Mantids frequently wait on flowers to seize other insects that visit the flowers. **d.** This mantis from the forests of Panama is concealed by its green color and leaflike form. The green color is not from chlorophyll, but is created by insect-produced pigments. **e.** Among the leaves on the forest floor in Malaya the mantid, *Deroplatys* sp., is especially hard to distinguish. **f.** Even the salt flats of the western United States have a distinctive mantid, *Litaneutra* sp., that resembles the dry dusty habitat in which it stalks its prey.

7
SOCIAL RELATIONSHIPS

Winged reproductives of African termites emerging from nest. The nuptials and wingless workers and soldiers are swarming from a fissure at the base of a termite mound in West Africa.

In this chapter we will be concerned with the cooperative relationships among members of the same species that are called **social** (Color photos 4a–4f) and are beyond relationships directly involved in sexual behavior. Social interactions form the basis for important evolutionary strategies in diverse animals, both invertebrate and vertebrate, including humans. Only about 2 percent of insect species are social: termites (2200 species), ants (8800 species), vespid hornets, yellowjackets, and paper wasps (860 species), sphecid wasps (at least

one species, neotropical *Microstigmus comes*), and the social bees (1000 species, of which the European honey bee is only one).

STEPS FROM SOLITARY TO EUSOCIAL BEHAVIOR

The evolution of social behavior has attracted the attention of many scientists, including Charles Darwin, who sought to explain how workers that did not reproduce could be favored by natural selection. Others have asked: Did the behavior evolve from the relationships of parents to their offspring, or as interactions among adults of the same generation, or some combination? In the process of development, or **sociogenesis,** of an individual colony, how do members of a colony change in behavior, caste, or physical location over time? To address these questions, we first need to define an array of behaviors that lead from solitary to true social behavior. These behaviors can be considered from two points of view: phylogenetically (i.e., for a given lineage, what are the steps in the evolution of social behavior over geologic time?) and sociogenetically (i.e., for a given group of individuals, what are the steps in the development of social behavior over the life of a colony?).

Solitary Behavior

The great majority of insects are **solitary**; that is, interactions among adults are largely limited to sexual behavior and competition, and adult-offspring contacts are confined to activities associated with oviposition. For example, the sexes of Phasmatodea meet during copulation and remain joined for long periods, sometimes for more than a day. Later the female, aloft in vegetation, drops her eggs to the soil litter without further care. The more advanced solitary species of insects exhibit some maternal care (i.e., preparation for the safety of offspring), even though they do not directly interact with the immatures. Many species of bees and wasps, for example, are solitary. They construct individual chambers or "cells" in earth or plant materials which are then provisioned with food for their larvae. This behavior is called **mass provisioning,** because the adult does not return to add food once the egg has hatched.

Subsocial Behavior

Adults of **subsocial** insects protect and/or feed their own offspring for some period of time after hatching, but the parent leaves or dies before the offspring become adults. Some provide food to the offspring in a behavior called **progressive provisioning.** Careful observation has revealed a surprising number of species that exhibit various degrees of maternal and/or paternal care. Among the subsocial insects are some species of the following: cockroaches, crickets, earwigs (Fig. 7.1), mantids, jumping plant lice, webspinners, thrips, 13 families of true bugs (Color photo 2b), treehoppers, 9 families of beetles, certain bees and wasps, plus some other arthropods such as spiders. Protection of young during critical early stages has obvious advantages to insects as well as to some amphibians, reptiles, and fish, and to all birds and mammals. The subsocial parent-offspring relationship is one possible phylogenetic or sociogenetic route to social behavior.

Parasocial Behavior

A different possible route from solitary to social behavior, at least in the Hymenoptera, is via interactions of adults of the same generation. This behavior, broadly termed **parasocial,** includes communal, quasi-social, and semisocial behaviors. A simple type of parasocial relationship is the **communal** colony or aggregation of a group of female bees or wasps. The adult females inhabit the same nest burrow, but each constructs and provisions cells for her own eggs. The frequent presence of at least one of the females in the communal burrow provides mutualistic defense for the colony against natural enemies. This relationship can be carried another step: a female in the communal burrow may cooperatively make and provision brood cells with other females while still laying eggs herself. Such cooperation, called **quasi-social behavior,** probably occurs early in nests that later exhibit a reproductive division of labor among the females.

The most advanced stage of parasocial behavior is called **semisocial.** Some females are mated and lay eggs, thus serving as queens. Others of the same age and size may mate or not, but they

Figure 7.1   Female earwig, *Forficula auricularia* (Forficulidae), exhibits subsocial behavior by protecting her eggs.

fail to lay eggs and they become workerlike in behavior. Thus a reproductive division of labor is established among females of the same generation.

Eusocial Behavior

True social or **eusocial behavior** evolved once in the orthopteroids to produce the termites and possibly up to a dozen times in the Hymenoptera. Each case represents an independent lineage of social evolution from essentially solitary ancestors. Colonies of eusocial insects are distinguished by three traits:

- Members cooperate in caring for the young.
- At least two adult generations overlap in the same colony so that the offspring aid the parents in the colony's work.
- The labor of reproduction is divided between reproductive individuals (queens and males or, in termites, kings) and nonreproductive workers.

Pheromones play a major role in the interactions of individuals in a colony. For example, colony odors allow members to distinguish between nest mates and individuals from other colonies. In **kin recognition,** individual pheromones that are probably blends of hydrocarbons in the epicuticular wax allow members to recognize close relatives within a colony. Many, but not all, social insects exchange liquids from the mouth or anus among members of a colony. This arrangement, called **trophallaxis,** is important in the distribution of food and social pheromones in the colony.

Primitively Eusocial Insects

Some species of sweat bees (Halictidae), some allodapine bees (Anthophoridae), bumble bees, and some social wasps are called **primitively eusocial.** These insects have queens and workers that are similar in appearance (queens may average larger in size) and differ mainly in behavior and the development of the reproductive organs. Colonies have fewer individuals and often last only a year. Queens often spend the winter apart from the colony and establish new colonies by themselves. Among these bees and wasps and their close relatives can be found a full range of behavior from solitary to eusocial. Certain species even exhibit these behaviors from one nest to another, revealing that eusocial behavior does not necessarily require a long progression through evolutionary steps.

Highly Eusocial Insects

Termites, ants, some social wasps, honey bees, and stingless bees are considered to be **highly eusocial.** They have large colonies that last for more than a year and often many years. The distinct castes are easily distinguished on sight, and their members display different behaviors or **caste polyethism.** Individual members do not live for long as independent insects. The highly eusocial insects have convergently evolved extraordinary abilities in a number of features, of which many species exhibit the following:

- Coordinated defense of the colony
- Use of pheromones in communication, recognition, and behavior
- Use of pheromones in control of postembryonic development and physiology of adults
- Regulation of the numbers of individuals in different castes and sexes
- Discovery and efficient exploitation of resources: food, water, nest sites and materials to construct nests
- Maintenance of a sanitary, stable environment inside the nests (nest homeostasis)
- Transfer of the nest and accumulated resources to the next generation

In many respects the colony is like an organism with individuals rather than cells specialized for reproduction and other functions. It has been argued that highly eusocial colonies represent a **superorganism,** or level of biological organization beyond the individual organism. As a result of their superior competitive traits, highly eusocial insects and especially ants and termites dominate small organisms in most terrestrial habitats and make up a major fraction of the total animal biomass (Wilson, 1990).

GEOLOGICAL HISTORY OF SOCIAL BEHAVIOR

The evolution of social behavior began in the Mesozoic Era (Table 15.3). Fossils of social termites, ants, wasps, and bees are all found first in the Cretaceous Period. The oldest are termites: *Cretatermes carpenteri* from Labrador, dated at about 97 million years ago, is considered to be relatively advanced among modern Hodotermitidae. Another termite of the same family, *Meiatermes araripena,* from Brazil, dates from 110 million years ago. The ants are the second oldest: two workers of a primitive ant, *Sphecomyrma freyi,* dating from about 80 million years ago, were found preserved in amber in New Jersey. On the basis of comparative anatomy, ants seemed to be just beginning to differentiate from wasps in the Cretaceous, while termites were already diversifying. A fossil nest of a social wasp has been found in Cretaceous deposits in Utah. Finally, what is clearly the worker of a stingless bee in a modern genus, *Trigona prisca* (Meliponinae, Apidae), was found in amber in New Jersey, dating from 96 to 74 million years ago. The fossil record indicates that distinct castes and presumably highly eusocial behavior were established early in these lineages and, once fixed, have remained largely unchanged up to modern times. Among living species of bees and wasps of the primitively eusocial type, social evolution continues in a dynamic state and behavior is remarkably flexible.

EVOLUTION OF EUSOCIAL BEHAVIOR

The critical transition from subsocial or parasocial behavior to eusocial behavior involves the appearance of a nonreproductive caste. The sterile individuals are said to be altruistic because they sacrifice their reproductive ability to benefit the survival of the colony. For altruism to evolve, a shift must occur from individual organisms as the units of natural selection to individual colonies as the units of selection. This was the solution proposed by Darwin. In modern terms, selection must act on **inclusive fitness,** that is, the sum of an individual's fitness plus the effect of the individual on the fitness of relatives other than direct descendants. Altruistic behavior is not limited to eusocial insects: the "defender morphs" of the polyembryonic

Copidosoma (Encyrtidae, Hymenoptera; see Chaps. 4 and 11) and the "soldiers" of certain aphids (see Chap. 34) are also examples of altruistic behavior.

Mechanisms that would facilitate the evolution of eusocial behavior are mutualism, parental manipulation, and kin selection. These are not mutually exclusive and probably act jointly in some groups. **Mutualism** is the pattern of behavior seen as responsible for the higher survival of individuals living in groups with common defense versus living alone. In **parental manipulation,** the queen suppresses reproduction by her daughter workers or other female nestmates through behavioral dominance, sterilizing pheromones, or partial starvation. Queens usually have better developed ovaries and more active corpora allata than the subordinate workers. Some workers may still lay a few eggs, which may be eaten by the queen. As a genetic trait, parental manipulation could be favored by selection, provided the workers do not leave the nest to reproduce elsewhere. For the workers to stay, the risk of failure in founding a new nest must be greater than the security of staying in the colony as a worker, laying fewer eggs, or waiting for the queen to die. The evolution of eusocial behavior by parental manipulation would be enhanced if the worker helped to raise closely related sisters (see below).

The third mechanism, **kin selection,** acts on the inclusive fitness of a colony. This is best explained by considering the parent-offspring relationships in a family of Hymenoptera. Recall that reproduction in this order is by haplodiploidy. Unfertilized eggs develop into males; fertilized eggs yield females. The haploid male receives one of each of the paired chromosomes from the mother (resulting, incidentally, in a situation where a son has a grandfather, but no father). The mother therefore shares half her genes with each son. The son will produce sperm without meiosis so that all his sperm will have the same set of chromosomes. The diploid daughters have half their chromosomes from the mother and half from the father. Accordingly, the mother shares half her genes with each daughter. Of special importance is the genetic relationship among sisters sired by a single father. They all share the same set of paternal chromosomes (as a result, half the chromosomes among sisters are the same). Among the maternal chromosomes, because the chromosomes are segregated randomly during meiosis, the sisters are likely on the average to share half the mother's contribution (as a result, one-fourth of the chromosomes among sisters are likely to be the same). On the average, therefore, sisters share three-fourths of their genes in common. The average fraction of genes shared is called the coefficient of relationship.

For genes for altruistic traits to increase under kin selection, the altruist must actually promote the survival of genes like its own by increasing the numbers of closely related individuals that will reproduce. Altruistic sisters (related at three-fourths) that live with their mother and increase the survival of their reproductive sisters by caring for them, defending the colony against enemies, and foraging for food for the colony are increasing altruistic genes more effectively than they would have done if they cared equally for their own daughters (related at one-half).

On the other hand, the genetic relationship is not favorable in semisocial groups, even when sisters are involved, because an altruistic sister would be helping the production of nieces (related at three-eighths) rather than her own offspring (related at one-half). Evidence exists that unrelated females may sometimes establish semisocial groups, thus further lowering the coefficient of relationship. Similarly, if the female parent of a eusocial group is inseminated by several males or if several females contribute female offspring, the average genetic relationship among the group of female offspring is rapidly lowered. It remains to be shown that local inbreeding might raise the general level of genetic relationship to the point of again yielding favorable coefficients. For insects like the termites, in which both sexes are diploid, inbreeding might also facilitate the evolution of altruism.

BIOLOGY OF TERMITES

All Isoptera are eusocial. Termites differ from the social Hymenoptera discussed further on in three important features: (1) termites live in colonies with castes of both sexes; (2) both sexes are diploid; and (3) these insects have a hemimetabolous life history in which the young instars may participate in the work of the colony. Almost certainly termites evolved along the subsocial route from cockroaches, with which they share many anatomical and physiological similarities. The step from subsocial to eusocial may have been aided by high degrees of genetic relatedness within a colony as a result of inbreeding.

A present-day cockroach, *Cryptocercus punctulatus* (Cryptocercidae, Blattodea), may give us some clues to the behavior of the termites' ancestors. The roaches exist as subsocial groups in cavities in decaying wood. They derive nourishment from cellulose by means of flagellate protozoa that live symbiotically in the roaches' intestine. Flagellates are transmitted among roaches and to the next generation by anal trophallaxis. This bond between generations requires a subsocial life, but not necessarily a eusocial life. Although the circumstances under which true sociality first arose in termites are not clear, it seems probable that termites evolved from a common ancestor with roaches during the early Mesozoic Era (Table 15.3) and were the first eusocial insects.

Castes of Termites

The castes of termites (Figs. 7.2, 24.1, 24.2; Color photo 4e) fall into four categories.

1. **Primary reproductives** are the only individuals that are fully winged after the last molt. They correspond to the alate, imaginal instar of other insects. Correlated with flight are a sclerotized and often pigmented body, compound eyes, and frequently a pair of ocelli. After the nuptial flight, the wings break at preformed lines of weakness near the base, leaving characteristic wing stumps. After insemination, the female's body undergoes extensive changes leading to sustained egg production. The swollen abdomen of queen Termitidae, for example, may reach 9 cm in length. The male remains small. Usually a colony has only one functional primary reproductive of each sex, and both are relatively long-lived in comparison to the sterile castes.

2. **Supplementary reproductives** are imaginal individuals capable of producing eggs and sperm but exhibiting partly inhibited development. In comparison to primary forms, the pale body is less sclerotized and the compound eyes are reduced. Functional wings are not developed at the last molt, but individuals with wing pads are termed secondary reproductives, and those without wing pads are called tertiary reproductives. A dead or declining primary reproductive will be replaced by one or more supplementary reproductives of the appropriate sex.

3. **Workers** are the most numerous members of a termite colony. They are usually of either sex and are wingless and sterile (Figs. 7.2, 24.2). They perform all the duties of foraging, nest construction, and tending the young. The mandibles are not exceptional but are suitable for gnawing and are strongly muscled. Most workers are pale and sightless, but tropical Hodotermitidae, which forage during the day for grasses, are pigmented and possess compound eyes. Some scientists limit the worker caste to a more or less distinctive group of older instars that have lost the capacity to develop further into soldiers or reproductives. Such worker castes are found almost entirely in the advanced family Termitidae. The workers are already different after the first molt and may be divided into subcastes, with one sex predominating in each. The "workers" of more primitive families are properly termed pseudergates or "false workers." These older nymphs without wing pads may later differentiate into soldiers or reproductives, or they may remain "arrested" and workerlike throughout their remaining instars. Pseudergates may also be derived from other partly differentiated castes by regressive molts resulting in loss

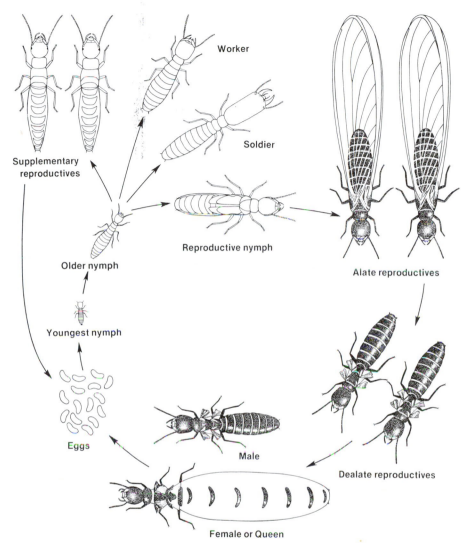

Figure 7.2 Diagram of life histories of castes in a termite colony such as that of *Reticulitermes hesperus*. [Redrawn from Kofoid (ed.), 1934; copyright 1934 by the Regents of the University of California; reprinted by permission of the University of California Press.]

of wing pads. Soldiers and sexually functioning reproductives are final instars and do not regress to pseudergates.

4. **Soldiers** are workers with special development of the head for defense. One type found in some species has enlarged mandibles. A second type, known as the nasute, has a nozzlelike frontal

projection from which is secreted a defensive fluid. In some species the soldiers have both large mandibles and the nasute head. Soldiers cluster about openings when a nest is damaged and engage intruding ants or other enemies. Soldiers may be of both sexes, but in advanced termites, soldiers may be all females or all males, depending on the taxonomic group.

Caste Determination in Termites

How is caste determined? Early investigations sought to distinguish between intrinsic or genetic mechanisms versus extrinsic or environmental influences. The research of Luscher (1961) on a relatively primitive termite, *Kalotermes flavicollis* (Kalotermitidae), points conclusively to the latter explanation: eggs are equipotent, capable of developing into any caste depending on the nest environment. Yet in the Termitidae, early caste differentiation and sex-specific castes suggest an additional genetic basis.

Determination of the caste of an individual *K. flavicollis* involves the familiar hormones of insect development plus pheromones from caste-determined members of the colony. Not unexpectedly, ecdysone promotes differentiation from pseudergate to the reproductive caste. On the other hand, whereas low doses of juvenile hormone in a pseudergate result in no further differentiation, high doses oddly promote differentiation into soldiers. Judging by the volume of the corpora allata just following a molt, the pseudergate at this time has the lowest level of juvenile hormone and is most likely to differentiate toward a reproductive if ecdysone is present. Proportions of the hormones are presumably regulated by the brain, which in turn is responsive to pheromones ingested by anal feeding from other termites. The pheromones have not been chemically identified, but experimental evidence points to several possible agents: (1) a substance produced by the queen that inhibits female pseudergates from becoming functional queens, possibly by suppressing their secretion of ecdysone after molts, (2) two substances produced by the male reproductive or king (viz., a corresponding male inhibitory substance and another substance that stimulates female pseudergates to become functional queens), (3) a substance produced by soldiers that inhibits pseudergates from becoming soldiers, and (4) a recognition pheromone peculiar to queens that allows queens to recognize each other and fight when more than one is present in the colony (similarly, kings recognize each other by their own recognition phero-

mone and fight). Male or female supplementary reproductives become active when the inhibitory influence of the reproductive male or female, respectively, is removed. In other termite species, both reproductives are known to stimulate soldier production. Thus, recruitment into castes may be both stimulated and inhibited by the presence of individuals already caste-determined. Excess numbers are destroyed by cannibalism.

Life History of Subterranean Termites

To illustrate the life cycle of a colony of termites, let us examine the rhinotermitid *Reticulitermes hesperus,* a common and especially destructive species in the western United States (Fig. 7.2; Kofoid et al., 1934). Other species of *Reticulitermes* are found in almost every state and are equally important as structural pests. These are known as subterranean termites because colonies are nearly always in contact with damp earth by means of closed tunnels. When decaying logs are rolled, pieces of lumber turned over, or, worst of all, when part of a house collapses, the presence of *Reticulitermes* is revealed by clusters of hundreds of white insects the size of rice grains. New colonies are established by a pair of primary reproductives (male and female). After the autumn swarming flight, the reproductives flutter to the ground and, by flexing the abdomen, break off their wings at the lines of weakness near the wing base. Females raise the abdomen and release a sex attractant. Males, running on the ground, orient to the odor and locate females. Each pair then moves off in tandem, with the female in the lead. On finding a crevice under or in a piece of wood, the two enlarge the cavity and plaster the walls with feces and wood bits. Copulation follows the first efforts at nest construction and is repeated at intervals for the rest of the lives of the reproductives. Initially fewer than 10 eggs are laid and carefully cleaned to prevent mold. The young nymphs are fed from the mouth or anus of the parents. By the second instar, they have acquired intestinal protozoa. Growth of the population is slow, and members are long-lived. Workers are estimated to live 3 to 5 years, and reproductives probably live longer. The first clearly differentiated castes are apterous, nonreproductive workers and soldiers. After no less than 3 or 4 years under favorable conditions, the colony produces its first large numbers of new primary reproductives. Fully winged in the seventh instar, they remain in the nest until early autumn rains soak the soil. Then on a bright warm day, they emerge by the thousands. During a weak flight of 200 m or so, they mingle with reproductives from other colonies in the vicinity.

Large colonies may also reproduce by a budding process, whereby a part of the constituency becomes isolated and develops its own supplementary reproductives. Outlying food sources, such as houses, may become rapidly infested in this way. Supplementary reproductives of the appropriate sex also replace primary reproductives that die or lose physiological control of the colony.

BIOLOGY OF SOCIAL HYMENOPTERA

Eusociality has emerged repeatedly in Hymenoptera, probably once leading to Formicidae, or ants; once or twice in Vespidae, or hornets and paper wasps; once in Sphecidae (the neotropical *Microstigmus comes*); and possibly up to eight times in Apoidea, or bees, of which honey bees, stingless bees, bumble bees, some allodapine bees, and sweat bees are eusocial. Social behavior in this order is an activity among adult females. As holometabolous insects, the larvae are helpless and rarely contribute to colony welfare. Males are usually winged and function solely in reproduction.

Why are aculeate (stinging) Hymenoptera predisposed to evolve social behavior? The search for possible factors among the presocial Hymenoptera has not yielded a simple explanation, because other insects have some of the same features. However, four features in combination may have uniquely set the stage for recurring sociality:

- **Morphology:** Adult aculeates are strong fliers, with versatile mandibulate mouthparts useful in nest construction, food manipulation, and brood care. Female aculeates have the ovipositor modified into a effective weapon of defense.

- **Behavior:** Nest building and various levels of presocial behavior exist, including parental manipulation and mutualism among long-lived females.

- **Genetics:** The haplodiploid reproductive system creates in a family genetic relationships that favor evolution of altruistic behavior among sisters living with their mother.

- **Geological age and statistical opportunity:** Aculeates date from the Cretaceous Period and number now about 50,000 species. Thus there has been ample chance for eusociality to arise.

BIOLOGY OF ANTS

The life histories of different species of ants are extraordinarily varied (Fig. 7.3). As an extreme example, species are known in which reproduction is by thelytokous parthenogenesis, queens are absent, males very rare, and unfertilized workers reproduce. In general, however, most colonies have reproductive queens and males and nonreproductive workers (Color photos 4c, 6e). Virgin queens usually mate with males from other colonies in the area. Depending on the taxonomic group, essentially two kinds of mating behavior exist. In the **female-calling syndrome,** virgin females on ground or vegetation release a sex pheromone from the pygidial gland, between the sixth and seventh abdominal tergites. Males are attracted and copulation ensues. In the **male-aggregation syndrome,** flying males form swarms in the air, usually above conspicuous landmarks like trees. Females fly into the swarms, and copulation ensues.

Colony growth can be divided into three stages. The **founding stage** begins with the end of the nuptial flight. The virgin queen mates with one or more males, which die without returning to the colony. The queen breaks off her wings, finds a suitable nest site in soil or plant material, and

Figure 7.3 Examples of ants: **a**, *Crematogaster* ants feed on honeydew from aphids on the underside a willow leaf. The ants may protect the aphids from enemies, but in this case two kinds of predator have escaped attention: a white-spotted coccinellid beetle larva and, at lower right, a fly larva, *Leucopis* (Chamaemyiidae). **b**, Leaf-cutter ants, *Atta* sp., cut pieces of leaves and carry them to their underground nest. The leaves are masticated to serve as a cultural medium for the fungus that the ants eat.

constructs the first cell. The first workers are reared by the queen and as adults are quite small. Depending on the species, the queen may forage for some food outside the nest (known as partially claustral founding type) or remain entirely in the nest (fully claustral founding type), drawing on her own tissues (flight muscles and fat body) to produce eggs and feed the larvae. Feeding is by special salivary secretions or by trophic eggs (eggs fed to larvae). In some species, young queens are accepted back into the parental nest.

In the second, or **ergonomic stage,** of colony growth the first brood of workers reaches adulthood. The queen begins to devote herself to egg laying, leaving foraging for food, enlargement of the nest, and brood care to the workers. The population of workers grows in number, and the average size of individuals increases. New castes may appear among the workers. At this stage, all activities in the nest are devoted to colony growth.

After a period of one season or up to 5 or more years, depending on the species, the colony enters the final, or **reproductive stage.** Part of the colony's resources are now invested in new queens and males. These leave the colony and the colony activities revert to the ergonomic stage until the next reproductive stage is reached.

Colonies of ants are perennial, but nest sites may be changed from time to time. Some species, such as the army ants and driver ants, are more or less constantly on the move. The number of eggs per queen per year varies from 400 in one species to 50 million in the African driver ants. The time from egg to adult varies with the species, on the order of 1 to 2 months. Adult queens are long-lived; some have been estimated to live for 18 to 29 years. Adult workers may live 1 to 2 years although most probably have much shorter lives. Males have a life span much shorter than workers or queens.

Worker ants lavish care on all the immature stages, from egg to larva to pupa. Workers lick, clean, and carry the immatures about the nest, and remove their waste products. They feed the larvae by regurgitation, by glandular secretions, and by placing the larvae near other kinds of food that has been prepared for them by cutting up or chewing. Some workers are capable of laying trophic eggs that are fed to larvae.

Larvae of some species are not entirely helpless, but contribute to the colony's welfare. In some species, larvae produce their own secretions that are consumed by the workers. It has been speculated that the larvae are able to digest certain foods and share the digested food with the workers. Larvae of weaver ants help in the construction of arboreal nests. Worker ants pull leaves together, then hold the larvae at the leaf edges, where they secrete silk to bind the leaves together.

Males and Castes of Ants

1. **Males** are generally considered to be a sexual form, but not a caste, because they are all alike and function only in reproduction. They are winged, with a fully developed thorax and fully developed compound eyes and ocelli. In some species they undergo some degeneration of internal structures (fat body depleted, corpora allata and midgut degenerated) prior to the nuptial flight and become "sexual missiles." In most colonies they are fed by workers and do no work. In some species, however, males are long-lived and exchange food with workers, thus displaying some, but not all, the behaviors of a caste.

2. **Queens** are the female reproductive caste in the colony. They have a fully developed thorax with flight muscles and wings, compound eyes, and ocelli. The wings are broken off after the nuptial flight and the flight muscles degenerate. Queens have a fully developed reproductive system with a spermatheca capable of storing sperm for the duration of the queen's life. Depending on the species, a colony may have a single queen (monogyny) or multiple queens (polygyny).

3. **Workers** are sterile females with reduced or no ovarioles, reduced or no eyes, no wings or flight muscles, and an otherwise simplified thorax. Workers are divided into subcastes according to size: minor, media, and major. When majors function mainly in defense of the colony they are called soldiers. Workers of different sizes also differ in the proportions of some body parts such as the head and mandibles. These grow relatively larger in relation to the rest of the body in large ants, resulting in ants suited for defense or grinding of seeds with mandibles. Large workers in the nests of the honeypot ants, *Myrmecocystus,* and some other species become so engorged with honey or water in the crop that they are almost immobile. Called "repletes," they are confined to the nest and serve as storage vessels.

Caste Determination

Caste determination is virtually always environmental rather than genetic. In rare instances, single alleles have been shown to determine some special reproductive types in one species. The determination of caste can be viewed in the following way. In the absence of modifying influences, a diploid female egg will develop into a queen. The anlagen or imaginal disks, of the larva are divided into two sets: (1) a dorsal group, including wing buds, incipient gonads, and ocellar buds, and (2) a ventral group, including leg buds, mouthparts, and central nervous system. In the morphogenesis of a queen, both sets develop with equal vigor. However, an interwoven set of environmental and/or maternal factors may inhibit or fail to promote (presumably via the endocrine system) the complete development of the dorsal set. Such an individual becomes a worker.

Certain experiments indicate that pheromones secreted by a functional queen inhibit development of more queens among her brood. Nutrition of larvae also probably varies both in quantity and quality as the colony grows. A large foraging force may provide optimum food, or the queen's inhibitory pheromones may be diluted, or both. In any event, new queens are produced from large colonies. Other environmental influences are chilling temperatures. Brood produced in the fall at temperate latitudes become dormant and later chilled by winter's cold. Dormancy is later broken by the warmth of spring, and the surviving larvae are able to grow rapidly, metamorphosing into queens. Maternal effects include the size of eggs laid by the queen. During periods of high egg production, eggs average smaller and produce mostly workers. Younger queens also tend to lay eggs that result in workers.

Life History of the Argentine Ant

The Argentine ant, *Linepithema humile* (formerly known as *Iridomyrmex humilis*) is a familiar species that differs in several respects from the normal life history of ants. Commercial traffic from South America distributed the ants to countries in the Northern and Southern Hemispheres within latitudes of 30° to 36°. In North America the first colonies were seen in 1891 in New Orleans, Louisiana. They are now widely distributed in the southern United States and in California, where they first appeared in 1905. Native ants are regularly displaced by Argentine ants, often by physical combat.

Nectar and honeydew from plant-sucking Homoptera are more than 99 percent of the diet of the Argentine ants. This search for sweet fluids has earned the species a place among the important household and agricultural pests. Massive numbers may invade well-kept kitchens or restaurants for food, often in late summer and after heavy rains. Beehives are entered and honey is taken from the combs. Mealybugs, scale insects, and other honeydew-producing Homoptera are tended and protected by the ants. The homopterous pests of citrus and other crops are protected by ants from attack by predators and parasitoids, thus greatly reducing the effectiveness of biological control programs.

Argentine ants occupy a wide range of habitats, provided sufficient moisture is available in soil, logs, or under cover objects. The semi-nomadic colonies do not have clearly defined limits or individuality. In a given area many nests are excavated in soil at favorable sites and near food sources, but these are parts of an extended colony. Workers move freely from one nest to another. A nest is abandoned when the food is depleted or when flooding or drying renders the site unsuitable. When the general area is unfavorable, occasionally a single large nest is made or the colony moves 100 m or more to a new area. Argentine ants exploit food resources in their territory by a rapid, highly organized exploration by workers. A column of ants extends from the nest, then spreads out into a fan-shaped front in which the individuals search randomly. Once food is discovered, a trail pheromone (Z-9-hexadecenal) from a sternal gland leads new recruits to the food until it has been consumed.

New colonies are established by colony fission or "budding" rather than the usual aerial dispersal of young queens to new sites. Reproductive males and new queens are produced in the spring, but these usually mate entirely within the nest. The inseminated queens shed their wings and begin laying eggs. During the growing season the number of queens in a colony may comprise as much as 3.5 percent of the biomass. In autumn, the worker population decreases. Workers then attack queens and dismember them, reducing the number of queens in the colony before the next year's cycle.

BIOLOGY OF SOCIAL VESPIDAE

The family Vespidae includes six subfamilies, of which most eusocial species are in two subfamilies: Polistinae (800 species in 29 genera; cosmopolitan) and Vespinae (60 species in 4 genera; Holarctic and Oriental tropics). Examples of social genera are *Vespa* and *Vespula* (Fig. 1.6), and the largest genus, *Polistes*. The subfamily Stenogastrinae (50 species in 6 genera; Oriental tropics) also has at least some eusocial species, but these are poorly studied.

The wasps prey on small invertebrates, scavenge on carrion, and also feed eagerly on sugary solutions such as nectar, honeydew, and fruit juices. Both males and females are winged and strong fliers. In some species, the queens and workers are similar in size and appearance, while in others the queens are larger than the workers, especially early in the season. Nests are constructed in cavities or exposed and with paper made of plant or wood fibers (Color photo 4d). Brood is reared in combs of hexagonal cells that open downward.

The typical life history in the Temperate Zone begins with inseminated females (queens) overwintering in protected places and founding new nests in spring either alone or in groups. The queen builds the first nest and feeds the larvae. The first brood usually comprises all females that remain on the nest as workers. The queen then devotes herself to laying eggs, and the workers build the nest and feed the larvae of subsequent broods. The last brood of the season includes males and females. On maturity, these leave the nest and mate. On the death of the queen, brood production ceases and the nest population declines as the workers and males die. The newly inseminated females overwinter, living on fat reserves and in a reproductive diapause. Nests may have a single queen (haplometrosis) or several queens (pleometrosis) that either founded the colony together initially (often sisters) or have joined an existing colony. Nest sharing by closely related females is common. In the Tropics, nests may be perennial and new nests may be formed by swarming of wasps, either a single queen and workers or multiple queens and workers.

BIOLOGY OF SOCIAL BEES

The Apoidea include about 25,000 to 30,000 species in over 4000 genera. Most are solitary in behavior, but perhaps 1000 species are eusocial. The most advanced social bees are the honey bees (*Apis* in Apidae; Color photos 4a, 4b, 5b) and tropical stingless bees (*Trigona* and *Melipona* in Apidae) that regularly form perennial colonies. These establish new nests by large flying swarms of workers plus the queen, because the queen cannot live apart from the colony. Less advanced are bumble bees (*Bombus* in Apidae), the social sweat bees (*Halictus* and *Lasioglossum* in Halictidae), and some allodapine bees of Africa, Asia, and Australia (*Allodape* and *Exoneura* in Anthophoridae) that have smaller colonies, usually lasting for one season, and females that spend part of the year not in colonies.

The life history of bumble bees is similar to that of social wasps. The inseminated queen overwinters in a protected place. In the spring she makes a nest in a preexisting cavity, constructs rounded wax cells for brood and food storage, and forages outside the nest for pollen and nectar. The larvae are fed progressively. The first brood consists of small females that function as workers. They take over the duties of nest building, foraging, and defense, leaving the queen to lay eggs. Early in the season there may be additional broods of female workers, but toward the end of the season both males and new, large queens are produced. These mate and the males die. After the founding queen and the workers die, the species survives the winter as the young inseminated queens.

Sweat bees are similar in life history with inseminated females overwintering, often in the old nest, and founding new nests in the spring. Nests are burrows in soil with individual cells mass provisioned for brood. Mostly females are produced early in the season and both sexes toward the end of the season. These mate, the males and old females die, and the young inseminated females overwinter. Behavioral interactions among the females are enormously varied, depending on the species, the history of the individual nest, and the length of the growing season. Within a single species, the populations in a region of short growing season (higher latitudes or altitudes) may live as solitary bees, while those in a region with a longer growing season have several generations and eusocial colonies. Furthermore, examples are known within a single species at one location where nests have solitary females and other nests have groups of females displaying various degrees of social behavior, including semisocial and eusocial. Females have the capacity to function as solitary bees, as workers, or as queens, and to change behavior during their adult life. A worker can replace the queen if she dies. The flexible behavior of sweat bees contrasts sharply with the rigid castes of highly social insects like the honey bee.

Allodapine bees have a different annual cycle: males and females overwinter together, and mating is at the beginning of the growing season. One or more females excavate a burrow in the soft pith of a dead stem. No cells are constructed; rather, the larvae are dumped together at the bottom of the burrow and all are fed progressively by the female or females. A reproductive division of labor may occur even when only two adult females occupy a nest. As with the sweat bees, a great range of behavior has been observed within a single species: solitary (subsocial), semisocial, and eusocial.

The stingless bees (Apidae, Meliponinae) found in tropical regions of the world are so named because the ovipositor is atrophied and does not function as a sting. The bees nevertheless defend themselves with mass attacks on intruders, biting and secreting toxic chemicals. Colonies are highly eusocial with large nests built of wax, resins, and other materials. Each colony has a single queen, many workers, and males. Depending on the species, nest populations range in number from 300 to 180,000. Brood is reared in horizontal combs of resin and wax in which the cells open upward. Food is stored in wax pots. The brood cells are mass-provisioned, so workers do not directly contact the developing larvae. Worker bees of *Trigona,* when returning to the nest from a food source in the forest, deposit droplets of odorous secretion from the mandibular glands on vegetation at intervals of 2 to 3 m. This "odor trail" guides new foragers to the site. Odor trails are especially efficient in marking surface routes over broken terrain or in the three dimensions of a tropical forest. Female caste determination in *Trigona* is apparently trophic: males and workers are reared in smaller cells and receive less food; the larger queens are reared in larger cells and receive more food. In *Melipona,* males, workers, and queens are reared in cells of the same size. It is likely that a certain proportion of females are genetically determined to develop into queens, provided the food is adequate; otherwise they develop into workers, like all other females.

Life History of the Honey Bee

The best-known and most useful social insect is the honey bee, *Apis mellifera.* Possibly it has been studied more than any other insect. The native home of the honey bee is in the Old World, where it has been exploited throughout human history. Colonies in natural cavities, such as a hollow tree, or in hives made by human beings, were killed to extract stored honey and wax. Only relatively recently, in the 1500s, did European beekeepers devise a hive with movable wooden frames that permits the removal of individual combs without destruction of the entire colony (Fig. 1.7). Apiculture is now the most widely practiced agricultural activity in the world. The role of the honey bee as a pollinator of crops is becoming increasingly important in North America. During the earlier stages in the development of agricultural lands, adjacent wild areas were usually inhabited by native pollinators. Under modern intensive cultivation, these pollinators have been eliminated, and honey bees now are regularly transported from crop to crop for this function.

The common honey bee is one of four distinct species or species complexes in the genus *Apis.* All are properly called honey bees because they are eusocial and construct wax combs for rearing brood and storing honey. All except *A. mellifera* are confined to the eastern Palearctic and Oriental regions. *Apis mellifera* originally inhabited western Asia, Europe, and Africa south of the Sahara Desert.

Races of European Honey Bees

Several distinctive races of honey bees evolved in Europe during the Pleistocene Epoch, when advancing glaciers fragmented their distribution. The bees survived in separate refugia in what are now Spain, Italy, Austria, and the Caucasus. The respective races are *A. m. mellifera,* the German, Spanish, or English honey bee; *A. m. ligustica,* the Italian honey bee; *A. m. carnica,* the Carniolan honey bee; and *A. m. caucasica,* the Caucasian honey bee. Beekeepers in these temperate regions have selected for productivity and ease of management, with the result that European bees are preferred in agriculture.

When colonies of European settlers were established abroad, honey bees were taken from the home countries. In the New World, *A. m. mellifera* was the race initially introduced. As communication and transportation increased, beekeepers exchanged queens and colonies around the world in an effort to improve productivity in the new climates. The Italian honey bee became especially favored and now is the predominant commercial honey bee in North America.

African Honey Bees

In Africa, *A. m. scutellata* evolved under quite different conditions. Here it was beset by a variety of natural enemies, including human beings, and existed in a tropical climate with dry seasons. The African honey bee is consequently highly aggressive in the defense of its nests and able to undertake migrations to more favorable places. The colonies also reproduce more frequently than European bees. The high productivity in the tropics has attracted bee breeders on several occasions to make hybrids with European stocks. These experiments went largely unnoticed until a recent episode had calamitous results.

European bees in tropical South America are unsatisfactory honey producers. In 1956 African honey bees were taken to southern Brazil for the breeding of a strain combining their higher productivity in the tropics with desirable features of European bees. In 1957 protective screens that confined the African queens were mistakenly removed and 26 queens with workers escaped. Hybrids were formed between the African bees and the resident European bees. In the Old World, however, *A. m. scutellata* and the European races had evolved apart for at least 2000 years, separated by the Sahara Desert. As a result, free cross-mating between the races was reduced, owing to the apparent preference of reproductives for their own race. Such hybrids as were formed perpetuated mostly African characteristics, hence the name "Africanized" for the bees in the New World (Fig. 7.4).

The Africanized bees directly compete for food and nest sites, even invading the hives of European bees and displacing them. Large numbers of wild colonies were also established in small cavities not suitable for European bees. As the Africanized bees migrated annually into new areas of Brazil, it became evident that the European bees were being completely replaced by a much higher population of Africanized colonies. Beekeeping became difficult or impossible. Reports appeared in the world press of deaths and injury to humans and domestic animals. Although some minor differences in the chemistry of bee venom may exist, the danger is created by the large numbers of bees that quickly sting and pursue a victim. Since 1957, Africanized bees spread throughout most of tropical South America, up

Figure 7.4 Workers of Africanized honey bees, *Apis mellifera scutellata* (Apidae, Hymenoptera), from Brazil, feeding on sweets spilled on a rock in a picnic area.

through Central America to Mexico, and reaching the United States at Brownsville, Texas, in October 1990. The bees have now been identified in the southern parts of the states from Texas to California. Southward in Argentina, the Africanized bees appear to be unable to survive the cold winters of the temperate climate. A similar temperature barrier may prevent northward invasion in the United States beyond the southern and southwestern states.

In southern Brazil, large numbers of Italian queens were distributed to beekeepers in a program to reduce the aggressive behavior of Africanized bees by further hybridization and by killing aggressive colonies. This measure has been apparently successful in producing bees that are manageable and still productive. The beekeeping industry has recovered in this area. The genetic approach offers the best means of controlling the undesirable traits of the bees, but over vast regions of tropical America such a tactic is impractical.

Males and Female Castes of Honey Bees

Both sexes of honey bees are winged, with well-developed compound eyes and ocelli.

1. **Males** or **drones,** are produced during the spring and summer seasons. They are distinguished by larger eyes and their larger body size. Drones are fed throughout their 4- to 5-week lives by the worker bees and are unable to extract nectar from flowers. Aerial excursions are taken from the hive in search of queens, with which they mate in the air. The male genitalia are uniquely constructed to evert violently and be torn off during copulation. Successful drones die after mating. At summer's end, unsuccessful drones are expelled from the hive, also to die.

2. The **queen** is the functional female reproductive in the colony (Color photo 4a). A queen is slightly larger than a worker and has a longer abdomen that extends well beyond the folded wing tips. From their mandibular glands is secreted the **queen pheromone,** which is a mixture of 9-keto-

(E)-2-decenoic acid, 9-hydroxy-(E)-2-decenoic acid, and lesser amounts of other components. Possibly another pheromone is produced from tergal glands, but the chemical identity of such a compound is unknown. The pheromone or pheromones have multiple functions: inhibition of the development of the ovaries of workers, inhibition of the construction special wax cells for rearing new queens, attraction of workers to the queen in the hive and during swarming, and attraction of males during aerial mating. The absence of the queen in a colony or swarm is quickly detected by the absence of her pheromones. The queen never forages for food, but does leave the hive for one or more mating flights early in her life and during periodic swarming. Ordinarily, only one queen is present at a time in a colony. She produces all the eggs that are fertilized and most of the unfertilized eggs. These may total 600,000 over the 2 to 3 years of her normal life span.

3. **Workers** are females whose reproductive organs are atrophied to the extent that copulation is prevented, although they can occasionally produce eggs (Color photo 4b). Some of the unfertilized, or drone, eggs in a colony come from this source. In an exceptional race of honey bees, restricted to the Cape region of South Africa, workers lay eggs that develop by thelytokous parthenogenesis into females. The sterile workers perform all the foraging, defense, wax secretion and comb construction, brood care and feeding, and regulation of the physical environment, and also locate new sites during swarming. Colonies normally contain 60,000 to 100,000 workers. During the active foraging seasons, a worker may live 5 to 6 weeks. Workers reared in the fall and overwintering may live 6 months or more.

Caste Determination in Honey Bees

The vertical comb is made of wax secreted by glands on the abdominal sterna of workers. Using their mandibles, the bees shape the scalelike pieces of wax into cells of hexagonal cross section. Each comb is composed of two layers of cells, which open horizontally on each side (Color photos 4a, 4b). Human engineers have imitated this design for creating strong, lightweight structures. The bees' cells are used as a nursery or for the storage of honey, or they are packed with lumps of pollen. Most of the cells are closely similar in size. In the brood area the queen lays a single, fertilized egg in each cell to produce workers. When food is plentiful, larger cells near the margin of the comb receive unfertilized eggs to produce drones. New queens are reared in much larger cells, which are round in cross section and are specially constructed to hang from the surface or lower edge of the comb (Fig. 4.6). The brood cells remain open while the larvae are fed progressively by the workers. The determination of caste among the females is directly related to differences in food.

For the first 2 days, drone and worker larvae are fed a mixture of glandular secretions and crop fluids from workers functioning as nurse bees. This "bee milk" contains clear hypopharyngeal gland secretions plus clear crop fluids, and a lesser amount of white mandibular gland secretions and some hypopharyngeal secretions. The white secretions diminish and disappear in the food on the third day, and afterward the larvae are fed "bee bread," consisting of the clear fluids variously mixed with pollen and honey. A female larva destined to become a queen is fed throughout her life on "royal jelly," which is composed of equal parts of the clear and white foods. These secretions include basic nutrients such as sugars, amino acids, proteins, nucleic acids, vitamins, and cholesterol, plus a fatty acid, *(E)*-10-hydroxy-2-decenoic acid. The quantity, quality, and timing of the food, not a queen-determining pheromone, are the factors that divide worker-destined and queen-destined larvae at an early age. With high-quality food, the neurosecretory system of the queen larva becomes active earlier than in the worker larva, leading to higher titers of juvenile hormone content in the queen. For the queen, this difference in hormone results in a higher respiratory rate as a larva, protection of her ovaries from degradation, and

initiation of vitellogenin synthesis in the pupal instar. Intermediates between queens and workers are rarely found in nature but can be experimentally produced.

Once the larva has reached full size, it defecates for the first time. A cocoon is spun that separates the larva from its feces. The workers cap the cell with wax. The larva now transforms to a pupa and then to an adult inside the closed cell. After the final molt, the new bee chews its way out of the cell to join the colony. Three weeks are required for growth from egg to adult.

Life of the Worker Honey Bee

A worker bee spends about 3 more weeks inside the hive before becoming a forager, or field bee. The bee's behavior and the tasks performed in the hive change as the bee ages. This continuum is called **age polyethism** and is probably regulated by juvenile hormone acting as a behavioral primer. Much time is spent resting or patrolling the comb. Young worker bees first clean the hive and feed larvae with secretions from their hypopharyngeal and mandibular glands, which are well developed at this time. With age these glands normally decline in activity and change secretions. The hypopharyngeal gland later secretes invertase, which converts the sucrose in nectar to the glucose and fructose of honey, and the mandibular gland secretes an alarm pheromone, 2-heptanone, when the worker later serves as a guard or forager. As the nursing function decreases, the sternal wax glands may become active if comb building is prevalent in the hive. Short flights from the hive begin at about this time. After the wax glands decline, activity shifts to the entrance, where the worker may guard the hive before finally becoming a forager. For the remaining 1 to 3 weeks of her life, she collects nectar and pollen from flowers (see "Learning and Communication in Honey bees," Chap. 6; Color photo 5b). Plant resins are also collected and used to seal openings in the hive. Beekeepers call the resins **propolis** or bee glue.

Annual Cycle of a Colony

The annual cycle begins when the spring flowers provide abundant food. The queen begins laying fertilized eggs that greatly augment the worker force. Normally the increased food supply, stored provisions, and high population inside the crowded hive lead to swarming. The inhibitory effect of queen pheromone declines, and workers construct the special cells in which new queens are reared. Egg production by the old queen declines, as does foraging by the workers. Before the new queens emerge, the old queen, part of the worker force, and some drones leave the hive. The mass flight, or **swarm,** is spectacular. The bees may periodically settle in a cluster on a tree limb or other object. Scout worker bees fly from the swarm in search of suitable new quarters. The scouts return and dance on the cluster, indicating the distance and direction to the favored site. Once the swarm has occupied the new hive, wax combs are constructed, food is collected, and the old queen begins to restore the worker population. In the original hive, the first new queens to emerge fight physically for the role as queen. Pupal queens are destroyed by the workers. The survivor flies from the hive several times over several days and may be inseminated by a drone each time. The sperm are stored in her spermatheca for use during the remainder of her life. Her new colony has the benefit of existing combs and stored food. The colony may produce swarms more than once during favorable seasons. In the fall when flowers no longer supply enough food, brood rearing declines and drones are expelled. More propolis is added to reduce cold drafts. The colony clusters together on combs filled with stored honey during cold weather. The bees remain alert through winter, using honey as a heat-giving energy source.

SELECTED REFERENCES

General

Andersson, 1984 (evolution of eusociality); Billen, 1992 (general); Breed et al., 1982 (general), 1989 (genetics); M.J.W. Eberhard, 1975 (kin selection); Engels and Buschinger, 1990 (castes and reproduction); Fletcher and Ross, 1985 (reproduction); Hermann, 1979–1982 (comprehensive); Hunt and Nalepa, 1994 (nutrition); Robinson, 1992 (division of labor); Tallamy and Wood, 1986 (subsocial insects); Wilson, 1971 (comprehensive), 1990 (success of eusocial insects).

Termites

Emerson, 1967 (fossil); Krishna, 1990 (fossil); Krishna and Weesner, 1969a, 1969b (comprehensive); Skaife, 1961 (general).

Ants

Holldobler and Wilson, 1990 (comprehensive); Shattuck, 1992 (scientific name of Argentine ant); Wilson et al., 1967 (fossil).

Wasps

Evans and Eberhard, 1970 (general biology); Ito, 1993 (communal aggregations); Jeanne, 1980 (evolution of social behavior); Ross and Matthews, 1991 (comprehensive); Spradberry, 1973 (comprehensive); Wenzel, 1990 (fossil).

Bees

Alford, 1975 (bumble bees); Dadant, 1975 (apiculture); Crane, 1990 (apiculture); Free, 1987 (pheromones); Menzel and Mercer, 1987 (neurobiology of honey bees); Michener, 1969, 1974 (comprehensive); Michener and Grimaldi, 1988 (fossil); Moritz and Southwick, 1992 (superorganism); Needham et al., 1988 (Africanized bees); Rembold, 1991 (caste determination); Robinson, 1996 (queen pheromone); Spivak et al., 1991 (Africanized bees); Winston, 1987 (honey bees).

See Also:

Additional information and references can be found in the chapters on Isoptera and Hymenoptera (Chaps. 24 and 46).

PART TWO

Insect Ecology

8
POPULATION BIOLOGY

A population of monarch butterflies. Each autumn monarch butterflies, *Danaus plexippus* (Nymphalidae), migrate long distances to sites in coastal California and the central Mexican highlands. There, hanging on tree limbs, the butterflies spend the winter before migrating north in the spring.

The diversity, abundance, and rapid development of insects have long attracted the attention of ecologists. Insects generally are more easily captured and handled than birds, mammals, or other vertebrate animals, facilitating their study. Most importantly, insects play key roles in determining the distribution of plants and other animals. Biological control of introduced species of weeds by introducing and establishing insect herbivores of the weeds is dramatic proof of the kind of impact insects can have on plant abundance and distribution (Chap. 14). Insect pollinators are essential to many species of flowering plants (Chap. 10). The trypanosomes transmitted by tsetse flies, which cause serious animal diseases, can dramatically influence which species of animals are able to survive within a region (Chap. 13). Economic and medical incentives to understand what causes numbers of insect pests or vectors to increase or decline reinforce a basic scientific interest in the biology of populations. What underlying factors and processes determine the distribution, growth, and the decline of insect populations? **Communities** consist of assemblages of populations that jointly occupy a defined location. Together with the physical environment, communities form ecosystems. Thus understanding how populations interact with other populations and the environment is an important part of larger scale ecological studies.

173

Populations are defined for purposes of research or description. For example, one population might be all the insects of a given species in a square meter of soil to a depth of 10 cms. Another population could be defined as all the individuals of a given species in a particular forest. Generally, a population is defined by the phenotype or genotype of its members and the location and duration of the study. Because of the impossibility of counting or measuring all members of most populations, one must sample a population to determine its attributes. Thus population studies extensively employ statistical measures and methods for description and analysis. The property most commonly measured or estimated in populations is population density. Studies of insect distribution and dispersal depend on measures of density within a location. **Population dynamics** is the study of changes in population size (usually inferred from measurements of population density) over time. Other attributes of populations may be important, depending on the purpose of a particular study: distributions of size, age, sex or phenotype (e.g., winged or wingless), genotype (possession of a particular sequence of DNA), or behavior.

Insect populations can have major ecological, social, and economic influences. Termites are examples of insects that have produced a variety of important effects. Their damage to buildings costs billions of dollars per year to repair or prevent. As the major herbivores in some tropical savannahs, termites are important in determining the types and amounts of plants that occur in those ecosystems. Termites also produce relatively huge amounts of methane and other gases as a consequence of their abundance, widespread occurrence, and consumption of plants. The collective contribution of termite flatulence to the composition of atmospheric gases and long-term climate change provoked by a greenhouse effect is controversial. Disagreements center on the accuracy of estimates of termite numbers worldwide and the rate that termite colonies digest plant materials and produce respiratory gases. These estimates range widely among investigators (Berenbaum, 1995), so the analysis of the contributions of insect metabolism to the atmospheric greenhouse effect requires realistic estimates of how many termites are alive and how much and what types of gases they generate.

Losses of food and property to insect pests also can affect human economies and social welfare. Most field studies aimed at improving management of agricultural pests require estimates of densities of pest populations and often of populations of the natural enemies of pests. Monitoring the relationship between pest population size relative to crop growth, pest damage, and the status of natural enemies of pests can improve pest control and avoid needless applications of pesticides (Chap. 14). The identification of endangered species of insects requires data to demonstrate their rarity. Even more important, plans to improve the survival of these species must be preceded by studies and projections of their population biology.

POPULATIONS OVER TIME: DYNAMICS

The ecological interactions that determine the distribution and abundance of insects have been the focus of many population studies. **Natural control** describes all the factors that cause populations to change in size. Insect species tend to have a **characteristic abundance.** This term refers to the approximate upper and lower limits of densities at which insects typically occur, and values may vary by two or three orders of magnitude (100 to 1000 times). Outbreaks represent major increases of densities of species that are many times their characteristic abundance. Examples include plagues of migratory locusts or outbreaks of caterpillars that completely defoliate large areas of forests.

To gain insight into the processes in population change, biologists have constructed mathematical models. We will briefly describe three simple models to illustrate how insights can be gained from such models, even though field populations may not behave in a closely similar way. We then discuss how life tables provide another approach to studying population behavior, and we conclude with a discussion of how sampling methods are used to measure changes in populations.

Exponential Growth Model

The earliest theoretical studies of population changes emphasized population growth. Malthus' theory of the geometric nature of increase of human populations heavily influenced Charles Darwin's theory of natural selection. One of the earliest mathematical models of population growth was the **exponential growth** model (Fig. 8.1a). This model is based on the notion that the rate of change of population growth should be proportional to the population size. That is, the larger the population, the larger the rate of growth. The differential (from the calculus) dN/dt expresses the rate of change of population of size N per unit of time t. The growth constant r represents the proportional relationship between rate of growth (dN/dt) and population size (N). Thus $dN/dt = rN$. We use calculus to derive another equation to express population size at any time t, $N_t = N_0 e^{rt}$, where N_0 is the population size at time 0 and e is the base of natural logarithms with exponents r and t. N_t is exponentially related to the product of the growth constant and time, $r \times t$. If the growth constant takes into account both births and deaths, $r = b - d$, where b is the birth rate and d is the death rate. The rate of growth would be negative where rates of death exceed birth rates.

The exponential growth model is useful in estimating the theoretical potential of populations to grow without restraints such as disease, competition, predation, weather, or other physical limitations. The term r_m represents the growth constant under specified conditions. As a hypothetical example, suppose that 10 asexually reproducing aphids have an r_m of 1.0 per week per aphid. After 4 weeks, $N_t = 10e^{(1)(4)} = 10 \times 54.6 = 546$. If the rate of growth per individual is doubled (i.e., $r_m = 2$), the number of aphids after 4 weeks will be $10e^8 = 29,810$ or about 55 times greater than when r_m was 1.

Logistic Growth Model

By incorporating ideas about how competition for food, space, and other resources might influence growth, the basic exponential growth model just discussed can be modified so that the rate of population growth can respond to competition for limited space and resources. In this second model, called the **logistic growth** model, r is defined as a variable where

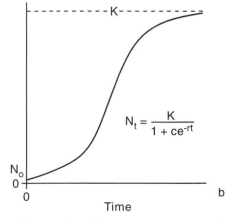

Figure 8.1 Exponential and logistic models of population growth of an initial population (N_0): **a**, exponential growth; **b**, logistic growth to carrying capacity (K, dashed line).

$$r = r_m \frac{K - N}{K}$$

and K is defined as the **carrying capacity** of a population. K represents the maximum stable population size for the amount of food, space, and other essential needs. Here, r_m is defined as the intrinsic rate of increase of a given population. The differential equation

$$\frac{dN}{dt} = r_m \frac{K - N}{K} N$$

can be solved for the population size (N) at any time (t) as

$$N_t = \frac{K}{1 + ce^{-r_m t}}$$

where the constant $c = \dfrac{K - N_0}{N_0}$

As shown in Figure 8.1b, N increases exponentially at first, when the population size is low, but begins to decrease as the carrying capacity is approached because the term e^{-rmt} becomes extremely small when t is large (much time has passed). In the logistic model, N can never exceed K. In reality, populations often exceed the carrying capacity, and the subsequent depletion of a common resource such as food or space leads to rapid decreases or "crashes" in population density. If new resources become available (e.g., if damaged plants grow back), the population can recover. Neither the exponential model nor the logistic model accounts for disease, predation, or parasitism. The emphasis is on innate growth ability and external resources, mainly food.

The logistic model of population growth has provided much of the basic framework and rhetoric for characterizing life history patterns or "strategies" of various insects. Insects that rapidly exploit new habitats by quickly reproducing in an exponential manner are considered to be *r*-selected, referring to the assumption that natural selection has resulted in populations that occupy new habitats and reproduce rapidly. Such insects are able to quickly colonize new resources. Aphids are

good examples of *r*-selected insects because in many phases of their life cycle they disperse on air currents, mature rapidly, and give birth to live young without having to mate (Chap. 4). It is not unusual for aphids to overwhelm a host plant so quickly that the surviving aphids starve when the plant collapses. Either declining plants or overcrowding by aphids triggers wing formation in subsequent generations of aphids, permitting the alates to disperse to colonize new plants in advance of the host plant's decline. Aphids have adapted to rapid exponential growth, followed by a collapse of the local population through competition or dispersal.

Migratory locusts, discussed in Chapter 4, are other insects whose behavior and dispersal are strongly influenced by population density. The phase polymorphism of the migratory desert locust (*Schistocerca gregaria*) depends on how densely the immature stages are crowded. Migrants appear only after the gregarious phase has been induced by crowding above a density that can persist without some locusts starving from competition (i.e., exceeding the carrying capacity). Rather than oscillating about the carrying capacity of the local environment, populations of both aphids and migratory locusts tend to fluctuate widely between extreme abundance and scarcity.

K-selected species are those that are adapted to stable habitats and environments in which population densities remain relatively stable from year to year. The populations of such species presumably remain near the carrying capacity (K). The carrying capacity is rarely constant but varies with time, as resources increase and decrease. Many tropical insect species that are adapted to specializing on one or a few species of plants or animals are examples of *K*-selected insects. Tsetse flies, for example, occur in relatively low densities and have very low rates of viviparous reproduction (Chap. 4). Populations remain relatively stable over time because the adults are long-lived and adapted to changing habitats as seasonal changes occur. Most insects have characteristics somewhere between extreme *r*-strategists and extreme *K*-strategists.

INTERSPECIFIC INTERACTIONS

Numerous experiments and observations indicate that natural enemies such as predators and parasitoids are major influences on insect populations. The role of predators, parasitoids, and pathogens in controlling insect pest populations is an important component of pest management programs and is covered in greater detail in Chapter 14. Experience with field studies and experiments demonstrate that natural enemies and diseases can regulate populations at an equilibrium level or characteristic abundance much lower than that dictated by the environmental carrying capacity. The population dynamics of interdependent predator-prey populations have been the subjects of numerous experimental studies and attempts to model the interactions mathematically. The first attempts to model predator-prey population dynamics, made separately by Lotka and by Voltera, led to the conclusion that predator and prey abundances should oscillate. Field studies commonly document out-of-phase oscillations of predator and prey populations (Fig. 8.2). Predator populations

typically lag prey populations both when prey numbers are increasing and when they are decreasing.

Populations of predators or parasites can respond to changing prey or host densities in several ways. Natural enemies can increase their production of offspring when food in the form of prey are more abundant. This **numerical response** is one explanation of why predator populations begin to increase only after growth in prey populations has occurred (Fig. 8.2). A second type of change in predator populations, called a **functional response,** occurs when predators increase their rate of killing a particular prey species (Fig. 8.3). The nature of the functional response may vary from one predator-prey combination to another; Figure 8.3 illustrates two types (called types II and III) of functional response. The increased efficiency of predation may result from learning or simply from the greater ease of encountering prey that are more closely aggregated. Functional responses in predators increase the number of prey consumed and may lead to positive numerical responses. Temperature may greatly affect functional re-

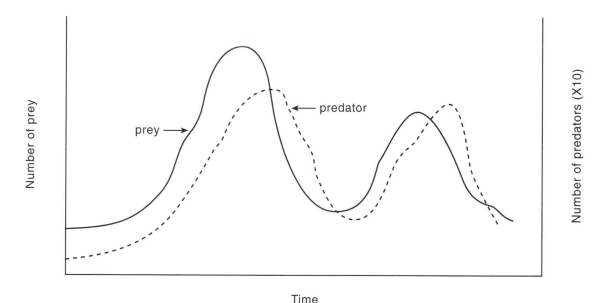

Time

Figure 8.2   Population densities of prey (solid line) typically begin to increase and to peak before predator populations (dashed line). The ratio of prey to predators is usually on the order of 10.1.

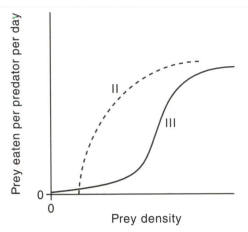

Figure 8.3 Two types (II, III) of functional response of predators to various prey densities.

sponses. For example, ladybird beetle larvae may eat fewer aphids during cooler weather.

Immigration, the movement of predators into an area, increases predator density. Some predators use the sex pheromones of their prey to locate prey, as discussed for predatory beetles (Cleridae) that feed on bark beetles (Chap. 6). Adult lacewings (Chrysopidae, Hemerobiidae) are attracted to aromatic amino acids such as tryptophan that are present in aphid honeydew. This behavior can be exploited to quickly attract lacewings to pestiferous populations of aphids by spraying aphid-infested crops with food sprays that contain the odors. The sprays provide the added bonus of increasing lacewing egg production because the adult lacewings that consume the food spray lay more eggs as a result of the improved nutrition.

The population models described above are only a few of the many types of models that can be applied to the analysis of insect populations. The simplest models take into account single populations or two populations. In reality, we usually find that multiple species of predators, parasitoids, and pathogens attack a single insect species. Most predators and many parasitoids typically feed on more than one species of prey. Factors that tend to change population density and intensify as population density varies are called **density-**dependent factors. Natural enemies that exhibit numerical or functional responses are one type of extrinsic, density-dependent mortality factor acting on prey populations. Intraspecific competition for food or shelter is an example of an intrinsic factor that influences population changes in a density dependent manner. For example, blow fly larvae (Calliphoridae) can quickly overpopulate a fresh animal carcass. If competition among the blow fly maggots is especially severe, only a few will survive to become adults, and those that do may be smaller and lay fewer eggs because of food scarcity due to competition with other larvae.

Aphids illustrate another type of intrinsic, density-dependent reaction to competition. As already mentioned, crowding of aphids on a plant induces the formation of wings in adult aphids. This increases the chances that the surviving aphids will escape overcrowded plants to occupy more favorable conditions. The trade-off is that winged (alate) aphids produce fewer offspring than do wingless (apterous) aphids. After an alate aphid has alighted on a plant to reproduce, its wing muscles begin to break down, converting internal resources from flight muscles to reproduction. If the conditions for growth are favorable, the offspring of alates will be apterae, which have higher rates of fecundity than do alates. Even among apterae, reproduction rates vary according to aphid density. In experiments that controlled the initial numbers of apterous bean aphids, reproduction was at a maximum when started with eight aphids per plant. Fewer starting aphids produced fewer offspring initially. After several cycles of reproduction, however, the plants that started with eight aphids were much more heavily crowded and had lower growth rates than on plants that started with fewer aphids (Way, 1968).

Competition for food and space is most severe among members of the same species because they have the most similar requirements. When different species have closely similar requirements, one species will tend to displace the others. This concept is called the **competitive exclusion principle.**

An especially well-documented example of

competitive exclusion was the interaction of three species of wasps in the genus *Aphytis*. These parasitoids were imported to control the California red scale (*Aonidiella aurantii*), which was a pest of citrus in California. Like the red scale itself, the parasitoid *Aphytis chrysomphali* was introduced inadvertently to southern California. It soon spread throughout southern California and effectively controlled populations of the scale in milder climate areas near the ocean. In 1946 entomologists in California intentionally introduced a second species, *A. lingnanensis,* from China. By 1958, *A. lingnanesis* had replaced *A. chrysomphali* throughout most of its former range in California. In the mid-1950s, *Aphytis melinus* was imported from southwest Asia in hopes of improving scale control in the warmer interior areas of California, where *A. lingnanensis* was not very effective in reducing red scale levels to noninjurious levels. Within 4 years after its first release, *A. melinus* had displaced *A. lingnanensis* and made up over 90 percent of the total population of *Aphytis* in the interior areas. *A. chrysomphali* remained in very low numbers near coastal areas, where *A. lingnanensis* still predominated (Fig. 8.4). These results and other field experiments with these species clearly demonstrate that species interactions can reduce the potential range of one or more of the species involved.

Territorial behavior limits the growth of some insect populations. For example, the cotton bollworm, *Helicoverpa zea* (Noctuidae), and numerous other herbivorous insects are cannibalistic under crowded conditions. Only the larger and more aggressive caterpillars survive direct encounters. Ovipositing apple maggots, *Rhagoletis pomonella* (Tephritidae), mark apples by dragging their ovipositors across the surface near the oviposition scar where the egg was inserted. In so doing, they deposit a pheromone that inhibits other females from laying eggs on the same fruit. Small fruits receive fewer eggs from the same or other females and thus avoid excessive larval competition. The oviposition behavior of parasitic Hymenoptera is even more complex. The number and sex of the eggs that a female will deposit into a single host

often depend on the size of the host and the number of eggs already present in the host, even if they were deposited by another female. In very small host insects, such as those in the egg stage, no more than a single parasitoid may be able to successfully develop from a single host, and multiple parasitism could lead to fatal competition for all parasitoids within the egg. Moreover, adult males of some species that parasitize insect eggs will emerge from the host eggs before females emerge and will guard the remaining eggs in an egg mass. The male not only mates with the females that emerge from the other host eggs but also wards off other males that arrive in search of mates.

LIFE TABLES

Another approach to studying changes in insect populations over time and the factors responsible for these changes is the construction of **life tables.** Stage-specific changes in birth and mortality are recorded for specified populations over a generation, or **cohort,** of a species in a particular habitat. A simple hypothetical example is given in Table 8.1. Beginning with the numbers of eggs at the start of a generation (l_x), life tables record how many individuals are lost in each stage (d_x) and the causes of mortality (d_xF). The numbers of insects that are lost through emigration or gained by immigration into the habitat may also be recorded. The mortality rate (q_x) is calculated as d_x/l_x. The final step in constructing a life table is to account for the number of offspring produced (m_x) by the surviving adults. In the hypothetical example illustrated in Table 8.1, Only 10 adults lived to produce an average of 100 eggs each. This would exactly replace the starting population of 1000 eggs. The ratio (R_0) of the number of replacement female offspring to the number of parental females at the start of the generation indicates whether the population has grown or shrunk from one generation to the next. In our simple example, $R_0 = 1$. In real populations, R_0 fluctuates above and below $R_0 = 1$.

Figure 8.4 Competitive displacement as of 1961 among three species of *Aphytis* parasitoids imported for control of California red scale. *Aphytis chrysomphali* occupied the entire citrus area in southern California before 1948. After that date, *A. lignanensis* replaced it. Then *A. melinus,* introduced in 1956, displaced *A. lignanensis* with little overlap except for near the ocean. By 1961, *A. chrysomphali* was extremely rare. (Data combined from DeBach and Sundby, 1963.)

Note that not all causes of mortality or loss in a life table may be known. The factor that causes the greatest amount of mortality may not necessarily have the greatest impact on changes in population density from one generation to another. For example, 80 to 90 percent of the first-instar larvae in a population may succumb to adverse heat and humidity, but this mortality may be relatively constant from year to year. Other factors such as predation or disease may explain most of the variation in population growth among locations or years. The identification of the **key factors** that are most important in explaining or predicting population changes from one generation to the next requires analytical methods (e.g., regression analysis) beyond the scope of our discussion here. Identifying key factors helps to understand why populations of some insects periodically reach very high levels in some seasons and remain low in most others. Obviously general conclusions from life table data require studies from several seasons and locations.

The basic life table approach emphasizes mortality factors. Other effects of habitat quality on female behavior and reproductive success may be incorporated. Table 8.2 illustrates life tables for two cohorts of the shoot-galling sawfly, *Euura lasiolepis* (Tenthredinidae), on willow shoots (Prezler and Price, 1988). Instead of beginning to follow a cohort of eggs laid on the plant, the study began with the numbers of eggs in female ovaries. A major difference between the two cohorts was the amount of water available to the plants in which the female sawflies deposited eggs. Willows with high water produced longer and more succulent stems, while the low-water plants were shorter and tougher. Observations and experiments showed that greater numbers of eggs were laid and larval survival was higher on the longer stems of the high-water plants. The galls develop in meristematic tissue, and fast-growing stems aid successful gall formation. The eggs in shoots grown in high-water conditions have lower mortality (d_x). Thus plant quality plays a major role in

determining the potential population size of this sawfly.

POPULATIONS AND THE PHYSICAL ENVIRONMENT

The physical environment provides a framework for population behavior. The incredible diversity of insects' morphological, physiological, and behavioral attributes attests to the many ways they use to adapt to their environment. Chapters 5 and 6 describe ways by which various insects cope with day length, changing seasons, and extremes of moisture and temperature. The physical environment, in concert with the biotic components of ecosystems, determines the distributions and abundance of insects. Physical factors may restrict the occurrence of insects to limits narrower than those imposed by the distributions of their food sources. For example, insects that specifically feed on only one species of plant may not occur throughout the geographic range of the plant. The localities where species can survive from year to year are often determined by physical factors such as extreme temperatures that limit survival or by adequate temperatures required for development. The total of all physical and biotic requirements for a species' continued existence is called its ecological or species **niche**. The **habitat,** or place in which an insect lives, is only part of its

TABLE 8.1
Life Table for a Hypothetical Insect Population

Age class or stage, x	Survivorship, l_x	Mortality, d_x	Mortality factor, d_xF	Mortality rate, q_x	Average fecundity, m_x
Eggs	1000	400	Weather	0.40	0
		100	Parasitism	0.10	
Larvae	500	300	Weather	0.60	0
		100	Parasitism	0.20	
Pupae	100	25	Predation	0.25	0
		25	Unknown	0.25	
Prereproductive adults	50	25	Unknown	0.5	0
Reproductives	25	15	Unknown	0.6	100

TABLE 8.2
Life Tables for Populations of the Sawfly *Euura lasiolepis* on Experimental Plants Given a High-Water Treatment or a Low-Water Treatment

Stage, (x)	Percentage alive at start of stage, lx	Mortality factor, dx	Number dying during x, dxF	Mortality rate, qx
High-water treatment				
Eggs in females	100	Failure to initiate gall	0.0	0.0
	100	Egg retention	<u>27.5</u>	<u>27.5</u>
		Total maternal response	27.5	27.5
Eggs in galls	72.5		0.0	0.0
First instar	72.5			
		Parasitism	0.0	0.0
		Host effects	<u>9.9</u>	<u>13.6</u>
		Total	9.9	13.6
Second instar	62.6	Host effects	0.0	0.0
Last instar	62.6			
Low water treatment				
Eggs in females	100	Failure to initiate gall	62.1	62.1
	37.9	Egg retention	<u>23.0</u>	<u>60.8</u>
		Total maternal response	85.1	85.1
Eggs in galls	14.9		0.0	0.0
First instar	14.9			
		Parasitism	0.0	0.0
		Host effects	<u>8.1</u>	<u>54.5</u>
		Total	8.1	54.5
Second instar	6.8	Host effects	0.3	3.9
Last instar	6.5			

[1]*Source:* Data from Prezler and Price (1988).

multidimensional niche, which can include the range of permissible temperatures and humidity, the total amount of heat (temperature over time) available during the season of development, and other factors. The **realized niche** is the portion of the niche that a species actually occupies. Competition with or exploitation by other species reduces the amount of the potential niche used by a species. As an example, recall the competitive displacement among *Aphytis* parasitoids of the California red scale and reconsider Figure 8.4. The realized niche of *Aphytis chrysomphali* was larger before the introduction of two other species of *Aphytis* reduced the geographic range occupied by *A. chrysomphali*.

Weather plays a major role in shaping changes in insect populations. Migrating locusts (*Schistocerca gregaria*) are carried by air masses (winds) in the intertropical convergence zone. The meeting of opposing air masses not only determines where migrating locusts will land but also regularly produces rains to promote the plant growth needed by successive generations of locusts that emerge from the eggs laid by migrant adults. Soil moisture, also strongly influences the success of locust egg hatch. Too much moisture suffocates the developing eggs. Too little moisture reduces the growth of plants on which the emerging hoppers feed. Plant growth responds directly to soil fertility, soil moisture, and weather. Insects such as aphids, sawflies (Tenthredinidae), and gall gnats (Cecidomyiidae) are especially dependent on succulent new plant growth.

Precipitation directly affects aquatic insects. Heavy flows of streams or flooding can scour away many aquatic habitats. Droughts can dry up some aquatic habitats and eliminate the insects that depend on them. Physical impacts on populations are often considered to be **density-independent factors,** whose intensity does not vary with population density. Some external factors such as weather may incorrectly appear to be density-independent. For example, weather events that catastrophically decrease population density may act more severely on individuals stressed by competition in crowded populations, and sometimes only the fraction of a population that can occupy limited shelters survives adverse weather conditions.

POPULATIONS AND TEMPERATURE

Although population growth is usually described with respect to time, another dimension, temperature, affects population growth. The growth rates of insects depend strongly on temperature. To illustrate, we examine data for the development of eggs of the fruit fly, *Drosophila melanogaster* (Drosophilidae). Plotting the time (*H*) required to complete a developmental stage at various constant temperatures shows that development generally speeds up as temperature increases (Fig. 8.5a). The data lie close to a curve that is similar in shape to a hyperbola. This means that over most of the range of temperatures normally experienced during growth, a plot of the **development rate** (simply the reciprocal of development time, or 1/*H*) approximates a straight line. In our example, the development rate deviates significantly from a straight line (Fig. 8.5b) only at temperatures below 15°C or above 27.5°C. We can fit most of the development rate data reasonably well to the simple equation of a straight line,

$$\frac{1}{H} = aT + b$$

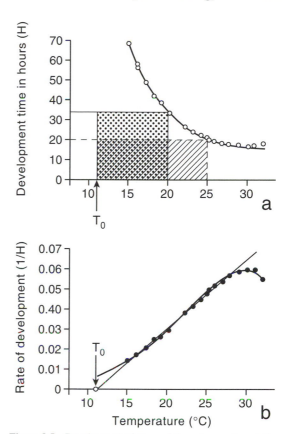

Figure 8.5   Development of *Drosophila melanogaster* relative to temperature: **a,** development time (*H*) at various constant temperatures; **b,** development rates (1/*H*) calculated from the data in **a;** developmental zero temperature (*T*$_0$) is intersection of straight line that best fits data on development rates with horizontal axis (zero development rate). Hatched area represents the physiological time requirement in degree-hours for 25°C (slanted lines) and 20°C (dots).

where *a* (the slope of the line) and *b* (the vertical axis intercept) are coefficients that are characteristic of the growth rate of the particular stage of development.

We estimate the temperature at which development rate is zero by projecting the straight-line portion of the development-temperature equation until it intersects the temperature axis. In our example this is about 11°C.

Another useful mathematical property of the similarity of the hyperbolic shape of the development time curve is convenient for making a simple model. The rectangular area formed by the difference between the threshold development temperature T_0 and any constant temperature T_t on one side and by the development time for the actual temperature experienced is an approximately constant amount. Expressed as an equation,

$$\text{temperature-time combination required} = (T_t - T_0)\,(H)$$

Therefore, we can calculate the time the egg will mature if we know when the egg was laid and the temperature-time requirement, and have a record of temperature.

To illustrate with our example of the fruit fly egg data: at 20°C the temperature difference is 9°C and the development time is 33.3 hours, giving a time-temperature requirement of 9°C × 33.3 hours = 300 degree-hours. At 25°C the requirement is 296 degree-hours. If the variability or errors in this approach of a few hours is acceptable, we can estimate within a few hours the average time required for the egg stage. Each subsequent stage in the life cycle may have its own development rate. Physiological time, measured as the product of time and temperature, typically is expressed as **degree-days.** In our example at 25°C, 296 degree-hours equals 12.33 degree-days.

In reality, the relationship between development rate and temperature is only approximately linear over a certain range of temperature (Fig. 8.5b). The actual development threshold temperature is lower than that estimated by the linear temperature/development rate equation. At high temperatures the development rate decreases and eventually drops to zero (no growth or death). Development at an average fluctuating temperature may be faster or slower than at the same constant temperatures from which the relationships are derived experimentally. The temperatures experienced by insects living a few centimeters beneath the surface of damp soil may be very different from those of insects on the surface during a hot, sunny day. Both temperatures could be very different from that recorded by a shaded thermometer. Finally, insects can heat to temperatures above ambient, as discussed in Chapter 5. Bees produce metabolic heat that warms their developing larvae during cool spring weather. Other insects absorb solar radiation to heat up above air temperature.

Despite its shortcomings and inaccuracies, the just described development model can provide useful estimates of how temperature affects development time. More complex models can account for nonlinear development responses to temperature or can be modified to reflect the effects of fluctuating temperatures. Estimates using a development model can be used to track the development of pest populations relative to temperature. The timing of applying pest control measures based on physiological time (degree-day models) is more efficient than simple calendar time. The same type of model can be applied to estimate the time of death of a corpse invaded by insects. If the relationships of development to temperature are known for the various stages of carrion-feeding flies or other insects in the corpse, the time of death can be calculated from historical temperature data. The initial invasion of the corpse by egg-laying insects usually is assumed to occur soon after death, but this event depends on the characteristics of the insect species involved and the season.

SAMPLING

Characteristics of populations, such as density, must be estimated from data collected in variable and complex environments. In **sampling** we collect data on a small number of individuals within a population and make inferences about the entire population from the sample data. Sampling differs from a census, which is gathering information on every member of the population. A census is practical only in limited laboratory studies. Reliable and accurate sampling data are the first requirement for field studies of insect abundance and distribution. Insects are particularly good subjects for population biology studies because their populations are often large and cycle quickly, but the same characteristics may present challenges in designing a sampling program to collect the necessary data.

Statistical considerations are an integral part of any sampling effort. Biological data typically are highly variable. Indeed, sampling seeks to document and interpret the significance of variability in population characteristics. Statistics help to estimate the optimal size of each sample, the numbers of samples to be taken, and the spatial patterns to be used in their collection. Decisions regarding the reliability of conclusions drawn from sampling data rely on statistical tests. An appropriate text or reference on biological applications of statistics should be consulted to appreciate or learn the usefulness and importance of methods of statistical inference.

A first requirement in selecting or applying any sampling effort is to know something of the biology of the species to be sampled. What are the life cycle stages? Where do they occur? What are the main food sources? Where are the eggs or offspring deposited? As sampling data are collected, more may be learned of key biological traits and sampling methods modified accordingly. The amount of biological information available for a species may determine how sampling data should be interpreted. Measuring environmental variables may be important to applying sampling data. For example, applying temperature-based development models to data from field populations requires relevant and accurate temperature data. Sampling data for life tables ideally estimate not only the abundance of living insects but also what killed the other insects in the same cohort.

Sampling methods can be categorized as **absolute sampling,** where every individual within a volume of habitat is counted, or as **relative sampling,** where the sample provides a relative index of how many individuals are present within the sampled habitat. An absolute sampling method may be as simple as counting all the insects on a plant organ such as a fruit or a branch. This method is accurate if the insects to be counted are immobile or unlikely to escape or be overlooked while the sample is taken. More elaborate procedures include quickly enclosing a unit of habitat within a device that prevents insects from escaping and counting the insects in the cage. Suction traps

may provide absolute samples of the aerial density of weak-flying insects such as aphids.

An indirect but absolute method is the **mark-release-recapture** technique. Individuals within the population are captured and marked by means of an easily recognized powder, paint, or other marking method that does not alter the insects' longevity or behavior. The marked insects are released, and then the population is sampled again after the released individuals have dispersed into the habitat. If the collection, redistribution, and recapture of marked and released individuals (n_r) is random with respect to the rest of the population and other assumptions of the method are satisfied, then the ratio of marked individuals that are recaptured (n_m) to the total number of marked and unmarked individuals ($n_m + n_u$) that are captured should be the same as the ratio of the number of marked individuals that were released (n_r) to the actual population size (N). The simplest estimate of the population size is

$$N = n_r \frac{n_u + n_m}{n_m}$$

Absolute sampling provides data to estimate actual population size. Relative sampling methods are faster and easier to use, but they provide only an index or relative measure of population densities or activity. For example, an investigator may be satisfied in knowing that population A is about twice as dense as population B, without having to know the actual size of either population. Relative measures of pest density on a crop may be related to the amount of damage and thus used as guidelines for implementing pest control measures.

The most common type of relative sampling is the use of a net or suction device to capture insects from vegetation. A sweep net jars insects from the vegetation, and they fall into the net as it is swept through the habitat. Such methods are not absolute measures because plant characteristics and density, temperature, moisture, collection technique, and other factors strongly influence the percentage of insects in the sampled habitat that are collected. Because heavier insects or those with a weaker grasp are more easily dislodged by sweeping, not all insects are equally likely to be captured by this method. Insects that bore into plant tissues may be missed entirely.

Traps that attract insects with pheromones, baits, or visual images provide relative measures of the combination of insect abundance and activity. Insects may be attracted to particular colors or shapes. For example, yellow attracts many leafhopper (Cicadellidae) species, and yellow panels coated with a sticky substance can monitor the flight activity of these insects. Mosquito biting rates can be estimated by quickly removing and counting mosquitoes that settle on people in a given time span. Tsetse flies in Africa were formerly surveyed by counting the flies that landed on cattle as they were led along on a tether. Traps to monitor tsetse flies are baited with the key compounds present in animal breath, which the flies use as cues to find hosts for blood meals. Some sampling methods measure the presence of individual insects through indirect evidence such as oviposition or feeding wounds, frass pellets, excrement, or exuviae.

Suction traps that capture flying insects provide an absolute sample of aerial planktonic insects. The volume of air collected per time period provides an estimate of aerial density over time. These methods have been used to calculate the aerial densities of aphids and other small arthropods over time, for example. Temperature and time of day are critical to interpreting these results. Most mosquitoes typically fly at dusk or during a particular time of night. Bees, wasps, and winged termites fly during daylight hours. The source of "haze" detected by weather radar in the midwestern United States during clear late summer afternoons was determined by aerial sampling with specially equipped helicopters to be clouds of insects (mainly aphids) arising from crops.

SELECTED REFERENCES

Population biology is treated by Begon et al. (1996), Cappuccino and Price (1995), Horn (1988), Price (1984) and Atkins (1978); competitive displacement by DeBach and Sunby (1963) and DeBach and Rosen (1991); sampling by Dent (1991), Horn (1988), Southwood (1979), and Pedigo (1989); pheromones by Prokopy (1991) and Metcalf and Metcalf (1993); life tables by Carey (1993), Horn (1988), Price (1984), and Prezler and Price (1988).

9

DIVERSITY AND ADAPTATIONS OF INSECTS IN SELECTED HABITATS

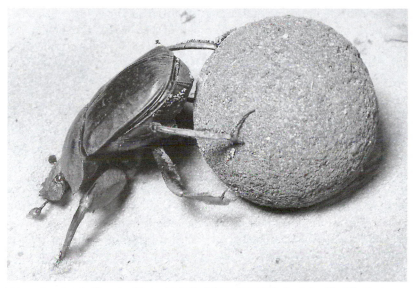

African dung beetle (Scarabaeidae) with ball of dung that will be buried as food for the larvae. Africa has many large herbivores and an associated fauna of dung beetles. The native fauna of dung beetles in Australia are adapted to the pellets of kangaroos and small animals. With the development of the cattle industry in Australia, methods were sought to recycle the dung that covered pastures and supported dung-breeding flies. A solution that met with limited success was to import dung beetles from Africa.

Ecology is the study of the interactions between organisms and the physical, chemical, and biotic factors of their environment. Chapters 10 through 13 deal with the biotic environment, namely, interactions of insects with plants, other insects, vertebrates, and microbes. In this chapter we focus on certain **habitats,** or places where insects live: soil, caves, deserts, high altitudes, and fresh and salt water. We will discuss some of the physical and chemical features of these environments, the kinds of insect present, and their special adaptations for survival. Insects in these habitats are frequently scavengers, and at the end of the chapter we will list the variety of foods eaten.

Soil and freshwater are home to insects of many kinds and also are vital natural resources for humans. Insect scavengers play an important role in the health of these resources by acting as decomposers and recycling nutrients. Unfortunately, soil and freshwater are easily contaminated by pesticides or pollutants, with the result that great numbers of these beneficial insects are destroyed. Because they are so sensitive, certain insect species, by their presence, abundance, or absence, can serve as biological monitors of environmental quality.

THE SOIL HABITAT

Soil is the medium that connects all habitats of the land, from the intertidal to the alpine. It is also a distinct realm and one on which green plants and ultimately all forms of terrestrial life depend. Thousands of kinds of soil have been identified, each the product of different circumstances during formation. In general, the process of soil formation leads to horizontal layers that are readily visible in a cross section such as a road cut. On the soil surface is usually a duff, or **litter** of whole or partly decayed leaves and other plant parts. The nature of the litter varies with moisture and the covering vegetation (i.e., grassland, coniferous forest, deciduous forest, or lichens and mosses). Beneath the loose litter is the soil's uppermost layer, or **topsoil,** which is darkened by the accumulation of a complex, organic substance called humus. Most plant roots and soil organisms occur in the topsoil. Below the humus-rich zones, the **subsoil** is paler in color, being only slightly enriched from above and having a higher content of clay. The topsoil and subsoil make up the true soil. Beneath the topsoil and subsoil may be a layer of varying thickness called the soil parent material, which grades below into the unaltered rock of the earth's crust.

To sustain a community of living plants and animals, the nutrients locked in dead plants and animals and animal waste must be released by decomposition. This recycling of nutrients is accomplished primarily by bacteria and fungi. Arthropods and other animals indirectly influence the process by feeding on the microbes and thereby regulating their populations. Arthropods also contribute directly by feeding on plant and animal waste at the soil surface. These materials are mechanically and chemically broken down, often with the aid of microorganisms in the gut, and feces are deposited in their burrows deeper in the soil. Their excavations also mix the surface layer between litter and soil. Microbes in the moister regions below then act again on the partly decomposed fragments, now with greater surface area, to produce humus. Arthropods are especially important in recycling of nutrients in grasslands and deserts.

Cracks, old root channels, and tiny cavities (up to 1 mm in size) between soil particles create pore spaces in soil. The total pore space of topsoil is usually about a third of the soil volume, but in some places may exceed half the volume. At greater depths, the soil is more compact and has smaller pore spaces. The pore spaces are filled partly by air and partly by water held by capillary forces, in proportions depending on the nature of the soil particles, rainfall, groundwater, and drainage. During periods of flooding, insects in the soil obtain oxygen from air trapped in their burrows or pore spaces. Soil air is usually saturated with water vapor except near the soil surface. In well-drained topsoil, oxygen in soil air is replenished by gaseous diffusion from the atmosphere. Soil water derives its oxygen from soil air. Oxygen, however, diffuses much more slowly in water and can be depleted by bacterial decomposition of organic matter. When this happens, anaerobic conditions are created in as short a distance as 1 mm from a water/air interface. Deeper in the soil, gaseous diffusion from the atmosphere decreases. Here the respiration of organisms reduces oxygen and increases the carbon dioxide content of soil air.

If the vegetation overhead is not too dense, sunlight penetrates the litter, open burrows, and superficial soil crevices. Below the surface, at depths of only a few centimeters at most, is permanent darkness. In deserts, grasslands, alpine regions, and deciduous forests after the leaves have fallen, the surface is largely exposed to the sun's radiation. Thus heated by the sun, the surface temperature of soil rapidly climbs well above that of either the air or the soil at slight depths beneath. Surface temperatures above 50°C are not infrequent even in the Temperate Zone. Owing to the low thermal conductivity of soil, the heating and cooling of deeper layers lags behind the fluctuations in temperature of the air, litter, and the soil surface. Heat acquired during day penetrates into the soil quite slowly. A high moisture content in soil will reduce the amplitude of these daily fluctuations in temperature. "Cover objects," such as large rocks or logs resting on the surface, shelter the soil beneath from extreme temperatures and maintain a high humidity by preventing evaporation. The spaces beneath cover objects provide insects a place to escape from predators and environmental extremes, especially in grasslands. When looking for insects under rocks and logs, it is a good practice to return such cover objects to their former position so that other insects can find shelter.

Insects and Other Small Organisms of the Litter and Soil

The litter of plant debris, the soil surface, and the soil itself are inhabited by at least some members of all orders of hexapodous arthropods except the Ephemeroptera. Also found here are many small organisms: annelids, bacteria, fungi, mollusks, nematodes, protozoa, and other terrestrial arthropods (arachnids, centipedes, isopods, millipedes, pauropods, symphylans). Indeed, in a Brazilian rain forest, about three-fourths of the total animal biomass was found to be small, soil-dwelling animals. In this and most other terrestrial habitats, social ants and termites dominate the animal community in terms of number of individuals and biomass (Wilson, 1990; Color photos 4c, 4e, 4f, 6e).

After termites and ants, the next in abundance in soil are Acarina, Collembola, all stages of Coleoptera, immature Diptera, and immature Lepidoptera, especially pupae. The following insects and small arthropods, although generally less numerous, spend their entire lives in litter, in soil, or in crevices under cover objects: Archeognatha; some species of Blattodea, Dermaptera, Embiidina, Grylloblattodea, Orthoptera, Thysanura, and Zoraptera; and all species of the arthropod orders Diplura, Pauropoda, Protura, and Symphyla. Some other orders are represented in soil at least by the immature stages that feed as scavengers or predators, or on algae, fungi, or mosses: Coleoptera, Mantodea, Mecoptera, Neuroptera, Psocoptera, Raphidioptera, Siphonaptera and Thysanoptera. Parasitoids and parasites, including Strepsiptera, attack other litter insects. Phthiraptera are found as parasites on ground dwelling birds and mammals.

Some species of insects that are ordinarily considered to be aquatic can be found in terrestrial habitats such as pockets of decayed leaves, moist litter, or soil away from the water. The exceptions are immatures of some species of Odonata, Plecoptera, and Trichoptera. The larvae of the damselfly *Megalagrion peles* (Odonata) of Hawaii regularly live in damp leaf mold caught in leaf axils. Several species of stoneflies in the Southern Hemisphere live far from water on cold, wet mountains. The larvae of the caddisfly *Enoicyla pusilla* (Trichoptera) inhabit forest litter in Europe.

Immobile stages of insects are commonly spent on the ground. Eggs are often deposited here: Phasmatodea drop eggs in litter, and Acrididae lay eggs in soil. Lepidoptera pupae and the resting "pupae" of Thysanoptera are often found in litter. Periods of quiescence or diapause also may be spent among dead leaves. Aggregations of diapausing adults of the lady beetle, *Hippodamia convergens* (Coccinellidae), are found on the ground in Sierra Nevada of California and higher parts of the Coast Ranges from June to February (Color photo 3d).

Adaptations of Insects in Litter and Soil

Insects that move about on the surface of the litter or soil are exposed to predation by visual hunters such as birds. Such insects may have the following characteristics: well-developed compound eyes;

circadian rhythms in behavior, often with nocturnal activity; and the body pigmented, with concealing patterns. Insects at the surface also are generally resistant to desiccation. Respiration is by tracheae, and the spiracles are usually fitted with closing devices that conserve water.

The active stages of insects and hexapodous arthropods in litter and crevices in soil often are somewhat dorsoventrally compressed and parallel-sided or anteriorly narrow, a form that facilitates movement among varied obstacles. Most are agile and possess well-developed means for locomotion on ground. Escape by jumping is often a character-istic (Archeognatha, Collembola, Orthoptera, and Schizopteridae in Hemiptera). The antennae and tactile sense organs are usually well developed.

Large, membranous wings are apparently an encumbrance in narrow passageways. Except for mating and dispersal, flight is an unnecessary in-vestment of energy and structure. The Embiidina spin silken runways to move out from cover ob-jects into the litter. The wings of male webspinners, if present, are flexible and bend freely inside the tunnels. The Apterygota, Collembola, Diplura, and Protura are, of course, wingless, and the immature pterygotes have small wing pads at most. Among adult pterygotes some have protective forewings (Coleoptera, Dermaptera, and Orthoptera); others fail to develop wings (some species in the orders Dermaptera, Coleoptera, Orthoptera, and Psocopt-era; all species of Grylloblattodea; and the sterile castes of Isoptera and Formicidae); or discard them after mating and dispersal (Zoraptera, reproductive castes of Isoptera and Formicidae).

Females of most soil-dwelling arthropods other than insects acquire sperm indirectly from the male without copulation. The male produces a spermatophore that is placed on the ground or a special web and later picked up by the female. Thanks to the generally moist environment, the sperm are not in danger of desiccation for at least a short period. Among the insects and related arthropods this behavior is known in the primitive orders Archeognatha, Collembola, Diplura, and Thysanura. Ancient arthropods were aquatic, and sperm was transferred in water without copulation.

During the early evolution of terrestrial arthropods from aquatic forms, litter and soil evidently pro-vided a moist place where sperm could continue to be transferred indirectly. Other insect orders (Pterygota) transfer sperm directly from the male genitalia to the female reproductive tract through copulation, thus protecting the sperm and allowing a fully terrestrial existence. Odonata are an excep-tion; males place the spermatophore in a secondary organ on their bodies (Fig. 21.2).

Subterranean Insects and other Organisms

The high humidity of soil air, films of soil water, and moderate temperature of soil at greater depths create conditions approaching an aquatic environ-ment. Many invertebrate phyla are represented in soil by minute aquatic forms that are found else-where in freshwater. Protozoa, for example, are the most numerous microscopic animals in moist soil. With increasing depth below the soil surface the variety of arthropod taxa abruptly declines. In the soil air of pore spaces live species of the minute arthropods Acarina, Collembola, Diplura, Pauropoda, Protura, and Symphyla. The under-ground nests of ants and termites are found here, each populated with large numbers of individuals. Other insects deep in soil are some species of Coleoptera, Diptera, Hemiptera, Hymenoptera (certain bees and wasps), and Orthoptera. Of these, the Coleoptera are the most numerous in species. All stages of the life cycle of Coleoptera can be found in soil (Fig. 9.1c,d), but generally only the immature stages of Diptera are present. Many species of bees and predatory wasps excavate sim-ple burrows in the earth, stocking them with pro-visions for their young. Roots are the food of nymphal Cicadidae (Fig. 9.1a) and certain other subterranean Hemiptera. Orthoptera are repre-sented by the omnivorous Stenopelmatinae (Gryl-lacrididae) and by mole crickets (Gryllotalpinae, Gryllidae; Fig. 9.1b) which feed on roots but also take insect larvae.

Soil-dwelling termites and ants make deep nest galleries, often into the subsoil and below.

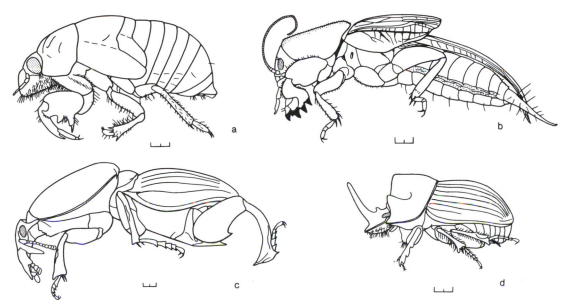

Figure 9.1 Examples of soil insects: **a**, cicada nymph, *Okanagana* sp. (Cicadidae); **b**, mole cricket, *Gryllotalpa* sp. (Gryllotalpinae, Gryllidae); **c**, unusual soil-dwelling cerambycid beetle from Brazil, *Hypocephalus armatus;* **d**, dung beetle, *Copris lugubris* (Scarabaeidae). Scales equal 2 mm.

Their extensive excavations and the organic matter brought into the nest contribute substantially to soil development and drainage. In the tropics and some arid regions, they are of primary importance as soil builders.

Large earthen mounds built by termites in Africa, South America, and Australia are conspicuous and long-lasting features of the landscape (Fig. 9.2). In southern Africa, mounds still occupied by termites have been carbon-dated at 4000 years old. In warm grasslands termites are among the primary decomposers of plant litter and animal dung. In some areas, termites are so numerous that they influence the regional types of soil and vegetation. The mounds are rich in organic matter from the termite's feces and in minerals from soil excavated below. Abandoned nests resist decay, and thus nutrients are locked up for many years. Moreover, in living colonies carbon is lost to the atmosphere as carbon dioxide and as methane gas produced by intestinal symbionts.

Subterranean nests of *Atta* ants in the Ameri-can tropics may be enormous and reach 3 to 6 m in depth. Earth brought to the surface by ants in one nest weighed 44 tons. In prairies of North America, excavations by ants are less impressive but significant in nutrient dynamics, soil drainage, and vegetation.

Adaptations of Subterranean Insects

An important factor limiting insects deeper in soil is the obvious difficulty in moving about. Unlike small arthropods that live in pore spaces, most soil insects are not small enough to freely negotiate interconnecting pore spaces for any distance. They must either tunnel by pushing aside particles and squeezing through, or excavate the soil in front and deposit it behind. The physical resistance of the soil, therefore, raises the energetic cost of locomotion and reduces the distance to which an individual can hunt for mates or food. Sight and windborne odors are also eliminated as aids in the

Figure 9.2 The large size of this mound made by fungus-growing termites, *Macrotermes* sp., exemplifies the diligence, long life, and importance of termite colonies in the African savanna. The mounds persist for many years with successive dynasties of queens.

search. For this reason, mating of pterygote insects is usually accomplished above ground. Except for the roots of green plants, insects must find their food in the form of other soil organisms, dead or alive. Active excavators such as ants, termites, silphid beetles, larval cicindelids, wasps, and bees obtain their food from energy-rich sources above ground.

Reduction in wings is a characteristic both of insects that live deeper in soil and of those that live in litter. Burrowing insects are generally round in cross section but are greatly varied in shape otherwise. "Tunnelers" have reduced or no legs and are of two kinds. Larval Elateridae (Coleoptera) and Therevidae (Diptera) are smooth, stiff-bodied, and slender. By sinuous movements of the trunk, they force their hard, tapered heads forward through the substrate. In contrast, larval Tipulidae

and Bibionidae (Diptera) are soft-bodied and use peristaltic movements to penetrate soil. "Excavators" may be fairly thick-bodied and armored, often exhibiting conspicuous molelike modifications of the head and forelegs for digging. Such fossorial adaptations are seen in Gryllotalpinae (Orthoptera), Cydnidae and nymphal Cicadidae (Hemiptera), and Scarabaeidae (*Copris, Geotrupes;* Coleoptera), to list only a few (Fig. 9.1). Worker ants and termites are not so conspicuously equipped for digging, but by their immense numbers and coordinated activity, they are able to remove and carry away immense quantities of particles, using their strong mandibles.

In the absence of light, insects and the hexapodous arthropods that live deeper in the soil tend to have reduced or no compound eyes, antennae variably developed (even absent: Protura); well-developed tactile sense organs, and pale pigmentation of the integument without colored patterns. They are usually repelled by light, higher temperatures, and humidities less than 90 percent. The upper limits of tolerance to temperature are relatively low, but some soil insects can remain active a few degrees above 0°C. Rhythmic behavior has been poorly studied. Vertical migration in response to daily temperature fluctuations may occur if the soil is exposed to solar radiation. Resistance to desiccation is low, probably because the epicuticle becomes abraded by soil particles or is naturally permeable to water. Some soil insects and arthropods respire cutaneously. Those that normally inhabit the deeper subsoil are tolerant of higher levels of carbon dioxide, at least for several days (some Collembola and larvae of the beetles Elateridae and Scarabaeidae).

THE CAVE HABITAT

In addition to tiny pore spaces in soil, larger cavities that exist much deeper below the soil layers are inhabited by insects and other organisms. Such cavities range from networks of air-filled spaces in rock, measuring a few millimeters in width, to giant caverns. Both terrestrial and aquatic habitats exist here because cavities below the water table are filled with still or running water. The full extent of the **cavernicolous,** or cave-dwelling, biota remains unknown because only a minority of cavities have openings to the surface, and still fewer are large enough for a person enter. Where access is limited, samples of the underground aquatic fauna can be taken from wells and springs.

Underground cavities are commonly found in three geologic situations. (1) Classic caverns exist where limestone is dissolved by carbonic acid, which is produced when water is charged with carbon dioxide. An extensive area with underground streams, sinkholes, and caves is called a karst topography. These systems may be millions of years old. (2) Volcanic deposits have cavities formed by gas bubbles or by gaps in the lava flow, or by lava flowing in covered channels that may leave large empty tubes many kilometers in length. These are generally a few thousand years old. (3) Cavities or fissures also are found in broken rock strata or talus and are of variable ages.

Insects of the Cave Entrance

The entrance to a cave is a transition zone between the outer world and the cave beyond. The humidity is often higher, the temperature less varied, and partial darkness prevails. Some animals here are only temporary residents; others have adapted in various degrees to life in caves. Cave-dwelling insects, which are often the same species normally found in moist litter, dark crevices, or the soil outside, include representatives from many orders and families. Flying insects, especially flies (Culicidae, Mycetophilidae), seek shelter at entrances during the day. The famous New Zealand glow-worm is the cavernicolous larva of the fungus gnat, *Arachnocampa luminosa* (Mycetophilidae). These luminous predators catch small flies by secreting sticky threads that hang from the ceiling of the cave. Bodies of water within the entrance to a cave may have the usual aquatic insects. However, some species of Trichoptera and Ephe-

meroptera are rare outside caves and may be less pigmented than their normal relatives.

Insects of Deep Caves

Beyond the entrance, food must come ultimately from the surface. The absence of green plants except for their roots largely limits cave-dwelling insects to scavenging, parasitism, or predation. Phytophagous insects do exist where roots are available. Otherwise, new organic matter must be washed in, or deposited as waste by organisms that enter and leave. Bats attract several taxa of parasites (Polyctenidae, Streblidae, and Nycteribiidae; some Siphonaptera and Cimicidae) and the accumulated feces of bats (called guano) attract scavengers (Arixeniina, Dermaptera; Blattodea; Coleoptera; Diptera, etc.).

The environment deep in a cave is stable and usually has the following conditions: darkness is total; temperature is constant and near the average annual surface temperature; the air is saturated with water vapor; the walls are covered with water; and oxygen is lower and carbon dioxide higher than in the atmosphere above ground. Even though they may be terrestrial animals, inhabitants of deep caves live in an essentially aquatic environment with the saturated atmosphere and wet walls. Animals that are adapted to continuous life in caves and reproduce there are called **troglobites.**

In continental caves, terrestrial hexapodous troglobites are frequently Collembola (Entomobryidae), Diplura (Campodeidae), Orthoptera (cave crickets, Rhaphidophoridae, Fig. 9.3a), and especially Coleoptera (Pselaphidae; trechine Carabidae; Leiodidae). Icy lava tubes of Idaho have leiodid beetles appropriately named *Glacicavicola bathyscioides* (Leiodidae, Fig. 9.3b).

Terrestrial troglobite arthropods exhibit special modifications. On the average, individuals of cave species tend to be larger than their close relatives outside. Cave crickets and trechine Carabidae have more slender bodies and longer appendages. Antennae are generally longer, sometimes exceeding the length of the body in Collembola.

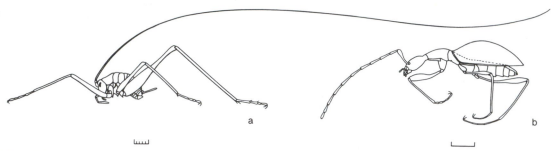

Figure 9.3 Examples of cave insects: **a**, cave cricket, *Tropidischia xanthostoma* (Gryllacrididae; scale equals 5 mm); **b**, blind cave beetle, *Glacicavicola bathyscioides* (Leiodidae; scale equals 1 mm).

Beetles may have erect, long hairs scattered over the body that are probably tactile. Most troglobites are repelled by light. The tracheal system and spiracles of trechine carabids are rudimentary; respiration probably takes place through the membranous abdominal terga. Resistance to desiccation and higher temperature is low. Long periods without food are tolerated, and growth is slow.

Activity and rest in bathysciine beetles do not follow a circadian rhythm, but are associated with temperature fluctuations. Some bathysciine beetles that live far from the entrance, where food is rare, have larvae that do not feed. The female lays one egg at a time; each is large and full of yolk. Shortly after hatching, the larva builds a clay capsule and remains inside 5 to 6 months, then pupates. The adult searches widely for food, in a manner not possible for a young larva. Adults live for about 3 years.

In the Hawaiian Islands, an amazing fauna has been found in lava tubes, including lycosid spiders, amphipods, centipedes, millipedes, predatory earwigs (*Anisolabris,* Carcinophoridae, Dermaptera), omnivorous crickets (*Thaumatogryllus* and *Caconemobius,* Gryllidae, Orthoptera), predatory bugs (*Nesidiolestes,* Reduviidae, Hemiptera), and scavenger bugs (*Speleovelia,* Mesoveliidae, Hemiptera). Phytophagous planthoppers (*Oliarus,* Cixiidae, Hemiptera) and caterpillars (*Schrankia,* Noctuidae, Lepidoptera) feed on roots hanging from the roof. The cixiid bugs have special tarsal claws and empodia for crawling on wet walls.

Many of these insects are flightless and have reduced or no eyes and the pale bodies of true cave animals, yet some species may have evolved from surface-dwelling ancestors in less than 100,000 years.

Aquatic troglobite insects have been discovered in underground waters throughout the world, including aquatic Coleoptera (some species of Dryopidae, Dytiscidae, Elmidae, Hydrophilidae, and Noteridae) and Hemiptera (Nepidae). The beetles are distinguished by reduced or no eyes, reduced pigmentation, thin, soft exoskeleton, and reduced or no hind wings. The waterscorpion (Nepidae, Hemiptera) from Romania is also blind.

Evolution of Cave Faunas

The restriction of troglobite insects to caves has stimulated research into their evolutionary origin and geographic distribution. No single explanation of their evolution is adequate for all species, but there is a common element in their history: initially they were adapted to life in cool, moist, dark places like leaf litter or crevices in soil. This **preadaptation** permitted them to enter caves, where they exploited new food resources and were less threatened by natural enemies and competition. In Hawaii, for example, evolution of cave dwelling is believed to be a continual process as species take advantage of new food resources in caves. The closest relatives of Hawaiian troglobites are species living nearby on the surface. In contrast, most troglobites in continental North America inhabit caves where colder climates and glaciation occurred in the Pleistocene Epoch (0.01 to 1.64 million years ago). Under these circumstance, insects became adapted to cold, humid, forest litter near the edges of glaciers. At that time glaciers extended to lower latitudes and altitudes. When the glaciers receded, some cool-adapted insects found refuge in caves, while the remainder of the fauna moved to higher latitudes and altitudes or became extinct. Therefore, the closest relatives of the troglobites are now far away or extinct. The scarcity of food and lower temperatures of caves in the Temperate Zone places strong selection pressure on efficient use of food at a low metabolic rate. In separate, isolated cave systems, insects independently evolved economizing adaptations: reductions of eyes, wings, flight muscles, pigmentation, and, in some beetles, number of offspring.

THE DESERT HABITAT

In contrast to life in well-watered regions of the earth, deserts have low, often unpredictable rainfall and long droughts, dry winds and abrasive sandstorms, sparse vegetation and barren sandy or rocky soil, and periodic deficiencies of nutrients; often, too, they are extremely hot with intense solar radiation. Despite this harsh environment, insects have become among the most successful of **deserticolous** animals. This is largely due to water conserving features such as a waxy epicuticle, closable spiracles, and the production of uric acid by the excretory system. In addition, they avoid daily and seasonal environmental extremes by their behavior and life history. Larval growth and activity of adult insects are often correlated with the rainy season in deserts.

Certain insects are unusually tolerant of heat, perhaps because they have enzymes that are not denatured at higher temperatures or epicuticular waxes that do not melt. However, they can use evaporative cooling for only short periods because the loss of water would be too great for organisms of such small body size. The most extreme case is the ant *Ocymyrmex barbiger,* which forages in the sun at temperatures up to 67°C on the surface of the Sahara Desert. Single workers move quickly in search of heat-killed arthropods before returning to the safety of their subterranean nest.

Many insects depend on the sandy substrate for protection against lethal temperatures and desiccation. During the hot season many forage openly only at night, spending the day in burrows or simply immersed in sand. Surface temperatures as high as 72°C and humidity as low as 10 percent are avoided by the desert cockroach, *Arenivaga investigata* (Polyphagidae, Blattodea). At depths of 30 to 45 cm in sand, the roach inhabits a zone where daily temperatures are well below the upper lethal limit (about 45.5 to 48.5°C) and humidity may reach above 82 percent, a level at which the roach can actually absorb water vapor. Like some tenebrionid beetles, the roach literally swims in sand. The body is biconvex in cross section, oval, and smooth, its lateral margins are sharp, resembling those of an aquatic beetle. Insects that run on the surface during the day may have long legs that lift the body well above the hot surface (some species of Mutillidae, Formicidae, and Tenebrionidae).

Tenebrionid beetles thrive in deserts as scavengers (Fig. 9.4). Here they are important decomposers of plant litter because bacteria and fungi are limited by aridity. At night, adults of *Onymacris unguicularis* stand on sand dunes in the fog, head down, and drink the water that condenses on their bodies and runs to their mouth. Although most desert tenebrionids are flightless, they retain the elytra, which are fused together to enclose an air space over the abdomen. Air is released by the abdominal spiracles into the space and escapes only at the tip of the abdomen, thus reducing the loss of water vapor.

THE HIGH-ALTITUDE HABITAT

Terrestrial animals and plants are generally confined to the dense, oxygen-rich atmosphere of low elevations. By comparison, the **highland,** or high-

Figure 9.4 Tenebrionid beetles of the Namib Desert in Africa. **a**, Species active by day run fast on the hot surface of the desert. **b**, These dune beetles are active at night and escape the heat of day by burrowing in sand. The beetles are disk-shaped and covered with fine hairs.

altitude environment of mountains is extremely inhospitable to life. It is the highest habitable region on earth and borders on outer space. Nevertheless, relative to other animals, arthropods have been among the most successful in living permanently at high altitudes. The highest elevations at which arthropods have been found is 6700 m, and the prize goes to the spiders. Vertebrates are poorly represented at high altitudes and perhaps relief from their predation has permitted the arthropods to flourish.

The highland is generally the treeless area above timberline at altitudes above about 2000 m.

An important difference between the lowland and the highland is that the latter has a progressively thinner atmosphere. Correlated with the decrease in atmospheric pressure are decreased air temperature and a decreased content in air of oxygen, carbon dioxide, and water vapor. With fair weather, the thin transparent atmosphere transmits intense solar radiation (especially ultraviolet) in daytime and allows rapid loss of radiation in the shade and at night. Over a 24-hour period, the temperature at the soil surface fluctuates over a wide range. The air can be exceedingly dry, so that evaporation is high. The growing season is short in summer and frost is common. At still higher altitudes, the average air temperature never rises above freezing and snow remains on the ground throughout the year.

Among the permanent animal residents, or **hypsobionts,** of the highlands are both terrestrial and aquatic insects. In the Himalayas near the highest elevations inhabited by arthropods, Collembola are dominant. At altitudes up to 6300 m, they occur in enormous numbers on snow and ice, in glacial waters, on rocks, and plants that form flat cushions. Feeding on wind-borne pollen and spores, the Collembola are at the base of a food pyramid that supports insect predators and scavengers, especially on snow. Collembola have been seen crawling about at −10°C.

In cold glacial torrents and glacial lakes at lower elevations are found the immature instars of Ephemeroptera, Plecoptera, aquatic Coleoptera (Dytiscidae, Hydrophilidae), and aquatic Diptera (Blephariceridae, Chironomidae, Deuterophlebiidae, and Simuliidae). The adults that emerge and mate later become food for scavenging insects and birds. Terrestrial insects include Orthoptera (Acrididae), Dermaptera, Lepidoptera (Papilionoidea, Geometroidea, Noctuoidea), and many Diptera (Syrphidae, Stratiomyidae, Empididae, Rhagionidae, Asilidae, Dolichopodidae, Calliphoridae, Anthomyiidae, Muscidae, and Tachinidae). Active on snow are the wingless snow tipulid, *Chionea* (Tipulidae, Diptera), and the snow "flea," *Boreus* (Boreidae, Mecoptera; Fig. 9.5). The small order Grylloblattodea is found primarily in highland environments, usually under rocks near snow. Archeognatha have been found basking on rocks at altitudes up to 5800 m in the Himalayas. The most important order in terms of the numbers of different genera and species is the Coleoptera, and especially the Carabidae. Other common beetle families are Staphylinidae, Histeridae, Tenebrionidae, Chrysomelidae, and Curculionidae.

Unlike birds and mammals, which must have adequate oxygen to generate body warmth, the most specialized hypsobiont insects have evolved a metabolism that functions normally at low temperatures in an oxygen-deficient atmosphere. They are closely associated with snow, which enables them to avoid the drying effect of higher temperatures. Active only in a narrow range of temperatures near 0°C, they die quickly if exposed to temperatures slightly above normal. For example, aquatic instars are active in water between −1.5 and 5°C, but they die if held in a human hand.

In winter, inactive hypsobiont insects are able to survive prolonged exposure to low temperature. Some are physiologically cold-hardy (see Chap. 5), but even without this mechanism, insects sheltered under insulating snow or protected in crevices can survive a few degrees below freezing without harm. As the short summer season progresses and patches of snow melt, terrestrial insects exhibit tolerance to higher temperatures and move 20 to 30 m away from the edge of the snow onto warmer soil. Here their development accelerates and the life history is completed.

In comparison with related lowland species, hypsobiont insects are usually more darkly pigmented, with black, brown, or reddish tones. Dense melanic pigments presumably prevent injury of soft tissues by the strong ultraviolet radiation. The melanism also aids in absorbing warmth from the sun. Other common features of hypsobiont insects in comparison to lowland forms are univoltine life history, slower growth and smaller body sizes of adults, mainly ground-dwelling habits, accompanied by a reduction or loss of wings, and increased density of body hairs and scales.

Most terrestrial insects in the highland are predators or scavengers. Much of their food is

Figure 9.5 A snow flea or wingless scorpion fly, *Boreus* sp. (Boreidae), is active on snow in winter at higher elevations. (Photo by H. V. Daly.)

transported from lower elevations by wind. Insects and hexapodous arthropods of many kinds, plus pollen, spores, and bits of vegetable matter, are deposited in this manner on the soil, rocks, or snow at high altitudes. Newly arrived living insects are numbed by the cold and become prey to the resident insects and birds. Plant life is generally sparse, but flowering plants are visited by flies, moths, and butterflies. Bees and bee-pollinated plants are usually absent.

THE AQUATIC HABITAT

In this section we discuss insects that spend at least the immature stages of their lives floating or submerged in water. All aquatic insects evolved from terrestrial ancestors and nearly all still have significant phases of their lives in terrestrial or aerial habitats. Aquatic species are a minor fraction of all insects, probably numbering no more than 3 to 5 percent of all species. Yet these insects are taxonomically diverse (Table 9.1), and fascinating in structure (Fig. 9.6) and biology. Some of them, such as mosquitoes, are of extreme importance in public health.

The limited number is probably the result of the limited amount of freshwater habitat in comparison with the land surface. On a global scale the total extent of freshwater is remarkably small. A few insects are truly marine. Some others live in the intertidal zone, in brackish coastal waters or saline lakes, or in hot springs (ephydrid flies in Iceland at 47°C). Most aquatic insects, however, are restricted to freshwater, where they are among the most important organisms in the aquatic ecosystem. As herbivores, predators, scavengers, and even parasitoids, insects are consumers in the food chain and, in turn, are food for fish, birds, and amphibians. Insects are important as decomposers of tree leaves and other plant material that fall into running water.

The most important limitation for submerged insects is a supply of oxygen. Oxygen readily dissolves in water, but at concentrations far less than that of air. Once dissolved, its movement in still water by diffusion alone is exceedingly slow, hence currents and turbulence in water are critical to the distribution of oxygen throughout the habitat. Respiration of aquatic insects is discussed in Chapter 5.

Aquatic insects also must regulate the balance of water and salt in their body fluids. The waxy epicuticle, evolved to prevent water loss in air, now operates in reverse to prevent an influx of water. Excretory organs are able to produce urine more dilute than hemolymph, thus preserving valuable salts. Ions may be absorbed from water by active transport through special areas of the cuticle, for example, the rectal gills of mosquito larvae. With abundant water for dilution, some species excrete nitrogenous waste in the form of ammonia.

Marine Insects

Whereas insects and other hexapodous arthropods constitute nearly three-quarters of the earth's animal species, they are greatly reduced in number of species in the earth's largest habitat, the ocean. By no means are insects scarce as individuals on the coasts or even far out at sea. The beach or littoral zone has abundant insect life, especially Coleoptera and Diptera. Each marine species has overcome certain physical barriers such as :

- **Tidal submergence.** Examples: the collembolan *Anurida maritima;* insects in several orders, including the bug *Aepophilus bonairei* (Saldidae, Hemiptera) of England and Europe.
- **Wave action.** Examples: especially midges of the genus *Clunio* (Chironomidae, Diptera) and the barnacle-eating larvae of the fly *Oedoparena glauca* (Dryomyzidae, Diptera).
- **Salinity.** Examples: especially salt-marsh mosquitoes (Culicidae, Diptera), shore flies (Ephydridae, Diptera), water boatman *Trichocorixa* sp. (Corixidae, Hemiptera), and, in Australia, the intertidal rockpool caddisfly larva of *Philanisus* sp. (Trichoptera).
- **Depth.** Examples: the midge, *Chironomus oceanicus* (Chironomidae, Diptera) dredged from 36 m, and another midge *Pontomyia* sp., from 30 m.
- **Life offshore.** Examples: five species of the pelagic water striders, *Halobates* sp. (Gerridae, Hemiptera), which occur hundreds of miles from land.

Of these barriers, perhaps the most limiting is the inability to respire for long deep beneath the surface, where oxygen may be deficient. No marine insect is known to spend its entire life underwater. Insects in general also have not been successful in colonizing deep freshwater lakes. Chironomid fly larvae have been found below 1300 m in Lake Baikal of Russia, but the life history is not completed there. Only the peculiar stonefly *Utacapnia* sp. (Capniidae, Plecoptera), in Lake Tahoe, California, can exist indefinitely in deep, still water.

TABLE 9.1
Taxa of Aquatic Insects[1]

Immature stages in freshwater

Ephemeroptera (all)
Odonata (nearly all)
Plecoptera (nearly all)
Megaloptera (all)[2]
Neuroptera
  Sisyridae[2]
Coleoptera
  Helodidae
  Limnichidae (some)
  Psephenidae[3]
  Ptilodactylidae
Diptera[3]
  Anthomyiidae *(Limnophora)*
  Blephariceridae
  Ceratopogonidae
  Chironomidae
  Culicidae

Deuterophlebiidae
Dixidae
Dolichopodidae *(Dolichopus)*[4]
Empididae (Clinocerinae,
  Hemerodromiinae)
Ephydridae
Ptychopteridae[4]
Phoridae *(Diploneura)*
Psychodidae[4]
Rhagionidae *(Atherix)*
Sarcophagidae *(Sarcophaga)*
Sciomyzidae
Scopeumatidae *(Hydromyza)*
Simuliidae
Stratiomyidae (Stratiomyinae,
  Adoxomyiinae)
Syrphidae (Sericomyiinae,

  Eristalinae)
Tabanidae[4]
Tanyderidae *(Protoplasa)*[4]
Thaumaleidae
Tipulidae (Tipulinae, Limoniinae)
Lepidoptera
  Pyralidae (Nymphulinae,
    Acentropus)[3]
Trichoptera (nearly all)[3]
Hymenoptera[3,5]
  Braconidae *(Chorebus)*
  Diapriidae *(Trichopria)*
  Ichneumonidae *(Asilops)*
  Mymaridae *(Caraphractus)*
  Scelionidae *(Tiphodytes)*
  Trichogrammatidae
    (Hydrophylita)

Immature and adult stages in freshwater

Collembola[6]
  Entomobryidae *(Sinella)*
  Hypogastruridae *(Xenylla)*
  Isotomidae *(Isotomurus)*
  Onychiuridae *(Onychiurus)*
  Poduridae *(Podura aquatica)*
  Sminthuridae *(Sminthurinus)*
Orthoptera
  Acrididae *(Marellia
    remipes)*[6]
  Tridactylidae[7]
Hemiptera (Heteroptera)
  Belostomatidae
  Corixidae
  Dipsocoridae[7]

Gelastocoridae[7]
Gerridae[6]
Hebridae[7]
Hydrometridae[6]
Macroveliidae[7]
Mesoveliidae[6,7]
Naucoridae
Nepidae
Notonectidae
Ochteridae[7]
Pleidae
Saldidae[7]
Veliidae[6]
Coleoptera[2]
  Amphizoidae

Chrysomelidae (Donaciinae)[3]
Curculionidae (Hydronomini,
  Erirrhinini, Bagouini)
Dytiscidae
Elmidae
Gyrinidae (adults)[6]
Haliplidae
Heteroceridae *(Heterocerus)*[4]
Hydraenidae (larvae)[7]
Hydrophilidae (most)
Hydroscaphidae
Limnichidae (some)
Noteridae[3]
Sphaeriidae[4]

Adult stage in freshwater

Coleoptera
  Dryopidae

Insects of the open ocean

Hemiptera (Heteroptera)
  Gerridae *(Halobates)*[6]

Diptera
  Chironomidae *(Pontomyia
    natans* adults[3])

(cont.)

199

TABLE 9.1 (continued)

Intertidal marine insects

Collembola
  Entomobryidae *(Entomobrya)*
  Hypogastruridae *(Anurida maritima)*
  Isotomidae *(Isotoma)*
Hemiptera (Heteroptera)
  Corixidae *(Trichocorixa)*
  Gerridae *(Rheumatobates)*
  Omaniidae
  Saldidae *(Aepophilus)*
Hemiptera (Homoptera)
  Aphididae *(Pemphigus)*
  Psyllidae *(Aphalara pulchella)*
Coleoptera
  Carabidae *(Thalassostrechus)*
  Chrysomelidae *(Haemonius)*

Curculionidae *(Emphyastes)*
Heteroceridae *(Heterocerus)*[4]
Hydraenidae *(Ochthebius)*
Hydrophilidae *(Cercyon)*
Limnichidae *(Hyphalus)*
Melyridae *(Endeodes)*
Salpingidae *(Aegialites)*
Staphylinidae (Aleocharinae, *Bledius, Thinopinus*)
Diptera
  Anthomyiidae *(Fucellia)*
  Canaceidae
  Ceratopogonidae
  Chironomidae (Clunioninae)
  Coelopidae

Culicidae
Dolichopodidae *(Aphrosylus)*
Dryomyzidae *(Oedoparena)*
Ephydridae
Helcomyzidae
Heleomyzidae *(Anorostoma)*
Tabanidae
Tipulidae *(Limonia)*
Trichoptera
  Limnephilidae *(Limnephilus affinus)*
  Philanisidae *(Philanisus)*
Hymenoptera[5]
  Encyrtidae *(Psyllaephagus)*
  Eulophidae *(Tetrastichus)*

---

[1]This list of taxa is intended to be representative, not exhaustive. The listing of a family without qualification does not necessarily mean that all species in the family have the same habits. Well-known or exceptional taxa are given in parentheses.
Source: Summarized from Usinger (1956a), R. T. Smith and Carlton (1975), and Cheng, (1976).
[2]Pupae terrestrial.
[3]Pupae aquatic.
[4]In saturated soil or semiliquid media.
[5]Adults enter water to oviposit on aquatic hosts.
[6]On water surface.
[7]On shore near water, occasionally enter water.

Added to the physical barriers is the intense biological competition from marine arthropods and other organisms. During the Paleozoic Era (Cambrian through Permian Periods, 570 to 245 million years ago: see Table 15.3), while insects flourished on land, the marine fauna was diversifying and occupying the habitats of the sea. The oceans now seem effectively closed to insects.

Insects of Lotic Water

Freshwater habitats can be divided into running, or **lotic** water, and still, or **lentic** water. Although the total volume of lotic water in the world is less than that of lentic water, the great river drainages exist for a million years or more, providing a continuously habitable environment for aquatic organisms over a wide geographic area. Because major drainages are more or less permanent and are otherwise favorable, insects have evolved the greatest variety of taxa in lotic waters. Indeed most kinds of aquatic insect are represented, and some occur only here. This is especially true in the rapidly flowing, shallow, rocky streams of the headwaters, where erosional processes predominate, oxygen concentration is high, and the mean monthly temperature seldom exceeds 20°C.

Water seeping into a stream from the ground or issuing from springs is commonly cooler than air in summer and warmer than air in winter. This difference may be sustained for some distance if the stream is shaded in summer or insulated by a snow cover in winter. Although such waters are subject to daily and seasonal fluctuations, the range of variation is usually less than that of the shallow regions of lakes. Furthermore, cooler water absorbs much more oxygen than warmer water (e.g., water saturated at 4°C holds 1.5 times the amount at 22.5°C). Turbulence over the rocky bed maintains dissolved oxygen at the point of saturation and distributes it to the depth of the stream. In summer, algae and other aquatic plants thrive in clear, shallow water. In autumn, deciduous trees along streams drop leaves that provide nutrients into the water. Consequently, with food, oxygen and moderate temperatures, insects are able to feed and grow throughout the year, even in winter in temperate regions.

The immatures of Ephemeroptera, Plecoptera, Trichoptera, and Megaloptera are much more numerous in streams than in lakes. Among rocks in swift, cool waters are also larvae of the fly families Blephariceridae, Simuliidae, and Deuterophlebiidae, and the beetles Dryopidae, Elmidae, Ptilodactylidae, and Psephenidae. Larvae of the odonates Agrionidae, Cordulegastridae, and Gomphidae also favor running water. Among aquatic holometabolous insects, Trichoptera and some Diptera remain in water as pupae while Megaloptera, aquatic Coleoptera, and some other Diptera leave the water to pupate in soil or litter.

Adaptations of Insects in Lotic Water

Organisms associated with the bottom of a body of water are often referred to as the benthic community or **benthos.** In strongly flowing streams,

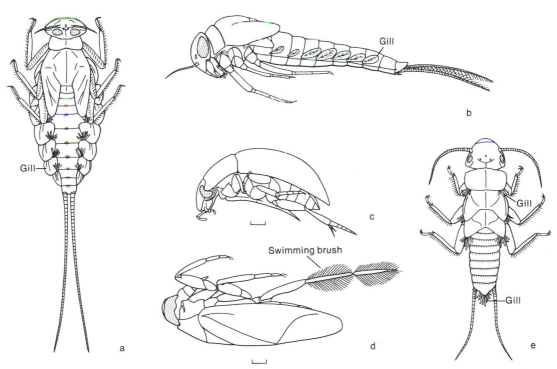

Figure 9.6 Examples of aquatic insects: **a**, mayfly naiad, *Epeorus* sp. (Heptageniidae); **b**, mayfly naiad, *Ameletus* sp. (Siphlonuridae); **c**, water beetle, *Tropisternus ellipticus* (Hydrophilidae); **d**, backswimmer, *Notonecta undulata* (Notonectidae); **e**, stonefly naiad, *Acroneuria pacifica* (Perlidae). Scales equal 1 mm.

an ever-present hazard to benthic insects is to be swept off the bottom and drift downstream, exposed to predators. The number of insects drifting usually rises at night, because many are nocturnally active. Drifting may involve only short distances and is not always entirely deleterious, since it serves to relieve competition for space and food and to effect a more even distribution of the population downstream. Evidence is inconclusive that adults tend to fly upstream to lay their eggs and recolonize the uppermost waters. The eggs of mayflies (Ephemeroptera), stoneflies (Plecoptera), and some dragonflies (Odonata) are equipped to sink rapidly, and firmly attach to the bottom, thus remaining near the ovipositional site.

Insects that inhabit swift water are generally prevented from visiting the surface for air. Their respiratory structures are tracheal gills or plastrons (Chap. 5) that function while bathed in oxygen-rich water. Larvae of some mayflies and stoneflies that normally reside in still pools create their own currents. The abdominal gills of mayflies have an undulating beat that circulates the surrounding water. Similarly, stoneflies pump up and down on their legs at a rate proportional to the oxygen need. Insects living in the swiftest waters do not exhibit such respiratory movements and depend entirely on the current to bring oxygen. They die quickly when removed from their natural environment, even when placed in still water of the same temperature and oxygen saturation, because the oxygen supply becomes depleted in the proximity of the gills and is not renewed by currents.

A number of similar morphological adaptations evolved independently by various insects in torrential waters. The body may be dorsoventrally compressed so that the insect can creep beneath stones or in crevices out of the current. Larvae of the mayflies, Heptageniidae (Figs. 9.6a, 20.4c; Color photo 2f), and larval beetles of Psephenidae are thus flattened (Fig. 39.10c). Small size also permits insects to enter small spaces protected from the current. Nearly all adult beetles in swift water are small.

Various attachment devices allow insects to cling to the substrate. The beetles Elmidae and Dryopidae have the last tarsal segment and claws enlarged for anchoring themselves. In addition, the larvae of Elmidae have prehensile hooks near the anus. Stout tarsal claws and anal hooks also are found on the free-living caddis larvae of Rhyacophilidae (Trichoptera, Fig. 45.2a) and the hellgrammites or larvae of Corydalidae (Megaloptera). Lacking true legs, the aquatic larvae of the fly families Simuliidae, Deuterophlebiidae, Blephariceridae, Chironomidae, Rhagionidae, Empididae, Syrphidae, and Ephydridae, and *Limnophora* (Anthomyiidae), have developed leglike prolegs for attachment. The prolegs are usually fitted with fine hooks. Suckers are developed near the anus of larval simuliids and on the ventral surface of *Maruina* (Psychodidae) and Blephariceridae. Frictional resistance to detachment is developed by ventral gills or hairy pads on the mayfly naiads *Rhithrogena* (Heptageniidae), *Ephemerella doddsi,* and *E. pelosa* (Ephemerellidae), the African *Dicercomyzan* (Tricorythidae), and the larval beetles of Psephenidae (Fig. 39.10c). Silk and sticky secretions are used for attachment by some Chironomidae and the caterpillars of Pyralidae (Lepidoptera). Larval simuliids move about with their prolegs on silken mats. *Hydropsyche* (Trichoptera; Fig. 45.3e) spin nets in the current and harvest waterborne food. Most Trichoptera attach their pupal cases to rocks with silk. The stone cases built by Trichoptera in streams are usually composed of heavier grains and a few heavy pebbles (Fig. 45.4). This provides ballast, ensuring that the caddis larva quickly sinks into still water if dislodged.

Few insects actively swim against the current. The streamlined, minnowlike mayflies *Ameletus* (Fig. 9.6b), *Isonychia, Centroptilus,* and some *Baetis* move in and out of the current with powerful, up-and-down movements of their caudal filaments. On the surface of riffles are the bugs *Rhagovelia* (Veliidae, Hemiptera; Fig. 34.9). Leaves packed against obstructions and wood debris have a special fauna of mayflies, caddisflies, and stoneflies. The stout-limbed *Ephemerella* (Ephemerellidae, Fig. 20.2a) is commonly found here.

Insects in crevices can utilize the rich supply of oxygen and food of fast running streams without the immediate danger of the current. Flattened mayflies are also found in the quiet water beneath rocks. Deeper still, in the sand or gravel soils under the stream that are saturated with water is the **hyporheic zone.** Here are found tiny young instars of insects, the narrow-bodied stoneflies and mayflies, and crane fly larvae, *Antocha* and *Hexatoma* (Tipulidae, Diptera). When the stream is dry, some immature aquatic insects can survive buried in the streambed.

The lower region of a drainage, or where it flows through relatively flat land, is characterized by slower flow, greater depth, occasional low oxygen concentrations, and monthly mean temperatures close to that of air and rising above 20°C. Depositional processes now predominate, so that the bed of the river is formed by fine sediments of sand and mud.

Fewer kinds of insects are specialized for life here, but some species are extremely abundant. Larvae of the burrowing mayflies Ephemeridae (Fig. 20.4b) excavate U-shaped burrows in silt. Undulating movements of their feathery gills circulate water in the burrow, bringing in organic matter, which is eaten, and oxygen. Other mayflies, such as Caenidae, Tricorythidae, and some Ephemerellidae, inhabit silty bottoms of streams and rivers. These have the functional gills kept clean of particles by the thick uppermost gill. The body surfaces of some bottom-dwelling mayflies, stoneflies, and dragonflies are covered by a protective layer of fine hairs. Larvae of chironomid midges are often abundant in soft, mucky bottoms, where they build tubes.

Insects of Lentic Water

Lakes are relatively short-lived in comparison to rivers because they normally fill with sediments during prolonged wet periods or evaporate during prolonged dry periods. Most lakes existing today originated in the Pleistocene Epoch (0.01 to 1.64 million years ago) and are less than 25,000 years old. Only a few large, deep lakes are much older. Lake Baikal of Russia and Lake Tanganyika in Africa probably existed in the Tertiary Period (1.64 to 65 million years ago).

In contrast to lotic water, the insect life of lentic water is influenced more by temperature and oxygen supply than by water current. Temperatures in shallow, marginal waters of lakes or small bodies of water such as ponds tend to fluctuate with air temperature. Large bodies of still water will warm slowly in summer, becoming stratified into an upper, warmer layer and deeper, colder water that may be 4°C near the bottom. In winter, lakes cool slowly. Insects in lakes are protected against freezing because water reaches maximum density at 4°C and is less dense at lower or higher temperatures. Surface water cooled to 4°C sinks. Ice forms at the surface after deeper water is uniformly 4°C and the colder, lighter water is at the top.

Biological communities of lakes can be divided into three zones: the **littoral zone,** or near-shore, shallow waters where light reaches the bottom; the **limnetic zone,** or offshore, open water to the depth of light penetration; and the **profundal zone,** or deep, offshore water below the penetration of light.

Oxygen is absorbed at the water's surface and is produced by photosynthetic plants in the littoral and limnetic zones. In summer the oxygen supply of the littoral zone may exceed saturation, but the concentration is inversely proportional to temperature. The supply varies daily because the oxygen consumed by plants and animals at night is not replaced. In the deep profundal zone, oxygen is often depleted by bacterial respiration.

Lakes that are poor in nutrients are called **oligotrophic.** Young, deep lakes are characteristically of this type. Precipitation and runoff from the watershed carry dissolved substances into the lake, promoting the growth of free-floating algae in sunny, open water. The gradual enrichment of water with nutrients, especially nitrogen and phosphorus, is called **eutrophication.** Continued addition leads to the production of organic matter in excess of decomposition. The decomposing organisms in deep water deplete the water of its oxygen, leaving the organic matter to accumulate with other sediments. Older lakes, rich in nutrients and shallow, are called **eutrophic.** When lakes receive large amounts of acid humic substances from the watershed, the water becomes brownish and poor in variety of aquatic life. Organic matter accumulates from plants in the shallow marginal waters. Such lakes or bogs are called **dystrophic.** Ultimately both eutrophic and dystrophic lakes fill completely with organic matter, are overgrown by vegetation, and disappear.

Slender fly larvae dominate the insect life of the deeper, open limnetic and profundal zones and the bottom beneath. The thin sediments of oligotrophic lakes are characterized by colorless midge larvae of *Tanytarsus* (Chironomidae). On the surface of the bottom ooze of eutrophic lakes, where oxygen is often depleted, the predominant or only insects are red, hemoglobin-containing larvae of *Chironomus.* These are able to survive the irregular oxygen supply and continue to function as important decomposers. The open waters of eutrophic lakes are inhabited by the phantom midge larvae, *Chaoborus* (Culicidae). As one of the few free-floating insects, the predatory *Chaoborus* spend the day near the bottom and swim toward the surface at night. The phantom larva is nearly transparent—hence the name. Gas sacs are visible at each end and act as organs of equilibrium. The volume of gas is controlled by the nervous system. When the sacs are compressed, the larva sinks; when they are expanded, it rises.

Adult midges emerge synchronously in enormous numbers from productive lakes throughout the world. *Chaoborus edulis* forms spectacular clouds seen at great distances over Lake Malawi in Africa. The emergence of *C. astictopus,* the Clear Lake gnat of Clear Lake, California, creates an annual nuisance to resort owners. When carried over land by winds, the masses of chironomids and *Chaoborus* remove significant quantities of organic matter from the lake ecosystem.

The littoral zone of lakes and ponds contains more different kinds of insect than deeper, open waters. Certain areas approximate the physical and chemical conditions of running water. On the wave-washed, rocky shores of oligotrophic lakes we find some of the same mayflies, stoneflies,

and caddisflies that are in streams. Stoneflies are sensitive to low oxygen and are rare in lakes except in such places. The numbers and variety of insects in shallow water increase with eutrophication. Mayflies, caddisflies, dragonflies, and damselflies live among submerged plants or on the bottom. These gilled insects depend on high levels of dissolved oxygen for respiration and do not rise to the surface for air.

The quiet, warm waters of shallow bays have a greater daily variation in dissolved oxygen. Here are found the many Hemiptera, Coleoptera, and Diptera that breath air, either directly or in air bubbles. On the surface are water striders (Gerridae and Veliidae, Hemiptera), whirligig beetles (Gyrinidae), Collembola, and marsh treaders (Hydrometridae, Hemiptera).

Shallow bodies of water may be created each year by rain and snow melt, lasting but a few weeks or months before the water is lost through evaporation or absorption into the soil. Such bodies of water formed in the spring are called **vernal lakes** or **ponds.** Insect life ordinarily teems in these temporary habitats because predatory fish are absent. Even the small volumes held in tree holes, cavities at the bases of leaves of plants such as the tropical bromeliads, and hoofprints are sufficient for some insects to complete their life. With warm temperature, growth is usually quite rapid. For example, the yellow fever mosquito, *Aedes aegypti,* completes its development in less than 2 weeks.

INSECT SCAVENGERS

Many, but by no means all, insects that live in soil and water are scavengers. Perhaps the word "scavenger" is poorly chosen. Does it indicate an animal that eats whatever comes along, one that eats mostly dead things, or one whose diet consists of items so small and miscellaneous that an accounting is difficult? Here we intend all these meanings when we call certain insects scavengers. The term is broad enough to convey the diversity of food that some insects are known to eat while

TABLE 9.2
Major Taxa of Scavenging and Fungivorous Insects [1]

Omnivorous, generally saprophagous, or microphytic insects

Collembola	Coleoptera	Scirtidae [2,3]
Diplura	Cantharidae	Staphylinoidea (many)
Campodeidae	Carabidae (some)	Tenebrionidae (many)
Protura (?)	Chrysomelidae [2]	Mecoptera [2]
Archeognatha	Cleroidea (many)	Bittacidae
Thysanura	Colydiidae (many)	Choristidae
Ephemeroptera [2,3]	Cryptophagidae	Panorpidae
Blattodea	Cucujidae	Diptera [2]
Isoptera (some)	Dryopidae	Anthomyiidae (some)
Grylloblattodea	Elmidae	Nematocera (many) [3]
Dermaptera (some)	Hydrophiloidea	Stratiomyidae
Plecoptera [2,3]	Hydroscaphidae [3]	Syrphidae (*Volucella*)
Orthoptera	Lathridiidae	Siphonaptera [2]
Gryllidae	Limnichidae	Trichoptera (many) [2,3]
Zoraptera	Nitidulidae	Hymenoptera (Formicidae)
Psocoptera	Psephenidae [2,3]	

Phytosaprophagous insects

Isoptera (some)	Scarabaeidae	Helcomyzidae
Embiidina	Silphidae (some)	Heleomyzidae
Coleoptera	Staphylinidae (some)	Lauxaniidae
Anthicidae	Tenebrionidae	Lonchopteridae
Cerylonidae	Throscidae	Muscidae
Corylophidae	Diptera [2]	Mycetophilidae
Hydrophilidae	Ceratopogonidae (*Forcipomyia*)	Sciaridae
Monommidae	Chloropidae	Stratiomyidae
Nitidulidae	Coelopidae	Syrphidae
Nosodendridae	Dolichopodidae	Tipulidae
Ptiliidae	Dryomyzidae	Trichoptera (many) [2,3]
Ptinidae		

Scatophagous and zoosaprophagous insects

Thysanura (Lepismatidae) [4]	Staphylinidae	Muscidae
Isoptera (some)	Tenebrionidae (Opatrinae;	Psychodidae
Hemiptera (Heteroptera) [4]	Diaperinae, Phalerini)	Sarcophagidae
Gerridae	Trogidae	Scatopsidae
Mesoveliidae	Mecoptera (Panorpidae) [4,5]	Sepsidae
Coleoptera	Diptera [2]	Sphaeroceridae
Cleridae (Corynetinae)	Anthomyiidae	Syrphidae
Dermestidae	Bibionidae	Lepidoptera [2]
Gyrinidae [4,5]	Calliphoridae	Pyralidae (*Bradypodicola*)
Hydrophilidae	Dolichopodidae (some)	Tineidae
Nitidulidae	Heleomyzidae	Hymenoptera (Formicidae)
Scarabaeidae	Lauxaniidae	
Silphidae		

(cont.)

TABLE 9.2 *(continued)*

Xylophagous insects in dead wood

Blattodea (Cryptocercidae)
Isoptera
Coleoptera
   Anobiidae[2]
   Archostemata
   Bostrichidae[2]
   Brentidae[2]
   Buprestidae[2]
   Cephaloidae[2]
   Cerambycidae[2]

Cucujidae[2]
Curculionidae *(Cossonus)*
Elateridae (some)[2]
Eucnemidae[2]
Lucanidae[2]
Lyctidae (all)[2]
Lymexylidae[2]
Mordellidae[2]
Oedemeridae[2]
Passalidae

Pyrochroidae[2]
Rhysodidae
Scarabaeidae[2]
Scolytidae
Scraptiidae[2]
Tenebrionidae[2]
Diptera[2]
   Tipulidae
   Xylophagidae

Insects infesting stored plant products: grains, nuts, seeds, dried fruits, flour, and other foods

Thysanura (Lepismatidae)
Psocoptera (Liposcelidae) .
Coleoptera (Polyphaga)
   Anobiidae
   Anthicidae[6]
   Bostrichidae
   Bruchidae
   Colydiidae
   Cryptophagidae[6]
   Cucujidae
   Curculionidae

Dermestidae
Erotylidae[6]
Lathridiidae[6]
Mycetophagidae[6]
Nitidulidae[6]
Trogositidae
Ptinidae
Tenebrionidae
Diptera
   Piophilidae[2]

Lepidoptera[2]
   Cosmopterygidae
   Gelechiidae
   Oecophoridae
   Pyralidae
   Tineidae
   Tortricidae
Hymenoptera
   Eurytomidae[2]
   Formicidae

Fungivorous insects

Isoptera (Macrotermitinae)
Zoraptera
Hemiptera
   Aradidae
   Derbidae
Thysanoptera (some)
Coleoptera (Polyphaga)
   Anthribidae
   Cucujoidea (many)
   Derodontidae
   Leiodidae (some)

Nosodendridae (some)
Trogositidae (a few)
Ptiliidae
Scaphidiidae
Scarabaeidae (a few)
Scolytidae (a few)
Tenebrionidae (many)
Diptera[2]
   Cecidomyiidae (some)
   Ceratopogonidae
   Drosophilidae

Mycetophilidae
Platypezidae
Sciaridae
Lepidoptera[2]
   Lycaenidae (a few)
   Lyonetiidae
   Oecophoridae
   Tineidae
Hymenoptera
   Formicidae *(Atta)*
   Siricidae[2]

[1] This list of taxa is intended to be representative, not exhaustive. The listing of a family without qualification does not necessarily mean that all species in the family have the same habits. Unless otherwise indicated, both the immature and adult stages are believed to have the same food habits. Well-known or exceptional taxa are given in parentheses.

[2] Feeding on this food mainly by immature stages.

[3] Algae are probably important in diet.

[4] Food is mainly dead insects.

[5] Feeding on this food mainly by adults.

[6] Fungi are important in diet.

also concealing our ignorance of the exact diets of others.

A complex terminology has developed to describe the diets of scavengers. Some are **omnivorous,** eating dead or living plants or animals. **Saprophagous** scavengers feed on dead organisms and have been subdivided into **xylophagous** forms, boring into and feeding on sound or decaying wood; **phytosaprophagous** forms, feeding on decaying vegetable matter; **scatophagous** or **coprophagous** forms, feeding on feces or dung; and **zoosaprophagous** or **necrophagous** forms, feeding on dead animals. **Detritivores** feed on particles of plant or animal matter plus the associated microbiota of fungi, bacteria, protozoa, and other tiny invertebrates. The food of saprophagous insects, therefore, is not all dead matter, because microbes flourish in such places and are nutritious.

Microphytic insects selectively eat small living organisms: bacteria, yeasts, fungi, algae, diatoms, lichens, spores, and loose pollen. Animal foods include Protozoa, and small invertebrates and their eggs. **Fungivorous** insects are adapted to feeding exclusively on living and dead fungi. Aquatic **filter feeders** strain microbes and organic matter from water with specialized brushes on their mouth parts or webs. Thus among the insects loosely called scavengers are those with specific choices of food. Major taxa of scavenging and fungivorous insects are listed in Table 9.2.

Saprophagous insects have unity in an ecological sense because they function in the complex world of the decomposers (Fig. 9.7). These should be counted among our most beneficial insects. Intact bodies of dead vascular plants and vertebrates, smaller dead organisms, and excrement are progressively disintegrated. The organic nutrients in streams and soils are derived from plant material that is decomposed in part by insects and other hexapodous arthropods. Scavenging species are usually specific to either plant or animal matter. Tissue is ingested and partially or completely digested, often with the aid of symbionts or unusual enzymes such as cellulase, keratinase, or collagenase. The feces of scavengers are further decom-

Figure 9.7 Insects as decomposers in water and on soil. **a,** Dead leaves are recycled in streams by aquatic insects and microbes. This mayfly larva (Heptageniidae) is resting on a leaf that is in the process of disintegration by insects and microbes. (Photo by H. V. Daly.) **b,** Many insects not normally considered to be scavengers visit excrement for moisture and other substances. Included here, on hyena dung, are butterflies, flies, bees, and wasps as well as scavenging scarab beetles.

posed by other scavengers or microbes, thus aiding in nutrient recycling. Associated with this community are insect predators and parasitoids that prey on the scavengers.

In terrestrial environments the scavenger species appear at a dead animal or plant in a predictable sequence as characteristic chemicals are emitted from the decaying organic material at each stage of disintegration. The kind of scavenger associated with a human corpse can provide estimates of the time of death in criminal investigations. The field of **forensic entomology** makes use of this and other evidence from insects in matters of law.

SELECTED REFERENCES

Insects and Pollution

Heliövaara and Väisänen, 1993.

Soil Insects

Kevan, 1962 (text); Kuhnelt, 1961 (text); R. F. Lawrence, 1953 (text); Lee and Wood, 1971 (termites); J. C. Moore and Walter, 1988 (regulation of populations); J. M. Moore and Picker, 1991 (ancient termite mound); Seastedt, 1984 (decomposition); Wallwork, 1970 (text); Wilson, 1990 (ants).

Biology of Collembola

Butcher et al., 1971; Hale, 1965; Christiansen, 1964.

Desert Insects

Cloudsley-Thompson, 1975; Edney, 1974; Louw and Seely, 1982.

Cave Insects

Barr, 1967, 1974; Culver, 1982; Decu et al.,1994 (Nepidae); Holsinger, 1988 (evolution); Howarth, 1983 (ecology); R. W. Mitchell, 1969 (tropical versus temperate); Vandel, 1965 (text).

Highland Insects

Edwards, 1987; Mani, 1962, 1968; Mani and Giddings, 1980.

Aquatic Ecology and Aquatic Insects, comprehensive treatments

Allan, 1995 (streams); Amos, 1967 (ponds); Cheng, 1976 (marine insects); Hynes, 1970 (streams); J. L. Lloyd, 1921 (caddisfly larvae); Merritt and Cummins, 1996 (text); Niering, 1966 (marsh); Resh and Rosenberg, 1984 (text); R. I. Smith and Carlton, 1975 (marine insects); Usinger, 1956a (California species), 1967 (rivers and streams); Ward, 1992 (text); D. D. Williams, 1987 (temporary waters); D. D. Williams and Feltmate, 1992 (text).

Aquatic Ecology and Physiological Adaptations, reviews

N. H. Anderson and Sedell, 1979 (detritus processing); Brittain, 1982 (mayflies); H. P. Brown, 1987 (riffle beetles); Cheng, 1985 (*Halobates*); Corbet, 1980 (Odonata); Cummins, 1973 (trophic relations); G. F. Edmunds and McCafferty, 1988 (mayfly subimago); Fremling, 1960 (*Hexagenia*); Hynes, 1976 (stoneflies); Mackay and Wiggins, 1979 (ecology of caddisflies); Morse, 1984 (caddisflies); Neumann, 1976 (intertidal chironomids); Oliver, 1971 (chironomids); Pinder, 1986 (chironomids); Pritchard, 1983 (tipulids); Spence and Andersen, 1994 (gerrids); Wallace and Merritt, 1980 (filter feeding); Ward and Stanford, 1982 (thermal responses); Weaver and Morse, 1986 (caddisflies).

Insect Pests of Stored Food

Gorham (1991).

Forensic Entomology

Catts and Goff, 1992; K. G. V. Smith, 1986.

See Also:

For references on specific soil and aquatic taxa, consult the appropriate chapters on insect orders.

10
INSECTS AND VASCULAR PLANTS

Acrocerid flies, such as this *Eulonchus* sp., frequently suck nectar of *Mimulus* flowers within which they may also spend the night. While the fly is immobilized by cold, part of the flower can be cut away to reveal the pollen-laden fly and the flower's reproductive structures. The stamens are positioned over the backs of insect visitors while the hairy, two-lipped stigma juts out from the mouth of the flower. This flower has not yet shed its pollen or received pollen from another plant. Later, when a visitor arrives from another *Mimulus* plant, pollen will be brushed onto the stigma and thus cause the lip to quickly close. As the visitor crawls deeply into the flower, pollen will be deposited on its back and when it withdraws, the closed stigma will not be receptive, thus eliminating the risk of undesirable self-pollination.

Approximately half the species of insects are **phytophagous**; that is, they feed on green plants (Color photos 5a–f). Insects in the six largest orders (Fig. 1.1) derive much or most of their food from plants. Insects now are the most important primary consumers of land plants, usually ex-ceeding vertebrate herbivores and competing directly with humans. The impact of insect feeding is checked by natural enemies and ameliorated by varied plant defenses. Not all relationships, however, are injurious to plants. The dominant plants today, the angiosperms or flowering plants, owe their origin to the feeding behavior of Meso-zoic insects that functioned as pollinators.

Phytophagy has evolved repeatedly from other food habits, usually scavenging. In the sections that follow, we will examine the relationships of insects with plant hosts. These relationships, both mutualistic and detrimental to plants, are envisioned as coevolutionary: that is, they involve the interaction of two or more kinds of organisms in which evolutionary changes in one affect the evolution of the others.

FLOWER-VISITING INSECTS

Evolutionary Origin of Insects as Pollinators

Insects may well have served to transport spores among early plants during the Paleozoic Era (570 to 245 million years ago; see Table 15.3). Here we are concerned with the relationships between insects and the angiosperms that probably began in the Mesozoic Era (245 to 65 million years ago). The earliest fossil evidence of flowering plants is leaves and pollen found in strata about 124 million years old during the Cretaceous Period of the Mesozoic Era. The evolution of angiosperms was so rapid that an astonishing 67 families are represented by the end of the Cretaceous Period (65 million years ago).

Early phytophagous insects doubtless fed on the reproductive organs, pollen, and seeds, as well as the vegetative parts of plants. The first step in the mutualistic association of insects and plants probably took place when insects accidentally carried pollen from male to female organs of the same species of plant. This transfer, called **pollination,** had previously required large quantities of wind-borne pollen, to ensure that a few grains would land by chance in the right place. In exchange for food, insects became the agents, or **pollen vectors,** that allowed small amounts of pollen to be carried quickly and placed precisely on the female organs.

The first pollen vectors may have included terrestrial or flying reptiles or early birds, but the most important were probably the flying insects. Among these, the Coleoptera are thought to be the most significant. They were well diversified in the Mesozoic Era. Interest in them is heightened by the number of primitive angiosperms today that are beetle-pollinated. However, beetles may consume the ovules of the plants that they pollinate (Color photo 5c). For this reason, certain flower structures such as carpels can be explained as defensive measures initially evolved against the powerful, chewing jaws of these insects.

The next steps in the evolution of mutualism included features of plants that attracted insects by taking advantage of their mosaic vision and color perception, and the important roles of chemicals in their behavior. A broken outline and colored petals aided visual recognition of flowers. Odors attracted insects at a distance. The first floral odors may have imitated odors of fruits or decay that appealed to scavenging beetles. Primitive flowers visited by beetles generally lack floral nectar.

Meanwhile, insects also were probably attracted to plants by the sweet honeydew produced by phloem-feeding Homoptera. The evolution of nectaries allowed the plant to locate a similar attractant near the reproductive structures and to regulate its quantity and content. Nectar is an aqueous fluid rich in sugars (commonly sucrose, glucose, and fructose) plus nutrients attractive to pollinators: amino acids, proteins, and lipids. Thus, both water and food are provided. Other substances include ascorbic acid, possibly serving as an antioxidant, and alkaloids, which might be toxic to certain unwanted flower visitors. The secretion of nectar inside the flower was an incentive to actively flying insects in need of carbohydrate fuel such as Lepidoptera, Diptera, and Hymenoptera. Indeed, the abundant supply of nectar as food must have contributed to the success of these orders. Floral nectar differs in composition from both honeydew and extrafloral nectar (from nectaries elsewhere on plants) in ways that suggest it is secreted expressly for the food needs of favored pollinators. Advanced families of plants have higher concentrations of amino acids than primitive families, and butterfly-pollinated flowers have higher concentrations than bee-pollinated flowers. This discrepancy is associated with the inability of most butterflies to ingest protein rich pollen, whereas bees obtain their amino acids from pollen. *Heliconius* butterflies, however, collect pollen on their galeae and digest it there, sucking up a solution of free amino acids. Flies that frequent protein-rich dung are attracted to the nectars of fly-pollinated flowers that are high in amino acids.

The most common pollinators are the adults of the endopterygote orders Coleoptera (Color photo 5c), Lepidoptera (Color photo 5a), Diptera (Chapter photo), and Hymenoptera (Color photo 5b; Table 10.1). Their searches for mates, oviposition sites, and plant or animal food are aided by a strong flight apparatus, highly developed senses, and, in some groups, learning ability. The same attributes aid pollinators as they search for and remember flowers.

Behavior of Flower-Visiting Insects

Individual insects that visit flowers of the same plant species during a single flight or longer period are said to be **flower-constant.** When all individuals of an insect species are restricted to visiting a single species of plant for pollen, the insect is said to be **monolectic.** If several, possibly related, plant species are visited by individuals of the insect species, the insect is said to be **oligolectic.** If many plant species are visited, the term **polylectic** applies. An individual bee that is a member of a polylectic species may also be flower-constant during successive visits. Flower constancy is benefi-

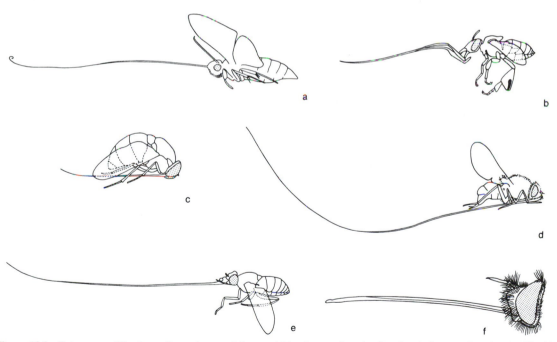

Figure 10.1 Extreme modifications of mouthparts of flower-visiting insects (length of proboscis in parentheses): **a,** sphingid moth, *Manduca quinquemaculata*—proboscis (12 cm) is coiled under head when not in use; **b,** male apid bee, *Euglossa asarophora*—proboscis (3 cm) is folded between legs when not in use; **c,** acrocerid fly, *Lasia kletti*—proboscis (2.4 cm) is swung forward in use; **d,** nemestrinid fly, *Megistorrhynchus longirostris*—proboscis (8.4 cm) is swung forward in use; **e,** tabanid fly, *Philoliche longirostris*—proboscis (5.8 cm) is carried in flight as shown; **f,** head of bombyliid fly, *Bombylius lancifer*—proboscis (9.7 mm) is carried as shown.

TABLE 10.1
Major Taxa of Terrestrial Phytophagous Insects [1]

External feeders on foliage, stems, roots, fruits, and/or seeds

Exposed feeders
Isoptera (some)
Dermaptera (some)
Plecoptera (some)[2]
Orthoptera
  Acrididae
  Eumastacidae
  Gryllidae
  Gryllotalpidae
  Tanaoceridae
  Tettigoniidae (some)
Phasmatodea (all)
Hemiptera (Heteroptera)
  Berytidae
  Coreidae
  Cydnidae
  Largidae
  Lygaeidae[3]
  Miridae
  Pentatomidae (most)
  Piesmatidae
  Pyrrhocoridae[3]
  Tingidae
Hemiptera (all Homoptera)
Thysanoptera (most)

Coleoptera (Adephaga)
  Carabidae (some)
Coleoptera (Polyphaga)
  Anthicidae (some)[2]
  Anthribidae (some)[2]
  Byrrhidae (some)[2]
  Byturidae
  Cantharidae (some)[2]
  Cerambycidae (some)[2]
  Chrysomelidae
  Coccinellidae *(Epilachna)*
  Curculionidae
  Elateridae (some)[2]
  Meloidae (some)[2]
  Scarabaeidae (Melolonthinae,
    Rutelinae)[2]
  Tenebrionidae (some)
Lepidoptera[4]
  Bombycoidea
  Geometroidea
  Hesperoidea
  Noctuoidea
  Notodontoidea
  Papilionoidea
  Sphingoidea
Hymenoptera (Symphyta)[4]
  Tenthredinoidea

Leafrollers and shelter makers[4]
Coleoptera
  Attelabidae
Lepidoptera
  Gelechiidae
  Gracilaridae
  Lasiocampidae
  Pyralidae
  Tortricidae (Tortricinae)
  Yponomeutidae
Hymenoptera (Symphyta)
  Megalodontidae
  Pamphiliidae
Casebearers[4]
Coleoptera (Polyphaga)
  Chrysomelidae (Clytrinae,
    Cryptocephalinae)
Lepidoptera
  Coleophoridae
  Incurvariidae
  Psychidae
  Tineidae
Open galls
Thysanoptera (some)
Hemiptera (Homoptera)
  Aphididae (some)

Internal feeders on foliage, stems, and/or roots[4]

Borers
Coleoptera (Polyphaga)
  Brentidae
  Buprestidae
  Cerambycidae (some)
  Curculionidae
  Languriidae
  Scolytidae
    (including Platypodidae)
Diptera
  Agromyzidae
  Anthomyiidae
  Chloropidae
  Ephydridae
Lepidoptera
  Cossidae
  Hepialidae
  Noctuidae
  Pyralidae
  Sesiidae
  Tortricidae (Olethreutinae)
Hymenoptera (Symphyta)
  Cephidae
  Siricidae (some)
  Syntexidae
  Xiphydriidae

Leafminers
Coleoptera
  Buprestidae
  Chrysomelidae
  Curculionidae
Diptera
  Agromyzidae
  Anthomyiidae
  Cecidomyiidae
  Drosophilidae
  Ephydridae
  Lauxaniidae
  Psilidae
  Sciaridae
  Syrphidae
  Tephritidae
Lepidoptera
  Coleophoridae
  Cosmopterygidae
  Cycnodiidae
  Elachistidae (some)
  Eriocraniidae
  Gracilariidae
  Helodinidae
  Heliozelidae
  Incurvariidae
  Lyonetiidae

Nepticulidae
Noctuidae *(Bellura)*
Opostegidae
Pyralidae
Tischeriidae
Tortricidae (a few Oleuthreutinae)
Yponomeutidae (Argyresthiinae)
Hymenoptera
  Argidae
  Tenthredinidae
Closed galls
Coleoptera
  Buprestidae
  Cerambycidae
  Curculionidae
Lepidoptera
  Cosmopterygidae
  Gelechiidae
  Tortricidae (Olethreutinae)
Diptera
  Agromyzidae
  Cecidomyiidae
Hymenoptera
  Cynipidae
  Eurytomidae
  Tenthredinidae

TABLE 10.1 (*continued*)

Flower feeders

Flower tissue feeders[2]
Coleoptera (Polyphaga)
  Anthribidae
  Buprestidae
  Cantharidae
  Cerambycidae
  Chrysomelidae
  Curculionidae
  Elateridae
  Meloidae
  Melyridae
  Nitidulidae (some species;
    both adults and larvae)
  Scarabaeidae
Lepidoptera
  Lycaenidae (some)[4]

Pollen feeders
Collembola (some)
Blattodea (some)
Dermaptera (some)
Plecoptera (some)[2]
Hemiptera (Heteroptera)
  Anthocoridae

Miridae
Thysanoptera (some)
Coleoptera (Polyphaga)
  Cephaloidae[2]
  Meloidae[2]
  Melyridae[2]
  Mordellidae[2]
  Nitidulidae (some)
  Oedemerdae[2]
  Ostomidae[2]
  Phalacridae[2]
  Tenebrionidae
    (especially Alleculinae)[2]
Diptera (probably many)[2]
  Anthomyiidae
  Bibionidae
  Bombyliidae
  Calliphoridae
  Mycetophilidae
  Muscidae
  Scatopsidae

Syrphidae
Tachinidae
Lepidoptera[2]
  Micropterygidae
  Nymphalidae *(Heliconius)*[5]
Hymenoptera
  Apoidea[6]
  Vespidae (Masarinae)[6]
  Xyelidae[4]

Nectar feeders[2]
Neuroptera
  Chrysopidae
Mecoptera
  Panorpidae *(Panorpa)*
Diptera (many)
Coleoptera
  Meloidae (Nemognathini)
Lepidoptera (most)
Trichoptera (some)
Hymenoptera (most)[6]

Internal feeders in fruits and/or seeds on living plants[3]

Coleoptera (Polyphaga)
  Bruchidae[4]
  Byturidae[4]
  Curculionidae[4]
  Nitidulidae (some)[4]
Diptera[4]
  Cecidomyiidae *(Contarinia)*

Tephritidae *(Ceratitis, Rhagoletis)*
Lepidoptera[4]
  Gelechiidae *(Pectinophora)*
  Incurvariidae *(Tegeticula)*
  Lycaenidae *(Strymon)*
  Noctuidae *(Heliothis, Helicoverpa)*

Nolidae *(Celama)*
Pyralidae *(Ostrinia)*
Tortricidae *(Cydia)*
Hymenoptera[4]
  Agaonidae *(Blastophaga)*
  Eurytomidae *(Bruchophagus)*

---

[1]This list of taxa is intended to be representative, not exhaustive. The listing of a family without qualification does not necessarily mean that all species in the family have the same habits. Unless otherwise indicated, both the immature and adult stages are believed to have the same food habits. Well-known or exceptional taxa are given in parentheses.

[2]Feeding is mainly by adults.

[3]Feeding is mainly on seeds.

[4]Feeding is mainly by larvae.

[5]*Heliconius* adults ingest nutrients from pollen that has been dissolved in nectar.

[6]Adult females of Apoidea and Vespidae (Masarinae) also store pollen and nectar in their nests as food for their larvae.

cial to both plants and insects. It is advantageous for a plant to attract flower-constant visitors because they are the most effective cross-pollinators. It is advantageous to insects to become temporary or permanent specialists because by learning to recognize a given flower and operate its floral mechanism, they reduce competition for food and also forage more efficiently.

Mutual Adaptations of Flowers and Their Insect Visitors

In a general way, the flower size, shape, position of reproductive parts, color patterns, odor, and nectar composition, as well as the time of flowering, can be matched with the sizes, anatomies, diets, sensory physiology, rhythmic activity, and foraging behaviors of the appropriate pollinators. Even among related species of plants, different species may depend on quite different kinds of pollinator. This specificity attracts effective pollinators and tends to reduce losses of pollen and nectar to nonpollinating visitors. The reward given to pollinators is thereby more closely regulated to promote cross-pollination.

Commonly nectar is situated deep within a floral tube so that casual visitors are unable to reach it. Elongation of the mouthparts into a sucking tube is a frequent adaptation among specialized flower-visiting insects. The elongation is achieved in many ways and often independently (Fig. 10.1). Curiously, the Hemiptera, with their long sucking beaks, never became regular flower visitors for nectar.

Apoidea as Flower Visitors
The largest group of efficient pollinators are the bees, or Apoidea. About 30,000 species are known in the world. Most species are solitary in behavior: that is, the females make and provision their own nests. Bees evolved from visually hunting, predaceous wasps that frequently visited flowers for nectar. Both sexes of bees take nectar as flight fuel. Females eat pollen as a source of the protein used in producing their eggs. Most females, except the cleptoparasites and social queens, also collect pollen and nectar for their larvae. This food is stored in the nest as provisions, or, in certain social species, the pollen is fed directly to the larvae. Some bee-pollinated plants, especially in the Neotropics, offer energy-rich fatty oils in addition to, or instead of, pollen and nectar.

The adaptations of bees that are associated with flower visitation are plumose or featherlike hairs, special pollen-transporting devices, modifications of the tongue, or glossa, for extracting nectar, and a diet of pollen and nectar. Correlated with their social behavior, honey bees have a highly developed system for foraging on flowers that involves communication (Chap. 6). The species of bees that visit oil flowers have special brushes to collect and transport lipids.

A few bee species are monolectic, but most bees are oligolectic or polylectic. Among the latter are honey bees, which visit an enormous variety of flowers, including those designed to attract insects other than bees. During times of food scarcity, honey bees collect honeydew or fruit juices as nectar substitutes. So compelling is their foraging behavior that when pollen is not available, honey bees have been observed to collect flour or even inert dust.

Certain primitive bees, such as *Hylaeus* (Colletidae), are relatively hairless, like wasps. They eat pollen and later regurgitate it with nectar while preparing the nest provisions. But most bees have abundant, plumose hairs that retain pollen grains brushed on the body during flower visits. Female bees methodically groom themselves and pack the pollen into special devices for transport to the nest. These receptacles are of two kinds: (1) pollen brushes, or **scopae,** of long, dense hairs on the hind legs of most bees or on the underside of the abdomen in Megachilidae, and (2) pollen baskets, or **corbiculae,** which are created by a circle of stiff hairs on the outer surface of the hind tibiae of honey bees (Color photo 5b), bumble bees, and their relatives in the family Apidae. Andrenidae, Colletidae, and Halictidae have short tongues suited to take nectar from exposed nectaries or flowers the bee can bodily enter. These are considered to be less specialized than the long-tongued

Megachilidae, Anthophoridae, and Apidae, which can reach nectar hidden in the inner recesses of specialized flowers. The expandable crop is used to carry nectar back to the nest and also functions on outbound flights as a fuel tank.

"Bee flowers" characteristically open at certain times during the day, emitting sweet or aromatic odors and presenting their pollen and nectar. Petals of bright blue, purple, yellow, and other colors within the bee's range of color vision are common. Red, which is invisible to bees, is an uncommon color of bee flowers. Patterns reflecting ultraviolet light are seen by the bee but are invisible to us. Separate petals that create a broken outline are suited to detection as a mosaic image by the bee's compound eyes. The two-lipped form of certain bee flowers, such as those of legumes or mints, provides a landing platform. This places the bee in a position favoring access to the food and the plant's reproductive structures. Distinctive stripes or spots serve as nectar guides, which orient the bee to the food.

Diptera and Lepidoptera as Flower Visitors

Two other large groups of pollinating insects show conspicuous modification for flower visiting: certain Diptera and most Lepidoptera (Color photo 5a). In both orders, the special modifications are mainly elongation of the mouthparts to reach hidden nectar. Although the mouthparts of flies are variously modified for bloodsucking or sponging, the general ability seems to be retained to ingest nectar and small particles such as pollen grains. Certain species in each of several families of flies have developed exceptionally long mouthparts for probing deep flowers. Some species with long mouthparts are found in Bombyliidae, Apioceridae, Nemestrinidae, Acroceridae, and Tabanidae (Fig. 10.1e). Shorter, but distinctly specialized mouthparts are also seen on species of *Rhingia* (Syrphidae), Conopidae, and Tachinidae. The long-tongued flies visit flowers of the same kinds visited by bees.

The so-called "fly flowers" are mainly adapted to less specialized, short-tongued insects that nor-mally feed on fluids from dead animals, feces, or plant juices. The flowers depend more on odor than their appearance to attract these insects. The shallow flowers are often white or dull-colored; the nectar is exposed; and the smell is often musty or rank.

The mouthparts of butterflies and moths are suited only to sucking fluids. Most species take nectar, but some moths do not feed at all as adults. The longest tongue of any insect is the 22.5 cm proboscis of the Madagascar hawk moth, *Xanthopan morgani praedicta* (Sphingidae). The moth is apparently the sole pollinator of the orchid *Angraecum sesquipedale,* a plant with nectar situated in an equally long tube and accessible only to this moth.

Flowers attractive to butterflies open during the day, have sweet odors, and often have nectar at the base of a deep tube. The flowers are erect and have a horizontal surface for landing. Red is visible to butterflies and is a common color of "butterfly flowers," such as carnations or the butterfly weed, *Asclepias tuberosa* (plant family Asclepiadaceae). "Moth flowers" bloom in the evening or night; most of them have heavy, sweet scents; and they are often a highly visible pale or white color. Hawk moths (Sphingidae) characteristically hover in front of a flower and take nectar by extending their long proboscis. Flowers visited by hawk moths are horizontal or drooping, with the reproductive parts situated to contact the hovering moth.

Coleoptera as Flower Visitors

The Coleoptera that visit flowers are active fliers that frequent open, sunny places, in contrast to their terrestrial, cryptic relatives. Beetles tend to linger in flowers, feeding on pollen and flower parts with their powerful mouthparts, and sometimes taking nectar (Color photo 5c). Elongation of the mouthparts has occurred in only a few cases. The heads of the cerambycids *Strangalia* sp. and *Cyphonotida laevicollis* (Fig. 10.2b) are prognathous, with the anterior portions somewhat elongated. The most spectacular adaptations of this sort in the Coleoptera are the greatly elongated galeae of *Nemognatha* (Meloidae), which form a sucking tube (Fig. 10.2a).

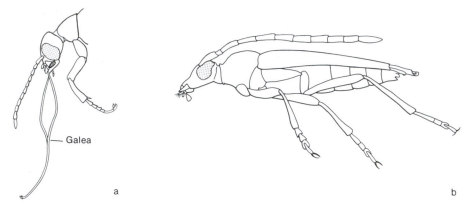

Figure 10.2   Adaptations of flower-visiting Coleoptera: **a**, meloid beetle, *Nemognatha* sp., with elongate galeae 7.5 mm long; **b**, cerambycid beetle, *Cyphonotida laevicollis*, with prognathous elongate head.

Other Species as Flower Visitors

Our final examples of insect pollinators are those in which pollen and nectar are not the prime incentives for the insects. We discuss bees, wasps, and moths that seek out other features of certain plant species.

Orchids and Hymenoptera

Pseudocopulation involves certain orchid species, such as the Mediterranean *Ophrys,* that have flowers somewhat similar to the shape of a female bee and odors that attract male bees. Depending on the species of orchid, the plant emits a chemical that is the same as the sex pheromone produced by females of a certain species of bee. Male bees of the same species are attracted by the odor and attempt to copulate with the flower. In the course of this behavior the males pick up the adhesive pollinium (a packet of pollen) from the orchid. As they move from flower to flower they pollinate the orchid. Similarly, in Australia, *Cryptostylis* orchids attract male ichneumonid wasps.

Another improbable relationship is that of the orchid-visiting male euglossine bees of the American Tropics. Males of *Euglossa* (Fig. 10.1b) and *Eulaema* are spectacular bees with bright metallic green or blue colors. They are extremely fast flyers and travel long distances in the forest. Nectar is taken from various flowers as flight fuel, but orchids are visited to obtain odors. The floral odors of each orchid species are created by blends of volatile compounds such as cineole, benzyl acetate, eugenol, methyl salicylate, methyl cinnamate, skatole, and vanillin. Males of each species of bee are attracted to certain odor blends. On arrival at an orchid, the males secrete lipids from labial glands onto the odor-producing surfaces and pick up the mixture with brushes on their forelegs. The lipids and fragrances are then transferred to special cavities in their large hind tibiae. During their contact with the orchid, the bees fall into a trap device in the flower. In the course of their escape, a pollinium is attached to a specific place on their body. At their next visit to an orchid of the same species, the pollinium is removed by the orchid, thus achieving pollination. Males are also attracted to chemicals in rotting wood. The bees' use of the odors is not clear. Investigators have speculated that the chemicals are metabolized by the males or converted to attractants for either males or females or both sexes.

Figs and Fig Wasps

Another set of examples of plants and their pollinators illustrates obligate mutualism, in which host specificity is compounded by the specificity of the larvae that feed on the plant tissues: the fig wasps and yucca moths lay their eggs in the ovaries of their hosts. Initially they probably were only accidental pollinators. Seed production, however, is advantageous to both the plant and the seed-infesting insects. The plants and their pollinators now are completely interdependent, but, as we will see, open to exploitation by closely related insects.

The largest group of highly specialized pollinators are the tiny wasps of the family Agaonidae, which pollinate figs. The genus *Ficus* (plant family Moraceae), found in tropical regions, includes about 900 species. Virtually every species of wasp is confined to a single species of fig. The fig

"fruit," or synconium, that we eat, is actually an inflorescence composed of an enlarged, cup-shaped receptacle that encloses many small, unisexual florets. The small opening of the synconium is called the ostiole and is guarded by scales to prevent entry of nonpollinating insects.

Pollination of the wild caprifig, *Ficus carica,* by the wasp *Blastophaga psenes* takes place as follows (Fig. 10.3). The plant has florets of two kinds: pollen-producing male florets (staminate flowers), located near the ostiole, and seed-producing female florets with short styles (pistillate flowers), located at the base of the synconium. Each wasp larva feeds inside a pistillate floret, causing a tiny gall and destroying the seed. At

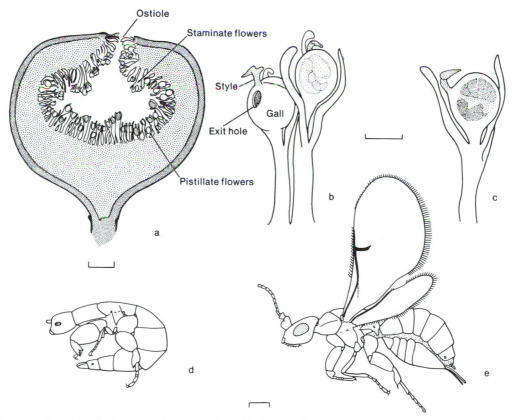

Figure 10.3 Pollination of caprifig by *Blastophaga* wasps (Agaonidae): **a**, caprifig at time of emergence of wasps (scale equals 5 mm); **b**, gall flower with male wasp inside and another with exit hole; **c**, gall flower with female wasp inside (scale for b and c equals 1 mm); **d**, male of *Blastophaga;* **e**, female of *Blastophaga* (scale for d and e equals 0.25 mm).

maturity, the adult males are flightless and have reduced legs, eyes, and antennae (Fig. 10.3d). They chew their way out of their galls first (Fig. 10.3b) and seek the galls containing adult female wasps. The male bores a hole in the gall, copulates with the female by means of an extensible abdomen, and then dies. The inseminated female emerges from her gall (Fig. 10.3c,e) and makes her way toward the ostiole. She passes through the staminate florets and becomes covered with pollen. Some of the pollen enters folds of the pleural intersegmental membranes between the terga and sterna of the abdomen. The female's body shrinks slightly, probably because of loss of water, and pollen in the folds is trapped. She leaves the fig, flies to another fig in the proper state of development, and attempts to enter it. The process of squeezing through the ostiole pulls off her wings and antennal flagella (Fig. 10.3a). Once inside, she penetrates deeply in the fig to reach the female florets. There she inserts her ovipositor through the short style to the ovarian region, where she lays an egg. In the moist fig, her body is rehydrated and swells, releasing the trapped pollen from the membranous folds of her abdomen. As she moves about, pollen is dusted on the female stigmas. After laying her eggs, she dies, still in the fig.

The fruit of the caprifig is generally not edible, but agricultural varieties have been selected that are edible. When figs were introduced in California in the late 1800s, the Smyrna variety failed to produce fruit. An enterprising grower traveled to Turkey to learn the secrets of successful fig culture. After some difficulties, including an epidemic of plague, he returned with the knowledge that special pollination is necessary. The Smyrna variety has only female florets with long styles. The wasps and suitable pollen must be obtained from the caprifig, in which the normal cycle can be completed. Fruits of the caprifig, containing pollen-laden females of *Blastophaga,* are hung in perforated bags among the limbs of the Smyrna fig trees. The *Blastophaga* emerge and enter the Smyrna figs. They are unable to lay eggs because their ovipositor is too short for the long-styled florets. The fig is nevertheless pollinated and normally ripens with seeds into an edible fruit. Some varieties of figs do not require pollination but, if pollinated, yield larger fruits with better flavor.

The behavior of *Blastophaga* inside the caprifig is considered primitive in comparison to most agaonid wasps. In other species the females have special containers in which pollen is packed and later actively placed on the stigmas. Some species of agaonid wasps have evolved a "cuckoo" or "cheater" relationship, ovipositing in the flowers without pollinating the stigmas.

Yucca and Yucca Moths

Another group of host-specific pollinators are the yucca moths, *Tegeticula* and *Parategeticula* sp. (Incurvariidae). All species of *Yucca* (plant family Liliaceae) are American in origin, but they have been introduced elsewhere in the world. More than two dozen species of *Yucca* east of the Rockies and the Mojave Desert are pollinated by what was considered one moth species, *T. yuccasella* (Riley, 1892). In the West, *Yucca brevifolia* is pollinated by *T. synthetica, Y. whipplei* by *T. maculata* (Powell and Mackie, 1966), and *Y. schottii* and *Y. elephantipes* by *Parategeticula pollenifera,* as well as *Tegeticula yuccasella* (Davis, 1967; Powell, 1984).

The eastern moths are active at night, the time at which flower scent is also strongest. The female *T. yuccasella* enters the white flower, climbs up the stamens to the anthers, and gathers the pollen in a ball (Fig. 10.4a). The pollen mass from one to four anthers is carried under her head, clasped by a prehensile elongation of the maxillary palpi and the bases of the forelegs. She then flies to another flower in the proper state of ovarian development. After inspecting the ovary, she bores in with her sclerotized, elongate ovipositor and lays an egg (Fig. 10.4b). Climbing the style, she packs some pollen on the stigma. This behavior is often repeated after each egg is laid. On the average, one egg is inserted into each of three compartments of the ovary. A few of the many seeds in pollinated flowers serve as food for the moth larvae. Unpollinated flowers do not develop seeds. When fully fed, the larvae leave the seedpod (Fig.

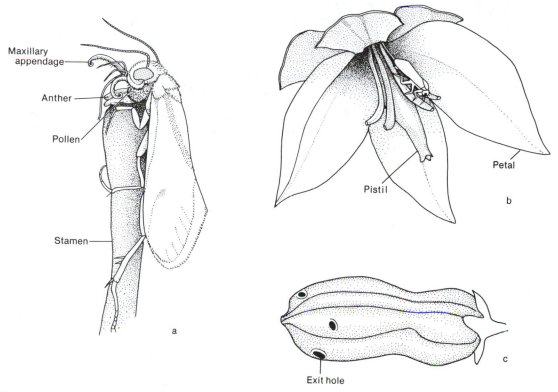

Maxillary
appendage

Anther

Pollen

Stamen

Petal

Pistil

b

a

Exit hole

c

Figure 10.4   Pollination of *Yucca* sp. by *Tegeticula* moths (Incurvariidae): **a**, female of *Tegeticula yuccasella* collecting pollen from anther; **b**, position of female moth during oviposition in ovary of *Yucca*; **c**, mature pod of *Yucca* showing emergence holes. (Redrawn from Riley, 1892.)

10.4c) and pupate in the ground. Emergence of the adults is timed to coincide with the flowering season.

Only the genera *Tegeticula* and *Parategeticula* include species with the specialized maxillary palps that are associated with pollination. Related moths of the genus *Prodoxus* are "cheaters" that do not pollinate their host. Females lay eggs in the inflorescence, and the larvae feed on sterile tissues of the fruit, scape, or leaves. Recently, certain populations of *Tegeticula yuccasella* were discovered to be distinct species that lacked functional palps for pollination and were cheaters on yuccas pollinated by their relatives. In this case, the cheaters feed on seeds, causing a heavy cost to the host plant and destabilizing the mutualistic relationship.

PHYTOPHAGOUS INSECTS AND THEIR MODES OF FEEDING

External and Internal Feeding

An adequate classification of phytophagous feeding habits would be overly complex for our purposes. Here we shall divide insects by their taxonomic groupings and by their general mode of feeding (Table 10.1).

The mouthparts and other characteristics of the immature and adult insects should be kept in mind when considering each group and its food habits. Phytophagous insects can be broadly divided into **external feeders,** whose body is outside the plant (Fig. 10.5; Color photos 2b, 2c, 5d, 5e), and **internal feeders,** those whose body is

Figure 10.5 Phytophagous exopterygotes: **a**, cicadellid leafhopper, *Graphocephala atropunctata;* **b**, pseudococcid mealybug, *Pseudococcus fragilis*. (Photos by H. V. Daly.)

physically inside the plant (Fig. 10.6b; Color photo 5f). The first major category, the external feeders, include a wide variety of insects, both exopterygotes and endopterygotes. During the life history of some species, especially small moths, the larvae may feed first inside, then outside.

External feeders may be freely visible to predators and parasitoids. Such **exposed feeders** are usually protectively colored and patterned if they are large enough to be edible by vertebrates (Color photos 1c, 1d, 2c). The manner of feeding is characteristic of the species. Many caterpillars and sawfly larvae feed along the leaf edge. Surface feeders with chewing mouthparts may ingest whole pieces of leaf, leaving holes or notches. Others remove most of the photosynthetic tissues, leaving a delicate skeletonized vascular network

(Color photo 5e). The cellulose-rich feces are called **frass.** Exposed feeders simply drop frass from their feeding stations.

The feeding injuries of leaf-feeding Hemiptera are often recognized as small, discolored spots where the inner palisade and spongy mesophyll cells are broken down and emptied. The insect's saliva sometimes has physiological effects on the leaf that are also injurious. Thysanoptera also remove the contents of the inner leaf cells, resulting in silvery air spaces inside the leaf. Damaged leaves and stems may wither, discolor, or fall prematurely.

Some external feeders gain protection by feeding on roots or within enfolding leaves. Those that are able to spin silk may bind together leaves in a sheltering cluster. The **leafrollers** are caterpillars

Figure 10.6 Phytophagous endopterygotes on apple: **a**, leaf mine of *Nepticula* sp. (Nepticulidae) on leaf at left, caterpillar of *Halisidota maculata* (Arctiidae) at right; **b**, gallery of codling moth larva, *Cydia pomonella* (Tortricidae). (Photos by H. V. Daly.)

and sawfly larvae that roll or fold leaves to create a tube. The larvae hide in the tube, feeding at its edge or on nearby leaves. Some moth species pupate in the roll. Adult attelabid weevils cut the leaf bearing their egg and the leaf forms a roll naturally. The weevil larva feeds inside. **Casebearers** attach bits of plants or frass in a tough silk bag and hide inside for protection.

The second major category, the internal feeders, are completely surrounded by living tissues. They are exclusively endopterygotes and usually the larval stages of the life history. **Leafminers** are small larvae that eat some or all of the mesophyll tissues between the outer layers of leaf blades or needles (Color photo 5f). No adult insects are leafminers. Fossil leafmines have been recovered from the Mid-Cretaceous Period at 97 million years ago.

Female insects select the host and lay their eggs on the leaf surface, or inserted in the mesophyll, or nearby on the plant. Usually mature leaves are mined. The larvae feed in a manner characteristic of the species, leaving a "signature" that is sometimes beautifully serpentine. Larvae of Diptera are cylindrical but soft-bodied and adaptable to the restricted mine. The other larval miners are highly flattened, usually colorless, legless, and with a flattened, wedgelike head that slopes forward. Linear mines begin as a tiny channel and progressively widen as the larva grows in size and appetite (Fig. 10.6a). The direction of the mine may be altered when vascular bundles are encountered. Blotch mines are created when the larva feeds in various directions, eating both vascular and mesophyll tissues. The frass is often deposited in the mine or ejected through an opening. Lepidoptera may lay frass inside in a continuous line (Color photo 5f) or pack it at one end or plaster it randomly on the mine floor, while Diptera lay two rows. Hymenoptera scatter the frass about the cavity or in piles. Some miners pupate in the mine, while other species leave to form a cocoon elsewhere.

Plant **borers** are insects that burrow in living or dead plant tissues other than leaves. The xylophagous scavengers such as long-horned beetles (Cerambycidae) are borers in dead plants. Their life span is usually long because of the lower nutrient value of the food, the variable and sometimes low moisture content, and the stable environment. Most plant-boring larvae tunnel into living buds, stems, roots, fruits (Fig. 10.6b), seeds, nuts, or grains. Life spans of these borers may be short, not only because the food value and moisture are higher, but also because the living plant tissues change rapidly over time.

Although this is a heterogeneous group, the larvae share several similarities arising from their sheltered mode of life. The bodies are often cylindrical, pale, legless, and with bumps or rough areas of skin that provide traction against the burrow walls. The mouthparts, mounted on a retractable head capsule, excavate tissue bit by bit. Antennae are short and retractable into grooves. These specializations give borers access to food rich in nutrients: newly developing leaves, cambium tissue, the flesh of fruits, and endosperm of seeds. Frass is often packed behind the larva as it advances.

Galls

Certain secretions of insects and mites stimulate abnormal development of growing plant tissue, resulting in misshapen leaves or in swellings called **galls.** These are usually initiated in the spring and early summer, when the meristematic tissue is active. Aphids feeding on the underside of a new leaf can cause the leaf to curl around them and enclose a favorably high humidity. Homoptera, Thysanoptera, and mites create galls by their feeding secretions. The gall is typically a hollow cavity that may remain open to the exterior and consequently is called an **open gall.** A group of insects or mites or several generations may live inside the gall, feeding and taking shelter.

The galls of endopterygote insects are usually, but not always, occupied by a single larva. Either the ovipositing female or the larva or both seem to be responsible for the biochemical stimulants. The galls do not have a permanent opening and are called **closed galls.** The host plant responds with a growth of specific color and design such that the insect can be identified by the gall it makes. The larva is surrounded by moisture and food and is largely protected from natural enemies. Some wasp parasitoids, however, are able to lay their eggs in or near the gall-making larvae so that insects of several kinds may emerge from a single gall. Closed galls are made by species in the orders Coleoptera, Lepidoptera, and especially Diptera and Hymenoptera (Fig. 10.7).

HOST SPECIFICITY OF PHYTOPHAGOUS INSECTS

Some exposed feeders accept a great variety of plants as food. The larvae of the gypsy moth, *Lymantria dispar* (Lymantriidae; Fig. 44.13i), are known to feed on 458 species of plants in the United States (Leonard, 1974). Insects that feed on many species of plants from different families are called **polyphagous. Oligophagous** insects feed on fewer species of plants, often related to one another as genera in a family or having certain similar biochemical constituents. **Monophagous** insects feed on one or more species in a single plant genus. Insects with the more restricted range of hosts are usually the leafminers, borers, and gall makers, which are surrounded by plant tissues, or the insects that have adapted to toxic plants. The early plant-feeding Neoptera are assumed to have been polyphagous, exposed feeders, now exemplified by Orthoptera. The oligophagous and monophagous taxa evolved as specialists, with occasional reversions to polyphagous habits (Dethier, 1954).

RESISTANCE OF PLANTS TO INSECT ATTACK

What factors favor the evolution of host specificity? To answer this question, we must consider in detail the nature of insect/plant relationships. Part of the explanation lies, on the one hand, in the

Figure 10.7 Cynipid gall. **a,** female gall wasp, *Andricus* sp. (Cynipidae), selects live oak for oviposition; **b,** appearance of a mature gall of the summer generation of wasps; **c,** wasp larvae develop inside the gall.

defense strategies of plants against attack. We have seen that all parts of a plant may be eaten: roots, stems, leaves, sap, flower parts, fruits, and seeds. Plant growth is affected by loss of food when photosynthetic tissue is eaten or phloem sap is drained. Damage to roots and xylem deprives the plant of water and minerals. Chemicals in an insect's saliva or secreted by the ovipositor may have general or specific effects on the host's physiology. Injured cambium or meristematic tissue results in abnormal growth. These influences lead to stunting, lowered production of seeds, or death. Damage to flowers, fruits, and seeds directly affects the population growth of plants. Phytophagous insects also transmit viruses and microorganisms that cause plant diseases. Yet, despite all the hungry insects, plants still seem to flourish.

In the stepwise coevolution of insects and plants, the plant evolves a protective measure; then insects evolve effective means to overcome the plant's defense, which in turn selects for new plant defenses, and so on. In the context of this reciprocal interaction, a plant that is less damaged because of heritable characteristics is said to be **resistant.** Three basic kinds of resistance have been identified by Painter (1958):

1. Resistant plants may be passed over by an insect seeking an oviposition site, shelter, or food. That is, either some chemical or physical feature of the plant is lacking, so that the plant fails to attract insects, or some feature is present that is repellent. This condition is called **antixenosis** (formerly termed nonpreference).

2. Resistant plants may adversely affect the biology of the insect. Physical or chemical properties of the plant may result in the insect's early death, abnormal development, decreased fecundity, or other deleterious conditions. For example, toxins, repellents, copious sap or pitch, or tissue that is nutritionally inadequate for insects could reduce or eliminate attack. This kind of resistance is called **antibiosis.**

3. Resistant plants may be **tolerant** and survive even when infested by insects at levels that kill or injure susceptible plants. Rapid compensa-

tion for or replacement of lost parts, excess production of seeds, rapid wound healing, and detoxification of insect salivary toxins are some of the possible characteristics that enable plants to tolerate damage.

The antixenosis and antibiosis properties prevent or reduce attack. These properties vary (as do other genetic traits) from plant species to species, geographically with the ranges of species, locally depending on ecology, and from individual to individual. For example, the oleoresins or pitch of conifers traps, expels, or is toxic to boring insects in needles or bark. The variation in physical and chemical composition of oleoresins gives almost every tree a unique chemical individuality (Hanover, 1975). For example, individual trees have been shown to vary greatly in susceptibility to scale insects.

Angiosperms have so-called **secondary metabolic products,** or metabolites such as essential oils, amines, alkaloids, terpenoids, phenolics, and glycosides whose primary function seems to be in chemical defense, although some may be otherwise integrated into the plant's metabolism (Ehrlich and Raven, 1967; Schoonhoven, 1972; Whittaker and Feeny, 1971). These chemicals, incidentally, also give us spices and flavorings. Secondary metabolites are another example of allelochemicals (Chap. 6). The toxic, biochemical shield partially protects flowering plants against herbivores, both insect and vertebrate. The chemicals may be localized in the plant to deter attack on vulnerable tissues. For example, the pyrethrum plant, *Chrysanthemum cinerariaefolium* (plant family Asteraceae or Compositae), has insecticidal pyrethrins concentrated in the seeds of the flowers (Chap. 14).

Insects have evolved a variety of physiological countermeasures to detoxify plant secondary compounds. For example, microsomal oxidases, esterases, glutathion transferases, exposide hydrolases, and other catabolic enzymes break down toxins into harmless products. One hypothesis is that herbivorous insect species develop resistance to man-made insecticides at such a high rate because herbivores already have adaptations such as catabolic enzymes to overcome plant defensive chemicals (Gordon, 1961). Changes in some of the same enzyme systems that insects use to detoxify plant chemicals also are responsible for evolved resistance to insecticides. As evidence of yet another step in coevolution, the pyrethrum plant mentioned above also has the compound sesamin, which is an inhibitor of insect oxidases.

Chemical substances in plants that reduce or prevent feeding are called **antifeeding compounds.** The substances vary greatly in chemistry and include many secondary metabolites. These operate as allomones; that is, they are beneficial to the plant or emitter. Injury to plant cells caused by insect feeding is believed to promptly activate production of certain antifeeding compounds at the site or even some distance from the site. Of special interest here are the **protease inhibitors** that are found in leaves and especially in seeds and tubers, like the potato. These polypeptide or phenolic compounds in plants are specific for common protease enzymes of insects. The inhibitors function to prevent digestion, such that the insect's gut becomes full and the nervous feedback from the distended gut stops further feeding. Protease inhibitors may function in concert with other allelochemicals to resist insect attack.

Alkaloids in some grasses that deter feeding in insects come from an unexpected source: endophytic fungi that live within the plant. For example, the endophytic fungus *Acremonium lolii* grows intercellularly in rye (*Lolium* sp.) and produces insect-repellent alkaloids. The toxins cause poisoning in sheep and cattle but also prevent damage to the grasses by the Argentine stem weevil, *Listronotus bonariensis,* and other insects. By cultivation and selection of endophyte strains, plants can be reinfected with fungal strains that repel serious insect pests but are not toxic to cattle. The endophyte toxins are concentrated in the basal leaf sheath, where insect feeding would be most damaging. The endophyte-plant-insect relationship involves a three-way coevolution.

Plant defensive compounds also include **insect growth regulators.** A wide variety of organisms, including plants, have been found to contain compounds exerting the same physiological effects of some of the developmental hormones of insects (Chap. 4). Extracts with juvenile hormone activity have been isolated from vertebrates, bacteria, yeast, protozoa, and plants, especially the balsam fir. These are called juvenile hormone mimics or **juvenoids.** Both ecdysone and 20-hydroxyecdysone, as well as over 65 compounds with similar structure, have been isolated from 111 plant families. Ferns and yews are the major sources of the active substances, called **phytoecdysteroids.**

The significance of the hormonelike compounds in plants is not fully clear; evidence is inconsistent that the natural compounds disrupt normal growth of insects. In the case of the balsam fir, paper towels made from the tree contain the active factor called juvabione. European linden bugs, *Pyrrhocoris apterus* (Pyrrhocoridae), fail to mature when reared in cages lined with the towels. The bugleweed of Kenya, *Ajuga remota* (plant family Lamiaceae or Labiatae), has phytoecdysteroids that interfere with normal metamorphosis. Ferns are relatively free of insect attack, but mulberry leaves, also rich in ecdysone, are eaten by silkworms without effect. Only trace amounts of hormone are present in some plants, causing no harm, and some insects are known to break down the chemicals when ingested.

Growth regulators of a different kind, called juvenile hormone antagonists, have been extracted from *Ageratum houstonianum* (plant family Asteraceae or Compositae). Named precocene I and II, the compounds cause atrophy of the corpora allata and precocious metamorphosis in hemimetabolous insects. Our last example is the chitin inhibitor plumbagin, which has been found in *Plumbago* (plant family Plumbaginaceae). Insects die at ecdysis because the cuticle lacks mechanical strength. Investigation of natural insect growth regulators from plants has led to the development of several novel pesticides.

Plant-feeding insects must compete with other herbivores for food. The resistance factors evolved by plants further limit the number of kinds of plant available to eat. Insects that are able to survive the antibiotic properties and develop a preference for a resistant plant will acquire not only food but also some relief from interspecific competition. Those able to bore into plant tissues must tolerate immersion in the host's chemical environment and physical constraints; but here also are food and relief from competition, plus escape from certain enemies and protection against desiccation, freezing, and other environmental hazards. Some insects store toxic plant substances and thereby acquire a defense for themselves against vertebrate predators (Rothschild, 1973; Chap. 12). Thus a number of advantages accrue to the monophagous or oligophagous insect that is adapted to the special conditions of life associated with plant hosts of one or a few kinds.

In specializing on plants toxic to other competing species, insects resistant to the plant's chemicals use the antibiotic chemicals as cues to locate and identify suitable hosts. For example, plants of the cucumber family (Cucurbitaceae) possess a variety of terpenoid compounds called cucurbitacins. Various species of chrysomelid beetles have specialized on certain cucurbit species and are not harmed by cucurbitacins. The chemicals even serve as cues to find the hosts and stimulate feeding.

The defensive mechanisms of plants and the host specificity of insects have evolved in response to essentially two periods of contact (Beck, 1965): (1) the period of oviposition, during which the female seeks suitable host plants on which to lay her eggs, and (2) the period of feeding, during which the immature or adult insects eat the plant. Oviposition behavior involves detection and orientation to a host plant at a distance, the search for specific sites in the plant, and, finally, the deposition of eggs, often followed by dispersal. The behavior is a series of complex events that involve the insect's sensory receptors and integration of physiological responses. Any heritable physical or chemical feature of the plant that reduces the numbers of eggs, and thus ultimately the number of feeding insects, will be selectively favored. This is resistance of the antixenosis type and is achieved either by deterring or failing to attract ovipositing females.

FEEDING BEHAVIOR OF PHYTOPHAGOUS INSECTS

Beck (1965) identified four steps in the feeding behavior of insects on plants: (1) host recognition and orientation; (2) initial biting or piercing of the plant; (3) maintenance of feeding; and (4) cessation of feeding, usually followed by dispersal. The behavioral response at each step depends on releasing stimuli provided by the plant and on the insect's response thresholds, which vary with its physiological state. A plant that lacked the appropriate releasing stimuli or that discouraged feeding at some step would present resistance of the antixenosis type.

Physical and chemical stimuli have been classified according to the response they elicit from insects (Beck, 1965; Dethier et al., 1960). During orientation to a host plant at a distance, certain stimuli, like volatile chemicals, may act positively as **attractants** or negatively as **repellents.** When in close contact with the plant, positive stimuli may stop further locomotion (i.e., be **arrestants**), or act as repellents to hasten the insect's departure. At the initiation of feeding, positive stimuli are called **incitants** and negative stimuli are **suppressants.** Feeding is maintained by **phagostimulants** and may be terminated by high concentrations of antifeeding compounds, also known as feeding deterrents.

The sensory apparatus and orientation behavior of insects are finely tuned to the characteristics of the desired host plants. The sensory receptors on the antennae and maxillae of caterpillars are chiefly involved in discrimination. When such receptors are experimentally removed, oligophagous caterpillars often accept as food a broader range of plant species. Although the female usually selects the host plant when she lays her eggs, the caterpillar must select the parts of the plant to eat, avoiding concentrations of toxins and finding the richest food.

Secondary plant substances that, as allomones, are repellent to most insects are, in fact, often the feeding stimulants for the appropriate monophagous insects. The same compound can be an attractant or feeding incitant, thus serving as a kairomone and benefiting the perceiver. The attraction of certain chrysomelid beetles to cucurbitacins in cucurbits was mentioned above. The terpenoid gossypol, produced by *Gossypium* or cotton plants, serves as a repellent to the noctuid moths *Helicoverpa zea,* and *Heliothis virescens* and to the meloid beetle *Epicauta,* but is an incitant for the boll weevil, *Anthonomus grandis.* For insects deciding to feed on a plant, the absence of antifeedants is apparently more important than the presence of feeding incitants. Young needles of pine with antifeedants are not consumed by sawfly larvae, but older needles, with reduced concentrations of antifeedants, are consumed. Nutrients, including sugars, amino acids, phospholipids, and ascorbic acid, can also be stimulants to certain insects.

SELECTED REFERENCES

Pollination, General

Buchmann, 1987 (oil flowers); Dafni, 1984 (mimicry); Eickwort and Ginsberg, 1980 (behavior of bees): Free, 1993 (agricultural crops); Kevan and Baker, 1983; Proctor and Yeo, 1973; Roubik, 1988 (tropical bees); Willemstein, 1987.

Pollination of Orchids by Bees

Dressler, 1982 (biology of bees); O'Toole and Raw, 1991 (general descripton with illustrations); Whitten et al., 1989 (labial glands of bees), 1993 (attraction of bees to rotting wood).

Pollination of Figs by Wasps

Galil and Neeman, 1977; Wiebes, 1979.

Pollination of Yucca by Moths

Pellmyr et al., 1996 (cheater yucca moths); Powell, 1992.

Phytophagous Insects

Barbosa and Letourneau, 1988 (novel aspects); Bernays, 1989 (interactions); Bernays and Chapman, 1994 (host plant selection); Chapman and de Boer, 1995 (regulation of feeding); Gordon, 1961 (role of nutrition in resistance to pesticides); Hespenheide, 1991 (leafminers); Juniper and Southwood, 1986 (plant surface); Labandeira et al., 1994 (fossil history); Norris, 1986 (antifeeding compounds); Powell, 1980 (larval food of microlepidoptera); Price et al., 1994 (gall insects); Rosenthal, 1986 (chemical defenses of plants); C. M. Smith, 1989 (plant resistance to insects); Strong et al., 1984 (community ecology); M. A. J. Williams, 1994 (galls).

See Also:

For additional discussion of specific taxa, consult chapters on insect orders. Insects that feed on dead plants and wood are discussed as xylophagous scavengers (Table 9.2) in Chapter 9, their nutrition in Chapter 5, and their symbionts in Chapter 13.

11
ENTOMOPHAGOUS INSECTS

Mantids wait for prey and strike quickly. This African mantid is eating a young katydid.

In this chapter we discuss insects that kill or injure one or more other invertebrates before completing their life cycle. Most of these carnivorous insects feed on other insects and are said to be **entomophagous** (Fig. 1.3c,d; Color photo 6). Snails, earthworms, spiders (Fig. 11.1a), millipedes (Fig. 11.1b), mites, and other terrestrial and freshwater invertebrates are also eaten by insects. Such animal bodies, with their high nitrogen and water contents, are among the most nutritious of foods eaten by insects. By feeding on other insects and invertebrates in their communities, entomophagous insects contribute to the control or regulation of the density of individuals in the prey popula-

tions. **Applied biological control,** discussed in Chapters 8 and 14, makes use of the regulatory action of entomophagous insects to lower the density of pest organisms below a level that causes economic damage. In this chapter we discuss the life histories, behavior, and evolution of entomophagous insects.

TYPES OF ENTOMOPHAGOUS INSECTS

For simplicity in discussing entomophagous insects, we provide several definitions: victims are called **prey** if killed directly, **hosts** if fed upon

Figure 11.1   Insect predators of other arthropods: **a**, With its sting this female *Pepsis* sp. (Pompilidae) wasp will paralyze the tarantula. The spider will serve as immobilized food for the wasp's larvae in an underground burrow. If the spider had been killed, it would soon spoil and be unsuitable. **b**, An adult, but larviform, female of the beetle, *Zarhipis integrepennis* (Phengodidae), is eating a millipede. The much smaller adult male has wings and bipectinate antennae.

in ecology to designate an organism that obtains nutrients from another organism, causing harm but not necessarily death. Unfortunately, in entomological literature a parasitoid may be called a parasite and may be said to parasitize an insect host even when the host dies. Furthermore, terms like "hyperparasite" or "superparasite" are used in reference to parasitoids. We prefer to distinguish between parasitoids and true parasites, and, further on, to define the terms hyperparasitoid and superparasitoid. Second, the behavior of sphecid wasps does not easily fit our definitions. The female wasps prepare nests that they provision with paralyzed insects as living food for their larvae. Should the victims be called hosts or prey? Wasps of some species place only one victim in a nest (Color photo 6f), while other wasps place two or more victims. Should the former be called parasitoids and the latter, predators? To avoid confusion within this single family of related wasps, we simply call all these wasps predators and their victims, prey.

Parasitoid insects are intermediate between predators and parasites in the sense that they live at first parasitically at the host's expense, but ultimately kill the host. From an ecological viewpoint, predators and parasitoids both act to eliminate individuals in the prey or host population, whereas true parasites permit the host to continue functioning in the community, though often at a suboptimal level.

while still living. The entomophages also can be defined according to their mode of feeding:

- **Predators** kill and consume more than one prey organism to reach maturity.
- **Parasitoids** require only one host to reach maturity, but ultimately kill the host.
- **True parasites** feed on one or more hosts, but do not normally kill the host.

Although these definitions will apply to most insects without confusion, some problems should be noted. First, the word "parasite" is widely used

Cleptoparasites and Other Kinds of Parasitic Behavior

Before further discussion of predators, parasitoids, and parasites, some additional kinds of behavior should be mentioned in which an insect consumes food collected at the energetic expense of the host or otherwise causes harm. Although the insect may not be directly entomophagous, the net result is a reduction in the number of host offspring and an increase in the offspring of the insect that benefits. **Cleptoparasites,** or "cuckoo" parasites, lay their eggs in the nests of other species, in the manner of cuckoo birds. The cleptoparasitic parent or the larva may kill the egg or larva of the host immediately, or the host larva may die of starvation after the cleptoparasite larva eats the nest provisions. In the former instance, the young cleptoparasites often possess enlarged mandibles suited for attack. For example, among the megachilid bees, the genus *Coelioxys* is cleptoparasitic on the related genus *Megachile,* which provisions a nest with pollen and nectar. A **social parasite** is a female that enters the nest of a social host and takes over the role of the queen. The social parasite's offspring are fed by host workers at the expense of host offspring. Females of the apid genus *Psithyrus* are social parasites of their close relatives the bumble bees, genus *Bombus.* Parasites that spend much of their life in their host's nests and consume food are known as **inquilines.**

Slavery is practiced by certain ants. Worker pupae of another species are taken by slave-making workers. The resulting adult slaves then become members of the colony and do most of the work. Members of social species may also rob one another. For example, honey bees from a strong colony may enter the nest of a weaker colony and remove stored food.

Phoresy is the transport of an insect on or physically inside the body of another insect. The insect that provides the transportation is usually not harmed. When the passenger is an adult female parasitoid or predator, however, she usually remains aboard until the host lays its eggs. The passenger then oviposits on the host's eggs. This practice occurs frequently in the wasp family Scelionidae. The first instar larvae of meloid and rhiphiphorid beetles, stylopid parasites (Strepsiptera), and eucharitid wasps are commonly phoretic on adults of their favored hosts. By this means the larvae gain entrance to the nests of the hosts.

RESISTANCE OF HOSTS OR PREY TO ATTACK

Prey insects may avoid or prevent attack by entomophagous insects by actively biting, stinging, or kicking. Enemies may be excluded by tough cocoons, hard puparia, or nests of mud or leaves, or by the inaccessible position of the insect in burrows in soil or plant tissues. Some still succumb to parasitoids and predators that are suitably equipped with strong mandibles or long, powerful ovipositors.

Exposed insects may violently resist attack or quickly escape. When approached by an enemy, an aphid may drop from the plant, walk away, or both kick and secrete oily droplets from its siphunculi. The odor of the siphuncular secretion alarms other aphids, and they drop off or walk away. Larvae and pupae of many Lepidoptera or Coleoptera thrash about when contacted. Larvae may bite, or regurgitate or secrete defensive fluids. One of the advantages of social behavior is mutual defense against insect enemies. Asian honey bees defend their colonies against vespid wasps in a remarkable behavior: a mass of worker bees covers each invader for up to 20 minutes, by which time the bees' metabolic heat has raised the wasps' temperature to a lethal 46 to 48°C.

It seems possible that some of the defensive strategies to avoid predation by vertebrates (Chap. 12) are also effective against insect enemies. Certain aphids are distasteful and are avoided by insect predators after an initial contact. The gustatory senses and learning ability of the predator are therefore important to the success of this defense. The role of protective coloration and behavior, however, is generally less effective because many insect predators, unlike vertebrate predators, do not depend on vision to find prey.

Fourteen orders of insects are known to encapsulate (see Chap. 5) foreign bodies, endoparasitic worms, and parasitoids. Among the parasitoids, larvae of certain species definitely stimulate encapsulation in hosts that are not the normal hosts: maggots of Tachinidae and some, but not all, wasp larvae of Ichneumonidae, Braconidae, Chalcidoidea (Encyrtidae, Eulophidae), Proctotrupoidea, and Cynipoidea. Yet in their normal hosts, these parasitoids usually escape encapsulation. How do they succeed?

Among wasps, some attack only eggs. By rapidly completing their growth before the host's hemocytes develop, they avoid encapsulation altogether. Rapid destruction of a larval host also allows a parasitoid to avoid encapsulation because the host dies quickly. If the parasitoid lingers for a while and allows the host larva to grow, a larger food supply becomes available, but at a greater risk of encapsulation. Certain parasitoids prevent encapsulation by inserting their eggs precisely into an organ of the host where the endoparasite will be separated from the host's hemolymph by a layer of connective tissue. Others first invade the alimentary canal and are sheltered from the host's hemocytes. When the host is suitably larger, the parasitoid feeds voraciously and so debilitates the host that it is unable to react in time. Some parasitoids are initially invested in cellular membranes, the cells of which become dissociated and enlarged, and circulate in the host's hemolymph. Apparently the cells absorb nutrients and similarly act to reduce the host's ability to muster an encapsulation.

Parasitoids in the Ichneumonoidea have viral DNA (called polydnaviruses) incorporated into their own chromosomes. The gene products of the virus disarm the ability of the host caterpillar to encapsulate the egg and the wasp larva (see Chap. 13; Whitfield, 1990).

PREDACEOUS INSECTS

Predators are usually larger in body size than other entomophagous insects. The need for more than one prey organism makes predators dependent on prey populations of higher density than other entomophagous insects. Prey are usually smaller in body size than the predator but proportional to the predator's size. In other words, larger predators take larger prey and smaller predators take smaller prey. These generalizations about relative body size have some exceptions: dragonflies are relatively large predators, yet they mainly feed on small prey. Wasps with stings (Fig. 11.1a) and ants in groups (Color photo 6e) can kill prey larger than themselves. Predators usually eat their prey immediately unless, as in some wasps, the prey

are stored for later consumption by larvae.

The same terminology used in describing the choice of foods by phytophagous insects can be applied here. Monophagous predators feed on one species of prey; oligophagous predators take several species of prey; and polyphagous predators take many species of prey. The last type are the "generalists." They take individuals of prey species largely in proportion to the relative abundance of the different prey species. Thus polyphagous predators, by continually shifting to the most abundant species of prey, tend to regulate populations of prey species in the community. Furthermore, if mobile, polyphagous predators are able to thrive in disturbed communities, where prey species vary in abundance as ecological succession proceeds. At the other extreme, monophagous predators are density-dependent on one species of prey and may regulate the prey at lower levels than polyphagous predators. Monophagous predators tend to be associated with undisturbed communities where their host maintains continuous, stable populations.

Review of Table 11.1 leads to the conclusion that a majority of predaceous taxa are carnivorous in all feeding stages. This does not mean they are always exclusively carnivorous; exact diets vary species by species. Adults of some of these visit flowers for energy-rich nectar, as do certain adults of taxa that are predaceous only as immatures. Scavenging, honeydew, symbionts, and plant foods also may supplement a predator's diet.

Some of the complexities of diet are revealed in the following example. Hagen and his associates (1970) carefully examined the diets of lacewings of the genus *Chrysopa* (Chrysopidae). Adults of about half the species have larger mandibles and are predaceous like the larvae. Adults of other species have shorter mandibles and feed only on honeydew and pollen. The foreguts of the second group contain yeast symbionts and are supplied with larger tracheal trunks for increased respiration. By this means the essential amino acids are acquired, even though the adult diet lacks animal prey. Predaceous adults of the first group lack the yeast and tracheal modifications. Females of both groups of *Chrysopa* require substantial amounts of

TABLE 11.1
Major Taxa of Entomophagous Insects [1]

Predaceous as both immatures and adults

Collembola
  Entomobryidae
  Isotomidae *(Isotoma)*
Diplura
  Japygidae
Odonata (all)
Mantodea (all)
Dermaptera
  Arixeniina (probably)
  Forficulina (most)
Orthoptera
  Acrididae
  Gryllacrididae
  Gryllidae *(Oecanthus)*
  Tettigoniidae *(Conocephalus)*
Psocoptera
  Atropidae
  Caeciliidae *(Caecilius)*
  Liposcelidae
Zoraptera
Hemiptera (Heteroptera)
  Anthocoridae *(Anthocoris)*
  Belostomatidae
  Corixidae (some)
  Enicocephalidae
  Gelastocoridae
  Gerridae (occasionally)
  Hebridae
  Largidae *(Euryopthalmus)*

Lygaeidae *(Geocoris)*
Miridae (some)
Nabidae
Naucoridae
Nepidae
Notonectidae
Ochteridae
Pentatomidae (some)
Phymatidae
Pyrrhocoridae *(Dindymus)*
Reduviidae (except Triatominae)
Saldidae (occasionally)
Veliidae (occasionally)
Thysanoptera
  Aeolothripidae *(Aeolothrips)*
  Phlaeothripidae *(Leptothrips)*
  Thripidae *(Scolothrips)*
Neuroptera (most)
Raphidioptera (all)
Coleoptera (Adephaga)
  Amphizoidae
  Carabidae (most)
  Cicindelidae
  Dytiscidae
  Gyrinidae
  Paussidae
Coleoptera (Polyphaga)
  Anthicidae *(Anthicus)*
  Anthribidae *(Brachytarsus)*

Cantharidae (many)
Cleridae
Coccinellidae (most)
Cucujidae *(Cryptolestes)*
Elateridae (some)
Histeridae
Lampyridae
Melyridae (many)
Nitidulidae *(Cybocephalus)*
Nosodendridae
Orthoperidae
Phengodidae
Pselaphidae
Scydmaenidae
Silphidae *(Xylodrepa)*
Staphylinidae (many)
Trogositidae
Diptera (Brachycera)
  Asilidae
  Dolichopodidae
  Empididae
Diptera (Cyclorrhapha)
  Anthomyiidae *(Hylemya)*
Hymenoptera (Apocrita)
  Chalcidoidea
  Chrysididae
  Formicidae
  Ichneumonoidea
  Vespidae

Predaceous primarily as immatures

Ephemeroptera
  Siphlonuridae *(Isonychia)*
Plecoptera (most Setipalpia)
Megaloptera (all)
Neuroptera
  Chrysopidae *(Chrysoperla)*
Coleoptera (Polyphaga)
  Dermestidae *(Thaumaglossa)*
  Hydrophilidae (most)
  Lycidae (some)
  Meloidae
  Rhipiphoridae (most)
Diptera (Nematocera)
  Cecidomyiidae *(Aphidoletes)*
  Ceratopogonidae
  Culicidae *(Chaoborus,*
    Megarhinus)
  Chironomidae (Tanypodinae)
  Mycetophilidae *(Arachnocampa,*
    Platyura)
  Tipulidae (Hexatomiini)

Diptera (Brachycera)
  Bombyliidae (most)
  Mydiidae
  Rhagionidae
  Tabanidae (most)
  Therevidae
  Xylophagidae
Diptera (Cyclorrhapha)
  Anthomyiidae (some)
  Calliphoridae *(Stomorhina)*
  Chamaemyiidae
  Chloropidae *(Siphonella)*
  Drosophilidae *(Gitonides)*
  Lonchaeidae *(Lonchaea)*
  Otitidae *(Elassogaster)*
  Phoridae *(Syneura)*
  Sarcophagidae
    (Sarcophaga,
    Miltogrammini)
  Sciomyzidae
  Syrphidae (most)

Lepidoptera
  Blastobasidae *(Holcocera)*
  Cosmopterygidae *(Stathmopoda)*
  Cyclotornidae
  Heliodinidae
  Lycaenidae (some)
  Lyonetiidae *(Ereunetis)*
  Noctuidae *(Eublemma)*
  Psychidae *(Platoeceticus)*
  Pyralidae *(Laetilia, Macrotheca)*
  Sesiidae *(Synanthedon)*
  Tineidae *(Dicymolomia)*
  Tortricidae (some *Tortrix)*
Trichoptera
  Glossosomatidae
  Hydrobiosidae
  Hydropsychidae *(Hydropsyche)*
  Polycentropodidae
  Rhyacophilidae
Hymenoptera (Apocrita)
  Sphecoidea (some)

(cont.)

TABLE 11.1 (*continued*)

Predaceous primarily as adults

Coleoptera (Polyphaga)
  Cerambycidae *(Elytroleptus)*
  Scarabaeidae
   (Cremastocheilus)
Mecoptera
  Bittacidae

Diptera
  Anthomyiidae (some)
  Blephariceridae
  Calliphoridae *(Bengalia)*
  Ceratopogonidae (some)
  Empididae

Hymenoptera (most Symphyta)
Hymenoptera (Apocrita)
  Bethylidae
  Dryinidae
  Mutillidae
  Tiphiidae

Parasitoids

Coleoptera
  Carabidae *(Lebia, Brachinus)*
  Colydiidae *(Deretaphrus)*
  Meloidae
  Rhipiceridae
  Rhipiphoridae
  Staphylinidae (some Aleocharinae)
Diptera
  Acroceridae
  Cryptochetidae *(Cryptochetum)*
  Anthomyiidae
  Conopidae

Nemestrinidae
Phoridae *(Megaselia)*
Pipunculidae
Pyrgotidae
Tachinidae
Lepidoptera
  Cosmopterygidae
   (Euclemensia, Antequera)
  Cyclotornidae
  Epipyropidae
Hymenoptera
  Bethyloidea

Chalcidoidea
Chrysidoidea
Evanioidea
Figitidae
Ilbaliidae
Ichneumonoidea
Megalyroidea
Pompiloidea
Proctotrupoidea
Scolioidea
Sphecoidea (some)
Trigonaloidea

Insect ectoparasites of insects

Diptera
  Ceratopogonidae (some)
  Chironomidae *(Symbiocladius)*

Hymenoptera
  Scelionidae *(Rielia)*

True insect endoparasites of insects

Strepsiptera (all)

Hypermetamorphic taxa

Neuroptera
  Mantispidae
Coleoptera
  Carabidae *(Lebia, Brachinus)*
  Colydiidae (a few)
  Drilidae
  Meloidae
  Rhipiceridae *(Sandalus)*
  Rhipiphoridae

Staphylinidae
  (Aleocharinae)
Strepsiptera
Diptera
  Acroceridae
  Bombyliidae
  Calliphoridae (some)
  Nemestrinidae
  Tachinidae (some)

Lepidoptera
  Cyclotornidae
  Epipyropidae
Hymenoptera
  Chalcidoidea (some)
  Ichneumonoidea (some)
  Proctotrupoidea
  Trigonaloidea

[1]This list of taxa is intended to be representative, not exhaustive. The listing of a family without qualification does not necessarily mean that all species in the family have the same habits. Unless otherwise indicated, both the immature and adult stages are believed to have the same food habits. Well-known or exceptional taxa are given in parentheses.

Source: Modified from Askew, 1971; Balduf, 1935; Clausen, 1940; Hagen et al., 1976; Hagen, 1987.

food before eggs are produced. As a consequence, females lay eggs only near abundant food sources that will later supply the lacewing larvae. An advantage to the honeydew-feeding species is the variety of insects in addition to aphids that are attracted to honeydew and may serve as prey of the larvae.

In contrast to other entomophagous insects, predators of both sexes must repeatedly find and subdue prey. Predators may be active during day and night. The eggs of predaceous insects are usually deposited by the females in close proximity to or at least in the vicinity of suitable prey. In this way the adult's well-developed compound eyes, chemoreceptive organs, and ability to fly are used to search for prey-rich habitats within which the less well-equipped immatures can forage. Plants of a certain height, odors of honeydew, odors of decay, and even the pheromones of prey are attractive to searching predators. The adult may eat prey or other attractive foods when they are found, or only lay eggs near the prey.

Many predatory insects utilize extraoral digestion. This allows insects with piercing-sucking mouthparts like the Heteroptera to inject powerful protease enzymes into the prey and later suck out the liquefied tissues. The predaceous larvae of Neuroptera and some Coleoptera have mouthparts for piercing-sucking and injecting enzymes that have been modified from the mandibulate type. Some larval and adult Diptera, with their specialized mouthparts, also feed in this manner. Extraoral digestion offers several advantages to predators: relatively larger prey can be subdued by smaller predators; concentrated nutrients can be extracted more quickly and completely from the prey; and the ingested food contains no bulky indigestible pieces to be passed through the gut.

Strategies of Insect Predators

Several kinds of strategy are used by insect predators in finding and capturing prey: (1) **random searching,** (2) **hunting,** (3) **ambush** or **"sit and wait",** and (4) **trapping.**

Insects of the first type roam in the appropriate microhabitat and seize prey after physical contact. Their orientation to objects in the microhabitat and their movements may increase the probability of encountering prey. For example, the predator may patrol leaf edges, veins, or stems and eat insect eggs or sedentary Homoptera. Random searchers may have either monophagous or polyphagous habits. The prey is accepted if proper incitants are detected by receptors on the forelegs, mouthparts, or antennae. The victim is then devoured, using mouthparts that are usually specially adapted for predation. After an initial contact or meal, further searching often involves more frequent turns so that the predator stays in an area of previous success. While seemingly inefficient, the random searchers are probably the most common type of insect predator and include species highly effective in regulating prey populations.

Most predatory Coleoptera are random searchers (Color photo 6a). Predatory carabids are distinguished by long, sharply hooked mandibles that contrast with the broad, blunt jaws of plant-feeding relatives. Coccinellid beetles, syrphid larvae, and neuropteran predators that feed on aphids also search at random. The mandibles of predaceous coccinellids may be incisors with one or two apical teeth and a basal tooth, or they may be small, with ducts for sucking prey juices. Each sickle-shaped jaw of a neuropteran larva is formed by the mandible and maxilla locked together to create a tube through which body fluids of prey are sucked (Figs. 38.1c, 38.2c; Color photo 6d).

The Lepidoptera are almost exclusively phytophagous. Caterpillars of certain moths retain their close association with plants but attack other phytophagous insects such as coccids and leafhoppers as well as mites. In Hawaii, 18 species of geometrid caterpillars in the genus *Eupithecia* have evolved into ambush predators of small insects. They appear like ordinary inchworms perched on a leaf and will wait for long periods until an unsuspecting victim approaches. When the prey appears, it is quickly seized. Caterpillars of the lycaenid butterflies may be phytophagous or partly or wholly predaceous on ant larvae and pupae or aphids. Under crowded conditions, larvae of noctuid moths become cannibals.

The relatively small size and high nutritive value of insect eggs make them vulnerable to predation by insects of varied food habits. Scavenging, phytophagous, predatory, or parasitoid insects may consume eggs. Collectively they are called **egg predators.** These predators are of special ecological importance because the prey never function in the community. The large clustered eggs of Orthoptera are a special case, since they are subject to frequent attack by many predatory larvae: clerids, meloids, bombyliids, rhagionids, anthomyiids, calliphorids, otitids, phorids, sarcophagids, and eurytomids. The adult female of the predatory species finds the eggs and lays her own eggs; the larvae then feed essentially as random searchers among the mass of eggs of the host.

The hunting insects differ from random searchers by using sight or other stimuli to orient to prey at a distance. Visual hunters have enlarged compound eyes with overlapping fields of vision that permit distance perception. Mandibles may be sharply toothed, and the legs may be strong and spiny for seizing elusive prey. Strong fliers often carry their prey to perching places or nests. Conspicuous examples of such aerial hunters are the adults of dragonflies, asilid flies, and the aculeate or stinging wasps. Stinging wasps are able to subdue prey that are physically larger than themselves, such as katydids or spiders, by injection of paralyzing chemicals with their sting. Dragonflies such as *Anax* (Aeshnidae) fly almost continuously, "hawking" for prey, while *Libellula* (Libellulidae) remain perched until prey approaches, then quickly dash in pursuit. Asilids commonly perch and await flying prey, then return to the perch after the victim is caught (Fig. 2.10c). Adult cicindelid beetles pursue their prey by running fast on open ground while the larvae ambush prey from burrows.

Stimuli other than visual may be used by hunting predators. Kairomones may attract predators to prey (Chap. 6). Although blind, the famous army ants (genus *Eciton*) of the New World Tropics are enormously successful predators. Odor and movement of prey are detected by chemical and tactile receptors while the colony is on one of its periodic raids. Prey animals many times larger than the individual worker ants are attacked by massive swarms. Cooperative foraging by "packs" of ants of other kinds also subdues large prey.

The backswimmers (Notonectidae) are voracious aquatic predators. They flush and stalk prey by sight and are able to detect the vibrations of prey movements with receptors situated along the forelegs. The water striders (Veliidae) also perceive prey by sight and vibrations.

Insects that ambush prey conserve energy by simply waiting for prey to approach within striking distance. Reduviid and phymatid bugs often remain motionless on flowers awaiting visitors. Such predators are exposed to predation by vertebrates and have either concealing or warning coloration (Chap. 12). Concealing coloration may also prevent detection by alert prey (Color photos 8a–f). Raptorial or clasping forelegs are frequently characteristic of insects that ambush, especially among the Hemiptera (Fig. 34.2a–c; Color photo 6b). Some reduviids aid prey capture by smearing sticky secretions or plant resins on their legs. The beaks of predatory bugs are stout and inject saliva that contains proteolytic enzymes. This is why their bites are painful to humans, in contrast to the mild bites of bugs that are parasites of humans. Some reduviids also have a potent toxin that quickly renders prey helpless. When the tissues of the victims have been digested, the resultant fluid is sucked by the bug.

The praying mantis is a familiar predator that strikes prey from ambush (Chapter photo). The coordination of the depth-perceiving vision, mobile head and prothorax, and toothed, raptorial forelegs has been carefully analyzed. The complete strike, timed by high-speed photography, takes 50 to 70 milliseconds. A fly or cockroach would require about 45 to 65 milliseconds to respond if startled by the first movement of the mantis—alas, too slow to escape the strike already in motion. The dragonfly naiad similarly grasps prey with quick strikes of its extensible labium. Although tiger beetle larvae (Cicindelidae) have stemmata as their primary visual organs, the larvae apparently have keen vision because they are able to seize prey that pass near the entrance to their burrows in soil.

Only a few kinds of insect trap prey. The "ant lion" larvae of *Myrmeleon* (Myrmeleontidae, Neuroptera) excavate conical pits in fine, loose sand (Fig. 38.2d). The larva waits motionless and buried at the bottom until a small insect ventures into the pit. The ant lion flicks sand toward the victim to cause the unstable sand to slide down. Once the prey has been seized, the ant lion extracts the body fluids with its sickle-shaped jaws. Larvae of the fly *Vermileo* (Rhagionidae) also construct pits and trap prey. Wheeler (1930) aptly describes these insects as "demons of the dust."

Larvae of the New Zealand glowworm, *Arachnocampa luminosa* (Mycetophilidae) live in moist caves. They spin slimy webs that dangle from the ceiling. The larvae glow in the darkness, attracting small flies that become entangled in the sticky webs. Larval mycetophilids of the genus *Platyura* also secrete webs that entangle prey and contain a toxic fluid. Even though silk is secreted by various insects, webs are surprisingly rarely used to trap prey in the manner of spiders. The aquatic caddisfly larvae of the Hydropsychidae spin webs that filter small prey and other food from passing currents. Though they do not spin their own webs, a group of slender reduviid bugs, the emesines, wait in ambush at spider webs to attack trapped prey.

PARASITOID INSECTS

Only the larvae of holometabolous insects, especially Diptera and Hymenoptera, are parasitoids. Their unique attributes combine certain features of parasites and predators. In addition to their requirement for only one host, parasitoids differ from predators in the following features: (1) the host is larger than the parasitoid, (2) parasitoids are frequently host-specific, attacking one or several related host species, (3) a lower density of host population will sustain a parasitoid population; and (4) the victim is usually searched for and selected by the diurnally active, adult female. A

species of parasitoid may be specific not only in the choice of host species, but also in the choice of the life stages of the host to be attacked. Eggs or young or both are most frequently attacked, but pupae and sometimes adult hosts are also eaten.

Types of Parasitoid Insect

Parasitoids may feed externally as **ectoparasitoids** or internally as **endoparasitoids.** Hosts openly exposed to other predators are usually attacked by endoparasitoids that conceal their presence by being inside the host. Hosts in protected situations, such as leaf mines, galls, or nests, are attacked by either endoparasitoids or ectoparasitoids. When the larva of a species characteristically develops in the ratio of one to a host, the species is termed a **solitary parasitoid.** When several larvae of the same species normally develop in a single host, the species is called a **gregarious parasitoid.** If the host is a phytophagous insect, the parasitoid is called a **primary parasitoid.** Parasitoids that attack other parasitoids are called **hyperparasitoids** or secondary parasitoids. **Multiple parasitoidism** occurs when two or more species of primary parasitoids attack one host individual. **Superparasitoidism** occurs when a host is attacked by more larvae of the same species than can reach maturity in the one host.

Adult parasitoids are free-living, and most of them are winged except for female velvet ants (Mutillidae) and certain other scolioid wasps. Nectar and honeydew are frequently consumed for energy to fly. Some parasitoid wasps feed on the body fluids that are released from the host's body when it is punctured by the parasitoid's ovipositor. This is called **host feeding.** In some instances, the host is inside a cocoon or for some other reason can be reached only by the parasitoid's ovipositor. A tube of coagulated host's hemolymph, or possibly of a secretion produced by the parasitoid, forms around the ovipositor. When the ovipositor is withdrawn, the parasitoid is able to suck the host's hemolymph as it wells up in the tube (Fig. 46.3c).

How Parasitoids Find and Attack Hosts

The adult parasitoid searches first for the habitat in which the appropriate hosts exist. Primary parasitoids are often attracted mainly by the plants that shelter favored insects hosts. In this behavior the parasitoids respond like phytophagous insects to shapes, colors, and odors of plants, yet they do not feed on green plant tissues. Their search movements may be essentially random or somewhat systematic in response to stimuli detected at a distance. Parasitoids are also known to be attracted by the sex pheromones or defensive secretions of their preferred hosts. Once near a potential host, female Hymenoptera examine suitable individuals on which to oviposit, using the tactile and olfactory senses of their antennae. Female *Trichogramma* (Trichogrammatidae, Chalcidoidea) walk back and forth on a host's egg while drumming with their antennae. By this means they determine the acceptability of the host, the number of eggs to lay, and often the sex of the host egg. The wasp then inserts her ovipositor through the egg chorion. Olfactory receptors at the tip of the ovipositor respond to the egg contents. At this point, the female may reject the host and withdraw the ovipositor without oviposition. Otherwise, she will lay an appropriate number of eggs.

The ovipositor of Hymenoptera plays an important role in locating and selecting the host, sometimes in paralyzing it, and in delivering the egg to the host (Figs. 46.3, 46.10). Chemoreceptors on the ovipositor tip respond to a variety of chlorides and aliphatic alcohols, and to hydrochloric acid, in much the same way as taste receptors on the anterior appendages. These receptors function in the final discrimination of hosts before the egg is laid. Hosts inside tough cocoons or sclerotized puparia, in leaf mines, or even deep in burrows in solid wood can be reached by parasitoids equipped with ovipositors of suitable length and strength. By alternate movements of the sharply ridged valvulae, parasitoids can drill through resistant substrates. Receptors at the tip sense the presence and suitability of the host. The highly elastic eggs then pass down the minute channel inside the ovipositor by becoming greatly elongated. After deposition in the host, eggs of some species increase in size by as much as a thousand times. Some wasps temporarily or permanently paralyze or kill the host before ovipositing. For example, the venom of *Bracon hebetor* (Braconidae) causes permanent paralysis of host caterpillars when diluted up to 1 in 200 million parts of host hemolymph.

Female Diptera lack the well-developed antennae of wasps and the needlelike ovipositor for precisely inserting eggs in hosts. Consequently, some flies stick their eggs on the bodies of their hosts, while others depend on first-instar larvae to actually locate hosts after the eggs have been laid in the proper habitat.

Insects that have two or more successive larval instars specialized for different modes of life are **hypermetamorphic** (see taxa listed in Table 11.1; Fig. 46.5). Among the hypermetamorphic parasitoids, the first-instar larva is specialized for active host finding and the subsequent instars specialized for feeding.

Superparasitoidism and multiple parasitoidism may be advantageous to the parasitoids in that the host's encapsulation reaction is dissipated. The resulting competition among larvae also may be disadvantageous. First-instar larvae of certain wasps are specialized to compete successfully in such situations. Some have large sickle-shaped mandibles and kill competitors; others release inhibitory toxins or, if larger, deprive smaller competitors of oxygen. Competition is also commonly reduced by the behavior of the female wasp. For example, *Trichogramma* (Trichogrammatidae, Chalcidoidea) reject hosts on which other females have walked. Some wasps seem to be able, by means of sensilla on their ovipositors, to select host individuals that have not been attacked.

Endoparasitoids are immersed in the body tissues and fluids of their hosts. Many synchronize their development with that of the host by responding to the host's hormones. Some obtain food in the normal manner via the mouth, but others, especially those in eggs, are greatly simplified and absorb food through the outer integument. Endoparasitoid wasp larvae commonly exchange gases through their outer integument, with or without a tracheal system, and rarely with open spiracles. Respiration may be aided by caudal filaments, anal vesicles, or an everted hindgut. Endoparasitoid fly larvae usually have at least a metapneustic tracheal system (Chap. 5). Some perforate the host's trachea or integument and obtain atmospheric air. Others take advantage of the host's encapsulation reaction and mold a respiratory "funnel" or tube of host tissue connected to a trachea or the integument.

Reproductive Biology of Parasitoid Hymenoptera

Among the parasitoids, Hymenoptera have several noteworthy features of reproductive biology. Recall that sex is determined in this order by a haplodiploid mechanism. The ratio of sexes is often unbalanced in favor of females. It is a peculiarity of Aphelinidae (Chalcidoidea) that the female may select the host according to the sex of the egg to be laid. Furthermore, the larva's feeding behavior may be influenced by its sex; males may be regular hyperparasitoids. Some wasps are exclusively parthenogenetic, in contrast to other entomophagous insects that are rarely so.

The unusual phenomenon of polyembryony (Chap. 4) is found only among parasitoids (Fig. 11.2b) and parasites; it occurs in some endoparasitoid Hymenoptera and a few Strepsiptera. As many as 1500 embryos in a single caterpillar have been counted after oviposition by one female of *Litomastix* (Encyrtidae). The most remarkable case of self-defense among parasitoids is the polyembryonic wasp *Copidosomopsis* (Encyrtidae; Cruz, 1981), in which larvae of two kinds develop from one egg. The first larvae to appear are "defender morphs" and attack any other parasitoid in the host. They fail to pupate and thus sacrifice themselves in favor of the normal feeding larvae that emerge later and complete development to adult wasps.

Figure 11.2 Parasitic and parasitoid insects. **a**, Parasitic fly, *Forcipomyia* sp. (Ceratopogonidae) sucking hemolymph of a stick insect (Phasmatodea) in Peru. An engorged fly firmly attaches to the host like a tick. **b**, Emergence of most of 789 brood produced by the polyembryonic wasp parasitoid, *Pentalitomastix plethoricus* (Encyrtidae), from caterpillar of navel orangeworm, *Paramyelois transitella* (Pyralidae). (b, photo courtesy of and with permission of Frank E. Skinner.)

PARASITIC INSECTS

True parasitic relationships between one insect and another are surprisingly rare. As we have seen, the usual result of attack is the death of the victim. Entomophagous parasites may be divided into ectoparasites and endoparasites.

Adult biting midges of the genus *Forcipomyia* (Ceratopogonidae) and several other genera are ectoparasites that take blood from both vertebrates and large insects such as walkingsticks. When feeding on the latter, the flies puncture the wing veins or intersegmental membranes (Fig. 11.2a). The flies visit the host only during feeding.

The small wasps of the family Scelionidae lay their eggs among the freshly laid eggs of host insects. Females of several genera have been observed attached phoretically to the bodies of female hosts. After the host's eggs have been laid, the wasp immediately oviposits. Young winged females of the European *Mantibaria manticida* (Scelionidae) search for and attach themselves to mantids of both sexes. Female hosts are more frequently selected. Once attached by their mandibles, the wasps shed their wings and await the host's oviposition, which may not take place for several months. In the interim, the wasp feeds as an ectoparasite on the mantid's hemolymph. When the mantid's ootheca has been deposited and is still soft, the wasp leaves the mantid to lay its own eggs. The wasp is said then to attempt to return to its host's body.

An unusual instance of ectoparasitic behavior is provided by the aquatic larvae of the midge *Symbiocladius* (Chironomidae). The larvae attach themselves behind the wing buds of mayfly naiads and feed.

Life History and Hosts of Strepsiptera

Strepsiptera are the only entomophagous insects that are true endoparasites. Numbering about 525 species, they probably evolved in the Mesozoic Era as the sister group to Coleoptera. Fossils have been found in Cretaceous amber. Hosts include insects in the orders Thysanura, Blattodea, Mantodea, Orthoptera, Hemiptera, and Diptera, but most of them are aculeate Hymenoptera. The infested hosts are said to be "stylopized," because the name of a common genus is *Stylops* (Fig. 40.2d).

The life history of *Stylops pacifica* (Stylopidae) was studied by Linsley and MacSwain (1957) near Berkeley, California. The hosts are two bee species in the genus *Andrena* (Andrenidae). On bright warm days in February and early March, adult bees emerge from their burrows in soil with puparia protruding from the body, commonly between the fourth and fifth terga of the abdomen

(Fig. 11.3a). As many as 16 percent are parasitized, and their next step is to visit the buttercup *Ranunculus* 11.3c). The male *Stylops* is winged and less than 3 mm long (Fig. 11.3b). On emergence from the puparium, he immediately begins a rapid flight in search of a female. The latter is hardly more than a sac of reproductive organs inside the puparium. Females release a sex attractant while the host flies from flower to flower. Males fly upwind, tracking the odor, until the female is located on the dorsum of a feeding bee. The male *Stylops* lands and inserts the aedeagus by puncturing the female's puparium (Fig. 11.3c; Chapter 40 photo). Once inseminated, the female ceases to release the attractant. Males mate but once and probably die on the day they emerge. Except during copulation, their intense flight activity never ceases, and they rarely feed. The tarsi even lack claws for clinging to objects. Linsley and MacSwain were able to capture the rare males by placing cages containing bees with virgin female parasites among flowers.

The first-instar larvae are triungulins, and they develop from the fertilized eggs while still inside the female's body. During the next 30 to 40 days, the female dies and the larvae move into a median brood passage in preparation for their exit. A large female *Stylops* probably produces 9000 to 10,000 triungulins. On warm days while the host is visiting flowers, the active larvae emerge through the ruptured puparium (Fig. 11.3d). Parasitized hosts move more rapidly among flowers than normal hosts. The triungulins are brushed singly or several at a time onto the flowers.

In contrast to the triungulins of Meloidae, Rhipiphoridae, and other Strepsiptera, those of *Stylops pacifica* do not readily attach themselves to new hosts, but are ingested with nectar by bees (Fig. 11.3e). The bee returns to its nest and prepares a ball of pollen mixed with regurgitated nectar. In this way the triungulin is deposited in a nest cell with the bee's egg (Fig. 11.3f). The triungulin penetrates the egg, transforming as it does into the second instar, and becomes an endoparasite (Fig. 11.3g).

As the host feeds and grows, the *Stylops* larva feeds on the host's nonvital tissues and passes

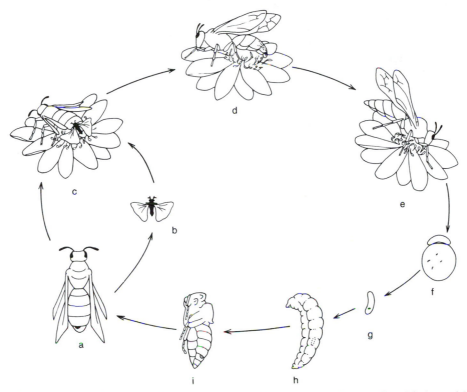

Figure 11.3 Life history of *Stylops pacific* (Strepsiptera). See text for explanation. (Redrawn from Linsley and MacSwain, 1957; published 1957 by the Regents of the University of California; reprinted by permission of the University of California Press.)

through an unknown number of instars (Fig. 11.3h). When the bee pupates, the *Stylops* protrudes the anterior portion of its body through the intertergal membrane and also pupates (Fig. 11.3i). The last larval skin is not shed, but forms a tanned puparium.

Strepsiptera have no known natural enemies. Major causes of mortality are the initial losses of triungulins that fail to find a female host of the proper species, and superparasitism. Linsley and MacSwain noted that usually only the larger female bees with a solitary female parasite lived long enough for the triungulins to escape. Smaller male hosts, those with several *Stylops,* or those with *Stylops* plus nematode parasites, usually died before the cycle was completed.

Bees that survive parasitism are variously af-

fected. The genitalia of both sexes may be reduced. Secondary sexual characters, especially in the female, may be shifted toward the opposite sex. Stylopized females may have more malelike yellow color and reduced pollen-collecting hairs on the legs.

SELECTED REFERENCES

Comprehensive Treatments of Entomophagous Insects

Askew, 1971 (parasites and parasitoids); Balduf, 1935, 1939 (general); Clausen, 1940 (general); Godfray, 1994 (parasitoids); Hagen, 1987 (nutritional ecology); Hagen et al., 1976 (predators); Holldobler and Wilson, 1990 (ants as predators, slavery); New,

1991b (predators); Price, 1974 (evolutionary strategies); Wheeler, 1930 (ant and worm lions).

Short Articles on Related Topics

Bay, 1974 (aquatic predators); Clausen, 1976 (phoresy); Cohen, 1995 (extraoral digestion); Dethier 1947 (chemoreceptors on wasp ovipositor); Kerkut and Gilbert, 1985 (general physiology, chemoreception, encapsulation); Linsley and MacSwain, 1957 (Strepsiptera life history); Montgomery, 1983 (predaceous caterpillars); Salt, 1970 (cellular defense reactions).

See Also:

For additional discussion of specific entomophagous taxa, consult chapters on insect orders.

12
INSECTS AND VERTEBRATES

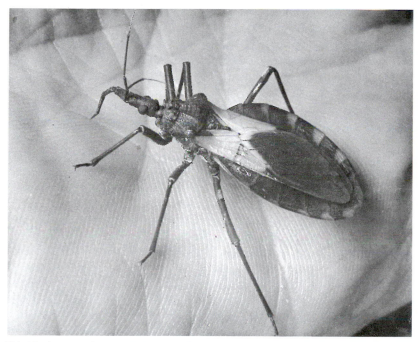

This Mexican parasitic bug, *Triatoma phyllosoma pallidipennis* (Reduviidae) is sucking human blood. *Triatoma* are vectors of Chagas' disease, which is caused by trypanosomes. The fully engorged bugs defecate while on the host, and the infected liquid feces is accidentally rubbed by the host into wounds made by the feeding bug or mucous membranes, especially around the eyes.

Insects have coexisted with freshwater and terrestrial vertebrates throughout most of their evolutionary history. In the course of this long association, feeding relationships have evolved such that certain insects are parasitic on vertebrates and certain vertebrates are predatory on insects. The orders Phthiraptera and Siphonaptera, and various families and species of Dermaptera, Hemiptera (suborder Heteroptera), Lepidoptera, Coleoptera, and Diptera are dependent on the tissues or blood of vertebrates as food during part or all of their lives. As you will learn in the next chapter, this parasitic behavior also has made possible the transmission of certain microbes and helminths among vertebrate hosts, including some of the most important diseases of humans.

In the geological past, the abundance of insects as food for vertebrates played a key role in the early evolution of reptiles, birds, mammals, and the primates. Judging by the structure of fossils, the early forms of each group appear to have been mainly insectivores. Today insects continue to be an important source of nutritious food for vertebrates. Many are specialists on insects.

Many familiar birds in urban and rural areas feed on insects and in some instances significantly reduce pest populations. The annual spring migration to northern forests and grasslands is timed to feast on the emergence of edible insects. Some freshwater fish, like trout, include insects as a regular part of their diet. The mosquito fish, *Gambusia affinis,* is a valuable agent for biological control of mosquitoes in ponds and irrigated rice fields. By at least AD 200, human anglers had learned to imitate insects by making "flies" of feathers or hair. Thus the popular recreation of fly-fishing takes advantage of the predatory behavior of fish on insects.

INSECT PREDATORS OF VERTEBRATES

The difference in size between vertebrates and insects protects most vertebrates from insect predators. Some large insect predators, however, are able to catch and kill small vertebrates. Large dragonfly naiads, dytiscid larvae (Fig. 39.5b), and belostomatids capture small fish or amphibians. These aquatic predators are unwelcome in home aquaria and fish hatcheries. Large praying mantises may kill young birds. Chicks and young birds are also killed by ants as small as Argentine ants. Ants probably initially seek moisture, but soon also take blood. Poultry farms have reported losses of chicks killed by ants.

INSECT PARASITES OF VERTEBRATES

The larger size of vertebrates prevents predation on them by most insects. All terrestrial vertebrates, however, are subject to attack by parasites. Only vertebrates that remain partly or completely submerged in water throughout their lives—fish, whales, and sea cows—are usually exempt, but mosquitoes have been seen biting fish at low tide. Other marine vertebrates, such as seals and marine birds, including penguins, periodically leave the water and are hosts to lice and occasionally other kinds of parasite. For example, mosquitoes transmit filarial worms to sea lions.

Insect parasites of vertebrates are commonly attracted to carbon dioxide as they search for hosts. Knowledge of this phenomenon can be used to capture parasites in traps baited with dry ice. Both carbon dioxide and vaporized lactic acid cause flying mosquitoes to orient toward the source. When these parasites are close to the host, the temperature gradient appears to stimulate probing by them and all other bloodsucking insects. A component of blood, probably nucleotides, serves to stimulate feeding.

Blood is the usual food and often is ingested with the aid of anticoagulants in the saliva. Other parasites subsist on sebaceous (oil) secretions, mucus, pus, lachrymal (tear) secretions, epidermis, skin debris, hair or feathers, or internal tissues. In some instances, the eating of foreign matter, fungi, or microbes on the host's skin is probably beneficial, but most parasites are detrimental to their hosts to some degree. Nevertheless, the health and survival of the host are essential to the parasite.

On an evolutionary timescale, the host's reaction to a new parasite is usually severe, while long-established parasites have evolved in such a way as to minimize irritation. The bite of a bloodsucking insect is often apparently painless to its normal host, but among other hosts the reaction may be swift. Occasionally hosts die as a direct result of parasitism. Excessive numbers of bloodsucking insects may kill a host by exsanguination or toxic effects, or internal parasites may destroy vital organs. Debilitating or fatal pathogens may be transmitted, or secondary infections may develop in wounds.

TYPES OF PARASITIC INSECT

Some terms will be needed for our discussion of insect parasitism. Parasites located on the outside of the host's body are **ectoparasites;** those inside the body are **endoparasites. Continuous** parasites remain on the host's body throughout the parasite's life. **Transitory** parasites spend a part of their life on the host and part elsewhere as free-living insects. **Intermittent,** or **temporary,** parasites visit the host only to feed. **Facultative** parasites are insects that normally complete their life without parasitism but are able to survive as parasites under certain circumstances. **Obligatory** parasites require a host to complete their life.

Parasites have evolved repeatedly from different lineages during the long association of insects and vertebrates as the codominant terrestrial animals. For our purposes, we divide parasites into three broad categories based on their general adaptation to parasitism:

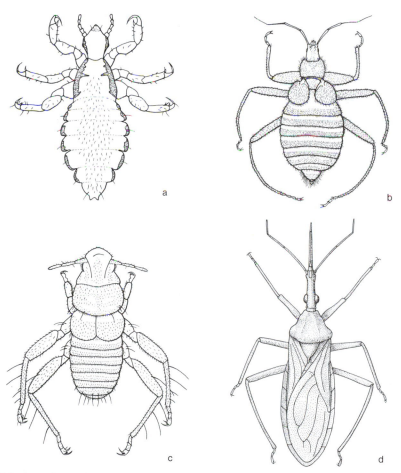

Figure 12.1 Parasitic hemipteroids: **a**, human body louse, *Pediculus humanus* (Pediculidae); **b**, primitive cimicid bat bug from Texas cave, *Primicimex cavernis;* **c**, polyctenid bat bug, *Hesperoctenes fumarius;* **d**, bloodsucking reduviid, *Rhodnius prolixus.*

- **Crawling ectoparasites** are apterous or sometimes winged; they cling to or crawl on their hosts, at least during feeding.
- **Flying ectoparasites** are winged, active fliers; they alight at or near the site of feeding.
- **Myiasis-producing Diptera** are tissue- or blood-feeding larvae.

Crawling Ectoparasites

Crawling ectoparasites are the most taxonomically diverse (Table 12.1). They exhibit a number of convergent adaptations associated with life in the hair or feathers of their hosts. In most species both sexes are parasitic. Adults are more or less flattened (Figs. 12.1, 12.2) and with a tough, leathery cuticle. Siphonaptera are laterally compressed for movement through hair (Figs. 12.2d, 43.1a); the others are dorsoventrally flattened and manage equally well in hair or feathers. The adults of continuous parasites and Siphonaptera have generally reduced sensory and locomotory apparatus:

- short antennae;
- compound eyes reduced (absent in Hemimerina,

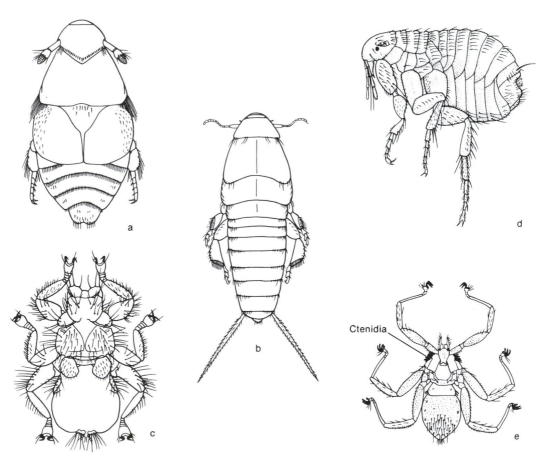

Figure 12.2 Various parasites of vertebrates: **a**, beaver beetle, *Platypsylla castoris;* **b**, parasitic dermapteran, *Hemimerus talpoides;* **c**, streblid bat fly; **d**, oriental rat flea, *Xenopsylla cheopis,* a vector of plague; **e**, nycteribiid bat fly.

TABLE 12.1
Crawling Ectoparasites

---

1. Continuous or apparently continuous ectoparasites

Dermaptera
  Suborder Hemimerina, 10 species, eat epidermis and skin debris of southern African rats of the genus *Cricetomys* (Fig. 12.2b).
Hemiptera
  Polyctenidae, or bat bugs, 6 species in the Old World and 12 in the New World, suck blood of bats (Fig. 12.1c).
Phthiraptera
  Amblycera and Ischnocera, the biting lice (Fig. 33.4a,b), include 2800 species, most of which feed on feathers, skin debris, and sometimes blood of birds, but some species in each suborder parasitize mammals.
  Rhynchophthirina, or elephant lice, include only *Haematomyzus elephantis* on African and Asian elephants, and *H. hopkinsi* on African warthogs.
  Anoplura, or sucking lice, include over 400 species, which suck blood of many mammals. Species attacking humans are the human body louse, *Pediculus humanus* (Fig. 12.1a), and the pubic louse, *Pthirus pubis* (Fig. 33.4d).
Diptera
  Hippoboscidae, or louse flies, include about 100 species, which suck blood from birds and large mammals.
  Streblidae, or bat flies, suck blood of bats in both the New and Old Worlds (Fig. 12.2c).

Nycteribiidae, or spiderlike bat flies, 195 species, suck blood of bats, mostly in the Malaysian region (Fig. 12.2e).
Coleoptera
  Platypsyllidae, *Platypsylla castoris,* or beaver beetles (Fig. 12.2a), feed on skin debris of beavers in Europe and North America, and the relict rodent *Aplodontia rufa,* of coastal Northwestern United States.
  Staphylinidae, tribe Amblyopinini, 41 species, feed mostly on rodents and also on marsupials in Central and South America; one species feeds on a rat from Tasmania. Parasites become embedded in skin, where they may suck blood, but specimens have also been found in bat guano.
Lepidoptera
  Pyralidae, sloth moths: *Bradypodicola hahneli* on the three-toed sloth, *Bradypus* spp.; *Cryptoses choloepi* on the two-toed sloth, *Choloepus hoffmanni;* and *Bradypophila garbei* on *Bradypus marmoratus.* Adults are phoretic on sloths. It is not known whether they feed on sebaceous secretions or on the algae that grow in the grooves of the sloth's hair. When the sloth descends to the forest floor to defecate, the female moth leaves the host and oviposits in the feces. The larvae develop in the feces and later, as adults, find their way to a sloth.

2. Transitory ectoparasites

Siphonaptera, or fleas, including nearly 1800 species, are ectoparasitic bloodsuckers in the adult stage only. The larvae live as scavengers in the host's nest or in ground litter of the host's habitat. Hosts are mostly mammals that nest in dens, holes, or caves. Fleas are not known to parasitize flying lemurs, primates (humans are an exception), aardvarks, elephants, or most horses and their allies. Birds are also hosts, mostly seabirds and perching birds, especially swallows. *Pulex irritans,* the so-called human flea, normally attacks pigs. *Tunga penetrans,* the burrowing flea, burrows under toenails of humans. *Xenopsylla cheopis* (Fig. 12.2d), the oriental rat flea, and others also bite humans among other hosts.

Coleoptera
  Scarabaeidae, adults of *Uroxys gorgon* and *Trichillum brachyporum* live in the hair of the South American sloth, *Bradypus,* and the larvae live in dung. Adults of species of *Onthophagus* (= *Macropocris*) cling to hairs near the anus of kangaroos and wallabies, feeding on secretions and awaiting the host's feces, which are buried by the adult beetle as food for the larvae.

(cont.)

TABLE 12.1 (*continued*)

3. Intermittent, or temporary, ectoparasites

Hemiptera

Cimicidae or bed bugs (Fig. 12.1b), including 74 species, suck blood from hosts in all stages but visit the host only to feed. Twelve genera feed only on bats, nine genera feed only on birds, and only *Cimex* has some species feeding on bats and others on birds. *Cimex hemipterus* and *C. lectularius* (Fig. 34.7b) are human bed bugs that also feed on chickens and bats.

Reduviidae, subfamily Triatominae, or conenose bugs, including about 80 species in the New World and 7 in the Old World, feed on blood in all stages and parasitize lizards, birds, and mammals. *Rhodnius prolixus* and some other species feed on humans (Fig. 12.1d).

Dermaptera

Suborder Arixeniina are sometimes regarded as ectoparasites because of their close association with bat caves and bats and because they share some of the morphological peculiarities of parasites. Actual parasitism, however, has not been demonstrated.

Fig. 12.2b; Polyctenidae, Fig. 12.1c; some Anoplura; and Siphonaptera);

- ocelli absent in endopterygote parasites (present in some Hippoboscidae);

- wings absent (except in some Hippoboscidae and Streblidae, Fig. 12.2c, in which the wings may be well-developed, vestigial, or absent);

- legs usually short, more or less stout, and with well-developed tarsal claws, which may be single and chelate on mammalian hosts (Nycteribiidae have long legs, Fig. 12.2e);

- stiff, posteriorly directed hairs or scales, or a comblike ctenidium (ctenidium present in Polyctenidae, Fig. 12.1c; Siphonaptera, Fig. 43.1a; Nycteribiidae, Fig. 12.2e; and *Platypsylla,* Fig. 12.2a); and

- where the life history is known, mating and oviposition usually takes place on or near the host.

The temporary parasites Cimicidae (Figs. 12.1b, 34.7b) and Triatominae (Reduviidae, Fig. 12.1d) move to and from their hosts. Consequently they retain long antennae, compound eyes, and walking legs. The Cimicidae are otherwise apterous, flattened, and have small eyes and no ocelli. Nymphs of Triatominae are crawling ectoparasites,

but the adults are strong fliers and have well-developed eyes and ocelli.

Flying Ectoparasites

All flying ectoparasites are intermittent parasites in the adult stage. Most species are Diptera, but also included here are some odd Lepidoptera. Four groups are:

1. **Mandibulate biting flies.** The parasites are females with blood-sucking mouthparts that include functional mandibles (Figs. 2.10b, 42.3b). They attack a wide range of vertebrate hosts, both warm- and cold-blooded. Species in all families attack humans. Blood meals are usually required for egg maturation and production. Species that can produce the first clutch of eggs without a blood meal are termed **autogenous.** The immature stages are free-living and usually aquatic or associated with damp soil. Larvae of tabanids and snipe flies are also found on dry litter or soil. Families of Diptera included are Psychodidae, or sand flies; Ceratopogonidae, or biting midges; Simuliidae, or black flies; Culicidae, or mosquitoes (Fig. 2.10b); Tabanidae, or horse flies and deer flies; and Rhagionidae, or snipe flies.

2. **Bloodsucking muscoid flies.** Both sexes suck blood. Most feed on mammals, but *Glossina palpalis* prefers crocodiles and monitor lizards. Humans are bitten by various species of *Glossina* (Fig. 2.10d) and *Stomoxys*. Larvae of *Stomoxys* and *Haematobia* live in manure, but *Glossina* is viviparous, producing a mature larva that soon pupates. Families of Diptera included are Muscidae (*Stomoxys calcitrans,* or stable flies and their allies; *Haematobia irritans,* or horn fly; *H. exigua,* the buffalo fly of Australia; and *H. stimulans,* the cattle fly of Europe) and Glossinidae (about 20 species of *Glossina,* or tsetse flies, in Africa).

3. **Eye gnats and flies.** These insects are attracted to the eyes of humans and other mammals, where they feed on lachrymal and sebaceous secretions, pus, and sometimes blood. Families of Diptera included are Chloropidae (*Hippelates* spp.) and Muscidae (*Musca* spp.).

4. **Eye moths** (fruit-puncturing moths). The moths have a sharp galea with which they puncture fruit and suck the juice. They are also attracted to fluids in the eyes of mammals. The noctuid moth, *Calpe eustrigata,* is able to puncture mammalian skin and take blood. Eye moths include about 23 species from Africa and Southeast Asia and in the following families of Lepidoptera: Noctuidae, Pyralidae, and Geometridae.

Myiasis-Producing Diptera

Larvae of certain flies infest the bodies of animals, including humans, and produce a disease called myiasis. Pupation is usually on the soil, and adults of the flies are free-living. Four groups of myiasis fly have been recognized (James and Harwood, 1969):

1. Larvae of many species that produce **enteric myiasis** and **pseudomyiasis** have been reported in alimentary canals of vertebrates. Pseudomyiasis involves fly larvae that are ingested accidentally or enter through the anus and are able to survive for varying periods in the intestine. The larvae may eat the ingested food of the host or, less frequently, tissues of the intestine. True enteric myiasis is created by the larvae of horse bot flies, *Gasterophilus* (Oestridae, Gasterophilinae), which live attached to the stomach wall while feeding on the wall and absorbing the host's food. Included are the families Muscidae, Calliphoridae, Sarcophagidae, and Oestridae.

2. Larvae of many species that produce **facultative myiasis** normally live as scavengers in dead animals. Adult females, attracted by odors of decomposition, lay eggs on the carcass. Wounds or deep, soiled wool of living sheep also attract ovipositing females. The resulting larvae become parasites in such wounds on sheep. Included are the families Muscidae, Calliphoridae, and Sarcophagidae.

3. Larvae that produce **obligatory myiasis** regularly infest living hosts. The primary screwworm, *Cochliomyia hominivorax* (Calliphoridae), of the New World, the Old World screwworm, *Chrysomyia bezziana* (Calliphoridae), and *Wohlfahrtia magnifica* (Sarcophagidae), also of the Old World, enter the body at wounds or through mucous membranes and create local infestations. Larvae of the cattle grubs *Hypoderma lineatum* and *H. bovis* (Oestridae) penetrate unbroken skin, migrate extensively in the host's body, and return to complete their development in tumors beneath the skin before dropping from the host to pupate. The female of the human bot, *Dermatobia hominis* (Oestridae, Cuterebrinae), attaches her eggs to another fly, such as a blow fly, tabanid, or mosquito. While the carrier insect is feeding, the bot larva hatches and drops to the host's skin, burrows into the subcutaneous tissues, and remains in the same place until fully fed. Mature larvae drop from the host to pupate. These flies occur in Latin America and attack other animals as well as humans. Included are the families Calliphoridae, Sarcophagidae, and Oestridae.

4. **Bloodsucking larvae** that are temporary feeders include the Congo floor maggot, *Auchmeromyia luteola,* which lives in huts and feeds on

sleeping persons; and various species of *Protocalliphora* in the Old and New Worlds which feed on nestling birds. All species are in the family Calliphoridae.

EVOLUTIONARY EFFECTS OF PREDATION BY INSECTIVOROUS VERTEBRATES

Insects are a regular part of the diet of many vertebrate species: freshwater fish, salamanders, frogs, toads, lizards, turtles, snakes, birds, and mammals, including bats. Some human cultures also regularly include insects in their diet. The effectiveness of these predators in catching insects is aided by some of the following: (1) physical agility; (2) keen senses of hearing, smell, taste, and/or vision (including color vision by some fish, some reptiles, most birds, and primates); (3) hunting strategies; and (4) the capacity to learn from experience. Insects are subject to predation in aquatic, terrestrial, and aerial habitats, and during day and night. To survive, exposed insects have evolved special patterns of coloration, behavior, and defensive properties to take advantage of what the predator fails to see or hear, cannot capture, or will not eat. Similar protective strategies can be found repeatedly among different insect taxa. This is because the insects are subjected to selection by predators with similar behavior and senses. The convergent evolution of protective strategies by insects provides some of the best opportunities to study natural selection in action. There has been much speculation, but too few instances of the protective strategies have been analyzed with critical experiments.

For our discussion, we first divide insect prey into creatures that are **palatable** (edible or acceptable) to a given kind of predator and those that are **unpalatable** (inedible or rejected). This distinction must be made with qualifications. Some individual insects will be relatively more or less palatable than others, and individual predators will vary in accepting prey depending on hunger and previous experience. Hard or spiny bodies, sticky secretions,

stings, stinging hairs, and distasteful or toxic chemicals are defenses that tend to render an insect unpalatable. Even a low frequency of unpalatable individuals however, will be an advantage favored by selection in the prey population. Unfortunately we do not know the palatability of most insects relative to their main predators, but observations indicate that most species of insects are edible by a wide variety of insectivores.

Palatable Insects

It is assumed that relatively palatable species evolve either an effective means of escape, or a disguise, or both. Another option that will be discussed further on is **Batesian mimicry,** in which a palatable species avoids attack by resembling an unpalatable species.

The most effective means of survival would be to avoid detection altogether. This can be accomplished by a combination of **cryptic behavior** and **cryptic coloration** (also called camouflage or concealing coloration). Insects with cryptic coloration are overlooked by predators unless the latter learn to seek them. A low density of the species of insect or of other species with similar protective strategies therefore prevents accidental detection and reinforcement of the predator's search image. The coloration may be one or a combination of the following: (1) **background resemblance,** such that the insect either closely matches the background against which it is seen, or **special resemblance** of an inanimate object of no interest to the predator such as a flower, twig, stone, or bird dropping (Fig. 12.3; Color photos 7e, 8a–f); (2) **obliterative shading,** in which the rounded shape of the insect, such as a caterpillar, is obscured by having the illuminated surface of the body more darkly pigmented and the shaded surface lighter in color; (3) **shadow elimination** by checkered borders or marginal fringes of hairs, which reduce the contrast along shaded edges next to the substrate; and (4) **disruptive coloration** of contrasting patterns and colors, to alter the perception of the

Figure 12.3 Examples of concealing coloration: **a**, A katydid (Tettigoniidae) that resembles a leaf. The katydid is active at night, remaining still among green leaves in the daytime. **b**, The Peruvian deadleaf butterfly, *Anaea* sp. (Nymphalidae), has a false stem and midrib line.

true body outline or contour. These patterns are usually accompanied by one or a combination of special behavioral traits in which the insect (1) selects an appropriate background and orients its body in a definite position relative to the background; (2) adopts a special prostrate posture to eliminate shadows or more closely resemble the object imitated; (3) rests motionless even when closely approached; and (4) when attack is imminent, escapes quickly, silently, and erratically.

Cryptic coloration and behavior are the rule among palatable insects exposed to diurnal, visu-ally hunting predators. Aquatic nymphs and larvae may be complexly patterned to resemble aquatic vegetation, detritus, or the variegated gravels or other substrates of the bottom of the body of water. Immature and adult insects that inhabit plants, whether as predators or herbivores, are often green (commonly created by the pigment insectoverdin) or colored or sculptured to resemble leaves, stems, or bark. The larger insects of litter, soil, sand, or gravel may closely match their surroundings. The color patterns are usually characteristic of the instar and may change from instar to instar, but usually do not change rapidly by physiological processes in the manner of chameleon lizards. The walkingstick, *Carausius* (Phasmatidae), however, does become darker or paler depending on various stimuli.

Jumping, dropping from vegetation, and flying are the most common means of escape for cryp-tically colored insects that have been discovered. When approached, leafhoppers on a stem will move behind the stem out of sight. Some species do the opposite, becoming motionless to avoid detection or feign death. Large mantids and Or-thoptera may adopt a threatening posture that de-ters attack. These behaviors are often aided by coloration that diverts the predator's strike or atten-tion. **Eyespots** may be of two main types: small, dark spots, with or without a light mark simulating a reflection from the "eye"; and large spots with concentric rings depicting a colored iris and dark pupil marked with a reflected highlight. The first are thought to imitate an insect's eye, and the second are unmistakable representations of a ter-restrial vertebrate's eye (Fig. 12.4). The small eyespots may be situated at the posterior end of the insect next to antenna-like processes, to imitate a head. Examples are hairstreak butterflies (Lycae-nidae, Fig. 44.11g), which have tails that resemble antennae on the caudal part of their wings, and, at the bases of the tails, spots that resemble eyes. When perched, the butterfly moves its wings to wiggle the false antennae. This ruse presumably directs a predator's attention to the wrong end of the body and permits the insect to escape in the opposite direction with minimal damage. Small

Figure 12.4 Example of special resemblance: the neotropical nymphalid owl butterfly, *Caligo* sp., has large eyespots on the hind wings. Seen from the side, the pattern resembles the eye of an *Anolis* lizard. False eyes startle and frighten would-be predators.

spots near the outer wing margin of other butterflies, such as the satyrs (Satyrinae, Fig. 44.11i) or the buckeye (Nymphalidae, Fig. 1.5f), presumably attract the bird's peck away from the vital body and give the insect a chance to free itself. Evidence that such butterflies do escape is provided by collected specimens that have notched margins or V-shaped bill marks on their wings.

Larger eyespots have been shown to frighten vertebrate predators. Cryptically colored insects may hide their eyespots and suddenly display them when contacted by a predator. The nymphalid butterfly *Caligo* (Fig. 12.4) and the Io moth, *Automeris* (Saturniidae; Color photo 7f), have conspicuous eyespots.

Unexpected displays of bright **flash colors** are similarly designed to startle enemies at the moment of escape. Underwing moths, *Catocala* (Noctuidae, Fig. 44.13b), have conspicuous red or yellow bands on the hind wings. The colors are suddenly revealed and visible during the moth's erratic flight, but abruptly disappear when the moth alights and resumes a cryptic posture. Similarly, the upper surfaces of the wings of *Morpho* butterflies are brilliant blue, while the lower surfaces are cryptically colored. In flight the butterfly is conspicuous, but at rest with its wings held together, only the cryptic pattern is visible.

Industrial Melanism

The best-documented study of the evolution of cryptic coloration concerns industrial melanism, the evolution of dark or melanic populations of insects in industrially polluted areas. Prior to the Industrial Revolution in England, trunks of trees such as oak were naturally light in color and often encrusted with light-colored lichens. Resting against this background, the pepper moth, *Biston betularia* (Geometridae), is virtually invisible. With the development of smoke-producing industries in the latter half of the eighteenth century, soot accumulated on vegetation in the areas surrounding and downwind from the factories. Lichens were killed and the bark of trees was stained black. The light, or *typica,* form of the pepper moth was conspicuous against this darkened background.

In 1848 a dark, or melanic, form of the moth, called *carbonaria,* was first recorded at Manchester in the polluted district. At that time the melanic form probably was not more than 1 percent of the population, but by 1898 it was estimated to be 95 percent.

Melanic forms of over a hundred of the 780 species of larger moths in England have also increased in frequency during the last 150 years.

Such dark forms of moths are becoming evident elsewhere in industrial areas throughout the world's Temperate Zone, including North America. Industrial melanics are not known in tropical regions, possibly because the heat-absorbing dark color is a disadvantage in warmer climates. The melanics are always members of species that are concealingly colored. They rest during the day on such objects as lichen-encrusted bark, rocks, or fallen logs. In the majority of species, the switch from predominantly light forms to dark forms does not involve intermediate shades and is usually controlled genetically by a single dominant gene.

Kettlewell (1973) carefully investigated various explanations for the rapid spread of the *carbonaria* form of the pepper moth. Mutagenic effects of the pollutants were discounted by experiments; the melanic forms must arise in populations by normal, low rates of mutation. Except where the melanics are favored by selection, the frequency remains quite low. However, migratory melanic males may fly into unpolluted environments. In industrially blackened areas, the resting melanic forms are less visible to insectivorous birds. Observation in the field and experiments verified that birds of several species selectively eat the conspicuous form (*typica* against dark trunks in polluted districts and *carbonaria* against light trunks in unpolluted districts) and overlook the concealed forms.

Thus in a relatively brief period, the cryptic patterns of many species of moth over wide areas shifted to a dark, patternless design as a result of bird predation in polluted areas. This dynamic relationship is a sensitive monitor of pollution. With improved pollution controls now in England and northern Europe, nonmelanic moths have reappeared in districts where only melanic forms had been observed. In West Kirby, England, melanic forms were 93 percent of the population in 1959 and 29.6 percent in 1989; in Holland, the change was from 60 to 70 percent in 1969 to less than 10 percent in 1988.

Studies of other melanic insects have shown that different insects are affected differently by pollution. Melanic forms of the ladybird beetle, *Adalia bipunctata* (Coccinellidae), also increased and then decreased in Britain in correlation with the level of smoke pollution. These beetles are believed to be unpalatable, so the changes in the frequency of the melanic forms cannot be explained by selective predation. It is speculated that the melanic beetles are more efficient in heat absorption than nonmelanics when the sun's radiation is partly blocked by smoke.

Unpalatable Insects

The coloration and behavior of unpalatable species often contrast strikingly with those of cryptic palatable species. The strategy here is twofold: (1) to create an unpleasant, *but not lethal,* ordeal for the vertebrate predator by means of poisonous or nauseating secretions, repulsive odors, effective stings or bites, or spiny surfaces; and (2) to be easily recognizable so that an experienced predator probably will not attack again. Vertebrate predators are relatively long-lived and are capable of learning, so once they have experienced an unpleasant insect, they avoid future encounters. As a result, the population of the unpalatable insect suffers less mortality. Unpalatable species usually have warning or **aposematic coloration** (Color photos 3b, 3d, 7a) and **aposematic behavior,** including one or more of the following: (1) bold, simple patterns of contrasting black and red, orange, or yellow, which are highly visible to vertebrates with color vision and also contrast against the predominantly green color of the landscape (green is the complementary color of red); (2) gregarious behavior that leads to local, dense aggregations (Chapter 8 photo; Color photo 3d); (3) free exposure when walking or resting and slow, conspicuous flight in full view of predators; (4) characteristic movements, buzzing sounds, or smells that warn against attack; and (5) when attacked, only sluggish efforts to escape, repulsive taste (Color photo 7c), and a durable body often permitting survival without fatal injury. For example, beetles of the family Lycidae aggregate, are sluggish, and have leathery, resistant, black and orange bodies. The colorful caterpillars of Papilionidae, when disturbed, release defensive odors from brightly colored **osmeteria** (Color photo 7d). Many beetles with defensive secretions are dark or black (Color photo 6a). Against pale soil or sand, the black color is visible during both day and night.

A naive predator has ample opportunity to sample an aposematic species and, even without color vision, has a number of cues by which to remember the experience. Bird predators are also known to learn vicariously by witnessing another predator's response. Learning can be reinforced by merely seeing an aposematic insect again without touching it. Not all the cues may be necessary to elicit an avoidance reaction. It is not known to what extent a predator's response includes some innate avoidance of aposematic colors or behavior.

Aposematic coloration is oddly missing among most insects that live in water, even though some aquatic insects are unpalatable and some fish are known to have color vision and learning ability.

A few cases have been reported of insects with venoms or toxic chemicals in their bodies that do kill vertebrates. These would seem to be exceptions to the rule that vertebrate predators survive their experience with unpalatable insects. In the first case, stings of honey bees cause some human fatalities each year. The venom is a complex mixture of enzymes, proteins, amines, and amino acids. Some persons are hypersensitive to bee stings and may sucumb after a single sting. These rare events are an unfortunate by-product of human physiology. On the other hand, a mass attack by Africanized bees (Fig. 7.4) results in large numbers of stings and invasion of the mouth and nose by bees, causing death of humans and domestic animals. The attacks by Africanized bees evolved in response to the persistence of their natural enemies in Africa such as the honey badger, *Mellivora capensis* (Mustelidae, Carnivora). African bees do not distinguish between tenacious enemies and defenseless animals.

In a second case, cantharid beetles contain the chemical cantharidin, which may be fatal to horses that eat the cantharids in pasture grasses or in dried hay. The beetles are normal aposematic insects. However, horses are not a natural enemy of the beetles; rather, they appear to be unusually susceptible to cantharidin and must be considered accidental victims.

In the third case, certain species of colorful African beetles of the genus *Diamphidia* (Chrysomelidae) and related genera have a protein in their bodies called diamphotoxin. When injected in the blood of mammals, the protein causes hemolysis and is ranked among the most toxic substances known. The !Kung people stir the body contents of mature larvae or pupae with certain plant ingredients and dry the mixture on the shafts of arrows just behind the arrowhead. To prevent injury to the hunter, the arrowhead is not coated. A well-placed arrow will kill an animal as large as mature giraffe in a few hours. How the beetles acquire the poisonous protein is unknown, and its function in the insect's life remains a mystery. It is possible that when eaten by a vertebrate the poison is detoxified by digestion and no longer lethal.

Mimicry

The sensory and learning abilities of relatively long-lived vertebrate predators have allowed aposematic species to survive. Furthermore, there is evidence that predators need only one unpleasant experience to generalize their learning to avoid insects resembling the species responsible for the first bad episode. This quickness to generalize also gives some advantage to insects that are in the process of evolving aposematic coloration and behavior but have not acquired the full set of advertising characteristics. The predator's ability to generalize (or failure to discriminate) is also the basis for the evolution of two types of mimicry: (1) **Müllerian mimicry,** named for Fritz Müller (1879) and involving two or more unpalatable and aposematic species, and (2) the previously mentioned **Batesian mimicry,** named for Henry W. Bates (1862) and involving one or more unpalatable species that serve as **models** and one or more palatable species that **mimic** the model's coloration and behavior (compare Color photos 7a and 7b). Models and mimics with a common pattern of coloration are often not closely related taxonomically and have clearly acquired their similarity through convergent evolution.

The mutual resemblance of Müllerian mimics confers protection on all members of the association in a given area. Because all are unpalatable, and predators can generalize, the resemblance among the species need not be exact, although it is often astonishingly precise. Naive predators presumably attempt to eat mimics in proportion to the relative abundance of each species, and each species suffers fewer losses in the process of educating predators when several mimics are together. Examples are found among the neotropical butterflies, especially the genus *Heliconius* (Nymphalidae) and the nymphalid subfamily Ithomiinae (Cover photo); the monarch, queen, and viceroy butterflies (Fig. 12.5); net-winged beetles (Lycidae) and certain other insects; and the convergent similarity of many ants, wasps, and bees, the females of which are equipped with stings.

Batesian mimics are avoided by predators familiar with the unpalatable model (Color photo 3a). An argument can be made that the populations of mimics must remain less numerous than the model's population. Otherwise a naive predator will be more likely to eat palatable mimics. Generally, Batesian mimics are rarer than their models (in some, only the female is the mimic), but recent studies show that strongly unpalatable models may confer protection even when outnumbered. Apparently tasting a model is truly a memorable experience for the predator! The resemblance of Batesian mimics to their models is often quite close. Examples include neotropical moths and katydids that resemble wasps; the syrphid flies and other insects that resemble wasps and bees (Chapter 16 photo); and long-horned beetles (Cerambycidae), leaf beetles (Chrysomelidae), and moths that resemble lycid beetles.

On careful analysis of mimetic associations, the distinction between Müllerian and Batesian becomes more difficult to make. Lincoln P. Brower and his associates (1964, 1968) studied the mimetic relationships of the monarch butterfly, *Danaus plexippus* (Nymphalidae) (Fig. 12.5a; Color photo 3b), and the queen butterfly, *Danaus gilippus* (Fig. 12.5b). The larval food plants of both species are milkweed, mostly in the genus *Asclepias*. Two species of milkweed in the southeastern United States, *A. curassavica* and *A. humistrata*, contain cardiac glycosides (cardenolides). In birds and cattle, these chemical compounds cause vomiting

Figure 12.5 Multiple mimicry among nymphalid butterflies: **a**, monarch butterfly, *Danaus plexippus;* **b**, queen butterfly, *Danaus gilippus;* **c**, viceroy butterfly, *Limenitis archippus*. See text for explanation. (Photos from the University of California Scientific Photography Laboratory.)

or, in larger amounts, death. Cattle in Costa Rica learn to avoid eating *A. curassavica;* hence the plant is protected. Monarch larvae feed on these milkweeds and acquire various amounts of cardiac glycosides, which are retained in the body of the adult butterfly. Insectivorous birds vomit after eating a sufficiently toxic adult monarch and will avoid pursuing other monarchs when sighted. Thus, not only are emetic monarchs (those that cause vomiting) effectively unpalatable and aposematic, but also the aposematic larvae on the poisonous milkweed are undisturbed by grazing cattle and birds.

Not all milkweeds, however, contain enough cardiac glycosides to cause emesis, and some lack the drugs altogether. Monarchs reared on these plants are palatable in laboratory tests with naive birds but rejected by birds previously fed emetic monarchs. The natural population of monarchs in an area may include both palatable and unpalatable individuals, yet the predators learn to avoid most of both types. Even as low as 25 percent unpalatable monarchs theoretically will gain protection for 75 percent of a population. Brower has called this kind of intraspecific relationship **automimicry**. A further complexity was recently revealed: as unpalatable adult monarchs age, the concentration of cardiac glycosides decreases; young adult monarchs are Müllerian mimics, old monarchs are automimics.

The queen butterfly also feeds on toxic milkweeds and resembles the monarch. Where these species occur together, they are usually considered Müllerian mimics. In Trinidad, however, Brower found both monarch and queen larvae feeding on both poisonous and nonpoisonous species of milkweed. Adult monarchs were 65 percent emetic, but only 15 percent of queen adults were emetic. The mimetic relationships here include automimicry between palatable and unpalatable members of each species, Batesian mimicry between palatable members of one species and unpalatable members of the other, and Müllerian mimicry between unpalatable members of both species.

For many years the viceroy butterfly, *Limenitis archippus* (Fig. 12.5c), was considered to be a palatable Batesian mimic of the monarch butterfly (Fig. 12.5a). The larvae feed on nontoxic plants such as willows *(Salix),* populars *(Populus),* and some kinds of fruit tree. Nevertheless, a high percentage of viceroys have been found to be unpalatable to birds. The viceroy now joins the monarch and queen butterflies as members of a Müllerian and Batesian mimicry complex.

Batesian mimicry may be thought of as involving coevolutionary interactions similar to those occurring in host-parasite complexes. The mimic, of course, plays the role of the parasite. Its strategy is to take advantage of the model without destroying it. The model gains nothing and faces the danger that a "credibility gap" will develop in its potential predators. For at some point, if the mimics get too common, most predators will associate only pleasant experiences with the aposematic pattern of the model. Such a development, of course, ruins the game for both model and mimic, since a conspicuous pattern now means "tastes good" to the local insectivores. One would expect that the model would evolve a pattern different from that of the mimic at a maximum rate, everything else being equal. It is to the mimic's advantage to maintain a maximum of resemblance to the model, until that critical point mentioned above is reached. Then the advantage becomes a disadvantage—the mimic is conspicuously patterned, but predators now associate that pattern with tastefulness. As a result, selection would tend to move the mimic away from the model into a more cryptic pattern.

In all cases it obviously is of advantage to the Batesian mimic to become distasteful if it can do so without sacrificing too much physiologically. In butterflies, at least, it appears that the usual source of noxious compounds is plant biochemicals, and thus food plant relationships must play a large role in the evolutionary dynamics of any given situation. Several different routes to evolve distastefulness are open to a butterfly. If it eats a food plant that does not produce an appropriate compound, it may switch food plants. If it is feeding on a plant with an appropriate compound or its precursors, it may evolve the ability to use the compound or synthesize a noxious compound from precursors. Finally the food plant of the butterfly may evolve an appropriate compound, which then may be picked up by the butterfly. In the latter case the mimetic butterfly would be involved in a complex of "selection races," involving the model, the food plant, and predators. As the food plant becomes more and more toxic, the butterfly must find ways of "breaking even" (by avoiding poisoning) or "winning" (by turning the poison to its own advantage). Predators may simultaneously be undergoing selection for ability to discriminate between model and mimic, and for "resistance" to the noxious properties of the model. Of course the presence or strength of such selection will depend on many variables. For instance, in some cases butterflies in a single population may make up such a small proportion of the targets of a single predator that selective influence on the predator will be negligible.

SELECTED REFERENCES

Behavior and Ecology of Insect Parasites of Vertebrates

James and Harwood, 1969; A. G. Marshall, 1981, 1987.

See Also:

For references on specific parasitic taxa, consult the appropriate chapters in Part Three.

Defense, Adaptive Coloration, and Mimicry

Alonso-Mejia and Brower, 1994 (monarch butterflies); Bates, 1862 (classic); Blest, 1957 (eyespots); Brower et al., 1971 (incipient mimicry); K. S. Brown, 1981 *(Heliconius);* Dettner and Liepert, 1994 (chemical mimicry); Edmunds and Dudgeon, 1991 (cryptic mantid); Ehrlich, 1970 (coevolution); D. L. Evans and Schmidt, 1990 (insect defense); Müller, 1879 (classic); Pasteels et al., 1983 (chemical defenses); Pasteur, 1982 (types of mimicry); Platt, 1983 (viceroy and other admiral butterflies, hybridization); Ritland, 1994 (queen butterfly), 1995 (viceroy butterfly); Ritland and Brower, 1991 (viceroy butterfly); Rothschild, 1985 (toxic plants and insects); Silberglied and Eisner, 1968 (mimicry of Hymenoptera); Turner, 1981 *(Heliconius).*

Industrial Melanism

Berry, 1990; Bishop and Cook, 1975; Brakefield, 1987, 1990; Brakefield and Lees, 1987; Clarke et al., 1990; Kettlewell, 1973.

13
INSECTS AND MICROBES

The Plague (La Peste) by the artist Pierre Mignard, 1610–1695. The scene is created from Biblical accounts and Thucydides' description of the plague in Athens. The painting depicts symptoms of swollen lymph nodes (buboes, being lanced in woman in left foreground), insanity (delirious victim being restrained from leaving house at left), and unquenchable thirst (victims kneeling at fountain in background). The sulfurous fumes poured by the angel in the sky and smoke from the torch and brazier are for disinfection or purification. (Reproduced by permission of Instituto Amatller de Arte Hispáñico, M. A. S. Ampliciones y Reproducciones, Barcelona, Spain.)

A diversity of microorganisms populated the earth long before higher animals appeared. These microbes altered the environmental conditions to make possible the oxygen-based respiration of plants and animals. The antiquity, variety, and abundance of microbes have led to the evolution of many unique insect-microbe relationships. The term "microorganism," as used here, simply refers to taxonomically diverse organisms that are difficult or impossible to see without magnification. Microbes may be unicellular organisms such as bacteria (prokaryotes), yeast, and protozoans, multicellular organisms such as fungi and nematodes, or may be viruses. Viruses are noncellular microbes consisting of a nucleic acid genome of RNA or DNA and a protein shell or capsid, some-

TABLE 13.1
Insect-microbe Relationships

Effect or role of microbe	Examples
Food	Fungus-feeding ants and termites, aquatic filter feeders, detrivores, many others
Aids digestion of food	Termites, beetles, wasps, others
Facultative symbiosis	Fruit flies (Tephritidae) and bacteria; bacteria in aphids
Obligate symbiosis	Beetles and mycangial fungi; aphids and symbiotic prokaryotes, polydnaviruses in parasitoid wasps
Affects sex ratio of insect offspring	Bacteria in parasitoid wasps
Limits reproductive compatibility of insect	Bacteria and spiroplasmas in vinegar flies, parasitic wasps
Insect pathogens	Nematodes, fungi, protozoa, bacteria, viruses
Insect-vectored pathogens of plants or animals	Nematodes, fungi, protozoa, bacteria, viruses

times with host-derived membranes. Viruses use the machinery of the host cell to transport, replicate, and assemble the viral genome and capsid.

Pathogens are microbes that harm their insect hosts. Although pathogenic relationships are the exception rather than the rule, pathogens are often important in regulating insect populations. At the other extreme, some bacteria, fungi, and viruses have evolved with their insect hosts to create mutually beneficial relationships to such an extent that neither partner can survive without the other. Selected examples listed in Table 13.1 illustrate the wide spectrum of insect-microbe relationships.

MICROBES AS FOOD

Insects that feed on living or dead organic matter consume decay microorganisms. Aquatic filter feeders such as the larvae of mosquitoes and black flies consume planktonic microorganisms along with organic debris. Among the particles ingested, microbes often contain the greatest concentrations of protein, carbohydrates, and other nutrients. Habitat and environmental conditions determine the types and concentrations of microbes present, but microorganisms typically account for no more than about 4 to 25 percent of the nutritive value of decaying plant matter. Associated microbes improve the food value of detritus by serving as sources of concentrated digestible nutrients and by trans-

forming complex organic substrates into metabolites that are more readily digested or assimilated. Some microbes may serve as either food or infectious pathogens. For example, the protozoan *Lambornella clarki* is free-living in the water in tree hole cavities, where it feeds on organic detritus (Fig. 13.1). The free-living stages (trophonts) are food for mosquito larvae (Fig. 13.1a). Under certain conditions, the protozoan parasitizes mosquitoes. It first forms an invasion cyst (Fig. 13.1b) on the host mosquito larva *(Aedes sierrensis)* that penetrates the mosquito's cuticle. The protozoan then multiplies and eventually kills the mosquito larva. Adult female mosquitoes that survive infection are castrated but introduce *L. clarki* into tree holes while attempting to oviposit (Washburn et al., 1988).

Termites in the family Termitidae (Chap. 24) and attine ants (Chap. 46) may cultivate fungi in "gardens" within the nest. Leaf-cutter ant workers cut leaves or petals from a variety of plants, return them to their underground nest, and arrange the harvested and processed plant materials on a nest to fuel a fungal garden. Cultivation of fungi by ants allows the ants to exploit a wide variety of plant species that might otherwise be toxic or indigestible. Some ants and fungi evolved from a common ancestral relationship, while other associations arose independently (Chapela et al., 1994).

Ambrosia beetles, the common name of beetles in several families (Scolytidae, Platypodidae, Lymexylidae), feed on fungi that grow along the

Figure 13.1 The protozoan *Lambornella clarki*: **a**, oval tro-phont and round theront stages that may be eaten and digested by mosquitoes; **b**, three cuticular cysts of *L. clarki* on the abdominal segments of a mosquito larva *(Aedes sierrensis)*. (Photo courtesy of and with permission of J. O. Washburn.)

galleries that the beetles excavate in wood. Adult females carry fungal spores in specialized cavities called **mycangia,** behind the head, along the thorax or at the bases of the coxae. Some beetles lack mycangia and transport spores that adhere to sticky secretions on their bodies. Female wood wasps in the genus *Sirex* (Siricidae) harbor spores of fungi in the genus *Amylosternum* in mucus-filled pouches along the female genital tract. Female wasps introduce the fungal spores during egg laying. The mucus secreted by the wasp with its eggs aids in fungal survival and reproduction. The fungus develops in the wood surrounding the egg and provides nourishment for the wasp larvae.

SYMBIOSES

The original and most general definition of **symbiosis** is "a close association or living together of two dissimilar organisms." More current usage of the term tends to describe relationships that are mutually beneficial. More specifically, such a relationship is called a **mutualistic** symbiosis. Examples are the protozoan and bacterial **symbionts** (or **symbiotes**) of some termite species that digest cellulose for their termite **hosts.** Killing the termites' intestinal symbiotes slowly starves the termites unless they can replenish their intestinal microflora. **Obligate symbiosis** describes a relationship in which both symbiote and host are required for the growth or reproduction of either host or symbiote. Where the host can grow or reproduce without the symbiote, the relationship is a **facultative symbiosis.** Symbiotes that cause no harm or benefit to their hosts are called **commensals.**

The role of microorganisms in aiding food digestion has been closely studied in termites, but we can expect that many insects in diverse orders use microbes for this purpose (Martin, 1987). The role of microbes in aiding insect digestion varies widely among insect groups. The consistent occurrence of bacteria, even in structures apparently specialized for containing bacteria, cannot be interpreted as proof of a beneficial role. The heads of fruit flies (Tephritidae, Diptera) contain a bulbous diverticulum of the esophagus that houses bacteria. Experimental investigations of the role of gut-associated bacteria in fruit flies failed to find any differences in growth and survival of fruit flies on a sterile diet compared to flies reared on a diet that included bacteria commonly found in the flies' guts. Although a variety of bacterial species can colonize the esophageal bulb, individual flies typically harbor only one type of bacteria. Many heteropteran species have caecal pouches of the midgut that contain distinctive bacteria. Newly emerged nymphs immediately feed on the surface of the eggshell, presumably to acquire the proper gut microflora. If the egg surface is sterilized before hatching, the nymphs that emerge do not grow or develop.

Symbionts that live within the host insect's cells are called **endosymbiotes**; those that live only outside cells are **ectosymbiotes.** Aphids (Aphididae), leafhoppers (Cicadellidae), scales (Coccoidea), and other homopterans house obligate endosymbionts in specialized cells, the **mycetocytes** (called **bacteriocytes** if the symbiote is bacterial). The mycetocytes may be loosely aggregated; or a specialized organ, the **mycetome** (or bacteriome), may house the mycetocytes. In some aphids, **facultative symbiotes** inhabit the sheath cells that surround the mycetome as well as the hemolymph. The term "facultative" describes the nonobligate nature of the association, meaning that not all aphids in a species have them. The intracellular habitat of the symbionts probably protects the symbionts from destruction by insect immune responses such as phagocytosis and encapsulation. Molecular studies have shown that primary symbiotes of aphids have genes for enzymes for the synthesis of some essential amino acids. Symbiotes are transmitted from mother to offspring by infecting the oocytes (or embryos in parthenogenic reproduction). The symbionts invade the ovariole or lateral oviduct to colonize eggs. Symbiotes of the human louse periodically migrate within the hemocoel from the mycetome to the lateral oviduct (Fig. 13.2). They penetrate the oviduct to invade the egg before the chorion is deposited. Circulating phagocytes consume any remaining symbiotes that fail to penetrate the oviduct. The developing embryo develops mycetomes, which the symbionts invade intracellularly. Many leafhoppers, membracids, and other hemipterans have at least two different types of symbiote that travel to the ovary within cells that detach from the mycetome, perhaps to evade attack by the insect immune system. Some hemipteran species harbor as many as five morphologically distinct types of symbiote (Buchner, 1965). Fungi and yeast are symbiotes found within caecal pouches in wood-feeding beetles such as anobiids and cerambycids. Endosymbiotic yeast inhabit the hemocoels of planthoppers and subsocial aphids.

Polydnaviruses are mutualist symbiotes of some species of braconid and ichneumonid wasp

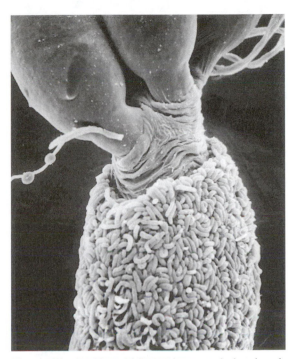

Figure 13.2 Symbiotes of human louse attached to lateral oviduct of female louse prior to invading eggs. The louse's phagocytes remove the symbiotes that fail to bore into the ovary. (Photos courtesy of and with permission of Mark Eberle.)

parasitoids. The viruses protect developing wasp larvae by suppressing immune responses of the host caterpillars. The viral DNA is integrated into the chromosomal DNA of the wasps and replicates exclusively in the ovaries of female wasps. When a female wasp lays an egg within a caterpillar, she also introduces viral particles, wasp venom, calyx fluid from the ovary, and proteins secreted in the ovary. Each element may act independently or synergistically in suppressing both humoral and cellular defensive reactions of the host. Although the polydnaviruses do not replicate in the parasitized caterpillars, viral genes are transcribed to produce messenger RNAs that are translated into proteins within the host caterpillar. It is thought that a viral protein may mimic host proteins to prevent encapsulation by the host caterpillar's hemocytes.

SEX RATIO ORGANISMS AND REPRODUCTIVE INCOMPATIBILITY

Prokaryotes that alter the ratio of male to female offspring are **sex ratio organisms.** A spiroplasma parasite of some species of *Drosophila* causes the offspring of infected female fruit flies to be all or mostly females. Male embryos die early during embryogenesis in infected mothers. A bacterium called "son killer" has the same effect on the parasitic wasp *Nasonia vitripennis.* Parasitoid wasps (Chalcidoidea) that harbor bacteria in the genus *Wolbachia* produce only females (thelotoky). Antibiotic or heat treatments that eliminate the bacteria cause the insects to produce both sexes. The parthenogenesis bacteria prevent chromosomes in unfertilized eggs from segregating. In all these cases, parents transmit sex ratio microorganisms to their offspring. Because the transmission usually is only from mother to offspring, the production of only female offspring is thought to maximize the production of the sex ratio organisms as long as the continued existence of the host insect is not threatened by the excessive elimination of males.

Bacteria can cause **reproductive incompatibility** between members of the same or closely related insect species. The presence of eggs fertilized by males from mothers that do not harbor the same microorganism prevents maturation of embryos. Eliminating the bacteria from insects by means of heat or antibiotics restores reproductive compatibility. Typically, the incompatibility is unidirectional, but it can be bidirectional. Incompatible crosses produce sterile males or no progeny, but the reciprocal crosses produce normal offspring. The associated prokaryotes invade the egg during oogenesis. Incompatibility bacteria interfere with the incorporation of the paternal chromosome in fertilized eggs. For example, transovarially transmitted bacteria confer reproductive incompatibility between the alfalfa weevils *Hypera postica* and *H. bruneipennis* (Curculionidae). Similar relationships have been demonstrated for bacteria found in *Culex pipiens* mosquitoes, *Drosophila* fruit flies, planthoppers, and numerous other insects. Reproductive incompatibility may favor speciation of the host insects by isolating otherwise reproductively compatible populations of insects.

INSECT PATHOGENS

Viruses, bacteria, fungi, protozoans, and nematodes are the principal pathogens of insects. Some of these are facultative pathogens that cause no harm unless they enter an insect's hemocoel or shift to a pathogenic form. For example, the bacterium *Serratia marcescens* can persist in soil without insects but will invade the hemocoel of insects through wounds. Once in the hemocoel, it is usually fatal to the infected insect. Most facultative pathogens enter the insect hemocoel after being ingested. Obligate parasites of insects multiply only within an insect host.

Viral Pathogens

Numerous types of virus infect insects (Table 13.2). The most important viruses for biological control are the **baculoviruses** (Baculoviridae), **entomopoxviruses** (Poxviridae), and **reoviruses** (Reoviridae). Baculoviruses include nuclear polyhedrosis viruses (NPVs) and granulosis viruses (GVs). Virus particles (virions) of baculoviruses of both types have double-stranded DNA and are embedded in a protective matrix of virally encoded protein. The combined virions and matrix are called an occlusion body. In the NPVs the protein matrix contains large numbers of embedded virions. The occlusion bodies of GVs contain a single virion but may be almost as large as those of NPVs. The baculoviruses have been found exclusively in arthropods and are believed not to infect vertebrates. They are found mostly in endopterygotes, especially the Lepidoptera. Viral occlusion bodies protect baculoviruses from deteriorating in the environment. Being sensitive to ultraviolet radiation, however, they persist for longest periods in soil and environments shielded from light.

Symptoms of viral diseases vary with the

TABLE 13.2
Selected Insect Viruses

Representative insect viruses	Nucleic acid; occlusions[1]	Taxa of principal insect hosts
Baculoviruses	ds DNA	
Nuclear polyhedrosis viruses (NPV)	Large polyhedra with numerous virions	Numerous orders; most in Lepidoptera
Autographa californica NPV		Noctuidae
Nonoccluded baculoviruses	No occlusion bodies	Numerous orders
Granulosis viruses (GV)	Single virion in an granular occlusion	Lepidoptera, Hymenoptera
Sawfly GV		Tenthridinidae
Reoviruses	ds RNA, isometric particles, usually in inclusion bodies	Lepidoptera, Diptera, Coleoptera, Neuroptera, Hymenoptera, Hemiptera
Rice dwarf virus		
Silkworm cytoplasmic polyhedrosis virus (CPV)	Transovarial transmission	Leafhopper vectors, rice
Entomopoxviruses	Inclusion bodies in cytoplasm	Lepidoptera
	ds DNA; spheroid or spindle inclusion body	Lepidoptera, Diptera, Coleoptera, Orthoptera, Hymenoptera
Polydnaviruses	ds DNA	Ichneumonidae, Braconidae
Rhabdoviruses	ss RNA, rod-shaped capsid	
Potato yellow dwarf virus		Leafhopper vectors, plants
Picornaviruses	ss RNA, no envelope	
Cricket paralysis virus		Orthoptera, Lepidoptera, others

[1]ds, double-stranded; ss, single stranded.

causal virus, the developmental stage of the host insect that is infected, the dosage of virus, and environmental factors, especially temperature. External symptoms of baculovirus diseases generally do not appear for the first 2 to 5 days after ingestion of virus by caterpillars. Infected insects typically begin to show changes in color and pallor several days after infection. Diseased larvae are less active and feed less. Some viral infections may prolong larval life by interfering with ecdysone metabolism, an effect that favors the virus in prolonging viral reproduction. Very young caterpillars may die within a few days of infection, but later instars of some species may live for weeks. Just prior to death, the integument of a baculovirus-infected caterpillar becomes fragile and easily torn. At death, the body contents are largely liquefied. The spread of virus to other caterpillars may also be aided by the disease-induced behavior displayed by some insect species whose larvae climb to the tops of plants just before dying. The virus-laden contents of the insects killed by the virus disease can then spread more widely to other insects feeding below. Granulosis viruses of sawflies (Tenthredinidae) that appear to infect only the midgut cells trigger viral epidemics that are sufficient to keep populations of forest pests below damaging levels for many years. To improve the usefulness of baculoviruses as biological pesticides, genetic engineering is being applied to NPVs to increase the speed at which they kill insects after infection and to increase viral host ranges.

The reoviruses have double-stranded RNA genomes enclosed in icosohedral viral coats. One type of reovirus, called cytoplasmic polyhedrosis virus (CPV), has been reported from over 250 species of insects, mostly Lepidoptera and Diptera. The polyhedra of CPVs, like the occlusion bodies of baculoviruses, protect the virions from environmental degradation. Some reoviruses replicate in both plants and insects and may be pathogens of hosts of both kinds.

The entomopoxviruses are a subfamily within the large virus family Poxviridae. They are structurally similar to poxviruses of vertebrates. These include the causal viruses of familiar human diseases such as smallpox and fowlpox (chicken pox), but the entomopoxviruses have restricted host ranges and do not infect vertebrates. Like the baculoviruses, they contain a double-stranded DNA genome.

Bacterial Pathogens

The relatively low prevalence of bacterial diseases in insects testifies to the effectiveness of insects' antibacterial defenses. Numerous bacterial species are facultative pathogens capable of rapidly invading wounded or freshly killed insects. The bacterial pathogens of insects most commonly used for insect control are species of the genus *Bacillus*. The spore-forming stage of *Bacillus thuringiensis (Bt)* contains crystalline proteins that are toxic to certain insect species. For example, the *B. thuringiensis* subspecies *kurstaki* kills numerous species of Lepidoptera. Other strains of *Bt* are used to control larval mosquitoes and black flies in aquatic environments *(Bt* subspecies *israeliensis),* or chrysomelid beetles *(Bt* subspecies *san diego).* Over 40 types of toxin protein have been identified by their chemical properties and insect host range. After ingestion by the host insect, the toxin crystals dissociate in the proper chemical environment (pH is especially important), bind to the midgut epithelium, and disrupt the midgut cells.

Bacillus popilliae is more host-specific than *Bt* and can play an important role in the natural regulation of its host insects, scarab beetles (Scarabaeidae, Coleoptera). Ingestion by root-feeding beetle grubs activates the spores of *B. popilliae* to multiply within the beetle's gut and invade the hemocoel, causing a condition that is called "milky disease" because infected grubs have a pale-colored abdomen. The large numbers of *B. popilliae* spores produced in an infected grub can survive in soil for years. Unlike *Bt,* however, *B. popilliae* cannot be fermented on an artificial medium for commercial production of biological insecticides. It multiplies only within host insects. In contrast to *Bt, B. popilliae* can set in motion an epidemic of milky disease that can persist in soil for years. The bacterium *Bacillus sphaericus* also produces crystalline proteins toxic to larvae of mosquitoes and other flies. It is less host-specific than *B. popilliae* but also cycles within natural populations, whereas the *Bt* subspecies *israeliensis* does not multiply or persist in significant numbers in natural environments after application for insect control. Another host-specific bacillus, *B. larvae,* causes American foulbrood disease of honey bees. The environmentally resistant spores survive for long periods in abandoned nests or hives of honey bee colonies destroyed by the disease.

Rickettsiae are bacteria that are intracellular parasites of animals, including insects. Many rickettsiae in insects cause disease. For example, *Rickettsiella melolonthae* causes the lethargy disease of scarab beetles (Scarabaeidae). Infected grubs turn an abnormally blue color and tend to rise to the soil surface rather than burrowing deeper, as healthy grubs normally do. This behavior may aid the dispersal of the pathogen to younger larvae, which are more numerous in the upper layers of soil.

Mollicutes are bacteria that lack any traces of a cell wall. As a result, they usually are spherical or amorphous but may be pleomorphic (taking on many shapes). Commonly known as "mycoplasmas," the mollicutes are the smallest free-living organisms. They are common and widespread in insects, but only a few are known to be pathogens. This may be partly because of the difficulty of detecting them microscopically or of culturing them. Some species of mollicute in the genus *Spiroplasma* have filamentous cells coiled in a helix. Some *Spiroplasma* species are plant pathogens transmitted by leafhoppers (Cicadellidae) (Table 13.3) and are also pathogenic to some of their insect vectors. *Spiroplasma apis* causes May disease (maladie de mai) of honey bees and may occur widely in flower nectaries during spring months. Most species of spiroplasmas and other mollicutes, perhaps numbering into the millions of species (Whitcomb, 1981), are associated with insects. They are seldom noticed, however, because they are barely discernible by ordinary light microscopy,

Protozoan Pathogens

Protozoa are a taxonomically diverse group of eukaryotic, single-celled microorganisms classified as a subkingdom in the kingdom Protista. They have subcellular organelles, a membrane-bound nucleus, and other features common to multicellular organisms. Protozoans commonly inhabit insect guts.

Only a small fraction of protozoan species are insect or animal pathogens, but protozoa transmitted by insect vectors cause important human and animal diseases such as malaria, African sleeping sickness, and Chagas' disease. One of the earliest studies of a protozoan disease of insects was Pasteur's investigation of pébrine disease of the silkworm. Although the French microbiologist never established the identity of the causal pathogen *(Nosema bombycis),* his systematic studies showed pébrine to be an infectious disease. Sanitary methods Pasteur developed for disease prevention continue to be used to this day. His early investigations of pébrine and other silkworm diseases led to his famous later studies of important human diseases such as rabies and cholera.

Most protozoan pathogens of insects are not highly virulent, instead causing chronic infections that do not kill the host insect. Some of the most common protozoan inhabitants of lepidopterous caterpillars are assumed to be nonpathogenic or only mild pathogens because they appear to universally infect some species that are very prolific pests. The infective stages of protozoans are usually spores or cysts. For example, the ciliates *Lambornella* spp. penetrate the integument through invasion cysts attached to the cuticle (Fig. 13.1). Vertical transmission (from parent to offspring) of protozoan parasites is common, either through invading the ovary (**transovarial**) or by contaminating the egg surface. All members of phylum Microspora, commonly called microsporidia, are obligate pathogens that multiply only in living cells. They are the most important group of insect pathogenic protozoans. *Nosema apis* causes serious, chronic disease in honey bees. Infected worker bees age more quickly and are sluggish. The hypopharyngeal glands of diseased bees atrophy, so these workers do not attend the queen bee, perhaps explaining the rarity of the nosema disease among queens.

Fungal Pathogens

Fungi are perhaps the most common insect pathogens. Agostino Bassi's nineteenth-century experiments with a fungal disease of silkworms first demonstrated the role of a microorganism as a cause of an animal disease. The evidence produced by Bassi and others disputed the then prevalent theory of spontaneous generation of insects and microbes. Spores, the infective stage of fungal pathogens, germinate on the insect cuticle. The developing fungal extensions, called **hyphae,** penetrate the cuticle to invade the hemocoel. Death ensues from fungal toxins or the conversion of insect tissues into fungal growth (Fig. 13.3b,c). Some fungal diseases kill the host insect while it is attached to plants or other external substrates. Grasshoppers, flies, and other insects characteristically climb to the tops of plant stems or other elevated locations to die after infection by fungi. This action facilitates the dispersal of the spores released from the cadaver.

The usefulness of fungal pathogens as microbial pesticides for controlling insect pests has been limited by the fungi's requirements for free moisture or high humidity and suitable temperatures for spore production (sporulation) or spore germination. Some fungi that are pathogenic to insects

Figure 13.3 Insect pathogens: **a**, an aquatic larva of the fly, *Chaoborus astictopus* (Chaoboridae), ruptured by infection of the coccidian protozoan, *Adelina* sp.; **b**, an adult of the weevil, *Metamasius hemipterus* (Curculionidae), infected with the fungus *Beauveria bassiana*, showing white mycelia emerging from between the segments; **c**, a nymph of the cicada, *Diceroprocta apache* (Cicadidae), infected with the fungus *Cordyceps sobolifera*, showing large, mature spore-producing structures at left; **d**, an adult mosquito, *Anopheles funestus* (Culicidae), with the large nematode *Empidomeris cozii* emerging from its body. (Photos courtesy of and with permission of Gerard M. Thomas and George O. Poinar, Jr.)

have never been grown in artificial media. Species in the fungal genera *Beauvaria, Entomophthora, Erynia, Metarhizium,* and *Nomuraea* are especially prominent pathogens for pest control. *Verticillium lecanii* is applied to control aphids and other homopterous pests in greenhouses.

Nematode Pathogens

Nematodes are small to microscopic, unsegmented worms that occupy a wide variety of habitats in soil, water, plants, and animals. Some nematode species are important insect pathogens; others are human or animal pathogens that are transmitted by insect vectors. Nematodes may be either facultative or obligate parasites. Some species of nematode infect their insect hosts passively through ingestion; others actively seek out and penetrate the host integument to enter the hemocoel.

Mermithid nematodes are striking parasites because of their large size relative to the host insect when they emerge (Fig. 13.3d). Early stages of the nematodes penetrate insects externally or after ingestion by the host. In an acceptable host, the worms increase in size through a series of molts. Their emergence from the insect kills the host. The nematodes form a cyst or resting stage in insects that are not suitable for development. Mermithids are frequently found in aquatic habitats.

Steinernematid and heterorhabditid nematodes (Fig. 13.4) are used for biological control because they are highly virulent, have a wide insect host range, can be produced commercially, and retain viability after prolonged storage. These nematodes have an environmentally resistant juvenile stage (dauer stage) that can be applied to soil or locations moist enough for nematode survival and movement. The nematodes enter and penetrate insects through the alimentary tract or enter spiracles and penetrate the tracheae to reach the hemocoel. Once in the hemocoel, the nematodes defecate to release bacteria (*Xenorhabdus* spp.) that are symbiotic in the nematode gut. The bacteria rapidly multiply and kill the insect host, while producing a powerful antibiotic that excludes other bacteria from colonizing the dying or dead host. The nematodes consume as food the bacteria and decaying insect host to develop into mature adults, mate, and lay eggs. Juvenile nematodes develop into adults or remain in the environmentally resistant dauer stage for as long as 5 years.

Figure 13.4 Insect pathogens: *Steinernema carpocapsae* nematodes emerging from dissected body of a parasitized caterpillar, *Spodoptera exigua* (Noctuidae). (Reproduced with permission from Tanada and Kaya, 1993.)

INSECTS AS VECTORS OF MICROBES

Insect **vectors** transport and transmit microorganisms to other organisms. Transmission may be the consequence of simply transporting a microbe to a site where the microbe can invade a suitable host. For example, numerous flower-visiting species of insects in several orders may be vectors of the bacterium *Erwinia amylovora,* which causes fire blight disease of apple and pear. The vectors carry the bacterium as an external contaminant from flower to flower. More often, insects that are vectors have special physical or physiological characteristics and behaviors that increase the chances of successful transmission of pathogens. The bacterium *Pantoea (Erwinia) stewartii* causes Stewart's wilt of maize and can be recovered from insects of many types, but only flea beetles (Chrysomelidae, Halticinae) infect plants through their feeding wounds. Similarly, any insect that feeds on the blood of a person with malaria may acquire the malarial parasite, but only certain mosquito (Culicidae) species in the genus *Anopheles* are vectors, that is, capable of transmitting the parasite to another person.

Vector Characteristics

Most ways of characterizing pathogen-vector relationships emphasize how the efficiency of vector transmission changes from species to species and over time. In addition, we may categorize the mechanisms by which insects transmit pathogens. **Vector specificity** refers to the degree of exclusivity of transmission between a particular pathogen and its vectors. For example, the beet curly top virus has only one known vector, the beet leafhopper, *Circulifer tenellus* (Cicadellidae). In contrast, many different aphid species (Aphididae), can transmit watermelon mosaic virus. **Transmission efficiency** or **vector competence** is a measure of how frequently an infectious vector successfully transmits a pathogen to a host. The brown citrus aphid, *Toxoptera citricida* (Aphididae), is a very efficient vector of citrus tristeza virus. In contrast, the melon aphid, *Aphis gossypii,* is a very inefficient vector of this virus; each new infection requires many thousands of aphids. Yet when it is extremely abundant, the melon aphid can be important in the spread of tristeza disease.

In the sections that follow, we discuss insect transmission of pathogens of plants and animals: viruses, prokaryotes, protozoans, nematodes, and fungi. In each of these pathogen categories we describe additional types of vector-pathogen relationship. Table 13.3 lists the pathogens, vectors, and transmission characteristics of some important plant and animal diseases that represent a diversity of vectors, vector-borne pathogens, and pathogen-vector relationships.

Virus Vectors

Most plant viruses have mobile insect, acarine, or nematode vectors because plants are not mobile and viruses do not disperse on their own. Some species of beetles and thrips are important vectors of plant viruses, but the most important groups of insect vectors of plant viruses are in the Hemiptera. The needlelike mouthparts and feeding behavior of sucking insects makes them ideal for transmitting viruses. Among the Hemiptera, aphids (Aphididae), leafhoppers (Cicadellidae), whiteflies (Aleyrodidae), and mealybugs (Pseudococcidae) have the greatest number of species that are plant virus vectors. Sucking, blood-feeding insects such as mosquitoes (Culicidae), black flies (Simuliidae), kissing bugs (Reduviidae), and sucking lice (suborder Anoplura) are important vectors of animal pathogens.

Changes in transmission efficiency over time (Fig. 13.5) are clues to the method or mechanism of transmission and may be features of the transmission process that are extremely important in trying to prevent the spread of particular diseases. The length of time required between a vector's acquisition of a pathogen, usually by feeding on an infected host, and its first successful inoculation of another host is the **latent period** (or extrinsic incubation period), usually estimated as an average or median value. Because not every insect capable of transmitting a pathogen will do so at every opportunity, it is not possible to determine absolutely that there is no latent period, only that such a phase is very short. Another dynamic characteristic of vector transmission of pathogens is the length of time for which a vector is able to continue to transmit after it has become infectious. Many aphid-borne viruses are **nonpersistent**; their aphid vectors are capable of transmitting to only one or a few plants in succession. The maximum efficiency of aphid acquisition of nonpersistently transmitted plant viruses occurs within seconds of feeding on virus-infected plants. Likewise, inoculation efficiency is at a maximum within seconds or minutes of beginning to feed on a new host. Molecules produced from the transcription of viral genes serve as "helper factors" that apparently aid aphid transmission. Viruses with defective helper factor genes are not aphid-transmissible, even though they may still be readily transmitted by mechanical inoculation in the laboratory. Mosquitoes or rabbit fleas transmit myxoma viruses nonpersistently to rabbits. The transmission process requires interrupted feeding, a pattern in which the mosquito or flea vectors do not take a long rest after a full blood meal.

The latent period is an important consideration in trying to control the spread of vector-borne diseases. Insecticides cannot prevent the spread of nonpersistent, aphid-transmitted viruses because aphids transmit so quickly after acquisition (no latent period). The key to control is usually to reduce the number of infected plants within the crop to prevent aphids from picking up the virus and transmitting it to nearby plants. Where aphids fly into the crop from a distance, repelling the landing aphids with reflective mulches or interfering with virus transmission with oil sprays on plant surfaces reduces virus spread. For pathogens transmitted only after a considerable latent period, insecticidal control of vectors often can reduce disease spread.

Circulative pathogens must enter the hemocoel of their vectors to be transmitted. **Propagative** pathogens multiply within their vectors, requiring a latent period of days to weeks, depending on temperature. Both circulative and propagative vi-

TABLE 13.3
Selected Pathogens Transmitted by Insect Vectors

Disease	Pathogen	Vector	Transmission
Viruses			
Animal			
Yellow fever	Togavirus	Mosquito (Aëdes aegypti)	Circulative, propagative
Bluetongue of cattle	Reovirus	Biting midges (Ceratopogonidae)	Circulative, propagative
Myxomatosis of rabbits	Myxomavirus	Mosquitoes (Culicidae) fleas	Nonpersistent, nonpropagative
Plant			
Barley yellow dwarf	Luteovirus	Aphids	Circulative, nonpropagative
Watermelon mosaic	Potyvirus	Aphids	Nonpersistent, nonpropagative
Squash mosaic	Comovirus	Leaf beetles (Chrysomelidae, Chrysomelinae)	Noncirculative, semi-persistent (hours to days), nonpropagative
Lettuce infectious yellows	Geminivirus (ds DNA)	Whiteflies (Aleyrodidae, Bemesia spp.)	
Bacteria			
Animal			
Plague	*Yersinia pestis*	Fleas	Noncirculative, persistent, propagative in gut
Epidemic typhus	*Rickettsia prowazekii*	Human body louse (Pediculus humanus)	Noncirculative, propagative in gut cells only
Relapsing fever	*Borrelia recurrentis* (spirochaete)	Human body louse (Pediculus humanus)	Noncirculative; propagative, but dies in hindgut
Tularemia	*Franciscella tularensis*	Horse flies	Nonpersistent on mouthparts
Plant			
Fire blight of pome fruit trees	*Erwinia amylovora*	Flying, flower-visiting insects	External contaminant on vectors
Stewart's wilt of corn (maize)	*Pantoea stewartii*	Flea beetles	Noncirculative, persistent
Pierce's disease of grapevines, citrus variegated chlorosis, others	*Xylella fastidiosa*	Xylem-feeding Hemiptera (Cicadellidae: Cicadellinae, Cercopidae)	Noncirculative, persistent, propagative in foregut
Pear decline (PD)	PD phytoplasma	Psyllid (Cacopsylla pyricola)	Circulative, propagative
Elm yellows (EY)	EY phytoplasma	Leafhoppers (Deltacephalinae)	Circulative, propagative
Corn (maize) stunt	*Spiroplasma kunkelii*	Leafhoppers	Circulative, propagative

(cont.)

TABLE 13.3 (*continued*)

Disease	Pathogen	Vector	Transmission
Protozoa			
Animal			
Chagas' disease	*Trypanosoma cruzi*	Triatomine reduviids	Propagative in gut, feces infect wound
African sleeping sickness	*Trypanosoma* spp.	Tsetse flies (*Glossina* spp.)	Circulative, propagative
Malarias	*Plasmodium* spp.	*Anopheles* mosquitoes	Cyclodevelopmental
Leishmanias	*Leishmania* spp.	Sand flies (Psychodidae: *Phlebotomus* spp.)	Cyclodevelopmental in gut and foregut
Plants			
Oil palm hartrot	*Phytomonas staheli*	Pentatomidae	Cyclopropagative
Milkweed trypanosomes	*Phytomonas emassiana*	Milkweed bugs (Lygaeidae)	Cyclopropagative
Nematodes			
Animals			
Filariasis (Brugian and Bancroftian)	*Wuchereria bancrofti, Brugia malayi*	Mosquitoes (*Culex* spp.)	Cyclodevelopmental
Onchocerciasis (river blindness)	*Onchocerca volvulus*	Black flies (Simuliidae)	Cyclodevelopmental
Canine heartworm	*Dirofilaria immitis*	Mosquitoes	Cyclodevelopmental
Plant			
Pine wilt	*Bursaphelenchus xylophilus*	Pine sawyers (Cerambycidae)	Carried in beetle tracheae
Fungi			
Plant			
Dutch elm disease	*Ophiostoma ulmi*	Elm bark beetles (Scolytidae)	Sticky spores carried externally, invade feeding wounds
Oak wilt	*Ceratocystis fagacearum*	Sap beetles (Nitiduidae)	Vectors carry spores to fresh wounds
Endosepsis of fig	*Fusarium moniliformae*	Fig wasps (Agaonidae)	Carried externally during pollination (Chap. 10)

ruses can be transmitted for extended periods, often for the life of the vector. The time required for a circulative and propagative virus to enter the hemocoel and pass into the salivary glands constitutes a latent period of hours to days. If the virus must multiply within the vector before it can be transmitted, the required latent period may be days to weeks. Yellow fever was the first **arbovirus** (**ar**thropod-**bo**rne virus) that allowed clear experimental demonstration that an insect (the mosquito *Aedes aegypti*) transmitted the causal agent. This discovery led within a short time to good control of yellow fever because the vector occurred most commonly near human habitations

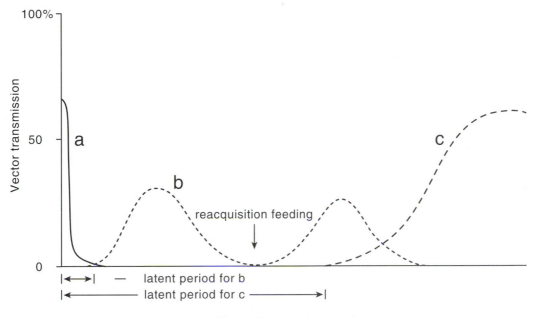

Figure 13.5 Changes in transmission efficiency over time after acquisition: **a**, nonpersistent (hours) transmission; **b**, persistent (days), circulative, nonpropagative transmission with the vectors feeding for a second time on an infected host (arrow); **c**, persistent (weeks), circulative, propagative transmission. The time between acquisition and first transmission is the latent period for **b** and **c**.

and because people with yellow fever served as sources from which *A. aegypti* could acquire the virus. Reducing mosquito populations and preventing mosquitoes from biting infected persons reduced yellow fever transmission.

Some other virus diseases are **zoonotic** diseases (or **zoonoses**), meaning that infected animals are primary vertebrate hosts from which vectors acquire the virus. For example, humans contract jungle yellow fever in parts of Brazil where populations of monkeys harbor the virus in forest canopies. Humans become infected when they contact mosquito vectors displaced by logging or other human incursions. Mosquito-transmitted western equine encephalitis virus occurs mainly in populations of wild birds, in which the disease is usually mild or even symptomless. When bird populations leave, the remaining mosquito vectors fly beyond their customary range to infect unusual hosts such as horses and humans, causing severe illness.

Vectors of Prokaryotic Pathogens

Bacteria transmitted by insect vectors have caused important human diseases such as bubonic plague and epidemic louse-borne typhus. Flea vectors maintain the causal bacterium of plague, *Yersinia pestis,* in wild rodent populations in Asia-Oceania, Africa, and the Americas. Fleas acquire the bacterium from feeding on infected animals. The plague bacterium multiplies rapidly within the flea's crop to form a bacterial mass in the proventriculus that blocks the flea's digestive tract. Infected fleas repeatedly attempt to regurgitate during further feeding to clear their blocked foreguts. This action introduces bacteria into feeding wounds in the vertebrate host. Maintenance of the zoonotic cycle of plague in rodents requires that the immune reactions of the rodent population be neither too high nor too low. Highly resistant individuals or species of rodents clear the bacteria from the bloodstream and become immune so quickly that the plague bacteria cannot survive in a rodent population. On the other hand, animals or vectors that die too quickly from plague also extinguish the reservoir of infected animals necessary for the bacteria to persist in the field. Humans contract the disease from infectious fleas or by contact with the blood of infected animals.

The plague bacteria spread through the circulatory system of the vertebrate host and cause swellings (buboes) of the lymph nodes. At this stage, timely treatment with antibiotics is highly successful in curing the disease. However, if the bacteria reach the lungs, the resulting pneumonic phase of the disease is highly infectious for other humans who inhale droplets expelled by coughing. Pneumonic plague is highly virulent and can kill infected people if antibiotic therapy is not begun before symptoms of fever and coughing appear. Plague caused influential major pandemics throughout history. The Black Plague killed over 25 million people in Europe during the fourteenth century, with major historical and lasting social and cultural impacts. European art from the seventeenth century (Chapter photo) depicts the terror of plague epidemics described by ancient historians.

The bacterial pathogen of epidemic typhus, *Rickettsia prowazekii,* multiplies in the cells that line the gut of *Pediculus humanus* (Pediculidae), the human body louse vector, rapidly killing the lice vectors. The transmission process is crude but effective. Victims scratch crushed lice or their feces into wounds caused by the irritating bites of the lice. Epidemic typhus occurs as a devastating disease of human populations under stressful conditions such as war or natural disasters that make regular bathing and changes of clothing impossible. Under these conditions body lice thrive, and symptomless human carriers of the rickettsias revert to releasing rickettsia into their own bloodstreams to be acquired by feeding body lice. Typhus, not bullets or cannon fire, was the major decimator of Napoleon's army that invaded Russia in 1812. The typhus epidemic along the eastern front in World War I had major impacts on Russian and Serbian armies, leading to their retreat and surrender.

Flea beetles (Chrysomelidae) transmit the bacterium *Pantoea (Erwinia) stewartii* to maize through feeding wounds, causing Stewart's wilt of maize. Various strains of the bacterium *Xylella fastidiosa* cause debilitating or lethal diseases of grape, peach, citrus, and numerous tree species in the Americas. The bacterium exclusively inhabits the xylem (water-conducting) tissues of plants. Vectors are xylem-feeding insects in several hemipteran families (Cicadellidae, Cicadellinae; Cercopidae, Cicadidae). Both bacterium and vectors are extreme organisms. The xylem-feeding vectors have the highest feeding rates of any terrestrial animal, consuming hundreds of times their weight daily. The bacterium thrives in the high-velocity (10 to 50 cm/s), nutritionally dilute stream of ingested sap in the vectors' foreguts. Vectors transmit the bacterium either immediately or after a very short latent period but retain their infectivity indefinitely unless they molt. These observations point to the foregut as the critical site for bacterial retention and release during transmission.

Phytoplasmas and spiroplasmas are small bacteria that lack a cell wall (class Mollicutes) and occupy the sugar-rich phloem tissues of plants. These plant pathogens usually complete the complex process required for vector transmission only in certain species of phloem-feeding leafhopper (Cicadelloidea), planthopper (Fulgoroidea), or psyllid (Psyllidae) vectors. Transmission requires a latent period of several weeks while the phytoplasmas or spiroplasmas multiply and pass from the gut to the hemocoel and eventually to the vector's salivary secretions, where they enter new plant hosts during the feeding of their insect vectors. Each of these steps represents a hurdle the pathogen must overcome, and completion of the series results in the high degree of vector specificity and interspecific differences in vector competency of these pathogens.

Vectors of Protozoans

Globally, human diseases caused by protozoans are among the most important diseases borne by insect vectors. Once a serious problem in subtropical regions, malaria now is confined mostly to the Tropics but is still the world's most important infectious disease. Mosquitoes in the genus *Anopheles* are the sole vectors of four species of *Plasmodium* that cause human malarias. Malarial parasites are specific to a particular vertebrate host. The basic malarial cycle was first observed in bird malarias. Malarias of birds, lizards, monkeys, and other animals have similar cycles but do not affect humans. Bird malarias accidentally introduced into Hawaii along with exotic birds may have eliminated some avian species native to the islands.

The developmental cycle of the malarial parasite is complex in both mosquito and vertebrate hosts. A simplified summary of the transmission cycle in both hosts is illustrated in Figure 13.6. Malaria has a **cyclopropagative** transmission process, meaning that the *Plasmodium* parasites change in form and function as they invade and multiply within various organs of the vector. Adult female mosquitoes acquire two forms of the malarial parasite, micro- and macrogametes, which are contained in infected red blood cells (erythrocytes) of the vertebrate host. In the midgut of the mosquito the microgametes quickly divide into forms that appear to have many external flagella. The "flagella" actually are separate haploid sexual cells that separate (exflagellate) and rapidly seek out and fuse with macrogametes to form a zygote that develops into a form called an **ookinete.** The motile ookinete burrows into the mosquito's gut epithelium to form an **oocyst.** The growing oocyst divides meiotically to produce thousands of haploid **sporozoites.** The speed of this process is temperature-dependent and does not continue below 16°C. The spindle-shaped sporozoites enter the host mosquito's hemocoel and migrate to the salivary glands. Mosquitoes inject the sporozoites into animal hosts during subsequent blood meals. In the bloodstream, the sporozoites reach the liver and eventually form large cells, called **schizonts,** each of which contains thousands of **merozoites.** The growing merozoites rupture liver cells. Some merozoites continue to infect other liver cells; some invade the red blood cells. In the erythro-

cytes, the merozoites transform into **trophozoites.** During subsequent cycles within the erythrocytes, trophozoites become schizonts that in turn produce merozoites and repeat the erythrocytic cycle. The synchronous release of merozoites from the red blood cells provokes the host's immune response: the characteristic fever and chills of a malaria attack. The periodicity of the erythrocyte schizogenous cycles are 48 hours for tertian ("third-day") malaria caused by *Plasmodium falciparum* and 72 hours for quartan ("fourth-day") malaria *(Plasmodium malariae).*

Numerous environmental factors govern the type and rate of transmission of malaria: minimum and average temperatures, rainfall and water accumulation leading to mosquito breeding success, the survival rate, the biting rate on humans and vector competence of mosquitoes, the level of immunity or chemotherapy in the human population, and the resistance of the malarial parasites to preventive medicines. All these factors are taken into account in devising strategies to prevent or control malaria. Mosquito vectors have become resistant to insecticides as a result of excessive insecticide treatments to prevent malaria transmission. In many parts of the world *Plasmodium* species are resistant to the medicines taken to prevent or cure malaria.

Flagellated protozoans cause two other major human diseases in the Tropics: African sleeping sickness and Chagas' disease. Tsetse flies (Glossinidae: *Glossina* spp.) are vectors of protozoan species in the genus *Trypanosoma,* which cause **trypanosomiasis** in animals and humans in Africa. Infection causes the disease known as nagana in domestic animals such as cattle and horses and as African sleeping sickness in humans. The pathogens are zoonotic in wild animals, which trypanosome infections do not adversely affect. The cyclopropagative pathogen changes morphologically as it migrates and multiplies in moving from the blood meal, through the gut, and eventually to the salivary glands. The tsetse fly has been called the "guardian of Africa" because trypanosomiasis excluded the use of domestic animals and limited

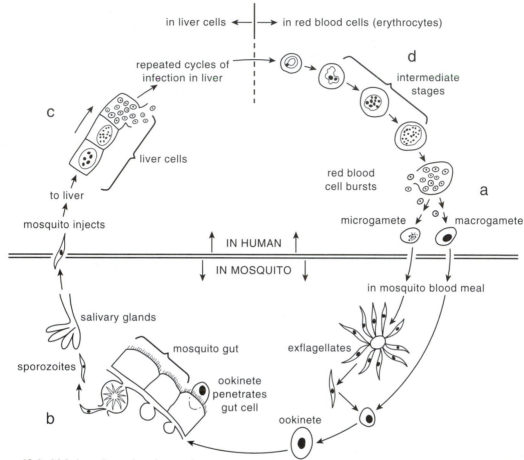

Figure 13.6 Malaria cycle: **a**, the micro- and macrogametes taken up in the mosquito's blood meal combine sexually in the mosquito's gut and the resulting zygote forms an ookinete, which penetrates the midgut cells, developing into an oocyst; **b**, the sporozoites released from the burst oocyst penetrate the salivary glands to be injected with saliva into the human bloodstream when the mosquito feeds; **c**, the sporozoites infect the liver cells in a series of cycles; **d**, eventually infecting the red blood cells, where they go through several intermediate stages to become haploid merozoites that burst the human erythrocytes and develop into micro- and macrogametes.

human habitation of much of tropical Africa, preserving the wildlife that were tolerant to these parasites. This exclusion of domestic beasts of burden because of nagana required greater human labor instead and undoubtedly contributed to the institutionalization of human slavery in central Africa.

Triatomine bugs (conenose bugs or kissing bugs) in the family Reduviidae are vectors of

Trypanosoma cruzi, the pathogen of Chagas' disease. This disease is zoonotic, with the trypanosomes infecting domestic animals such as chickens, dogs, rats, and other wild or peridomestic animals. The principal vector, *Rhodnius prolixus* (Reduviidae), typically defecates on the host immediately after a blood meal. Scratching the feeding wound introduces the pathogen from the vector's feces into the host's bloodstream. Chagas' disease seldom kills its victims directly. It produces symptoms of chronic lethargy, irritability, and increased susceptibility to other diseases. Combined with poor nutrition and other diseases of poverty, it lowers life expectancy while causing major human suffering. Sand flies (Psychodidae) transmit protozoans in the genus *Leishmania.* The parasites are cyclopropagative within the alimentary tract of the sand flies and are regurgitated by the flies during feeding to infect the vertebrate host. The resultant diseases cause internal or external swellings or ulcers in humans and other animals. The diseases are also zoonotic in wild and domestic animals.

Insects as Vectors of Nematode Pathogens

Filariases are nematode-caused diseases with mosquito vectors. Large numbers of juvenile filarial worms **(microfilariae)** accumulate in the vertebrate host's lymphatic system to produce swellings that may eventually grossly distort the limbs, breasts, or groin to produce a condition called elephantiasis. Unlike previously discussed examples of vector-borne pathogens, where disease results from a single infection, the severe Bancroftian *(Wuchereria bancrofti)* and Brugian *(Brugia malayi)* forms of filariasis result only after many thousands of individual infections from infective mosquitoes. The filarial worm bores through the soft cuticle of the labium of the mosquito vector and crawls onto the host during mosquito feeding. The tiny (1.5 mm), eyeless worm must find and enter the feeding wound before it clots. Once in the host's bloodstream, the worm undergoes another molt to sexual maturity and seeks a mate. Mating encounters are likely to be successful only if a sufficient concentration of adult worms is present in the host. Mated female worms produce thousands of eggs, which then hatch into microscopic microfilariae. The microfilariae circulate in the blood and lymphatic system. At dusk, when *Culex* mosquitoes are most likely to feed, the tiny worms migrate to the peripheral circulation in capillaries beneath the skin to maximize the chances of being acquired by a feeding mosquito.

For this reason, blood tests for filarial infections must be taken during early evening. In parts of the South Pacific where day-feeding *Aedes polynesiensis* is the predominant vector, the microfilariae appear in the peripheral capillaries of humans most frequently during afternoon. Similar circadian synchrony of vector feeding with the appearance of microfilariae in the peripheral blood is common for many other worm-vector associations. The worms taken up by mosquitoes penetrate the gut wall of the mosquito and undergo development changes in the flight muscles, finally reaching the mosquito's head. Here they await a chance to exit during the mosquito's next blood meal.

Cyclodevelopmental transmission describes developmental changes of parasites without multiplication within the vector. Onchocerciasis is a filarial disease that can lead to blindness when microfilariae invade the eye tissues and die. The black fly (Simuliidae) vectors breed in rivers in parts of Africa and South America, thus the disease's common name "river blindness." The pathogen *Onchocerca volvulus* has a complex cyclical development in the adult black flies of the genus *Simulium.*

The nematode *Bursaphelenchus xylophilus* causes pine wilt disease by invading the woody tissues of pine trees. The nematodes crawl into the tracheae of adult pine sawyer beetles (Cerambycidae) (Fig. 13.7) while the beetles bore galleries in nematode-infested trees. After the beetles have moved to different trees to feed or excavate egg-laying tunnels, the nematodes exit their beetle vectors to infect the tree. Other species of wood-boring beetles transport fungal-feeding rhabditoid nematodes externally. The nematodes feed on fungi or decaying matter in the beetle galleries but cause no direct harm to their beetle vectors or plants.

Insects as Intermediate Hosts of Flatworms and Tapeworms

Insects may serve as intermediate hosts of tapeworms (Order Cestoda) and helminth worms of other types, such as flukes and flatworms (Phylum

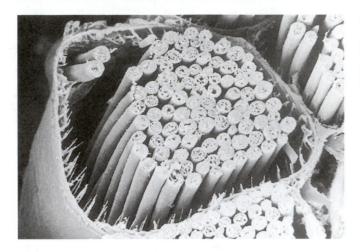

Figure 13.7 The pine wilt nematode, *Bursaphelenchus xyllphilus*, in the tracheae of sawyer beetles (Cerambycidae). The beetles bore galleries in the diseased trees and transport the nematodes to other trees. (Photograph courtesy of and with permission of M. Linit.)

Platyhelminthes). The insects are not truly vectors for these worms because they do not initiate infection of another host. Instead, the tapeworms infect animals that eat the insects containing the worms. For example, larval fleas or lice consume eggs of the dog tapeworm, *Dipylidium caninum.* Dogs or cats acquire the still viable tapeworms by consuming the lice or adult fleas. Flukes in the genus *Prosthogonimus* emerge from snail hosts and seek out dragonfly naiads. The flukes enter the anus of the naiad when it takes in water for respiration. Birds that eat the infected naiads or adult dragonflies become infected with the worms.

Vectors of Fungi

Fungi transmitted by insect vectors cause important plant diseases. Bark beetles (Scolytidae) are the main vectors of fungal plant pathogens in the genera *Ceratocystis* and *Ophiostoma.* Bark beetles such as *Scolytus multistriatus* (Scolytidae) are the principal vectors of the fungus *Ophiostoma ulmi,* which causes Dutch elm disease. The sticky spores of the Dutch elm fungus line the galleries beneath the bark of diseased elms and adhere to the bodies of adult bark beetles that emerge from the galleries. The adults then disperse to new elm trees and introduce spores into feeding wounds made by the beetles on healthy elm twigs. Trees weakened by Dutch elm disease are especially attractive to egg-laying adult beetles and are more susceptible to bark beetle attack, thus accelerating the spread of both the disease-causing fungus and bark beetles.

In a similar fashion, various bark beetle species spread fungi of the genus *Ceratocystis* to cause blue stain disease of conifers. Sap beetles (Nitidulidae) pick up spores of the oak wilt fungus, *Ceratocystis fagacerum,* from sap that oozes from fungal cushions found beneath the bark of diseased trees. Subsequently, the beetles may introduce the spores to fresh wounds on uninfected trees. The *Amylosternum* yeasts mentioned earlier in the discussion of symbiotic relationships are plant pathogens only in conjunction with the oviposition wounds made by their *Sirex* wasp hosts and vectors. In this situation the vector is also a direct component of the resulting plant disease.

SELECTED REFERENCES

General insect pathology is reviewed by Beckage et al. (1993), Cantwell (1974), Steinhaus (1949), and Tanada and Kaya (1993); insect nematology by Gaugler and Kaya (1990) and Poinar (1979, 1983); general relationships of microbes with insects by

Steinhaus (1946); insect symbiosis with microorganisms by P. L. Baumann et al. (1995), Buchner (1965), Campbell (1989), Douglas (1989, 1994), and Margulis and Fester (1991); symbiotes in lice by Eberle and McLean (1983); polydnaviruses by Lavine and Beckage (1995). Insects and fungi are discussed by Martin (1987) and Wilding et al. (1989). Vectors of plant pathogens are treated by Carter (1973), Linit (1988), and Purcell (1989); animal disease vectors by Dunn (1969), Hoeppli (1969), James and Harwood (1969), Lehane (1991), and Snow (1974). The series *Advances in Vector Research* (from 1983) provides up-to-date reviews on vectors of animal and plant pathogens.

14
PEST MANAGEMENT

Caterpillars of the western grape leafskeletonizer, *Harrisina brillians* (Zygaenidae), consuming a grape leaf. Populations of this pest are usually adequately controlled by natural enemies (primarily a baculovirus and a tachinid parasitoid fly), but rare outbreaks require insecticidal treatment. (Photograph courtesy of and with permission of Diane Stark.)

Any insect that interferes with human welfare or aesthetics can be considered to be a pest. Even beneficial insects are pests in certain circumstances. For example, honey bees provide valuable services as pollinators (Chaps. 1 and 7), but their stings can provoke severe allergic reactions. The public image of the aggressive Africanized honey bee pursuing and stinging people looms larger than its service in pollination and honey production. Ladybird beetles (Coccinellidae) are widely recognized for their helpful role as insect predators, but a larval ladybird beetle crawling about fresh produce is automatically considered to be a pest by most shoppers.

Pest management draws on and integrates many of the facts and principles of insect biology covered in earlier chapters. In this chapter we discuss approaches to preventing pest damage to agricultural crops, forests, and commercial products (including structures). Many of the principles outlined here apply to control of medically important pests (Chaps. 12 and 13). Modern methods of pest management emphasize interactions of insects with the environment. Pest management methods, whether based on pesticides or agronomic practices such as plowing or other forms of tillage, can have major impacts on the environment. Pest management uses tools and ideas from population biology (Chap. 8). The evolution of plant attributes that prevent or reduce insect herbivory (Chap. 10) provides a basis for breeding plants that resist insect pests.

PEST DAMAGE

Direct pests cause immediate injury or disturbance (Fig. 14.1). Biting flies or insects that feed on marketable fruits such as apples or on furniture are examples. **Cosmetic damage** is an injury that is largely or totally aesthetic. Thrips feeding on rinds of oranges or skins of grapes create noticeable scars on the fruit (Fig. 14.2a) but do not alter its nutritional value or flavor. Thrips' scarring blemishes grapes as fresh fruit but is inconsequential for grapes used to make raisins or wine.

Indirect pests create conditions that lead to damage by other factors. Thrips can cause immense damage to fruit or flower crops as vectors of tomato spotted wilt virus even when their numbers are so low that direct injury by the thrips' feeding is not noticeable. Honeydew excreted by scale insects (Coccoidea) or whiteflies (Aleyrodidae) causes indirect cosmetic injury to plants by supporting the unsightly growth of otherwise harmless sooty mold fungi on leaves or fruit (Fig. 14.2b). If the sooty mold layer is thick and widespread on most leaves, photosynthesis can be significantly reduced because of shading to the leaves. Honeydew can cause direct injury if it falls on cotton fibers or another plant product that directly affects the usefulness of the crop.

Complete or total control of insect pests is usually not attainable except for special cases such as protecting valuable historical documents or artifacts from insect feeding. Biological control of pests by natural enemies or disease requires that some level of the pest population persist; otherwise the natural enemies will starve or emigrate. The concept of an **economic injury level** (EIL), at which the amount of damage equals the cost of preventing damage, guides decisions about what level of damage should trigger control actions (Fig. 14.3). Because pest (and plant) populations are not static but are constantly changing, control action should be taken at an **economic threshold** (ET) of pest density. This is defined as the level of pest density at which action must be taken to prevent the economic injury level from being reached.

The characteristic abundance or general equilibrium level of population density for various insects may vary enormously from one pest species to another. The codling moth, *Cydia pomonella* (Tortricoidea), is a serious pest of apples at much lower densities than caterpillar pests that attack

Figure 14.1 Direct pest damage to cotton fruit by the cotton bollworm, *Helicoverpa zea* (Noctuidae). (Photograph courtesy of and with permission of University of California at Davis, IPM Education and Publications.)

Figure 14.2 Cosmetic injury: **a**, thrips scarring of grapes; **b**, rhododendron leaf at top shows sooty mold growing on sugary excrement (honeydew) from whiteflies; undamaged leaf is below. (a, photograph courtesy of and with permission of J. K. Clark, University of California at Davis, IPM Education and Publications; b, photograph by A. H. Purcell.)

Various combinations of economic and ecological modeling have produced methods to objectively calculate EILs. In general, the EIL should equal the cost (C) of control on a per production unit basis (e.g., dollars per hectare) divided by the product of the market value (in dollars per kilogram) of the crop (V), the yield loss caused per pest (D, in kilograms per insect), and the decimal fraction of pests reduced by the control used (K):

$$EIL = \frac{C}{VDK}$$

Changes in the independent variables V, D, and K with time or other variables may further complicate the calculation of EIL. The effects of insect density on yield loss may vary considerably with time. For example, pests that attack fruit may be tolerated in higher levels before bloom time or after a vulnerable stage in crop maturity has passed.

The relationship of the typical or characteristic densities of pest populations to the EIL determines whether an insect is never a pest (populations always remain below the EIL), is a minor or occasional pest (populations occasionally or rarely

forage crops such as alfalfa. Much higher densities of caterpillars can be tolerated in alfalfa hay without concern for economic damage. The EIL is inversely related to the unit value of the crop or commodity to be protected. For high-value crops such as apples or other fresh fruits, small percentages of insect infestation can provide economic justification for control measures. Only very substantial damage to low-value crops such as hay would meet or exceed the cost of insecticide applications. Changes in the cost of control have the opposite effect. As control measures become more expensive, the EIL increases because a greater amount of damage must be sustained to equal or exceed the cost of control.

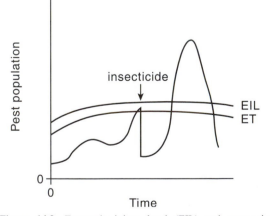

Figure 14.3 Economic injury level (EIL) and economic threshold (ET) for pest density. The pest population (unlabeled curve) treated with insecticide (arrow) dropped below the ET, then rebounded above the EIL. The variation of the EIL over time depends on changes in the crop's vulnerability to pests and to other factors.

exceeding the EIL), or is a major pest (populations always near or above the EIL). Pests whose numbers either usually or always exceed the EIL are **key pests** if pest management programs for a particular crop must always contend with their control. The codling moth is a key pest of apples and pears in North America and Europe because its damage to fruit is always above the EIL. If insecticides are not used to control this moth, other control measures are needed (e.g., application of pheromones to disrupt mating, application of insect viruses and natural enemies). Using insecticides may disrupt natural enemies that usually keep densities of occasional pests below the EIL. Using noninsecticidal control methods may allow pests that are normally suppressed by insecticides to reach damaging levels.

To use an economic threshold to trigger the timing of pest control actions, pest populations must be monitored by sampling (Chap. 8). The reliability and expense of sampling and the predictability of changes in pest population have to be taken into account in devising threshold levels. Economic thresholds also require an understanding of how pest populations relate to damage. The EIL may change as the crop matures. For example, lygus bugs, *Lygus* spp. (Miridae), feeding on the flower buds of cotton plants can exceed the EIL early in the plant's fruiting cycle because their feeding causes the eventual loss of cotton fruits (bolls). Higher densities of damage by lygus bugs feeding later in the season may actually increase cotton yields by eliminating buds that would not mature, but would divert some of the plant's reserves from fruits that could be expected to ripen.

A BRIEF HISTORY OF PEST CONTROL

In the preinsecticide era, traditional agricultural practices intentionally or unintentionally incorporated a variety of pest control methods. A degree of plant resistance resulted from the greater survival and seed production of plants that better resisted local pests and diseases and were adapted to local climates. Natural enemies and diseases of pest insects provided a degree of natural control. In regions where the characteristic abundance of key pests always exceeded levels at which the crop could be profitably produced, farmers simply did not attempt to grow the crop. Early farmers often had little appreciation or understanding of how insects and mites damaged plants, apart from the most obvious cases of damage to foliage or fruit. As agriculture increasingly applied scientific methods, pests and their damage came under greater scrutiny. The result was new information on pest identification, life histories, natural enemies, and diseases. Some pests could be avoided or their damage minimized by manipulating normal agricultural practices such as the date of planting or harvesting.

The first recorded intentional use of natural enemies for pest control was in 324 BC in China, where farmers placed colonies of predaceous ants in citrus orchards to control pests. In this example, native insects were used to control native pests, a practice that has continued to the present.

Scientists realized that exotic insects often became pests or worse pests after arriving in a new area without the natural enemies that had coevolved with them. In their place of origin, these insects were often held in check by various natural enemies. Freed from regulation by these natural enemies, the exotic insects became abundant enough to reach pest status. Logically, such exotic pests should be controlled by importing specific natural enemies from the region where both the pest and the natural enemies originated. The first spectacular success with this approach, called **classical biological control,** was the control of cottony cushion scale on citrus in 1889. But to most farmers interested in pest control, biological methods were ignored or underutilized for decades afterward. Early insecticides such as soaps, oils, and arsenic compounds were widely used where damage justified the expense.

Beginning in the 1940s, the development of synthetic pesticides rapidly led to widespread and intensive use of insecticides to control agricultural and medical pests. The new chemical pesticides had many advantages. They were effective, fast-acting, and inexpensive compared to pest damage, and they offered a choice of control action for individual farmers. Negative aspects of pesticides that offset these advantages were soon recognized. Insects and mites that had been only occasional or minor pests, or never were considered to be pests because of their rarity, began to appear in damaging numbers following treatments with some insecticides. Before pesticide treatments, these **secondary pests** had been limited to nondamaging levels by natural enemies, but they reached abnormally high populations after pesticide treatment reduced populations of key predators and parasitoids (Fig. 14.4).

A related effect was **pest resurgence,** the rapid increase of pest populations to even higher than normal levels following pesticide applications (Fig. 14.4). Again, this result could be attributed to disruption of preexisting levels of natural enemies. In both phenomena, pest insects were less sensitive to pesticides than were natural enemies. In a few cases, notably spider mites (Tetranychi-

dae, Order Acari), pesticides even stimulated or increased pest reproduction.

Both pest resurgence and secondary pest outbreaks tended to intensify when the same insecticides were used repeatedly. By comparing the insecticide sensitivity of field populations of pests to the sensitivity of reference populations not exposed to insecticides, entomologists documented the evolution of genetically controlled **resistance** to insecticides. Resistance is now a major problem for pest management in many crops worldwide.

Other problems in using pesticides were recognized later. Investigations of the fate in the environment of the insecticide DDD, chemically similar to the better known insecticide DDT (**di**chloro**d**iphenyl**t**richloroethane), showed that concentrations of DDD in animal tissues increased at higher levels in the food chain. Clear Lake, California, was treated three times over 3 years with DDD. Chemical analyses detected only trace amounts of the chemical, about 0.01 part per million (ppm), in the lake's water immediately after

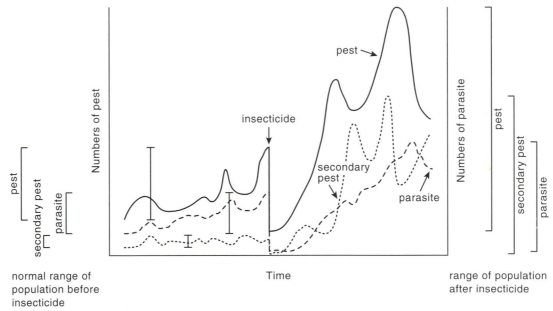

Figure 14.4 Resurgence of pest population (solid curve) due to insecticide treatment (arrow) severely reduces a key parasite population (dashed curve). Insects that are habitually below injurious levels (dotted curve) can become secondary pests when released from control by natural enemies eliminated by insecticides. Brackets indicate the characteristic abundance (normal range) of the three populations.

treatment. Algae had about 5 ppm, small fish 40 to 250 ppm, larger fish 340 to 2500 ppm, and fish-eating birds an average of 1600 ppm (Rudd, 1964). This **biological amplification** was especially prominent in DDT and chemically related pesticides that are chemically stable for years, whether in soil, plants, or animals. Furthermore, being soluble in fat and insoluble in water, DDT tends to accumulate in fatty tissues of animals that eat other plants or animals that bear its residues. Predators such as fish-eating birds had high levels of DDT because their prey were predatory fish that also had amplified DDT levels. Experiments demonstrated that these high levels of DDT interfered with the reproduction of some birds. In the landmark book *Silent Spring,* Rachel Carson documented or suggested numerous adverse side effects of pesticides on nontargeted organisms such as pollinating insects and earthworms. Data increasingly suggested that some pesticides had chronic effects on the health of mammals. The **oncogenic** (causing cancer) and **teratogenic** (causing developmental defects in fetuses) potential of pesticide residues on food had obvious implications for human and animal health.

Heightened public concern and political reactions to pesticides continue to motivate pest management approaches that reduce or eliminate pesticide applications. Rather than relying solely on insecticides, pest management practices increasingly integrate a variety of control methods for economical pest control while minimizing adverse environmental or social impacts. This approach is called **integrated pest management** (IPM).

Integrated pest management programs take a broad perspective of how insects, weeds, and diseases fit into the ecology of a cropping system. Fungicides to control plant diseases may upset fungal diseases of insects that would otherwise adequately control a key pest. Weed control may eliminate sources of pollen and nectar required by adult predators or parasitoids, or it may eliminate nonpest insects that serve as prey to increase populations of key predators or parasitoids. If crop yields are relatively constant over a range of crop densities, planting crops at higher initial densities may allow for some plant mortality or damage due to pests without affecting yields. Plowing may kill some important pests but increase soil erosion.

Good pest management requires an understanding of the overall ecology of the crop and of adjacent crops or ecosystems from which harmful or beneficial insects may immigrate. Economic consequences of weeds, nematodes, pathogens, and insect pests should be incorporated into management decisions and practices. Agronomic practices (selection of the crop variety, fertilization, irrigation, etc.) attempt to optimize crop yields and make the growing of the crop as economical as possible by allowing large areas to be planted, maintained, and harvested as a single unit. Unfortunately, a uniform stand of plants of the same age and genotype under optimum growth conditions provides near ideal conditions for the explosive growth of most pest populations.

PEST MANAGEMENT METHODS

Integrated pest management aims to coordinate a variety of pest control actions to optimize the advantages and reduce the disadvantages of any one method. Some methods are inherently incompatible. For example, insecticides often upset biological control of pests by killing natural enemies. The choice and dosage of insecticide and the timing of application may reduce interference with natural control to acceptable levels. Some pest control methods are **complementary,** meaning that their effects are compatible or even may be additive when used together. Other combinations of control methods may be synergistic; that is, the combined effects are greater than the sum of the different methods used separately. For example, the simultaneous use of biological control and the planting of crop varieties that are resistant to pests often results in a reduction in pest populations greater than the sum of the reductions that would have been obtained by using both methods separately. Here we briefly describe the major pest control methods and the advantages and disadvantages of each type.

Insecticides

Development of synthetic insecticides in the 1940s rapidly led to their widespread use for pest control. Insecticides effectively and rapidly reduce pest numbers, allow individual choice of insecticide and timing of use, can be applied with relative ease, and provide high economic returns if used properly. For almost 20 years, most other methods of pest control were ignored or neglected, but the serious disadvantages we have already discussed soon became evident. Most species of pest insects have developed resistance to one or more insecticides, requiring more frequent or heavier applications. The consequences of these adverse effects often led to a "pesticide treadmill." This metaphor describes what happens when pest resurgence, secondary pest outbreaks, or resistance requires greater use of pesticides, but the increased use only makes the problems worse. Use of insecticides must be regulated because of the hazards and adverse environmental impacts posed by these substances. Moreover, insecticide manufacturers and applicators are subject to civil suits for unsafe use or adverse effects from applications. As a result of regulation and legal concerns, the cost of developing new insecticides continues to climb. Total development costs for a new insecticide exceeded $50 to $100 million in 1990. When resistance shortens the marketable life span of a new insecticide, commercial efforts to develop new products are further discouraged.

Despite the problems posed by insecticides, their use is a cornerstone of modern agriculture because no equally effective alternative methods of control have been found for many pest problems. Many adverse effects of insecticides result from misuse or abuse. Proper choices of insecticide and dosage are critical to pest management programs to minimize adverse effects.

Insecticides can be classified in several ways. Depending on the route of entry, insecticides are classified as **contact poisons** absorbed through the exoskeleton, **stomach poisons** that act after being ingested, and **fumigants** that enter the insect through the tracheal system. Insecticides can also be classified by their mode of action. Already mentioned, for example, are nervous system poisons that inhibit acetylcholinesterases (Chap. 6) and the chitin synthesis inhibitors and insect growth regulators (Chaps. 4, 10). The following sections describe some of the distinguishing properties of various categories of insecticides grouped by similarities of origin or chemical structure.

Inorganic Chemicals

The earliest insecticides were inorganic chemicals. **Arsenic** salts such as lead arsenate and copper acetoarsenite were easily produced with the industrial technology of the nineteenth century and effectively killed a wide range of chewing pests. Because of their direct toxicity to some plants (**phytotoxicity**), their risk as toxicants for humans and animals, and their long persistence in the environment, arsenic compounds are rarely used for insect control. The legacies of their overuse many decades ago are persistently high levels of arsenic in some soils, even after more than 100 years. **Cryolite** (sodium fluoroaluminate) can be mined directly or manufactured. It is much less toxic to vertebrates than many current insecticides and is selective in that it acts as a stomach poison to kill chewing insects such as caterpillars and grasshoppers. **Soaps** made as salts of fatty acids and **oils** of plant, animal or petroleum origin continue to be used as insecticides. They require direct contact to kill insects and have little residual activity. **Silica gels** and boric acid, which are relatively nontoxic to humans, are used to kill household roaches. These materials absorb wax from the cuticle of roaches, killing the insects by increasing water loss from the cuticle.

Botanicals

Plants that contain chemicals to reduce insect herbivory are sources of natural insecticides (Chap. 10). One of the earliest insecticides consisted of pulverized flowers of the pyrethrum daisy, *Pyrethrum cineariaefolium.* The active chemicals in the flower extracts, collectively called **pyrethrum,** are chemically classified as pyrethrins and cinerins. Insecticides incorporating pyrethrum are still manufactured because of their low mammalian toxicity and because they are made from natural sources rather than being synthesized. They provide extremely quick knockdown of insects. For this reason and for the sake of their low risk to vertebrates, they are widely used for pests of households or of domestic animals. Piperonyl butoxide is commonly added as a **synergist** to pyrethroid insecticides. By itself, piperonyl butoxide is scarcely toxic, but in mixtures, it multiplies the toxicity of pyrethrins to insects manyfold.

Nicotine is a pyrrolidine alkaloid present in high concentrations (2 to 14 percent) in tobacco species and some other plants. It was first used for aphid control in 1763. Nicotine interferes with insect nerve transmission by reducing the action of acetylcholinesterase and disturbing the postsynaptic receptors of the central nervous system. Awareness of its dangerous toxicity to mammals has greatly reduced its use. Other insecticides manufactured as natural products from plants are **ryanodine,** from the roots and stems of the plant *Ryania speciosa,* and **rotenones,** from a variety of tropical plants that were used originally by native peoples for fish poisons. Botanical insecticides are generally too expensive for large-scale use in agriculture but are applied for rapid pest control by organic farmers, or in dairies or other facilities where synthetic chemicals are not legally permitted. The tropical neem tree, *Azadirachta indica,* has long been a source of crude insecticides produced from infusions of its fruits soaked in water. **Azadirachtin** is the principal active component among the numerous antifeedant, repellent, and insecticidal compounds in neem extracts. This liminoid chemical can kill insects by disrupting normal development but is used mainly for its antifeedant or insect-repellent properties.

Synthetic Organic Insecticides

The synthetics are by far the most widely used chemicals for insect and mite pest control. Most **organochlorine** compounds (e.g., DDT, chlordane, lindane, dieldrin) have been banned in most developed countries because of the problems of biological amplification and environmental persistence discussed above. Some organochlorine insecticides are reserved for emergency control of medically important vector insects. Organochlorine compounds interfere with sodium channel function of the central nervous system.

Organophosphate insecticides were developed during World War II as a by-product of nerve gas research. These compounds are derivatives of phosphoric acid. They phosphorylate acetylcholinesterase, the enzyme that breaks down acetylcholine during the process of synaptic nerve transmission. As a result, acetylcholine accumulates at synapses, hindering normal nerve transmission. Because they break down rapidly in most environments, organophosphates do not persist as pollutants or interfere with the activity of natural enemies beyond initial mortality. Plants can take up and translocate some organophosphate compounds. This **systemic action** is especially effective against sap-feeding insects.

Carbamate insecticides are esters of carbamic acid. They have short (days) to medium (weeks) persistence after application. They inhibit acetylcholinesterase by competing with the enzyme for active sites. Carbaryl is a commonly used carbamate with low mammalian toxicity. Aldicarb is a highly toxic carbamate that is also systemic in plants.

The synthetic analogues of botanical pyrethrins called **pyrethroids** were developed as commercial insecticides. Although they may have chemical structures that appear rather different from natural pyrethrins, the chemically active sites of pyrethroids are similar in activity to natural pyrethrins. Their mode of action is similar to pyrethrins in disturbing the normal function of sodium channels in depolarization of the nerve cell membrane during nerve signal transmission. Pyrethroids have become widely used because they act very quickly at low dosages and have low mammalian toxicity. In addition, they persist in an active form for longer periods than natural pyrethrins, which break down quickly in sunlight. Synthetic pyrethroids are among the most expensive insecticides because they are the most recent to be commercialized, and the proprietors of the formulas have not yet recovered the costs of development.

Insect Growth Regulators

A variety of chemicals that affect insect growth perpetuate immature stages of development, making these agents the most appropriate for insects such as fleas and mosquitoes, which are damaging only in the adult stage. These chemicals are nontoxic and break down rapidly after application, especially in ultraviolet light. Insect growth regulators are selective in their effects because each analogue is effective against a limited range of insect families. **Chitin synthesis inhibitors** such as diflubenzuron interfere with development of the cuticle, usually leading to death during or immediately after molting.

Microbial Insecticides

Preparations of insect pathogens can be applied much in the same ways (dusts, sprays, baits, etc.) as chemical insecticides. Preparations of spore-forming bacteria such as *Bacillus thuringiensis* (*Bt*) (Chap. 13) have been widely marketed. The toxicity of *Bt* applications is short-lived, lasting only a few days at most. The numerous strains of *B. thuringiensis* differ in the spectrum of insects that are poisoned. For example, strains effective against larvae of some Lepidoptera are not active against beetle or mosquito larvae. Insecticidal preparations of *Nosema* pathogens of grasshoppers or *Bacillus popillae* for the Japanese beetle (*Popillia japonica*) are inoculative; they multiply within infected insects and survive for prolonged periods (months to years) before new inoculative applications are needed. The bacterium *Serratia entomophila* is produced commercially and applied to the soil as an inoculative control for root-feeding larvae of the scarab beetle pest *Costelytra zealandica*. This measure controls the beetle grubs adequately for at least several years.

Baculoviruses, especially nuclear polyhedrosis viruses (NPV), are produced commercially for control of certain species of caterpillars. They are expensive but particularly useful where poisonous insecticides cannot be used because of hazards to people or animals. Baculoviruses infect arthropods only and are considered to be safe to mammals. Baculoviruses break down rapidly in ultraviolet light after application, however, and the diseases they cause require several to many days to kill the targeted pest. Many viruses cause the host insect to cease feeding within hours of ingesting the lethal agent, so that the slowness of killing the pest is not a problem. The limited host range of baculovirus insecticides can be a disadvantage if there are numerous pests to be controlled.

Fungi have been tried as microbial pesticides since the late 1800s, with inconsistent success. Fungal pathogens must have humidity, surface moisture, and temperature within narrow limits to germinate and infect insects or to sporulate. Commercial preparations of the fungi *Verticillium lecanii* and *Hirsutella thompsonii* are marketed for controlling some pests in greenhouses.

Nematodes of the families Steinernematidae and Heterorhabditidae can be mass-produced along with their accompanying bacteria (Chap. 13) and will remain viable in storage for many months. These nematodes attack and kill a wide range of insects but require suitable moisture to survive for long out of soil.

Formulation and Application

Insecticides that have the same active toxicants may perform differently because of other ingredients, called **adjuvants,** with which they are packaged. A variety of chemicals, usually named on the insecticide's label as "inert ingredients," can be used to dilute the concentration of active toxin in the final mixture. Dilution makes packaged insecticides safer and easier to ship, store, and apply. Talc, powdered clay, or other finely pulverized carriers provide a surface to which the active insecticide can be adsorbed for application as a dust or wettable powder. Other adjuvants serve various purposes. Emulsifiers disperse water-insoluble insecticides to form an emulsion when the product is added to water, and such mixtures can be easily sprayed in large quantities of water. Spreaders reduce the surface tension of droplets on waxy plant surfaces and cause the droplet to spread. Stickers help it to remain firmly attached to plant surfaces. Buffers maintain a desired level of acidity or alkalinity that may be essential for the chemical stability of the insecticide.

The manner in which insecticides are applied may affect their performance. Since only a small amount of an insecticide applied to a crop reaches a location that is likely to affect targeted pests, the method of delivery is critical for effectiveness. Insecticides that kill pests only by contact (e.g., soaps or oils) require thorough coverage and will not affect insects that are sequestered (within plants, under soil, etc.). The concentration, droplet size distribution, and sprayer velocity of the pesticide application can affect the final coverage of the insecticide on the targeted crop. Some insecticides are encapsulated in minute granules for gradual release to soil or plants.

Insecticide Toxicity

Comparing the toxicity of different insecticides to a particular insect or to different populations of insects, as in studies of insecticide resistance, re-quires standardized measures of toxicity. Individuals within any population of insects will vary in their susceptibility to a uniform dosage of any insecticide. The mortality that results from varying the dose of insecticide administered to individual insects produces results such as those depicted in Figure 14.5a. Plotting dosage on a logarithmic scale and mortality on a probit scale (normalized percentages) results in the data approximately fitting a straight line (Fig. 14.5b) to simplify mathematical computations and comparisons. The line that best fits the data points (by linear regression) can be described as an equation from which we can estimate the dosage that kills a given percentage of the test insects. There is no probit for 100 percent or 0 percent because probits approach these limits asymptotically. The most common statistic from dosage-mortality experiments is the median lethal dosage or **LD_{50}**. This is the estimated dosage that will kill half the treated insects. The median toxicity is less variable from one experiment to another than are other levels of toxicity, such as the LD_{95} (dosage lethal to 95 percent of test organisms). Where it is difficult to deliver the same dosage per insect, concentrations of insecticide are varied

Figure 14.5 Results of experimental administration of various dosages of insecticides: **a**, mortality plotted on linear scale; **b**, plotted on probit scale and dosage plotted on a logarithmic scale. R represents a pouplation resistant to the insecticide; S represents a population 10 times more susceptible. The dose that kills half of the test population (LD_{50}) is shown by the dashed line.

to estimate the LC_{50} (concentration of insecticide that kills 50 percent of the insects). The toxicity must be measured separately for oral (ingestion) or dermal (external) applications.

The relative toxicity (LD_{50}) of insecticides to mammals is expressed as milligrams of pure insecticide per kilogram of body weight or as an LC_{50}. Hazards of toxicity to humans and animals are estimated from tests using small mammals such as rats or rabbits. Table 14.1 shows the LD_{50} for legally defined toxicity categories and some examples of widely used insecticides in each category.

Insecticide Resistance

The rapid evolution of resistance to insecticides in pest populations is a clear example of natural selection, Darwin's "survival of the fittest." When populations of insects are exposed to an insecticide, usually only a few individuals may tolerate the exposure without dying. If the basis of their tolerance is genetic, a larger percentage of genes for resistance will occur in the next generation. After repeated cycles of selection for resistant individuals, most of the population may have genes for resistance (Fig. 14.6). Not only insects, of course, but all other pest or pathogen groups se-

lected by chemical treatments—mites, nematodes, weeds, fungi, protozoa, and bacteria—have evolved resistance to chemicals. Resistance appears to occur much less frequently in populations of natural enemies. Resistance will develop most rapidly where all individuals in the population are exposed to high doses that leave only a few resistant survivors. If the resistance allele is dominant, resistance may develop rapidly. The frequency in Africa of *Anopheles gambiae* mosquitoes that were homozygous (RR) or heterozygous (Rr) for dieldrin resistance ranged from 0.04 percent to 6 percent before spraying. After a single spray cycle using dieldrin insecticide, the frequency of individuals with R alleles increased to 86 percent. Resistance alleles that are recessive reduce the speed with which resistance appears in a population under selection. Resistance based on more than one gene (polygenic) will develop more slowly than will monogenic resistance. Once resistance has developed in a population, it may persist for many generations without further selection. Reversion, or the decrease in frequency of resistance alleles in the absence of selection, has been documented.

Resistance to DDT was first noted in houseflies in Sweden in 1946, not long after it first began to be used. The numbers of insecticide-resistant arthropod species have steadily grown from less than 50 recorded as resistant in 1955 to

TABLE 14.1.
Toxicity Categories and Selected Insecticides

Toxicity category	Oral/dermal LD_{50} (mg toxin/kg body weight)	Required signal words on label	Representative insecticides
Highly toxic	<50 oral and/or <200 dermal	Danger—Poison	Parathion, aldicarb, nicotine sulfate
Moderately toxic	50–500 oral and/or 200–2000 dermal	Warning	Rotenone, diazinon
Slightly toxic	500–5,000 oral and/or 2,000–20,000 dermal	Caution	Malathion, carbaryl, acephate, chlorpyrifos
Low toxicity	>5,000 oral and/or >20,000 dermal	Caution	Permethrin

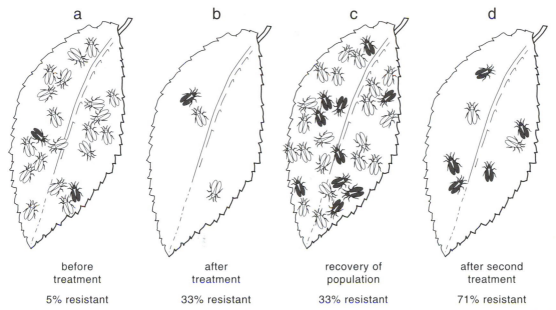

a	b	c	d
before treatment	after treatment	recovery of population	after second treatment
5% resistant	33% resistant	33% resistant	71% resistant

Figure 14.6 Selection for resistance to insecticide: **a**, Starting with a pest population in which genetically resistant individuals (dark) are greatly outnumbered by susceptible individuals (light), insecticides differentially kill susceptibles, producing **b**, a higher proportion of resistant individuals whose offspring, **c**, are more likely to be genetically resistant. **d**, After a second treatment, the percentage of resistant individuals increases again.

481 in 1989 (Georghiou and Lagunes-Tejeda, 1991). The diamondback moth, *Plutella xylostella* (Yponomeutidae), became resistant to the biological pesticide *Bacillus thuringiensis* within 3 years of heavy use in Hawaii and in Japan. More than half of the resistant pest species have **multiple resistance** (i.e., resistance to insecticides in more than one chemical class). This characteristic usually reflects the development of resistance to a succession of insecticides as first one, then another insecticide is adopted following the appearance of resistance to each in a series of chemicals. Sometimes resistance to one insecticide confers resistance to a related insecticide to which the species has never been exposed. This phenomenon, called **cross-resistance,** is possible because a single mechanism such as a detoxifying enzyme may act against insecticides that share the same mode of action. For example, resistance to one organophosphate compound may confer resistance to another organophosphate. Cross-resistance to insecticides in other chemical groupings also occurs.

Research into the basis of resistance has revealed numerous biochemical, physiological, and behavioral mechanisms. Enzymes responsible for detoxification in resistant insects include various esterases, mixed microsomal oxidases, glutathione transferases, and others. Resistance to pyrethroids may be caused by reduced sensitivity of the sodium channels of nerve axons, which in turn is due to a reduction in the number of receptor sites for the insecticide. Altered acetylcholinesterases may be less susceptible to inhibition by carbamate and organophosphorus insecticides. Resistance to toxins of *Bacillus thuringiensis* appears to be due to changes in receptor sites on midgut cells, where the toxins attach. Amplifying the amount of enzymes that break down pesticides by having multiple copies of the gene for the enzyme may increase resistance dramatically (Devonshire and Field, 1991). Resistant *Culex* mosquitoes may have hundreds of copies of an esterase gene.

One proposed explanation for the rapid appearance of resistance to insecticides in herbivorous insects is that their natural adaptations to detoxify a variety of plant defensive chemicals also enable them to detoxify synthetic pesticides (Chap. 10). A counterargument to this hypothesis is that mosquitoes and other insects that do not feed on plants are among the best examples of insects to have rapidly become resistant. Resistance is rarely observed in insect predators and parasitoids, thus compounding the problems posed by the development of resistance in pest species. By intensive selection and breeding in the laboratory, resistance to specific pesticides has been incorporated into predators and parasitoids. Establishing resistant natural enemies in field populations may reduce the likelihood of pest resurgence and thus reduce the number of pesticide treatments needed (Hoy and Herzog, 1985).

Resistance Management

Resistance to insecticides is one of the most pressing problems in pest management. The commercialization of new chemicals requires years of expensive and lengthy development and safety testing. Older insecticides have been rendered obsolete by resistance, and pests are becoming resistant to new insecticides. The recently observed leveling of the increase in the numbers of resistant species is not cause for optimism. It simply means that most major pests have become resistant. The problem is especially acute in controlling mosquitoes and other insect vectors of human pathogens, where insecticides have been the only effective or the most environmentally acceptable control measure.

Preventing or slowing the development of resistance in pest populations depends on understanding the factors that lead to its evolution and modifying practices to avoid or minimize these factors. Ideally, other pest control methods should replace insecticides except in emergencies. Tolerating higher levels of damage, especially where it is largely cosmetic, reduces the need for control. Reducing the number of insecticide applications reduces the number of selection cycles. The simplest

and most broadly applicable approach is to eliminate unnecessary applications by basing control decisions on careful monitoring of pest populations and a good understanding of the relationship of pest numbers to crop damage. Economic thresholds are of no value if populations are not monitored regularly. Economic injury levels that are unrealistically low increase insecticide use.

Resistance develops most rapidly where all stages and all members of a population are exposed to an insecticide. For example, insecticides to kill larval mosquitoes often are applied over large areas. This approach exposes most of the population to selective pressure for resistance. Treatments aimed at killing adult mosquitoes near human habitations exerts selective pressure for resistance on a much smaller fraction of the mosquito population. The latter method may be as effective as larval treatments in preventing mosquito transmission of human pathogens if humans are the main sources for mosquitoes to acquire the pathogens. Monophagous pests that feed on frequently treated crops, such as the cotton boll weevil, *Anthonomus grandis* (Curculionidae), will develop resistance more quickly than will polyphagous pests. Examples of the latter are grasshoppers that feed and reproduce in surrounding habitats where they are not exposed to insecticides. Insecticide-treated populations that frequently interbreed with unexposed populations will develop resistance more slowly. Short generation times and numerous offspring per generation speed the development of resistance.

Lower doses and chemicals that break down rapidly after use help to avoid resistance. Persistent insecticidal residues can continue to select for resistance long after applications are made. Applying insecticides as a rotation or as a mixture of different chemicals with different modes of action can delay the onset of resistance.

Biological Control

Recall that interspecific competition, predators or parasitoids are important constraints on the growth of insect populations (Chap. 8). Secondary pests, for example, are organisms that become pests when insecticides eliminate natural enemies. De-Bach (1964) defined **biological control** as "the action of predators, parasites and pathogens in maintaining another organism's density at a lower average than would occur in their absence." Biological control avoids hazards to health or the environment associated with insecticides. Where effective biological control can be established, it is generally more stable and far more economical than chemical methods. As of 1992, 517 projects in 196 islands or countries had been declared successful in reducing the population densities of 421 pest species. The severest shortcoming of biological control may be the difficulty of implementing it in certain situations, such as crops grown for only a few months. Natural enemies may not increase to useful levels quickly enough to be effective in such rapidly changing environments. Another caution in the use of biological control is to avoid effects on nontargeted species that might be driven to extinction by importing certain natural enemies. Indeed, some native insect species have vanished from island ecosystems for this reason.

Biological control is not cheap, but the benefits far outweigh investments in its research and implementation because such measures may continue for many years without any further inputs. Only a few successful major projects are needed to return more than the amount invested in all biological control projects.

Four approaches are used for biological control: **importation** of natural enemies to control exotic pests (**classical biological control**), **conservation** of existing natural control agents, **augmentation** of natural enemies by releasing mass-reared natural enemies, and **inundation** of pest populations by releases of mass-reared agents.

Classical Biological Control

The best-known type of biological control is the importation and establishment of natural enemies to control pests that become established in new areas away from their area of origin. The first success with this method, implemented in Califor-

nia in the 1880s, controlled the cottony-cushion scale, *Icerya purchasi* (Margarodidae), by introducing a coccinellid, the vedalia beetle, *Rodolia cardinalis*. A key to this first success was determining that the cottony-cushion scale came from Australia and identifying its specific natural enemies. Entomologists found the scale to be rare in Australia and discovered that vedalia beetles and certain parasitoid flies, *Cryptochetum iceryae* (Cryptochetidae), attacked the scale. These natural enemies were carried to California, reared, and released. Success was rapid and spectacular. Moreover, both natural enemies still keep the cottony cushion scale well below economic thresholds. The vedalia beetle predominates in warmer areas, the fly in cooler locations.

Classical biological control has become far more complicated today. Molecular as well as traditional methods are used to identify pests and parasitoids. More elaborate steps are taken in quarantine to confirm the identity of natural enemies, their freedom from disease and hyperparasitoids, and their suitability for release. Possible adverse effects on native nonpest insects that might be eliminated by imported natural enemies must be considered. Generally the most host-specific natural enemies are most effective for biological control. A notable exception was control of the endemic coconut moth, *Levuana iridescens* (Zygaenidae), in Fiji by the parasitoid tachinid fly, *Bessa remota*. This fly was imported from Malaysia, where it attacked other moth species. In Fiji, biological control by the tachinid fly made the coconut moth extinct.

Biological control has succeeded in controlling only some exotic pests, despite major efforts in cases such as the gypsy moth, *Lymantria dispar* (Lymantriidae), in North America, where more than 60 species of natural enemies have been introduced. Controversy continues to surround the question of whether importing only one or a few natural enemies is more effective than introducing a larger complex of parasitoids and predators. For example, hyperparasitoids and superparasitoids may upset effective control of a targeted pest by other parasitoids.

The more detail given to study of the targeted pests and its natural enemies, the greater chances of success with biological control. The walnut aphid, *Chromaphis juglandicola* (Aphididae), is native to the Old World and was not controlled satisfactorily in central California by introductions of a parasitoid wasp from France, *Trioxys pallidus* (Chalcidoidea). The parasitoids collected from France were effective only in cooler localities near the ocean. Later collections of this species from hot, arid regions of Iran quickly brought the walnut aphid under excellent biological control. The ability of the Iranian strains to withstand the hotter, drier interior areas of California was the key difference.

There are fewer cases of successful biological control of exotic weeds than of insect pests. On the other hand, a higher percentage of biological control attempts against exotic weeds have succeeded, perhaps because more detailed biological studies were prompted by concerns that herbivores imported to control weeds might also damage desirable plants. Exotic plants that have no host-specific herbivores are usually the worst weeds.

The first step in classical biological control of weeds is to learn where the weed originated and discover its natural enemies. Next, the weed's ecology, including natural enemies and diseases, must be studied in detail. This includes testing potential biological control agents to ensure that useful plant species are not attacked. Finally, the most promising species of natural enemies are imported and screened under quarantine to ensure that the imported organisms themselves are free of parasitoids and diseases before mass rearing and release. Biological control of weeds has worked best against perennial rangeland or pasture weeds rather than against weeds in frequently disturbed environments such as annual croplands. This is because of the combined effect on the target weed of the herbivore and competition from surrounding plants.

Outstanding examples of biological control of weeds are the control on two continents of the rangeland weed *Hypericum perforatum* and the aquatic weed *Salvinia molesta*. *Hypericum perfor-atum* is native to Europe, where it was known as St. John's wort. Australian researchers identified several beetle herbivores specific to *H. perforatum* and established them in Australia. Two of these, *Chrysolina hyperica* and *C. quadrigemina* (Chrysomelidae), were imported and released in California. Ever since the 1950s, these imported insects have effectively controlled this weed in the western United States, where it is known as Klamath weed.

Salvinia molesta is an aquatic weed that had choked lakes and other waterways after its accidental introduction throughout Asia, Africa, and Australasia. Initial attempts at classical biological control failed because of the misidentification of the weed as a look-alike but nonpest species, *Salvinia auriculata,* and its herbivore weevil, *Cyrtobagous singularis* (Curculionidae). Further research established the true identity of the exotic weed as *Salvinia molesta* but misidentified another weevil, now named *C. salvinae,* as a specific herbivore of the weed. After the taxonomic problems had been clarified, *C. salvinae* was introduced and established. One more detail, however, remained before the project could succeed. *Salvinia molesta* thrives best on nitrogen-rich waters. Fertilizing the targeted lakes with nitrogen increased the food quality of the plants for the weevils. In turn, weevil populations increased to high levels that devastated the remaining weeds.

Conservation

A strategy to enhance the effectiveness of established natural enemies is to minimize disturbances to existing natural control. The most common approach is to eliminate or severely restrict the use of insecticides. A second method is to preserve or provide habitats or refuges from which natural enemies can move into nearby crops to attack pests. Leaving strips of alfalfa among cotton fields in California provides reservoirs of predators and parasitoids of key pests. Harvesting only half the alfalfa strip at any time further ensures a continuing habitat for natural enemies. This method has not been widely adopted because it requires higher costs for cutting alfalfa and closer monitoring of pest and natural enemy populations, and because pesticides applied to the cotton crop are difficult to keep out of adjacent alfalfa. Furthermore, a small amount of damage to cotton from pests that move in from the alfalfa strips must be tolerated, and growers have traditionally tried to ensure extremely low damage levels by using pesticides.

Augmentation

Releasing additional natural enemies of a species that is already established may be useful when unusual environmental conditions eliminate some natural enemies or keep them below an effective level. Where unusually bad weather has eliminated a natural enemy, inoculative releases of individuals of this species may reestablish it. Augmentative releases also may increase natural enemies at a particularly critical phase in the growth of the crop to be protected or during a vulnerable phase of the target pest's life cycle. Inoculative applications of the bacterial pathogen *Bacillus popillae* against Japanese beetles (*Popillia japonica*) often must be repeated after a few years to achieve effective levels of beetle control (Chap. 13).

Inundative Releases

Releasing large numbers of natural enemies can be similar to the use of insecticides, as discussed in connection with microbial pesticides. This strategy is used where establishing stable biological control is not feasible. Greenhouse pests such as aphids or whiteflies (Aleyrodidae) may be controlled by releasing predaceous fungus gnats (Mycetophilidae), beetles (Coccinellidae), parasitoids (Chalcidoidea), or fungal sprays (*Verticillium lecanii*). The rapid, short growing cycle in the highly artificial greenhouse environment makes long-term control with inoculative releases an unrealistic goal. Most microbial pesticides do not remain at effective levels in the environment after application. Viral or bacterial pesticides kill their targeted hosts rapidly, leaving few infected hosts to provide inoculum to infect additional hosts over a prolonged period. Exceptions may be the granulosis viruses of sawflies (Tenthredinidae) and the western grape leafskeletonizer, *Harissina brillans* (Zy-

gaenidae). These granulosis viruses have effectively controlled their host insects for years after the initial application. The viruses appear to be limited to cells lining the midgut, and some virus-infected larvae survive to the adult stage, with the virus carrying over on eggs as a surface contaminant to the next generation or from one season to the next.

Host Plant Resistance

Plant resistance to insects is a heritable set of characteristics of a plant that make it less susceptible to damage by insects. You may recall three categories that describe mechanisms of plant resistance to insects discussed in Chapter 10: **antibiosis** adversely affects herbivore longevity or fecundity; **antixenosis** (formerly called nonpreference) deters insect feeding or oviposition; and **tolerance** minimizes loss of yield despite insect damage. These are relative properties in comparison to other plant genotypes that have a lesser degree of resistance and are thus said to be more susceptible.

Plant resistance is one of the cheapest and most easily applied methods of pest management. It is compatible or even synergistic with other methods of pest control, except for a few cases in which plant resistance also harms important natural enemies. The major direct costs are for breeding and testing new varieties (or cultivars). Indirect costs are the lower yields or quality of some resistant varieties. Plant resistance is especially desirable for crops with such a low unit value that insecticides are too expensive to use. For this reason, plant resistance is used most extensively in grain and forage crops. One major problem is the long time required to develop some resistant varieties. Another problem is the development of insect populations that overcome plant resistance.

Conventional plant breeding by controlled crosses of selected parents has identified specific genes for resistance to insects. In most cases the specific mechanism of resistance is unknown, but this is no impediment to using plant resistance for pest management. When a breeder can identify specific genes for resistance (R genes), it is relatively easy to incorporate them into new varieties. A major problem is the potential appearance of resistant races of pests after cultivars with the new gene combinations have become widely used. **Vertical resistance** was defined (van der Plank, 1963) by plant pathologists to describe resistance to which new populations (termed "races") of plant pathogens quickly formed. The **gene-for-gene theory** postulates that for every resistance gene in a host plant, there is a matching gene that confers susceptibility in a plant parasite (whether of plants or of animals). There are abundant examples of fungal pathogens of plants that fit gene-for-gene relationships. Fewer gene-for-gene relationships have been noted in breeding for plant resistance to insects. One proposed example for insects is resistance in wheat to the cecidomyiid pest, the Hessian fly, *Mayetiola destructor.* Populations of the Hessian fly that overcame the resistance appeared in response to the widespread deployment of R genes. Such populations with distinct abilities to overcome resistance have been termed **biotypes,** but the definition or usage of the term biotype varies widely. With respect to plant resistance, biotypes are defined by the plant cultivars the insect does or does not infest. There are now at least 10 biotypes of Hessian fly defined by their reactions to combinations of four alleles in wheat for resistance to Hessian fly.

Laboratory experiments with the brown planthopper, *Nilaparvata lugens* (Delphacidae), demonstrated how biotypes might evolve in this pest. The planthopper biotypes were reared on rice cultivars that were resistant at the start of the experiment. After 10 to 15 generations, each biotype was reproducing about as well on the originally resistant cultivars as on other cultivars formerly classified as susceptible (Claridge and Den Hollander, 1982). This microevolution of changes of the brown planthopper in response to plant resistance is analogous to laboratory selection for insecticide resistance. In both types of selection, survival increased gradually from one generation to the next after each cycle of selection.

Durability is a major goal for resistance breeding. Repeated selection among high-yielding cultivars for resistance is aimed at achieving long-lasting or **horizontal resistance,** defined as resistance that avoids the development of pest biotypes. One of the earliest examples of using plant resistance was the adoption of wild grapevines, *Vitis rupestris* and other wild grapevines from North America as rootstocks to counteract the grape phylloxera, *Daktulosphaira vitifoliae* (Phylloxeridae). This approach was instantly successful and has endured at least 140 years in France. Because no biotypes that overcome the resistance of wild grape species have appeared over this long period, we assume that the resistance is horizontal. Vertical resistance has been identified only on hybrids of wild grape species crossed with commercial grape, *Vitis vinifera.* Biotypes of phylloxera have appeared on hybrid varieties; thus such resistance is vertical. The replacement of hybrid rootstocks with traditional selections from wild grape appears to have resolved this biotype problem by employing the horizontal resistance of the purely wild species, but only after projected losses exceeding a billion dollars over 10 years in California alone.

Effective plant resistance is a still elusive goal for some major pests of important crops. The long time required for the repeated selection and testing necessary to develop resistant cultivars is the main barrier for breeding resistant fruit trees. Breeding trials may require 3 to 10 years for plants to produce fruit for a single selection cycle. A major advantage of plant resistance is that it can be combined with other methods to produce acceptable results where either method alone would not. For example, resistance to an aphid pest of wheat and other grain crops, the greenbug, *Schizaphis graminum* (Aphididae), is adequately controlled by a parasitoid in combination with a resistant variety, but either method alone is not adequate for control (Fig. 14.7). By reducing the rate at which pests can multiply, plant resistance may allow predation and parasitism to adequately control pests. Even where resistant varieties do not lower pest densities enough to avoid the need for insecticides, they may delay or reduce the number of pesticide applications.

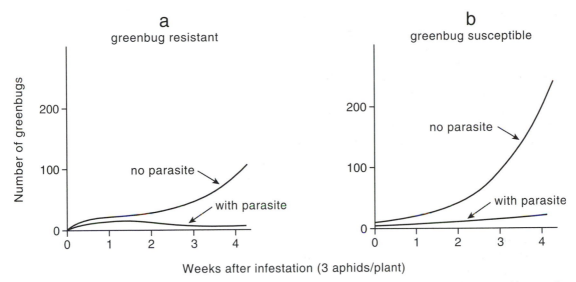

Figure 14.7 Effects on populations of the greenbug, *Schizaphis graminum* (Aphididae), of combining a parasitic wasp for biological control with plant resistance: **a**, on a greenbug-resistant variety (Will barley); **b**, on a susceptible variety (Rogers barley). The parasite reduces greenbug populations on the resistant variety to lower levels than on the susceptible variety. (Redrawn from data by Starks et al., 1972.)

Early human selection that domesticated plants for cultivation intentionally selected for rapid growth, gigantism of fruits, taste, reliable germination, and ease of harvest. Natural selection increased crop resistance to climatic unpredictability, pests, diseases, and weeds with little conscious direction from humans. Breeding governed by understanding of genetic principles rapidly improved yields, adaptability, and resistance of major crops to multiple pests and diseases. About the same time, other developments in soil and fertilizer technology and chemicals for disease and pest control decreased farmers' dependence on inherent plant resistance to pests. The goal of decreasing pesticide use has renewed interest in plant resistance. Wild relatives and progenitors of domesticated plants have provided important sources of resistance genes, but these often have led to impermanent, vertical resistance.

Applications of molecular genetics have begun a new era in crop breeding, with breeders transferring genes among organisms as dissimilar as plants, bacteria, and animals. Among the first uses of the new technology was the incorporation into plant genomes of genes controlling the production of insecticidal toxins produced by *Bacillus thuringiensis* (Chap. 13). Even before this technology was deployed, strategies were being devised to deal with the likelihood that pests would develop resistance to *Bt* toxins. In contrast to pesticide applications made only when necessary, putting *Bt* toxin genes into plants leads to constant toxin production by plants. This is very likely to accelerate the development of pest resistance to *Bt* toxins. Possible strategies to retard resistance include combining multiple toxin genes in the same plants, planting mixtures of various genotypes, or mixing in seeds of plants without *Bt* genes to ensure that some proportion of the pest population will not be exposed to *Bt* toxins. Another strategy would be to deploy different toxin genes sequentially as resistance to first one, then another toxin develops. Incorporating antixenosis with antibiosis (*Bt* toxins) should delay the development of resistance by repelling pests and thus reducing their exposure to the toxins.

New methods of gene transfer in plant breeding put an increased premium on understanding pest resistance mechanisms in plants so that the genes controlling these characteristics can be identified to transfer into crop plants. Since plant-produced toxins that provide resistance to insect herbivores may be distasteful or toxic to humans and animals, whether incorporated into new cultivars by conventional or molecular methods of breeding, possible adverse health effects should be considered when plant resistance strategies are assessed.

Cultural Controls

Manipulations of normal agricultural practices to reduce pest damage are classified as **cultural control** techniques. The most common types of cultural control are tillage, timing of harvest and planting, crop rotation, sanitation, and water management.

Tillage, or cultivation, is the mechanical manipulation of soil, chiefly for weed control, drainage, and disruption of impenetrable soil layers. Plowing, perhaps one of the best-known methods of cultivation, inverts a layer of soil to bury weeds and crop residues, giving crop seedlings a fast start over competing weeds. The main disadvantages are that plowing, followed by hoeing or disking for weed control, requires large amounts of fuel and leaves soils vulnerable to high rates of erosion. Shredding and deep plowing to bury cotton plant residues after harvest destroy overwintering larvae of the pink bollworm, *Pectinophora gossypiella* (Gelechiidae), but these practices are useful only if applied in a region-wide program. Otherwise, surviving pink bollworms in unplowed fields will develop into adults and migrate into adjacent fields the following spring. Plowing usually does not kill soil-inhabiting pests that are mobile in soils. The widespread increase in using herbicides (pesticides that control weeds) has largely replaced tillage in some crops, reducing erosion, soil compaction, and costs of machinery and fuel. No-till farming has reduced some pests and increased others compared to traditional tillage, requiring changes in pest management tactics.

The timing of planting and harvesting is a long-established method for pest management. For example, wheat planted after the Hessian fly's brief period of egg laying in the autumn escapes most infestation by this pest. Crops vulnerable to some virus diseases can be planted after the main flights of specific insect vectors. Delaying the planting of crops until after spring migratory flights of the beet leafhopper, *Circulifer tenellus* (Cicadellidae), reduces losses to the curly top virus transmitted by this insect. Similarly, the timing of harvest can avoid pest injury. Adult moths of the navel orange worm, *Paramyelois transitella* (Pyralidae), lay eggs on ripe almonds in trees or on the ground. Damage from larvae that hatch from the eggs increases steadily after the nuts have matured, especially when they are on the ground. Early harvest minimizes damage. Potatoes grown for use as seed are harvested in many regions as soon as possible following increases in flights of the green peach aphid, *Myzus persicae,* which is the main vector of potato leaf roll virus. Infection with the virus greatly reduces the value of a seed crop but can be tolerated at higher levels in main crops grown for food. Early-maturing cultivars of cotton allow the profitable harvest of cotton crop in regions where late-season pest explosions lead to the pesticide treadmill syndrome.

Crop rotation varies the place rather than the time that crops are planted. This measure may be effective against crop-specific pests that have soil-inhabiting stages between crop cycles. For example, populations of corn rootworm, *Diabrotica virgifera* (Chrysomelidae), that build up in soils in maize (corn) fields do little or no damage to soybeans that are planted in the same fields the next year. Maize fields planted after soybeans have reduced numbers of rootworms because few rootworms develop in soybeans. Crop rotation has additional benefits. It may reduce crop-specific pathogens, weeds, and nematodes in the soil and help to maintain soil fertility.

Sanitation is the removal of weeds or crop residues that might lead to pest problems. Burning or shredding pruned branches from fruit trees or grapevines destroys borers (Bostrichidae, Anobiidae, other Coleoptera) that overwinter in prunings to emerge during the spring and damage remaining branches. Knocking down and disking under the "sticktight" nuts remaining in winter-dormant almond trees destroys naval orangeworm larvae that overwinter in the old nuts. Some pests build up populations faster in fields with certain weeds. For example, Russian thistles growing in sugar beet fields lead to denser populations of beet leafhopper, *Circulifer tenellus* (Cicadellidae) and beet webworms, *Loxostege sticticalis* (Pyralidae). Weeds next to croplands may harbor stinkbugs (Pentatomidae) or lygus bugs, *Lygus* spp. (Miridae), which move in to damage fruit crops. On the other hand, outlying weeds may be important reservoirs and sources of nectar and pollen for natural enemies.

Water management by draining or tillage to avoid standing water is a major tactic to reduce mosquito populations by eliminating breeding habitats. The adverse environmental consequences of draining wetlands call for new ideas for mosquito control. Flooding dormant grapevines for more than a month during the winter effectively controls the grape phylloxera, but prolonged flooding cannot be tolerated by most crops.

Pheromones and Attractants

Research on the effects of external chemicals on insect behavior has led to the use of semiochemicals (Chap. 6) for pest management. Chemical attractants are used to monitor pest abundance and activity. Food lures, aggregation pheromones, and sex pheromones attract insects in measurable numbers where densities are too low to monitor by direct sampling methods. Pheromones are especially valuable for detecting initial insect outbreaks or evaluating the effectiveness of quarantine programs in excluding insects. For some pests, traps baited with sex pheromones can indicate when mating flights begin, helping to determine when to apply control efforts.

Mass-trapping of pests attracted to traps has not been widely successful as a method of pest con-

trol. Low-density pests such as tsetse flies, *Glossina* spp. (Glossinidae), can be controlled in limited areas with inexpensive and simple traps baited with attractants. The most attractive components and combinations of chemicals were derived from analyzing the components of cattle breath and their effects on tsetse fly behavior. Other studies have demonstrated the feasibility of mass-trapping with aggregation pheromones to control bark beetles (Scolytidae) (Chap. 6) and sex pheromones for the grape berry moth, *Endopiza viteana* (Tortricidae), but cost/benefit ratios have been uneconomic.

Mating disruption of pests using sex pheromones has succeeded in controlling the pink bollworm in cotton and the oriental fruit moth, *Grapholita molesta* (Tortricidae), in peaches and plums. Synthesized blends of volatile sex pheromones can be delivered by emitters dispersed throughout the crop. In cotton the emitters are segments of thin, hollow fibers dispensed by aircraft or ground equipment. In orchards, the emitters are hollow plastic tubes tied onto branches. High levels of widespread pheromones in the crop inhibit normal mating or reproduction. The exact mechanism of inhibition is unknown. Males may be unable to locate female pheromone plumes within the high background concentrations of sex pheromone. The effects of high pheromone backgrounds on females are less clear.

Using pheromones avoids the environmental damage and health risks of applying toxic chemicals and may enhance biological control. Unfortunately, many important pests are not equipped with pheromones that act at long distances. Moreover, timing and coverage of pheromone application are critical for success, and the method controls only a single pest species. Thus a pest species targeted for control with pheromones should be a key pest.

Genetic Manipulation

Mass releases of insects that can mate normally but do not reproduce have been used increasingly to attempt to eliminate certain pests from entire regions. The landmark first success of the **sterile**

insect release method was with the screwworm fly, *Cochliomyia hominivorax* (Calliphoridae), a serious pest of livestock. This fly occurs in very low population densities during winter months. Mass-reared flies were sterilized by radiation and the males distributed by aircraft over infested areas to overwhelm natural populations. The populations of screwworm decreased exponentially with each release of the same numbers of sterile flies because the ratio of sterile to fertile flies increased after each release. In contrast, insecticides kill a relatively constant proportion of insects with each treatment. The screwworm fly was eradicated first from the island of Curaçao. Eventually coordinated control and releases of sterile males eliminated the pest from the United States and northern Mexico. The program continues in Mexico to keep screwworm flies from reestablishing north of Central America.

For sterile releases to succeed, the number of offspring produced from matings after each release must drop enough to ensure that the ensuing adult generation decreases in size with each release. The ratio of sterile to fertile insects required to decrease populations depends on many factors. Examples of insect characteristics that tend to facilitate the sterile insect release method are single instead of multiple matings by females, methods to economically separate mass-reared males from females, a low rate of increase of natural populations, and competitive mating performance of sterilized compared to normal insects. Lowering the target pest population before release increases the sterile/fertile ratio after release. For this reason, sterile insects are first released when populations are just recovering from normal seasonal lows. Often the targeted pest population is reduced by insecticides or other controls before releases. Using combinations of control methods to lower populations, even pests that can reach high peak populations, such as the cotton boll weevil, the codling moth, and fruit flies (Tephritidae), have been eradicated from prescribed regions by means of sterile releases. Because the ultimate goal of sterile releases is to eliminate an insect species from a particular region, this approach must be coordinated over the entire target area, necessitating massive investments for research and implementation. As a result, only programs that will produce economic and social benefits valued at least several times the total costs of control have been attempted.

The release of pest insects altered with introduced genes to decrease targeted insect populations has not been successful for pest management. Natural selection quickly selects against genes that make an insect population less competitive. A more realistic but still speculative goal would be to change the pest status of targeted insects, rather than their population density. For example, reducing the ability of key populations of mosquitoes to transmit virus would greatly reduce the damage or risk caused by the mosquitoes without the need to try to change population size. A mechanism to accomplish this goal must increase the numbers of insects with replacement genes relative to numbers of wild-type insects.

COMPARISON OF INTEGRATED PEST MANAGEMENT FOR COTTON AND ALFALFA

Two major crops, alfalfa and cotton, serve to illustrate how pest management methods have been integrated in different ways. We mention only major highlights. It is important to realize that the pest complexes and methods for managing them vary greatly among different regions. Optimum methods to control pink bollworm in Louisiana may not be relevant to California or to Sudan, but comparisons of problems and approaches to solving them in different regions are instructive.

IPM in Alfalfa

Alfalfa is a widely grown perennial crop used for animal feed. It is broadly tolerant of damage by herbivores and diseases. Cosmetic damage or the presence of insects in alfalfa hay is of little concern to the consumers (usually livestock). The low unit value of alfalfa hay compared to most crops requires that pest damage must be relatively extensive to justify the cost of insecticide treatments. The widespread use of the crop as feed for dairy cattle further restricts the choices of insecticides and posttreatment intervals to prevent chemical residues from entering dairy products. Moreover, many farmers who grow alfalfa specialize in another crop or in dairy farming and raise alfalfa mainly as a rotation crop or as an adjunct to their main activity. They are not likely to use sophisticated pest management schemes or complicated monitoring methods.

Over a thousand insect species are recorded from alfalfa fields. The key pests in the United States are the alfalfa weevil, *Hypera postica* (Curculionidae), and the potato leafhopper, *Emposaca fabae* (Cicadellidae), in eastern North America. In western North America the Egyptian alfalfa weevil, *H. bruneipennis,* is the key pest. Caterpillars of several noctuid species and aphids such as the pea aphid, *Acyrthosiphon pisum,* are occasional pests. In hot desert areas of the western United States, the spotted alfalfa aphid, *Therioaphis maculata,* was a key pest, but plant resistance and biological control have relegated it to the status of occasional pest. Alfalfa seed production has additional pests. Computer-based models of alfalfa growth relative to weather, pests, disease, and management (irrigation, fertilization, time of cutting, etc.) and models for pest and natural enemy populations can be used to assist management decisions and establish control guidelines.

The weevils are controlled by combinations of methods. Early first harvests decrease survival of the first generation of larvae in late winter or early spring and are least disruptive to natural enemies. A complex of parasitoid wasps imported from the Old World seems to be fairly successful against alfalfa weevils in most circumstances after the first cutting. The relatively high EIL of alfalfa provides for a stable supply of insect prey to maintain natural enemy levels. Alfalfa varieties with multiple resistance to pests are available and often popular. Insecticides can generally be reserved for the rapid and unexpected appearance of pests above the economic threshold.

Potato leafhoppers are severe pests only in the eastern and midwestern United States, and EILs have been established and refined in these areas. Adult leafhoppers migrate from southern states early in the growing season, so biological control of potato leafhopper is difficult to establish. Research showed that EILs for potato leafhopper are lower than had been believed because feeding damage reduces the protein content of the hay even before visible damage can be seen. Monitoring and establishing EILs for this pest have increased the use of insecticides to control it. This is counter to the general result that monitoring for IPM reduces insecticide use by eliminating unnecessary treatments. Insecticides for IPM should be selected and timed to cause the least disruption to key natural enemies. In addition to concerns about pesticide residues in dairy cattle feed, care must be taken to avoid killing honey bees that visit alfalfa blossoms.

Alfalfa IPM has generally succeeded in integrating multiple control methods for effective pest control with minimal environmental impacts. The crop is exotic to North America, and new pest species appear regularly in the United States. Crop and pest models continue to improve and incorporate more variables and objectives, so alfalfa IPM should continue to adapt and improve.

Cotton IPM

Cotton is a perennial tropical plant that is grown as an annual crop in tropical and subtropical regions in both arid and high rainfall climates around the world. Vast areas of cotton are planted, and the crop has a major role in world commerce. In the United States, cotton accounts for about 40 percent of all pesticides applied, but this figure can fluctuate widely from year to year. Pesticides are used by practically all cotton growers, but the number of applications varies widely with region and among individual producers. Insecticides remain a pivotal control method for cotton pest management.

The key pest in the eastern United States is

the cotton boll weevil, *Anthonomus grandis.* It lays its eggs inside the flower buds (squares) or young fruits (bolls), which soon fall to the ground. The weevil has from two to seven generations per year, depending on temperatures. Generations of entomologists have devoted careers to studying its biology and control. The boll weevil is native to Central America. Natural enemies have been imported and established, but biological control alone is not effective in keeping weevil densities below an economic threshold. Economic thresholds vary from about 12 to 25 percent of buds damaged by oviposition or feeding. High rates of insecticides are used wherever the weevil is a problem; from 3 to 13 applications per year is typical. The noctuids—the cotton bollworm, *Helicoverpa zea,* and the tobacco budworm, *Heliothis virescens*—are secondary pests that result from pesticide applications. Whiteflies (Aleyrodidae) are the most recent secondary pests induced by the use of the latest insecticides (synthetic pyrethroids). These secondary pest outbreaks, documented repeatedly in diverse regions throughout the Americas, indicate that natural (unintentional) control by natural enemies and diseases keeps populations of many insects below damaging levels.

Pink bollworm is a key pest in hot, arid cotton-growing regions. Cultural methods (shredding the crop and burying it by plowing) and pheromone techniques (mating confusion) can effectively control pink bollworm. Growers often bypass pheromone control of pink bollworm in favor of less expensive insecticidal controls that have to be applied anyway to control other pests.

Pest-resistant genetic traits and cultivars of cotton are available but little used commercially. Instead, the emphasis in cotton breeding is on increasing yields and lint quality by genetically improving agronomic traits. New varieties of cotton have been genetically engineered to incorporate *Bt* toxin genes, which are effective against lepidopterous pests. Rapid evolution of pest resistance to these genes is a major concern even before the release of the new varieties.

Rapidly maturing varieties of cotton are widely used for cotton pest management. The use of short-season cultivars in cotton IPM is a cultural method. The most damaging populations of key cotton pests typically occur late in the growing season. Early-maturing cultivars can ripen bolls before these damaging population densities appear. They were developed originally for resistance to root diseases and nematodes, but their benefits for insect pest management soon became evident. Fast-maturing cultivars are the cornerstone of IPM programs where heavy late season insecticide use would be unprofitable. By maturing the crop early, growers of short-season cotton varieties forego the potentially higher yields of longer maturing varieties but gain two advantages for pest control. The first is that the crop escapes attack by late season pests. Second, the crop-free period following harvest is prolonged, increasing the natural mortality of crop-specific pests such as boll weevil and pink bollworm. This longer crop-free period results in fewer pests that live to colonize the following year's crop.

Models of cotton growth responses to weather and farming practices such as irrigation and time of planting are widely used by growers for planning and management. In conjunction with computer-based models and field experiments, EILs have been established for key and occasional pests by region. Adherence to these guidelines varies widely, however, among pest management consultants, growers, and regions. Monitoring cotton pests to determine the need for treatment has been practiced for about a hundred years. Sampling to estimate pest abundance is now coupled to crop models that project crop growth and damage levels relative to control options.

Because insecticides have been applied to cotton so often and for so many years and still account for so much of the total amount of insecticide applied today, management of insecticide resistance is a priority for cotton IPM. In Australia, resistance levels of key pests to insecticides in use are monitored along with pest densities. Although participation is voluntary, almost all Australian growers take part in regional resistance management programs that base the choice of insecticide to be applied throughout the region on resistance monitoring. The choice of insecticides to be used seeks to maximally retard the development of resistance in key pests.

Cotton production in the Sudan, as in many other developing countries, depends almost entirely on insecticides for pest management. The crop is crucial to the Sudan's export earnings, and widespread crop failures because of pest damage are not economically or politically acceptable. Growers apply up to 30 applications of broad-spectrum insecticides per crop cycle. Aircraft are used to treat large areas at one time to accommodate the many small farms and to get around the lack of ground equipment. This application tactic has produced spectacular levels of resistance and outbreaks of secondary pests such as whiteflies (Aleyrodidae). Because of the heavy reliance on chemicals, the growers' reaction to resistance is to

increase the dosages and variety of chemicals. Because organochlorine and organophosphate insecticides can be produced cheaply, they are most often used, with dire environmental consequences. Cultural methods using planting dates and destruction of crop residues are widely practiced but inadequate by themselves. Interest in using plant resistance to pests is high, but the use of plant resistance is modest.

One of the largest and most impressive but also expensive developments in cotton IPM is the program to eradicate boll weevil from the southern United States. The program began in the Atlantic states (Virginia, Carolinas) and is scheduled to progressively eliminate weevil populations toward the southwestern United States. The eradication program aims to lower early spring populations of boll weevils by coordinated use of insecticides. Then pheromones are used to attract males to bait

TABLE 14.2.
Comparisons of IPM in Cotton and Alfalfa

	Alfalfa	Cotton
Crop features		
Crop duration	3–5 years	Annual (120–210 days)
Mode of weed control	Rigorous in establishing new crop, then minor	Rigorous use of herbicides, some cultivation
Control methods		
Insecticides	Not unusual to never apply; typically used for key pest over economic threshold; crop use is sensitive to residues	Use is universal, 3–13 times per year, depending on grower, region, and year
Biological control	Effective against several important exotic pests	Little success with imported natural enemies
Natural control	Extensive and important	Insecticides induce secondary pests; occasional pests usually suppressed
Cultural control	Early cutting for weevils; companion or mixed plantings partly effective	Short season cultivars central to some programs; post-harvest crop destruction, planting dates complement control with insecticides
Plant resistance	Widespread use against key pests; effective in combination with other methods; crop cultivars are heterozygous	Available but not used (unless short season cottons considered)
Pheromones	Not used because of expense and large number of pest species	Useful against two major pests

containing a chemical sterilant. Eggs from females that mated with sterile males do not hatch.

To summarize, cotton pest management has advanced from sole reliance on chemicals to an IPM system that is based on combining cultural and pheromone controls with insecticides. Chemicals remain the primary management approach; even the most promising developments of new transgenic varieties use genes for insecticide production by the plant. Nonchemical methods focus on delaying and reducing the number of pesticide applications. Plant and pest models are increasingly used in IPM decision making, but the use of resistant varieties will probably be postponed unless insecticide resistance eliminates chemicals as an option.

Table 14.2 contrasts some of the major differences between IPM in cotton and alfalfa.

SELECTED REFERENCES

General pest management methods, strategies, and principles are covered by Burn et al. (1987), Dent (1991), Horn (1988), Metcalf and Luckmann (1982), Metcalf and Metcalf (1993), Pedigo (1989), and Pfadt (1985); biological control by DeBach and Rosen (1991) and Hoy and Herzog (1985); plant resistance by Painter (1951) and C. M. Smith (1989); insecticides by Dent (1991), Hassell (1990), Metcalf and Metcalf (1993), and Ware (1989); insecticide resistance by Devonshire and Field (1991), Metcalf and Metcalf (1993), and Rousch and McKenzie (1987); pheromones by David and Birch (1989); cotton pest management by Frisbie et al. (1989), Horn (1988), Luttrell (1994), and Luttrell et al. (1994); alfalfa pest management by Pfadt (1985) and Flint (1981).

PART THREE

Insect Diversity

15
AN EVOLUTIONARY PERSPECTIVE OF THE INSECTS

This parasitic wasp (Ichneumonidae) has been preserved as a fossil in fine-grained volcanic shale, called paper shale, from the Miocene Epoch (23.3 to 5.2 million years ago) in west-central Nevada. When the shale is split into sheets, the wasp is visible as "counterpart impressions" on both surfaces. Such fossils indicate that most, if not all, Tertiary insects belong to orders, families, and even genera alive today.

The phylum Arthropoda is split into major subdivisions that reflect basic differences in structural organization (Table 15.1). The insects belong to the subphylum Labiata, which is characterized by mouthparts consisting of the labrum, a pair of mandibles, a pair of maxillae, and a pair of second maxillae fused into a labium, which forms the floor of the oral cavity. Although there are differences in detail of the mouthpart structures among the classes of Labiata, their close relationship is confirmed by a general similarity in body organization.

In all Labiata the head bears a single pair of antennae and the mandibles never bear palps, as they do in Crustacea. The segmental nature of the trunk is fundamentally similar in all Labiata, with dorsal terga, lateral pleura, and ventral sterna, the last producing invaginations or apophyses into the hemocoel for the purpose of muscle attachment. The alimentary canal consists of fore-, mid-, and hindgut divisions with a peritrophic membrane and Malpighian tubules. Respiration is by a tracheal system, usually with two spiracles per segment, and circulation is by means of a dorsal vessel with segmental contractile chambers.

TABLE 15.1
Conspectus of the Higher Classification of the Phylum Arthropoda

Subphylum Trilobita—trilobites
Subphylum Chelicerata
    Class Merostomata—horseshoe crabs and eurypterids
    Class Pycnogonida—sea spiders
    Class Arachnida—spiders, mites, scorpions, phalangids, etc.
Subphylum Crustacea—crabs, shrimp, lobsters, etc.
Subphylum Labiata
    Superclass Myriapoda
        Class Diplopoda—millipedes
        Class Chilopoda—centipedes
        Class Pauropoda—pauropods
        Class Symphyla—garden centipedes
    Superclass Hexapoda
        Class Parainsecta—Protura and Collembola
        Class Entognatha—Diplura
        Class Insecta—insects

In Crustacea, Arachnida, and Pycnogonida the alimentary canal typically expands into a large, glandular digestive and storage organ in the trunk region, and respiration is by book lungs or gills. In addition, in the last three subphyla the head and thorax are fused into an integral body region or **tagma,** the cephalothorax. In the Labiata the head is always an independent and freely moving tagma. The differentiation of the arthropod body into distinct regions specialized for different functions is **tagmosis.** This is one of the most important developments in the evolution of the Arthropoda and has produced the distinctive body plans of the various arthropod phyla and classes. Within the Labiata two patterns of tagmosis are evident. In the Myriapoda the body is differentiated into head and trunk. In the Hexapoda the trunk is further differentiated into the thorax and abdomen, the former specialized for locomotion, the latter for housing the intestine and reproductive organs. This organizational change led first to the shortening of the body and to the characteristic hexapod gait, and eventually to the evolution of wings in pterygote insects. The general organizational features of the hexapod body as proposed by Kukalová-Peck (1987) are shown in Figure 15.1.

Until recently all six-legged arthropods have been included in the Insecta. However, the hexapodous condition represents an **evolutionary grade** and may have been evolved more than once. In other words, six-leggedness is a level of structural organization that could easily have evolved in different groups of myriapods that lost all but three pairs of legs. Common grades of body organization are frequently shared by distantly related organisms adapted for similar modes of life. For example, in both birds and bats the forelimbs are modified as wings (though in different manners). In arthropods the six-legged body may have evolved three times from different multilegged ancestors— once in the Insecta, and again in the Parainsecta and in the Entognatha (Fig. 15.2; Table 15.1). Together these three taxa are referred to as the Hexapoda, but the monophyly of Hexapoda is debatable.

Independent evolution of the Parainsecta and Insecta is strongly indicated by fundamental differences in head structure. First, all antennal segments of Parainsecta are musculate, while in the Insecta only the basal two segments bear muscle insertions. More importantly, the Parainsecta have the mandibles and maxillae deeply retracted into pouches in the head capsule (Fig. 17.1c), the tips being protruded during feeding. This condition is termed **entognathy** or endognathy. The entognathous hexapods have **monocondylic** mandibles, adapted for externally triturating food particles into minute pieces before ingestion. In all Insecta, except Archaeognatha, mandibles are **dicondylic,** articulated with the head capsule at two points. These mandibles move transversely and are adapted for biting off and grinding food particles. The elongate styliform or bladelike mandibles of the Parainsecta lack the basal grinding lobe of Insecta. Differences in mouthpart structure among the entognathous taxa suggest, however, that entognathy may not represent a primitive divergence from the ectognathous hexapods, but is a condition that has arisen independently several times in taxa that primarily inhabit soil and litter. Entognathy may be favored in soil interstitial habitats because the enclosed mouthparts are protected from abrasion as the animals force their way through constrictions.

Ecologically, as well as in some morphological features, the Parainsecta resemble myriapods more than insects. Like myriapods, the Parainsecta are primarily inhabitants of litter and the upper soil layers, requiring high relative humidity or sources of free water. Direct evidence of their former multilegged condition is provided by their styliform, rudimentary abdominal appendages, which function as skids in supporting the abdomen. Sperm transfer is indirect, by means of a spermatophore that is attached to the substrate and later picked up by the female, with or without direct contact with the male. Some of these primitive features are also characteristic of apterygote insects, as discussed below. Diplura have often been classified as Parainsecta because the mouthparts are elongate and partly enclosed by the head capsule, and the mandibles are monocondylic. The mandibles of Diplura, however, have a distinct incisor lobe and the maxillae are relatively unmodified compared to those of Protura and Collembola.

Figure 15.1 Diagram of a hypothetical, primitive, winged insect, showing serial homologies of the appendage segments on the head, thorax, and abdomen. Some basal leg segments are incorporated into the sides of the body as follows: head (epicoxa); thorax (epicoxa, subcoxa); and abdomen (epicoxa, subcoxa, trochanter). Lobes between some segments of the appendages are called exites (outer lobes) and endites (inner lobes). Leg segments are: 1, epicoxa; 2, subcoxa; 3, coxa; 4, trochanter; 5, prefemur; 6, femur; 7, patella; 8, tibia; 9, basitarsus; 10, tarsus (primitively with two subsegments); 11, posttarsus (primitively with two claws). Ecx sidelobe, a lobe of the first leg segment or epicoxa; T, tergum. (Redrawn from Kukalová-Peck, 1987. Reproduced by permission of *Canadian Journal of Zoology.*)

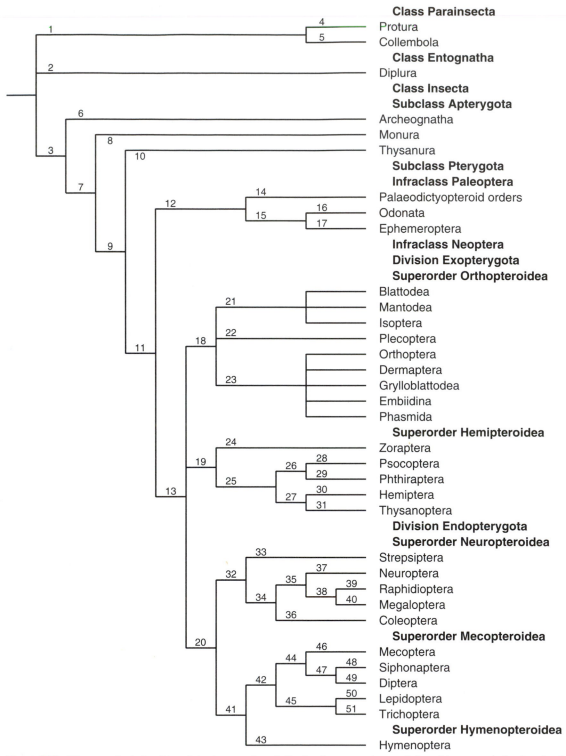

Figure 15.2 Diagram of relationships of the insect orders, indicating important characteristics of insects and closely related hexapodous arthropods.

Important characteristics of insects and closely related hexapodous arthropods.

1. development anamorphic (paedogenetic in Collembola); antennae musculate or absent; mouthparts entognathous
2. development epimorphic; antennae musculate; mouthparts entognathous
3. development epimorphic; antennae flagellate; mouthparts ectognathous
4. antennae and tentorium absent; development anamorphic
5. abdomen with 6 segments (apparently paedogenetic) and 3 sets of specialized appendages
6. mandibles monocondylic (single articulation)
7. mandibles dicondylic (2 articulations)
8. abdominal appendages with claws
9. abdominal appendages usually absent; lacking claws if present
10. primitively wingless
11. primitively winged (wings may be secondarily lost)
12. paleopterous
13. neopterous
14. mouthparts suctorial; nymphs terrestrial
15. mouthparts mandibulate; nymphs aquatic
16. indirect flight muscles very small; males with unique secondary copulatory structures
17. indirect flight muscles large; subimago and imago functionally winged
18. mouthparts mandibulate; Malpighian tubules numerous; central nervous system diffuse (numerous separate ganglia); development exopterygote
19. mouthparts modified, suctorial in most; 4–6 Malpighian tubules; central nervous system concentrated (posterior ganglia fused); development exopterygote
20. mouthparts, Malpighian tubules and central nervous system variable; development endopterygote
21. dorsal longitudinal muscles very small; corporotentorium perforated; immatures terrestrial; eggs in specialized ootheca (lost in Isoptera)
22. dorsal longitudinal muscles large; corporotentorium imperforate; immatures aquatic; ootheca absent
23. dorsal longitudinal muscles large; corporotentorium imperforate; immatures terrestrial; ootheca absent
24. cerci present; nymphs with ocelli; mouthparts simply mandibulate
25. cerci absent; nymphs without ocelli; mouthparts specialized
26. mandible retained and laciniae modified as rods, or all mouthparts highly modified (Anoplura)

27. mandibles and laciniae modified as stylets
28. free-living
29. ectoparasitic
30. mouthparts symmetrical; both mandibles modified as stylets; labium modified as sheath
31. mouthparts asymmetrical; right mandible absent; labium unmodified
32. head with gula or pregular bridge; larvae campodeiform or highly modified
33. legs without trochantins; immatures endoparasitic in insects
34. legs with trochantins; immatures never endoparasitic in insects (but some are endoparasitoids)
35. forewings membranous
36. forewings modified as elytra
37. larvae with incomplete intestine and suctorial mouthparts
38. larvae with complete intestine and chewing mouthparts
39. larvae terrestrial
40. larvae aquatic
41. head without gula (pregular bridge may be present); larvae usually eruciform or apodous; larvae frequently with abdominal legs
42. adult with rostrate (modified as rostrum or proboscis) mouthparts; 4–6 Malpighian tubules
43. adult with mandibulate mouthparts; Malpighian tubules numerous
44. larvae usually apodous (exception, Mecoptera); labial glands seldom producing silk; wing membrane bare or with few, unmodified setae
45. larvae eruciform, frequently with abdominal legs; silk production by labial glands; adults with wing membrane covered with hairs or scales
46. adults with 4 wings; larvae with thoracic legs and usually with abdominal prolegs
47. adults with 2 wings or wings absent; larvae apodous
48. adults apterous; ectoparasites living on mammals and birds
49. adults almost always with 1 pair wings; usually free living; when ectoparasitic seldom living on host
50. adults with mouthparts developed as coiled proboscis (occasionally atrophied); wings covered with scales; larvae almost always terrestrial
51. adults with mouthparts atrophied; wings covered with hairs; larvae aquatic

Based on these characters, Diplura have traditionally been considered the entognathous order most closely related to Insecta. In a fossil dipluran from the Pennsylvanian Period of Illinois, the mouthparts are clearly ectognathous and large compound eyes were present (Fig. 18.2). These characters suggest that many of the similarities between Diplura and Parainsecta have resulted from convergence in form due to similar litter and soil environments and that Diplura alone should be recognized as the separate class, Entognatha. Other characters of Diplura, such as the musculate antennae, are clearly primitive features shared with Myriapoda, indicating the evolutionary origin of the insects.

Protura and Collembola differ from Diplura as well as Insecta in basic features. Protura lack antennae, and they grow by a process of **anamorphosis,** whereby abdominal segments are added at the time of molting. The newly eclosed nymphs have nine segments, while mature individuals have twelve segments. Anamorphosis is characteristic of myriapods. In contrast, insects (and other hexapods) grow by **epimorphosis,** in which the number of segments remains constant. Collembola are unique among hexapods in their six-segmented abdomen, also found in early instars of some myriapods. It has been suggested that the six-segmented abdomen indicates that adult Collembola are paedogenetically derived from Myriapoda. Eggs of Collembola are **microlecithal** (small-yolked) and develop by complete or **holo-** blastic cleavage. Eggs of other hexapods divide by **meroblastic,** or superficial, cleavage because of the large yolk mass.

The great differences among the various entognathous orders indicate a very early divergence. Fossil Collembola are recorded from the Devonian Period (Shear et al., 1984), but the only other fossil representing this group is the dipluran mentioned above. This is probably because the small, soft bodies of Parainsecta and Entognatha do not lend themselves to preservation. The extant orders of the primitive hexapods perhaps represent remnants of a much more diverse Paleozoic fauna.

ORIGIN OF WINGS

Wings have contributed more to the success of insects than any other anatomical structures; yet the historical origin of wings remains largely a mystery. The earliest insect fossils that have been discovered, from the Pennsylvanian Period, were already winged. The primitively wingless hexapods, from which we might have gained insight into the origin of flight, are poorly represented in the living fauna. Thus the body structures that developed into wings, the steps in the evolution, and the ecological circumstances that favored wings, are debatable. Figure 15.3 shows what a primitive apterygote might have looked like (see further discussion, "Exite Origin of Wings").

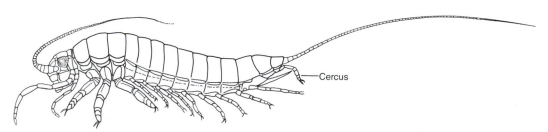

Figure 15.3 Reconstruction of a primitive apterous insect, Monura *(Dasyleptus),* from the Pennsylvanian Period (= Upper Carboniferous; 323 to 290 million years ago) of Illinois. Length without antennae or terminal filament is about 35 mm. Note that the cerci retain the shape of legs. Not shown are the dense bristles that covered the body. (Redrawn from Kukalová-Peck, 1987. Reproduced by permission of *Canadian Journal of Zoology.*)

Predaptations for Flight

Before discussing two theories of wing origin, let us examine certain general deductions about the origin of pterygotes and their early environment. Wings evidently evolved only once, because the veins and articulatory sclerites at the wing base can be homologized among all pterygotes. Furthermore, wings probably evolved after the apterygotes had acquired the characteristics of resistance to desiccation, breathing of air by means of tracheae, and six-leggedness. These attributes may be considered to be preadaptations for flight. These features, that is, evolved in response to one set of environmental conditions and provided the basis for the evolution of new features under another set of conditions. Here adaptations to terrestrial life and locomotion have provided the preadaptations for aerial life and locomotion.

Protection against desiccation is obviously needed by small organisms that are freely exposed to the drying effect of air currents. The assumption that the tracheal system preceded wing evolution is supported by the association of that system with wing ontogeny. At one time tracheae were even thought to influence the pattern of venation, but this is incorrect. The veins arise as blood-filled extensions of the hemocoel between the layers of wing epidermis. Nevertheless, tracheae supply oxygen within the wing and presumably antedate the wing in evolutionary origin.

The six-legged condition probably also evolved before the wings. The reasoning is that wings are restricted to the thoracic segments because only those segments had the skeletal and muscular arrangements associated with walking legs. Certain flight muscles were originally leg muscles, and the enlarged pleura that first evolved to support the legs were modified as a stiff fulcrum for the wings. Of the three segments, the mesothorax and metathorax supported the more powerful legs and possibly were also in the most favorable position aerodynamically to evolve the flight mechanism.

Fossils indicate that the early pterygotes may have molted more than once after the wings became functional, thus perpetuating the apterygote pattern of life history in which adults molt repeatedly. Today only the Ephemeroptera have two instars with wings. All other pterygotes possess functional wings only in the final instar. Judging by the occasional failure of living insects properly to shed the cuticle enclosing the wings, one suspects that repeated molting of wings large enough to fly would involve a decided risk during growth. Flight is also expensive energetically and would divert nutritional resources away from growth in the immature stages. The strategy of delaying reproduction and dispersal until the last instar has been successful in the insects. This style of life history combines the advantages of flight and offsets the hazards of wing molting. Flight is advantageous in promoting sexual outcrossing among unrelated mates, thus reducing inbreeding; in dispersing to new habitats; in escaping enemies; and in locating specific feeding and oviposition sites. It is no coincidence that three of the largest, most diverse orders of insects, Lepidoptera, Diptera, and Hymenoptera, are primarily aerial in habit.

Ecological Considerations

Turning to the probable circumstances under which flight evolved, we should note that by the middle of the Devonian Period some plants reached 6 m or more in height. Later, in the Mississippian and Pennsylvanian Periods (=Carboniferous Period), trees of the coal swamps reached 41 m in height. Food in the form of spores, pollen, seeds, or vegetative parts might have provided incentives for early apterygotes to adopt arboreal habits. Arboreal herbivores also may have been joined by predatory relatives and arachnids. Therefore, the appearance of wings, which allowed these early insects to escape predators and to move from plant to plant, was a major evolutionary development regardless of any other function wings might once have served.

The distribution of vegetation and bodies of freshwater was probably patchy and seasonal in the Ordovician and Devonian. Dispersal by air would have been a distinct asset, both for terrestrial forms as well as for those that had become secondarily aquatic. It has been emphasized by Rainey (1965) that dispersal by wind is especially advantageous in arid regions. Thermal updrafts over heated ground lift objects aloft. The prevailing winds tend to converge at low-pressure areas where rainfall is likely. Thus wind movement is toward areas of renewed plant growth and freshwater.

Flight is not essential for passive airborne dispersal, but organisms must be quite small to benefit from passive dispersal. Glick (1939) collected wingless arthropods by airplane at heights up to 4500 m. Included in his catch were mites, spiders, Thysanura, Collembola, Siphonaptera, ants, and immature instars of Hemiptera, Orthoptera, Coleoptera, Lepidoptera, and Diptera. The possession of wings, however, permits insects both large and small to reach altitudes of favorable wind currents, and to remain there longer, thus greatly increasing the distances traveled. Wings are, of course, important for local movements because they enable insects to search for mates and select suitable feeding or breeding sites.

All theories of wing origin must somehow explain how selection would favor intermediate steps of wing development before the wing could function in flight. To lift the insect in air and provide a forward thrust, the wings must be thin, stiff, and large relative to the body; they also must be articulated at the bases and moved by muscles in a complex propulsive stroke.

The time from the appearance of fossil land plants in the Devonian Period to the first fossils of winged insects in the Pennsylvanian Period is 100 million years. Perhaps it is not surprising that arthropods could become fully terrestrial and then aerial during that time. A much shorter period in the Cenozoic Era was sufficient to evolve humans from shrewlike prosimians.

Paranotal Origin of Wings

According to one theory, flight first evolved in an arboreal insect. If dislodged by wind or while escaping predators, a wingless insect would fall to the ground unchecked. Although probably not harmed, it would have to regain its footing before escaping further. The addition of thin, lateral expansions on the terga, called paranotal lobes, might create a stable orientation or attitude during falling so that the insect would land on its feet for a quick getaway. Calculations have shown that some degree of attitudinal control is given when even small, fixed wings are added to models, provided the models are no smaller than 1 cm in length. Paranotal lobes also might have functioned in other ways, such as giving increased lateral protection from predators when the insect was flattened against a substrate.

Larger paranotal lobes would facilitate gliding to the ground or from plant to plant. The development of a suitable basal hinge on the thoracic nota would permit the paranotal lobes to flap when the thorax was deformed by muscular contractions. The final refinements would include modification of the basal hinge, muscle attachments, and nervous system to allow the wing angle to be varied during each stroke.

The anatomical evidence for this theory is based mainly on the presence of broad thoracic nota in Thysanura and Archeognatha, and winglike prothoracic lobes on some fossil Palaeodictyoptera, Ephemeroptera, and Protorthoptera. The prothoracic lobes of some fossils appear to be articulated and to have well-developed venation (Fig. 15.4b). Presumably they indicate what the wings were like in an incipient stage of evolution.

Exite Origin of Wings

Another theory proposes that flight originated in an insect with an aquatic, gill-bearing nymph. The lateral abdominal gills of certain extant Ephemeroptera bear some resemblance to wings, being movable, thin, and membranous and having a ramifying pattern of tracheae. The similarity is quite striking in nymphs of several orders of fossil insects from the Paleozoic (Kukalová-Peck, 1983) (Fig. 15.4a). These fossils have wings on the meso-

Figure 15.4   Reconstruction of Paleozoic insects by J. Kukalová-Peck; **a**, naiad of *Protereisma* sp., a mayfly from the Permian Period (290 to 245 million years ago); **b**, *Homaloneura joannae*, a palaeodictyopteran from the Pennsylvanian Period (= Upper Carboniferous Period; 323 to 290 million years ago). (Courtesy of and with permission of J. Kukalová-Peck.)

thorax and metathorax that are too small for flight, but are curiously held outstretched and curved obliquely backward from the body. The winglets have fairly distinct venation, which seems to be homologous to that of the wings and what appears to be a movable basal articulation.

The similarity of the winglets and abdominal gills in the Paleozoic mayfly nymphs reinforces the idea that the gills and wings are serially homologous. By beating the gills, living mayfly nymphs create currents of water over the body to replenish oxygen. Sclerotized gills or gill plates also function to some extent in locomotion. The articulated winglets of Permian nymphs may have moved in the same fashion and for the same purposes. More powerful strokes might have propelled the nymph forward. Thus the winglets might have functioned like fins, stiffened by venation, and moved in a propulsive stroke by muscles.

In another Paleozoic order, the Palaeodictyoptera, nymphs are believed to have been terrestrial. The flattened abdominal flaps may have functioned as covers for the spiracles, reducing water loss. Flight could have originated when the spiracle covers became enlarged to allow gliding in nymphs that fell from emergent vegetation. Palaeodictyoptera appear to be among the most primitive flying insects, suggesting that the aquatic habits of nymphal Ephemeroptera and modifications of the spiracle covers as gills were probably evolved subsequent to the origin of flight. This is important, since it much easier to imagine flight arising in a terrestrial, rather than an aquatic ancestor. The prothoracic winglets, which were placed too far forward to be aerodynamically useful, were fused to the pronotal margin and then lost in modern insects. Various authors have proposed that the original function of articulated notal flaps was in courtship display (Alexander and Brown, 1963), in thermoregulation (Kingsolver and Koehl, 1985), in sculling along the surface of water, as in some winter stoneflies (Marden and Kramer, 1994; but see Will, 1995), and in other behaviors (Kingsolver and Koehl, 1994).

If wings and gills are serially homologous, what familiar structures of the arthropod body do they represent? Opponents of the gill theory, such as Snodgrass (1958), believed that the abdominal gills were serially homologous to the thoracic legs. Since legs are already present on the thoracic segments, the wings must have evolved from something else, such as paranotal lobes. Wigglesworth (1976), however, argued in favor of the gill theory by homologizing gills with the basal lobes of the arthropod limb, the **exites.** If derived from exites, both wings and legs could be present in the thorax.

Kukalová-Peck (1983, 1987) has shown that exites existed on the leg bases of numerous Paleozoic insects, both winged and wingless. The styli of modern Parainsecta and Apterygota have been interpreted as endite lobes, and it appears that early insects had a variety of endite and exite processes on the basal (precoxal and subcoxal) segments of their legs. The fossils illustrated by Kukalová-Peck further show that the Paleozoic order Monura (Fig. 15.3) had well-developed abdominal legs homologous with those on the thorax. The earlier presumption of Manton (1972, 1973) and others that the insect leg is primitively uniramous and thus basically different from the legs of other arthropods appears to be incorrect. The primitive structure of the myriapod trunk segment probably incorporated both exites and endites of various function (Fig. 15.1).

An important aspect of the exite theory of wing origin is that wings are thereby derived from a structural feature that already had the required basal articulation and musculature. Complex structures are almost always evolved from precursory features that are preadapted for the necessary modifications. If wings were developed from fixed paranotal lobes, it is difficult to explain how the basal hinge and musculature could have appeared completely *de novo*.

CLASS INSECTA

As restricted here, the class Insecta may be defined as the hexapodous arthropods that at some time in their life show the following features: head, five-segmented with ectognathous mouthparts comprising mandibles without palps; maxillae and labium both with palps; antennae with intrinsic muscles in first two segments; thorax, three-segmented, with one pair of legs per segment; legs, six-segmented (coxa, trochanter, femur, tibia, tarsus, pretarsus); abdomen, primitively eleven-segmented with cerci; respiration through tracheal system with spiracles on last two thoracic and first eight abdominal segments; excretion by Malpighian tubules; postembryonic development epimorphic. In addition, the vast majority of insects are characterized by the possession of compound eyes and wings in the adult stage, though these are absent in some primitive forms and secondarily lost in many specialized ones.

By conservative estimates the class Insecta encompasses at least 876,000 extant species, classified in 32 orders and about 939 families (Table 1.1). These numbers dwarf those in any other class of animals. Indeed, each of the four largest orders of insects (Coleoptera, Lepidoptera, Hymenoptera, Diptera) exceeds the next largest phylum of animals (Mollusca) by a considerable margin. The family Curculionidae (weevils), with an estimated 60,000 species, is by far the largest family of animals, larger than all nonarthropod phyla except Mollusca. Species known only from fossils represent at least 12 additional orders, which arose and radiated in the late Paleozoic and early Mesozoic Eras.

The morphological and ecological diversity within the Insecta is enormous. As detailed elsewhere in this book, insects occupy essentially every conceivable terrestrial habitat and have extensively colonized freshwaters. Only marine situations are relatively devoid of insects, although several thousand species inhabit the sea/land interface presented by intertidal regions, and a few Hemiptera live on the surface of open seas far from land. Insects include scavengers, herbivores, predators, and parasites. Many species are narrowly specialized for highly restricted foods, substrates, or activity periods, with consequent morphological modifications. The immensity of diversity within the Insecta and the attendant complexity of their phylogenetic relationships have resisted attempts to develop an entirely consistent and comprehensive evolutionary history. Nevertheless, the broad outlines of insect evolution have been repeatedly confirmed by independent lines of investigation. Areas of uncertainty or dispute are indicated where appropriate in the following account.

The classification proposed by Handlirsch (1908, 1926) has formed the framework for most modern classifications (Brues et al., 1954; Hennig, 1953; Mackerras, 1970; O. W. Richards and Davies, 1977; Rohdendorf, 1969; Wille, 1960), although many minor differences are evident. Kristensen (1991) and Kukalová-Peck (1991) provide excellent reviews of ideas regarding hexapod phylogeny.

Perhaps the most significant advance over Handlirsch's classification is the recognition of the major differences in flight mechanism between the paleopterous and neopterous orders, as first set forth by Martynov (1925). More recently, Tiegs and Manton (1958), Manton (1964, 1972, 1973), and Anderson (1973), among others, have shown conclusively that the entognathous hexapods are fundamentally different from Insecta, as detailed above. The classification adopted here is shown in Table 1.1 and Fig. 15.2.

SUBCLASS APTERYGOTA

The apterygotes are similar in body organization to pterygote (winged) insects, but primitively lack wings. These insects have ectognathous mouthparts and intrinsic musculature in only the basal two antennal segments, and they have compound eyes and ocelli. The long, slender cerci and median filament recur in Ephemeroptera (mayflies), and the body form of Apterygota strongly resembles that of Ephemeroptera nymphs. Like the Entognatha, apterygotes have rudimentary abdominal appendages, employ indirect insemination, and

molt throughout life. These are primitive features that have been retained in both hexapod lines of evolution; they do not indicate close relationship.

Two levels of organization are evident within the Apterygota. In the Archeognatha the elongate monocondylic mandible has the incisor lobe distinct from the molar lobe (Fig. 19.1b), a primitive condition similar to that in many Crustacea. In Thysanura the dicondylic mandible has the incisor and molar lobes much closer, as in pterygote insects. Archeognatha have large, compound eyes, as in Pterygota. Extant Thysanura have the eyes reduced to a few lateral facets, but relatively large compound eyes are present in some fossils. The Archeognatha and Thysanura also show differences in the endocrine system (Watson, 1965), in the structure of the spermatozoa (Wingstrand, 1973), and in the abdominal musculature (Birket-Smith, 1974). The order Monura, known from abundant Paleozoic fossils, appears to be more primitive than extant Apterygota and has been interpreted as ancestral to winged insects and other Apterygota.

The order Monura (Fig. 15.3) superficially resembled Archeognatha but had a full set of abdominal appendages, each with paired terminal claws, and appeared to have segmental divisions on the head. The lateral caudal filaments were in the form of typical cerci, but were also tipped by claws. Monura, as well as some Paleozoic species of Archeognatha and Thysanura achieved body sizes of 50 mm or more.

SUBCLASS PTERYGOTA

The Pterygota are primitively winged insects with the mesothoracic and metathoracic segments enlarged and bearing wings, or secondarily wingless; mandibles primitively dicondylic, adapted for chewing, or highly modified; abdomen primitively with 11 segments; the anterior 10 segments without appendages and the eleventh segment frequently with cerci.

Pterygota differ biologically from Apterygota

in employing **direct insemination** via copulation and in molting only until sexual maturity. No Pterygota possess abdominal appendages similar to those of apterygotes, but the abdominal gills of mayflies are probably exites of abdominal appendages. The abdominal appendages found on some pterygote larvae may be secondarily derived structures.

Infraclass Paleoptera

Pterygota comprise two major divisions, differing in their wing hinging mechanism and flight musculature. In the Paleoptera the wings cannot be flexed over the back at rest, whereas in the Neoptera wing flexing is allowed by the configuration of axillary sclerites and folds in the wing base. Wing flexing was necessary to permit winged insects to reenter the protected environments provided by litter and by subcortical (under bark) and other restricted situations. Some paleopterous insects, such as the suborder Zygoptera (Odonata), simulate wing flexing through the steep backward tilt of the pterothorax (Chap. 21).

The extant orders of Paleoptera exemplify many primitive features. In Odonata (dragonflies) the male stores the spermatozoa in a secondary copulatory organ on the second abdominal sternite. The female is responsible for completing sperm transfer, as in Apterygota. Possibly this behavior represents a modified indirect fertilization. Furthermore, in Odonata the flight musculature is entirely **direct** (attached directly to the wing bases) rather than primarily **indirect** (attached to the thoracic terga) as in Ephemeroptera and nearly all Neoptera. Notably primitive features of Ephemeroptera include the paired male external genitalia of some species and abdominal gills which, in at least some cases, appear to be homologous with the wings. The winged but sexually immature subimaginal instar of mayflies is unique, but suggests the indeterminate molting of Apterygota.

Extant Paleoptera are restricted to the two relatively small orders Ephemeroptera and Odonata, both aquatic as naiads. However, during the Paleozoic Era the Paleoptera radiated extensively, producing a much larger, more diverse fauna than at present. Two major evolutionary lineages existed. The palaeodictyopteroid lineage comprised a very diverse assemblage characterized by prothoracic winglets, long caudal filaments (cerci), and hypognathous mouthparts adapted for piercing and sucking. The clypeus was strongly domed, apparently to house a cibarial pump. Members of this lineage are classified as four orders: Palaeodictyoptera, Diaphanopterodea, Megasecoptera, and Permothemistida. Some of these species had wingspreads of 50 cm or more. It has been suggested that the piercing mouthparts were used to reach into the conelike fruiting structures of primitive plants such as lycopods and seed ferns. Other species were probably predators, while some very small, delicate Diaphanopterodea had the body form of mosquitoes and may have been ectoparasites. The Permothemistida had the hind wings reduced, as in modern mayflies. Palaeodictyopteroid nymphs were terrestrial and some were probably capable of flight, judging from the size of their wings (Shear and Kukalová-Peck, 1990). No members of the palaeodictyopteroid assemblage survived later than the Permian Period.

The other major lineage of paleopterous insects includes the extant Ephemeroptera and the dragonfly-like Protodonata. Paleozoic mayflies were generally similar to modern forms, but often were much larger (to 45 cm wingspan). Some of the fossils show mouthparts that appear to have been functional in the adults. Many of the Paleozoic forms had prothoracic winglets, and in some the forewings and hind wings were subequal. Naiads were aquatic. Protodonata were generally similar to Odonata in body form. Like Odonata they had powerful chewing mandibles and strong legs used for grasping prey. As in Odonata, nymphs had a prehensile labium or mask, and were predators. Older nymphs had large wings and probably could fly. Protodonata lacked the specialized secondary genitalia of dragonflies and differed in several features of wing venation. This order contained the largest known insects, with wingspans up to 71 cm in *Meganeura americana*.

Infraclass Neoptera

The Neoptera comprise about 99 percent of all insects. They present an exceedingly diverse assemblage but share the ability to flex the wings over the back by means of a pleural muscle inserting on the third axillary sclerite. For almost every other character, the Neoptera are variable, or at least subject to exceptions. Neopterous insects may be classified in two major divisions, the Exopterygota and Endopterygota, based on modes of growth and development. In Endopterygota the wings and other presumptive adult structures develop as internal buds, or **anlagen,** in the immature (larva), which usually differs from the adult in many features. In Exopterygota the immature (nymph) is generally similar to the adult except in the size of the wings, which develop as external pads. Some Hemiptera and Thysanoptera are physiologically holometabolous, with a "pupal" instar characterized by histolysis of muscles, alimentary canal, and other structures. However, the ontogeny of the wings is exclusively external in these species, which are highly evolved members of their evolutionary lineages. Thus, holometabolism represents an evolutionary grade that developed independently in Exopterygota and Endopterygota. For these reasons, Exopterygota and Endopterygota are preferable to Hemimetabola and Holometabola as taxonomic names.

Division Exopterygota

The Exopterygota comprise two major lineages, designated here as the superorders Orthopteroidea and Hemipteroidea. These differ in numerous major morphological characteristics (Table 15.2), the most important involving the structure of the central nervous system, the number of Malpighian tubules, and wing venation. In all these features, as well as in the structure of the mouthparts, the Hemipteroidea are specialized compared to the

TABLE 15.2.
Major Evolutionary Trends in Exopterygote Insects

Orthopteroid
Nymphs with ocelli
Chewing mouthparts; gular region well developed
Antennae long, multiarticulate
Hind wings dominate in flight ("posteromotoria")
Tarsomeres variable in number
Cerci usually large, multiarticulate
Malpighian tubules numerous
Central nervous system with separate ganglia in thorax and abdomen
Herbivores, predators, very few parasites; some social, subsocial (termites, cockroaches)

Hemipteroid
Nymphs lacking ocelli
Chewing mouthparts in most primitive groups; gradual development of piercing-sucking
  mouthparts in advanced groups
Tendency toward reduction in number of antennal segments
Forewings dominate in flight ("anteromotoria")
Three or fewer tarsomeres
Cerci absent (except in Zoraptera)
Four or six Malphighian tubules
Central nervous system strongly concentrated in thorax
Herbivores, predators, parasites; none social, few subsocial

more primitive Orthopteroidea. Fossils assignable to both the Orthopteroidea and the Hemipteroidea appear in the Pennsylvanian Period, when they coexisted with the very diverse paleopterous fauna. The division Exopterygota is considered by some authorities to be an evolutionary grade rather than a monophyletic lineage, and they reject it as a formal taxonomic name.

SUPERORDER ORTHOPTEROIDEA.   The orthopteroid orders are characterized by chewing mouthparts, long multiarticulate antennae, multiarticulate cerci, and numerous Malpighian tubules, as well as other primitive features. One possible derived orthopteroid feature is the enlarged anal lobe of the hind wing.

Undoubtedly the orthopteroid insects represent the most primitive Neoptera. Yet the interrelationships of the orthopteroid orders are poorly understood because of the mosaic distribution of primitive and specialized characteristics. For example, Dermaptera are the only Neoptera that possess paired external genitalia (paired penes of some species), but they display highly specialized hind-wing venation (to permit folding beneath the specialized forewings) and strongly sclerotized cerci modified as forceps. Morphological features of Embiidina are strongly specialized for life within self-constructed silken tubes, and Phasmatodea for camouflage. In contrast, nearly all features of the Grylloblattodea and Plecoptera are so generalized that their relationships with other orders are difficult to determine.

Not only are the relationships among the various orthopteroid orders difficult to specify, but their relationship to the Hemipteroidea and the Endopterygota remains a matter of contention. The derived characters linking the orthopteroid orders could have evolved independently more than once, and Orthopteroidea may not represent a monophyletic assemblage (Kristensen, 1975, 1991).

Within the Orthopteroidea, the Blattodea (cockroaches), Mantodea (mantids), and Isoptera (termites) form a group of clearly related orders, sometimes combined as the single order Dictyoptera. All Dictyoptera share several unique morphological features (proventricular armature, wing venation pattern, structure of female genitalia, perforated corporotentorium), and Blattodea and Isoptera share additional structural and biological features (see Chaps. 22, 24; Table 24.1). Phylogenetically the termites and mantids might be considered to be specialized cockroaches.

Among the remaining Orthopteroidea, Plecoptera are noteworthy for their primitive morphological features, including tracheal gills in nymphs, distinct anapleurite and coxopleurite in some nymphs, and multiarticulate cerci. Orthoptera show a long, separate evolution, with recognizable fossils as early as the Pennsylvanian Period. Their closest affinities are probably with the Phasmatodea. In the Zoraptera are combined features of both the Orthopteroidea (chewing mouthparts, cerci) and the Hemipteroidea (six Malpighian tubules, concentrated central nervous system, two-segmented tarsi, and reduced wing venation). Based on wing venation, they have been placed near the Blattodea (Kukalová-Peck and Peck, 1993), but their relationships remain uncertain.

SUPERORDER HEMIPTEROIDEA.   The most important feature in interrelating the hemipteroid orders is the configuration of the mouthparts. With the exception of Zoraptera, all Hemipteroidea have the clypeus enlarged to house the cibarial pumping structure and the lacinia modified as a sclerotized, styliform organ. In other mouthparts much variation exists. In one evolutionary line leading through Psocoptera (book lice), to Phthiraptera (lice), the lacinia is a sturdy rod that is applied to the substrate as a brace to stabilize the head while the mandibles scrape food particles. The hypopharynx and pharynx are modified as a mortar and pestle for crushing. Some Phthiraptera (suborder Anoplura) have the mouthparts further modified to suck blood, but their general similarity in numerous characters to more primitive lice (suborders Amblycera and Ischnocera) and Psocoptera leaves little doubt of their evolutionary origin.

In the evolutionary line leading to Thysanoptera (thrips) and Hemiptera (bugs), both mandibles and maxillae are modified as styliform piercing organs. The more primitive arrangement occurs in Thysanoptera, where the maxilla and single mandible are relatively short, thick blades, guided by the conical labrum and labium. As in Hemiptera the maxillary stylets are attached to the stipes by a special lever arm (Chaps. 34, 35). In Hemiptera the two mandibular and two maxillary stylets interlock to form a highly efficient piercing-sucking tube, which is enclosed by the troughlike labium. A few Thysanoptera and Hemiptera are physiologically holometabolous, as mentioned above. Their highly specialized mouthparts, concentrated central nervous system, and other features preclude them from being ancestors of the Endopterygota.

Compared to the Endopterygota, the exopterygote orders display limited biological diversity. Most Orthopteroidea live on the surface of ground, on vegetation, or in litter, while the great majority of Hemipteroidea inhabit foliage. Specialized modes of life, such as ectoparasitism in Phthiraptera and Hemiptera, or social behavior in Isoptera, are exceptional. There are no parasitoids, no endoparasites (see Chaps. 11, 12), and very few aerial species. Exopterygota scarcely participate in exploiting pollen and nectar, which are very important factors in the diversification of the Endopterygota. In general, exopterygotes utilize the same food resources in all instars (exceptions include cicadas) and are almost all external feeders. Inability to feed internally has barred them from a multitude of niches occupied by endopterygote larvae. For example, except for termites (Isoptera), almost no exopterygotes bore or mine living or dead wood, foliage, fruits, or seeds. By external feeding only, Hemiptera have exploited the vascular tissues of woody plants, an important habitat for endopterygotes, especially Coleoptera.

Most extant orders of Hemipteroidea are known as fossils from the Permian (Phthiraptera are known only from the Recent Epoch, less than 10,000 years ago), but the large group of Pennsylvanian and Permian fossils previously known as Protorthoptera actually represent a diverse fauna of primitive hemipteroids (Kukalová-Peck and Brauckmann, 1992). These are now assigned to the orders Caloneurodea, Glosselytrodea, and Blat-

tinopsodea. Members of these three orders are the most common Paleozoic insect fossils, suggesting that the hemipteroids, which constitute about 8 percent of the present fauna, were formerly more abundant.

Division Endopterygota

About 85 percent of the extant species of insects, representing immense taxonomic and biological diversity, are endopterygotes. While the evolution of holometabolism is not entirely clear, its chief adaptive value seems to be in allowing adults and larvae to utilize different resources. Resource division is usually accompanied by morphological divergence. This divergence is extreme in Diptera and most Hymenoptera, where the legless, grublike larvae frequently lack eyes and have the antennae and mouthparts reduced to papillae. Differences between adults and larvae are least in Megaloptera and Raphidioptera, which are among the most primitive endopterygote orders. Besides sharing holometabolous development, the Endopterygota also flex the wings along the jugal fold, in contrast to flexion along the anal fold in Exopterygota.

The origin of the Endopterygota is obscure. As noted, holometabolous development independently evolved in the Hemipteroidea, but there are no other Exopterygota that suggest a tendency toward holometabolism. The most primitive endopterygotes—Mecoptera, Megaloptera, and Raphidioptera—are extremely generalized, with little hint of what their ancestors may have been. Their wing venation has been compared to that of Palaeodictyoptera, but the resemblance is apparently convergent. Venational similarities to Plecoptera (Hamilton, 1972) are possibly more indicative of the true relationships of the Endopterygota. Alternatively, Endopterygta and Hemipteroidea may be sister taxa (Kukalová-Peck, 1991; Kukalová-Peck and Brauckmann, 1992).

The endopterygote orders form three evolutionary lineages, recognized as the superorders Neuropteroidea, Mecopteroidea, and Hymenopteroidea (Table 1.1). The more specialized orders, such as Coleoptera, Diptera, or Lepidoptera, clearly belong to one lineage or the other, but the primitive orders (Megaloptera, Mecoptera) are similar in body plan and evidently arose from a common ancestor.

SUPERORDER NEUROPTEROIDEA. The most primitive Neuropteroidea are certainly the Megaloptera and Raphidioptera. The pupae of both groups are capable of limited movements and have functional (decticous) mandibles that are used in defense. Although the larvae of Megaloptera are aquatic and those of Raphidioptera terrestrial, the morphological similarity of these neuropteroids is so great that they are sometimes classified as a single order. They are distinct as fossils since the Jurassic Period, however, and are here recognized as separate orders in accordance with recent practice.

Adult Neuroptera are similar to Megaloptera, but the larvae are highly specialized in having the mandibles and maxillae adapted for piercing and sucking, and in having the midgut end blindly. In addition, fossil evidence shows a long period of evolution independent of Megaloptera, with great diversification in the Permian and Cretaceous Periods, justifying recognition as a separate order.

Coleoptera and Strepsiptera are so highly specialized that their exact relationships are uncertain. Larvae of some aquatic Coleoptera are exceedingly similar to immature Megaloptera, and fossils such as *Tshekardocoleus* may represent intermediates between Megaloptera and Coleoptera. Mickoleit (1973) has shown that the coleopteran ovipositor is homologous with that in other neuropteroids.

The Strepsiptera are narrowly specialized for endoparasitism, and are morphologically distinct from other neuropteroids. Like Coleoptera, they fly using the hind wings (posteromotoria). First-instar larvae (triungulins) of Strepsiptera are similar to triungulins of the parasitoid beetle families Meloidae and Rhipiphoridae.

SUPERORDER MECOPTEROIDEA. Extant Mecoptera number no more than a few hundred species, but fossils record a large, diverse fauna in the late Paleozoic and Mesozoic Periods. Like Megaloptera, pupae of Mecoptera are decticous. Mecoptera have larvae with compound eyes, a feature shared with some primitive Lepidoptera. Most modern Mecoptera are characterized by the elongation of the lower face and mouthparts, producing a strong resemblance to primitive Diptera (Figs. 41.1b, 42.3a). Likewise, wing venation in primitive flies is similar to that of Mecoptera, and some fossils are difficult to place in either order. Modern Diptera are characterized by the reduction of the hind wings to halteres, but metathoracic wing pads are visible in some *Drosophila* pupae, and certain fossil Diptera may have been four-winged.

Siphonapteran (flea) adults are extremely specialized as ectoparasites of mammals and birds, but the larvae are similar to larvae of primitive Diptera, from which they presumably evolved. The former winged condition of Siphonaptera is revealed by pupal wing pads in some species.

Lepidoptera and Trichoptera are generally similar in wing venation, mouth-part morphology, and larval body form, though their larvae are terrestrial and aquatic, respectively. The most primitive Lepidoptera (suborder Zeugloptera) and all Trichoptera have decticous pupal mandibles, and Zeugloptera have compound larval eyes, as in Mecoptera. Lepidoptera are generally characterized by the elongation of the apposed maxillary galeae as a sucking proboscis. In Zeugloptera the maxillae are unmodified, and in Eriocraniidae the proboscis is formed by relatively short galeae, apposed only during feeding.

SUPERORDER HYMENOPTEROIDEA. Relationships of Hymenoptera are uncertain. They possess numerous Malpighian tubules, differentiating them from all other Endopterygota, which typically have four or six tubules. Hymenopteran wing venation is highly distinctive, with fusions of many veins to produce a few large cells. The eruciform larvae of primitive Hymenoptera share many characteristics with larvae of Lepidoptera and/or Mecoptera, including abdominal pseudopodial legs, a single tarsal claw, and labial silk glands. Probably the Hymenoptera diverged relatively early from the mecopteroid lineage.

THE GEOLOGICAL RECORD OF INSECTS

Our knowledge of insects that lived in the past is based on traces or remains preserved as fossils in the earth's sedimentary rocks (Table 15.3). The hard parts of insects are resistant to decay, and the chances are great that at least the wings will be preserved if insects are quickly immersed in a protective, fine-grained medium such as mud or volcanic ash. Unfortunately, through the ages terrestrial deposits on the surface are usually destroyed by the erosive action of water or wind. Other deposits become too deeply buried to be discovered. Today the places where insect fossils can be found are limited in number. Still, the fossil record of insects is surprisingly extensive, and throughout most of their evolutionary history the diversity of fossilized insects is greater than that of vertebrates (Labandeira and Sepkoski, 1993).

Fossil evidence can be used to establish the first appearance of a taxon and, in the case of extinct taxa, the last appearance. When contemplating the duration of their existence, remember that the organisms must have existed for some period prior to the earliest record we have found. Likewise, taxa now extinct undoubtedly persisted for some time after the latest fossils we happen to have found.

The chronological sequence in which the insect orders appear in the geological record is of limited value in determining their phylogenetic relationships. For example, the extant order Protura, though apparently of ancient origin, is unknown as a fossil, as is the suborder Mallophaga of the Phthiraptera. Zoraptera, probably originating in the Paleozoic Era, are known only from a single Tertiary Period fossil in Dominican amber.

Unfortunately evidence of the crucial steps leading to the origin of insects has not yet been found in the fossil record. Attention has been focused on strata of the Devonian Period (Table 15.3) because the first definite land plants have been found here. Fossils named *Eopterum devonicum* and *Eopteridium striatum* in these rocks were thought to be the oldest winged insects. They have now been identified not as insects but as the winglike tails of crustaceans. The first fossil Collembola of Devonian age were also controversial, but recently other fossils of Collembola and possibly of Archeognatha have been found.

Most insect fossils are found in sedimentary deposits, either coal beds or fine-grained siltstones. A difficulty with sedimentary fossils is that the anatomical detail necessary to critically analyze small, complex organisms like insects is obscured. Many of the fossils taken from coal beds show remarkable preservation of mouthparts, the wing base, and other important structures; the large size of many Paleozoic insects eases anatomical interpretation. For the great majority of small insects, especially those preserved in siltstones or sandstones, only gross features of anatomy are recognizable.

Many insects of the Cretaceous Period are preserved in hardened lumps of fossil plant resin called amber. The brownish, translucent substance often contains organic matter such as pollen, hair, leaves, or insects. Polished pieces are prized as jewelry. The resins do not decay or dissolve, but harden with time. Because of their light weight, lumps are carried by stream water to be deposited, eroded, and redeposited several times. Consequently, pieces may be much older than the rock in which they are finally embedded. Amber found with lignite or coal seams, however, probably was not carried far from the trees that produced the resin. Fossil inclusions in amber are remarkable for the detail of preservation, including setation, surface sculpturing, mouthparts, genitalia, and sometimes even soft tissues. DNA has been identified and partially sequenced for amber-preserved weevils from the Cretaceous Period and from other, less ancient insects as well.

EXTINCT ORDERS

The most common fossils are wings or fragments of wings. Many extinct orders have been described on the basis of unusual patterns of venation exhibited by fossil wings. As a result, the classification and the phylogenetic relationships of extinct insects are controversial. The first appearance of a modern insect order is also subject to argument

Geological Periods (duration in millions of years)

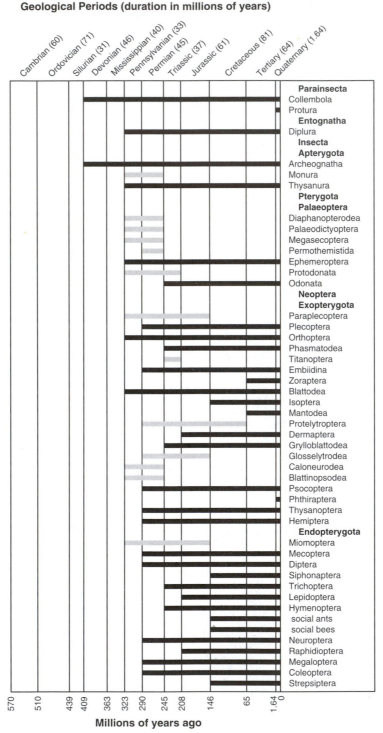

Cambrian (60)
Ordovician (71)
Silurian (31)
Devonian (46)
Mississippian (40)
Pennsylvanian (33)
Permian (45)
Triassic (37)
Jurassic (61)
Cretaceous (81)
Tertiary (64)
Quaternary (1.64)

Parainsecta
Collembola
Protura
Entognatha
Diplura
Insecta
Apterygota
Archeognatha
Monura
Thysanura
Pterygota
Palaeoptera
Diaphanopterodea
Palaeodictyoptera
Megasecoptera
Permothemistida
Ephemeroptera
Protodonata
Odonata
Neoptera
Exopterygota
Paraplecoptera
Plecoptera
Orthoptera
Phasmatodea
Titanoptera
Embiidina
Zoraptera
Blattodea
Isoptera
Mantodea
Protelytroptera
Dermaptera
Grylloblattodea
Glosselytrodea
Caloneurodea
Blattinopsodea
Psocoptera
Phthiraptera
Thysanoptera
Hemiptera
Endopterygota
Miomoptera
Mecoptera
Diptera
Siphonaptera
Trichoptera
Lepidoptera
Hymenoptera
 social ants
 social bees
Neuroptera
Raphidioptera
Megaloptera
Coleoptera
Strepsiptera

570 510 439 409 363 323 290 245 208 146 65 1.64 0

Millions of years ago

because the ancestral insects often do not closely resemble the modern forms. Recent treatments of fossil insects differ in methodology, higher categories, and number of orders, as well as in the families assigned to the orders. Carpenter's (1992) classification is traditional and evolutionary, while that of Kukalová-Peck (1991) is cladistic. For example, Kukalová-Peck does not recognize the Exopterygota or Orthopteroidea as valid taxa because they are evolutionary grades, not monophyletic groups. Here we list orders commonly mentioned in scientific literature.

SUBCLASS APTERYGOTA

Order Monura. Insects resembling Archeognatha, with ventral abdominal appendages and a long, caudal style and leglike lateral cerci. Some specimens exceed 30 mm in body length excluding the style. (Pennsylvanian-Permian; Fig. 15.3)

SUBCLASS PTERYGOTA

Infraclass Paleoptera

Order Protodonata (= Meganisoptera). Closely related to Odonata, but lacking the nodus, pterostigma, and arculus in the venation. *Meganeura* reached 71 cm in wingspread, but this was exceptional. Most were the size of large dragonflies. Nymphs had a labial mask similar to that of Odonata; presumably they were aquatic predators. Adults were predatory. (Pennsylvanian-Triassic)

Order Palaeodictyoptera. This order (Fig. 15.4b) and the next three are closely related. All possessed cerci (no caudal styles), external ovipositors, and long, piercing beaks of five stylets with the paired mandibles and maxillae and the hypopharynx supported in an elongate, troughlike labium. The beaks were held vertically beneath the head or slightly thrust forward, not directed to the rear as in Hemiptera. The postclypeus was swollen and may have housed a sucking pump. Nymphs of this order and of Megasecoptera and Diaphanopterodea were evidently terrestrial. The wing pads were free and curved obliquely backward. Nymphs had beaks like the adults, and all stages presumably had similar feeding habits. Many probably sucked the juices of plants, but some may also have probed crevices to reach the ovaries of primitive gymnosperm plants. Other species were probably predators, and some small mosquito-like forms may have been ectoparasites of vertebrates. Adults of this order reached 50 cm in wingspread, but most were smaller. Adults may have continued molting, as in Apterygota. The wings were often dark, with patterns of light spots or transverse bands. The hind wings were variable in size and shape, being larger than, equal to, or smaller than, the forewings. The pronotal lobes were large and sometimes exhibited a veined membrane. The abdominal terga had lateral lobes. Cerci were about twice as long as the abdomen. (Pennsylvanian-Permian)

Order Megasecoptera. Distinguished by long cerci and by distinctive wings that were long, nearly equal in size and shape, frequently elongate and petiolate at the base. Megasecoptera and Diaphanopterodea had simple or branched integumental processes on the body that were densely covered with setae. Some processes were quite long, even longer than the body, and formed fringelike rows on the tergites. (Pennsylvanian-Permian)

TABLE 15.3

Fossil record of the hexapods

The horizontal bars indicate that the oldest (earliest) fossils of an order were found in the period indicated toward the left side of the table. Shaded bars indicate that the order became extinct before the present and the last fossils were found in the period indicated toward the right side of the table. Example: Fossils of Monura have been found in the Pennsylvanian and Permian Periods. Solid bars indicate that the order survived to the present, but fossils may or may not have been found in the intervening periods. Geological periods are divided into three eras: Paleozoic Era (Cambrian to Permian Periods), Mesozoic Era (Triassic to Cretaceous Periods), and Cenozoic Era (Tertiary and Quaternary Periods). The Mississippian Period is also known as the Lower Carboniferous; the Pennsylvanian Period is known as the Upper Carboniferous. The distribution of insect fossils follows Kukalová-Peck (1991); the geological time periods follow Harland et al (1990).

Order Diaphanopterodea. Distinguished by a wing-flexing mechanism superficially like that of Neoptera. The nymphs were shrouded in filamentous projections of unknown function. (Pennsylvanian-Permian)

Order Permothemistida. Small with reduced prothoracic winglets and very small hind wings and short rostra. Nymphs unknown. (Permian)

Infraclass Neoptera

Division Exopterygota
Orders Paraplecoptera and Protoperlaria. Believed to be related to Grylloblattodea and Plecoptera. The anterior and posterior median veins are fused basally and the anal region of the hind wing is expanded. Mouthparts were mandibulate, prothoracic winglets were present, and moderately long cerci were present. In her classification, Kukalová-Peck (1991) includes fossils from the Jurassic in this lineage. Carpenter (1992) placed these insects in Protorthoptera. Paraplecoptera (Pennsylvanian to Jurassic); Protoperlaria (Pennsylvanian and Permian; not shown in Table 15.3)

Order Titanoptera. These were gigantic orthopteroids with raptorial forelegs and elongate mandibles suggesting predaceous habits. The forewings bore prominent fluted regions that may have functioned in sound production. Wingspans reached 36 cm. (Triassic)

Order Protorthoptera. This order previously contained a diversity of fossils, most of which have been reclassified as members of other orders of the Orthopteroidea or of other superorders. The Protorthoptera may eventually be eliminated as a formal taxonomic group. Carpenter (1992), however, placed families of the Paraplecoptera and Protoperlaria in the Protorthoptera. According to him, all families but one of uncertain classification were extinct at the end of the Permian. (Pennsylvanian to Permian; not shown in Table 15.3)

Order Protelytroptera. Small, robust insects superficially resembling beetles, but related to Blattodea and Dermaptera. Forewings formed into elytra; the broad hind wings folded beneath. Short cerci were present. (Permian to Cretaceous)

Orders Caloneurodea, Blattinopsodea, and Glosselytrodea. Small to large insects with domed clypeal region to accommodate sucking pump. The mandibles are variable, but often elongate and styliform. Maxillae were sometimes modified as a chisel-shaped organ, as in Psocoptera. Antennae were long, wings subequal. Short, annulate cerci were present. Carpenter (1992) considered the Glosselytrodea to be endopterygotes. Caloneurodea and Blattinopsodea (Pennsylvanian-Permian); Glosselytrodea (Permian-Jurassic)

Order Miomoptera. Small insects with mandibulate mouthparts, short cerci, and wings of equal size with simplified venation. In some Permian deposits they are more numerous than all other orders. The phylogenetic relationships of the Miomoptera are uncertain. According to some authors, they are endopterygotes and possibly ancestral to the Hymenoptera. (Pennsylvanian-Jurassic)

FOSSIL HISTORY OF INSECTS

Clear remains of a Collembolan, possible belonging to a family existing today, occur in the chert deposits of Devonian age near Rhynie, Scotland. Crowson (1970), who noted fossil-like traces of modern insects in the same rocks, suspects the insects died in crevices at more recent dates, after the rocks were formed. However, Collembolan fragments have also been found in the Devonian mudstone deposits near Gilboa, New York (Shear et al., 1984). The Gilboa material also includes fragments of cuticle that may be from an archeognathan (Table 15.3).

The earliest fossils that are unquestionably ancient and members of the Pterygota are found in the Pennsylvanian Period of the Paleozoic Era. The Pennsylvanian Period is also known as the Upper Carboniferous Period. A few isolated wings have been found near the base or oldest strata, but the best fossils begin to appear near the middle of the period, which was characterized by vast tropical swamps with luxuriant forests. This ancient vegetation has yielded extensive coal deposits. Two of the most important fossil localities are associated with such beds of coal. Fossiliferous shales at Commentry, France, contain diverse and well-preserved fossils. Insects from nearby coal forests evidently were buried in the fine sediments of stream deltas along the shore of a freshwater lake. Near Morris, Illinois, iron-bearing concretions or nodules that enclose fossil insects can be found along Mazon Creek. The ancient environment was a low, coastal plain with coal-producing forests.

Depending on how the orders are classified, the fossil record indicates that up to 18 orders of hexapods were present before or during the Pennsylvanian Period: Collembola, Diplura, Archeognatha, Monura, Thysanura, Palaeodictyoptera, Megasecoptera, Diaphanopterodea, Protodonata, Ephemeroptera, Paraplecoptera, Protoperlaria, Protorthoptera, Orthoptera, Blattodea, Caloneurodea, Blattinopsodea, and Miomoptera. Considerable insect evolution must have taken place before the Pennsylvanian because both the Paleoptera and Neoptera are represented, plus what some authorities believe are the first Endopterygota, namely, the Miomoptera. A possible endopterygotan larva also has been described from the Pennsylvanian. From this early fauna of hexapods, however, only the Collembola, Diplura, Archeognatha, Thysanura, Ephemeroptera, Orthoptera, and Blattodea survived until the present. Fossil wings of Blattodea are so common that the period is often called the "Age of Cockroaches." These insects were similar to modern cockroaches except that the wings were different in some features, their ovipositors were much longer, and they probably did not deposit oothecae. The Orthoptera had jumping hind legs, as in modern forms.

During the next period, the Permian, the climates of the world changed drastically. In the Northern Hemisphere increasing aridity is evidenced by great windblown deserts and thick salt deposits left by drying seas. Colder climates in the Southern Hemisphere were marked by several periods of glaciation.

Permian limestones of the Wellington Formation in Kansas and Oklahoma have abundant insect fossils. The strata measure some 213 m and include fossils of terrestrial plants, marine arthropods, and salt deposits. This is interpreted as a coastal swamp with intermittent freshwater habitats.

Associated with the changing climate and landscape of the Permian, the insect fauna also underwent major changes in composition. Ten new orders that have survived until the present made their appearance: Plecoptera, Embiidina, Hemiptera, Psocoptera, Thysanoptera, Neuroptera, Megaloptera (including a larva), Mecoptera, Diptera, and Coleoptera. Both the Orthopteroidea and Hemipteroidea are now represented, and the Endopterygota are definitely present as Neuropteroidea and Mecopteroidea. Three additional orders are also new: Permothemistida, Glosselytrodea, and Protelytroptera. With the appearance of insects of new kinds and the survival of a rich fauna from the Pennsylvanian, life of the Permian may have been more diverse in insect orders than even the present.

Permian Ephemeroptera had wings of equal size, and some had conspicuous mandibles. Nymphs are also found in Permian deposits (Fig. 15.4a). The mandibles, known from one specimen, are large and have well-developed teeth. The small wings have distinct veins and curve obliquely backward from the body. Gills were present on the first nine abdominal segments.

The end of the Permian was a time of massive extinction of Paleozoic hexapods that had thrived for at least 78 million years and probably much longer. Of the 14 ancient orders present in the Permian, the following were extinct by the end of the period: Monura, Diaphanopterodea, Palaeodictyoptera, Megasecoptera, Permothemistida, Caloneurodea, Blattinopsodea, Protoperlaria, and Protorthoptera. The remaining five ancient orders survived into the Mesozoic Era, later also to become extinct by the end of the era: Protodonata (extinct by the end of the Triassic Period); Paraplecoptera, Glosselytrodea, and Miomoptera (Jurassic); and Protelytroptera (Cretaceous).

The Mesozoic Era was a crucial time, as modern insects were evolving in association with flowering plants and mammals. After the various extinctions of the ancient orders that arose in the Paleozoic, the Mesozoic fauna took on a relatively familiar and modern appearance. Odonata, Phasmatodea, Grylloblattodea, Trichoptera, and Hymenoptera are first represented in the Triassic Period, plus the huge Titanoptera that existed only in the Triassic. During the Jurassic Period, fossils of Dermaptera, Raphidioptera, and Lepidoptera first appear. Also at this time, existing families of Odonata, Orthoptera, Diptera, and Hymenoptera can be recognized.

Cretaceous amber has been found in Alaska, Canada, Siberia, Lebanon, and New Jersey. The insects thus preserved include the earliest ants (worker caste), the earliest stingless bees (worker caste), Strepsiptera, and undisputed Collembola, plus Aphididae, parasitoid Hymenoptera, and Chironomidae. Mesozoic insect fossils have also been found in other types of preservation in Russia and

China. Several specimens of Siphonaptera have been described. Isoptera are known from Cretaceous deposits in Laborador. Thus, by the end of the Cretaceous, termites, ants, and bees had evolved eusocial behavior with worker castes, and entomophagous parasites and parasitoids were present. Most modern families had evolved and the fauna had essentially a modern aspect. The Cretaceous insect fauna survived into the Cenozoic Era, apparently without the extinctions suffered by the dinosaurs.

Of the several epochs of the Tertiary Period in the Cenozoic Era, the Oligocene is best known. Amber from the shores of the Baltic Sea is dated from the Eocene and Oligocene. Here for the first time are Mantodea, plus many representatives of modern orders. The species do not exist now, but the insects belong to modern types. The silverfish family Lepidotrichidae was thought extinct, but living relatives of those entrapped in Baltic amber

Figure 15.5 Insects preserved in Oligocene and Miocene amber from Chiapas, Mexico: **a** (left), fulgorid bug; **b** (right), stingless bee, *Trigona silacea* (Apidae). Note sensilla on antennae. (Photos courtesy of Joseph H. Peck, Jr.; b, reprinted by permission of *Journal of Paleontology*.)

were discovered in 1959 in the coastal forests of northern California. In general Baltic amber fossils represent distinct genera in modern families.

Oligocene and Miocene amber from Chiapas, Mexico, and from the Dominican Republic has yielded many insects (Fig. 15.5). Included are Zoraptera and stingless bees of the genus *Trigona* that differ only slightly from species living today in Central America. In general, fossils from Mexican and Dominican amber are very similar to modern species, and the greatest significance of these deposits is probably for paleoecology and biogeography.

Another important locality of Oligocene age is the Florissant Fossil Beds National Monument, near Colorado Springs, Colorado. A creek flowing through the area was dammed by lava and mudflows emanating from volcanoes about 24 km away. Fine ash from the volcanoes settled into the lake and buried a great variety of plants and animals that fell into the still water. Over a hundred species of higher plants and thousands of insect species have been described from these deposits. Half or more belong to genera existing today. Some, however, are no longer present in North America. *Glossina,* the tsetse fly, present as a fossil at Florissant, today is found only in tropical Africa.

Calcareous nodules containing insects have been found in the Mojave Desert of California, from deposits of the Miocene Epoch. Apparently a Miocene freshwater lake was surrounded by volcanoes. Many orders of insects are represented, but none for the first time. Some are beautifully preserved by replacement of the original organic matter by colloidal silica.

Asphalt or tar pools in southern California, dating from the Pleistocene Epoch of the Quaternary Period, have yielded abundant insect remains. Inside the skulls of the saber-toothed cat, *Smilodon,* are puparial shells of dipterous larvae that presumably were scavengers on animals trapped in the tar. Today adult aquatic insects, such as dragonflies and water beetles, are attracted to the waterlike reflections of the liquid tar and become entombed. The fossils taken from the asphalt pools

without exception represent species that occur in southern California today.

The oldest record of the suborder Anoplura of the Phthiraptera was obtained when lice were found on the carcass of a rodent frozen during the Pleistocene Epoch in Siberia (1.64 million to 10,000 years ago).

SELECTED REFERENCES

Phylogenetic Relationships

Anderson, 1973 (embryology and phylogeny of arthropods); Hamilton, 1972 (phylogeny of winged orders); Hinton, 1955a, 1958, 1963a (panorpoid orders); Istock, 1966 (evolution of metamorphosis); Kristensen, 1975 (comprehensive review of insect phylogeny), 1981, 1991; Kukalová-Peck, 1991 (summarizes extensive literature on phylogeny and fossil record), 1992; Manton, 1973; Remington, 1956 (entognathous orders); Sharov, 1966; E. L. Smith, 1969 (evolutionary development of external genitalia); Snodgrass 1950, 1958; Tiegs and Manton 1958; Tuxen 1959, 1970b (entognathous orders); Wille, 1960.

Origin of Wings

Paranotal lobe theory: Flower, 1964; Hamilton, 1971; Hinton, 1963b; Müller, 1879. Tracheal gill/exite theory: Oken, 1831; Wigglesworth 1976. Fossil evidence of wing origin: Kukalová-Peck, 1978, 1983, 1985; Wootton, 1976. Additional theories or general discussions: Alexander and Brown 1963; Brodsky, 1994; Kingsolver and Koehl, 1994; Marden and Kramer, 1994; Sharov, 1966; Snodgrass, 1958; Will, 1995.

Geological Record of Hexapoda

Carpenter, 1953, 1970, 1977, 1992 (reviews of insect fossil record); Elias, 1994 (Quaternary insects); Harland et al.,1990 (revised ages for geological time scale); Kukalová-Peck, 1987, 1991; Kukalová-Peck and Brauckmann, 1992 (morphological studies of Paleozoic insects); Labandeira et al.,1988 (Archeognatha); Martynova, 1961 (review); Michener and Grimaldi, 1988 (oldest bee); Plumstead 1963 (insect paleoecology); Poinar, 1992, 1993 (insects in amber); Smart and Hughes, 1973 (insect paleoecology); Smart et al., 1967 (critical dates for insect fossils); Wootton, 1981.

16
KEYS TO THE ORDERS OF HEXAPODA

Is this insect a wasp? Some insects superficially resemble other insects in a relationship called mimicry (Chap. 12). Careful use of a taxonomic key will identify this insect as a flower fly of the family Syrphidae (Diptera).

The keys in this book consist of series of numbered dichotomies or two-way choices. A pair of choices is called a **couplet.** Each half of a couplet leads either to a subsequent couplet (as indicated by the appropriate number) or to the name of a taxon (order, family), which is then identified.

Beginning with the first couplet, one works through the key, comparing the characteristics of the specimen at hand with the dichotomous choices. If a couplet lists more than one feature, the primary, or most diagnostic, character is compared first. Secondary characters that follow should be used to confirm identifications made with the primary character, or as alternatives when primary characters are missing or damaged. Numbers in parentheses indicate the couplet immediately preceding, so that the keys can be worked backward or forward. It is emphasized that any large group of organisms, such as the Insecta, contains exceptional species that do not fit keys. The keys in this book should identify nearly all the insects encountered during general collecting.

KEY TO THE CLASSES AND ORDERS OF COMMON

Hexapodous Arthropods

Body without wings, or with rudimentary or vestigial wings less than half body length   **Key A, p. 327**

Body with 1 or 2 pairs of wings at least half as long as body (wings may be modified as rigid covers over abdomen, or folded)   **Key B, p. 331**

KEY A

Wings Absent or Rudimentary

  1. Legs absent or reduced to unsegmented papillae shorter than one-fifth body width   **2**

      Legs with 4 to 5 distinct segments, almost always terminated by 1 or 3 claws   **3**

2(1). Mouthparts enclosed in a slender, tubular rostrum (Fig. 16.1a); antennae and eyes usually absent; body segmentation indistinct or absent; sessile plant feeders, frequently covered by a waxy or cottony shell (Fig. 34.16d)   **Hemiptera** (Chap. 34)

      Mouthparts mandibulate or internal, never enclosed in a tubular rostrum; antennae and eyes present or absent; body segmentation usually distinct; seldom with protective covering   **Legless endopterygote larvae, not keyed further**

3(1). Legs terminated by a single claw or without claws   **4**

      At least middle legs terminated by 2 claws   **16**

4(3). Head with large compound eyes almost always present laterally; ocelli frequently present on vertex   **13**

      Head with compound eyes absent or vestigial, stemmata (lateral ocelli) often present (Fig. 16.1b,d); ocelli absent from vertex   **5**

5(4). Mouthparts enclosed in a slender, tubular rostrum (Fig. 16.1a) antennae and eyes usually absent; body segmentation frequently indistinct or absent; plant feeders, frequently with body covered by a waxy or cottony shell (Fig. 34.16d)   **Hemiptera** (Chap. 34)

      Mouthparts mandibulate or concealed in head capsule; body segmentation rarely indistinct   **6**

6(5). Antennae with 2 or more segments   **8**

      Antennae absent   **7**

7(6). Tarsus terminating in single claw; lateral ocelli

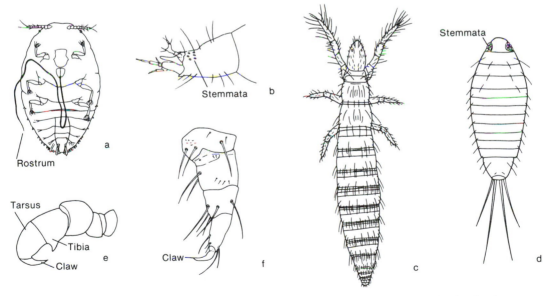

Figure 16.1 Key characters: **a**, a scale insect, Hemiptera *(Parlatorea)*; **b**, head of larval carabid beetle, showing stemmata (simple eyes); **c**, Protura (Acerentomidae); **d**, triungulin larva of Strepsiptera (Mengeidae); **e**, chelate tarsus of Phthiraptera (Anoplura); **f**, leg of endopterygote larva (Lepidoptera, Noctuidae). (a, modified from University of California Extension; b, modified from Melis, 1936; c, modified from Berlese, 1909.)

absent; body cylindrical, elongate (Fig. 16.1c); minute, pale arthropods found in soil   **Protura** (Chap. 17)
Tarsus without claw; stemmata (lateral ocelli) large, usually set in pigmented patches Fig. 16.1d); body fusiform (Fig. 16.1d); minute insects usually found in flowers, foliage.

       Strepsiptera (first instars, Chap. 40)

8(6). Abdomen with 6 segments; segments 1, 3, and 4 usually with medial unpaired appendages (Fig. 17.3a)   **Collembola** (Chap. 17)
Abdomen with 8 to 11 segments; without appendages or with paired unsegmented appendages on some segments   **9**

9(8). Tarsus and claw chelate (Fig. 16.1e); abdomen without appendages; ectoparasites with flattened body, thick, short, 3- to 5-segmented antennae   **Phthiraptera** (Chap. 33)
Tarsus and claw usually fused (Fig. 16.1f), very rarely chelate; abdomen with or without appendages; body shape extremely variable   **(Endopterygote orders, larvae) 10**

10(9). Abdomen with unsegmented, paired walking appendages on some preterminal segments; body caterpillar-shaped (Fig. 16.2a)   **11**
Abdomen without walking appendages on preterminal segments; not usually caterpillar-like   **12**

11(10). Abdominal appendages bearing rows, circles, or patches of short curved spines (crochets, Fig. 16.2b)   **Lepidoptera** (Chap. 44)

Abdominal appendages without crochets
       Hymenoptera (Chap. 46)

12(10). Abdomen with ventrally directed hooked prolegs on last segment (Fig. 45.2a); thoracic legs with trochanters 2-segmented; aquatic larvae usually dwelling in tubular cases
       Trichoptera (Chap. 45)
Abdomen without appendages or with dorsal or lateral appendages without terminal hooks; if prolegs are ventrally directed, then terminal hooks are absent   **Coleoptera** (Chap. 39)

13(4). Mouthparts enclosed in tubular rostrum (Fig. 16.1a); cerci absent   **Hemiptera** (Chap. 34)
Mouthparts mandibulate, never enclosed in a rostrum; cerci present or absent   **14**

14(13). Terminal abdominal segment with 2 or 3 long filaments; antennae multiarticulate, slender; long-legged aquatic insects, usually with leaflike abdominal gills
       Ephemeroptera (naiads) (Chap. 20)
Terminal abdominal segment without filaments; terrestrial; gills absent   **15**

15(14). Body caterpillar-shaped (Fig. 16.2a) or grublike (Fig. 16.2c), with short, thick legs; tarsi with 1 segment   **Mecoptera** (larvae) (Chap. 41)
Body slender, elongate, with long, slender legs; tarsi with 4 to 5 segments
       Mecoptera (adults) (Chap. 41)

16(3). Anterior legs with first tarsal segment globular, at least 3 times as thick and long as second segment   **Embiidina** (Chap. 28)

a

b

Crochets

c

d

Figure 16.2 Key characters: **a**, larva of Lepidoptera (Noctuidae); **b**, abdominal leg of larva of Lepidoptera (Noctuidae); **c**, scarabaeiform grub of Endopterygota (Coleoptera, Scarabaeidae); **d**, a katydid (Orthoptera). (a, b, modified from Melis, 1936; c, modified from Ritcher, 1944; d, modified from University of California Extension.)

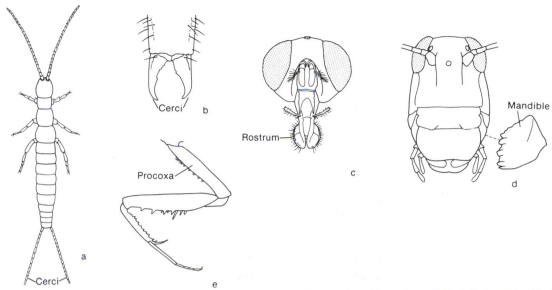

Figure 16.3 Key characters; **a**, Diplura, Campodeidae; **b**, forceps of Diplura, Japygidae; **c**, haustellum of Diptera (Muscidae); **d**, chewing mouthparts of a grasshopper (Orthoptera); **e**, raptorial leg of a mantid (Mantodea). (a, d, modified from University of California Extension; c, modified from Snodgrass, 1953.)

Anterior legs with tarsal segments about equal in thickness and length   **17**

17(16). Posterior legs modified for jumping, with femur greatly thickened (Fig. 16.2d)
                              Orthoptera (Chap. 29)

Posterior legs similar to middle pair   **18**

18(17). Last abdominal segment bearing cerci, either single-segmented (Figs. 16.3b, 16.4a) or multiarticulate (Fig. 16.3a)   **19**

Last abdominal segment without cerci   **32**

19(18). Last abdominal segment with median, multiarticulate filament; short, 1-segmented appendages present on at least abdominal segments 7 to 9   **20**

Last abdominal segment without median filament; appendages absent from segments 7 to 9   **21**

20(19). Head with large, contiguous compound eyes; maxillary palp with 7 segments; body cylindrical, arched   **Archeognatha** (Chap. 19)

Head with widely separated lateral ocelli or without eyes; maxillary palp with 5 segments; body usually flattened
                              Thysanura (Chap. 19)

21(19). Mouthparts modified as a rostrum or haustellum (Fig. 16.3c) or projecting beaklike below the head (Fig. 34.2e); labial palp with 2 segments   **22**

Mouthparts short, mandibulate (Fig. 16.3d), never forming a beak or haustellum; labial palp with 3 segments   **23**

22(21). Mouthparts modified as a rostrum or haustellum (Fig. 16.3c); metathorax frequently with halteres; antennae frequently with 5 or fewer segments   **Diptera** (Chap. 42)

Mouthparts with elongate mandibles and maxillae projecting beaklike (Fig. 41.1b); halteres absent; antennae with at least 12 segments   **Mecoptera** (Chap. 41)

23(21). Tarsi with 5 segments   **24**

Tarsi with 1 to 4 segments   **27**

24(23). Anterior legs raptorial (Fig. 16.3e)
                              Mantodea (Chap. 23)

Anterior legs not raptorial, fitted for walking   **25**

25(24). Head prognathous (mouthparts directed anteriorly)   **26**

Head hypognathous (mouthparts directed ventrally)   **Blattodea** (Chap. 22)

26(25). Cerci with 5 to 9 segments, long, flexible; body not modified for camouflage
                              Grylloblattodea (Chap. 25)

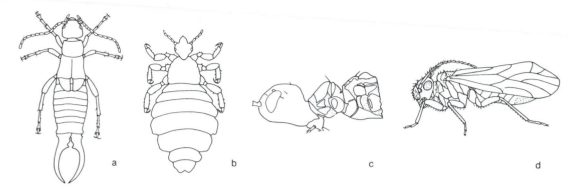

Figure 16.4 Key characters: **a**, an earwig (Dermaptera); **b**, a louse (Phthiraptera); **c**, lateral aspect of the anterior end of a thrips (Thysanoptera); **d**, a book louse (Psocoptera). (a,b, modified from University of California Extension; c, modified from Faure, 1949.)

Cerci with 1 segment, usually short; body almost always sticklike or leaflike
Phasmatodea (Chap. 30)

27(23). Tarsi with 1 to 2 segments **28**
Tarsi with 3 to 4 segments **29**

28(27). Cerci either multiarticulate (Fig. 16.3a) or 1-segmented, forceps-shaped (Fig. 16.3b); tarsi with 1 segment; soil or litter dwellers with long, parallel sided body (Fig. 16.3a)
Diplura (Chap. 18)
Cerci with 1 segment, short, not forceps shaped, tarsi with 2 segments; body stouter not parallel sided **Zoraptera** (Chap. 31)

29(27). Cerci forceps-shaped, strongly sclerotized (Fig. 16.4a) **Dermaptera** (Chap. 26)
Cerci not forceps-shaped **30**

30(29). Tarsi with 3 segments **31**
Tarsi with 4 segments **Isoptera** (Chap. 24)

31(30). Antennae more than half as long as body; labium small, without movable, apical teeth **41**
Antennae much less than half as long as body; labium jointed, with large, movable apical teeth (Fig. 21.5) **Odonata** (Chap. 21)

32(18). Tarsi with 5 segments **36**
Tarsi with 1 to 3 segments **33**

33(32). Mouthparts enclosed in a long, slender rostrum projecting beneath the head (Figs. 16.1a, 34.1b); maxillary and labial palps absent
Hemiptera (Chap. 34)
Mouthparts not in the form of a rostrum; maxillary and labial palps usually present **34**

34(33). Antennae longer than head, with at least 5 segments, usually with more than 10 segments **35**

Antennae shorter than head, with 3 to 7 segments **40**

35(34). Head cone-shaped, directed ventrally or posteriorly (Fig. 16.4c); antennae with 4 to 9 segments; body elongate, slender
Thysanoptera (Chap. 35)
Head not cone-shaped; antennae almost always with more than 12 segments; body stout (Fig. 16.4d) **Psocoptera** (Chap. 32)

36(32). Abdomen strongly constricted at base (Fig. 16.5a); antennae frequently elbowed (Fig. 16.5a) **Hymenoptera** (Chap. 46)
Abdomen not constricted at base; antennae not elbowed **37**

37(36). Body densely covered with scales or long hairs; mouthparts usually a coiled proboscis (Fig. 16.5e) (sometimes vestigial)
Lepidoptera (Chap. 44)
Body bare or sparsely covered with hairs, rarely scaled; mouthparts not a coiled proboscis **38**

38(37). Mouthparts a slender tube-shaped rostrum (Figs. 43.1b, 42.3b) or a haustellum (Fig. 16.3c); antennae usually with 3 or fewer segments **39**
Mouthparts mandibulate (Fig. 16.3d), never forming a rostrum or haustellum; antennae almost always with 9 to 11 segments
Coleoptera (Chap. 39)

39(38). Body strongly flattened laterally; thorax and head usually bearing large flattened, backwardly directed spines
Siphonaptera (Chap. 43)
Body cylindrical or flattened dorsoventrally; head and thorax not bearing special spines **Diptera** (Chap. 42)

40(34). Antennae usually concealed in grooves; ecto-parasites on birds and mammals; with flat-tened bodies, reduced eyes and pigmentation (Fig. 16.4b)   **Phthiraptera** (Chap. 33)

Antennae free; free-living aquatic naiads with long legs, large compound eyes and darkly pigmented bodies (Fig. 21.3)

Odonata (Chap. 21)

41(31). Cerci with numerous segments; aquatic naiads with gills usually present on legs, thorax, or abdomen   **Plecoptera** (Chap. 27)

Cerci with 1 segment; terrestrial, without gills   **Phasmatodea** (Chap. 30)

KEY B

Wings Present, Functional

1. Mesothoracic wings thick, strongly sclerotized or parchmentlike at least at base, or vesti-gial   **2**

Mesothoracic wings membranous, sometimes covered with scales; never vestigial   **10**

2(1). Mesothoracic wings vestigial, scalelike or club-shaped; hind wings large, fan-shaped   **9**

Mesothoracic wings covering about half or more of abdomen; never club-shaped or scalelike   **3**

3(2). Abdomen with large, strongly sclerotized for-ceps-shaped cerci (Fig. 16.4a); forewings short, leaving at least 3 abdominal segments exposed   **Dermaptera** (Chap. 26)

Abdomen with cerci absent or not forceps-shaped; forewings usually covering entire ab-domen   **4**

4(3). Mouthparts a slender, elongate rostrum, hinged at the base and projecting beneath head and backward (Figs. 16.1a, 34.1b)

Hemiptera (Chap. 34)

Mouthparts short, mandibulate (Fig. 16.3d); head sometimes prolonged as a beak (Figs. 39.19b, 41.1b) but never with a hinged base   **5**

5(4). Forewings without venation, usually strongly sclerotized and meeting at midline at rest (Fig. l6.5b); antennae rarely with more than 11 segments, frequently clubbed

Coleoptera (Chap. 39)

Forewings with extensive, reticulate venation; antennae usually with more than 12 seg-ments, filiform, never clubbed   **6**

6(5). Hind legs with femora greatly enlarged for jumping (Fig. 16.2d)

Orthoptera (Chap. 29)

Hind legs not modified, similar to middle legs   **7**

7(6). Anterior legs raptorial (Fig. 16.3e)

Mantodea (Chap. 23)

Anterior legs not raptorial, fitted for walking   **8**

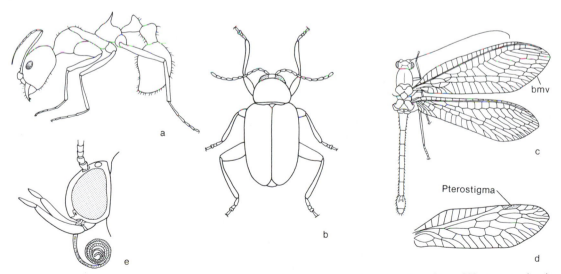

Figure 16.5 Key characters: **a**, an ant (Hymenoptera, Formicidae); **b**, a beetle (Coleoptera); **c**, wings of Neuroptera, showing bifurcating marginal veins (bmv); **d**, wing of a snakefly (Raphidioptera), showing pterostigma; **e**, head of a moth, showing coiled proboscis. (a,b, modified from University of California Extension; e, modified from Snodgrass, 1935.)

8(7). Head prognathous (mouthparts directed anteriorly); body sticklike or leaflike
Phasmatodea (Chap. 30)
Head hypognathous (mouthparts directed ventrally); body not sticklike or leaflike
Blattodea (Chap. 22)

9(2). Hind legs with femora enlarged for jumping (Fig. 16.2d); prothorax projecting posteriorly over wings and abdomen
Orthoptera (Chap. 29)
Hind legs not modified for jumping, similar to middle legs; prothorax small, not projecting over hindbody   **Strepsiptera** (Chap. 40)

10(1). One pair of wings present   **11**
Two pairs of wings   **13**

11(10). Abdomen bearing 1 to 3 long filaments on terminal segment; mouthparts vestigial   **12**
Abdomen not bearing long filaments; mouthparts rarely vestigial   **Diptera** (Chap. 42)

12(11). Antennae long, filiform; wing with a single vein, without cells
Hemiptera (Chap. 34)
Antennae short, bristle-shaped; wing with closed cells, usually with veins very numerous   **Ephemeroptera** (Chap. 20)

13(10). Abdomen bearing 2 to 3 long, terminal filaments   **21**
Abdomen without terminal filaments or with very short processes   **14**

14(13). Tarsi with 1 to 4 segments   **15**
Tarsi with 5 segments   **22**

15(14). Mouthparts a slender, tube-shaped rostrum projecting below the head (Figs. 16.1a, 34.1b)   **Hemiptera** (Chap. 34)
Mouthparts short, mandibulate, never an elongate projecting rostrum (Fig. 16.3d)   **16**

16(15). Forelegs with basal segment of tarsi globular, at least twice as thick as second segment
Embiidina (Chap. 28)
Forelegs with tarsal segments about equal in size   **17**

17(16). Antennae shorter than head, bristle-shaped; insects at least 1.5 cm long; abdomen long, slender; wings with extensive network of veins   **Odonata** (Chap. 21)
Antennae longer than head, filiform   **18**

18(l7). Tarsi with 4 segments   **Isoptera** (Chap. 24)
Tarsi with 2 to 3 segments   **19**

19(18). Last segment of abdomen bearing cerci; minute insects occurring in decaying wood

Zoraptera (Chap. 31)
Cerci absent   **20**

20(19). Wings linear, narrow, with no more than 2 veins; head cone-shaped, directed ventrally or posteriorly (Fig. 35.1a)
Thysanoptera (Chap. 35)
Wings oval with at least 4 longitudinal veins; head not cone-shaped
Psocoptera (Chap. 32)

21(13). Hind wings larger than forewings; mouthparts mandibulate, almost always well developed   **Plecoptera** (Chap. 27)
Forewings larger than hind wings, broadly triangular; mouthparts vestigial
Ephemeroptera (Chap. 20)

22(14). Forewings densely covered with hairs or scales   **23**
Wings bare, or with fringe of marginal hairs   **24**

23(22). Wings covered with scales; mouthparts usually a coiled proboscis (Fig. 16.5e)
Lepidoptera (Chap. 44)
Forewings covered with hairs; mouthparts mandibulate   **Trichoptera** (Chap. 45)

24(22). Forewings about 1.5 times longer than hind wings; forewings and hind wings usually markedly different in shape and venation; abdomen usually strongly constricted at base   **Hymenoptera** (Chap. 46)
Forewings and hind wings similar in size, shape, and venation; abdomen not constricted at base   **25**

25(24). Head not prolonged ventrally; wings usually with numerous crossveins in costal margin (exception, Coniopterygidae)   **26**
Head prolonged ventrally, beaklike (Fig. 41.1b); wings with 1 to 3 crossveins in costal margin   **Mecoptera** (Chap. 41)

26(25). Hind wings broader at base than forewings; veins not bifurcating near wing margin (Fig. 36.1c)   **27**
Hind-wing breadth at base less than or equal to forewing breadth; at least radial veins bifurcating just before wing margin (Fig. 16.5c) (exception: Coniopterygidae)
Neuroptera (Chap. 38)

27(26). Pronotum quadrate or nearly so; wings without pterostigma   **Megaloptera** (Chap. 36)
Pronotum at least 3 times longer than wide; wings with pterostigma (Fig. 16.5d)
Raphidioptera (Chap. 37)

17
THE PARAINSECTA: PROTURA AND COLLEMBOLA

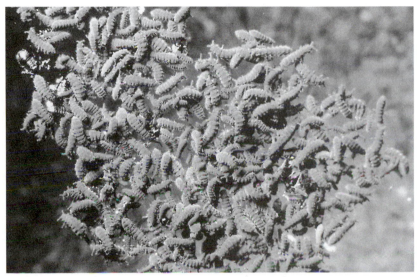

A mass of the aquatic collembolan, *Anurida maritima* (Hypogastruridae), supported by surface tension of a saline tide pool in Agadir, Morroco. This collembolan has a cosmopolitan distribution.

ORDER PROTURA

Very small to minute entognathous hexapods. Eyes, antennae absent; mandibles, maxillae styliform, retracted into invaginations in cranium; 1- or 2-segmented, non-ambulatory appendages present on abdominal segments 1 to 3; cerci absent; postembryonic development anamorphic, with 9 abdominal segments in first instar, 12 segments in adult.

Protura are weakly sclerotized, elongate, fusiform arthropods that occur exclusively in moist, decaying leaf litter, moss, soil, and similar situations.

About 400 species are known; all are easily recognized by the absence of antennae and the characteristic body shape with conical head. Because of their small size and cryptic habits, Protura are seldom encountered, but soil or litter extracts frequently contain large numbers of individuals. Eggs of *Eosentomon*, the only genus observed, are spherical and covered with either large conical tubercles or irregular polygonal plates. First-instar larvae, called prelarvae, eclose with 9 abdominal segments. Uneclosed prelarvae lie in a spiral position inside the eggs, with the head and posterior abdomen adjacent. No other details of embryology are known. The prelarva molts to a larva with 9 abdominal segments, the apical one different from that of the prelarva. The first-instar larva molts to a second-instar larva with 10 abdominal segments, followed by another molt to produce a subadult with 12 abdominal segments. The tenth segment is added between the eighth and ninth. The eleventh and twelfth segments are added on either side of the newly acquired tenth. There is no instar with 11 abdominal segments. A final molt produces the sexually mature adult, also with 12 abdominal segments. Occasional molts then occur throughout life.

Among hexapodous arthropods, Protura are unique in lacking antennae in all instars, but pseudoculi (Fig. 17.1b), present on the anterior cranium, may represent rudimentary antennae. The forelegs are usually held in front of the body, apparently functioning as sensory organs. Mouthparts of Protura are quite unlike those of insects but resemble those of Collembola. Mandibles and maxillae are in the form of elongate, simple blades, apparently used to triturate food particles. At rest the mouthparts are deeply retracted into invaginations in the head (Fig. 17.1c,d), with only the labrum visible externally. The only observation of

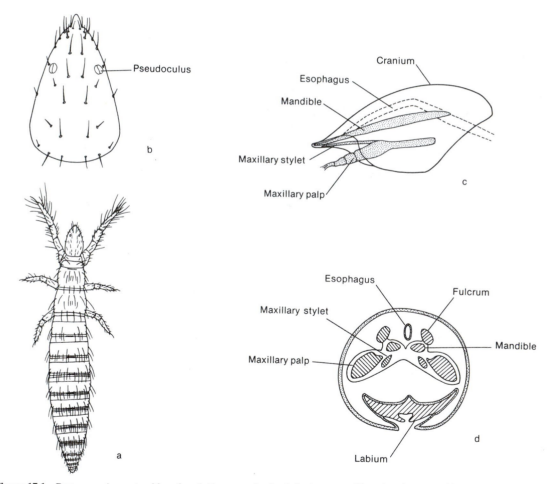

Figure 17.1   Protura: **a**, Acerentomidae, dorsal *(Acerentus barberi);* **b**, Acerentomidae, dorsal aspect of head *(Acerentomon);* **c**, longitudinal section of head, diagrammatic; **d**, transverse section of head, diagrammatic. (a, modified from Ewing, 1940; b, modified from Price, 1959.)

feeding has been on fungal mycorrhizae associated with tree roots. The Protura are the subject of a monograph by Tuxen (1964), whose classification is followed here. Their biology and ecology are summarized by Bernard and Tuxen (1987).

KEY TO THE FAMILIES OF

Protura

1. Spiracles present on mesothorax and metathorax; tracheae present   **Eosentomidae**
   Tracheal system and spiracles absent   **2**
2. Abdominal terga with transverse sutures and latero-tergites   **Acerentomidae**
   Abdominal terga without transverse sutures or latero-tergites   **Protentomidae**

ORDER COLLEMBOLA (SPRINGTAILS)

Entognathous hexapods characterized by 6-segmented abdomen with specialized, nonambulatory appendages on segments 1, 3, 4; antennae with 4 (rarely 6) musculate segments; cerci absent.

The name "springtail" refers to the ability of many species to jump using the abdominal appendages. Springtails range in length from less than 1 mm to more than 10 mm, averaging 2 to 3 mm. About 6000 species occur worldwide.

Several different body plans are characteristic of different families (Fig. 17.3), the most distinctive being the compact, globular form in Sminthuridae. Collembola occur in a wide variety of moist situations, especially soil and litter, where they are

Figure 17.3 Collembola: **a**, Sminthuridae *(Neosminthurus curvisetis);* **b**, Onychiuridae *(Onychiurus subtenuis).* (a, modified from Mills, 1934; b, modified from Folsom, 1917.)

frequently exceedingly abundant and are among the most important consumers in many soil ecosystems. In Arctic regions Collembola are particularly dominant in soil habitats. Densities of two to five individuals per cubic centimeter have been recorded, and under these conditions the soil may consist largely of collembolan fecal pellets. Collembola also inhabit fungi, lower levels of vegetation, the surface of slowly moving or stagnant water, snow or ice fields and, occasionally, buildings. Several species are abundant on seashores, ranging into the upper intertidal zone; a few forms occur in ant or termite nests. Collembola are readily wind-borne, and many species are extremely widespread, appearing even on coral atolls and oceanic islands. They have also been widely spread by commerce.

Mouthparts of Collembola are always entognathous, but the mandibles vary depending on the type of food consumed, rather than being uniformly styliform, as in Protura. Species feeding on plant material have a defined molar surface for grinding. Simple, piercing mandibles are found in species that feed on liquids such as juices of fungi.

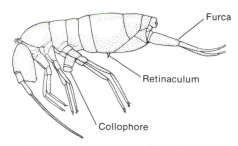

Figure 17.2 Collembola: Entomobryidae, *Tomocerus flavescens.* Entomobryidae usually have the fourth abdominal segment largest; *Tomocerus* is exceptional with the third segment largest. (Modified from Folsom, 1913.)

Collembola share a six-segmented abdomen with early instars of myriapods, which add segments after each molt until the adult segmentation is attained. Their reduced segmentation has been interpreted as evidence that Collembola represent a neotenic myriapod lineage. The arrangement and structure of abdominal appendages (Fig. 17.2) are unique among arthropods, but the structures are believed to be homologous to the abdominal legs of myriapods. The **collophore,** used in imbibing free water, is almost always present on the first abdominal segment. The **furca,** usually three-segmented, is held under tension beneath the abdomen by the **retinaculum** when at rest. When released it propels the animal up to 0.2 m through the air.

Most Collembola are bisexual, but parthenogenetic forms are known. Insemination is indirect, with males depositing stalked spermatophores on the substrate. The spermatophores are later picked up by the females, usually without direct participation by males. Old spermatophores may be consumed as fresh ones are deposited. In Sminthuridae, some species have developed elaborate courtship interactions involving contact between special sense organs on the frons. In *Sminthurides,* which live on the surface of water, the prehensile antennae of males are used to grasp the antennae of females. The much larger females carry the males about, periodically lowering them to the water surface where they deposit a spermatophore, which is then taken up by the female. Eggs, deposited singly or in clutches, are dropped onto the substrate. Those oviposited during unfavorable seasons may diapause for many months before hatching. Eggs of Collembola are unique among those of hexapodous arthropods, being **microlecithal** (lacking large yolk reserves) and developing by **holoblastic** or complete cleavage. The first four or five molts are accompanied by minor changes in body proportions, chaetotaxy, and color pattern. Sexual maturity is usually attained in the fifth or sixth instar, but occasional molts occur throughout life, with a total of 50 recorded for some species. Most springtails probably feed on decaying vegetation, but spores, pollen, fungal mycelia, or animal material are included in the diet of various species. Some species attack living plant material, occasionally becoming economically important, and a few are predators on microorganisms such as rotifers.

KEY TO THE FAMILIES OF

Collembola

1. Body elongate, cylindrical, or subcylindrical; first 6 postcephalic segments distinct (Figs. 17.2, 17.3b)   **2**
    Body globular; first 6 postcephalic segments fused (Fig. 17.3a)   **6**
2. Prothorax developed as distinct segment, bearing short setae (Fig. 17.3b)   **3**
    Prothorax reduced, indistinct, without setae (Fig. 17.2)   **5**
3. Head prognathous; eyes present or absent (as in Fig. 17.3b); third antennal segment often with large, complex sensilla   **4**
    Head hypognathous; eyes present; third antennal segment without unusually large sensilla
                                                           Poduridae
4. Eyes absent; third antennal segment with large, complex sensilla; body usually white
                                                           Onychiuridae
    Eyes present or absent; third antennal segment with simple sensilla   **Hypogastruridae**
5. Body with scales or clavate setae; fourth abdominal segment usually much longer than third
                                                           Entomobryidae
    Body with simple setae only; third and fourth abdominal segments subequal   **Isotomidae**
6. Eyes absent; antennae shorter than head
                                                           Neelidae
    Eyes present; antennae longer than head
                                                           Sminthuridae

Sminthuridae (Fig. 17.3a). Easily distinguished from other Collembola by the globose body, frequently covered with brightly colored scales. The collophore, retinaculum, and furca are usually present. The few agriculturally important Collembola are sminthurids.

Neelidae. Three genera of very small (< 1 mm) collembolans usually found in highly organic situations such as humus or rotten logs. Many of the species are unpigmented.

Onychiuridae (Fig. 17.3b). Usually unpigmented, lacking furca and retinaculum. Mostly inhabitants of soil interstices.

Poduridae. Includes only *Podura aquatica,* which is cosmopolitan on the surface of still, fresh water. The body is bluish to reddish and about 1.25 mm long.

Hypogastruridae. Body form variable; small, active species with furca and retinaculum present are similar to Onychiuridae; large (to 1 cm long), sluggish species superficially resemble immature Coccinellidae (Coleoptera). The color of hypogastrurids varies from creamy to greenish, bluish, or black. *Anurida maritima* occurs in marine intertidal situations. *Hypogastrura nivicola,* which occurs on snow and ice, is known as the snow flea.

Entomobryidae (Fig. 17.2). A large family, frequently encountered under bark, on the lower leaves of plants, or occasionally in buildings and other relatively dry situations, as well as in leaf litter and soil. Many species are closely covered by overlapping scales. Most entomobryids are easily separated from other families by the enlarged fourth abdominal segment.

Isotomidae. A large group of common species that occur in leaf litter, garden soil, or rotting vegetation. Body shape is similar to that of Entomobryidae, but simple setae are present rather than scales. Isotomids, like entomobryids, are often greenish, bluish, or purplish.

SELECTED REFERENCES

PROTURA

Biology and Systematics

Bernard and Tuxen, 1987 (biology, immatures); Ewing, 1940 (systematics); Kevan, 1962 (biology); Kuhnelt, 1961 (biology); Stehr, 1987 (immatures); Tuxen, 1964 (systematics), 1986 (New Zealand species); Wallwork, 1970 (biology).

COLLEMBOLA

Morphology

François et al., 1992 (cephalic anatomy).

Biology and Ecology

Butcher et al., 1971; Cassagnau, 1990 (biogeography and ecology); Christiansen, 1964, 1992; Christiansen and Bellinger, 1988 (maritime species), 1992 (Hawaiian species); Greenslade, 1991; Hale, 1965; Joosse, 1976 (excellent general discussions), 1983; Kevan, 1962; Kuhnelt, 1961; Merritt and Cummins, 1996 (aquatic species); Petersen, 1994 (ecology); Schaller, 1971 (reproductive behavior); Tichy, 1988 (walking); Waldorf, 1974 (sex pheromone); Wallwork, 1970; Witteveen et al., 1988 (life history of *Anurida maritima*).

Systematics

Maynard, 1951; Mills, 1934; W. R. Richards, 1968 (generic classification, world); Salmon, 1951, 1956; Scott, 1961 (pictorial key, nearctic genera); Stehr, 1987 (immatures).

18
THE ENTOGNATHA: DIPLURA

This dipluran, *Occasjapyx kofoidi* (Japygidae), was found in Potter Creek Cave in northern California. None of the cave japygids are especially adapted for cave life.

Entognathous hexapods with elongate body, short legs, slightly differentiated thoracic and abdominal segments. Eyes absent, antennae multiarticulate, flagellum musculate. Abdomen with 10 or 11 segments, appendages on segments 1 or 2 to 7; cerci developed as multiarticulate, moniliform appendages or as short, sclerotized forceps, bearing terminal gland openings.

Extant species of Diplura are functionally entognathous, with the mandibles and maxillae largely enclosed in an invagination on the ventral side of the head (Fig. 18.1b–d). They also differ from typical insects in lacking eyes. In the Pennsylvanian Period fossil *Testajapyx,* the mouthparts appear to be ectognathous, and well-developed compound eyes are present (Fig. 18.2; Kukalová-Peck, 1987). In these characters as well as in overall morphology Diplura are the most insectlike of the entognathous hexapods. Unlike insects, diplurans have antennae that are musculate rather than flagellate.

Like the other entognathous orders, Diplura inhabit leaf litter, soil interstices, moss, and other moist, cool situations. Typically, they are unpigmented and weakly sclerotized organisms, 5 to 15

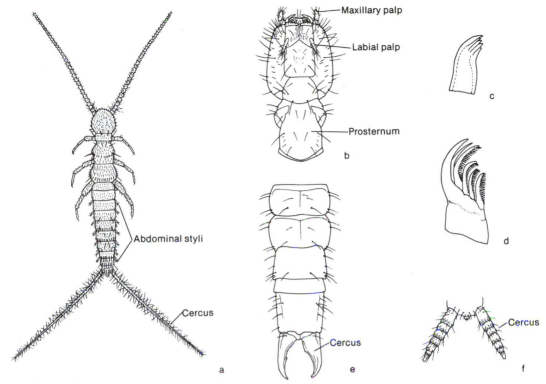

Figure 18.1 Diplura: **a**, Campodeidae *(Campodea montis);* **b**, Japygidae *(Japyx kofoidi),* ventral aspect of head; **c**, same apical portion of mandible; **d**, same, apical portion of lacinia; **e**, Japygidae *(Metajapyx steevsi),* abdomen; **f**, Projapygidae *(Anajapyx menkei),* cerci. (a, modified from Gardner, 1914; b, c, d, modified from Silvestri, 1928; e, modified from Smith and Bolton, 1964; f, modified from Smith, 1960.)

mm long, but *Anajapyx* (Japygidae) reaches 50 mm in length. Diplura are cosmopolitan, with the Campodeidae best represented in temperate regions, especially in the Northern Hemisphere. The Japygidae predominantly inhabit tropical, subtropical, and south temperate regions. About 800 species are known.

Known species are bisexual. Insemination is indirect, by means of a stalked spermatophore, which the males attach to the substrate. Early instars are similar to adults except for chaetotaxy, and metamorphosis is essentially absent. Molting continues throughout life, as in the Thysanura, but sexual maturity is attained when the setation is complete. Campodeidae apparently feed on decaying vegetation, while Japygidae are predatory, using the forceps to capture prey. Projapygidae appear to feed chiefly on mites.

Diplura were formerly placed in a single taxon with Thysanura, which are also primitively wingless with multiarticulate antennae and cerci. However, the antennae of Thysanura are flagellate (lacking intrinsic muscles in the flagellum). Imms (1936) showed that the Diplura are generally similar to the Symphyla in body organization. The Japygidae are unique among hexapodous arthropods in possessing four pairs of thoracic spiracles, with two pairs on each of the mesothorax and metathorax.

KEY TO THE FAMILIES OF

Diplura

1. Cerci multiarticulate, not sclerotized   **2**
    Cerci with 1 segment, sclerotized, forciculate (Fig. 18.1e)   **Japygidae**

Figure 18.2 Reconstruction of a primitive dipluran, *Testaja-pyx thomasi*, from the Pennsylvanian Period (= Upper Carbon-iferous Period; 323 to 290 million years ago) of Illinois. Length without antennae was 47.5 mm. Note the compound eyes (facets of ommatidia were rounded rather than hexagonal), segmented labial palp, comblike structure of the laciniae, exites on the legs and abdominal segments, and stout cerci. **a**, dorsal view of whole insect; **b**, ventral view of head. (Redrawn from Kukalová-Peck, 1987. Reproduced by permission of *Canadian Journal of Zoology*.)

2. Cerci less than half length of antennae (Fig. 18.1f), strongly moniliform, and bearing terminal gland openings   **Projapygidae**
   Cerci about as long as antennae, filiform or weakly moniliform, without terminal gland openings (Fig. 18.1a)   **Campodeidae**

Campodeidae (Fig. 18.1a). White delicate, active species mostly about 3 to 5 mm long. Abundant in leaf litter and soil throughout North America.

Japygidae. Body more compact than in Campode-idae, with terminal abdominal segment and forceps scle-rotized. Commonly ranging from 5 to 15 mm in length; in North America most abundant in Southwest and West.

Projapygidae. Body compact, as in Japygidae, but cerci multiarticulate. In North America occurring only in California.

SELECTED REFERENCES

Morphology

Barlet, 1989 (musculature and nervous system of tho-rax); Pages, 1989 (abdominal sclerites and append-ages); Rathman et al., 1988 (external morphology).

Biology

Kevan, 1962; Kuhnelt, 1961; Schaller, 1971 (reproduc-tive behavior); Wallwork, 1970.

Systematics and Evolution

Imms, 1936; Kukalová-Peck, 1987 (implications of fos-sils); Remington, 1956.

19
THE APTERYGOTA:
ARCHEOGNATHA AND THYSANURA

Bristletails or Machilidae have a characteristic humpbacked profile and stylets visible on the underside of the abdomen.

ORDER ARCHEOGNATHA (MACHILIDS, BRISTLETAILS, MICROCORYPHIANS)

Elongate, subcylindrical, with arched trunk and 3 multiarticulate terminal filaments; antennae multiarticulate; 3 ocelli and large, dorsally contiguous compound eyes (Fig. 19.1c); mouthparts partly retracted into head, mandibles (Fig. 19.1b) elongate, monocondylic; maxillary palp elongate, 7-segmented; 1-segmented appendages (styli and protrusible vesicles) present on abdominal segments 1 to 9; styli also present on meso- and metathoracic leg bases (Fig. 19.1a).

———

Superficially the Archeognatha are similar to the Thysanura, and the two are sometimes classified as a single order. Structurally the Archeognatha share certain features, especially mouthpart structures, with the entognathous hexapods, especially the Diplura, as well as showing similarities to the Thysanura and Pterygota (Chaps. 15, 18). The occurrence of archeognathan-like fossils as early as the Devonian Period (Labandeira et al., 1988) as well as abundant fossils of the related Monura (Chap. 15) seems to indicate that a much larger fauna of primitive, flightless insects existed during the Paleozoic Era (Sharov, 1966; Kukalová-Peck, 1987). The Archeognatha probably represent a relictual stock of such forms.

In contrast to the entognathous hexapods, the

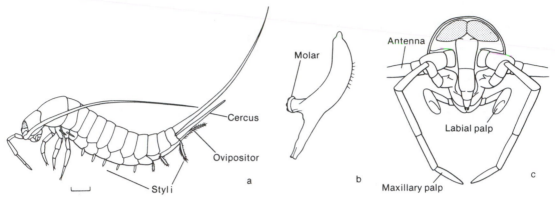

Figure 19.1 Apterygota, Archeognatha, Machilidae: **a**, *Machilis*, female (scale equals 1 mm); **b**, mandible of *Machilis;* **c**, head of *Machilis*. (b, modified from Snodgrass, 1950.)

Archeognatha have successfully adapted to a variety of habitats besides soil and litter. Machilids commonly occur in chaparral or dry coniferous woodlands, where they are dependent on localized sources of free water, which they imbibe through the eversible membranous vesicles adjacent to their abdominal appendages. Meinertellids inhabit extremely xeric habitats, including desert sand dunes, and require little if any free water.

Biology and life histories are known in detail for only a few species. Fertilization is indirect, as in the entognathous groups. During courtship the males of most genera spin a fine thread from the gonapophyses. One end of the thread is attached to the substrate, the other end held at the tip of the abdomen (Fig. 19.2). Several spermatophores are deposited on the thread as it is formed. These are later picked up by the female after a series of complicated courtship maneuvers. Considerable variation in details of courtship have been observed among various genera of Archeognatha (Sturm, 1992). In *Petrobius* the sperm droplet is placed directly on the ovipositor. Females mate during each instar, depositing clutches of up to 30 eggs in crevices.

Immatures resemble adults except for chaetotaxy and the lack of styli on the thorax. Scales appear on the body in the third instar. Food consists of lichens, algae, vegetable detritus, and probably dead arthropods. Sexual maturity is attained after about 8 to 10 molts over a period of up to 2 years. Sexually mature individuals continue to molt periodically throughout life. Life spans of 2 to 4 years have been recorded, with a total of as many as 66 molts. These are active, fast-running insects,

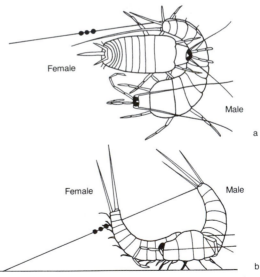

Figure 19.2 Mating positions of *Machilis:* **a**, viewed from above; **b**, viewed from the side. The male spins a thread fastened to the ground at one end and with the other end held by his genitalia above the female. Sperm drops (black dots) are suspended on the thread. (Redrawn from Schaller, 1971. Reproduced, with permission, from *Annual Review of Entomology*, Volume 16, 1971, by Annual Reviews Inc.)

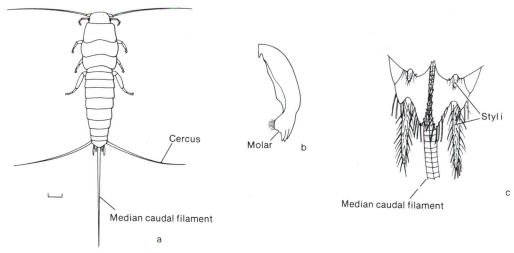

Figure 19.3   Apterygota, Thysanura, Lepismatidae: **a**, *Ctenolepisma lineata* (scale equals 1 mm); **b**, mandible of *Ctenolepisma;* **c**, abdominal apex of *Heterolepisma*. (b, modified from Snodgrass, 1950; c, modified from Folsom, 1924.)

capable of random escape jumps by suddenly arching the body and slamming down the abdomen. The abdominal appendages are used as skids to support the abdomen. About 350 species of Archeognatha have been described.

Meinertellidae. Abdominal sterna small; antennae without scales. Mostly small (<5 mm) diurnal species found in dry regions.

Machilidae. Abdominal sterna relatively large; antennae with scape and pedicel usually scaled. Larger (5 to 15 mm), mostly crepuscular species. Cosmopolitan. The commonly encountered North American archeognathans are all machilids.

ORDER THYSANURA (SILVERFISH)

Elongate to ovate, flattened insects with 3 multiarticulate terminal abdominal filaments; antennae multiarticulate; eyes of compound type but small, rudimentary, or absent; mouthparts mandibulate; mandibles short, semidicondylic (Fig. 19.3b), maxillary palps 5-segmented; 1-segmented appendages (Fig. 19.3c) absent or present on variable number of abdominal segments.

In the morphology of the mouthparts the Thysanura are comparable to primitive pterygote insects, such as mayflies, rather than the superficially similar Archeognatha. Structurally they are more variable than the Archeognatha, ranging from slender, elongate, free-living forms, to ovate, short-tailed types that frequent mammal burrows or ant nests. Thysanura are rapid runners but lack the jumping ability of archeognathans. Various species occur in many different habitats, including relatively hot, dry situations, including houses. *Ctenolepisma* is able to extract water through the rectum from atmospheres as low as 60 percent relative humidity.

Life histories of Thysanura are similar to those of Archeognatha, with complex courtship and indirect insemination, ametabolous metamorphosis, and surprisingly slow development and long life spans. *Tricholepidion gertschi* (Lepidotrichidae) (Fig. 19.4), the most primitive species, has a 2-year cycle. Survival of lepismatids over periods of 3 to 4 years is recorded, during which molting and reproduction continue. As in Archeognatha, mating occurs in each instar after sexual maturity. Most Thysanura are probably omnivorous. The household species occasionally damage books by consuming the starch in their bindings. The order includes about 370 species.

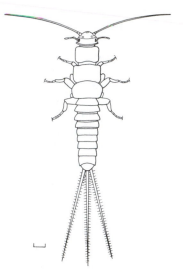

Figure 19.4   Apterygota, Thysanura, Lepidotrichidae, *Tricholepidion gertschi* (scale equals 1 mm).

KEY TO THE FAMILIES OF

Thysanura

1. Compound eyes present   **2**
   Compound eyes absent   **Nicoletiidae**
2. Body hairy; ocelli present; tarsi with 5 segments;
   exertile vesicles present on abdominal segments 2
   to 9   **Lepidotrichidae**
   Body scaly; ocelli absent; tarsi with 3 or 4 segments;
   exertile vesicles absent   **Lepismatidae**

Nicoletiidae. Mostly commensal forms, occurring in ant or termite nests or mammal burrows, or cavernicolous or subterranean. Eyes and ocelli are absent; body either elongate, slender, or broad, oval, with short terminal filaments. In the United States, restricted to the South and Southwest.

Lepidotrichidae. Body slender, elongate, moderate in size (12 to 15 mm); lateral and median ocelli present. *Tricholepidion* (Fig. 19.4), the only genus, has the body sparsely clothed with simple setae. The family was known only from Oligocene fossils until living insects were discovered in northwestern California in 1959.

Lepismatidae. Body slender, elongate, small to moderate in size; lateral ocelli absent, median ocellus

reduced as median frontal organ (Fig. 19.3). The body of lepismatids is always clothed in scales, differentiating them from Lepidotrichidae and most Nicoletiidae. Lepismatids occur in litter, rubbish, in dry areas beneath loose bark, in animal burrows and many other situations. All the frequently encountered silverfish belong to this family, including common household species such as *Lepismodes inquilinus* (the firebrat) and *Lepisma saccharina* (the silverfish).

SELECTED REFERENCES

ARCHEOGNATHA

Biology and Ecology

Benedetti, 1973; Joosse, 1976 (excellent general discussion); Schaller, 1971 (reproductive behavior); E. L. Smith, 1970; Sturm, 1992 (reproductive behavior).

Systematics and Evolution

Labandeira et al., 1988 (fossils); Remington, 1954, 1956; Sharov, 1966; Sturm and Bach de Roca, 1993 (overview of systematics).

THYSANURA

Morphology

Barnhart, 1961; Birket-Smith, 1974 (abdomen); Noble-Nesbitt, 1989 (spiracular mechanism); E. L. Smith, 1970; Wingstrand,1973 (sperm); Wygodzinsky, 1961.

Biology and Ecology

Delany, 1957, 1959; Edwards, 1992 (function of coxal vesicles); Joosse, 1976; Joosse and Verhoef, 1987; Schaller, 1971; E. L. Smith, 1970; Sturm, 1987 (reproductive behavior); Watson, 1965 (endocrine).

Systematics and Evolution

Remington, 1954 (generic classification); Slabaugh, 1940 (domestic species); Wygodzinsky, 1961 *(Tricholepidion),* 1972 (revision, North American species).

20
ORDER EPHEMEROPTERA (MAYFLIES)

Subimago of the mayfly, *Hexagenia limbata californica* (Ephemeridae, hanging from streamside vegetation). After a second ecdysis to the imaginal instar, the mayfly will have transparent wings.

Adult. Delicate paleopterous insects with elongate, subcylindrical body (Fig. 20.1a). Head with short, multiarticulate, filiform or setaceous antennae, 3 ocelli, and large, compound eyes; mouthparts mandibulate, vestigial. Mesothorax enlarged, supporting large, triangular forewings; hind wings small, rounded, or absent; venation extensive, with numerous crossveins (Fig. 20.3). Abdomen slender, elongate with 10 segments; tenth segment with long cerci and usually with median caudal filament.

Naiads. Extremely variable in body form (Fig. 20.4) but almost always with large compound eyes, short multiarticulate antennae; mouthparts mandibulate; tarsi 1-segmented, with single claw; 10-segmented abdomen with cerci and usually median caudal filament; gills present on at least some of anterior 7 abdominal segments.

Subimago. Similar to adult but with duller coloration and relatively shorter appendages.

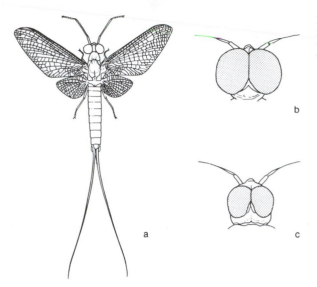

Figure 20.1   a, Adult of *Hexagenia bilineata* (Ephemeridae); **b**, *Homoneura dolani*, dorsal aspect of head of male; **c**, same of female. (a modified from Needham, 1917; b, c modified from Edmunds et al., 1958.)

Some Paleozoic fossils have the forewings and hind wings of similar size and venation, but all extant adult and subimaginal mayflies are easily recognized by the large, triangular forewings that are held vertically above the body at rest, and by the long abdominal filaments. Naiads are usually distinguished by the combination of three terminal filaments and abdominal gills. Forms with two terminal filaments (Fig. 20.4c) may superficially resemble nymphs of Plecoptera. Mayflies are exceedingly abundant in nearly all permanent freshwater habitats, where they constitute a basic food item for nearly all predators. Adults of many species, especially in temperate climates, tend to emerge synchronously, sometimes in extremely large numbers, and may be temporary nuisances near lakes or large streams.

The Ephemeroptera, together with the Odonata, are the only extant insects with a paleopterous flight mechanism. These two relatively small orders, sometimes classified as the infraclass Paleoptera, constitute the remnants of an extensive fauna which mostly became extinct at the end of the Mesozoic Era (see Chap. 15). In mayflies, although wing-flexing capability is lacking, the dorsal longitudinal thoracic muscles are relatively large, and flight is largely powered by these indi-

rect muscles. In Odonata, flight is powered entirely by direct musculature. The major differences in flight mechanism, as well as the grossly different body plans and biological adaptations, indicate the extreme antiquity of the paleopterous orders and their long period of independent evolution.

Mayflies are remarkable in possessing paired penes and oviducts opening through separate gonopores. Among other insects, the external genitalia are paired only in some male earwigs (Dermaptera). Mandibles of nymphs (Fig. 20.2b) are notable in their general similarity to those of Archeognatha (Fig. 19.1b), but are unique in having three points of articulation, one being a movable slider which affords unusual mobility (Kukalová-Peck, 1985). The sliding condyle can be moved along a troughlike socket, enabling the mandible to roll, much as in Archeognatha (Kukalová-Peck, 1985). Musculature and development of the nymphal gills show similarities to legs, suggesting possible homology with the abdominal appendages of the entognathous hexapods (see Chap. 15). In most other morphological features, mayflies are highly specialized, as described below.

Mayfly naiads molt to **subimagos** similar to the adults but sexually immature. Coloration of the subimago is usually subdued relative to the adult, the caudal filaments and tarsi are often shorter and the wing veins are microsetose (bare in imagos). Compared with adults, the subimagos are slow, clumsy fliers. All mayflies have subimaginal instars but in the family Palingeniidae the molt to the adult female instar is suppressed and the subimagos are reproductively functional. In Oligoneuridae the subimaginal cuticle is molted from the body but retained on the wings. Mayflies are the only extant insects in which a functionally winged form molts, but functionally winged immatures may have occurred in several orders that became extinct at the end of the Paleozoic (Kukalová-Peck, 1991).

Duration of the subimaginal instar is generally correlated with the length of adult life. In species that mate and reproduce the same night, the subimago usually lasts only a few minutes; in other species the adult may live several days, and the subimago usually persists for about 12 to 24 hours. In a few species the imaginal molt is partly or completely suppressed. Adults engage in mating flights, which may involve large, dense swarms, especially in nocturnally active, Temperate Zone species. The male swarms often occur in character-istic situations. Some species, for example, typically fly over emergent objects in streams, others in forest clearings near water. Time of day of swarming and, in temperate climates, season are also characteristic of species to a degree. In tropical areas seasonality is often absent. Nuptial flights or dances frequently entail characteristic movements, especially by males, which rapidly fly upward several meters, then slowly drift down. These patterns are assisted by the caudal filaments, which act as a counterbalance during the downward segment of the flight. In a few species solitary males patrol in search of females. Adults and subimagos have vestigial mouthparts and never feed. The midgut is enlarged as an air sac, increasing the efficiency of flight. The esophagus and hindgut are modified as valves, apparently for regulating and maintaining the amount of air in the midgut.

Before copulation, the male approaches the female from below, grasping her mesothorax with his elongate forelegs, which are provided with a reversible joint at the first tarsal segment. Also correlated with courtship is the enlargement of the eyes in males (Fig. 20.1b,c). In many genera the dorsal facets are larger than the ventral ones; in

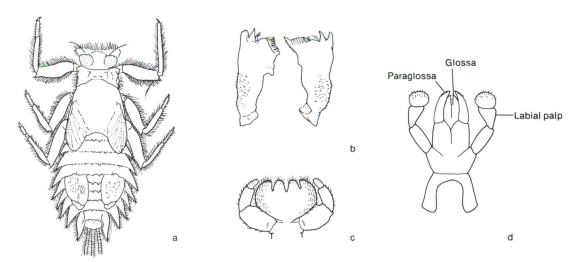

Figure 20.2   Structures of mayfly naiads: **a**, *Ephemerella hecuba,* dorsal view, caudal filaments removed (Ephemerellidae); **b**, mandibles, *E. hecuba;* **c**, labium, *E. hecuba;* **d**, labium, *Baetis* (Baetidae). (a, b, c modified from Allen and Edmunds, 1959; d modified from Murphy, 1922.)

some the eyes are divided into distinct dorsal and ventral lobes.

Several hundred to several thousand minute eggs are deposited, either in small batches or en masse, depending on the species. Females may oviposit while in flight or may descend below the water surface. Eggs are quite variable in shape and sculpturing among different genera and families, and are typically provided with coiled filaments that unroll when the eggs contact water, acting to attach the eggs to the substrate. Eggs have been characterized for many species and are of some taxonomic importance. Hatching occurs after one to several weeks or may follow diapause of several months. Development of immatures requires one to several years, with up to 27 instars. Most species are scavengers or herbivores; a few predaceous forms are known. Naiad morphology is generally correlated with habitat. Free-swimming, campode-iform or shrimplike nymphs (Figs. 9.6b, 20.4d) usually occupy small, standing bodies of water or sluggish streams. Flattened bodies with widely separated, laterally directed legs (Fig. 20.4a,c; Color photo 2f) are adaptations for clinging to stones or logs in fast-flowing streams. In bur-rowing forms, the legs or mandibles (Fig. 20.4b) may be specialized for digging, and the plumose gills for creating a current in the burrow, as in *Ephemera* and *Polymitarcys*. Species that live in water with high sediment content display a variety of structural modifications to protect the gills. In Tricorythidae, Neoephemeridae, and Caenidae, the gills on abdominal segment 2 are enlarged, form-ing a protective cover over the respiratory gills. In Baetiscidae and Prosopistomatidae the mesonotum is expanded as a carapace over the thorax and anterior abdominal segments and gills. Many Ephemerellidae and Heptageniidae are specialized for living in fast moving water. In these nymphs the body is broad and flat; the legs, which are robust and directed laterally, are adaptations that allow clinging to stones or submerged logs while presenting a low profile in the fast moving water. Some species occur in the spaces beneath stones, where the water movement is greatly reduced.

Classification of mayflies has been subject to numerous rearrangements, with great differ-ences in the number of families recognized. The classification adopted here follows Edmunds (1959) and Edmunds et al. (1976). Adult mayflies, probably because of their high degree of special-ization for a short adult life and aerial copulation, show only minor morphological differences among families. The key to adults is based largely on wing venation and must be used carefully. As reflected in their great morphological divergence, nymphs are highly specialized for living on or in different substrates, and usually can be easily keyed to family. The general geographic and habi-tat distribution of nymphs and the numbers of North American species in each family are listed in Table 20.1.

KEY TO THE NORTH AMERICAN FAMILIES OF

Ephemeroptera[1]

ADULTS

1. Forewings with veins MP_1 and CuA strongly divergent at base; vein MP_2 abruptly bent toward MP_1 at base (Fig. 20.3a,b)   **2**

   Forewings with veins MP_1 and CuA gradually divergent at base (Fig. 20.3d–f); vein MP_2 gradually bent toward MP_1 at base (Fig. 20.3d,e) or missing basally (Fig. 20.3f,h)   **6**

2(1). Basal portion of forewing with costal crossveins faint or incomplete; hind wing with veins MP_1 and MP_2 separating distally to middle of wing   **Neoephemeridae**

   Forewing with costal crossveins strong, never incomplete; hind wing with veins MP_1 and MP_2 separating proximally to middle of wing   **3**

3(2). Forewings with veins Sc and R_1 curving posteri-orly around apical margin of wing (Fig. 20.3a); middle and hind legs atrophied to trochanter   **Polymitarcyidae**

   Forewings with veins Sc and R_1 straight to apex, or slightly curved, ending anterior to apex (Fig. 20.3b); middle and hind legs with femur, tibia normal   **4**

[1] Names of the veins (Fig. 20.3b) are those of Edmunds et al. (1976): MA, medius anterior; MP, medius posterior; CuA, cubitus anterior (not cubitoanal); CuP, cubitus posterior.

TABLE 20.1
Characteristics of families of mayflies in North America.

Family	Approximate no. of species in North America	Distribution	Habitat	Substrate	Habits
Siphlonuridae	56	Widespread	Streams, rivers, lakes, ponds, swamps, temporary pools	Sandy or rocky or on vegetation or debris	Active swimmers; filter feeders
Oligoneuriidae	30	Southeast to southwest	Moving water, small to large streams, rapids	Shifting sand or under sticks, stones	Burrowing in sand or clinging to substrates; poor swimmers; filter feeders
Heptageniidae	132	Widespread	Mostly in moving water; also in ponds, rock pools, lake margins	Usually on stones; sometimes in vegetation or debris, or under stones, logs	Clinging to substrate with gill holdfast; poor swimmers
Ametropodidae	9	Northeast to southwest	Slow-moving streams	On vegetation, stones, or sand	Crawling and swimming
Baetidae	146	Widespread	Streams, rivers, ditches, lakes, ponds	Usually in vegetation; also in debris, under stones	Most actively swimming or climbing about on vegetation; negatively phototactic; herbivorous
Leptophlebiidae	70	Widespread	Mostly slow moving streams, backwashes, lake margins	Under stones, logs; in accumulations of leaves or other debris	Mostly crawling; negatively phototactic; herbivorous or omnivorous
Ephemerellidae	79	Widespread	Variable	On stones, logs, other objects	Climbing about substrate; poor swimmers
Tricorythidae	22	Widespread	Streams, rivers	On sticks, logs, branches, or on sand, gravel	Crawling about substrate; poor swimmers; herbivorous
Caenidae	18	Widespread	Slowly moving, quiet, or stagnant water	Sand or silt bottoms	Resting on bottom, partly covered by fine sand or silt; herbivorous or omnivorous
Neoephemeridae	4	Southeast to central	Slow- to fast-moving streams	On or in debris anchored in current; in vegetation	Crawling about substrate; poor swimmers

TABLE 20.1 (*Continued*)

Family	Approximate no. of species in North America	Distribution	Habitat	Substrate	Habits
Potamanthidae	8	East, midwest	Swift streams	On sandy or silty bottoms or under stones	Collectors, filterers
Behningiidae	1	Southeast	Large rivers	Loose sand	Burrowing in sand and swimming; predatory
Ephemeridae	13	Widespread	Streams, rivers, lakes	Sand, silt, or mud	Burrowing in substrate; probably ingest mud
Palingeniidae	2	Central and southeast	Large rivers	Clay banks	Burrowing
Polymitarcyidae	6	Widespread	Large rivers; swift streams	Sand, silt, or mud substrates	In burrows under partly embedded stones or in clay banks; filter feeders
Baetiscidae	12	Widespread	Small to moderate streams; lake margins	Sandy or gravelly substrates	Resting partially buried in sand or silt

4(3). Apical anterior portion of forewing with dense, interconnecting network of marginal crossveins (Fig. 20.3a); cubital intercalary veins straight, not connected with vein CuA (Fig. 20.3a) **Polymitarcyidae**

Radial portion of forewing with marginal crossveins sparsely interconnecting, not forming network; cubital intercalary veins sinuous, joining vein CuA (as in Fig. 20.3b) **5**

5(4). Vein A_1 forked **Potamanthidae**

Vein A_1 not forked (Fig. 20.3b) **15**

6(1). Lateral ocelli about one-tenth as large as compound eye **8**

Lateral ocelli about one-half as large as compound eye **7**

7(6). Forewing with numerous crossveins between veins R_1 and R_2 **Tricorythidae**

Forewing with 1 or very few crossveins between R_1 and R_2 (Fig. 20.3c) **Caenidae**

8(6). Forewing with venation reduced to 7 to 9 longitudinal veins; veins Sc and R_1 fused **Oligoneuriidae**

Forewing with at least 15 longitudinal veins (Fig. 20.3d–h); veins Sc and R_1 separate from base to apex **9**

9(8). Forewing with cubital intercalary veins present (Fig. 20.3d) **10**

Forewing with cubital intercalary veins absent **Baetiscidae**

10(9). Hind leg with tarsus 5-segmented **Heptageniidae**

Hind leg with tarsus 3- or 4-segmented; basal segment(s) fused with tibia **11**

11(10). Forewing with 1 or 2 long intercalary veins between veins MP_2 and CuA (Fig. 20.3e) **Ephemerellidae**

Forewing without long intercalaries between veins MP_2 and CuA (Fig. 20.3f,g,h) **12**

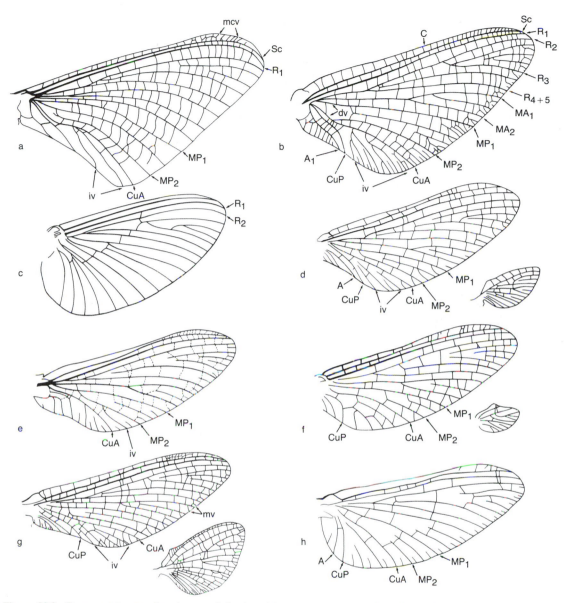

Figure 20.3 Representative mayfly wings: **a**, Polymitarcyidae *(Tortopus primus);* **b**, Ephemeridae *[Ephemera simulans]* **c**, Caenidae *(Brachycercus lacustris);* **d**, Heptageniidae *(Heptagenia maculipennis);* **e**, Ephemerellidae *(Ephemerella lutulenta);* **f**, Leptophlebiidae *(Thraulodes speciosus);* **g**, Ametropodidae *(Siphloplectron basale);* **h**, Baetidae *(Callibaetis fluctuans).* iv, intercalary veins; mv, marginal veinlets; mcv, marginal crossveins; dv, divergent veins; for names of veins, see note 1 on previous page (Modified from Burks, 1953.)

Figure 20.4 Representative mayfly naiads: **a**, Baetiscidae *(Baetisca columbiana),* caudal filaments removed; **b**, Ephemeridae *(Ephemera varia);* **c**, Heptageniidae *(Epeorus longimanus);* **d**, Baetidae *(Callibaetis coloradensis),* caudal filaments removed. (a, modified from Edmunds, 1960; b, modified from Needham, 1920; c, d, modified from Jensen, 1966.)

12(11). Forewing with vein CuP abruptly angled near midpoint (Fig. 20.3f)   **Leptophlebiidae**
Forewing with vein CuP straight or evenly curved   **13**

13(12). Forewing with 1 to 2 pairs of long, parallel cubital intercalary veins; marginal veinlets never free (Fig. 20.3g)   **Ametropodidae**
Forewing with several short cubital intercalary veins, or with marginal veinlets free, detached (Fig. 20.3h)   **14**

14(13). Hind wing very small or absent; forewing with vein MP_2 detached from MP_1 (Fig. 20.3h)   **Baetidae**
Hind wing large; forewing with vein MP_2 basally connected to MP_1 **Siphlonuridae**

15(5). Male with pronotum about three times as broad as long; female with caudal filaments shorter than body   **Palingeniidae**
Male with pronotum less than twice as broad as long; female with caudal filaments longer than body   **Ephemeridae**

NAIADS

1. Thorax with notum enlarged posteriorly, leaving only 4 or 5 abdominal tergites exposed, and concealing gills (Fig. 20.4a)   **Baetiscidae**

Thorax with notum smaller, leaving at least 7 abdominal tergites exposed; gills exposed (Fig. 20.4b–d)   **2**

2(1). Mandibles produced as long tusks projecting forward beyond the head (Fig. 20.4b)   **3**
Mandibles short, without tusks   **6**

3(2). Gills with fringed margins (Fig. 20.4b)   **4**
Gills bifurcate, with margins straight, unfringed (Fig. 20.5a)   **Leptophlebiidae**

4(3). Gills dorsal, held above abdomen; fore tibiae broad, flattened   **5**
Gills lateral, held at sides of abdomen; fore tibiae slender, cylindrical   **Potamanthidae**

5(4). Hind tibiae with apex produced into a sharp, angulate point, with a row of spines on the apical margin (Fig. 20.5b   **17**
Hind tibiae with apex rounded or truncate (Fig. 20.5c   **Polymitarcyidae**

6(2). Head with dense tufts of setae on anterior corners   **Behningiidae**[2]
Head without dense tufts of setae   **7**

7(6). Forelegs with dense row of long setae on inner margin; maxillae with tuft of gills at base   **8**

[2]Known only from naiads in North America.

Forelegs without setae arranged in dense row on inner margin; maxillae without gills **9**

8(7). Abdomen with dorsal gills on first segment; forelegs with gills on coxae

Siphlonuridae

Abdomen with ventral gills on first segment; forelegs without gills on coxae

Oligoneuriidae

9(7). Abdominal segment 2 with gills subquadrate, meeting at midline of abdomen (Fig. 20.5d **10**

Abdominal segment 2 with gills rounded or triangular, not meeting at midline of abdomen (Fig. 20.5e **11**

10(9). Mesonotum with rounded lobes projecting laterally from anterior corners; metathorax with distinct wing pads **Neoephemeridae**

Mesonotum without projecting lateral lobes; metathorax without wing pads

Caenidae

11(9). Gills absent from abdominal segment 1, or vestigial, threadlike; sometimes absent from segments 2 or 3 **12**

Gills present on at least abdominal segments 1 to 5 **13**

12(11). Gills present on abdominal segment 2, triangular or oval **Tricorythidae**

Gills absent on abdominal segment 2, sometimes absent from segment 3

Ephemerellidae

13(11). Body cylindrical; eyes or antennae located anteriorly or laterally (Fig. 20.4d) **15**

Body dorsoventrally flattened; both eyes and antennae located dorsally (Fig. 20.4c) **14**

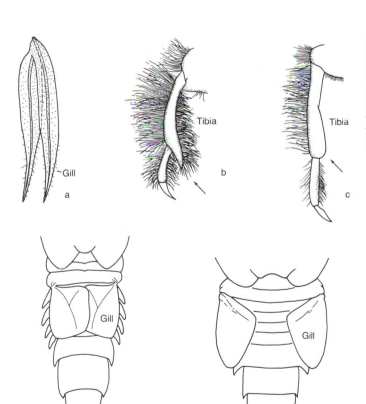

Figure 20.5 Characters of mayfly naiads. **a**, bifurcate gill of Leptophlebiidae; **b**, hind tibia of Ephemeridae; **c**, hind tibia of Polymitarcyidae; **d**, abdominal gills of Caenidae; **e**, abdominal gills of Tricorythidae. Arrows in b and c point to key characters of tibiae. (a–c, e, adapted from *An Introduction to the Aquatic Insects of North America,* 3rd ed., by R. W. Merritt and K. W. Cummins. Copyright © 1996 by Kendall/Hunt Publishing Company. Used with permission.)

14(13). Gills with 2 equal filaments; labial palps with 3 segments   **Leptophlebiidae**

Gills unbranched or with slender filament attached near base; labial palps with 2 segments   **Heptageniidae**

15(13). Claws on middle and hind legs long, slender, and without spines; claws on forelegs bifid or with several stout spines

                              Ametropodidae

Claws on all legs similar in structure   **16**

16(15). Labium with glossae and paraglossae long, slender (Fig. 20.2d); abdomen with posterolateral angles of segments 8 and 9 prolonged as flattened spines   **Siphlonuridae**

Labium with glossae and paraglossae short, broad (Fig.20.2c); abdomen seldom with segments 8 and 9 produced posteriorly (Fig. 20.4d)   **Baetidae**

17(5). Mandibular tusks round in cross section

                              Ephemeridae

Mandibular tusks with toothed, dorsal keel

                              Palingeniidae

SELECTED REFERENCES

General

Eatonia, 1954 to date (international journal); Needham et al., 1935; Peters and Peters, 1973.

Morphology

Edmunds and Traver, 1954.

Biology and Ecology

Alba-Tercedor and Sanchez-Ortega, 1991 (general); Berner, 1959 (nymphs); Berner and Pescador, 1988 (Florida species); Brinck 1957 (mating, reproduction); Britt, 1962 (detailed life histories of two species); Brittain, 1982; Edmunds, 1972; Edmunds and MaCafferty, 1988 (subimago); Elliott, 1968 (activity patterns of nymphs); Eriksen, 1966 (burrowing forms); Fremling, 1960 (biology of *Hexagenia*); Giberson and Mackay, 1991 (effect of acid streams); Heise and Flannagan, 1987 (life histories of *Hexagenia* and *Ephemera*); Hunt, 1950 (biology of *Hexagenia*); Ide, 1935 (effects of temperature); Jacobi and Benke, 1991 (snag-dwelling species); Lyman, 1955 (seasonal distribution); Sweeney et al., 1993 (use as bioassay organisms).

Systematics and Evolution

Berner 1950 (Florida species); Burks, 1953 (Illinois species); Day, 1956 (California species); Edmunds, 1959 (North America); Edmunds and Allen, 1987 (nymphs); Edmunds and Traver, 1954; Edmunds et al., 1963 (family classification of nymphs), 1976 (North and Central America); M.D. Hubbard, 1990; Koss, 1968 (eggs); Landa and Soldan, 1985; Leonard and Leonard, 1962 (Michigan species); Merritt and Cummins, 1996 (North America); Stehr, 1987 (immatures).

21
ORDER ODONATA
(DRAGONFLIES AND DAMSELFLIES)

This delicate damselfly, *Hetaerina americana* (Calopterygidae), is clinging to streamside vegetation. Unlike dragonflies, damselflies hold their wings upright, pressed together, when at rest.

Adult. Medium to large, paleopterous insects with large mobile head, enlarged thorax, two pairs of long, narrow, membranous wings, and slender, elongate abdomen (Fig. 21.1a,c). Head globular, hypognathous, constricted behind into a petiolate neck; compound eyes large, multifaceted; median, and two lateral ocelli present; antennae short, setaceous with 3 to 7 segments; mouthparts mandibulate. Prothorax reduced, mobile; posterior thoracic segments fused into rigid pterothorax with enlarged pleura, reduced sterna, terga; wings similar, with extensive network of veins, apical pterostigma near anterior margin. Abdomen of male with sterna 2 to 3 modified as complex copulatory organs; tergite 10 with unsegmented appendages.

Naiad. Body more robust than in adult, with less mobile head, smaller eyes, and longer antennae; mouthparts mandibulate, with labium modified as jointed, extensile grasping organ (Fig. 21.5); abdomen 10-segmented; respiration in Anisoptera by tracheate lamellae in rectum; Zygoptera with 3 external caudal gills or lamellae.

Their distinctive body configuration makes adult Odonata one of the most readily distinguished groups of insects. They differ from the superficially similar Myrmeleontidae (Neuroptera) in having setaceous antennae (clubbed in Myrmel-

355

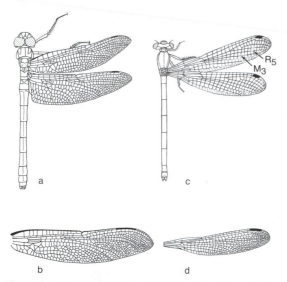

Figure 21.1   Adult dragonflies: **a**, Anisoptera, Macromiidae *(Macromia magnifica);* **b**, Petaluridae *(Tanypteryx hageni);* **c**, Zygoptera, Coenagrionidae *(Argia);* **d**, Lestidae *(Archilestes).* (Modified from Kennedy, 1915, 1917.)

eontidae) and in numerous other features. The unique prehensile labium, which is thrust rapidly forward to grab prey, differentiates the naiads from all other insects. All Odonata are predatory in all instars, mostly on organisms much smaller than themselves, and are probably important natural control agents of mosquitoes and other aquatic insects. Odonata are commonly associated with aquatic habitats, but as adults many dragonflies are highly vagile, straying many miles into deserts or other arid regions, where they may colonize animal watering troughs or other very limited sources of water. Two broad adaptive modes exist within the order. The more robust dragonflies (Anisoptera) are powerful, active, searching predators. The more frail damselflies (Zygoptera) spend more time perching and darting after suitable prey.

The flight mechanism is more primitive in Odonata than in any other insects. Not only is wing-flexing capability lacking, as in Ephemeroptera, but the flight muscles attach directly to the bases of the wings, with a consequent reduction in the size of the thoracic terga and sterna. The wing venation of extant Odonata is highly modified and very difficult to homologize with other insects. In some Paleozoic fossils, however, the venation is similar to that of contemporary mayflies (Riek and Kukalova-Peck, 1984). Despite their "primitive" thoracic structure, dragonflies are powerful, agile fliers. Aerial efficiency is increased by synchronization of the two pairs of wings (which move slightly out of phase). In the Zygoptera, wing flexing is simulated by the extreme backward tilt of the pterothorax.

Odonata are unique in the development of elaborate secondary male copulatory organs (Fig. 21.2b,c). Prior to courtship, the male transfers sperm from the primary genitalia on the ninth abdominal segment to the genital fossa on sternite 3 by curling the abdomen anteroventrally. The cerci are modified as strong clasping organs, used to hold the female by corresponding grooves on the prothorax (Zygoptera) or occipital region of the cranium (Anisoptera). While the courting partners are joined in tandem (Fig. 21.2a), the female bends her abdomen forward to the male genital fossa, where it is grasped by a second set of male claspers. Copulation and insemination then ensue in this "wheel position."

Other adaptive features facilitate efficient predation. In Anisoptera the nearly hemispherical eyes allow almost 360° vision. In Zygoptera the eyes are smaller but are placed at the lateral extremities of the transversely elongate head, maximizing the field of vision. The arrangement of the legs, with the posterior pairs successively longer, creates a basket for sieving prey from the air. The spinose femora and tibiae are useful in holding prey items, and the forward shift of leg attachments allows easy transfer of prey items to the mouth while in flight.

Naiads emerge from the water before the final molt, usually at dawn. Dispersal flights away from the place of emergence succeed metamorphosis in most species. These "maiden flights" may extend over distances of more than a kilometer in the stronger flying Anisoptera, or may cover only the few yards to riparian vegetation in the Zygoptera. The first few weeks of adult life, during which

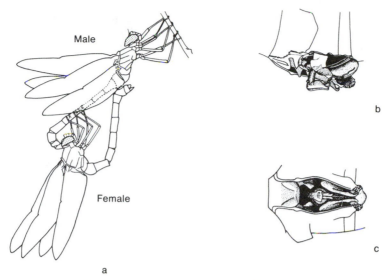

Male

Female

a

b

c

Figure 21.2   Copulation in Odonata: **a**, tandem pair of *Macromia magnifica* (Macromiidae) in "wheel position"; **b**, **c**, secondary male genitalia of *Macromia* (b, lateral aspect; c, ventral aspect). (Modified from Kennedy, 1915.)

sexual maturity and complete coloration develop, may be spent well away from aquatic habitats, but courtship and mating usually occur around water. Male Anisoptera frequently defend territories. Aeshnidae and Corduliidae patrol elongate areas over open water, sometimes flying almost incessantly. Libellulidae, Gomphidae, and Petaluridae usually protect a spot of open water by darting out from a central perch where they spend most of their time. Mid-air clashes are mostly between conspecific males, but other species may be investigated as well. Territoriality is less obvious but also important in Zygoptera (Paulson, 1974).

Both sexes of Odonata are polygamous, and males of some damselflies are able to remove sperm from previous matings before depositing their own sperm (Waage, 1979, 1984). To ensure sperm precedence, males of many Odonata maintain the tandem position for long periods after insemination, and in some species the male remains in tandem until the eggs have been deposited. In other species the male hovers near the female until after oviposition, driving away competing males. Zygoptera and some Anisoptera (Aeshnidae) oviposit in stems of vegetation; most

Anisoptera lack an ovipositor and deposit eggs freely or attach them to submerged vegetation.

Naiads (Fig. 21.3) of nearly all Odonata are aquatic, occupying a variety of situations from swift, cold streams to lakes, ponds and even estuaries. Respiration is by diffusion through the cuticle, enhanced in Zygoptera by the caudal lamellae, which function as gills. In Anisoptera water is

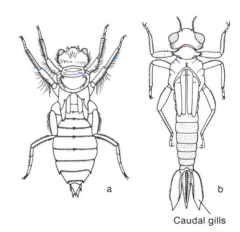

a

b

Caudal gills

Figure 21.3   Naiads of Odonata: **a**, Anisoptera, Libellulidae *(Libellula);* **b**, Zygoptera, Coenagrionidae *(Argia).*

rhythmically pumped in and out of the muscular rectum, where the highly convoluted surface increases gas exchange. Forceful contractions of the rectum are also used to propel the naiad forward in rapid spurts.

Naiads of a few species inhabit unusual situations such as tree holes, and some tropical Zygoptera live in bromeliads or hollow, water-filled, broken stems of plants. Naiads of most Petaluridae dig burrows in bogs. The bottom of the burrow, where daylight hours are spent, is below the water level, but the naiads may venture out of the water while hunting at night. A few other naiads, such as *Megalagrion* (Coenagrionidae) are essentially terrestrial, and many other genera are able to withstand prolonged periods out of water. This relative lack of dependence on water has been interpreted to indicate the Odonata were originally terrestrial as naiads, as in the palaeodictyopteroid orders (see Chap. 15).

Naiads are ambush predators, waiting concealed until prey is close enough to seize, or slowly creeping until within range of the prehensile labium. Development ranges from a few months in species that colonize temporary ponds to several years in larger species inhabiting permanent bodies of water. Naiad morphology is broadly correlated with microhabitat. Naiads of Zygoptera are typically slender and shrimplike (Fig. 21.3b). They usually frequent aquatic vegetation. Body form in the Anisoptera is variable. Relatively elongate, slender naiads (Aeshnidae, some Gomphidae, Corduliidae, Libellulidae; Fig. 21.3a) mostly frequent vegetation or surfaces of substrates. Burrowing forms tend to be shorter, broader, and often flattened (Petaluridae, Corduligastridae, most Gomphidae). Long-legged, flattened naiads (most Corduliidae, Libellulidae) usually sprawl on soft bottoms.

KEY TO THE SUBORDERS OF

Odonata

ADULTS

1. Hind wings with base broader than in forewings; venation dissimilar (Fig. 21.1a)   **Anisoptera**

Forewings and hind wings similar in shape and venation (Fig. 21.1c,d)   **2**
2. Eyes separated by more than their width in dorsal view (Fig. 21.1c)   **Zygoptera**

Eyes separated by less than their width in dorsal view (Oriental region)   **Anisozygoptera**

NYMPHS

1. Body slender with 3 caudal tracheal gills (Fig. 21.3b)   **Zygoptera**

Body stout, without external gills (Fig. 21.3a)   **2**
2. Antennae with 4 or 6 to 7 segments   **Anisoptera**

Antennae with 5 segments (Oriental region)   **Anisozygoptera**

KEY TO NORTH AMERICAN FAMILIES OF

Zygoptera

ADULTS

1. Wings with 10 or more antenodal crossveins (as in Fig. 21.4b)   **Calopterygidae**

Wings with 3 or fewer antenodal crossveins   **2**
2. Veins M_3 and R_5, arising closer to arculus than nodus (Fig. 21.1d)   **Lestidae**

M_3 and R_5 arising closer to nodus than to arculus (Fig. 21.1c)   **Coenagrionidae**[1]

NYMPHS

1. Median and lateral gills similar   **2**

Median gill bladelike, flattened; lateral gills triangular in cross section   **Calopterygidae**
2. Median lobe of labium grooved (Fig. 21.5c)   **Lestidae**

Median lobe of labium projecting, not grooved (Fig. 21.5d)   **Coenagrionidae**[2]

KEY TO NORTH AMERICAN FAMILIES OF

Anisoptera

ADULTS

1. Triangles similar in forewings and hind wings (Fig. 21.4a)   **2**

[1] Protoneuridae, with two uncommon species in southern Texas, will key here. Vein Cu_1 is absent or rudimentary and Cu_2 short in Protoneuridae. Both veins are well developed in Coenagrionidae.

[2] Protoneuridae key here. They differ from Coenagrionidae in having the proximal portion of the gills thickened and darkened.

Arculus

Nodus

acv

a

b

Figure 21.4 Venation at base of wings of dragonflies: **a**, Gomphidae *(Ophiogomphus);* **b**, Libellulidae *(Erythemis).* Arrow and stippled pattern indicate anal loops; arrow and vertical lines indicate triangles; acv, antenodal crossveins. (Modified from Needham and Westfall, 1955.)

Triangles dissimilar in forewings and hind wings (Fig. 21.4b)   **5**

2(1). Eyes broadly contiguous dorsally (Fig. 21.6a)
                                                              Aeshnidae

Eyes separated or barely touching dorsally (Fig. 21.6b)   **3**

3(2). Labium with median lobe deeply notched   **4**
Labium with median lobe not notched
                                                              Gomphidae

4(3). Pterostigma moderate, intersecting 4 to 5 crossveins   **Cordulegastridae**
Pterostigma elongate, intersecting 6 to 7 crossveins (Fig.21.1b)   **Petaluridae**

5(1). Eyes with posterior margin lobed (Fig. 21.6c)
                                                              6

Eyes with posterior margin straight (Fig. 21.6d)
                                                              Libellulidae

6(5). Hind wing with anal loop elongate, with two parallel rows of cells (as in Fig. 21.4b)
                                                              Corduliidae

Hind wing with anal loop short, rounded, with cells not arranged in 2 parallel rows (as in Fig. 21.4a)   **Macromiidae**

NAIADS

1. Labium strongly convex, projecting anteroventrally   **2**
Labium flat   **5**

2(1). Lateral lobes of labium with large, irregular, interlocking teeth (Fig. 21.5a)
                                                              Cordulegastridae

Lateral lobes of labium with regular dentition; teeth usually small, not interlocking (Fig. 21.5b)   **3**

3(2). Head with prominent, pyramidal horn or tubercle; squat naiads with very long legs
                                                              Macromiidae

Head bare or sometimes with small tubercle; body form variable   **4**

4(3). Cerci nearly as long as median anal spine (epiproct); lateral spines on abdominal segment 9 less than one-fourth length of segment
                                                              Corduliidae

Cerci usually about half to three-fourths length of epiproct; if longer, lateral spines of segment 9 more than one-fourth length of segment
                                                              Libellulidae

Movable hook
Labial palp

a

b

c

d

Figure 21.5 Ventral aspects of labia of naiads of Odonata: **a**, Cordulegastridae *(Cordulegaster);* **b**, Libellulidae *(Tarnetrum);* **c**, Lestidae *(Lestes);* **d**, Coenagrionidae *(Argia).* (a, b, modified from Musser, 1962; c, d, modified from Kennedy, 1915.)

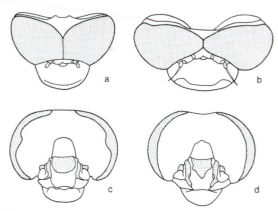

Figure 21.6 Anisoptera heads: **a**, Aeshnidae *(Anax junius);* **b**, Cordulegastridae *(Cordulegaster dorsalis);* **c**, Corduliidae *(Cordulia shurtleffi);* **d**, Libellulidae *(Libellula nodisticta);* **a**, **b**, dorsal views; **c**, **d**, posterior views.

5(1). Antennae with 4 segments; fore and middle tarsi with 2 segments **Gomphidae**

Antennae with 6 or 7 segments; tarsi with 3 segments **6**

6(5). Antennae with segments hairy, wider than long **Petaluridae**

Antennae with segments bare, longer than wide **Aeshnidae**

SUBORDER ZYGOPTERA

Coenagrionidae. The dominant family of Zygoptera, especially in temperate regions (Color photo 3e). Naiads inhabit a variety of situations, especially sluggish or static waters, where they clamber about on vegetation. **Protoneuridae** occur along streams.

Lestidae. Relatively large damselflies that frequent marshlands and swamps; bodies usually metallic blue, green, or bronze, wings transparent. Eastern North America to the Pacific Coast.

Calopterygidae. Primarily a tropical family, with only two genera in North America; bodies variably colored, wings usually pigmented, at least basally; naiads inhabit riffles in streams and creeks, eastern North America to Pacific Coast.

SUBORDER ANISOPTERA

Aeshnidae. Cosmopolitan; possibly the most familiar dragonflies to nonentomologists; adults strong-flying, sometimes migratory, and found far from water. Naiads occupy a variety of habitats, including brackish water.

Gomphidae. Cosmopolitan, medium to large in size, found most abundantly about moving water; the enlarged abdominal apex is a useful diagnostic character for adults; naiads usually burrow or hide in debris.

Petaluridae. Large, dull-colored, clear-winged species that mostly frequent bogs (where the naiads burrow) or, less frequently, streams; a relictual family with only about a dozen species, distributed mostly in temperate regions. Two species in North America.

Cordulegastridae. A small family of medium to large dragonflies, dark brown to black with yellow striping; adults frequent brooks and streams, where the naiads inhabit bottom debris. Holarctic and Indomalaysian, 25 species.

Corduliidae and **Macromiidae.** Widespread and abundant, especially in north temperate regions; adults mostly metallic blue or green with spots, bands; naiads usually occur in vegetation in a variety of aquatic habitats, including streams, ponds, and swamps.

Libellulidae. A dominant family, especially in the Tropics; in cross section the abdomen is triangular; cuticle usually nonmetallic, but often powdered with silvery pruinosity; wings frequently banded or clouded; nymphs stout, usually inhabiting vegetation or bottom debris in still water.

Suborder Anisozygoptera. Includes only two species from Japan and India, but Mesozoic fossils are numerous. Wing venation of adults is similar to Zygoptera, while the naiad resembles Anisoptera.

SELECTED REFERENCES

General

Odonatologica, 1972 to date (international journal).

Morphology

Ando, 1962 (embryology); Snodgrass, 1954 (naiad).

Biology and Ecology

Alcock, 1994 (postmating associations); Corbet, 1962, 1980 (general); Corbet et al., 1960, 1985 (general); Cordero and Miller, 1992 (mating behavior); Kormondy, 1961 (territoriality and dispersal); Mizuta, 1988a,b (adult ecology of *Ceriagrion*); Paulson, 1974 (reproductive isolation); Polcyn, 1994 (thermoregulation); Tol and Verdonk, 1988 (conservation); Waage, 1979, 1984 (reproduction); Wissinger, 1989a (population ecology of *Libellula*), 1989b (competition and predation among naiads).

Systematics and Evolution

Fraser, 1957 (higher classification); Garman, 1927 (Connecticut species); Gloyd and Wright, 1959 (North America); Merritt and Cummins, 1996; Needham and Westfall, 1955 (North America); Robert, 1962 (Quebec); Stehr, 1987 (immatures); E. M. Walker, 1953, 1958 (Canada and Alaska); E. M. Walker and Corbet, 1975 (Canada and Alaska); Wright and Petersen, 1944 (nymphs).

22
ORDER BLATTODEA (COCKROACHES)

The tropical roach *Phoraspis convexa* (Blaberidae) is active during the day on leaves in Amazonian forests where these insects feed on detritus, especially bird droppings. Their bright colors and full exposure to predators must indicate that they are inedible.

Small to large Exopterygota with broad, flattened body, strongly hypognathous head (Figs. 22.1). Mouthparts mandibulate, generalized; compound eyes large, ocelli developed as two "ocelliform spots"; antennae long, multiarticulate. Prothorax large, mobile, with shieldlike notum; mesothorax and metathorax similar, rectangular, slightly smaller than prothorax. Wings with longitudinal veins usually much branched, connected by numerous crossveins; forewing sclerotized as a protective **tegmen,** with anal lobe undifferentiated; hind wing membranous, with large, fan-shaped anal lobe, which is folded by longitudinal pleats at rest. Legs robust, long, often spiny; coxae very large, tarsi 5-segmented. Abdomen with 10 distinguishable segments; tergite 10 bearing cerci of 1 to many segments; tergites 5 to 6 and sometimes sternites 6 to 7 bearing openings of scent glands; ovipositor of 3 pairs of small internal valves; male genitalia complex, strongly asymmetrical (Fig. 22.3b).

Nymphs similar to adults except in development of wings, genitalia, and sometimes color and integumental texture.

The Blattodea constitutes a relatively small order at present, with about 4000 described species, but comprised the dominant group of Neoptera in the upper Carboniferous. About 30 species of cockroach are cosmopolitan inhabitants of man's dwellings, placing them among the most familiar insects to nonentomologists. Domiciliary species are commonly regarded with disgust because of their invasions of cupboards and pantries, which they spot with their feces and, depending on the species, may permeate with the musty-smelling secretions from their abdominal glands. Cockroaches have not been definitely implicated in transmission of human diseases, and they neither bite nor sting. Yet they must be considered of potential medical importance because of the large number of pathogenic microorganisms, including bacteria, viruses, protozoa, and parasitic worms, that have been isolated from their bodies or excrement. It may be pointed out that the same characteristics that adapt species such as *Periplaneta americana* (Blattidae) to life in dwellings also produce a hardy laboratory animal. Cockroaches have been important subjects for investigation of hormonal control of insect growth and development, and especially for the study of insecticidal mode of action and insecticide resistance.

Cockroaches are among the first neopterous insects to appear in the fossil record and are extremely generalized in most morphological features. The Blattodea are sometimes classified as a suborder of Orthoptera, with which they share many features, such as mandibulate mouthparts, a common pattern of wing venation, and the presence of cerci. However these groups differ in several profound characters, the most important probably being the mode of oviposition and the type of flight musculature.

All cockroaches produce characteristic oothecae (Fig. 22.1), consisting of a double-layered wrapper protecting two parallel rows of eggs. This trait extends to the Mantodea and the egg arrangement occurs in the primitive termite *Mastotermes* (Chap. 24), although the oothecal covering is absent. In contrast, Orthoptera lay eggs singly or in pods, never with an ootheca. In most neopterous insects, including Orthoptera, wing movements are powered almost entirely by indirect flight muscles. In Blattodea, Mantodea, and Isoptera, the indirect muscles are very small, the downstroke of the wing apparently being effected by basalar and subalar muscles (Snodgrass, 1958). These differences, together with the long period of separate evolution indicated by fossils, suggest that it is appropriate to recognize Blattodea as a separate order.

Most of the familiar household cockroaches are nocturnal, positively thigmotactic crevice dwellers. Tropical species, comprising the bulk of the world fauna, include diurnal species, often brightly colored, and sometimes arboreal. North American species of *Arenivaga* (Polyphagidae) are burrowers in sand or litter, and *Attaphila fungicola* (Polyphagidae), less than 3 mm long when mature, is a symbiont of the leaf-cutting ant *Atta fervens* in Texas. Some tropical species are known to be amphibious, and some in Australia are cavernicolous. Most species are solitary or simply gregarious, but *Cryptocercus punctulatus* (Cryptocerci-

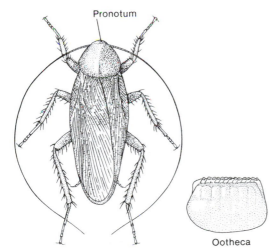

Pronotum

Ootheca

Figure 22.1 Blattidae, *Periplaneta americana*: adult female and ootheca. (Modified from Patton, 1931.)

dae), the wood roach, occurs subsocially in galleries in punky, rotting wood. In this species, as in termites, wood is digested with the aid of symbiotic flagellate protozoa, which are lost after each molt. The defaunated nymphs become reinfested by consuming fecal pellets of adults or older nymphs.

Sex pheromones have been isolated from several species and are probably present in all. Courtship by males, involving stridulation or posturing, normally precedes copulation. Males also produce a secretion from the abdominal tergites, which induces the female to mount the male just before copulation. After genitalic contact, the pair turn tail to tail. Copulation, during which a spermatophore is formed and transmitted by the male, is usually of considerable duration. Most species are bisexual, but some, including *Periplaneta,* may reproduce parthenogenetically.

Oothecae are quite variable in structure, enabling recognition of those of household species. Species also differ in the length of time the ootheca is retained by the mother, in its placement upon deposition, in the number produced by a single female, and in the number of eggs contained. For example, the American roach deposits an average of 50 oothecae per female, each containing about 12 to 14 eggs, while the German roach, *Blattella germanica* (Blattellidae), averages only five oothecae, each containing about 40 eggs. The American roach cements the oothecae to objects in the environment, while the German roach retains each ootheca until shortly before the eggs hatch. Cockroaches that deposit oothecae long before eclosion usually attempt some form of concealment. Some species tuck the ootheca into cracks or crevices, while in others dirt or debris may adhere to its outer surface.

Members of the family Blaberidae and a few Blattellidae are ovoviviparous or viviparous. In most ovoviviparous species an ootheca is formed, then retracted into a brood sac where the eggs develop. The oothecal wall is usually membranous rather than sclerotized. In a few genera no ootheca is formed, the eggs moving directly to the brood sac. In these ovoviviparous species the eggs are produced with sufficient yolk to complete development, needing only water from the mother. The blaberid genus *Diploptera* is strictly viviparous, producing small eggs with little yolk that depend on the mother for nutrition.

Embryonic and nymphal development is relatively slow, ranging from a few months in smaller species, to over a year in larger ones. Food of domestic species is extremely generalized, but many wild species are difficult to rear, especially in early instars. Adult roaches are often long-lived, some species surviving 4 years or more under laboratory conditions.

The North American fauna, numbering about 50 species, includes about a dozen important domiciliary species, including *Blattella germanica* (German roach*)* and *Supella longipalpa* (brown-banded roach), both species in Blattellidae, and *Blatta orientalis* (Oriental roach) (Fig. 22.2) and *Periplaneta americana* (Fig. 22.1) (American roach), both species in Blattidae. Only *Parcoblatta* (Blattellidae) among native roaches is of any economic significance. *Periplaneta americana,* the so-called American roach, apparently originated in Africa (Cornwell, 1968).

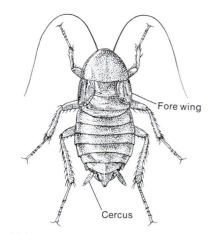

Figure 22.2   Blattidae, *Blatta orientalis*. (Modified from Patton, 1931.)

KEY TO COMMON NORTH AMERICAN
FAMILIES OF

Blattodea

1. Body length less than 3 mm; eyes very small;
   symbiotic in nests of *Atta* (Formicidae)
   Polyphagidae *(Attaphila)*
   Body length greater than 3 mm; eyes almost always large; not in ant nests   **2**
2(1). Middle and hind femora with numerous stout
   spines on posterior margin   **3**
   Middle and hind femora with one or two apical
   spines or lacking spines (hairs may be present)   **5**
3(2). Fore femora with posterior row of similar
   spines   **4**
   Fore femora with spines long and stout proximally,
   shorter and slender distally (Fig. 22.3a)
   Blattellidae
4(3). Female with last visible sternal plate divided by
   a longitudinal fissure (Fig. 22.3e); male styli
   slender, long, straight (Fig. 22.3d)
   Blattidae
   Female with last visible sternite not divided (Fig.
   22.3c); male with last sternite and styli usually
   asymmetrical or unequal in size (Fig.
   22.3b)   **Blattellidae**
5(2). Abdomen with seventh tergite and sixth sternite
   expanded as plates that conceal abdominal apex,
   including cerci; body elongate, parallel-
   sided   **Cryptocercidae**
   Abdomen with eighth to tenth segments and cerci
   exposed; body oval or broadly oval   **6**
6(5). Fore femora with 1 to 3 spines on posterior margin   **Blaberidae**
   Fore femora without spines on posterior margin   **7**
7(6). Frons flat; wings with anal region folded in fanlike
   pleats at rest; at least 16 mm long; pale
   green   **Blaberidae**
   Frons bulging, swollen; wings with anal region
   folded flat against preanal portion at rest (females frequently wingless); less than 16 mm
   long; brown, tan, or gray   **Polyphagidae**

Blattellidae. A large, worldwide family of small
to large, mostly slender-bodied, fast-moving roaches.
The most common introduced domiciliary pests *(Blattella, Supella)* and most of the roaches commonly

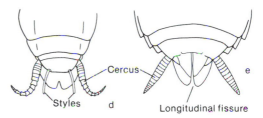

Figure 22.3 Taxonomic characters of Blattodea: **a**, anterior
leg, Blattellidae *(Blattella);* **b–e**, ventral aspects of external
genitalia; **b**, Blattellidae, male *(Blattella);* **c**, Blatellidae, female; **d**, Blattidae, male *(Periplaneta);* **e**, Blattidae, female.

encountered outdoors in North America are blattel-
lids.

Blattidae. Primarily tropical roaches of medium
to large size, represented in North America almost entirely by introduced domiciliary species *(Periplaneta,
Blatta, Neostylopyga).*

Polyphagidae. A relatively small family distributed mostly in arid and semiarid regions. North American genera *(Arenivaga, Eremoblatta, Compsodes)* are
almost entirely restricted to the arid southwest. Males
are winged, while females are apterous. *Attaphila texana,* which is commensal in the nests of leaf-cutter
ants is usually placed in Polyphagidae, sometimes in
Blattellidae.

Blaberidae. This large, primarily tropical family
contains ovoviviparous and viviparous species exclusively. North American representatives are tropical spe-

cies that have usually become established only near ports of entry (e.g., *Leucophaea,* the Madeira cockroach, in heated buildings in New York City) or in the subtropical areas of southern states *(Blaberus, Nauphotea, Panchlora).*

Cryptocercidae.[1] This is the most primitive family of Blattodea, with two species in China and one in North America. *Cryptocercus punctulata* occurs in moist, rotten wood in the Pacific Northwest and the Appalachians. The roaches occur as family groups of a male, a female and their offspring. Adults are apterous.

SELECTED REFERENCES

General

Cameron, 1961 *(Periplaneta);* Chopard, 1949a; Cornwell, 1968; Guthrie and Tindall, 1968; Huber, et al., 1990a (neurobiology); Scharrer, 1951 (wood roach, *Cryptocercus).*

Morphology

Bao and Robinson, 1990 (genitalia of *Blatta orientalis);* Guthrie and Tindall, 1968.

[1]*Cryptocercus* is placed in Polyphagidae in the recent classification of Grandocolas and Deleporte (1992).

Biology and Ecology

Cleveland et al., 1934 (wood roach, *Cryptocercus*); Engels and Buschinger, 1990; Fisk, 1987; Guthrie and Tindall, 1968; O'Neill et al., 1987 (social behavior in *Panesthia*); Roth, 1970 (reproduction); Roth and Hartman, 1967 (sound production); Roth and Willis, 1960 (biotic associations); Rugg and Rose, 1991 (biology of *Macropanesthia*); Schal et al., 1984.

Systematics and Evolution

Atkinson et al., 1991 (North American species); Gurney and Fisk, 1986 (keys to adults, nymphs and oothecae of household species); Helfer, 1987; McKittrick, 1964; Rehn, 1950 (North American genera), 1951 (higher classification); Roth, 1970; Roth and Hartman, 1967; Stehr, 1987 (immatures); Thorne, 1990, 1991 (use of symbionts in classification); Thorne and Carpenter, 1992 (higher classification and phylogeny); Vickery et al., 1985 (Canadian species).

23
ORDER MANTODEA (MANTIDS)

Female of *Litaneutria obscura* (Mantidae). This subapterous species frequents arid habitats.

Moderate-sized to very large Exopterygota adapted for predation. Head small, triangular, freely mobile on slender neck; mouthparts mandibulate; eyes large, lateral; ocelli usually 3 or none; antennae slender, filiform, multiarticulate. Prothorax narrow, elongate, strongly sclerotized; mesothorax and metathorax short, subquadrate; forewings narrow, usually cornified with reduced anal region; hind wings much broader, membranous, with large anal region; venation of largely unbranched longitudinal veins, usually with numerous crossveins; wing folding of longitudinal pleats; wings frequently reduced, especially in females; forelegs raptorial, with very long coxae; middle and hind legs long, slender, adapted for walking; tarsi 5-segmented. Abdomen with 10 visible segments, weakly sclerotized, slightly flattened dorsoventrally; segment 10 with variably segmented cerci; ovipositor short, mostly internal; male genitalia bilaterally asymmetrical. Nymphs similar to adults except in development of wings and genitalia.

Mantids are primarily tropical in distribution but are among the more familiar insects to nonentomologists everywhere because of the large size and striking appearance of many species. The common mantid pose, with the anterior part of the body elevated and the forelegs held together as if in prayer, has inspired folk beliefs since ancient times and the common name, praying mantis. In reality, this is the posture assumed for ambushing prey.

Mantids show a high morphological resemblance to cockroaches (Blattodea). Important similarities include the strong development of the direct pleurosternal flight muscles (see Chap. 22) and relatively small size of the indirect longitudinal muscles, five-segmented tarsi, multisegmented cerci, and strongly asymmetrical male genitalia. The eggs are always enclosed in an ootheca, which differs in details but is generally similar to that of cockroaches. These features strongly differentiate roaches and mantids from Orthoptera, with which they are sometimes classified as a single order. Nearly all the features that distinguish mantids from roaches are related to their predatory mode of life. Most obvious is the modification of the forelegs as strong grappling hooks for snaring arthropod prey (Fig. 23.1a). The elongate thorax and coxae increase the reach, while the sharp femoral and tibial teeth secure struggling victims. The long, slender posterior pairs of legs elevate the body, allowing unimpeded vision. The large eyes are situated far apart, as in other visually orienting predators such as damselflies, and the mobile head can be cocked in any direction. The cryptic coloration, usually in shades of green, brown, or gray, is a further adaptation for the sedentary mode of predation practiced by these insects (Color photo 8f).

In temperate regions adult mantids appear late in summer, when they are seen on vegetation, tree trunks, and other perches. As predators they are generalists, consuming any arthropod small enough to subdue, including smaller mantids. Larger species have occasionally been seen feeding on small frogs or lizards. Large mantids may be quite pugnacious, readily unfurling the hind wings, which may be brightly colored, and adopting threatening postures. All mantids are "sit-and-wait" predators that rely on camouflage to get within reach of prey, as mentioned above. Tropical species may be gaudily colored, with foliate appendages and specially textured or sculptured cuticle creating a strong overall resemblance to flowers, leaves, bark, or other plant parts (Color photos 8a–f).

Sex pheromones have been found in females of some species. Copulation may be preceded by courtship, allowing the male to reduce the chances of attack as he approaches the larger female. During copulation the male sits astride the female with his abdomen curled ventrally. Sperm is transferred by a spermatophore. In many species the female attacks and partly or wholly consumes the male during or immediately after copulation, but this may be partly an artifact caused by confined spaces of captivity.

Figure 23.1 Mantodea: **a**, *Stagomantis carolina,* male (scale equals 5 mm); **b**, ootheca, dorsal (diagrammatic); **c**, ootheca, transverse section (diagrammatic). (b, c, modified from Breland and Dobson, 1947.)

Gravid females become heavily swollen by the large mass of developing eggs. At the time of oviposition the female constructs a characteristic ootheca (Fig. 23.1b,c) from the frothy secretion from the accessory glands. This secretion quickly hardens into a protective envelope around the eggs, which are aligned in rows, four abreast. Structural details of oothecae, as well as site of attachment, vary greatly among species. Each female normally lays several to many egg masses. Females of a few species protect the ootheca until the nymphs have emerged. In temperate regions overwintering usually occurs in the egg stage.

As in cockroaches, the first-instar nymph is a nonfeeding stage that quickly molts. Succeeding instars vary in number among different species from 3 or 4 to 12. There is usually a single generation per year.

Although they are commonly considered beneficial insects and their oothecae are often sold in nurseries, mantids are generalist predators that consume desirable and undesirable species alike. In any case, natural populations seldom if ever become dense enough to have much effect on most pests.

A single family, **Mantidae,** occurs in North America. The most familiar species is the Carolina mantid, *Stagmomantis carolina.* Other conspicuous forms include the Chinese mantid, *Tenodera aridifolia,* and the praying mantid, *Mantis religiosa,* both introduced into the eastern United States. In the western states the small, wingless gray or brown species of *Litaneutria* are not infrequently encountered on the ground or on low vegetation. The world fauna comprises approximately 2000 species in eight families.

SELECTED REFERENCES

Chopard, 1938, 1949a (general); Crane, 1952 (defensive behavior); Edmunds and Dudgeon, 1991 (crypsis); Gurney, 1948a (North American species); Kevan, 1977 (higher classification); Preston-Mafham and Preston-Mafham, 1990 (general); Ramsay, 1990 (morphology, biology and New Zealand species); Reitze and Nentwig, 1991 (feeding ecology); Roeder, 1935 (sexual behavior); Stehr, 1987 (immatures); P. S. Terra, 1992 (maternal care); Thorne and Carpenter, 1992 (phylogeny); Vickery et al., 1985 (Canadian species); Yager, 1990 (auditory structures and function).

24
ORDER ISOPTERA (TERMITES)

Workers of the fungus-growing termite, *Macrotermes* sp. (Termitidae), foraging on leaf mulch in Tanzania.

Neopterous Exopterygota living socially in small to very large colonies. Individuals small to moderate in size, with subcylindrical, weakly sclerotized body (Figs. 24.1, 24.2). Head hypognathous or prognathous with mandibulate mouthparts, 10 to 32 segmented moniliform antennae; compound eyes and sometimes 2 ocelli present in reproductives, usually absent in soldiers and workers; frons almost always bearing conspicuous frontal pore or fontanelle, through which discharges frontal gland. Thorax with subequal segments, usually with lightly sclerotized tergites, membranous sterna; reproductives with 2 pairs of similar wings with simple, longitudinal venation or extensive network of cross-veins; wings broken at humeral suture and lost after dispersal flight; soldiers, workers, and nymphs without wings; legs short, adapted for walking. Abdomen with 10 similar segments and reduced eleventh segment; terminal segment bearing 1 to 8 segmented cerci.

Termites constitute a relatively small order of insects, with about 2300 described species. All termites are specialist feeders on cellulose, the most abundant organic compound in terrestrial habitats.

Because of this a large part of the stored energy in an ecosystem is available to them. Wood of living trees is usually avoided, but sound or decaying dead wood, twigs, fresh and dead grass, leaves, seeds, humus, and dung are consumed. Termites may be pests of agricultural crops and forestry. In tropical regions, termites are ecological analogues of earthworms in the Temperate Zone, playing a crucial role in soil development, drainage, and erosion.

North American species are of retiring habits: except for the reproductives, which disperse aerially prior to founding new colonies, they seldom venture out of their concealed galleries. Yet most nonentomologists are familiar with termites because of their nearly worldwide distribution and especially because of their proclivity to invade and consume man-made wooden or paper objects. They also damage dried food, fabrics, rubber, hides, wool, linoleum, and insulation materials. Termites are economically important everywhere except extremely cold regions, and in parts of the Tropics and Subtropics they may be among the most destructive arthropods. In northern Australia *Mastotermes* (Mastotermitidae) is considered to be the single most important insect pest. These termites form immense underground colonies from which they rapidly invade man-made structures as well as dead and living trees and crop plants. Aside from their economic importance, termites are of general biological interest because of their highly developed social behavior (see below and Chap. 7).

Termites are highly adapted for living in densely populated colonies, sociality apparently having evolved by the early Mesozoic Era or even before. Yet many characteristics, especially of the more primitive families, are shared with cockroaches (Blattodea). In particular, the primitive termite family Mastotermitidae shares many similarities with the roach family Cryptocercidae (Table 24.1). The Mastotermitidae is now represented by a single species, *Mastotermes darwiniensis,* in northern Australia, but *Mastotermes*-like fossils from the Mesozoic are widespread. The most convincing similarities, perhaps, are the shared features of wing structure, egg deposition, and intesti-

nal symbionts. The woodroach, *Cryptocercus,* harbors 21 to 22 species of hypermastigote and flagellate protozoa. All these are members of families that occur in primitive families of termites, including the shared genus *Trichonympha.* The nymphs of *Cryptocercus,* which are not unlike termite workers in general appearance, occur gregariously with adults in galleries in rotting wood. Another similarity, not listed in Table 24.1, is the presence of relatively large direct pleurosternal muscles and only small, poorly developed indirect dorsal longitudinal flight muscles in both roaches and termites (see also Chap. 22).

Termites, like the social Hymenoptera, are characterized by the development of morphologically differentiated **castes,** which perform different biological functions. Least modified morphologically are the **reproductives** (Fig. 24.1), which have relatively large eyes, unmodified mandibulate mouthparts, and a sclerotized body with distinct thoracic and abdominal tergites. Primary reproductives always bear wings initially, later breaking them at a basal fracture zone (humeral suture) by pressing the extended wings firmly against the substrate. Males do not change in external appearance after shedding their wings, but the abdomens of females gradually become swollen, or **physogastric** (Color photo 4e), from the enlargement of the ovaries. Physogastry is especially evident in the more specialized termites of the families Termitidae and Rhinotermitidae, in which the abdominal sclerites are eventually isolated as small islands on the bloated, membranous abdomen (Fig. 7.2). Many species, especially of the more primitive families, have the capacity to develop substitute kings and queens. These are neotenic individuals that never develop wings, do not become so physogastric as the primary reproductives, and differ in other minor morphological features.

The worker caste (Fig. 24.2a) is distinguished mostly by regressive characteristics. Compound eyes are usually absent, the number of antennal segments is usually smaller than in other castes, and the body is very lightly sclerotized except in a few species that actively forage outside their galleries. In Kalotermitidae and Hodotermitidae

TABLE 24.1
Comparison of selected characters of Isoptera and Blattodea

	Isoptera		Blattodea	
	Isoptera except Mastotermitidae	Mastotermitidae	Cryptocercidae, Panesthiidae	Blattodea except Cryptocercidae
No. of antennal segments	10–22	20–32	>32	>32
No. of tarsal segments	3–4	5	5	5
Wing venation	Reticulate in Kalotermitidae, Hodotermitidae, reduced in Termitidae	Reticulate	Reticulate	Reticulate
Hind wing anal lobe	Absent	Present	Present	Present
Humeral suture	Present in both wings	Present in forewings; absent in hind wings	Wings absent in Cryptocercidae; sutures present in Panesthiidae	Absent in both wings
Female external genitalia	Absent or vestigial	Short ovipositor with 3 pairs of valves	Short ovipositor with 3 pairs of valves	Short ovipositor with 3 pairs of valves
Egg deposition	Laid separately	Laid in pods	Contained in oothecae	Contained in oothecae
Intestinal symbiotes	Flagellate protozoa (spirochaetes in Termitidae)	Flagellate protozoa	Flagellate protozoa in Cryptocercidae; amoebae in Panesthiidae	Bacteria
Bacteriocytes	Absent	Present	Present	Present
Social organization	Eusocial	Eusocial	Primitively subsocial	Nonsocial

(such as *Zootermopsis* in North America) the worker caste is absent, being functionally replaced by nymphal reproductives, or **pseudergates.** Workers or pseudergates are the most abundant caste, performing foraging, care of young nymphs, feeding of soldiers and reproductives, and nest construction and maintenance.

Soldiers (Fig. 24.2b,c) are characterized by specialization of the head for defense of the colony, especially against ants. The mandibles are usually enlarged as long, slender, scissorlike blades, sometimes with grotesque serrations, ridging, or twisting. In some Termitidae the mandibles of soldiers are atrophied; the head is drawn into a nozzle-shaped projection bearing the opening of the frontal gland. These **nasute** soldiers deter invaders by spraying or exuding repellent or entangling substances. The proportion of soldiers varies from none in some species of Termitidae and Kalotermitidae to over 15 percent in *Nasutitermes* (Termitidae). In some species two classes of soldiers may occur, differing in structure as well as size.

Social organization in Isoptera is extremely complex and variable (see Chap. 7). In general, the more primitive species (Kalotermitidae, Hodotermitidae) excavate relatively small nests directly in logs, timbers, or other food material. For example, in *Incisitermes minor* (Kalotermitidae), which infests dry wood in California, mature colonies number only a few thousand individuals, including about 8 to 10 percent soldiers. Since these primitive termites do not construct passageways or for-

Figure 24.1 Adult female termite, Hodotermitidae *(Zootermopsis nevadensis).* (Modified from Banks, 1920.)

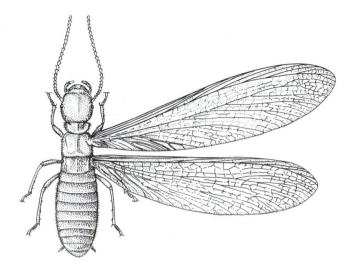

age away from the nest, the potential size and age of the colony are limited by the size of the initial food source. *Mastotermes,* generally considered to be the most primitive living termite, is exceptional in constructing massive, densely populated subterranean colonies.

Most members of the families Rhinotermitidae and Termitidae construct subterranean galleries from which the workers forage for food. The size of such colonies varies considerably, but frequently they are very large, containing millions of individuals. In many species the mature **termitarium** includes a portion that extends above the ground surface as a dome or spire, which may reach 7 m in height (Fig. 9.2; Color photo 4f). Such structures, which may be inhabited for 50 years or more, consist of an ensheathing hard layer of soil particles cemented together with excrement or glandular secretions, enclosing a honeycomb of living chambers (Fig. 24.3). Covered passageways are

Figure 24.2 Variation in body form of termites: **a**, worker, Rhinotermitidae *(Prorhinotermes simplex);* **b**, soldier, Rhinotermitidae *(Prorhinotermes simplex);* **c**, soldier, Termitidae *(Nasutitermes costaricensis)* (Modified from Banks, 1920.)

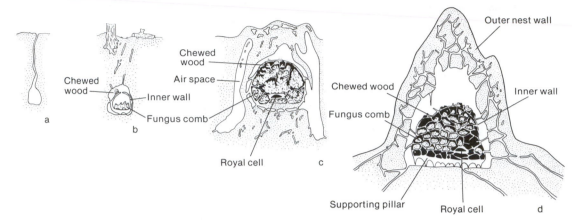

Figure 24.3 Development of the termitarium of *Macrotermes bellicosus* (Termitidae) of Africa: **a**, initial chamber excavated by founding couple; **b**, **c**, intermediate stages of development; **d**, structure of mature nest. (Modified from Grassé and Noirot, 1958.)

extended to food sources up to 100 m from the termitarium. In *Macrotermes, Nasutitermes* (Termitidae), and a few other genera, workers forage in the open at night or on cloudy days, collecting grass, pine needles, and similar materials. These are cut into short lengths and carried to storage chambers in the nest. Some members of the Termitidae construct enclosed, carton-type nests on trees, without any direct contact with the ground, and a few species ("dry wood termites") live in and consume the seasoned timbers in buildings without any contact with the soil. In general, however, some connection with the ground is necessary for the colony to obtain moisture.

Termitaria not only provide protection from predators but stabilize the temperature and humidity within the nest. The Old World subfamily Macrotermitinae (Termitidae) has capitalized on the homeostatic climate within the nest to grow fungi *(Termitomyces),* which are used as food, in a manner analogous to fungus culture by New World attine ants. The fungus is grown on spongy combs constructed of the termites' feces. The fungal combs are continually eaten, together with fresh woody material, and replaced. Lignin in the feces is broken down by the fungus, freeing additional cellulose, which the termites are able to utilize. In other species cellulose is obtained from

various sources, including wood or bark, dry grass or leaves, or highly organic soils. The dry dung of ungulates commonly harbors termites, particularly in arid regions. Actual digestion of the cellulose is accomplished by the microorganisms in the gut.

Worldwide, most of the economically important termites are members of the Termitidae. In parts of Africa and Asia damage to neglected structures occurs with remarkable rapidity. In temperate regions, where Termitidae are rare (14 species in North America), Rhinotermitidae cause more damage, principally by invading foundation timbers from their subterranean nests.

As in social Hymenoptera, termite reproductives disperse from established colonies. Dispersal occurs at characteristic seasons for each species and sometimes at characteristic times of day. For example, *Incisitermes minor,* discussed earlier, flies during warm dry periods, whereas *Reticulitermes hesperus* (Rhinotermitidae) flies after the first fall rains in California, usually early in the morning. Dispersal flights of a species are often simultaneous over a district and cover variable distances. A few species are efficient colonizers of islands and other isolated habitats, but sexuals of most species probably fly no more than a few hundred meters. After returning to earth, the unmated males and females dehisce their wings, then join in pairs that almost always exhibit tandem running, the male following the female (Chap. 7).

Founding pairs of reproductives initiate colonies in suitable locations in soil, logs, or other substrates. Colony growth is slow at first. After the first year, a typical colony may number only a dozen workers and the primary reproductives. The first soldiers appear in 2 or 3 years, and winged dispersal castes are not produced for 4 years or more. Eventually, in species with large colonies, egg production is almost continuous, and large, physogastric queens may lay 2000 to 3000 eggs per day. Individuals of all castes develop slowly and survive for considerable periods. The ultimate instar is usually achieved after 7 to 10 molts, requiring a year or more. Workers and soldiers live up to 4 years; reproductives may survive 15 years or more.

Termite colonies and those of ants are exploited by a variety of other organisms. Specialist vertebrate predators include Central and South American anteaters and armadillos; African aardvarks, aardwolves, bat-eared foxes, meercats, and pangolins; Indian sloth bears, and Australian echidnas and marsupial anteaters. Termites also are eagerly eaten by generalist predators such as birds, reptiles, frogs, and other insects, especially ants. Predators gorge themselves during the mass emergence of the reproductives from a colony, leaving few survivors to establish new nests. Humans also collect and eat the workers and reproductives.

Staphylinid beetles are particularly well represented among insect associates or **termitophiles** (Kistner, 1979, 1982). Some termitophiles consume eggs or young nymphs of their host, while others are nest scavengers. The latter frequently show considerable structural modification, with special secretory areas that are attractive to the termite; in others, the body is modified to resemble the contour or texture of the host.

Classification of Isoptera has attained a relatively mature state. New genera are infrequently discovered, and no new species have been described from North America since 1941. The family Termitidae is dominant, with about 1400 species, mostly tropical in distribution. Only 14 Termitidae species occur in North America, all restricted to the Southwest. The family Kalotermitidae contains 16 North American members, including several economically important species. Hodotermitidae are represented by three species and Rhinotermitidae by nine species, including *Reticulitermes,* the most destructive genus in North America. In most of the northern and central United States only a single species of *Reticulitermes* is present. Details of the type of damage caused by various termites and methods of control are found in texts dealing with economically important insects.

Workers are extremely similar and difficult to distinguish, and are not keyed here. All termites should be stored in liquid preservatives to prevent shriveling of their bodies.

KEY TO NORTH AMERICAN FAMILIES OF

Isoptera

1. Head with median dorsal pore or fontanelle; wings with 2 strong and several faint veins, without crossveining **2**

   Head without fontanelle; wings with at least 3 strong veins, usually with extensive crossveining **3**

2(1). Pronotum flat; basal portion of dehisced wing longer than pronotum ..... **Rhinotermitidae**

   Pronotum concave, saddle-shaped; basal portion of dehisced wing shorter than pronotum ..... **Termitidae**

3(1). Head with ocelli ..... **Kalotermitidae**

   Head without ocelli ..... **Hodotermitidae**[1]

SOLDIERS

1. Head produced anteriorly into a hornlike projection (Fig. 24.2c); mandibles vestigial ..... **Termitidae**

   Head without hornlike projection; mandibles large **2**

2(1). Mandibles with one or more large, marginal teeth **4**

   Mandibles without marginal teeth **3**

3(2). Head longer than broad, with median dorsal pore or fontanelle ..... **Rhinotermitidae**

   Head about as long as broad, without fontanelle ..... **Kalotermitidae**

[1]The only genus in North America, *Zootermopsis,* is placed in the family Termopsidae in some classifications.

4(2). Mandibles with 1 prominent marginal tooth; fontanelle present; head narrowed anteriorly

Termitidae

Mandibles with at least 2 prominent marginal teeth; fontanelle absent; head not narrowed anteriorly   **5**

5(4). Antennae with third segment enlarged, 2 to 4 times length of fourth segment

Kalotermitidae

Antennae with third segment subequal to fourth   **6**

6(5). Hind legs with femora swollen; antennae with at least 23 segments   **Hodotermitidae**[1]

Hind legs with femora slender; antennae with fewer than 23 segments   **Kalotermitidae**

SELECTED REFERENCES

General

Ebeling, 1968 (economic); Harris, 1961 (economic); Krishna and Weesner, 1969a,b; Skaife, 1961; Snyder, 1956, 1968 (bibliography).

Biology and Ecology

Breznak and Brune, 1994; Howse, 1970 (social behavior); Kistner, 1979, 1982 (termitophiles); Krishna and Weesner, 1969a,b; Luscher, 1961 (social behavior), 1984 (defense mechanisms); Prestwich, 1984 (chemical defense); Wilson, 1971 (social behavior and biology).

Systematics and Evolution

Desalle et al., 1992 (fossil DNA sequence); Emerson, 1952, 1955 (biogeography); Prestwich, 1983 (chemical systematics); Roonwal, 1962 (higher classification); Snyder, 1954 (North American species); Stehr, 1987 (immatures); Thorne, 1990, 1991 (use of symbionts in classification); Thorne and Carpenter, 1992 (phylogeny).

25
ORDER GRYLLOBLATTODEA (GRYLLOBLATTIDS)

Female of *Grylloblatta* sp. (Grylloblattidae) from 1280 m (4200 ft), Austin Pass, Mt. Baker, Washington. Grylloblattids occur in cool habitats, often at edges of glaciers and snow banks. They feed on organic detritus.

Elongate, cylindrical, or slightly flattened Exopterygota of moderate size (Fig. 25.1). Antennae multiarticulate, filiform, about as long as thorax; compound eyes reduced or absent, ocelli absent; mouthparts mandibulate, prognathous. Prothorax subquadrate, larger than mesothorax and metathorax; wings absent; legs long, slender, adapted for walking, tarsi 5-segmented. Abdomen 10-segmented, with filiform, 8- or 9-segmented cerci; female with straight, short ovipositor. Nymphs similar to adults.

Grylloblattids comprise one of the smallest orders of insects, with 16 species in four genera of the family **Grylloblattidae** occurring in boreal areas of North America and eastern Asia. Collections have been made from as low as 300 m elevation in central Japan and 500 m in northern California, but most specimens are found at high altitudes under cool, moist conditions, frequently near snowfields from central California to western Canada and Wyoming and in Japan, China, Korea, and

Figure 25.1 Grylloblattidae: *Grylloblatta campodeiformis,* male (scale equals 1 mm).

Siberia. Species in Japan and Korea have been found in caves. Activity occurs nocturnally or diurnally at low temperatures, and grylloblattids scavenge on the surface of snow or ice, as well as about snowmelt or moist surfaces in or near rockslides or under rotting logs. Food includes both plant and animal material. Prior to mating the male chases the female, then positions himself on her right side during copulation. The large, black eggs are deposited beneath stones, in rotten wood, or in moist soil, hatching only after a prolonged diapause of several months to about a year. Development is slow, with the eight recorded nymphal stadia requiring up to 7 years for completion. As development proceeds, segments are added to the antennae and cerci, and later instars gradually darken.

These generalized insects are of special interest because they combine features of the Orthoptera (external ovipositor; structure of the tentorium) and the Blattodea (five-segmented tarsi; multiarticulate cerci; asymmetrical male genitalia). It has been suggested that the presently reduced fauna represents the remnants of an evolutionary line that produced both the blattoid and orthopteroid orders, but some embryological and comparative anatomical evidence suggests that grylloblattids are a highly specialized lineage of Orthoptera (Ando, 1982). The Mesozoic *Blattogryllus* is the only possible fossil of this order.

SELECTED REFERENCES

Ando, 1982; N. Ford, 1926 (behavior); Gurney, 1948b (systematics); Kamp, 1963, 1973 (systematics); Stehr, 1987 (immatures); E. M. Walker, 1949 (morphology).

26
ORDER DERMAPTERA (EARWIGS)

Male of *Forficula auricularia* (Forficulidae). The forceps in this species show stong sexual dimorphism.

Neopterous Exopterygota of medium size; body elongate, slightly flattened in dorsoventral plane, with leathery cuticle (Fig. 26.1a,c). Head prognathous; compound eyes moderate or absent, ocelli absent; antennae multiarticulate, filiform; mouthparts mandibulate, adapted for chewing. Prothorax mobile, with large notum (except in wingless forms); cervical region of several separate sclerites; mesothorax small, mesothoracic wings short, elytriform, or absent; metathorax large, wings semicircular with distinctive radiating venation (Fig. 26.1b), or absent; legs adapted for walking, tarsi with 3 segments. Abdomen with 8 segments in female, 10 in male; cerci

modified as strongly sclerotized forceps, except in parasitic forms.

The forciplate cerci, terminating the flexible, telescoping abdomen, distinguish earwigs from all insects except Japygidae (Diplura), which are entognathous and blind and differ in numerous other features. In the elytriform forewings, Dermaptera resemble staphylinid beetles (Coleoptera), which, however, never have forceps. About 1800 species occur in all parts of the world but are most diverse in tropical and subtropical regions. In most temperate regions the familiar earwigs represent a few cosmopolitan, synanthropic species. Although they frequently become extremely abundant, most are of no more than nuisance importance. Earwigs are generalized in most aspects of their internal anatomy; they have unmodified, mandibulate mouthparts and unspecialized legs. The retention of three pairs of cervical sclerites is probably a primitive feature. In the more primitive families male genitalia are distally paired, as in Ephemeroptera, the only other insects with paired external genitalia.

With a few exceptions, the Dermaptera are quite uniform in most anatomical and biological features, being rather narrowly adapted for a cryptic life. For example, entry into crevices or cavities under stones, or logs, where nearly all Dermaptera take refuge by day, is facilitated by the smooth, leathery cuticle and the specialized wing structure. In the flying wings, the sclerotized, platelike pre-anal veins are crowded into the wing base (Fig. 26.1b). The membranous anal portion, with numerous, radiating veins, folds fanwise and transversely beneath the abbreviated mesothoracic wings, with only a sclerotized apical portion exposed. The mesothoracic wings are sclerotized, without defined veins. At rest they are held closed by rows of bristles on the metanotum.

Forceps (Fig. 26.1a,c) are used in defense, prey capture, and courtship. In most Dermaptera forceps are single-segmented throughout life, but are usually straighter, less strongly sclerotized, and setose in early instars. In *Diplatys* and *Bormansia* (suborder Forficulina) the cerci are multiarticulate until the last nymphal instar, in which they are unsegmented but straight, becoming forcipulate only after the ultimate molt. In *Arixenia* and *Hemimerus,* which are commensals or ectoparasites of vertebrates, the cerci are never forcipulate. The forceps of males are often larger and stouter than

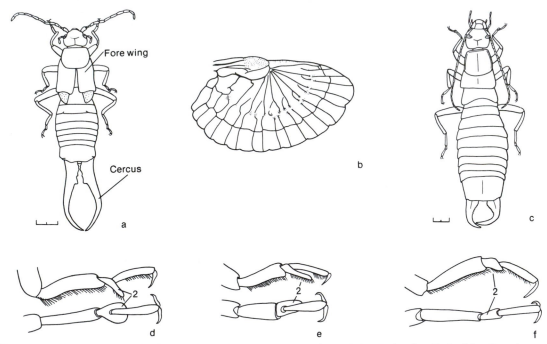

Figure 26.1 Dermaptera: **a**, Forficulidae, adult (*Forficula auricularia;* scale equals 2 mm); **b**, wing, Forficulidae *(Doru linearis);* **c**, Labiduridae *(Anisolabis maritimus,* male; scale equals 2 mm); **d**, tarsus, Forficulidae *(Forficula);* **e**, tarsus, Chelisochidae *(Chelisoches);* **f**, tarsus, Labiduridae *(Labidura).*

those of females, and sometimes more strongly curved or asymmetrical. Polymorphism in the size of the forceps, as well as in wing development is common.

Biological information is available primarily for a few cosmopolitan species but probably applies to most of the order. Most earwigs appear to be omnivorous, but some, such as *Anisolabis,* are reported to be predatory. Nearly all are nocturnal, retreating to protected shelters during the day. Nymphs and adults differ little in biological requirements. The order is notable for the development of maternal care of the young. Females oviposit clutches of 15 to 80 ovoid, delicate eggs in self-constructed burrows or retreats, then protect the eggs (Fig 7.1) and first-instar nymphs for several weeks, after which the mother becomes cannibalistic. Nymphs pass through four or five instars in various species. *Arixenia* and *Hemimerus* are parasites or nest associates of mammals. Both show reduction of sense organs, flattening of the body, and other morphological modifications associated with parasitism. In both genera embryonic development is internal, with first instar nymphs appearing viviparously.

KEY TO THE SUBORDERS AND NORTH AMERICAN FAMILIES OF

Dermaptera

1. Eyes vestigial or absent; cerci not forciculate
Suborders **Arixeniina** and **Hemimerina**
Eyes well developed; cerci forciculate, at least in adults   (Suborder **Forficulina**) **2**

2(1). Second tarsal segment produced longitudinally or laterally (Fig. 26.1d,e)   **3**
Second tarsal segment cylindrical, not produced (Fig. 26.1f)   **4**

3(2). Second tarsal segment expanded laterally (Fig. 26.1d)   **Forficulidae**
Second tarsal segment produced longitudinally below third (Fig. 26.1e), not expanded laterally   **Chelisochidae**

4(2). Tarsus with padlike arolium beneath claws (as in Fig. 34.2d)   **Pygidicranidae**
Tarsus without arolium   **5**

5(4). Antennae with 25 to 30 segments   **Labiduridae**

Antennae with 10 to 24 segments   **6**

6(5). Forewings meeting along entire inner margin   **Labiidae**
Forewings rounded, inner margins separated at base, or absent   **Carcinophoridae**

Pygidicranidae. This is considered the most primitive family of earwigs, with paired, subequal penis lobes on the male both flexed cephalad. Pygidicranidae is primarily a tropical family, with a single adventitious, representative, *Pyragropsis buscki* in southern Florida.

Labiduridae. Represented in North America by only a single introduced species, *Labidura riparia.* Labidurids have paired penis lobes, one pointing caudad, one cephalad. These medium to large (45 mm), often apterous earwigs occur in many habitats.

Carcinophoridae. This large, widespread family includes both winged and wingless species and predators as well as omnivores. Males have paired penes as in Labiduridae. Carcinophoridae is represented in North America by two genera, *Euborellia* and *Anisolabis.* *Anisolabis* has been introduced to both the Atlantic and Pacific coasts, where it is restricted to coastal strand situations, including the upper intertidal. At least some species of *Euborellia* are introduced. They are usually encountered in debris or litter. The largest known earwig (55 mm long) is an Australian carcinophorid.

Labiidae. A large, heterogeneous assemblage of primarily winged species distributed mostly in the Old World. *Labia* and *Prolabia* are common introduced representatives in North America. *Vostox,* encountered about rotting cacti and other succulent vegetation in the southwestern states, is extremely active and flies readily when disturbed.

Chelisochidae. Members of this small family are mostly paleotropical. One species, *Chelisoches morio,* introduced into California.

Forficulidae. A large, cosmopolitan family of winged species. *Forficula auricularia* (Fig. 26.1a) has been introduced into nearly all temperate regions, including North America, and is often a nuisance because of its large numbers. A single, median penis lobe is present in Forficulidae, Labiidae, and Chelisochidae.

SELECTED REFERENCES

General

Brindle, 1987; Chopard, 1949d; Guenther and Herter, 1974; Vickery et al., 1985.

Biology and Ecology

Briceno and Eberhard, 1994 (function of cerci); Eisner, 1960 (defensive behavior); Fulton, 1924; Guppy, 1950 (biology of *Anisolabis*).

Systematics and Evolution

Blatchley, 1920 (species of Northeastern states); Giles, 1963, 1974 (higher classification); Hebard, 1934 (Illinois species); Langston and Powell, 1975 (California species); Popham, 1965 (families and subfamilies, world); Stehr, 1987 (immatures); Vickery et al., 1985 (Canadian species).

27
ORDER PLECOPTERA (STONEFLIES)

Adult Plecoptera often rest on vegetation or rocks next to the streams and lakes in which their aquatic nymphs develop.

Neopterous Exopterygota with aquatic naiads, terrestrial adults.

Adults (Fig. 27.1a) with head hypognathous or prognathous with mandibulate mouthparts, bulging, lateral compound eyes, 2 to 3 ocelli, long, filamentous antennae with numerous segments. Thorax with 3 subequal segments composed of little-modified tergal, pleural, and sternal sclerites; wings membranous; forewing narrower than hind wing, with anal region much reduced; venation of longitudinal veins with complex network of cross-veins, or with crossveining absent, wing folding of longitudinal pleats; legs robust, long, with 3-segmented tarsi bearing 2 terminal claws. Abdomen weakly sclerotized, cylindrical or flattened, with 10 similar segments and cerci of 1 to many segments. Thorax and/or abdomen often with remnants of nymphal gills.

Naiads (Fig. 27.1b) similar to adults, but with gills usually larger or more extensive, cerci longer, wings and genitalia undeveloped.

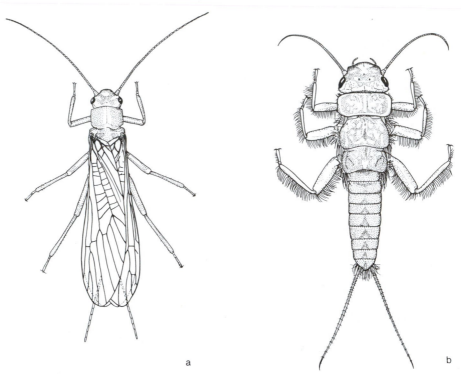

a   b

Figure 27.1 Representative adult and nymphal stoneflies: **a**, adult of *Isoperla confusa* (Perlodidae); **b**, naiad of *Acroneuria pacifica* (Perlidae). (a, modified from Frison, 1935; b, modified from Claassen, 1931.)

Plecoptera are rather obscure as adults, though some species are present throughout most of the year. The nymphs, which are familiar to fishermen and anyone who has turned stones in cool brooks, are important in food chains in aquatic ecosystems.

Stoneflies are extremely generalized in most features of their morphology. Their unmodified, mandibulate mouthparts (Fig. 27.2), with two-lobed maxilla and four-lobed labium, each with well-developed palps, are similar to those of Orthoptera. The multiarticulate antennae and cerci, unspecialized cursorial legs, and (in primitive forms) complex network of wing veins are shared, at least in part, with primitive insects such as Ephemeroptera and Grylloblattodea. Plecoptera are notable for the primitive structure of the thoracic pleura, which are not divided into episternum and epimeron in the prothorax. The single, crescent-shaped pleurite is suggestive of the subcoxal arc

of primitive insects such as silverfish. Venation is most primitive in the family Eustheniidae, presently restricted to Australia, New Zealand, and South America, but Pteronarcidae of the North American fauna show a relatively complete venation (Fig. 27.3a). In other families, especially Nemouridae, venation is variably reduced (as in Fig. 27.3b–d).

Nymphs (Fig. 27.1b) of most stoneflies respire through the general body surface, but many bear external tracheated gills. The gills vary from straplike, unbranched filaments to profusely branched tufts and may occur on the head, thorax, legs, or abdomen, or in the rectum. Differences in the distribution and structure of gills are of uncertain functional significance but are extremely useful in identification.

Adult stoneflies usually remain near the streams or lakes from which they emerged. Many species rest exposed on stones or vegetation, but others hide in cracks and crevices and may be encountered in such unexpected places as beneath slightly loose bark. Most species are weak, fluttery fliers, and many will run rather than fly to elude predators. Most stoneflies are diurnal, but some members of the suborder Setipalpia are nocturnal. Food of adults consists of algae, lichen, rotten wood, and foliage, though many species appear not to feed.

Emergence is highly seasonal for most species, and since the adult life span is usually only a few weeks, the composition of stonefly faunas is highly characteristic of time and location. Many species, especially of the family Capniidae, emerge during the coldest months, retaining normal abilities to crawl or fly, copulate and oviposit, even during periods of prolonged subfreezing temperatures (Sailer, 1950). Brachyptery, or winglessness, is not uncommon among these "winter stoneflies", and some species use their wings to scull along

the water surface (Marden and Kramer, 1994). The entire life cycle of *Utacapnia* is passed under water in Lake Tahoe, California (Jewett, 1963).

In some species mating is preceded by courtship involving drumming of the male abdomen against the substrate, sometimes producing an audible sound. Copulation, which may occur several times, takes place on substrates, with the male astride the female. Males of some species seek female nymphs, then immediately copulate with the freshly emerged adult before the exoskeleton has hardened.

Eggs, numbering from several hundred to several thousand per individual, are extruded into a ball temporarily carried on the venter of the abdomen, which is curled up so that the egg mass appears to be dorsal. Eggs are later dropped into water during flight, washed off by dipping the abdomen just beneath the surface, or deposited when the female crawls into the water, as in wingless species. The eggs of many species are

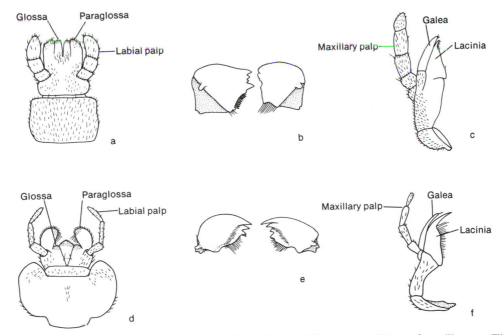

Figure 27.2 Taxonomic characters of stonefly naiads (left to right): **a, d**, labia; **b, e**, mandibles; **c, f**, maxillae, **a–c**, Filipalpia (Nemouridae, *Nemoura venosa*); **d–f**, Setipalpia (Perlidae, *Perlinella drymo*). (Modified from Claassen, 1931.)

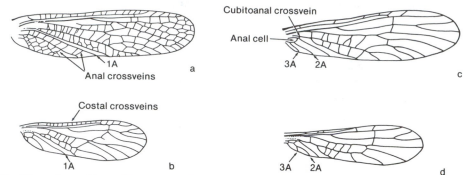

Figure 27.3 Wings of stoneflies: **a**, Pteronarcidae *(Pteronarcys proteus);* **b**, Peltoperlidae *(Peltoperla mariana);* **c**, Perlodidae *(Isoperla bilineata);* **d**, Chloroperlidae *(Utaperla sopladora).* (a, c, modified from Needham and Claassen, 1925; b, d, modified from Ricker, 1943, 1952.)

enclosed in a sticky material that anchors them to the substrate. Eclosion usually occurs after 2 to 3 weeks, or even several months, but in some species eggs are not deposited until they are ready to hatch. Species that inhabit seasonal streams may feature prolonged diapause so that the eggs hatch only under favorable conditions.

Nymphs develop slowly, requiring about a year to reach maturity in many species, and up to 3 to 4 years in others. Full growth requires numerous molts. From 12 to 24 instars, differing in the number of gills and segments in the antennae and cerci, are recorded for North American species; up to 33 instars have been recorded in other Northern Hemisphere species. Stoneflies are most abundant in cool streams with moderate or swift current, but they also occur around lake margins and in warm, sluggish streams. Stony, irregular bottoms support the most varied and densest fauna. In general, stonefly nymphs require well-oxygenated water and are among the first aquatic life to disappear in polluted streams. In New Zealand and nearby subantarctic islands a number of species have terrestrial naiads that inhabit cool, wet situations.

Most stonefly nymphs are primarily herbivorous, consuming submerged leaves, algae, diatoms, and similar items. Nymphs of Perlidae and Chloroperlidae are commonly predaceous, favoring may-

fly nymphs as prey. These feeding differences are reflected in mouthpart structure. Herbivorous species have short, stout mandibles with a pronounced molar lobe (Fig. 27.2b); predatory species have elongate, sickle-shaped mandibles with the molar lobe reduced or absent (Fig. 27.2e). A few days before adult emergence, feeding ceases, and shortly before the last molt the nymph crawls onto an emergent stone, log, or vegetation, often several feet from the water surface.

About 2000 species of Plecoptera have been described worldwide, with about 550 species in North America north of Mexico. The order is most abundant and diverse in cool temperate regions, and in North America is best represented in the north and in the western mountains, although other species are characteristic of the Great Plains, Midwest, and Southeast. The classification adopted here follows Ricker (1952, 1959).

Nymphs and adults should be preserved in 80 percent alcohol. The exuviae of the last nymphal instars may be softened in warm water to relax the gills and are nearly as useful as nymphs for identification. Nymphs are superficially similar to those of Ephemeroptera but differ by having no median caudal filament, two tarsal claws, and gills, if present, on the head, thorax, and abdomen, but not along the sides of the abdomen (cf. Figs. 27.1b and 20.4).

KEY TO NORTH AMERICAN FAMILIES OF

Plecoptera

ADULTS

1. Labium with glossae and paraglossae of approximately equal length (Fig. 27.2a) (suborder **Filipalpia**)   **2**

Labium with glossae much shorter than paraglossae (Fig. 27.2d) (suborder **Setipalpia**)   **7**

2(1). Basal segment of tarsus less than half the length of apical segment; sides of thorax with remnants of gills   **3**

Basal segment of tarsus more than half the length of apical; thorax without gill remnants   **4**

3(2). Gill remnants present on first 2 or 3 abdominal segments; 3 ocelli   **Pteronarcidae**

Abdomen without gill remnants; 2 ocelli   **Peltoperlidae**

4(2). Tarsus with second segment much shorter than first   **5**

Tarsus with first and second segments subequal   **Taeniopterygidae**

5(4). Cerci with one segment   **6**

Cerci with at least four segments   **Capniidae**

6(5). Wings flat over dorsum at rest; labial palp with apical segment globular, much larger than basal segments   **Nemouridae**

Wings at rest rolled down around sides of abdomen; labial palps with apical and preapical segments subequal   **Leuctridae**

7(1). Thoracic sterna or pleura with profusely branched gill remnants at corners; forewing with cubito-anal crossvein intersecting anal cell (as in Fig. 27.3b) or distad from it by no more than its own length   **Perlidae**

Thoracic segments without gills or with unbranched, straplike gills; forewing with cubito-anal crossvein distad from anal cell by its own length or more (Fig. 27.3c) or absent   **8**

8(7). Forewing with veins 2A and 3A arising from anal cell separately (Fig. 27.3c); gill remnants sometimes present on thorax or abdomen   **Perlodidae**

Forewing with veins 2A and 3A arising from anal cell as single vein (Fig. 27.3d); gill remnants absent   **Chloroperlidae**

NYMPHS

1. Abdomen with tufts of branched gills present on first two or three abdominal segments   **Pteronarcidae**

Abdomen without gills   **2**

2(1). Thorax with sternites produced posteriorly and overlapping; head flattened; body often cockroachlike   **Peltoperlidae**

Thorax without overlapping sternites; head not flattened; body not cockroachlike   **3**

3(2). Labium with glossae and paraglossae about equal in length (Fig. 27.2a); dorsal surface of body usually unpatterned   **4**

Labium with glossae much shorter than paraglossae (Fig. 27.2d); dorsum usually patterned   **7**

4(3). Hind legs extending beyond abdominal apex; wing pads oblique to longitudinal body axis   **5**

Hind legs not extending as far as abdominal apex; wing pads parallel to longitudinal body axis   **6**

5(4). Tarsus with first and second segments subequal   **Taeniopterygidae**

Tarsus with second segment much shorter than first   **Nemouridae**

6(4). Abdominal segments 8 and 9 annular, without membranous, ventrolateral fold separating tergal and sternal regions   **Leuctridae**

Abdominal segments 8 and 9 with membranous lateral folds separating tergal and sternal regions (as in anterior segments)   **Capniidae**

7(3). Thorax with branched gills ventrally and laterally   **Perlidae**

Thorax without branched gills   **8**

8(7). Cerci at least as long as abdomen; dorsum usually patterned   **Perlodidae**

Cerci less than three-fourths length of abdomen; dorsum seldom patterned   **Chloroperlidae**

SELECTED REFERENCES

General

Needham and Claassen, 1925.

Morphology

Wu, 1923.

Biology and Ecology

Alba-Tercedor and Sanchez-Ortega, 1991 (general); Dewalt and Stewart, 1995 (life histories); Freilich, 1991 (ecology of *Pteronarcys*); Holdsworth, 1941 (biology of *Pteronarcys*); Hynes, 1976 (general); Jewett, 1963; Membiela, 1990 (mating calls); Newcomer, 1918 (economic); Richardson and Gaufin, 1971 (feeding); Sailer, 1950 (low-temperature activity); Zwick, 1990 (upstream oviposition flights).

Systematics and Evolution

R. W. Baumann, 1987; Claassen, 1931 (naiads); Frison, 1935, 1942 (Illinois); Gaufin et al., 1966 (Utah); Harden and Mickel, 1952 (Minnesota); Illies, 1965 (phylogeny, biogeography); Jewett, 1956 (California), 1959 (Northwest); Merritt and Cummins, 1996 (North America); Needham and Claassen, 1925 (general); Ricker, 1959 (general); Stehr, 1987 (immatures); Stewart et al., 1988 (immatures; reprinted in 1993).

28
ORDER EMBIIDINA (EMBIOPTERA, WEBSPINNERS, EMBIIDS)

Male and silk gallery of the webspinner, *Antipaluria caribeana* (Clothodidae) from Venezuela. Adult males usually are winged, while females always are wingless.

Small to moderate-sized Exopterygota, with subcylindrical, elongate body modified for living in tubular galleries (Fig. 28.1a). Head prognathous with mandibulate mouthparts, 12 to 32 segmented antennae, kidney-shaped eyes; ocelli absent. Prothorax small, narrower than head; mesothorax and metathorax broad, flattened, and usually bearing wings in male but narrower, elongate, without wings in female; wings long, narrow, with similar, simple venation (Fig. 28.1c–e); legs short, robust; tarsi 3-segmented (Fig. 28.1b); forelegs with basal tarsomere enlarged, globular; hind legs with femur enlarged. Abdomen with 10 subequal segments, 2-segmented cerci on segment 10. Nymphs similar to adults except in development of wings and genitalia.

Embiidina are monotonously similar in body form but vary in size and coloration. Males of some species are attracted to lights, but most are obscure because of their small size, dull color, and retiring habits. This predominantly tropical order is poorly represented in the Temperate Zones, and except for introduced species is absent from islands, presumably because of the lack of wings in females.

Although the world fauna included fewer than 200 named species in 1991, it seems likely that as many as 2000 species remain undescribed (E. S. Ross, 1991). The U.S. fauna comprises eight native species and three others introduced from the Old World.

With the exception of dispersal by adult males, all instars of all species of Embiidina exclusively inhabit self-constructed silken galleries. This narrow specialization for a single niche is strongly reflected in the overall morphological uniformity of embiids. The most obvious modification involves the fore tarsi (Fig. 28.1a), which contain the silk-producing glands. Each swollen basitarsomere usually contains up to about 75 secretory units, exceptionally as many as 200, each connected to a hollow spinning bristle by a minute cuticular tubule. Each secretory unit consists of a syncytial layer of cytoplasm enclosing a storage chamber for the liquid silk. Silk production is involuntary, and since each tarsal bristle extrudes an individual fiber, a broad swath of silk is available continuously. The forelegs are rapidly shuttled back and forth, and the body is rotated as galleries are extended onto new food sources. Additional silk is continually added to old galleries, which eventually become opaque from multiple laminations of silk. As individuals grow, they progressively construct larger tunnels, so that an active embiid colony contains a maze of passageways of various sizes and ages.

Rapid backward movement within the narrow passageways is aided by the highly tactile cerci and strongly developed tibial depressor muscles of the swollen hind femora. In alate individuals, the wings are soft and flexible, allowing them to flex quickly over the head during backward motion. During flight the wings are stiffened by inflation of a blood sinus formed by the principal veins, especially the radial. The remaining veins are relatively weak, with few branches, and with almost no venational differences between fore and hind wings. In most other morphological features embiids are somewhat similar to earwigs (Dermaptera) and stoneflies (Plecoptera), indicating a very early divergence from the orthopteroid lineage.

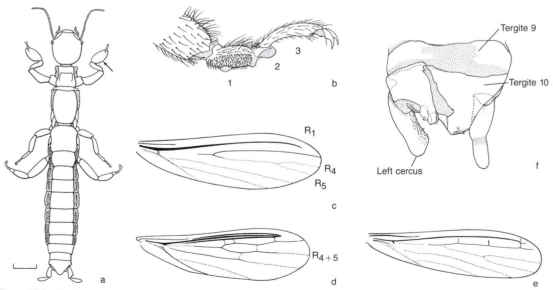

Figure 28.1   Embiidina: **a**, adult female of *Haploembia solieri* (Oligotomidae; scale equals 1 mm), with enlarged basitarsomere at arrow; **b**, hind tarsus of *H. solieri*; **c**, forewing of Teratembiidae *(Oligembia brevicauda)*; **d**, forewing of Anisembiidae *(Anisembia schwarzi)*; **e**, forewing of Teratembiidae *(Teratembia geniculata)*; **f**, abdominal terminalia of Anisembiidae *(Chelicera galapagensis)*. (b–f, modified from E. S. Ross, 1940, 1952, 1966.)

Embiidina are strongly gregarious, a typical colony consisting of one or more adult females with offspring of various numbers and sizes. A colony is usually centered around a secure retreat, such as a crevice beneath a stone or a crack in a log. Passageways are extended to close-by leaf litter, bark, lichen, moss, or other vegetable material that is suitable as food. In humid tropical habitats webways may run exposed over bark or other irregular surfaces, sometimes covering considerable areas. In North America, colonies are normally relatively small and concealed beneath stones, logs, or other shelters. In arid regions the galleries extend down cracks into cooler soil regions. Embiids are easily cultured in small containers of leaves or other habitat material.

Eggs are laid within the galleries and are sometimes covered by masticated cellulose. Eggs and newly emerged nymphs are briefly tended by the mother. There are four nymphal instars. The sexes are similar until wing buds appear in the last two instars of males. Asymmetry in male external genitalia appears only after the last molt. At maturity males leave the home colony for a short dispersal flight before entering another colony. They do not feed, and after mating they live only a short time, frequently being consumed by the females. The elongate, prognathous mandibles of male embiids are used to grasp the head of the female during mating. The highly complex, asymmetrical male cerci (Fig. 28.1f) of some species function as clasping organs during copulation. Parthenogenesis is recorded in a few species, and wingless males occur in others, especially those inhabiting arid regions.

In North America the order is restricted to the southern and southwestern states. The best differentiating characters are found in the wings, genitalia, and mouthparts of adult males. The following key is based on the work of E. S. Ross (1940, 1944) and C. Davis (1940).

KEY TO THE NORTH AMERICAN FAMILIES OF

Embiidina

ADULT MALES

1. Mandibles with apical teeth; left cercus with 2 segments; inner apical surface of cercus smooth   **2**

   Mandibles without apical teeth; left cercus with segments fused into a single member; inner apical surface of cercus lobed and minutely spiculate (Fig. 28.1f)   **Anisembiidae**

2(1). Wings with R_{4+5} unbranched (as in Fig. 28.1d)   **Oligotomidae**[1]

   Wings with R_{4+5} 2-branched (Fig. 28.1c,e)
       Teratembiidae

SELECTED REFERENCES

General

E. S. Ross, 1991.

Morphology

Barth, 1954 (tarsal silk glands); Mukerji, 1927.

Biology and Ecology

Choe, 1994c (communal behavior); Edgerly, 1988 (maternal behavior), 1994 (communal behavior); Mills, 1932; Mukerji, 1927; E. S. Ross, 1944, 1970, 1991.

Systematics

C. Davis, 1940; E. S. Ross, 1940 (North American species), 1944 (New World), 1957 (California species), 1970 (general), 1984a (North American species), 1984b (Mexican species); Stehr, 1987 (immatures).

---

[1]*Haploembia solieri,* common in the Southwest, has both parthenogenetic and bisexual races. In all instars the hind basitarsi have two ventral papillae. One papilla is present in all other North American embiids.

29
ORDER ORTHOPTERA (GRASSHOPPERS, CRICKETS, ETC.)

The devastating grasshopper, *Melanoplus devastator* (Acrididae), is an important pest of agriculture. It breeds in the grassy foothills of California and migrates into croplands.

Usually medium to large Exopterygota with subcylindrical, elongate body; hind legs almost always enlarged for jumping (Fig. 29.1). Head hypognathous, with compound eyes; ocelli present or absent; antennae multiarticulate; mouthparts mandibulate. Prothorax large, with shieldlike pronotum curving ventrally over pleural region; mesothorax small, with wings narrow, cornified; metathorax large, wings broad, with straight, longitudinal veins, numerous crossveins; wing folding by longitudinal pleats between veins; tarsi with 1 to 4 segments. Abdomen with first 8 or 9 segments annular, terminal 2

to 3 segments reduced; cerci with 1 segment. Nymphs similar to adults except in development of wings and genitalia.

———————

Many Orthoptera, especially tropical species, are bizarrely modified to mimic living or dead vegetation or other insects (Fig. 12.3a). In subterranean or myrmecophilous forms, wing loss may be accompanied by lack of differentiation of thoracic

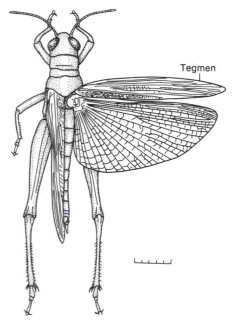

Tegmen

Figure 29.1 The grasshopper *Melanoplus cinereus*, Acrididae, dorsal aspect. Scale equals 5 mm.

and abdominal segments, and body proportions may differ markedly from surface-dwelling types. Yet nearly all Orthoptera are readily distinguished from other insects by the enlarged hind femora and characteristic grasshopper-like shape (Fig. 29.1). The Orthoptera are easily the dominant group of mandibulate hemimetabolous insects, with about 20,000 species distributed throughout the world. Most species are terrestrial herbivores, but various adaptations for predation or for subaquatic or subterranean life characterize many genera, tribes, or families.

A majority of Orthoptera conform to two broad adaptive modes, which correspond taxonomically to suborders. The **Caelifera** include mostly alert, diurnal forms with acute vision and hearing, which rely on jumping to escape from predators. Antennae and legs are relatively short, allowing unimpeded, rapid movement. These species, exemplified by the Acrididae (grasshoppers and locusts), are predominantly terrestrial (as opposed to arboreal). They are especially adapted to exposed situa-

tions in open habitats, such as savannah woodlands, steppes, and deserts. The great majority are herbivores; many have specialized dentition for consuming the tough foliage of monocotyledonous plants.

In contrast to the Caelifera, the **Ensifera** (katydids, crickets) tend to rely on crypsis to avoid predators. Most species are nocturnal, either retreating to protected niches during the day or escaping notice through camouflage or mimicry. Antennae and legs are long, sometimes extremely so; tactile responses are highly developed, though vision and hearing are also important. Movements tend to be slow and careful, but become abrupt and vigorous upon physical contact. Many Ensifera are omnivorous or predatory. Phytophagous species usually feed on stems or foliage of dicotyledonous plants, rather than grasses.

Nearly all Orthoptera defend themselves by kicking with the hind legs, which are usually sharply spurred on the tibiae. Most species will attempt to bite if held, and the larger ones can readily break the skin. Other defensive mechanisms include regurgitation of the crop contents or release of secretions from integumental defensive glands, often present in species with aposematic coloration (Color photo 7c). If seized by a predator, the hind legs are often autotomized at the femur; one-legged grasshoppers testify to the efficiency of this mechanism.

Hearing is acutely developed in most Orthoptera, and auditory stimuli are important in courtship in many species. In the Caelifera, auditory organs (if present) are on the first abdominal tergum, beneath the folded wings (Fig. 29.2b). Sound is produced by rubbing the hind femora against the lateral, vertical part of the forewings. In the Ensifera, auditory organs are located on the anterior tibiae (Fig. 29.2a). Stridulation is usually accomplished by vibrating the horizontal, overlapping parts of the forewings over one another (Fig. 6.8d,e). The wingless Stenopelmatidae tap the abdomen against the substrate to produce drumming patterns used in courtship.

The songs of Orthoptera are almost always species-specific and are often produced at a charac-

Figure 29.2   Auditory organs: **a**, Tettigoniidae *(Microcentrum californicum);* **b**, Acrididae *(Dissosteira carolina).*

teristic time of day. In general grasshoppers stridulate during the day, katydids at night, and crickets during both day and night. In some species singing may be synchronized and very loud. The crackling sounds produced by many acridids during flight may be designed to confuse birds.

Intraspecific polymorphism is common in Orthoptera, especially in the Caelifera. In the extreme case, distinct differences in behavior and physiology, as well as in color pattern and body proportions, may occur in response to population fluctuations and ecological conditions. Such phase transformation is very pronounced in periodically migratory grasshoppers such as the desert locust *(Schistocerca gregaria)* and occurs in species from all geographic regions (see Chap. 4). More typically, polymorphism involves only color, cuticular sculpturing, or wing length. The function in many aspects is obscure.

Nymphs are structurally similar to adults, except in the less-developed wings and genitalia, but may differ strikingly in coloration. Many common katydids, for example, are brightly banded as early instar nymphs, but solid green as adults. Immatures

of apterous species may be very difficult to distinguish from adults. A peculiarity shared only with Odonata is that the orientation of nymphal wing pads changes during development. In early instars the costal margin is lateral. Later, usually at the penultimate molt, the costal margin is turned inward; subsequent reversal to the lateral position occurs at the adult molt.

Sound communication is a frequent adjunct to courtship (see Chap. 6) and stridulatory mechanisms are frequently restricted to males. Copulation occurs with the male astride the female and may endure for several hours or longer. Orthopteran eggs are typically elongate or oval and relatively large. Eggs are usually inserted singly into plant tissue or soil substrates in the Ensifera. Burrowing species deposit the eggs in the burrow. In the Caelifera clutches of 10 to several hundred eggs are buried in shallow pits excavated by the ovipositor valves. Each pod of eggs is surrounded by a foamy waterproof matrix. Egg diapause is widespread in Orthoptera.

KEY TO THE NORTH AMERICAN FAMILIES OF

Orthoptera

  1. Antennae usually with many more than 30 segments (Suborder **Ensifera**)   **2**
      Antennae usually with many fewer than 30 segments (Suborder **Caelifera**)   **8**
2(1). Forelegs modified for digging, with femur and tibia broad and flat and tibia and tarsus strongly toothed   **Gryllotalpidae**
      Forelegs normal, adapted for walking   **3**
3(2). Hind coxae nearly contiguous ventrally; eyes very small; body dorsoventrally flattened
                                                  Myrmecophilidae
      Hind coxae widely separated ventrally; eyes large; body cylindrical or subcylindrical
                                                              4
4(3). Tarsi with 4 segments, at least on middle and posterior legs   **5**
      Tarsi with 3 segments   **Gryllidae**
5(4). Anterior tibiae with auditory organs (Fig. 29.2a), without ventral, articulated spines
                                                   Tettigoniidae
      Anterior tibiae without auditory organs, usually with ventral, articulated spines   **6**

6(5). Antennal bases separated by at least the length of the first antennal segment   7

Antennal bases contiguous or nearly so
Raphidophoridae

7(6). Tibiae narrow; tibial spurs small, inconspicuous   **Gryllacrididae**

Tibiae robust, flattened or club shaped; tibial spurs large, flattened   **Stenopelmatidae**

8(1). Tarsi with 3 segments   9

Tarsi with 1 or 2 segments   **Tridactylidae**

9(8). Pronotum extended backward to cover wings   **Tetrigidae**

Pronotum not prolonged backward over wings and abdomen   10

10(9). Antenna longer than fore femur   11

Antenna shorter than fore femur
Eumastacidae

11(10). Antennae at least half as long as body, longer than body in males   **Tanaoceridae**

Antennae less than half the length of body
Acrididae

SUBORDER ENSIFERA

Tettigoniidae (katydids). This variable family comprises approximately 5000 species predominantly from the Tropics. Most katydids are arboreal, winged insects, but apterous species, such as the Mormon cricket *(Anabrus simplex),* are common. Feeding habits of Tettigoniidae are diverse. While the majority are herbivorous, others are detritivores or predators. A few, including the Mormon cricket, are crop pests. Large, tropical Tettigoniidae may live several years and are often strikingly mimetic. Different species mimic ants and wasps as well as living or dead leaves, flowers, or lichen. Those resembling other insects often display unusual mimetic behavior. Most familiar North American species are relatively unmodified, green or brown, annual katydids (Fig. 29.3d).

Gryllidae (crickets; Fig. 29.3a). Approximately 1200 species, distributed worldwide. Most species live on the ground under logs or stones or in dense vegetation, but others are arboreal or subterranean. Gryllidae are often brachypterous, apterous, or polymorphic for wing length. Most species are omnivorous.

Gryllotalpidae (mole crickets). About 50 species worldwide, seven occurring in the United States. The antennae are much shorter than the body and the forelegs flat and platelike, adapted for burrowing in moist soil.

Myrmecophilidae. Small, flattened yellowish or brownish crickets that inhabit ant nests, feeding on secretions produced by the ants. Worldwide in distribution, with about 50 described species. Five of the six North

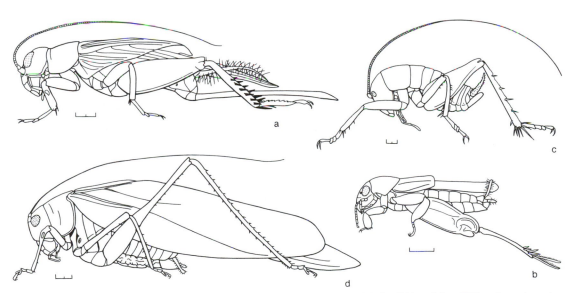

Figure 29.3 Representative Orthoptera: **a**, Gryllidae *(Acheta assimilis,* female); **b**, Tridactylidae *(Tridactylus minutus);* **c**, Raphidophoridae *(Ceuthophilus californicus);* **d**, Tettigoniidae *(Microcentrum californicum,* male). Scales equal 2 mm except that for b equals 1 mm.

American myrmecophilids occur in the West, where they are common in the nests of many different ants.

Raphidophoridae (cave crickets, camel crickets). North American representatives are apterous (Fig. 29.3c), but winged species occur in other areas. Most species are at least partly subterranean, and many are restricted to caves, tree holes, animal burrows, and similar situations. Most appear to be opportunistic scavengers; some are predaceous. There are about 600 species worldwide.

Stenopelmatidae (Jerusalem crickets, potato bugs). Stenopelmatidae number about 200 species of winged or wingless crickets worldwide. The North American fauna includes a few genera of flightless, heavy-bodied crickets that are found under stones or logs or burrowing in moist soil. They are restricted to the western states, especially of Pacific North America. The commonly encountered *Stenopelmatus,* lacking an obvious ovipositor, is also distinguished by the bicolored, banded abdomen. *Stenopelmatus* species usually turn onto their backs when molested, kicking the legs and moving the mandibles. They bite painfully if handled carelessly.

Gryllacrididae (leaf-rolling crickets). The single North American species, confined to the southeastern states, constructs shelters of rolled leaves tied with silk produced by glands that open into the mouth. Gryllacridids in other regions also use the secretions to stabilize burrows in the ground.

SUBORDER CAELIFERA
Acrididae (grasshoppers). The largest family of Orthoptera, with over 10,000 species. The great majority are typical "grasshoppers," with minor structural variation from the acridid body plan (Fig. 29.1). Many species are habitat-specific, and many have restricted diets, while others may be diversely adapted or polyphagous. Brachypterous species occur in several different lineages. Such familiar Orthoptera as migratory locusts, lubber grasshoppers, and band-winged grasshoppers are all acridids.

Tanaoceridae (long-horned grasshoppers). This family is restricted to the deserts of southwestern North America, where the few nocturnal species are found on sagebrush and other shrubs.

Eumastacidae (monkey grasshoppers). Worldwide Eumastacidae numbers more than 300 species of small to medium-sized, wingless grasshoppers that occur mostly in tropical or warm temperate regions. The 12 North American species are southwestern, occupying a variety of habitats from deserts to high mountains, where they are found at night on the tops of shrubby plants. In life the hind legs are held at right angles to the body.

Tetrigidae (pygmy locusts). A well-defined group of about 1000 species, mostly tropical or subtropical. Few are more than 15 mm long. Polymorphisms involving wing and pronotal development occur in many species, and many are grotesquely flattened and contorted. Many species are associated with moist situations and are proficient swimmers. Rain forest species are often encountered on mossy tree trunks. Tetrigids feed on algae or detritus.

Tridactylidae (pygmy mole crickets; Fig. 29.3b). At least 75 species of this specialized family occur worldwide. All frequent margins of ponds or streams. Eggs are laid in burrows in moist sand. Tridactylids are excellent swimmers and powerful jumpers. Food is apparently algae.

SELECTED REFERENCES

General

Acridia, 1972 to date (articles on grasshoppers and locusts); Beier, 1955; Chopard 1938, 1949c; Kevan, 1982; Uvarov 1966.

Morphology

Albrecht, 1953; Judd, 1948.

Biology and Ecology

Alexander, 1957, 1962 (sound production); Baccetti, 1987; Bailey and Rentz, 1990 (Tettigoniidae); J. A. Byers, 1991 (chemical ecology of locusts); Chapman and Joern, 1990 (grasshoppers); E. J. Clark, 1948 (general); Dethier, 1992 (sound production); Farrow, 1990 (locust migration); Fowler and Pinto, 1989 (mole cricket life history); Gwynne and Morris, 1983 (mating systems); F. Huber et al., 1990b (neurobiology of crickets); Iseley, 1944 (feeding specificity); Lomer and Prior, 1992 (biological control); T. E. Moore et al., 1989; Mulkern, 1967 (food selection); Preston-Mafham and Preston-Mafham, 1990.

Systematics and Evolution

Alexander, 1962; Ball et al., 1942 (Arizona species); Blatchley, 1920 (northeastern states); Brooks, 1958

(Canadian species); Chapco et al., 1994 (mitochondrial DNA of grasshoppers); Dirsh, 1975 (higher classification); Fischelson, 1985 (eastern Mediterranean species); Froeschner, 1954 (Iowa species); Hebard, 1934 (Illinois species); Helfer, 1987; Kevan, 1977, 1982; Merritt and Cummins, 1996 (aquatic species); Otte, 1981a,b (North American Acrididae); J. A. G. Rehn and Grant, 1961 (North American species); Richman, 1993 (New Mexico species); Sharov, 1971 (higher classification); Stehr, 1987 (immatures); Strohecker et al., 1968 (California species); Vickery et al., 1985 (Canadian species); T. J. Walker, 1963 (calling songs).

30
ORDER PHASMATODEA
(STICK INSECTS)

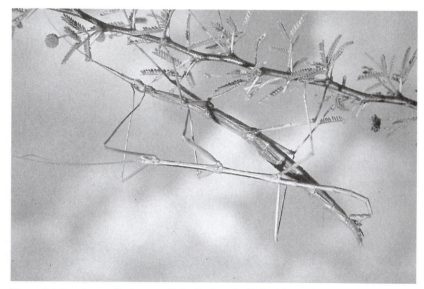

These stick insects are clinging to a limb of a "catclaw" shrub in southeastern Arizona. Males of the order are always much smaller than their mates. In southeast Asia and the Papuan region, many species resemble broad green leaves.

Cylindrical, elongate or flattened, leaflike Exopterygota, mostly of moderate to large size. Antennae multiarticulate, filiform or moniliform; compound eyes present, ocelli present in some winged males; mouthparts mandibulate, prognathous. Prothorax short, mesothorax and metathorax often elongate (winged species) or short (wingless species); pronotum without ventrolateral extensions covering pleural region; mesothoracic wings narrow, cornified; metathoracic wings broad, with straight longitudinal veins, numerous crossveins, anterior margin cornified, wing folding longitudinal, between veins; wing reduction or aptery common. Legs usually elongate, slender; tarsi with 5 segments. Abdomen with 10 (occasionally 11) apparent segments; terminal 2 or 3 sternites concealed beneath subgenital plate; cerci with 1 segment, external or concealed. Nymphs similar to adults except in development of wings and terminalia.

—————

Approximately 2500 species of phasmatids are known, predominantly from the Tropics, especially in the Oriental region. Walkingsticks appear to be relatively rare insects, even considering the extraordinary difficulty of finding them, and they seldom become economically important. Superficially similar to some Orthoptera, they differ in proventricular structure, in having asymmetrical male genitalia (symmetrical in Orthoptera), and in lacking wing pad reversal in the nymphs.

Nearly all phasmatids are modified to mimic vegetation of various sorts. Many species strikingly resemble either leaves or twigs, but others are generally similar to foliage without clearly conforming to either morphological type. Integumental sculpturing is diversely modified as thorns, tubercles, ridges, and other processes or textures that heighten the similarity to vegetation. Body form ranges from extremely slender to foliate, or only the appendages may be foliate. Resemblance to plant parts is further enhanced by posturing, frequently in grotesque, asymmetrical positions. Such postures may be maintained, even when the insects are disturbed, or they may drop from their purchase and feign death. A few species, such as *Carausius morosus,* are able to effect color changes by altering the dispersal of pigment granules in the epidermis. A few others, such as *Anisomorpha buprestoides* (Fig. 30.1b) of the southeastern United States, are aposematically colored. These active, agile insects secrete large quantities of a defensive substance (anisomorphal) in the prothoracic glands, readily spraying or oozing the milky product upon very slight disturbance. Yet another common defensive mechanism is limb autotomy, which takes place between the femur and trochanter. Partial leg regeneration may occur in nymphs.

All Phasmatodea are herbivores, typically leading solitary, sluggish lives on vegetation, especially trees and shrubs. Males are usually distinctly smaller than females. During copulation, which may be very prolonged, the male sits astride the female. In *Timema* the male spends nearly the entire adult instar riding on the larger female. Eggs, numbering about 100 to 1200 in different species, are deposited singly, either glued to vegetation or dropped to the ground. Females of *Anisomorpha* dig shallow pits in the sandy substrate,

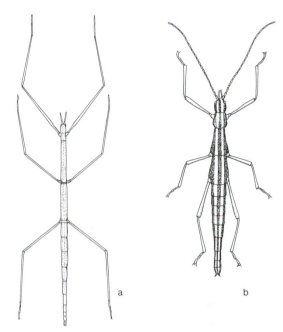

Figure 30.1 Representative Phasmatodea, Phasmatidae: **a,** *Parabacillus coloradus;* **b,** *Anisomorpha buprestoides.* (a, modified from Henderson and Levi, 1938; b, modified from Caudell, 1903.)

deposit a few eggs, and cover them. Phasmatid eggs are distinctive, with the thick, tough chorion often sculptured or patterned so that superficially the eggs often resemble seeds. Egg diapause is common and may endure more than a single season. Newly hatched nymphs climb onto convenient vegetation. They are generally similar to adults, both in appearance and behavior, but in a few species high population densities cause brightly colored nymphs and adults to develop, in a manner analogous to phase polymorphism in locusts. Unlike locusts there is no modification of behavior, and no increase in gregariousness.

Two families, both widespread, are recognized here. The **Phylliidae** include many leaflike species. There are no North American representatives. The **Phasmatidae** include species of diverse body form. Most North American species are sticklike, but *Anisomorpha* (southeastern states) and *Timema* (California) are fusiform with relatively short legs.

Timema is distinguished from all other genera in having the tarsi three segmented.

SELECTED REFERENCES

Bedford, 1978 (biology); Carlberg, 1987 (chemical defenses); Chopard, 1938, 1949b (general); Craighead, 1950 (biology of North American species); Littig, 1942 (morphology); Mazzini and Scali, 1987; Severin, 1911 (life history of *Diaphemeromera*); Stehr, 1987 (immatures); Vickery et al., 1985 (Canadian species).

31
ORDER ZORAPTERA

Zorotypus barberi (Zorotypidae) found under bark of a rotting log in Panama. Some adult individuals have wings that are later shed. All species of this small order are minute and very active.

Minute to small Exopterygota with mandibulate, chewing mouthparts, 9-segmented moniliform antennae, compound eyes, and ocelli present or absent. Prothorax large, subquadrate; mesothorax and metathorax subquadrate or transverse; wings with large pterostigma and venation reduced or absent; legs adapted for walking, with 2-segmented tarsi. Abdomen with 11 similar segments, unsegmented cerci on segment 11. Nymphs similar to adults except in development of wings and/or genitalia.

_____

These obscure insects resemble termites in general appearance and frequently occur about decaying wood. The order is known from the single family

Zorotypidae, which includes about 30 species, usually placed in a single genus, *Zorotypus*. The species are distributed sporadically in most parts of the world except Australia.

Their unspecialized, chewing mouthparts and the presence of cerci indicate a relationship of the Zoraptera with the orthopteroid orders. The male genitalia, which are asymmetrical, suggest affinities with cockroaches and termites. The Malpighian tubules number only six, a feature common to the hemipteroid orders. Termites, however, are unusual among the orthopteroids, in also having reduced numbers of Malpighian tubules (6 to 9). The central nervous system of Zoraptera is concentrated into two large thoracic and two abdominal ganglia, recalling the condition in the hemipteroid orders, and the wings, when present, have the venation extremely reduced, somewhat as in Psocoptera or Thysanoptera. As in hemipteroids, the hind wings are smaller than the forewings (Fig. 31.1b). According to Kukalova-Peck and Peck (1993), the basic pattern of venation can be derived from

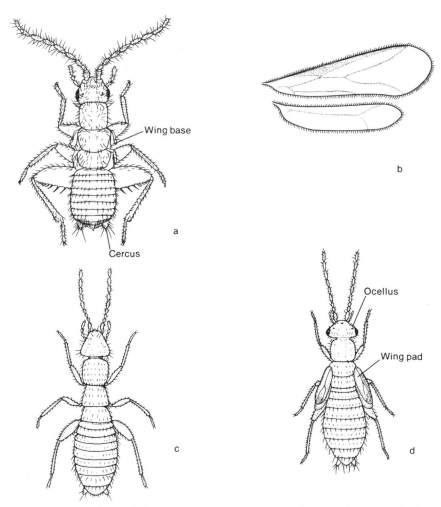

Figure 31.1   Zoraptera, Zorotypidae: **a**, dealate adult of *Zorotypus mexicanus;* **b**, wings of *Zorotypus hubbardi;* **c, d**, apterous adult and nymph of alate form of *Zorotypus hubbardi*. (a modified from Bolivary Peltain; b, c, d modified from Caudell, 1920.)

Female   Male

Figure 31.2 Mating position of *Zorotypus hubbardi* (Zorotypidae); male on right, female on left. (Modified from Shetlar, 1978.)

that of the Blattodea, and the dorsal longitudinal muscles are greatly reduced, again as in Blattodea, suggesting that similarities to the hemipteroids may be convergent.

Zoraptera occur in moist leaf litter, in rotting wood or sawdust, and in or about termite colonies. Species that have been studied are polymorphic, one adult form being largely unpigmented and lacking wings, compound eyes, and ocelli (Fig. 31.1c). Winged forms are more darkly pigmented, with small compound eyes and three ocelli (Fig. 31.1a). The significance of polymorphism in Zoraptera is uncertain, but the winged forms probably function in dispersal and may develop when the substrate becomes unsuitable because of drying or other deterioration. Food apparently consists mostly of fungi, though arthropod fragments have also been found in the alimentary canal.

In the only species observed, the female climbs onto the back of the male after a courtship dance. After coupling the female moves forward, turning the male onto its back and dragging it about in this position (Fig. 31.2). Eggs of Zoraptera are ovoid, without remarkable sculpturing or micropylar structures. Eclosion is by means of an egg burster on the head of the embryonic nymph. Polymorphism is evident in the nymphs (Fig. 31.1d), which develop external wing pads and have eyes and ocelli if destined to produce winged adults. The number of nymphal instars and other details of the life history and biology are unknown.

Two species occur in North America. *Zorotypus hubbardi* is widespread in the southeastern United States. *Zorotypus snyderi* is known from southernmost Florida and Jamaica.

SELECTED REFERENCES

Caudell, 1920 (biology); Choe, 1989, 1994a,b (mating behavior); Delamare-Deboutteville, 1948 (biology); Gurney, 1938 (systematics); Kukalová-Peck and Peck, 1993; Poinar, 1988 (fossil); Riegel, 1963 (distribution); Shetlar, 1978; Stehr, 1987 (immatures).

32
ORDER PSOCOPTERA (PSOCIDS, BOOKLICE)

A winged psocid on a leaf. Some species, especially as nymphs, form dense groups, often protected by a covering of silk produced by mouthpart glands. The tiny, wingless booklice are members of this order.

Minute to small Exopterygota with stout, soft body and large mobile head with swollen clypeus (Fig. 32.1). Mandibles of chewing type, maxillae with lacinia modified as a slender rod, detached from stipes (Fig. 32.2b); compound eyes usually large, convex; 3 ocelli present in winged, absent in wingless forms; antennae filiform, usually with 13 segments. Prothorax small, collarlike, especially in winged forms; mesothorax and metathorax subequal, sometimes with fused terga in wingless forms; forewings about 1.3 to 1.5 times longer than hind wings, with simplified venation; few crossveins or closed cells, pterostigma present; legs slender, adapted for walking, tarsi with 2 to 3 segments. Abdomen 9-segmented, without cerci. Nymphs similar to adults, except in the smaller number of antennal segments and ocelli, and undeveloped wings.

Figure 32.1 Representative adult Psocoptera: **a**, *Speleketor flocki;* **b**, *Psyllipsocus ramburi.* (Modified from Gurney, 1943.)

Because of their small size and retiring habits, Psocoptera frequently escape notice. However, they are exceedingly common animals in many habitats, numbering over 3000 species worldwide. Most psocids are scavengers, and they often occur in buildings, where they occasionally cause minor damage to stored materials. For example, *Liposcelis,* the common booklouse, sometimes infests insect collections or herbarium specimens, as well as invading dry, starchy foodstuffs.

Psocids are among the more generalized of hemipteroid insects, with chewing mandibles bearing well-developed molar and incisor lobes. The maxilla is specialized in having the detached lacinia modified as a stout, heavily sclerotized rod (Fig. 32.2b). Other maxillary structures are unmodified. The hypopharynx bears dorsally a concave **sitophore sclerite,** which is directly opposed by a cuticular knoblike structure on the dorsal wall of

the cibarium (Fig. 32.2a). During feeding, the rodlike laciniae are apparently shoved against the substrate as a brace to steady the head while the psocid scrapes up lichen, fungal hyphae, algae, or detritus with its mandibles. The sitophore apparatus provides an auxiliary grinding and crushing device. The hypopharynx also functions in taking up water vapor from moist atmospheres. It may be pointed out here that the lacinia and sitophore in biting lice (Phthiraptera) are highly similar to the structures just described and provide the best evidence of the origin of those highly specialized insects (see Chap. 33).

Psocids are frequently gregarious, mixed groups of winged or wingless adults and immatures of various sizes congregating in areas that are damp but not wet. Caves, hollow stumps, lichen-covered rock faces, and similar protected situations are favored by some species, and these frequently enter man-made structures. Other species frequent the foliage, twigs, and branches of trees and shrubs—hence the name "barklice." Fungi, especially bracket fungi on trees, leaf litter, and bird, mammal, wasp, and ant nests, provide suitable habitats for certain forms. Many species are rapid, nimble runners, often moving backward or forward with equal ease. At least some psocids have the ability to spin silk from labial glands. When the insects are numerous, entire branches or trees may be fairly covered with small canopies of the harmless, gossamer webs, which shelter psocid colonies.

Polymorphism is of frequent occurrence, involving either sex or both. Wingless forms usually lack ocelli and may differ greatly in thoracic structure from winged individuals. Minor differences in development of various body setae and specialized sense organs have also been described. Unlike the seasonal changes of Homoptera, polymorphism in most Psocoptera is not obviously related to environmental factors and is of uncertain significance. Parthenogenesis is frequent, males being extremely rare or unknown in some species; in others both asexual and sexual races are known.

Mating is typically preceded by courtship dances by the male, which later climbs over the

Figure 32.2 Oral structures: **a**, median section of head of *Ectopsocus briggsi;* **b**, maxilla of *Speleketor flocki.* (Modified from Snodgrass, 1944.)

female from back to front, backs under her, and copulates from the ventral position. After coupling, the pair turn end to end, as is common in hemipteroid insects. Eggs are deposited singly or in small clusters in crevices in irregular substrates, or on foliage. A few species are viviparous. Six nymphal instars are typical. Species that breed out of doors usually have one to three generations per year, depending on climate, while those inhabiting man-made structures have continuous generations. Most psocids are solitary, at least as adults, but aggregations often occur about suitable food. In a few species family groups of nymphs remain together until they become sexually mature, and aggrega-

tions of various instars are sometimes encountered beneath communal webs.

Identification of Psocoptera is difficult because of their small size, and slide-mounted specimens may be needed for examination of some of the important structures. Various authorities recognize different families, ranging from about 15 to 25 in number, and many genera have been repeatedly transferred among families. However, the great majority of commonly encountered species belong to a few families. The common booklouse, *Liposcelis divinatorius,* is a member of the Liposcelidae (suborder **Troctomorpha**). The suborder **Trogiomorpha** contains many short-winged or wingless forms, many frequenting caves. A few species are common in granaries, barns, and so on. Most of the common, winged species encountered outdoors are members of the suborder **Psocomorpha**.

KEY TO THE SUBORDERS OF

Psocoptera

ADULTS

1. Antennae with at least 20 segments; tarsi with 3 segments .... **Trogiomorpha**

   Antennae with 12 to 17 segments; tarsi 2 or 3 segmented **2**

2(1). Tarsi 3-segmented; antennae with some segments secondarily ringed; wing with pterostigma not thickened .... **Troctomorpha**

   Tarsi 2-segmented, or if 3-segmented, antennal segments not ringed; pterostigma thickened .... **Psocomorpha**

SELECTED REFERENCES

Baz, 1991, 1992 (biology and ecology); Cope, 1940 (morphology); Greenwood, 1988 (wing dimorphism); Gurney, 1950 (general); King and Ahmed, 1989 (sperm structure); Mockford, 1951 (Indiana species), 1957 (biology), 1993 (North American species); Mockford and Gurney, 1956 (Texas species); New, 1987 (biology); Pearman, 1928 (sound production); Rudolph, 1982 (water uptake); Smithers, 1990 (keys to genera); Smithers and Lienhard, 1992 (bibliography); Sommerman, 1943 (biology); Stehr, 1987 (immatures); Thornton, 1985 (ecology).

33
ORDER PHTHIRAPTERA (LICE)

A sucking louse from a Chinese water buffalo. Seen on a microscope slide, the tarsal claws are adaptations for clinging to the hair of active hosts. There are many louse species found on various mammals, including humans. The tiny eggs (called "nits") are firmly attached to hairs.

Small, wingless Exopterygota with the body strongly modified for ectoparasitism on vertebrates. Eyes reduced or absent, ocelli absent; antennae short, thick, 3- to 5-segmented; mouthparts mandibulate or highly modified for piercing and sucking. Thorax with segments poorly separated or fused, or with prothorax free, mobile; legs usually short, stout, often with tarsi modified, cheliform for clinging to host's pelage or plumage. Abdomen with 8 to 10 distinct segments, cerci absent. Nymphs similar to adults.

All members of this specialized order, numbering slightly over 3000 species, are permanent ectoparasites of birds and mammals, including domestic stock and fowls and man. Most species are scavengers of dried skin or other cutaneous debris. A relatively small number feed on blood, sebum, and other tissue fluids. All species can cause general irritation and may prove debilitating in severe infestations. The Anoplura are of special importance because of their ability to transmit epidemic typhus and other rickettsial diseases to humans; fortunately the prevalence of these diseases has decreased greatly since the advent of insecticides of low mammalian toxicity.

Lice are highly adapted to survive on the bodies of their hosts. The cuticle is tough and elastic, providing hydrostatic protection for the internal organs. Sensory structures are reduced or absent; antennae are short and can be tucked against the head; the body is usually streamlined or flattened. The tarsi are frequently shaped to cling to the pelage or plumage of the host, and in mammal-infesting species the tarsus and its single claw form a clamp for grasping hairs (Fig. 33.4b–d). Most of these modifications are characteristic of ectoparasitic insects from diverse orders and are evident in parasitic crustaceans as well.

The biting lice (suborders **Amblycera** and **Ischnocera**) have clearly evolved from psocid ancestors, which they closely resemble in mouthpart configuration. As in psocids, the maxillary laciniae are sclerotized rods, detached from the stipes, and the hypopharynx and cibarium are specialized as a sitophore apparatus (Fig. 33.1c,d). As in Psocoptera the hypopharynx also functions in uptake of atmospheric water vapor (Rudolph, 1983). The biting lice have normal mandibles, which are used to clip off the host's skin debris or feathers or, in some species, to pierce or lacerate the host's integument. In the sucking lice (suborder **Anoplura**), the functional mouthparts consist of three slender stylets of uncertain homology (Fig. 33.2a–c). At rest these are withdrawn into a pouch in the head capsule. The oral opening bears small denticles, which are anchored in the host's integument while the stylets are inserted. In the small suborder **Rhynchophthirina,** whose members are parasites of elephants and warthogs, the very small mandibles are attached apically on an elongate

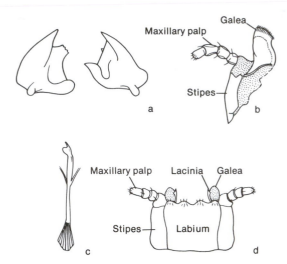

Figure 33.1 Mouthparts of Amblycera: **a**, mandibles of *Laemobothrion;* **b**, maxilla of *L. gypsis;* **c**, maxillary fork (lacinia) of *Ancistrona vagelli;* **d**, labium and maxillae of same. (Modified from Snodgrass, 1944.)

rostrum with the cutting edge external and the abductor muscles greatly enlarged (Fig. 33.3).

Phthiraptera are dependent on their hosts for food, shelter, and oviposition sites; most of them survive only a short time if isolated from hosts. All orders of birds and most orders of mammals serve as hosts. Lice are usually quite host-specific, and many species are further restricted to limited parts of the host's body. For example, some bird lice occupy the hollow quills of the large primary wing feathers, while others infest the chin, nape, or other parts of the body that are inaccessible to preening. Many mammal-infesting lice range widely over the hosts, whereas certain others are restricted to localized sites such as legs, tail, head, flippers, or even to the interdigital fossae of the feet, where the hosts have difficulty grooming. The species inhabiting humans occupy head or body hair, including pubic hair. Anoplura associated with marine mammals are unusual in being capable of surviving on land for long periods while their hosts remain at sea. The number of louse species occupying a host species varies from 1 to at least 15, but averaging about 2 to 6 for birds and 1 to 3 for mammals. The numbers of individuals on

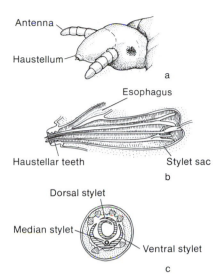

Figure 33.2 Mouthparts of Anoplura, diagrammatic: **a**, head with stylets protruding from oral opening; **b**, median section through head; **c**, transverse section near oral opening. (Modified from Furman and Catts, 1961.)

prepared slides. The key that follows includes suborders. Keys to families and genera appear in most medical entomology textbooks.

KEY TO THE SUBORDERS OF

Phthiraptera

1. Head prolonged anteriorly as a rostrum, with mandibles at apex   Suborder **Rhynchophthirina**
   Head not prolonged as a rostrum; small mandibles articulated ventrally on head capsule, or absent   **2**

2(1). Head relatively small, narrower than prothorax, sometimes fused with thorax; mandibles absent   Suborder **Anoplura** (Fig. 33.4c,d)
   Head relatively large, wider than prothorax, free, mobile mandibles adapted for chewing   **3**

3(2). Antennae capitate, 4-segmented; head with grooves for receiving antennae; mandibles horizontal; maxillary palpi present   Suborder **Amblycera** (Fig. 33.4b)
   Antennae filiform, 3- to 5-segmented; head without antennal grooves; mandibles vertical; maxillary palpi absent   Suborder **Ischnocera** (Fig. 33.4a)

a host individual may vary from none to hundreds or even thousands.

Most lice are bisexual, but males are frequently less common than females; a few species are known to be parthenogenetic. The number of eggs per individual varies from a few dozen to over 300 in different species, but usually only a few eggs are deposited each day. In most species the eggs are glued to hairs or feathers, or left in places inaccessible to preening and grooming, such as the interior of feather shafts. There are three nymphal instars, each usually molting after about 1 week.

No single classification of lice is generally accepted (see Haub, 1980; Kim and Ludwig, 1982). The suborders Amblycera, Ischnocera, and Rhynchophthirina are sometimes treated as a single order, Mallophaga, coordinate with the order Anoplura, but this classification has the disadvantage of concealing the structural diversity within the Mallophaga. The most recent authoritative classifications do not use the term Mallophaga. Several family classifications of Phthiraptera have been proposed. Those of Clay (1970) and von Keler (1969) are adopted here. Identification of families is frequently difficult without properly

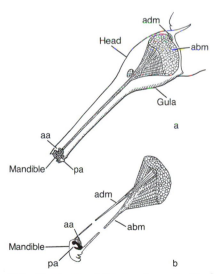

Figure 33.3 Mouthparts of *Haematomyzus* sp., Rhynchophthirina; **a**, side view of head; **b**, mandible and muscles. aa, anterior articulation of mandible; pa, posterior articulation of mandible; adm, adductor muscle of mandible; abm, abductor muscle of mandible. (Modified from Habu, 1972)

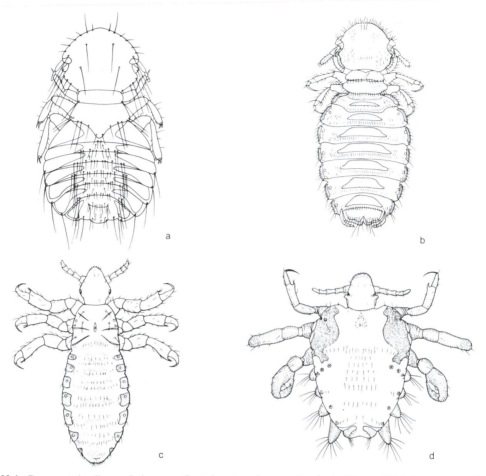

Figure 33.4 Representative lice: **a**, Ischnocera, *Goniodes gigas*, from poultry; **b**, Amblycera, *Trichodectes caprae,* from goat; **c**, **d**, Anoplura (c, the human body louse, *Pediculus humanus*); d, the pubic louse, *Pthirus pubis*). (a, modified from Emerson, 1956; b, modified from Patton, 1931; c, courtesy of D. Furman; d, modified from Ferris, 1951.)

Suborder Rhynchophthirina. A single family, Haematomyzidae, with two species restricted to elephants and the warthog, respectively. In the reduced mandible and vestigial maxillae and labium, as well as other characteristics, Rhynchophthirina are somewhat intermediate between the Amblycera and Ischnocera and the Anoplura.

Suborders Amblycera and **Ischnocera.** The largest and most diverse group of lice, with about 2500 species, mostly associated with birds. The two suborders differ in biological as well as morphological characteristics. The Amblycera have horizontally biting mandibles,

which in most species are used to scrape off feathers, loose skin, or other cutaneous debris. In a few genera the mandibles are modified for piercing integument. The mammal-infesting species (families Boopidae, Trimenoponidae, and Gyropidae) are almost entirely confined to marsupials and rodents from the Neotropics and Australasian region, but *Heterodoxus* infests dogs in many areas. Two other families (Menoponidae and Laemobothridae) occur on restricted groups of birds. The Ischnocera have vertically biting mandibles. The bird-infesting lice of this suborder (family Philopteridae) feed largely on feathers. This is the largest family, with specialized lineages infesting a diverse array of birds ranging from songbirds to marine and freshwater divers and waders, raptors, and galliform birds, including domestic fowl. Mammal-infesting species of Philopteridae infest placental mammals exclusively, feeding on dead or living skin. Bovicolidae infest a great diversity of ungulates and include species of economic importance. Trichodectidae infest carnivores and rodents and some primates. Finally, two families have very restricted hosts (Dasygonidae on hyraxes; Trichophilopteridae on lemurs).

Suborder Anoplura. About 250 bloodsucking species confined to placental mammals. The family **Pediculidae** specializes on primates. Included are two species that infest humans—*Pediculus humanus,* the body and head louse (Fig. 33.4c), and *Pthirus pubis,* the pubic or crab louse (Fig. 33.4d). The largest family of Anoplura is Hopleuridae, infesting rodents. The families Haematopinidae and Linognathidae occur mainly on pigs, ungulates, and hyraxes and include economically important species. Echinophthiriidae are restricted to Pinnipedia (seals, etc.) and Neolinognathidae to insectivores.

SELECTED REFERENCES

General

Askew, 1971; Buxton, 1947 (species that infest humans); Furman and Catts, 1961 (medically important species); Hellenthal and Price, 1991; Rothschild and Clay, 1952.

Morphology

Snodgrass, 1944 (mouthparts); Symmons, 1952 (head); Troester, 1990 (functional morphology of head of Anoplura).

Biology and Ecology

Busvine, 1948; K. C. Emerson, 1956 (ecology of species on chicken); Hopkins, 1949 (host associations); Martin, 1934 (life history of *Columbicola*); Matthysse, 1946 (biology of cattle lice); Murray, 1960, 1987; Trivedi et al., 1991 (distribution of species on poultry).

Systematics and Evolution

Barker, 1994 (classification and phylogeny); Busvine, 1948 (races of body louse); Clay, 1970; Durden and Musser, 1994 (host records, distribution, and world list); Ferris, 1931, 1951; Haub, 1980; von Keler, 1969; Kim et al., 1986; Lakshminarayana, 1985 (glossary of taxonomic terms); Lyal, 1985; Stehr, 1987 (immatures).

34
ORDER HEMIPTERA (BUGS, LEAFHOPPERS, ETC.)

A predatory bug, *Apiomerus* sp. (Reduviidae) with its soldier beetle prey. *Apiomerus* bugs collect sticky plant resins which are applied to setal brushes on the foretibiae and help hold struggling prey.

Minute to large Exopterygota with styliform suctorial mouthparts (Figs. 34.1, 34.2e–g). Head capsule strongly sclerotized, seldom with separate sclerites delimited by sutures; compound eyes usually large, ocelli present or absent; antennae short to long, filiform or setaceous, with 10 or fewer segments (exception, numerous segments in male Coccoidea); mandibles and maxillae extremely elongate, needlelike piercing stylets, ensheathed in elongate, tubular labium, which is jointed or unjointed; palps absent. Prothorax large, distinct, except in wingless forms; mesothorax represented dorsally by scutellum, usually large, triangular; metathorax small. Forewings larger than hind wings, frequently coriaceous, at least basally; flexed flat or obliquely over abdomen, usually without folds in wing membrane; legs usually adapted for walking, forelegs raptorial in many predatory taxa (Fig. 34.2a–c); tarsi with 3 or fewer segments, frequently with arolia, pulvilli, or other pretarsal structures (Fig. 34.2d). Abdomen commonly with anterior 1 to 2 segments reduced or absent, posterior 1 to 2 fused, reduced, associated with genitalia; cerci always absent.

The Hemiptera, with approximately 55,000 species, are the fifth largest insect order. Numerous fossils ascribed to the Hemiptera are known from the Permian Period, and it seems certain that the order arose earlier, in the Carboniferous Period. Most of the large group of fossils previously known as Protorthoptera appear to be forerunners of the hemipteroid orders (see Chap. 15) and show that these early hemipteroids had evolved the suctorial mouthparts characteristic of modern forms, making them admirably preadapted to feeding on the vascular tissue of higher plants, which diversified during the Mesozoic Era. Among the present hemipteran fauna, members of one suborder, the **Homoptera,** feed almost exclusively on the internal fluids of plants. Members of the Sternorrhyncha feed in the phloem tissue, whereas most Auchenorrhyncha feed in the xylem. Exceptions are the Fulgoroidea, which are phloem feeders, and the Cicadellidae, in which different subfamilies feed in phloem, xylem, or parenchymal tissue. Xylem and phloem feeders often produce a stylet sheath from salivary secretions. The sheath forms a tube around the stylets and remains in the plant tissue when the stylets are withdrawn. The other suborder, the **Heteroptera,** includes a minority of predatory and ectoparasitic species, but most species of the larger, more diverse families are principally phytophagous, feeding in parenchymal tissues.

Plant fluids do not provide a nutritionally complete diet, and most Hemiptera support symbiotic bacteria or yeasts in their guts. In Heteroptera the symbiotes may occur freely in the alimentary canal or may be housed in special gut diverticula. In Homoptera they are inside special cells or mycetocytes. The symbiosis is usually obligatory, and the microorganisms are transmitted transovarially from one generation of bugs to the next.

Suctorial plant feeding is the only major feeding niche that endopterygote insects have failed to invade. It seems likely that their complete exclusion has been due to the prior occupation by the Hemiptera, which differentiated before the large orders of Endopterygota.

Specialized styliform mouthparts are the most

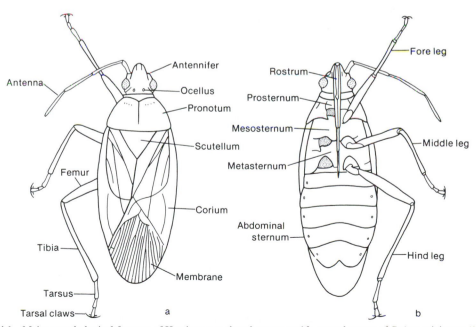

Figure 34.1   Major morphological features of Hemiptera: **a**, dorsal aspect, and **b**, ventral aspect of *Boisea trivittatus* (Coreidae).

obvious feature shared by all Hemiptera. The mandibles and maxillae have become highly specialized as a two-channeled piercing tube for delivering salivary secretions and taking up food (Fig. 34.2e–g). Proximally, the maxillary and mandibular stylets are hinged to a short arm (as in Fig. 35.1c), on which are inserted the protractor and retractor muscles. This specialized structure differentiates the Hemiptera and Thysanoptera from the other hemipteroid insects (see Chap. 15), in which the muscles insert directly on the stylets. The labium is developed as a stout, usually jointed sheath for the stylets and is normally the only portion of the rostrum visible externally. Most Hemiptera probe only a short distance into the host tissue, and the stylets are not greatly longer than the rostrum. Penetration is accomplished by repeated contraction and relaxation of the protractor muscles, necessitated by the small scope of

movement allowed by the protractor levers. In Aradidae, which feed on fungal mycelia (which are usually long and sinuous) the stylets are commonly longer than the body and are coiled in pouches within the head capsule. In Psylloidea and Coccoidea the stylets are coiled in the **crumena,** a pouch in the neck or anterior thorax. In predatory forms such as Reduviidae and Belostomatidae, where quick penetration of a potentially dangerous prey is important, the protractor muscles are very long, allowing the stylets to be inserted with a single muscle contraction.

The alimentary canal of Hemiptera is modified for uptake of liquid food. Salivary glands are universally present. Extraoral digestion is apparently widespread, and relatively large quantities of saliva are injected into the host's tissue, causing local necrosis in plants, or, more rarely, systemic disturbances. Injection of saliva may also be an

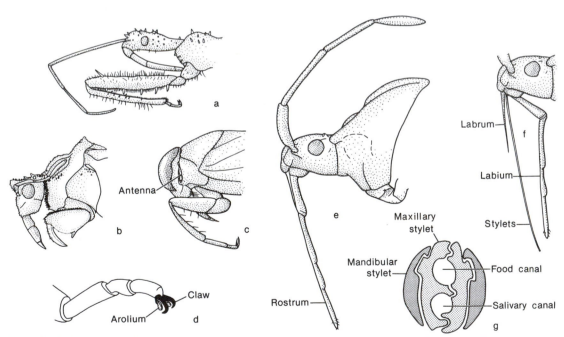

Figure 34.2 Legs and mouthparts of Heteroptera: **a,** raptorial forelegs of *Sinea* (Reduviidae); **b,** same of *Phymata granulosa* (Reduviidae); **c,** same of *Notonecta* (Notonectidae); **d,** arolium of pretarsus; **e,** rostrum of *Anasa tristis* (Coreidae); **f,** rostrum with labium removed to show feeding stylets; **g,** cross section of feeding stylets. (g, modified from Tower, 1914.)

important factor in transmission of microorganisms, especially plant viruses. In predatory species, saliva is highly toxic and paralytic, enabling relatively large prey to be quickly subdued. In many phytophagous Heteroptera, especially Lygaeidae, Pyrrhocoridae, and Pentatomidae, the midgut is differentiated as several distinct regions. These may include a distensible, anterior crop and a variable number of posterior caeca. The number and configuration of the caeca, which contain the digestive symbiotes, are characteristic of some families or genera.

Most Homoptera, with the major exceptions of the Fulgoroidea and some Sternorrhyncha, have the tubular posterior midgut coiled closely adjacent to the dilated, croplike anterior midgut. The Malpighian tubules may also be involved, forming a complex organ called a **filter chamber** (Chap. 5; Fig. 5.8c), whose function is apparently to convey excessive amounts of water to the hindgut. In Diaspididae (Coccoidea), no direct connection between midgut and hindgut exists, excess liquid being resorbed by the enlarged Malpighian tubules. In many Homoptera and some Heteroptera (Tingidae), surplus fluids, containing sugars, amino acids, and other small molecules, are excreted as honeydew, which forms the basis for commensal relationships with ants.

Flight in Hemiptera is powered primarily by the mesothoracic wings, which are usually longer than the metathoracic pair. This condition differentiates the Hemiptera from the orthopteroid orders, in which the anterior pair of wings functions primarily as a protective cover. Venation in Hemiptera is frequently reduced, especially in the smaller Homoptera, where only a few longitudinal veins support the wing membrane. In Heteroptera the wings usually conform closely to the outline of the body, and wing structure is frequently highly specialized. The anterior basal region, or **corium,** which covers the abdomen laterally (Fig. 34.4), is usually more heavily sclerotized. The corium may bear a distal line of weakness, the **cuneal fracture,** which allows the apex (membrane) of the wing to bend downward over the abdomen in repose. The arrangement of veins and closed cells in the corium and membrane is characteristic of families and is of great taxonomic utility.

Hemiptera commonly have pretarsal structures (Fig. 34.2d), important in maintaining a purchase on smooth, exposed surfaces of plants, or in holding prey. It is notable that in several other phytophagous orders special holdfast organs have evolved (e.g., crochets in larval Lepidoptera, Chap. 44). Most Heteroptera secrete aromatic compounds, apparently defensive in nature. Gland openings in nymphs are on the dorsum of the abdomen; at the last molt, new openings appear on the metapleural region, below the wings. Many Heteroptera that are well endowed with defensive secretions are aposematically colored, usually in red, orange, or yellow and black. Many Reduviidae are also brightly colored, as an advertisement of their potent bites rather than defensive secretions.

Sound production is widespread in Hemiptera (see also Chap. 6). In Homoptera sounds are produced by vibration of a pair of parchmentlike tymbals in the base of the abdomen (Fig. 6.8a–c). In Heteroptera sound is produced by several different stridulatory mechanisms. Reduviidae stridulate by rubbing the rostrum over the striate walls of the prosternal furrow, producing an audible rasping noise. Corixidae stridulate by drawing the anterior femora over the clypeal margin, and some Pentatomidae rub a comb on the inner surface of each wing over a striate area at the base of the abdomen. Other mechanisms involving the coxae and the abdominal venter occur in other families; clearly sound production has evolved independently several times in Hemiptera.

Hemiptera typically mate with the male initially in the dorsal position. In aquatic forms the male remains mounted on the female throughout, but in terrestrial forms the male dismounts after intromission and the pair remain connected end to end. In Cicadellidae mating usually occurs with the pair side by side. Cimicidae and some Nabidae and Anthocoridae are unusual in that the male punctures the body wall of the female, depositing the sperm directly into the hemocoel or into a special integumental pouch. The sperm then migrate to the genital tract, penetrate its wall, and enter the lumen, where fertilization occurs.

Most Hemiptera are hemimetabolous. Nymphs differ from adults in such details as numbers of antennal segments, presence of ocelli, and chaetotaxy, as well as wing development. Morphological differences among instars are usually gradual, but color changes may be abrupt and striking. For example, *Largus cinctus* (Largidae) is bright metallic blue with a red dorsal spot as a nymph, but molts to a brownish or blackish adult with a dull orange stripe on each flank. In some Homoptera developmental differences are as extreme as in the Endopterygota (see below).

Reproduction is usually bisexual, with one to several generations per year. In several families of Homoptera, however, differences among instars may be extreme, and cyclical or permanent parthenogenesis is common. These complications in reproductive strategy are discussed below. Most Hemiptera are oviparous, ovipositing either on or in plant tissue or on soil, stones, or other substrates. Eggs of Homoptera are usually of simple, ovoid structure, whereas those of Heteroptera are greatly differentiated among various families, frequently with opercula, spines, or appendages (Cobben, 1968).

In some classifications Homoptera and Hemiptera are considered to be separate orders. However, the differences between them are no greater than the differences between suborders of Coleoptera, Lepidoptera, or Hymenoptera. Furthermore, the superfamily Coleorrhyncha (Heteroptera) is intermediate in several important characters and in the past has been placed in the Homoptera. In addition, the monophyly of the Homoptera has been seriously questioned, and in some classifications Sternorrhyncha and Auchenorrhyncha are considered to be separate suborders, with Sternorrhyncha the sister to all other Hemiptera (Campbell et al., 1994; Campbell et al., 1995; Schaefer, 1996; Schuh and Slater, 1995). The classification followed here is traditional in recognizing Sternorrhyncha and Auchenorrhyncha as subdivisions of Homoptera.

KEY TO THE SUBORDERS OF

Hemiptera

1. Mesothoracic wings almost always held rooflike over abdomen, uniformly membranous; if wings absent, rostrum arising near posterior margin of head, without gula **Homoptera** (p. 426)
   Mesothoracic wings folded flat over abdomen; if held rooflike, basal portion thickened, apical portion membranous; if wings absent, rostrum arising anteriorly on head (Fig. 34.2a,b), with sclerotized gula **Heteroptera** (p. 416)

SUBORDER HETEROPTERA (TRUE BUGS)

Head horizontal, with base of rostrum clearly distinct from prosternum, although the distal segments may extend between the anterior coxae; antennae with 4 or 5 (occasionally fewer) segments; prothorax usually large, trapezoidal or rounded, frequently longer than broad; tarsi usually with 3 segments.

Whereas the Homoptera are exclusively terrestrial herbivores, the Heteroptera are believed to be primitively predaceous and include a diversity of ectoparasites and scavengers, as well as herbivores, with numerous amphibious and aquatic representatives. Heteroptera include a few viviparous species, but apparently none is parthenogenetic, in sharp contrast to the Homoptera. Phytophagous Heteroptera include many species that feed on the parenchyma of higher plants (Miridae, Coreidae, Pentatomidae, Tingidae, etc.) and some that are restricted to particular plant parts. For example, many Lygaeidae, Pyrrhocoridae, and Largidae feed preferentially or exclusively on seeds, sometimes from a very narrow host range. Many Cydnidae are subterranean, apparently feeding on roots. Other Heteroptera limit their feeding to certain plant taxa. For example, certain Miridae are restricted to ferns, others to specific genera or families of angiosperms. Peloridiidae feed only on liverworts and mosses and Aradidae feed exclusively on wood-rotting fungi.

Among terrestrial forms, the Reduviidae, Nabidae, and Anthocoridae are almost entirely predaceous (Color photo 6b), and many Miridae, Pentatomidae, and Lygaeidae, as well as occasional members of other families, are predatory. In addition many more species are facultatively predatory. Modes of predation range from "sit-and-wait" types (many Reduviidae) to active searchers, such as *Geocoris* (Lygaeidae). Some species, especially in the Pentatomidae, are host-specific, but most heteropteran predators prey on all suitably sized organisms in a given habitat. All aquatic Heteroptera are predators except some Corixidae, which feed on algae and diatoms. The numerous amphibious families are either scavengers or predators, mostly of much smaller, weaker organisms.

Two families of Heteroptera are exclusively ectoparasitic. Cimicidae inhabit bird and mammal nests, including houses, and suck blood from the occupants. Polyctenidae suck blood from bats. In addition, all members of the reduviid subfamily Triatominae are ectoparasites of mammals and birds.

Heteroptera are among the most successful insect colonizers of marine habitats. *Halobates* (Gerridae) occurs in all tropical seas, commonly hundreds or thousands of kilometers from land. *Rheumatobates* occurs in brackish mangrove swamps and *Hermatobates* around coral reefs. Many Saldidae frequent seacoasts, and some, such as *Aepophilus,* are intertidal, enduring long periods of submergence during high water.

Regardless of feeding habits, most Heteroptera are univoltine or bivoltine, typically with 5 instars. Overwintering is usually by the adult.

Approximately 40 families of Heteroptera are recognized. These are arranged in about 8 series and 15 superfamilies, but no classification is universally accepted, and several of the superfamilies are very heterogeneous and difficult to characterize. The suborder is frequently treated as three broad divisions, the Hydrocorisae, Amphibicorisae, and Geocorisae. As suggested by their names, these categories are ecologically distinct, but their morphological separation is difficult. The series **Coleorrhyncha** contains only the family Peloridiidae, with about 20 species distributed in Australia, New Zealand, southern South America, and a few intervening islands. Superficially the adults resemble tingids, with the reticulate wings held flat over the abdomen (Fig. 34.3). The alimentary canal lacks a filter chamber, as in the Heteroptera, but the rostrum is inserted posteriorly on the head, as in Homoptera. The lateral prothoracic region and wing pads of nymphs are demarcated by sutures (a characteristic of some fossil Homoptera) and superficially resemble the paranotal lobes of Palaeodictyoptera. All instars live and feed on mosses and liverworts. The morphological characteristics, as well as the discontinuous, circum-Antarctic distribution and restriction to primitive host plants, seem to indicate that the Peloridiidae are a relictual family, possibly the most primitive extant Heteroptera.

The following key includes North American families.

KEY TO THE FAMILIES OF THE SUBORDER

Heteroptera

1. Antennae very short, inserted ventrally, with no more than 1 or 2 segments visible from above (aquatic or subaquatic species)   **2**

   Antennae much longer than head, usually inserted dorsally, and with at least 3 segments visible from above   **9**

2(1). Anterior tarsus with a single flattened, paddle-shaped segment (Fig. 34.10c); rostrum short, with 1 or 2 segments

   Corixidae (p. 425)

   Anterior tarsus usually with 2 or 3 segments, never paddle-shaped; rostrum long, with 3 or 4 segments   **3**

3(2). Ocelli present   **4**

Figure 34.3 Peloridiidae; **a**, adult; **b**, nymph. (a, modified from Myers and China, 1929; b, modified from Helmsing and China, 1937.)

Ocelli absent   **5**

4(3). Antennae retractable into grooves beneath eyes; anterior femora thickened, raptorial
                                   Gelastocoridae (p. 426)

Head without grooves for antennae; anterior femora slender   **Ochteridae** (p. 426)

5(3). Abdomen with slender caudal respiratory appendages at least one-fourth length of body (Fig. 34.10b); tarsi with 1 segment
                                        Nepidae (p. 426)

Abdomen with respiratory appendages short or absent; tarsi with 2 segments   **6**

6(5). Body oval, flattened; anterior legs raptorial, with femora thickened and grooved to receive tibiae   **7**

Body strongly convex, arched dorsally, usually elongate; anterior femora not raptorial   **8**

7(6). Eyes bulging, prominent; wing membrane with reticulate veins; abdomen with short, flat apical breathing appendages
                                 Belostomatidae (p. 426)

Eyes molded to outline of head and thorax, not prominent; wing membrane without veins; abdomen without breathing appendages
                                     Naucoridae (p. 426)

8(6). Hind legs with tibia and tarsus flattened, oarlike, without tarsal claws; body length greater than 5 mm   **Notonectidae** (p. 425)

Hind legs with tibia and tarsus cylindrical, with 2 tarsal claws; body length less than 4 mm
                                        Pleidae (p. 425)

9(1). Eyes present   **10**

Eyes absent; body flattened; forewings short, hind wings absent
                                   Polyctenidae (p. 424)

10(9). Anterior legs with claws attached before apex of tarsi; apical tarsal segment longitudinally cleft   **11**

Anterior legs with claws attached apically; apical tarsal segment not cleft   **12**

11(10). Middle coxae close to hind coxae, remote from fore coxae; hind femora much longer than abdomen   **Gerridae** (p. 424)

Middle coxae about equidistant from fore and hind coxae; hind femora about as long as abdomen   **Veliidae** (p. 424)

12(10). Head linear, as long as thorax including scutellum; eyes distant from base of head
                                 Hydrometridae (p. 424)

Head shorter (usually much shorter) than thorax

including scutellum; eyes usually near base of head   **13**

13(12). Scutellum concealed by pronotum; ocelli absent; tarsi with 2 segments; hemelytra frequently with raised, reticulate venation (Fig. 34.5d)   **Tingidae** (p. 422)

Scutellum usually visible, frequently large; if concealed, ocelli present or tarsi with 3 segments   **14**

14(13). Body greatly flattened dorsoventrally, with texture and appearance of tree bark; forewings in repose usually much narrower than abdomen (Fig. 34.7c)   **Aradidae** (p. 422)

Body not textured like bark, usually not greatly flattened; forewings in repose usually about same width as abdomen   **15**

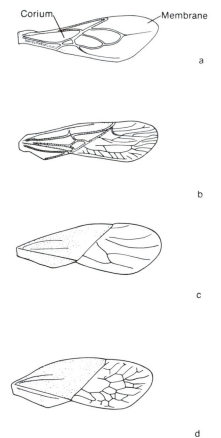

Figure 34.4 Wings of Heteroptera: **a**, Reduviidae (*Sinea diadema*); **b**, Nabidae (*Nabus alternatus*); **c**, Lygaeidae (*Lygaeus kalmi*); **d**, Largidae (*Largus*).

15(14). Antennae with basal 2 segments short, thick; apical segments slender, filamentous   **16**

Antennae with segments of subequal length and diameter or gradually tapering, or with only first segment short, thickened   **17**

16(15). Pronotum, wings, and head, including eyes, with long erect spines

Leptopodidae (p. 424)

Body not spiny

Dipsocoridae, Schizopteridae (p. 424)

17(15). Wings reduced to leathery flaps no longer than first abdominal segment, without veins; body ovoid, flattened, leathery (Fig. 34.7b)

Cimicidae (p. 424)

Wings usually large; if reduced, usually extending beyond first abdominal segment and with veins; body elongate, thick   **18**

18(17). Forewings with cuneus present (Fig. 34.5a,b)   **19**

Forewings without cuneus   **21**

19(18). Ocelli present; forewing with 0 to 1 basal cells in membrane (Fig. 34.5b)   **20**

Ocelli absent; forewing with 2 to 3 closed cells in membrane near cuneus (Fig. 34.5a)

Miridae (p. 423)

20(19). Rostrum with 4 segments; tarsi with 2 segments   **Microphysidae** (p. 423)

Rostrum and tarsi each with 3 segments

Anthocoridae (p. 423)

21(18). Antennae with 4 segments   **25**

Antennae with 5 segments   **22**

22(21). Scutellum large, triangular or U-shaped, reaching wing membrane (Fig. 34.7d)   **23**

Scutellum smaller, not reaching wing membrane (as in Fig. 34.6a–d)   **24**

23(22). Tibiae set with stout spines; apices of hind coxae fringed with stiff setae

Cydnidae (p. 421)

Tibiae not set with spines; hind coxae without fringe of stiff setae

Pentatomidae (p. 421)

24(22). Anterior femora thickened, raptorial (Fig. 34.2a); antennae with second segment about one-fourth length of third

Nabidae (p. 423)

Anterior femora slender; antennae with second segment about half the length of third

Hebridae (p. 424)

25(21). Ocelli present   **26**

Ocelli absent   **36**

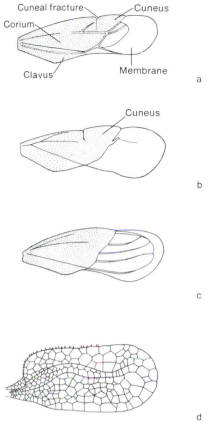

Figure 34.5 Wings of Heteroptera: **a**, Miridae *(Irbisia mollipes);* **b**, Anthocoridae *(Tetraphelps);* **c**, Saldidae *(Salda abdominalis);* **d**, Tingidae *(Corythuca obliqua).*

26(25). Middle and hind tarsi with 1 or 2 segments; size minute to small   **27**

Middle and hind tarsi with 3 segments; size small to large   **30**

27(26). Forewings entirely membranous; forelegs thickened, raptorial; body slender, elongate, linear   **Enicocephalidae** (p. 423)

Forewings with base thickened, coriaceous; forelegs slender; body not linear   **28**

28(27). Rostrum with 4 segments; corium shorter than wing   **29**

Rostrum with 3 segments; wing with corium extending to apex (1 species, Florida)

Thaumastocoridae (p. 424)

29(28). Body covered with velvety hairs; femora ex-

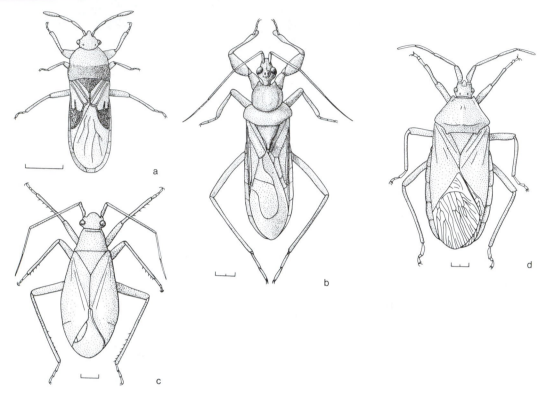

Figure 34.6 Representative Heteroptera: **a**, Lygaeidae (*Blissus leucopterus,* scale equals 1 mm); **b**, Reduviidae (*Triatoma protracta;* scale equals 2 mm); **c**, Miridae (*Irbisia mollipes;* scale equals 1 mm); **d**, Coreidae (*Anasa tristis;* scale equals 2 mm).

tending far beyond body margins
Hebridae (p. 424)

Body coarsely punctate on dorsum; never with velvety hair; femora extending about to body margins   **Piesmatidae** (p. 422)

30(26). Rostrum with 3 segments   **31**
Rostrum with 4 segments   **32**

31(30). Ocelli set between eyes; forewings with 4 to 5 elongate, parallel cells in membrane (Fig. 34.5c); head broader than long   **40**
Ocelli set behind eyes; forewings usually with 1 or 2 large cells in membrane (Fig. 34.4a); head longer than broad
Reduviidae (incl. **Phymatidae,** p. 422)

32(30). Rostrum with basal segment projecting, not capable of being apposed to ventral surface of head; tarsi without arolia
Nabidae (p. 423)
Rostrum capable of being apposed to ventral

surface of head and pronotum; tarsi with arolia (Fig. 34.2d)   **33**

33(32). Pronotum with posterior median lobe that covers scutellum; similar to Microveliidae in general appearance   **Macroveliidae** (p. 424)
Pronotum variable in shape, but without lobe covering scutellum   **34**

34(33). Forewings with 6 or fewer main longitudinal veins; if wings abbreviated, antennifers (tubercles on which antennae are inserted) located ventrolaterally   **35**
Forewings with at least 7 (usually many) main longitudinal veins (Fig. 34.6d); antennifers located dorsolaterally   **Coreidae** (incl. **Alydidae, Rhopalidae**) (p. 421)

35(34). Body elongate, slender, linear; antennae and legs extremely long, threadlike, with tips of femora and antennal scape slightly swollen (Fig. 34.7a)   **Berytidae** (p. 422)

Body usually stout; legs and antennae not extremely long and slender

Lygaeidae (p. 422)

36(25). Tarsi with 3 segments   **38**

Tarsi with 1 or 2 segments   **37**

37(36). Tarsi with 2 segments; rostrum retractable into groove between fore coxae

Reduviidae (p. 422)

Tarsi with 1 segment; rostrum not retractable into groove   **Nabidae** (p. 423)

38(36). Anterior femora swollen, bearing several teeth; forewing with 4 to 5 longitudinal veins in membrane (Fig. 34.4c)

Lygaeidae (p. 422)

Anterior femora slender, usually without teeth; forewing with numerous longitudinal veins in membrane (Fig. 34.4d)   **39**

39(38). Pronotum with upturned lateral margins

Pyrrhocoridae (p. 422)

Pronotum with lateral margins rounded

Largidae (p. 422)

40(31). Hemelytra with 4–5 closed cells in membrane (Fig. 34.5c)   **Saldidae** (p. 424)

Hemelytra without veins

Mesoveliidae (p. 424)

Cydnidae (including Thyreocoridae) (burrowing bugs and Negro bugs). Broadly oval, brown or black

bugs 3 to 10 mm long, sometimes with orange or yellow bars on the forewings. Ocelli present, antennae with five segments, tarsi with three segments; forelegs frequently modified for digging. Most species are probably phytophagous, feeding on either roots or foliage. As in pentatomids, thoracic repugnatorial glands are present. Worldwide in distribution.

Pentatomidae (including Scutellaridae) (shield bugs; Fig. 34.7d; Color photo 2b). Broad, usually triangular or trapezoidal in shape. Ocelli present, antennae with five segments; tarsi with three segments and arolia. Most species are phytophagous, including economically important pests on rice and crucifers; the subfamily Asopinae are predators, mostly of lepidopterous larvae. All instars produce aromatic repugnatorial secretions released through openings in the thorax (adults) or abdomen (nymphs). Some species are able to spray these secretions several centimeters. Pentatomids include a variety of brightly metallic-colored or grotesquely sculptured members, particularly in the Tropics, where the majority of the approximately 3000 described species occur.

Coreidae (squash bugs, etc., Fig. 34.6d; Color photo 5d). Small to large, robust, elongate bugs with trapezoidal prothorax. Ocelli present, antennae with four segments, inserted dorsolaterally on head, tarsi with three segments and arolia; wing membrane with numerous veins (Fig. 34.6d). All species are phytophagous,

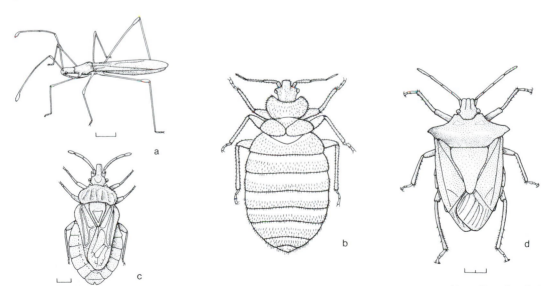

Figure 34.7 Representative Heteroptera: **a**, Berytidae (*Metapterus banksi;* scale equals 2 mm); **b**, Cimicidae (*Cimex lectularius);* **c**, Aradidae (*Aradus proboscideus;* scale equals 1 mm); **d**, Pentatomidae (*Euschistus conspersus;* scale equals 2 mm).

mostly attacking growing shoots. A few, such as *Anasa tristis* in North America, are economically important. Most produce repugnatorial secretions, which some large species are capable of spraying short distances. The subfamilies Alydinae and Rhopalinae are sometimes recognized as separate families. In Rhopalinae the openings of the scent glands are slitlike or concealed. The box elder bug, *Boisea trivittatus,* is a common species in most of North America (Fig. 34.1). In Alydinae the body is narrow and the head is nearly as broad as the prothorax (much broader in the other subfamilies). Approximately 2000 species, worldwide in distribution.

Lygaeidae (seed bugs). Oval or elongate bugs, mostly 2 to 15 mm in length. Ocelli present, antennae with four segments (with few exceptions), inserted ventrolaterally on head; tarsi with three segments and arolia, wing with 4 to 5 veins in membrane (Fig. 34.4c). *Lygaeus, Oncopeltus,* and other large, bright orange or red and black "milkweed bugs" are the most commonly noticed members of this family. The great majority of species are small, usually dull-colored, frequently ground- or litter-dwelling insects. Most feed on seeds, which are grasped by the swollen, toothed forelegs; some species occur in mammal nests, apparently feeding on stored seeds. *Geocoris* is a common predator in low vegetation, and chinch bugs *Blissus,* Fig. 34.6a) feed on stems and roots of grasses, including grain crops. Loss or shortening of wings is frequent in ground-dwelling lygaeids, which may mimic ants. There are over 2000 species, worldwide.

Pyrrhocoridae and **Largidae** (cotton stainers). Oval or elongate bugs, mostly more than 5 mm in length. Ocelli absent, antennae with four segments, tarsi with three segments and arolia; wing with anastomosing veins in membrane (Fig. 34.4d). Most species are bright orange, red, or yellow and black. Feeding is predominantly on fruiting structures of plants, and several species are economically important. Worldwide in distribution, but mostly tropical.

Piesmatidae. Small oval insects resembling Tingidae in the lacelike texture of the dorsal surface, but with the scutellum exposed. Ocelli present, antennae with four segments; tarsi with two segments and pulvilli. Phytophagous, widespread but uncommon.

Berytidae (stilt bugs; Fig. 34.7a). Slender, elongate, parallel-sided bugs, similar to lygaeids in most characteristics. All are primarily phytophagous but apparently require animal food such as insect eggs to mature. Most berytids appear to have broad host ranges, but *Jalysis wickhami* mostly frequents plants with sticky glandular surfaces, and other species are restricted to grasses. About 100 species, widespread.

Aradidae (flat bugs; Fig. 34.7c). Strongly flattened, oval in dorsal silhouette, with rough gray, black, or brown cuticle; mostly 5 to 10 mm long. Ocelli absent, antennae with four segments, usually thick, cylindrical; tarsi with two segments. Aradids feed on wood-rotting fungi, about which large numbers of all instars often congregate, either on the outer surface of trees or beneath loose bark. The feeding stylets are extremely elongate, frequently longer than the entire body, apparently as an adaptation for penetrating and following fungal mycelia. The elongate stylets also allow the feeding insects to remain camouflaged on bark adjoining fungal masses. Abbreviation or absence of wings is common in aradids, either in both sexes or in females only.

Tingidae (lace bugs). Small, dorsally flattened species with raised, reticulate venation (Fig. 34.5d); usually pale-colored. Ocelli absent, antennae with four segments, tarsi with two segments; pronotum usually covering scutellum, often crested or hood-shaped. Tingids are gregarious, phytophagous bugs that usually feed on undersides of leaves. Honeydew is secreted, and some species are tended by ants; a few form galls. Nymphs of tingids are often strikingly different from adults, with spines or other outgrowths from the dorsum. Over 750 species, widely distributed, with about 150 in North America.

Reduviidae (assassin bugs; Fig. 34.6b; Color photo 6b). Small to large; ocelli present (rarely absent); antennae with four or six to eight segments; rostrum with three segments, fitting into intercoxal stridulatory groove on prosternum; tarsi usually with three segments. Reduviidae is one of the largest families of Heteroptera, with over 3000 described species. These show an enormous variation in body form, size, color, and cuticular sculpturing, ranging from the gnatlike Emesinae to squat, robust types such as *Apiomerus.* Nearly all reduviids are predatory, but *Triatoma, Rhodnius* (subfamily Triatominae, Fig. 12.1d), and a few others are ectoparasites of mammals and birds and vectors of *Trypanosoma cruzi,* which causes Chagas' disease in humans. The forelegs are raptorial in all predaceous reduviids and may be greatly enlarged and specialized, as in *Phymata* (Fig. 34.2b), which ambushes prey from hiding places

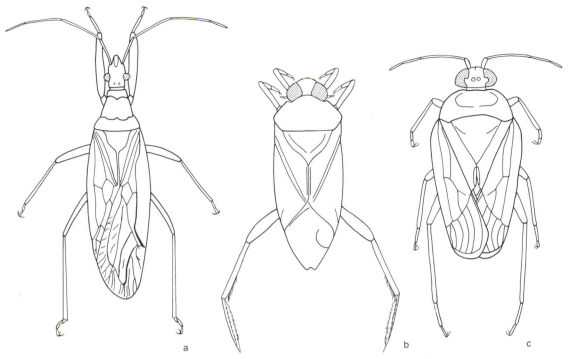

Figure 34.8   Representative Heteroptera: **a**, Nabidae; **b**, Notonectidae; **c**, Saldidae.

on flowers. Several of the larger reduviids, sometimes called kissing bugs, freely deliver painful bites if not handled gingerly.

Enicocephalidae. Small, reduvioid bugs with head enlarged, globose behind eyes, antennae and rostrum with four segments, tarsi with one to two segments; forewings entirely membranous. Most species frequent damp leaf litter or other debris and are predaceous. Adults form dense aerial swarms at dusk in the autumn. Widespread but uncommon.

Miridae (plant bugs; Fig. 34.6c). Delicate, oval or elongate bugs, mostly 2 to 10 mm long. Ocelli absent, antennae and rostrum with four segments; tarsi with three segments; forewings with distinct corium, clavus, and (usually) a large cuneus (Fig. 34.5a). Mostly phytophagous, feeding on nearly all higher plants, with a wide range in host specificity. *Lygus* is among a number of economically important species, feeding on many crops. Many mirids are predaceous, feeding on mites, aphids, and other small arthropods. Unlike most Hemiptera, many mirids overwinter as eggs. Over 5000 species, distributed worldwide.

Anthocoridae (minute pirate bugs). Compact, short-legged, 1 to 5 mm long. Ocelli present, antennae with four segments, rostrum with three segments; tarsi with three segments; forewings with incomplete cuneus (Fig. 34.5b). Most species are predators of foliage-inhabiting insects, but others occur on bark, in leaf litter, on fungus, and even in stored products, where they attack mites and small insects or insect eggs. Over 300 species, widespread.

Microphysidae. Similar in appearance to Anthocoridae, but tarsi with two segments, rostrum with four segments. Most species, including the single North American representative, *Mallochiola gagates,* inhabit moss and lichen on or about dead trees, and are probably predators. *Embiophila* (sometimes placed in the family Plokiophilidae) is a predator of Embiidina, whose tunnels it infests, and related species probably attack spiders in their webs. Predominantly an Old World family.

Nabidae (damsel bugs; Fig. 34.8a). Superficially similar to Reduviidae, but without prosternal groove for rostrum. Ocelli present or absent, antennae with four or five segments, rostrum with four segments. These predaceous bugs commonly inhabit vegetation and may be of some importance in controlling crop pests. Fertilization in some genera is directly into the hemocoel, as in Cimicidae. About 400 species, widely distributed.

Cimicidae (bedbugs; Fig. 34.7b). Oval, flattened ectoparasites, with forewings always reduced to short scalelike flaps. Ocelli absent, rostrum retractable into ventral groove, tarsi with three segments. Unlike lice, bedbugs contact their hosts only for periodic feeding, spending most of their time in nest material or adjoining cracks or crevices. Most are parasites of birds, some attack cave-dwelling bats, and one species, *Cimex lectularius*, is associated with humans. Fertilization is by a spermatophore, forcibly injected by the male, through a modified area of the female abdominal body wall (Carayon, 1966), a process termed **traumatic insemination.**

Polyctenidae. Highly specialized parasites of bats, mostly in tropical and subtropical regions. These bugs, which infest the bodies of their hosts, are highly modified for moving through fur (Fig. 12.1c). All species are viviparous, with only two nymphal instars. Two species occur in southwestern United States.

Thaumastocoridae. Small flattened bugs with broad head, eyes frequently on short, lateral stalks; tibiae usually with membranous apical appendage. Predominantly a tropical family. New World species appear to be restricted to palms, feeding on the unfolding leaves. The single U.S. species, *Xylastodoris luteolus*, occurs in Florida.

Dipsocoridae and **Schizopteridae.** Oval, predaceous bugs 1 to 1.5 mm long; recognized by the enlarged basal antennal segments. Most species occur in leaf litter, moss, and lichen. Dipsocoridae are probably predaceous; feeding habits of Schizopteridae are unknown. Members of both families jump as well as run when disturbed. Widely distributed but uncommon.

Saldidae (shore bugs; Fig. 34.8c). Oval, with broad head and large, prominent eyes. Ocelli present; rostrum with three segments, projecting below head, even in repose; tarsi with two segments; forewing with four or five long, parallel-sided cells in membrane (Fig. 34.5c). Active predatory bugs capable of rapid running, jumping, or flying. Saldids are widely distributed in wet habitats such as marshes and stream margins, including saline situations, but may sometimes be found far from water, especially during dry seasons. *Aepophilus* (Europe) and *Omania* (tropical) are intertidal.

Leptopodidae. These insects are native to tropical and subtropical regions of the Eastern Hemisphere, where they mostly inhabit dry areas along streams. One species, *Patapius spinosus*, introduced to California, occurs under debris in arid regions.

Gerridae (water striders). Slender, elongate insects specialized for skating on the surface film of still or moving water, using the greatly elongate middle and hind legs; rostrum with four segments. Gerrids are most abundant on still or slowly moving water, where they scavenge and attack small arthropods. *Halobates* occurs in marine situations, including the open seas at great distances from land. Other genera inhabit estuaries or other saline situations.

Veliidae. Similar in appearance and habits to Gerridae but with three-segmented rostrum and shorter middle and hind legs. *Velia* commonly frequents margins of quiet pools in streams; *Rhagovelia* (Fig. 34.9) inhabits riffles in fast-flowing streams, using the fanlike hemicircle of setae on the hind tarsi to swim against the current.

Mesoveliidae. Similar to Veliidae but with claws inserted apically. Mesoveliids are predators and scavengers frequenting seeps, margins of quiet pools along streams, or ponds. In this family, as well as the Veliidae and Gerridae, wing polymorphism is common, and wingless, short-winged, or fully winged individuals may be encountered in the same population.

Macroveliidae. A small family confined to the New World. A single species occurs in western North America, commonly about the margins of springs, seeps, and streams, and sometimes in cavities beneath logs or large stones or in other protected sites. Unlike Veliidae and Mesoveliidae, these bugs are incapable of skating on the surface film of water.

Hebridae. Small, stout-bodied, predaceous bugs; antennae with four or five segments, ocelli present, rostrum with three visible segments, retractable into groove between coxae. Hebrids are covered with hydrofuge pile that allows them to crawl beneath the water surface without wetting their cuticle. Widespread in marshes and at stream margins but uncommon.

Hydrometridae. Extremely elongate, linear bugs with threadlike legs and antennae. Hydrometrids are sluggish, slow-moving predators, inhabiting quiet pools, marshes, and swamps, mostly in tropical regions.

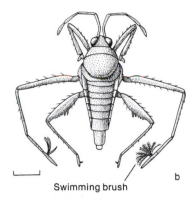

Swimming brush

Figure 34.9 Veliidae *(Rhagovelia distincta):* **a**, winged form; **b**, apterous form. Scale equals 1 mm.

Notonectidae (backswimmers; Figs. 9.6d, 34.8b). Stout, wedge-shaped bugs that swim on their backs, using the oarlike hind legs. Ocelli absent, antennae with four segments. Notonectids are free-swimming predators in open water or small pools or along stream and lake margins. Their bite is painful.

Pleidae. These insects are distinguished from notonectids by having three antennal segments, and by the partial fusion of the head and thorax. Both families are widely distributed, but pleids are much less commonly encountered.

Corixidae (water boatmen; Fig. 34.10c). Flattened, streamlined swimming bugs with fore and hind legs oarlike, without tarsal claws. Rostrum short, concealed; antennae with three or four segments. Corixidae occur in almost all still water, including salt ponds

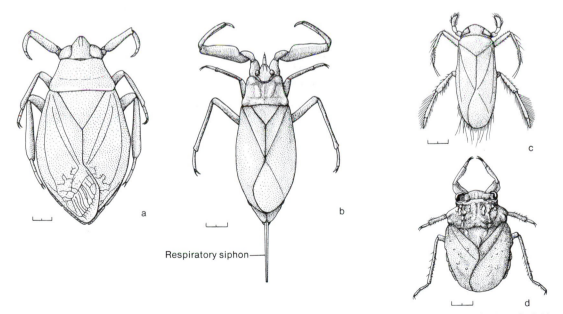

Respiratory siphon

Figure 34.10 Representative Heteroptera: **a**, Belostomatidae *(Belostoma flumineum);* **b**, Nepidae *(Nepa apiculata);* **c**, Corixidae *(Hesperocorixa laevigata);* **d**, Gelastocoridae *(Gelastocoris oculatus).* Scales equal 2 mm.

and brackish pools; they are less abundant in streams. Corixids include a few predators, but most use the paddlelike forelegs to scoop up bottom detritus containing algae, diatoms, and other microorganisms, from which they suck the protoplasm.

Nepidae (waterscorpions; Fig. 34.10b). Elongate, cylindrical, or flattened with long, slender legs; anterior legs raptorial; posterior pairs adapted for walking. Antennae with three segments, tarsi with one segment, without claws on anterior legs. Nepids are sluggish predators that ambush prey from camouflaged positions on aquatic vegetation or debris, and are only indifferent swimmers. They are widespread, but more abundant in tropical regions.

Belostomatidae (giant water bugs; Fig. 34.10a). Broad, flattened, oval bugs with the middle and hind legs flattened, fringed for swimming, the anterior pair raptorial; antennae with four segments. Belostomatids are moderate-sized to large (to about 100 mm long) free-swimming predators that kill small vertebrates as well as other arthropods and bite fiercely if handled freely. Females of *Belostoma* and *Abedus* have the peculiar habit of attaching their eggs to the backs of males, where they are carried until hatching. Widely distributed.

Naucoridae. Similar to belostomatids in general appearance, but more rounded and streamlined in dorsal aspect; anterior femora extremely thick; middle and posterior legs cylindrical, spiny. Naucorids frequent vegetation, bottom debris, undersides of stones, and similar concealed situations, in both running and still water, where they prey on a variety of small arthropods. Respiration is by means of a gas bubble, periodically renewed at the surface, or in some species by a plastron, which occurs in no other Hemiptera and enables these bugs to remain permanently submerged. Mostly tropical in distribution. North American species are mostly in the southwest.

Gelastocoridae (toad bugs; Fig. 34.10d). Squat, stout-bodied, with prominent, bulging eyes and raptorial forelegs. Gelastocorids are commonly associated with sandy or muddy banks. They actively search for prey, which they capture by sudden, short hops. As in damselflies, the widely spaced, hemispherical eyes allow vision in all directions, and probably provide binocular depth perception.

Ochteridae. Similar to Gelastocoridae, but with longer antennae, long, four-segmented rostrum, and anterior legs similar to posterior pairs. Ochterids are widespread but infrequently encountered. They occupy the same situations as saldids and gelastocorids.

SUBORDER HOMOPTERA (LEAFHOPPERS, PLANTHOPPERS, etc.)

Head deflexed, with rostrum closely appressed to prosternal region, its base extending between anterior coxae; antennae usually with 2 to 10 segments; prothorax usually small (exception: large in Membracidae), transverse or quadrangular; tarsi with 3 or fewer segments.

All Homoptera are phytophagous, but within the role of terrestrial herbivore they have become exceedingly diversified, both in body form and in details of life history.

Most species can be grouped into one of two broad categories on the basis of morphological and biological features. The superfamilies Cicadoidea, Fulgoroidea, Cercopoidea, and Cicadelloidea comprise the series[1] **Auchenorrhyncha.**[2] In this group the antennae are very short and bristlelike, the rostrum is clearly articulated with the head, and the ovipositor is bladelike, for inserting the eggs into plant tissue. The Auchenorrhyncha are mostly active insects that live exposed on plant surfaces and jump to escape predators. Exceptions include cicadas, which cannot jump, but readily fly as adults, and nymphal cercopoids, which live immersed in a self-produced coating of froth as immatures and have practically lost the ability to jump.

The Auchenorrhyncha have simple life cycles. Immatures generally resemble adults, except for the absence of wings, although nymphs of cicadas are specialized for subterranean life and are superficially quite different from adults. Nearly all Auchenorrhyncha are bisexual, and they never have alternating generations, as in the Sternorrhyncha, discussed below. Males of many species produce sound from a pair of tymbals in the base of the abdomen. In cicadas the tymbals are associated

[1] A series is a taxonomic category used in the Hemiptera for classifying groups between a suborder and a superfamily.
[2] Some molecular evidence (Sorensen et al., 1995) suggests that Auchenorrhyncha is not monophyletic, but other evidence (W. C. Wheeler et al., 1993) is contradictory. The classification applied here follows that of Carver et al. (1991).

with a resonating chamber. The chorusing noise produced by large aggregations of individuals may be almost deafening and probably functions to confuse birds and other predators, as well as in courtship. Sounds produced by cicadellids are too feeble to be heard by humans without amplification; their function in most species is apparently in courtship.

The superfamilies Psylloidea, Aphidoidea, Aleyrodoidea, and Coccoidea constitute the series **Sternorrhyncha.** These Homoptera have relatively long, filamentous antennae, not bristlelike as in the Auchenorrhyncha. The rostrum is inserted on the extreme posteroventral part of the head and is usually closely appressed to the thoracic sterna, from which it may appear to arise. Tarsi are one- or two-segmented. The ovipositor is usually undeveloped, and eggs are deposited on plant surfaces. Adult psyllids and aleyrodids are active jumping or flying insects, but the Sternorrhyncha are predominantly nonvagile, specialized for rapid feeding and reproduction. Parthenogenesis is widespread, and in Aphidoidea may be alternated with sexual reproduction at certain times of year. The development of complex life cycles in these groups is discussed in greater detail further on. Many Sternorrhyncha are notable for the tendency of the immature and adult stages to diverge morphologically. In aleyrodids and male coccoids the degree of differentiation is equivalent to that in endopterygote insects. The nonmotile "larvae," or neanides, with greatly reduced appendages and sense organs, are essentially a feeding stage. The motile adult is preceded by a nonfeeding "pupal" stage, and physiologically these insects are holometabolous.

Many Sternorrhyncha form commensal relationships with ants, which are attracted to their copious exudates of honeydew and to some extent protect the homopterans from predators and move them about their host plants. Ant-tended Homoptera often lack special protective devices such as heavy wax coats or sclerotized cuticle. They exude honeydew, rather than ejecting it, as do many free-living forms, and some ant-tended species have special devices for holding droplets of honeydew until it is taken by ants.

Ant-tended Homoptera tend to live in clusters. Alarm pheromones cause them to aggregate rather than disperse, and they allow themselves to be carried about by ants without struggling. Such relationships are most specialized with some aphids. For example, the corn root aphid, *Anuraphis maidiradicis,* is tended by ants (chiefly *Lasius*), in all instars. The eggs are collected in the fall and overwintered in the ant nests. The first-instar aphids are carried to the spring host (smartweed). Later generations are transferred to the summer host (corn). In many sternorrhynchous Homoptera, dermal glands secrete a protective cover of wax, either in the form of characteristic filaments or plates, or as a powdery deposit. In some Coccoidea the eggs are enmeshed in a flocculent envelope of wax filaments, and wax is a constituent of the scale of diaspidids and other armored scales.

Aphidoidea and Coccoidea require slide mounting or other special preparation for reliable identification to family. For that reason, families of these groups do not appear in the following key. Nearly all the commonly encountered Aphidoidea are members of the family Aphididae. Most Coccoidea encountered by nonspecialists are members of Diaspididae (armored scales) and Pseudococcidae (mealy bugs).

KEY TO SUPERFAMILES AND MAJOR DIVISIONS OF THE SUBORDER

Homoptera

ADULTS

1. Tarsi with 1 or 2 segments, or legs absent (series **Sternorrhyncha**)   **2**

   Tarsi with 3 segments (series **Auchenorrhyncha**)   **5**

2(1). Tarsi with 1 segment and 1 claw, or legs vestigial or absent; females larviform or gall-like or scalelike (Fig. 34.16c–e), males with mesothoracic wings and atrophied mouthparts (Fig. 34.16b)   **Coccoidea** (p. 433)

   Tarsi with 2 segments and 2 claws; body insectlike, with recognizable legs, antennae, and mouthparts; wings 4 in number, if present   **3**

3(2). Tarsal segments subequal in size and shape; antennae with 7 to 10 segments   **4**

Tarsi with first segment small, triangular, or absent; antennae with 1 to 6, usually with 5 or 6 segments **Aphidoidea** (p. 431)

4(3). Antennae with 7 segments; wings membranous, without sclerotized leading edge (Fig. 34.14e); body powdered with white wax
Aleyrodoidea (Aleyrodidae; p. 431)
Antennae usually with 10 segments; wings with sclerotized leading edge; body not powdered with wax **Psylloidea (Psyllidae**; p. 431)

5(1). Antennae inserted on frons, between eyes (Fig. 34.13a,b); pedicel subequal to scape in diameter; mesocoxae short, contiguous or close together **6**
Antennae inserted laterally, beneath eyes (Fig. 34.13c,d); pedicel enlarged, often globular; mesocoxae elongate, distant
Fulgoroidea (p. 429)

6(5). Head with 3 ocelli; anterior femora thickened; body more than 15 mm long
Cicadoidea (Cicadidae, p. 428)
Head with 2 ocelli, or ocelli absent; anterior femora slender; body rarely more than 10 mm long **7**

7(6). Metatibiae with 1 or 2 large spurs on shaft (Fig. 34.11a) metacoxae short, conical
Cercopoidea (Cercopidae, p. 430)
Metatibiae with numerous small spines on shaft (Fig. 34.11b); metacoxae transverse, platelike
(Cicadelloidea; p. 430) **8**

8(7). Scutellum exposed as a triangular sclerite (Fig. 34.13b) **9**
Scutellum hidden by pronotum, which extends backward usually about half the length of wings **10**

9(8). Hind tibia with one or several rows of small spines **Cicadellidae** (p. 431)
Hind tibia with one or two spines on shaft and circlet of spines at apex, or lacking spines
Aetalionidae (p. 431)

10(9). Pronotum extended backward as a narrow, straplike process that exposes the scutellum laterally **Aetalionidae** (p. 431)
Pronotum extended broadly backward to completely conceal the scutellum (Fig. 34.13a) **Membracidae** (p. 431)

Superfamily **Cicadoidea** (cicadas; Fig. 34.12; Color photo 2a). A single North American family, **Cicadidae,** of moderate to very large insects with membranous, transparent wings, sometimes banded or spotted.

Adult cicadas are familiar because of the high-volume sounds produced by the males (see Chap. 6; Fig. 6.8a–c), but because of the ventriloqual quality of their calls, they are never conspicuous. Most species are arboreal, and they are often gregarious. Females slash twigs with their ovipositors to insert their eggs, a habit that sometimes becomes economically damaging. Newly eclosed nymphs drop to the soil, where they lead subterranean lives feeding on roots. Most species require several years to mature, and often all the members of a single generation appear simultaneously. Usually, a generation matures each year, but in the "periodical cicadas" *(Magicicada)* of eastern North America, generations are restricted to certain years at a specific location. These species require either 13 or 17 years to mature, and consequently may unexpectedly appear after several years' absence.

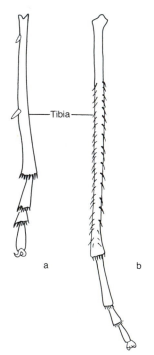

Figure 34.11 Metatibiae of Homoptera showing spurs and spines: **a,** Cercopidae; **b,** Cicadellidae.

Figure 34.12   A cicada, *Platypedia* sp. (Cicadidae) from California. Immature cicadas develop underground, sucking sap of roots. Adults of this genus audibly tap twigs with the edge of their wings. Males of other species make a loud, whining noise with special organs (tymbals) at the base of their abdomen.

At most localities the individuals emerging in any given year consist of two or three morphologically similar species with synchronized life cycles. The ecological and evolutionary significance of these synchronized mass emergences restricted to certain years is not entirely understood, but they may be a means of alternately "swamping" and starving potential predators. The large family Cicadidae is distributed throughout the world but is especially abundant and diverse in the Tropics and Subtropics. The Tettigarctidae, in which both sexes possess the apparatus for sound production, has two species endemic to Australia, but is known from the Northern Hemisphere as fossils.

Superfamily **Fulgoroidea** (fulgorids, planthoppers; Fig. 34.13c,d). Clypeus separated from frons by distinct frontoclypeal ridge, not extending posteriorly between eyes; antennae usually with second segment enlarged, with conspicuous sensory organs; 2 ocelli (occasionally 3) on sides of head near eyes; tegulae present on mesothorax.

Fulgoroidea range from small to large in size, and occur in a diverse and sometimes bizarre array of body forms. Many species (Cixiidae, Delphacidae,

Achilidae) superficially resemble Cicadellidae. Others have broad, mothlike wings (Flatidae) or have the wings reduced and strap shaped (Dictyopharidae, Issidae). Exaggeration of the proportions of various parts of the body, including head, wings, and legs, occurs in various species. The functions of such modifications are obscure.

Details of the biology and life history are unknown except for a very few fulgoroids. Most are phloem feeders, but some (Achilidae, Derbidae) feed on fungus. Most species probably spend their entire lives on foliage, but most Cixiidae apparently live underground as nymphs, feeding on grass roots, or in ant nests. Fulgoroids are primarily a tropical group, and most of the relatively few economically important species occur on sugarcane, coffee, and similar crops.

Recognition of fulgoroid families is rendered difficult by the great structural variation within many families, together with a repeated tendency toward convergence in external appearance, especially in the numerous flightless species from different families. About 20 families, many obscure or principally tropical, are recognized by most

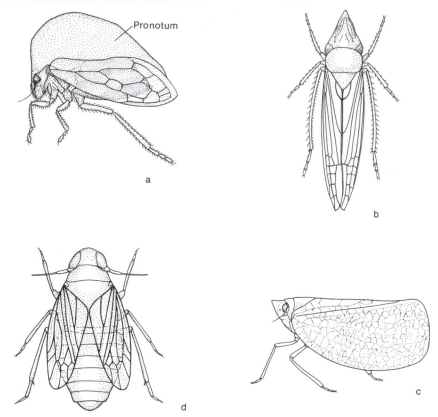

Figure 34.13   Representative Homoptera: **a**, Membracidae *(Stictocephala inermis);* **b**, Cicadellidae *(Draeculacephala mollipes);* **c, d**, Fulgoroidea; **c**, *Acanalonia bivittata* (Acanaloniidae); **d**, *Delphacodes campestris* (Delphacidae). (a, modified from Yothers, 1934; b–d, modified from Osborn, 1938.)

authorities. About half of these occur in North America. The interested student is referred to Kramer (1950) and Nault and Rodriguez (1985) for more complete discussions of the superfamily.

Superfamily **Cercopoidea** (spittlebugs or froghoppers). Adults superficially similar to Cicadellidae; best distinguished by the tibial characters given in the key.

The soft, whitish nymphs live within frothy masses of "spittle," which is produced by blowing air through a viscous mucosaccharide secreted by the Malpighian tubules and released through the anus. Air is expelled by contracting a cavity formed by the ventral extension of the abdominal tergites beneath the abdomen. The spittle serves both as a protective device and as a means of reducing

evaporation (Marshall, 1966). Nymphs of some Australian species construct hardened tubes or shells from the anal secretion, attaching these to twigs or branches. The tube in turn is filled with spittle. All instars are phloem feeders. A single family, **Cercopidae,** is usually recognized. Cercopids are cosmopolitan, but most species are tropical.

Superfamily **Cicadelloidea** (leafhoppers and treehoppers; Fig. 34.13a,b). Distinguished by the tibial and coxal characters listed in the key. Cicadelloids also differ from other superfamilies in venational characters and in having the tentorium reduced.

Cicadelloids feed in the phloem, xylem, or parenchymal tissues, almost always of higher plants.

Many species are cryptically colored, and some, especially membracids, are structurally modified to resemble thorns or other plant parts (Color photo 7e). Membracid nymphs are often gregarious and may be tended by ants. The great majority of Cicadelloids have simple life cycles with many generations per year. Cicadellidae use a wide array of host plants, including herbs, grasses, shrubs, and trees. In contrast, **Membracidae** and **Aetalionidae** feed mostly on woody vegetation. Adults, especially of Cicadellidae, are highly mobile and may migrate long distances, particularly when local food sources (such as ephemeral native plants) wither. In large part because of their tendency to move among hosts, cicadellids are extremely important as transmitters of plant diseases, especially viruses. The dominant family is **Cicadellidae,** with about 9000 species worldwide. Membracids include about 2200 species and Aetalionidae only about 50, with 3 species in North America.

Superfamily **Psylloidea** (psyllids; Fig. 34.14a,b). Minute to small Homoptera with long, bristlelike antennae, 3 ocelli; forewings with venation reduced by fusion of R, M, and Cu veins; metacoxae enlarged, elongate.

Psyllids often bear a striking resemblance to minute cicadas. As adults they are highly active, readily jumping and flying if disturbed. Nymphs are sluggish and are morphologically rather distinct from adults, with squat, flattened bodies. Many Australian species produce protective polysaccharide shells or **lerps**; North American species may be covered with honeydew, but they never produce hardened coverings. They mature on foliage, usually of woody plants. A few species cause the growth of galls. The life cycle is simple, with one to several bisexual generations per year. Psyllids are of considerable importance to specific crops because of disease transmission or toxic effects of the saliva which is injected into the hosts. **Psyllidae,** the single family, is cosmopolitan.

Superfamily **Aleyrodoidea** (whiteflies; Fig. 34.14c–e). A single family, **Aleyrodidae.** Adults similar in general size and body form to Psyllidae; antennae long, bristlelike; wings opaque with dusting of white wax, venation greatly reduced, with 2 or 3 longitudinal veins, no clavus; all instars with characteristic **vasiform orifice** on last abdominal tergite (Fig. 34.14d).

These minute insects are phloem feeders, usually on the undersides of the leaves of angiosperms. A few are known from ferns, none from gymnosperms. Eggs, attached by short pedicels to the undersurfaces of leaves, are frequently arranged in arcs or circles. First instars are flattened, ovoid, with short legs. The second and third instars (larvae) are sessile, with atrophied legs. The fourth instar ("pupa") is similar to the larva but feeds only briefly, then becomes inactive. Before the final molt, the developing adult organs become visible through the pupal cuticle. The pupal stage is frequently ornamented by wax plates or filaments in highly individualized patterns, and the taxonomy of Aleyrodidae is based mainly on these structures. Reproduction may be either sexual or asexual; in some species unfertilized eggs produce males, but parthenogenetic races of only females are known. Aleyrodids are primarily tropical, and the most economically important species in North America attack citrus or greenhouse plants.

Superfamily **Aphidoidea** (aphids; plant lice; Fig. 34.15). Soft-bodied, globular or flattened insects with abdominal segmentation indistinct; either winged or apterous as adults, forewings with 3 to 5 longitudinal veins and distinct clavus; cornicles usually present but sometimes reduced, cone-shaped or ring-shaped.

The Aphidoidea are remarkable for the development of highly complex life cycles, which may involve cyclical parthenogenesis, cyclomorphosis (seasonal change in body form), and alternation of hosts. These are apparently adaptations to survive in seasonal climates, and in highly temperate areas all three phenomena may occur in a single species. For example, in northern states, the woolly apple aphid (*Eriosoma lanigerum,* Fig. 34.15) overwinters as eggs on elm trees, where one or more spring generations of wingless, individuals mature. In early summer, winged individuals appear and migrate to apple trees and related plants such as

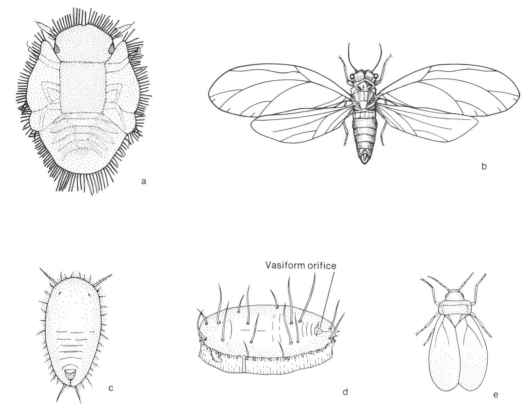

Figure 34.14 Psyllidae and Aleyrodidae: **a**, **b**, Psyllidae, nymph and adult, respectively *(Paratriazoa cockerelli).*; **c–e**, Aleyrodidae, nymph, pupa, and adult, respectively *(Trialeurodes vaporariorum).* (*a, b,* modified from Knowlton and Janes, 1931; c–e, modified from Lloyd, 1922.)

hawthorn, where an indefinite number of wingless generations follows. All these spring and summer generations are parthenogenetic and viviparous. In autumn, winged individuals migrate back to elm trees and parthenogenetically produce males and females, which reproduce sexually, each female laying a single egg. The annual cycle of *Rhopalosiphum padi* is illustrated in Figure 4.9.

Many species with alternating generations occupy hosts that are ephemeral annuals or otherwise unsuitable during part of the year. Production of sexual forms is predominantly controlled by changes in temperature and photoperiod, and under subtropical or artificial conditions sexual forms may never appear.

Additional morphs occur in certain species. In

Periphyllus californiensis, for example, the first-instar nymph exhibits specializations for oversummering: it is flattened, waxy, and hirsute. Several aphids of the subfamilies Hormaphidinae and Pemphiginae produce **soldier morphs,** which defend the colonies by gripping predators with their enlarged, raptorial forelegs and piercing them with the sharply pointed frontal horns or the feeding stylets. In some cases, constituting the only known instance of altruistic social behavior in Hemiptera, the soldiers do not molt or reproduce.

The influence and interaction of other factors involved in control of aphid polymorphism are discussed in greater detail by Hille Ris Lambers (1966) and Moran (1992).

Taxonomy of Aphidoidea is unsettled, with

one to nine families recognized by different specialists. The more important differences among families involve features of the life cycle, rather than morphological characters. **Aphididae** is the dominant family, with over 3000 species, the great majority from north temperate areas and very few from the Tropics. Two other small families are commonly recognized: **Adelgidae** cause conelike galls on conifers, with at least one generation occurring on spruce *(Picea);* **Phylloxeridae** occur on oak *(Quercus),* grape *(Vitus),* walnut *(Juglans),* and other plants, usually forming galls on at least one part of the plant. All but a few species have a single host. In the latter two families the cornicles are reduced or vestigial. Aphidoidea are of exceptional economic importance. Besides the direct effects caused by sucking plant sap, many species damage their hosts by injecting toxic saliva or transmitting diseases, especially viruses. For example, *Myzus persicae,* the peach aphid, vectors at least 100 different viruses.

Superfamily **Coccoidea** (scale insects; mealybugs; Fig. 34.16). Minute to small, sexually dimorphic Homoptera; mature females with ovoid to globular body, frequently with segmentation indistinct, eyes small, arranged in groups of simple lenses; antennae 1- to 9-segmented or vestigial; legs segmented, with single tarsal claw, or atrophied; rostrum 1- to 2-segmented, short; stylets extremely long, coiled in body; body frequently protected by waxy, cottony, or hard scalelike covering. Mature male usually gnatlike, with long,

multiarticulate antennae, lateral eyes, enlarged mesothorax bearing 1 pair of wings with 1 or 2 longitudinal veins; hind wings reduced to halteres.

Morphologically Coccoidea are among the most highly modified insects, being specialized for a sessile or nearly sessile life attached to their host plants. The first instars, or crawlers (Fig. 34.16a), are always motile and function in dispersal. **Pseudococcidae** (Fig. 34.16e) (mealybugs) possess functional legs in all instars, but in other families of scale insects at least one instar is nonmotile. The crawlers can survive up to several days without feeding, during which period surprisingly large distances may be traversed. If a spot suitable for feeding is located, the first instar inserts its mouthparts and, depending on the family, becomes sessile or nearly so after the first molt, with atrophied legs, antennae, and eyes. At this time the insect begins secreting the protective covering, or scale, which, in different species, varies enormously in hardness, texture, and shape. A few species produce no scale, protected instead by greatly thickened cuticle. Males pass through one more instar than females, which are apparently neotenic. The last two larval instars in males display external wing pads and other adult features, much as in endopterygote pupae, and the last instar does not feed. Adult males do not feed and usually survive only a few days. Reproduction is oviparous or viviparous. Oviparous forms usually cover the eggs with waxen oothecae or retain them beneath

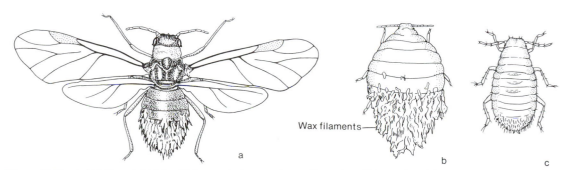

Figure 34.15   Aphidoidea, Aphididae (*Eriosoma lanigerum,* the woolly apple aphid). **a**, winged female; **b**, apterous female; **c**, male (Modified from Baker, 1915.)

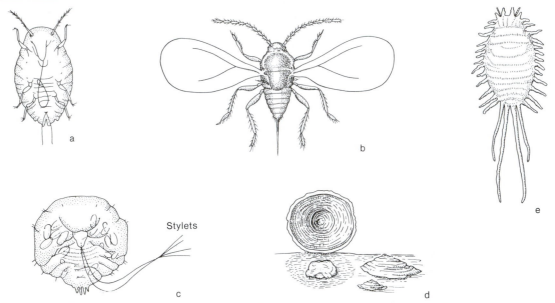

Figure 34.16 Coccoidea: **a–d**, various instars of the San Jose Scale, *Quadraspidiotus perniciosus* (Diaspididae); **a**, first-instar nymph or crawler; **b**, adult male; **c**, adult female, ventral, extracted from scale revealing eggs inside body; **d**, scales; **e**, *Pseudococcus longispinus* (Pseudococcidae). (a–d, modified from U.S. Department of Agriculture; e, modified from Green, 1896–1922.)

the mother's scale or body. Parthenogenesis occurs in various families, but its extent is uncertain. Some Margarodidae are hermaphroditic.

Various scale insects have one or more instars specialized beyond the description given above. In **Margarodidae,** intermediate instars of females are legless "cysts," capable of withstanding periods of up to several years without food or water. Some members of the **Eriococcidae** form galls, within which the insects mature. In some species the male and female galls are strongly differentiated, while in others the male remains within the parent gall. First instars of some species, such as the coccid *Walkeriana,* bear elongate filaments enabling them to travel on breezes, in the manner of a dandelion seed, and first-instar females of *Cystococcus* (Eriococcidae) cling to adult males as a means of dispersal.

It should be pointed out that although many coccoids are serious economic pests, several useful products, such as cochineal and other dyes, lac (used in shellac), and wax, have been derived from

certain species, and *Dactylopius* (Eriococcidae) has been used in biological control of prickly pear cactus *(Opuntia)*.

Many species of armored scales **(Diaspididae)** may be recognized by the characteristic scales they produce (Fig. 34.16d), but family and species classification and identification of most Coccoidea are extremely difficult, requiring removal of the scale, clearing, staining, and mounting on slides. About a dozen families are recognized by most authorities. For identification of the common economic species, the reader is referred to Metcalf and Metcalf (1993). The most complete taxonomic treatments of North American species are those of Ferris (1937–1955) and McKenzie (1967).

SELECTED REFERENCES

General

Blackman, 1974 (biology of aphids); Dolling, 1991; J. W. Evans, 1982 (Peloridioidea); Kosztarab, 1987

(Coccoidea), 1996 (Coccidae, northeastern North America); Miller and Kosztarab, 1979 (Coccoidea); Minks and Harrewijn, 1987 (aphids); Myers, 1929 (biology of cicadas); Nault and Rodriguez, 1985 (leafhoppers and planthoppers); Poisson and Pesson, 1951; Rosen, 1990 (armored scales); Schaefer, 1996; Schuh and Slater, 1995; Southwood and Leston, 1959 (Heteroptera); Van Emden, 1972.

Morphology

Ekblom, 1926 (Heteroptera); Hamilton, 1981 (head); Kramer, 1950 (Homoptera); Poisson and Pesson, 1951; Spooner, 1938 (head capsule); M. L. Williams and Kosztarab, 1972 (Coccoidea).

Biology and Ecology

Aldrich, 1988 (chemical ecology of Heteroptera); Alexander and Moore, 1962 (periodical cicadas); Baumann et al., 1995 (symbionts of aphids); Beardsley and Gonzalez, 1975 (armored scales); Blackman, 1974 (aphids); Blackman and Eastop, 1994 (tree-dwelling aphids); Buckley, 1987 (mutualism between Homoptera and ants); Byrne and Bellows, 1991 (whitefly biology); Cobben, 1978 (mouthpart structures and feeding strategies); Claridge, 1985a,b (sound production in Homoptera); Delong, 1971 (Cicadellidae); Dixon, 1985 (aphids); Hamilton, 1983 (biology and systematics of leafhoppers and planthoppers); Hille Ris Lambers, 1966 (aphid polymorphism); Houk and Griffiths, 1980 (intracellular symbionts); Kennedy and Stroyan, 1959 (aphids); McGavin and Preston-Mafham, 1993 (Heteroptera); N. C. E. Miller, 1956 (Heteroptera); Moran, 1992 (aphid life cycles); Mukhopadhyay, 1993 (chemical ecology); Myers, 1929 (cicadas); Ossiannilsson, 1949 (sound production in Homoptera); Pickett et al., 1992 (chemical ecology of aphids); Staddon, 1986 (scent glands); Way, 1963 (mutualism between Homoptera and ants); Stern and Foster, 1996 (soldier morph in aphids); K. S. Williams and Simon, 1995 (periodical cicadas); Wilson, 1971 (mutualism between Homoptera and ants).

Systematics and Evolution

Andersen and Weir, 1994 (evolution of sea skaters); Ben-Dov, 1993 (Coccoidea); Blatchley, 1926 (eastern North American species); W. E. Britton et al., 1923 (Connecticut species); A. R. Brooks and Kelton, 1967 (Canadian species); Campbell et al., 1994 (DNA and evolutionary origin of whiteflies); Campbell et al., 1995 (paraphyly of Homoptera); Cobben, 1968 (systematic and evolutionary importance of eggs), 1978 (mouthparts and feeding strategies); Cox, 1987 (Pseudococcidae); DeCoursey, 1971 (nymphs); J. W. Evans, 1963 (phylogeny of Homoptera); Ferris, 1937–1955 (scale insects of North America); Froeschner, 1960 (Cydnidae); Herring and Ashlock, 1971 (nymphs); Henry and Froeschner, 1988 (catalogue, North America); Hodgson, 1994 (genera of Coccidae); Hungerford, 1959 (aquatic species); Kosztarab and Kosztarab, 1988 (Coccoidea); Kosztarab et al., 1986 (Coccoidea); Kramer, 1950 (phylogeny of Homoptera); McKenzie, 1967 (Coccoidea of California); Merritt and Cummins, 1996 (Aquatic and semiaquatic forms); Nur, 1980 (chromosome systems in Coccoidea); Polhemus, 1996 (aquatic forms); Slater and Baranowski, 1978; C. F. Smith et al., 1992 (aphids of North Carolina); Sorensen et al., 1995 (cladistic relationships of Homoptera); Stannard, 1956 (phylogenetic relationships); Stehr, 1987 (immatures); Usinger, 1956b (aquatic species of California); W. C. Wheeler et al., 1993 (cladistic relationships); M. L. Williams and Kosztarab, 1972 (Coccoidea of Virginia).

35
ORDER THYSANOPTERA (THRIPS)

Cuban laurel thrips, *Gynaikothrips ficorum* (Phlaeothripidae) from San Jose, California, showing adults, immatures, and eggs (Photo courtesy of and with permission of Ronald E. Stecker).

Minute to small Exopterygota with slender, elongate body (Fig. 35.1a). Head elongate, hypognathous or opisthognathous; mouthparts asymmetrical, with maxillae and left mandible modified as piercing stylets; eyes prominent, round or kidney-shaped, with large, round facets; antennae 4- to 9-segmented, inserted anteriorly. Thoracic segments subequal or prothorax largest; mesothorax and metathorax fused; wings narrow, straplike, fringed with long setae; legs short, adapted for walking; tarsi 1- or 2-segmented with eversible, glandular vesicle at apex. Abdomen 11-segmented, terminating in a tubular apex or with a large, serrate ovipositor.

These highly distinctive insects are remarkable for their ubiquity, inhabiting all sorts of vegetation, as well as leaf litter, fungus, and the subcortical region of decaying trees. About 4500 species are known worldwide, with temperate and subtropical grasslands and woodlands supporting a particularly diverse fauna. Most foliage-inhabiting thrips imbibe plant juices released from cells pierced by the oral stylets. This method of feeding produces a characteristic mottling or silvering of the affected surfaces. Many phytophagous species preferentially infest the reproductive structures of plants,

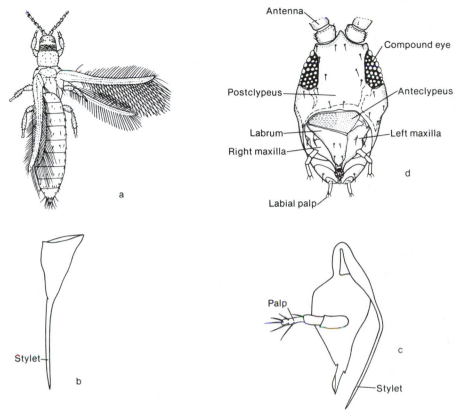

Figure 35.1 Thysanoptera: **a**, *Anaphothrips striata* (Thripidae), dorsal aspect; **b**, mandible, same; **c**, maxilla, same; **d**, frontal view of head, *Chirothrips hamatus*. (a–c, modified from Hinds, 1900; d, modified from Jones, 1954.)

and the most serious economic damage involves buds, flowers, and young fruits. Various species transmit bacterial, fungal, or viral diseases of plants, the most serious being spotted wilt of tomatoes and other plants. Many species are predators of aphids or other small, soft-bodied arthropods, especially eggs of mites and Lepidoptera. Some feed on fungal spores, and a few cause formation of galls. Flower-inhabiting thrips commonly suck the liquid contents from pollen grains, and the larger Tubulifera can ingest small pollen grains and fungal spores. Thrips occasionally create a nuisance by biting humans. The unique, asymmetrical mouthparts (Fig. 35.1d) of thrips share several general similarities with the specialized styliform mouthparts of Hemiptera. The single, awl-shaped mandible and the maxillary stylets

(Fig. 35.1b,c) slide through a conical guide formed by the labrum, labium, and external maxillary plates. Internally the maxillary stylets are attached to the maxillary plates by a slender lever bearing the attachments of the protractor and retractor muscles, as in the Hemiptera. In feeding, the oral cone is appressed to the food and the stylets are punched a short distance into the tissues. The exuding fluids are sucked into the alimentary canal by a cibarial pump.

The fringed wings are a specialization that appears repeatedly in very small insects, frequently in conjunction with reduction of the area of the wing membrane. Winged thrips, competent fliers, may be disseminated long distances by favorable air movements, and mass flights are recorded for several species. Wingless species often climb to the tops of grass stems and are readily dispersed by wind. The tarsus bears an adhesive vesicle that is protruded during walking, enabling thrips to traverse very smooth surfaces. These vesicles, apparently an adaptation for clinging to exposed plant surfaces, are unique to thrips, but analogous structures occur in a variety of phytophagous species in different orders. Most thrips are bisexual, but parthenogenesis is common, and males are rare or unknown in some species. Polymorphism is also of frequent occurrence, being most evident in the degree of development of the wings, which may be present, reduced, or absent, especially in males.

Female thrips are diploid, while males are produced from unfertilized eggs and are haploid. This type of parthenogenetic reproduction, termed **arrhenotoky,** is nearly universal in Hymenoptera (Chap. 46). Eggs are large in relation to body size. In different species, numbers of eggs vary from as few as two to several hundred. In the suborder **Tubulifera,** in which the ovipositor is not developed, eggs are deposited in cracks, in crevices, under bark, or exposed on foliage surfaces. In the suborder **Terebrantia,** the serrate, bladelike ovipositor is used to insert the eggs into plant tissue. A few species are ovoviviparous. Eggs hatch after about 2 to 20 days, depending on temperature. The newly emerged nymphs resemble adults except in the absence of wings and in the reduced number of antennal segments. Two feeding instars are followed by a "prepupal" or "propupal" instar and one "pupal" instar in the Terebrantia, or by one prepupal and two pupal instars in the Tubulifera (Fig. 35.2). During these transformational stages the mouthparts are nonfunctional and the insects are capable of only slow, restricted movement. The pupal stages are usually passed in a cell prepared by the preceding instar from soil or litter particles, but some species construct a silk cocoon or transform naked in cracks or crevices or on foliage. The prepupal and pupal stages are analogous to the pupae of endopterygote insects, and histolysis of various organs, especially head muscles, has been demonstrated in various species. Unlike metamorphosis in endopterygotes, external growth of wing pads in thrips occurs in more than one instar. Furthermore, their specialized mouthparts, reduced wing venation, and the organization of internal organs indicate that thrips represent a specialized derivative of hemipteroid insects, rather than primitive Endopterygota.

Many thrips pass through numerous generations annually, requiring as few as 10 days to develop from egg to adult. Others with a single generation spend most of the year in pupal diapause. Nymphs, pupae, or adults may overwinter in various species. Adults are commonly dark or

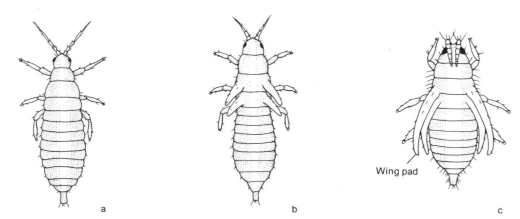

Figure 35.2   Developmental stages of Thysanoptera *(Hercothrips fasciatus):* **a**, last-instar larva; **b**, prepupa; **c**, pupa. (Modified from Bailey, 1932.)

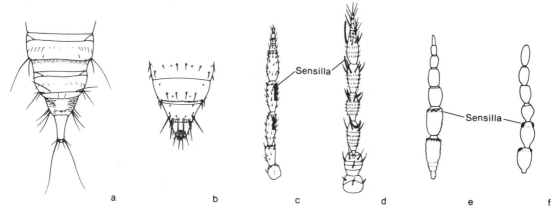

Figure 35.3 Taxonomic characters of Thysanoptera: **a**, suborder Tubulifera, abdominal apex (*Haplandothrips irretius*, Phlaeothripidae); **b**, suborder Terebrantia, abdominal apex (*Anaphothrips striata*, Thripidae). **c–f**, Antennae, showing arrangement of sensilla: **c**, Aeolothripidae *(Aeolothrips scabiosatibia);* **d**, Thripidae *(Anaphothrips striata);* **e**, Heterothripidae *(Heterothrips salicis);* **f**, Merothripidae *(Merothrips morgani).*

cryptically colored, but larvae are often bright red or yellow because their colored tissues show through the transparent cuticle.

Thrips commonly occur in aggregations on plants or fungi, but usually as members of the same generation. Social interactions among generations have been described in a species from Panama (Keister and Strates, 1984), in which nymphs engage in communal foraging. Australian species of *Kladothrips, Onchothrips,* and *Onychothrips* form galls which they inhabit and defend against other individuals using the enlarged, heavily spined forelegs (Crespi, 1992). In *Onychothrips tepperi* a few of the immature offspring inhabiting a gall have reduced wings and enlarged forelegs, suggesting that they defend their unmodified siblings against intruders.

KEY TO THE FAMILIES OF

Thysanoptera

  1. Terminal abdominal segment cylindrical, tubular (Fig. 35.3a); forewing (if present) with membrane smooth, glabrous (suborder **Tubulifera**)
  Phlaeothripidae
  Terminal abdominal segment rounded or conical (Fig. 35.3b); forewing (if present) with membrane bearing minute hairs (microtrichia) (suborder **Terebrantia**)   **2**

2(1). Antennae with longitudinal patches of sensilla on segments 3 to 4 (Fig. 35.3c); females with ovipositor curved upward   **Aeolothripidae**
  Antennae with round or transverse patches of sensilla on segments 3 to 4 (Fig. 35.3d–f); ovipositor curved downward or reduced   **3**
3(2). Antennae with sensilla on segments 3 to 4 arranged as simple or forked cones (Fig. 35.3d)
  Thripidae
  Antennae with sensilla on segments 3 to 4 as continuous or interrupted bands encircling apices of segments (Fig. 35.3e,f)   **4**
4(3). Antennae with 8 segments   **Merothripidae**
  Antennae with 9 segments   **Heterothripidae**

Most common thrips, including those of economic significance, are in the families Thripidae and Phlaeothripidae. Aeolothripidae are sometimes abundant on flowers or in litter. Most species have banded or maculated wings. Merothripidae and Heterothripidae are poorly represented in North America, a few species occurring in litter or on fungus and in flowers, respectively.

SELECTED REFERENCES

Ananthakrishnan, 1984, 1990, 1993 (bionomics, reproductive biology); Ananathakrishnan and Gopichandran, 1993 (chemical ecology); Ananathakrishnan and Raman, 1989 (thrips and galls), 1993 (ecology); Bailey, 1957 (systematics, California species); Crespi, 1987 (fighting behavior), 1992 (ecology of gall thrips); Jones, 1954 (morphology); Keister and Strates, 1984 (social behavior); T. Lewis, 1973 (comprehensive general work); Mound et al., 1980 (relationships among families; much general information); Palmer et al., 1989 (economic species); Sharov, 1975 (phylogenetic relationships); Stannard 1957 (systematics of Tubulifera), 1968 (Illinois species); Stehr, 1987 (immatures).

36
ORDER MEGALOPTERA (DOBSONFLIES, ALDERFLIES)

Adult *Sialis* sp. (Sialidae) on streamside vegetation. Adult *Sialis* are commonly called alder flies. The larvae are aquatic.

Moderate-sized to large Endopterygota.

Adults (Fig. 36.1a) with strong mandibles, maxillae, and labium, large compound eyes, ocelli present or absent. Thorax with 3 subequal segments; mesothorax and metathorax with large, subequal wings; metathoracic wings with pleated anal region folded over abdomen at rest.

Larva (Fig. 36.2) with large head, large prognathous mandibles, large sclerotized prothorax, smaller sclerotized mesothorax and metathorax; abdomen elongate, with segmented or unsegmented filaments on segments 1 to 7 or 8, sometimes with tufts of gills at the filament bases; terminal segment bearing elongate medial appendage or paired prolegs.

Pupa decticous, motile.

Megaloptera are morphologically similar to Neuroptera as adults, and in some classifications are included as a suborder of Neuroptera. The most reliable differentiating feature is the presence of an anal fold in the hind wing (Fig. 36.1c) (no folding in Neuroptera). In contrast, larvae of the two groups are well differentiated. Whereas nearly all neuropterous larvae have specialized suctorial mouthparts (Fig. 38.1c), in the Megaloptera mandibles and maxillae are unmodified and adapted for chewing. The alimentary canal is complete in Megaloptera (ending in the blind midgut in Neuroptera), and pupation occurs in an earthen cell (within a silk cocoon in Neuroptera). In general structure (Fig. 36.2), larvae of Megaloptera are more similar to adephagous beetle larvae than to Neuroptera, and separation from Gyrinidae and Dytiscidae is sometimes difficult.

Megaloptera are usually regarded as the most primitive endopterygote insects. They are most closely related to the Raphidioptera. There are about 300 species worldwide, distributed mostly in temperate regions.

Dobsonflies and alderflies are familiar insects around permanent water, being especially prevalent about cool, well-oxygenated streams. Adults are weak fliers, seldom straying more than a few hundred meters from water. Apparently they do not feed, despite their well-developed mandibles, which are greatly enlarged in males of some Corydalidae (Fig. 36.1b). Mating occurs on streamside vegetation or on the ground. Single females may produce as many as several thousand eggs, and communal egg deposition sites on vegetation or other objects overhanging water may be used by many individuals, resulting in large egg masses. Eggs hatch in 1 to 4 weeks, the newly emerged larvae dropping into the water.

Larvae shelter beneath stones, in sunken vegetation, or in other protected situations. All types of aquatic situation are utilized, but different species have distinct preferences for temperature, substrate, and other characteristics. Sialidae require permanent aquatic habitats, but some Corydalidae occupy intermittent streams in western North America, where they burrow into the substrate while the stream is dry. All megalopteran larvae are predaceous, attacking any aquatic organisms small enough to subdue, and in turn are consumed

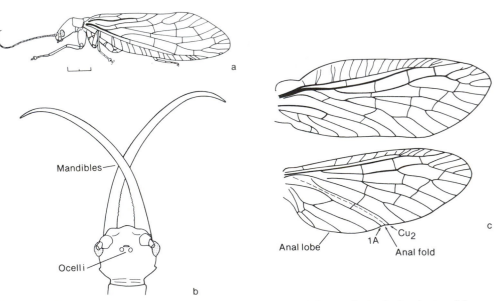

Figure 36.1 Adult Megaloptera: **a**, *Sialis californicus* (Sialidae; scale equals 2 mm); **b**, head of male *Corydalus*, antennae removed (Corydalidae); **c**, wings, *Sialis californicus*.

Figure 36.2   Larva of *Sialis* (Sialidae). Scale equals 2 mm.

by fish, frogs, and other predators. The tactile filaments along the sides of the abdomen are used in detecting prey and avoiding enemies. They may be lost without permanent detriment to the larva. There are numerous larval instars, and full growth apparently requires a year in Sialidae and up to five years in Corydalidae. The mature larvae leave the water to excavate a pupal chamber in moist sand or soft earth beneath stones or driftwood, where they may remain for several months in the larval stage before pupating. Pupae retain full mobility and can defend themselves with the mandibles.

KEY TO THE FAMILIES OF

Megaloptera

1. Adults with 3 ocelli (Fig. 36.1b), fourth tarsal segment cylindrical, wingspan 45 to 100 mm; larvae with 8 pairs of unsegmented abdominal filaments, prolegs on last abdominal segment
                                       Corydalidae

Adults without ocelli, fourth tarsal segment bilobed, wingspan 20 to 40 mm; larvae with 7 pairs of segmented abdominal filaments, terminal filament on last abdominal segment (Fig. 36.2)
                                       Sialidae

Corydalidae (dobsonflies). Widespread, with species throughout North America. Larvae occur predominantly in running water, especially fast-flowing streams. The wings are folded flat over the back at rest. Adults are partly nocturnal, commonly appearing about lights.

Sialidae   (alderflies). Widespread   in   North America, most diverse in the west. The wings are held at a steep angle over the back at rest, in the manner of lacewings. Larvae occur in ponds and sluggish watercourses, as well as in swiftly flowing streams. Adults diurnal.

SELECTED REFERENCES

Afzelius and Dallai, 1988, 1994 (spermatozoa); Chandler, 1956 (California species); Hayashi, 1988, 1989 (biology of larvae), 1992 (spermatophore production); Hayashi and Nakane, 1989 (radio tracking); Kelsey, 1954, 1957 (morphology); Mangan, 1994 (pupal ecology); Merritt and Cummins, 1996 (general); New and Theischinger, 1993 (general); Parfin, 1952 (Minnesota species); H. H. Ross, 1937b (systematics of *Sialis*); Stehr, 1991 (larvae); Whiting, 1991 *(Sialis),* 1994 (phylogeny and classification of Sialidae).

37
ORDER RAPHIDIOPTERA (SNAKEFLIES)

A female snakefly, *Raphidia* sp., (Raphidiidae), on a honeysuckle vine feeding on aphids. Their active larvae hunt prey under bark and in leaf litter. Their movements are like a snake, hence the name.

Medium-sized Endopterygota.

Adult (Fig. 37.1a) with strong mandibles, maxillae, and labium, prominent compound eyes, ocelli present or absent. Prothorax slender, elongate, necklike; mesothorax and metathorax subequal; wings similar in size and venation, with large pterostigma (Fig. 37.1c); female with long, slender ovipositor.

Larva (Fig. 37.1b) with large head, strong prognathous mandibles; prothorax sclerotized, slightly larger than membranous mesothorax and metathorax; abdomen fleshy, cylindrical.

Pupa decticous, motile.

In most features snakeflies resemble Megaloptera (Fig. 36.1). They are sometimes considered to be specialized terrestrial members of that order. Unlike the Megaloptera, which do not feed as adults, snakeflies are voracious predators throughout life, and the elongate head capsule and prothorax (Fig. 37.1a) are adaptations to increase mobility of the head, which is used to strike the prey in a snakelike fashion. Fossils similar to modern Raphidioptera date from the Jurassic Period, suggesting a very early divergence from the Megaloptera. The world fauna numbers about 175 species in two families, Raphidiidae and Inocellidae.

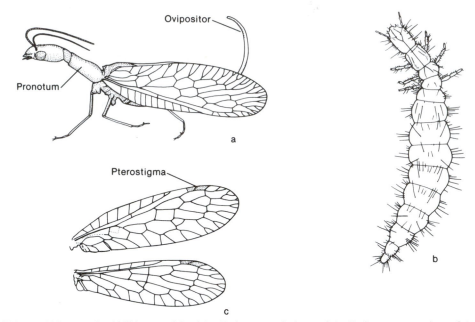

Figure 37.1   Raphidioptera, Raphidiidae: **a**, adult of *Agulla bracteata;* **b**, larva of *Agulla bracteata;* **c**, wings of *Agulla astuta.* (Modified from Woglum and McGregor, 1958, 1959.)

Raphidioptera predominantly inhabit woodlands, where the adults are encountered on foliage, flowers, tree trunks, or similar places. The male copulates from beneath the female, following a long courtship. Up to 800 eggs are inserted into crevices in bark or rotting wood in small groups or in clusters up to about 100. Larvae of Raphidiidae (Fig. 37.1b) commonly occur under loose bark and in porous rotten wood, leaf litter, and similar places; those of Inocellidae have been found only under loose bark. In addition to their legs, larvae use a pygidial adhesive organ to nimbly move in any direction and to maintain a purchase on vertical surfaces. As in Megaloptera, pupation is in a rough cell constructed by the last-instar larva, without a cocoon. The pupa retains full mobility and frequently moves to a second location before adult emergence. The life cycle normally requires two years.

Snakeflies occur throughout the temperate areas of the Northern Hemisphere. Species in the southern extremes of this range in Mexico, northern Africa, and northern India inhabit montane areas. Both families occur in North America, restricted to the region west of the Rocky Mountains. In **Raphidiidae** (Fig. 37.1) ocelli are present, antennal segments are basally constricted, and antennae are usually less than half as long as the body. In **Inocellidae** ocelli are absent, antennal segments are cylindrical, and antennae are usually about as long as the body.

SELECTED REFERENCES

Achtelig and Kristensen, 1973 (phylogenetic relationships); Afzelius and Dallai, 1988 (spermatozoa); Aspock, 1986 (general); Aspock and Aspock, 1975; Carpenter, 1936 (systematics of North American species); Ferris and Pennebaker, 1939 (morphology); Woglum and McGregor, 1958, 1959 (life history).

38
ORDER NEUROPTERA
(LACEWINGS, ANTLIONS, ETC.)

Adult antlion (Myrmeleontidae). These insects are commonly attracted to lights. Larvae live in dry sand or friable soil.

Small to large Endopterygota.

Adult with chewing mouthparts, large lateral eyes, ocelli present or absent; antennae multiarticulate, usually filiform or moniliform. Mesothorax and metathorax similar in structure, with subequal wings, usually with similar venation. Abdomen without cerci.

Larvae with clearly defined head capsule; mandibles and maxillae usually elongate, slender, modified for sucking (Fig. 38.1c); thoracic segments with walking legs with 1-segmented tarsus usually bearing 2 claws; abdomen frequently bearing adhesive disks on last 2 segments, without cerci.

Pupa exarate, decticous, enclosed in silken cocoon.

The Neuroptera comprise a small but highly variable order of predominantly predatory insects that display a mixture of primitive and specialized features. Permian fossils include several families of Neuroptera, and by the mid–Mesozoic Era, forms similar to most modern families had appeared. Neuroptera occur in all parts of the world, but many families show relict distributions. The fauna of Australia is especially rich, including several endemic families as well as a great variety of primitive forms from other families. Ithonidae are restricted to North America and Australia, and Polystoechotidae to North America and southern

South America. Most North American Neuroptera are green or various shades of gray or brown, sometimes with dark wing maculations; they are mostly rather obscure insects. Many Australian and oriental species are strikingly colored or patterned. The order numbers about 5000 species worldwide. The families Chrysopidae and Myrmeleontidae are dominant, each with about 2000 species.

Adult Neuroptera are soft-bodied insects of generalized body plan (Fig. 38.1a). The mouthparts are adapted for chewing, with strong mandibles and maxillae, and small labium. The thoracic structure and wing venation are especially generalized, compared with those of most other Endopterygota. In most species the thoracic segments are subequal, with the prothorax freely movable. Mantispidae, which have enlarged, raptorial forelegs, also have the prothorax elongate, further increasing their striking range. The wings (Fig. 38.3) are similar in size, shape, and venation, except in Coniopterygidae, in which the hind wings may be much reduced, and in Nemopteridae, in which the hind wings are very long and slender, usually expanded at the apex. In Ithonidae, as well as several exotic families, characteristic sensory structures (nyg-

mata) occur in the wing membrane. Wing coupling occurs in several families, especially those with strong flying members. In Hemerobiidae coupling is by a bristlelike frenulum on the base of the hind wing. Similar coupling occurs in some Chrysopidae and Mantispidae. The wings of Coniopterygidae are coupled by hamuli-like setae on the costal margin of the hind wings. Abdominal scent glands, usually situated on the first or fifth to seventh abdominal tergites, occur in several families, including Myrmeleontidae and Ascalaphidae. In Chrysopidae glands occur on the prothorax and offer some protection against vertebrate and invertebrate predators.

Immatures of Neuroptera (Figs. 38.1b, 38.2) are all predatory, frequently specialized for particular types of prey. A peculiarity of all species is the blindly ending midgut, a feature also found in most Hymenoptera. During pupation the alimentary canal becomes complete, and the accumulated fecal material is voided shortly after adult emergence.

In Polystoechotidae and Ithonidae, larval mandibles and maxillae are short and blunt. The larva of *Polystoechotes* is somewhat similar in general morphology to larvae of Megaloptera (Withy-

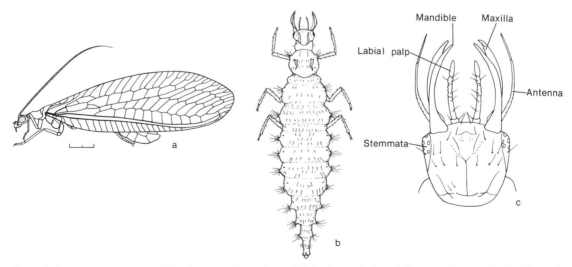

Figure 38.1 Neuroptera, Chrysopidae *(Chrysopa carnea);* **a**, adult (scale equals 2 mm); **b**, mature larva; **c**, head of larva. (b, modified from Tauber, 1974; c, modified from Withycombe, 1925.)

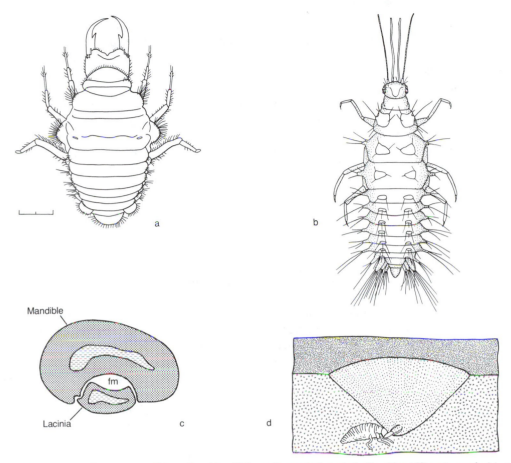

Figure 38.2   Larvae of Neuroptera: **a**, Myrmeleontidae (*Vella;* scale equals 2 mm); **b**, Sisyridae *(Climacia areolaris);* **c**, cross section of mandible and maxillary lacinia that enclose the food canal or meatus (fm); **d**, diagram of a larva in sand pit. The larva seizes small insects that fall into the pit and are unable to climb out. (b, modified from Parfin and Gurney, 1956; d, modified from Scholtz and Holm, 1985)

combe, 1925), and Polystoechotidae are probably among the most primitive extant Neuroptera. Larvae of Ithonidae are scarabaeiform in general shape, living in soil. It is unclear whether they are strictly carnivorous or feed at least in part on decaying plant material. All other neuropteran larvae are predatory, with specialized mouthparts. The mandibles are elongate, with a deep mesal groove that is covered by the maxilla to form an efficient piercing and sucking apparatus (Fig. 38.2c). Configuration of the larval body is related to mode of life. Chrysopidae (Fig. 38.1b), Hemerobiidae, and Coniopterygidae are freely roving predators of small, soft-bodied arthropods, especially Homoptera and mites. Their legs are relatively long with large empodia; the abdomen is slender, tapering, and prehensile, with a terminal adhesive disk, adaptations for clinging to vegetation.

Larvae of Osmylidae (Old World and South America) are similar in body form to Berothidae. They occur in forested situations, sometimes near streams, and they may be subaquatic. In contrast, in the Myrmeleontidae (Fig. 38.2a; Color photo 6d) and Ascalaphidae, the abdomen and thorax are either globular or broadly flattened and are relatively inflexible. The legs are short and in

Myrmeleontidae are adapted for backward movement and for digging. The head is broad and flat, with enormous, sickle-shaped jaws. These larvae are sedentary predators that ambush prey from concealed positions. Some Myrmeleontidae construct conical pitfalls in dry, friable soil, the larva buried at the bottom with open jaws (Fig. 38.2d), while most either lie in ambush at the surface or actively hunt on or just beneath the surface. Walking is either backward or forward depending on the species, but is backward in most pit-building forms.

Ascalaphid larvae sit motionless in litter or on vegetation, snapping the jaws closed on any suitably sized prey. Larvae that frequent exposed situations on vegetation sometimes bear specialized hooked or spatulate setae, which anchor a camouflaging layer of debris, including the dry husks of prey. This habit is shared with some chrysopid larvae. Larvae of Nemopteridae have a short bulbous abdomen, similar to that of Myrmeleontidae, but the prothorax forms a neck up to several times as long as the abdomen. These larvae frequent rock crevices and caves in the Old World and South America. The family Psychopsidae (Old World) is related to Chrysopidae, but the larvae have become adapted as ambush predators living under *Eucalyptus* bark and resemble myrmeleontid larvae in body form, with the head grossly enlarged and the body obese.

Larvae of Sisyridae (Fig. 38.2b) are superficially similar to those of Chrysopidae and Hemerobiidae but are specialized predators of freshwater sponges. They are distinguished by their straight, needlelike mandibles and jointed abdominal gills. Mantispid larvae are specialized parasitoids of spider egg masses or of larvae of vespid or sphecoid wasps or bees. The motile first instar, which is similar to the larvae of chrysopids, searches for a suitable host, sometimes attaching to adult hosts as a means of locating the immatures. After entering the spider egg sac or hymenopteran brood cell, the mantispid larva molts to the grublike second and third instars, with short, ineffective legs and reduced head capsule. All feeding and development occur in a single host cell or egg sac. Larvae of Berothidae are predators of immature ants and termites, in whose nests they live, using neurotoxic chemicals to overcome their much larger hosts (Johnson and Hagen, 1981; Brushwein, 1987). Eggs of berothids are placed on termite-infested logs. Like mantispids, larvae of berothids are hypermetamorphic, with the motile first instar entering the termite nest and the later, obese instars feeding and completing development.

Many adult Neuroptera are predators of a variety of insects; other species imbibe nectar, pollen, or the honeydew secreted by homopterous insects. Adults are relatively short-lived except for those that diapause. Most of them are weak, erratic fliers, but flight is strong and direct in many Ascalaphidae. Eggs are deposited in soil in ground-dwelling forms, or on foliage in arboreal types, being sessile (Hemerobiidae, Dilaridae, Coniopterygidae) or attached by a short or long stalk (Mantispidae, Chrysopidae, Berothidae).

Three larval instars are typical of most families, with four or five instars reported in Ithonidae. The length of larval life is correlated with type of prey and prey availability. In species that consume relatively abundant prey, such as Homoptera, full growth may be attained after only a few weeks. In Myrmeleontidae and Ascalaphidae, where hunting success is highly unpredictable, larvae are capable of ingesting very large meals, then fasting for many months if necessary. In such forms larval life may last several years.

All Neuroptera pupate within shelters of silk produced by the Malpighian tubules and spun out through the anus. In Hemerobiidae the cocoon is usually wispy, but Chrysopidae and soil-dwelling types may incorporate sand or debris to form a tough protective covering. In some families the cocoon is double-layered. The movable pupal mandibles are sharply toothed, serving to open the cocoon. It may be noted that pupal mandibles in Trichoptera and a few primitive Lepidoptera have the same function. Pupae of Neuroptera usually emerge from the cocoon, and may move about before eclosing.

KEY TO NORTH AMERICAN FAMILIES OF

Neuroptera

1. Wings with few veins and less than 10 closed cells (Fig. 38.3a); small to minute insects covered with whitish exudate

   Coniopterygidae

   Wings with numerous veins; more than 10 closed cells (usually very numerous); body without whitish exudate   **2**

2(1). Forelegs raptorial (Fig. 38.4b); prothorax elongate   **Mantispidae**

   Forelegs not raptorial; prothorax subquadrate or slightly longer than broad   **3**

3(2). Antennae filiform, moniliform, or pectinate (see Fig. 2.7)   **5**

   Antennae gradually or abruptly clubbed   **4**

4(3). Antennae nearly as long as body, with abrupt knob at apex   **Ascalaphidae**

   Antennae about as long as head and thorax, gradually thickened (Fig. 38.4a)

   Myrmeleontidae

5(3). Wings with sensory spots (nygmata) in membrane between branches of radial vein; hind wing abruptly narrowed at base

   Ithonidae

   Wings without sensory spots; hind wing gradually narrowed at base (Figs. 38.3b-d)   **6**

6(5). Antennae pectinate (see Fig. 2.7) (male), or with ovipositor exserted, as long as abdomen (female)   **Dilaridae**

   Antennae filiform or moniliform; ovipositor internal, not visible   **7**

7(6). Wings with numerous crossveins between veins R_1 and R_S, (Fig. 38.3d) or R_s arising from 2 to 3 stems (Fig. 38.3c)   **8**

   Wings with 1 to 4 crossveins between veins R_1 and R_S (as in Fig. 38.3b)   **9**

8(7). Forewing with some costal crossveins bifurcate (Fig. 38.3c); body and wings usually brown

   Hemerobiidae

   Forewing without bifurcate costal crossveins (Fig. 38.3d); body and wings green or brown

   Chrysopidae

9(7). Forewing with about 15 parallel branches of radial sector, wingspan at least 40 mm

   Polystoechotidae

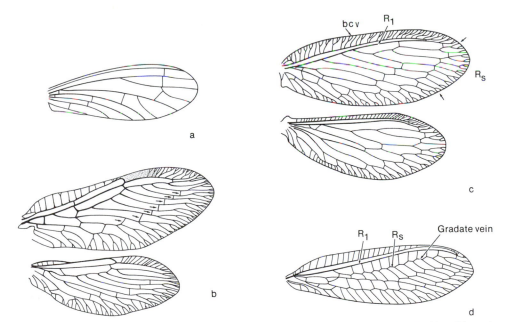

Figure 38.3 Wings of Neuroptera: **a**, forewing of Coniopterygidae *(Conwentzia hageni);* **b**, Mantispidae *(Plega signata);* **c**, Hemerobiidae *(Hemerobius ovalis);* **d**, forewing of Chrysopidae *(Chrysopa carnea).* bcv, bifurcate costal crossveins; extent of R_S marked by marginal arrows; gradate vein in b marked by arrows. (a, modified from Banks, 1907.)

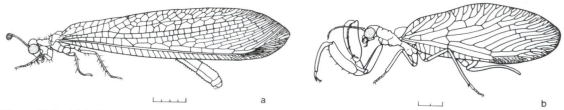

Figure 38.4 Adult Neuroptera: **a**, Myrmeleontidae (*Myrmeleon immaculatus;* scale equals 4 mm); **b**, Mantispidae (*Plega dactylota,* female; scale equals 2 mm).

Forewing with 4 to 7 branches of radial sector (as in Fig.38.3b); wingspan less than 30 mm

                                       10

10(9). Forewing with some costal crossveins bifurcate (as in Fig. 38.3b,c); gradate vein present (as in Fig. 38.3b)   **Berothidae**
Forewing without bifurcate crossveins (as in Fig. 38.3d); gradate vein absent   **Sisyridae**

Coniopterygidae. Small or minute insects, superficially resembling whiteflies (Aleyrodidae) because of the reduced wing venation (Fig. 38.3a) and the whitish powder that covers the body and wings. Coniopterygids are common, widespread insects but usually escape notice because of their small size. Adults may feign death when disturbed.

Ithonidae and **Polystoechotidae.** A single, rare ithonid, *Oliarces,* occurs in desert areas of southern California. Adults sometimes emerge in large evening swarms in spring or summer. The two North American polystoechotids are more widespread but also rare. Adults of both families are relatively large (30 to 70 mm), with broad wings and robust body.

Dilaridae. Small lacewings (wingspan, 6 to 12 mm) whose larvae hunt on tree trunks. Widely distributed, with 2 uncommon species of *Nallachius* in eastern North America. Females oviposit in bark crevices of dead trees, especially oak and tulip trees, where the larvae are probably predators of beetle larvae.

Berothidae. Small lacewings, superficially similar to Hemerobiidae, usually distinguished by scalloped outer wing margins. The North American fauna includes less than 15 uncommonly encountered species.

Sisyridae. Widespread but seldom encountered, being restricted by the occurrence of freshwater sponges, their only hosts. Eggs are deposited on foliage overhanging water, the newly emerged larvae (Fig. 38.2b) drop-

ping into the water. Pupation occurs on emergent vegetation or on other objects near the water. At least one of several species occurs in most regions of North America.

Hemerobiidae and **Chrysopidae** (lacewings; Fig. 38.1). The superficially similar adults are reliably distinguished by the presence or absence of bifurcate costal crossveins (Figs. 38.3c,d). Wings of Hemerobiidae are frequently hairy, those of Chrysopidae bare. These two families represent one of the dominant groups of Neuroptera, with many North American species.

Mantispidae (Fig. 38.4b). Distinguished from all other Neuroptera by the enlarged raptorial forelegs. Nocturnal species are gray or brown; diurnal species include bright green forms as well as others patterned in red, yellow, and black, apparently mimicking vespid wasps. Mantispidae are general predators, most abundant in tropical and subtropical regions.

Myrmeleontidae and **Ascalaphidae.** Differentiated from other Neuroptera by the elongate body form, similar to that of Odonata. Ascalaphidae are not familiar insects in most of North America, but Myrmeleontidae (antlions, Chapter photo, Figs. 38.2a, 38.4a) are among the commonest Neuroptera, especially in arid regions, where large numbers of the nocturnal adults often collect about lights.

SELECTED REFERENCES

Morphology

Ferris, 1940; Mickoleit, 1973 (ovipositor); Withycombe, 1925.

Biology and Ecology

Brushwein, 1987; Henry, 1989 (stridulation in lacewings); Johnson and Hagen, 1981 (Berothidae);

McKeown and Mincham 1948 (Mantispidae); New, 1975 (Chrysopidae); Redborg and Macleod, 1985 (Mantispidae); Oswald, 1993a,b (systematics of Psychopsidae and Hemerobiidae); R. C. Smith, 1922 (Chrysopidae); Stehr, 1991; W. M. Wheeler, 1930 (Myrmeleontidae); Withycombe. 1925.

nadian species); MacLeod and Adams, 1967 (Berothidae); Meinander, 1972 (Coniopterygidae); Merritt and Cummins, 1996 (Sisyridae); Mickoleit, 1973 (phylogeny); Peterson, 1951 (larvae); Rehn, 1939 (North American Mantispidae); Stehr, 1991 (immatures).

Systematics

Banks, 1927 (nearctic Myrmeleontidae); Carpenter, 1940 (general); Klimaszewski and Kevan, 1985 (Canadian species);

39
ORDER COLEOPTERA (BEETLES)

Tiger beetles are very alert, active predators on beaches and open flats. The metallic green *Cicindela oregona* (Cicindelidae) frequents patches of bare ground.

Adult. Minute to large Endopterygota. Compound eyes present (absent in many specialized forms), ocelli almost always absent; antennae almost always with 11 or fewer segments; mouthparts adapted for chewing, with recognizable mandibles, maxillae, and labium. Prothorax large, mobile, with ventral extensions of notum usually extending to vicinity of coxae; mesothorax small, fused with metathorax ventrally; mesonotum visible externally as scutellum; mesothoracic wings developed as sclerotized, rigid, elytra with specialized longitudinal venation; metathorax large (small in flightless forms), with strongly sclerotized sternum, weakly sclerotized tergum. Abdomen usually with sterna strongly sclerotized, terga membranous or weakly sclerotized

with 2 or more terminal segments usually reduced; cerci absent.

Larva. Body form extremely variable but almost always with sclerotized head capsule, antennae with 2 to 4 segments, mouthparts mandibulate; thoracic segments with 4 or 5 segmented legs bearing 1 or 2 terminal claws; abdomen without legs, but frequently bearing segmented or unsegmented, sclerotized **urogomphi** on terminal segments. Wood-boring or other internally feeding larvae sometimes have legs, sensory organs, and head capsule reduced.

Pupa. Adecticous, exarate, similar to adult in general form or rarely obtect.

_____

The Coleoptera, with more than 300,000 described species, constitute the largest order of insects. Five families of beetles number over 20,000 species; weevils (Curculionidae) top the list with at least 60,000 members, substantially more than most individual phyla of animals. As might be expected in such a large assemblage, Coleoptera are exceedingly variable ecologically and biologically. The majority of beetles are terrestrial herbivores, but several entire families and portions of others are predatory, frequently with highly specialized host ranges or life cycles. Either larvae or adults or both may be aquatic, with numerous freshwater and a few marine (intertidal) species known. In addition, beetles have exploited an extraordinarily diverse array of narrow, specialized niches, including seed predation; boring in leaves, stems, or other restricted plant parts; formation of galls or tumors on plants; and life inside ant or termite colonies. Cer-

tain relatively specialized habitats that have been extensively utilized, often to the near exclusion of other insects, are the subcortical region of woody plants; leaf litter and the top layers of soil; and fungi, including fungi in soil, subcortical habitats, dung, and carrion, as well as the fruiting bodies of mushrooms and shelf fungi. One major source of food that has been scarcely exploited by Coleoptera is nectar from flowering plants, probably because beetles are relatively inefficient fliers.

To a major extent, the evolutionary success of Coleoptera seems to stem from the protection from physical trauma that is provided by the high degree of sclerotization and body compaction. The most obviously affected structures are the forewings or elytra, which are modified as protective covers for the hind body (Fig. 39.1). Elytra are almost always present, even in wingless forms, but may be abbreviated (Staphylinidae and others), in which case

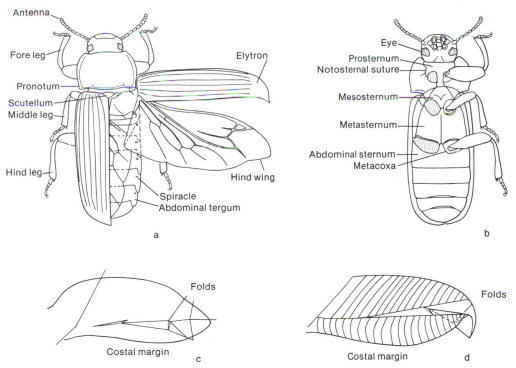

Figure 39.1   Anatomy of Coleoptera: **a**, dorsal aspect of *Tenebrio molitor* (Tenebrionidae), right elytron and hind wing extended; **b**, ventral aspect of same; **c**, diagram of folds in hind wing of same; **d**, folding of wing of same.

the abdominal terga are sclerotized. In most beetles the elytra form hard sheaths, molded to the shape of the abdomen and interlocking with one another along the midline at rest. The movement of the elytra is limited to opening and closing at times of flight. In flightless forms the elytra often interlock immovably along the midline and may dovetail with the abdominal sterna laterally.

Except during flight, the metathoracic wings are concealed beneath the elytra. This is accomplished by a complex series of foldings in the wing membrane (Figs. 39.1c, 39.3b), conforming the size and shape of the wing to the space available in the subelytral cavity. Wing venation is specialized to allow the necessary folding, with the few closed cells in the unfolded anal portion of the wing (Fig. 39.1a). Reduction in venation is common in species of small body size, which often have only a few longitudinal veins and no closed cells. Species of very small body size may also show reduction in the area of wing membrane, compensated by a fringe of long setae on the posterior margin, as in the family Ptiliidae. Flightlessness is relatively common in Coleoptera, and may only entail degeneration of the flight muscles following an initial dispersal flight, as in *Helichus* (Dryopidae), or may involve permanent shortening or complete loss of the wings and permanent atrophy of the associated muscles. In lineages that have long been flightless, the pterothorax becomes shortened, the metatergum is greatly reduced or lost, and the elytra are usually immovably attached along the midline and interlocked with the abdominal sternites, as in many desert-dwelling Curculionidae and Tenebrionidae. Wing polymorphism (reviewed by Thayer, 1992) is also of common occurrence in Coleoptera.

Body compaction may be increased in a variety of ways across families. Commonly the head is more or less deeply retracted into the thorax. In many families the coxae are enclosed in cavities in the sternal sclerites, reducing mobility but increasing the strength of the coxal articulations. Antennae and the distal joints of legs may also be withdrawn into concavities, and in strongly modified species the heavily armored body is practically without projections when the head and appendages are retracted.

The spiracles open into the subelytral cavity, with the exception of the mesothoracic pair, which is concealed between the prothorax and mesothorax. The enclosed position of the spiracles creates a moisture gradient, greatly enhancing water retention, and Coleoptera are usually a conspicuous element of arid habitats. Arid-adapted forms are often flightless, with elytra tightly interlocking with the abdominal sternites, further decreasing water loss. Spiracles may be provided with special closing mechanisms, and Malpighian tubules may be cryptonephric, additional means of increasing the efficiency of water retention.

In general, body form in Coleoptera reflects the mode of life. Subcortical species are often dorsoventrally flattened, sometimes extremely so, allowing entry between tightly adherent bark and wood. In these species the legs are laterally directed and are usually short. Species that burrow through substrates tend to be very compact, either circular or oval in cross section. Swimming forms are streamlined.

Many Coleoptera, especially those that lead exposed lives, store defensive substances, which are self-produced or obtained from plants. These may be contained in the hemolymph (Buprestidae, Coccinellidae, Endomychidae, Lycidae, Lampyridae, Meloidae, Oedemeridae, some Chrysomelidae) or held in reservoirs and forcibly ejected when needed (Carabidae, Rhysodidae, Cicindelidae, Dytiscidae, Gyrinidae, Silphidae, Staphylinidae, Tenebrionidae, some Cerambycidae). Compounds produced range from alkaloids to various organic acids, aldehydes, and quinones. The bombardier beetle, *Brachinus* (Carabidae), releases a gaseous mixture of quinones and steam, produced by an exothermic reaction in two chambered reservoirs located at the abdominal apex. Larvae of several of these families also produce defensive chemicals, sometimes different from those of the adults. The variation in structure and location, as well as the diversity of substances produced, makes it clear that chemical defensive mechanisms have arisen numerous times independently in Coleoptera.

Coleoptera include the most familiar bioluminescent organisms (fireflies, glowworms), in most of which luminescence is used in mate location and courtship when one sex is flightless (see Cantharoidea, below).

LIFE HISTORY

Sexes are generally similar except for minor differences in size or development of sense organs or appendages. Striking exceptions include many Lampyridae (fireflies) and related families, in which females are wingless and frequently larviform. In some Scarabaeidae (June beetles) and Lucanidae (stag beetles), the males bear elaborate horns or greatly enlarged mandibles, which are used in jousting over females in some species. Female-produced pheromones have been isolated from many beetles and are probably produced by the great majority. The pheromones have the usual function of attracting males, whose antennae are frequently larger or more elaborate than those of females, apparently to enhance pheromone detection.

Copulation almost always proceeds with the male mounted dorsally on the female and may be preceded by specialized courtship behavior, such as stridulation (Scolytidae) or palpation and antennation (Meloidae). Eggs are usually simple and ovoid, with relatively thin, unsculptured chorion, and oviposition takes place on or close by the larval food. Eggs are laid singly or in small batches in many Coleoptera, but egg masses are deposited by some (e.g., Coccinellidae). Numbers of eggs vary from one or two produced at a time in Ptiliidae and some other very small forms to over a thousand in Meloidae. Ovoviviparity occurs sporadically in several families, including Carabidae, Staphylinidae, Tenebrionidae, and Chrysomelidae. Parental care is not typical of Coleoptera, but pairs of sexton beetles (Silphidae) and dung-feeding Scarabaeidae prepare special larval chambers, stock them with appropriate food, and sometimes remain with the larvae until growth is partly com-

pleted. Adult Scolytidae excavate oviposition galleries in dead or living trees and remain for some time with their larvae, as do a few Staphylinidae and Carabidae in burrows excavated in soil or sand.

Immatures are usually of limited mobility, but many predatory larvae are active hunters; in parasitoid Coleoptera the highly mobile first instars (triungulins) function in locating the host. Body form among larvae reflects mode of life and is at least as variable as in adults. However, most coleopterous larvae can be classified as **campodeiform** (Fig. 39.5b), active, predatory, with prognathous head, long legs, and elongate urogomphi (Carabidae; Staphylinidae); **eruciform** (Fig. 39.17h), caterpillar-like, with cylindrical body, short legs, hypognathous head (Coccinellidae; Erotylidae); **scarabaeiform** (Fig. 39.9d), obese, C-shaped body with moderate legs; burrowing in soil or rotting wood (Scarabaeidae; Lucanidae); **cucujiform** (Fig. 39.17i), flattened or subcylindrical body, prognathous head, moderate legs, laterally directed; usually with well-developed urogomphi of diverse configuration (Cucujidae, Nitidulidae, Colydiidae); **apodous** (Fig. 39.19e), grublike, with legs, eyes, antennae, and urogomphi reduced or absent (Curculionidae, Bruchidae); or **clateriform** (Fig. 39.12b), elongate, cylindrical, with short legs and tough leathery cuticle; urogomphi present or absent; usually soil inhabiting (Elateridae, Tenebrionidae). Normally, all larval instars, which vary greatly in number, are similar, but hypermetamorphosis is the rule among parasitoid forms, with an initial campodeiform instar and later instars grublike. It should be stressed that many other, often highly distinctive body forms are associated with specialized modes of life; the larval types described above correspond to general modes of adaptation, not evolutionary relationships.

Pupation usually occurs in a cell constructed by the larva in the feeding substrate or nearby. Cocoons are uncommon but are formed by some Curculionidae as well as a few other families from silk produced by the Malpighian tubules. Limited pupal mobility is enabled by abdominal movements. Various parts of the body frequently bear tubercles, knobs, or setae characteristic of the pupa. Adjacent abdominal segments sometimes have opposable "gin traps," pointed sclerotized prominences supposedly used in defense. Pupae that occur on exposed surfaces (e.g., Coccinellidae) often remain within the tough larval cuticle. The pupal stage usually lasts 2 to 3 weeks; pupal diapause occurs in forms that pupate in protected cells (e.g., Cerambycidae) but seems to be relatively uncommon in Coleoptera, which diapause more frequently as larvae or adults.

The economic importance of Coleoptera proceeds almost entirely from their consumption of crops and other products valued by man. Scolytidae are probably the most important pests in coniferous forests, and, aside from termites, beetles (Anobiidae, Ptinidae, Dermestidae) are the most important pests in wooden structures. Coleoptera also include a multitude of species (many families) that feed on stored food products, and some of these will also consume woolen articles, preserved insects, or herbarium specimens. Beetles are essentially of no medical importance.

KEY TO THE SUBORDERS OF

Coleoptera

1. Pronotum with notopleural sutures (Fig. 39.2)   **2**
   Pronotum without notopleural sutures (Fig. 39.1b) [1]
                                   Polyphaga (p. 461)
2. Metacoxae dividing first visible abdominal sternite (Fig. 39.2)   **Adephaga** (p. 459)
   Metacoxae not dividing first visible abdominal sternite (Fig. 39.1b)   **3**
3. Antennae clubbed (Fig. 39.3c; two small families in North America, Sphaeriidae and Hydroscaphidae)
                                  Myxophaga (p. 459)
   Antennae filiform (Fig. 39.6a; two small families in North America, Cupedidae and Micromalthidae)
                                Archostemata (p. 458)

SUBORDER ARCHOSTEMATA
Archostemata, including only the families Cupesidae, Micromalthidae, and Omatidae, are most nota-

[1] *Micromalthus,* with one rare species in eastern North America, lacks notopleural sutures. It will key to Staphylinidae.

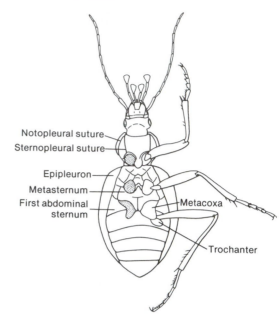

Figure 39.2   Ventral aspect of *Scaphinotus* (Carabidae).

Notopleural suture
Sternopleural suture
Epipleuron
Metasternum
First abdominal sternum
Metacoxa
Trochanter

bly distinguished from other beetles by having the wing membrane spirally rolled in repose, rather than folded (Fig. 39.3b). The presence of distinct pleural sclerites in the prothorax (Cupedidae) and closed cells in the cubitomedian area of the wing are primitive features shared with the Myxophaga and Adephaga.

Cupedidae are wood-inhabiting insects of moderate size (5 to 15 mm), which bore in decaying trunks of both conifers and angiosperms. They are seldom encountered but are locally abundant. Males of some species are attracted to chlorine bleaches, which apparently mimic a sex pheromone. Adults are characterized by the elongate, parallel-sided body, with reticulately sculptured elytra (Fig. 39.3a). Larvae are eruciform, with five-segmented legs with one or two claws. Only 25 species are known; five occur in North America. The earliest fossils definitely assignable to the Coleoptera, impressions of elytra from Permian beds in Asia, are very similar to elytra of extant Cupedidae. The genus *Tshekardocoleus,* described by Rohdendorf (1944), appears to have venation intermediate between that of Coleoptera and Megaloptera.

Figure 39.3   Coleoptera, Archostemata and Myxophaga; **a**, Archostemata, Cupedidae (*Priacma serrata;* scale equals 2 mm); **b**, rolled wing of *P. serrata;* **c**, Myxophaga, Hydroscaphidae (*Hydroscapha natans;* scale equals 0.25 mm).

The **Micromalthidae** are minute beetles (1 to 2 mm) known from the single species *Micromalthus debilis,* which occurs in rotting wood in eastern North America and has been introduced to several other continents. The life history of *Micromalthus* is extremely complex, with at least five distinct larval forms, including one that is neotenic and another that is ectoparasitic on the mother (Pringle, 1938). Females are diploid, males haploid.

Omatidae includes three Old World genera.

SUBORDER MYXOPHAGA

Four small families are included in the Myxophaga. The wings are folded basally, but rolled apically, and contain a closed cell (oblongum) in the cubitomedian area of the wing; the prothoracic notopleural sutures are distinct, and the mandible has a distinct molar lobe. These features suggest relationships to both Archostemata and Polyphaga. Larvae are ovate with broad, deflexed head, five-segmented legs with single claws, and tracheal gills on the abdomen. All species that have been studied are aquatic, apparently feeding on algae.

Two families occur in North America. **Hydroscaphidae** inhabit margins of small sluggish streams and springs, including hot springs, in western North America. Adults are 1 to 2 mm long, broad and flattened, with truncate elytra exposing two or three abdominal tergites (Fig. 39.3c). Pupae are pharate in the last larval cuticle and float on the surface film. Five species are known, with *Hydroscapha natans* the single North American representative. **Microsporidae** (= Sphaeriidae) are minute (0.5 to 1 mm long), highly convex beetles that occur in the interstices of wet gravel and debris along stream margins. The elytra completely cover the abdomen, differentiating microsporids from hydroscaphids. Two North American species are known, and 11 worldwide. **Torridincolidae,** with several genera in tropical regions of the New and Old Worlds, inhabit splash zones around cataracts and similar situations.

SUBORDER ADEPHAGA

These predominantly predatory beetles represent one of the two major evolutionary lines of Coleoptera, with two large, diverse families and several smaller, more specialized ones. Adephaga have distinct pleural sclerites in the prothorax (Fig. 39.2), and usually have a distinct oblongum in the wing, but fold the wing membrane in repose. They are distinguished from all other Coleoptera by the immovable metacoxae, which are fused to the first abdominal sternite, and by the enlarged trochanters of the hind legs (Fig. 39.2). In all Adephaga the molar lobe of the mandible is reduced or absent, a common modification in predatory species of Polyphaga as well. Other useful distinguishing features of Adephaga include large, round (not emarginate) eyes; long, filiform or moniliform antennae; abdomen with six visible sternites; tarsal segmentation 5-5-5. Larvae are campodeiform, with five-segmented legs with two claws.

The dominant family is the **Carabidae** (ground beetles; Fig. 39.5a; Color photo 6a) with about 20,000 species distributed throughout the world. Carabidae are most abundant in moist situations and tend to be nocturnal, but there are many exceptions. Size ranges from about 1 to 50 mm. Adults and larvae of most species are probably nonspecific predators, but diets are highly restricted in others (e.g., *Brachinus,* which attacks hydrophilid prepupae or pupae). About 2500 North American species are known.

Cicindelidae and Rhysodidae are closely related to Carabidae, where they are sometimes placed as subfamilies, although Rhysodidae superficially appear quite different. Fast-running, mostly diurnal predators, **Cicindelidae** (tiger beetles) (Chapter photo) are among the most agile flying beetles. Larvae are sessile predators that construct deep burrows in friable soil, usually near water. **Rhysodidae** are sluggish, heavily sclerotized beetles that inhabit moist, decaying wood.

Haliplidae inhabit slowly moving or standing water, especially the shallows along streams or pond margins. The strongly convex adults are distinguished by the large metacoxal plates, which conceal the bases of the hind legs. Larvae feed exclusively on algae. Adults mingle this diet with occasional small arthropod prey.

The Dytiscidae, Gyrinidae, and Amphizoidae are aquatic predators. **Dytiscidae** (Figs. 39.4, 39.5b,c) is the dominant family, with about 4000 species (330 in North America) occupying all freshwater habitats. Adults and larvae must periodically surface to renew their air supply, somewhat restricting them to shallow water. Adults carry an air bubble beneath the elytra, which they renew by breaking the surface film with the abdominal apex. Body size ranges from about 1 mm to more than 25 mm. Adults and larvae are predatory, the larger species able to kill tadpoles and small fish as well as invertebrates. Adults have typical chewing

mouthparts, while the larval mandibles are deeply grooved or hollow, allowing suctorial feeding.

Gyrinidae (whirligigs) are specialized for surface swimming, with the eyes divided, allowing simultaneous submarine and aerial vision. Adults tend to be highly gregarious, often congregating in dense swarms in eddies or quiet pools, where they gyrate rapidly and erratically on the water surface, sometimes diving to escape disturbances from above. Specialized, club-shaped antennae, which are sensitive to surface film disturbances, enable the beetles to avoid one another and to locate the floating insects on which they prey. Larvae frequent aquatic vegetation.

Amphizoidae include only five species, all but one from western North America. A single species occurs in Asia. Adults and larvae inhabit floating debris trapped in eddies or entangled in shoreline vegetation, mostly along cold, fast-flowing streams, but are unable to swim.

KEY TO THE FAMILIES OF THE SUBORDER

Adephaga

1. Metacoxae covering no more than first visible abdominal sternite   **2**

   Metacoxae expanded as plates covering at least first 3 sternites   **Haliplidae**

2(1). Hind coxae greatly enlarged, extending laterally to meet elytra (Fig. 39.4)   **3**

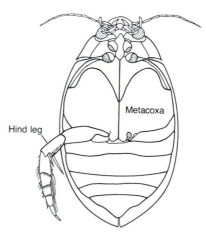

Figure 39.4 Ventral aspect of Dytiscidae, showing large hind coxae (metacoxae).

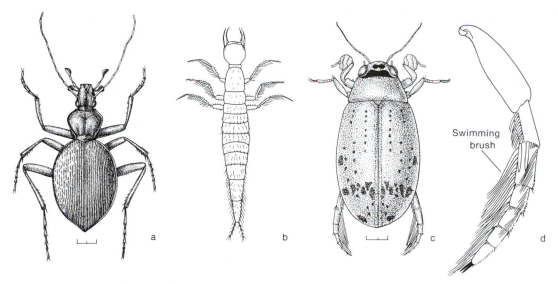

Swimming
brush

Figure 39.5   Representative Adephaga: **a**, Carabidae, adult *(Scaphinotus);* **b**, Dytiscidae, larva *(Hydaticus);* **c**, Dytiscidae *(Eretes sticticus);* **d**, hind leg of *E. sticticus.* (b, modified from Böving and Craighead, 1931.) Scales equal 2 mm.

Hind coxae not extending laterally as far as elytra
(Fig. 39.2)   **5**

3(2). Legs with long hairs fringing tibiae and tarsi (Fig.
39.5d)   **4**

Legs without long fringing hairs
                                                    Amphizoidae

4(3). Eyes divided into dorsal and ventral lobes
                                                       Gyrinidae

Eyes rounded or emarginate, not divided
                               Dytiscidae (including **Noteridae**)

5(2). Metasternum without transverse suture in front of
metacoxae; antennae filiform or serrate (Fig.
39.6a,e)   **6**

Metasternum with transverse suture just before
metacoxae; antennae thick, moniliform (Fig.
39.6b)   **Rhysodidae**

6(5). Clypeus extending laterally in front of antennal
insertions   **Cicindelidae**

Clypeus not extending laterally as far as antennal
insertions   **Carabidae**

SUBORDER POLYPHAGA

This is an exceedingly diverse assemblage, including more than 90 percent of the Coleoptera. The wings lack a closed cell (oblongum) in the cubito-median field and are always folded (not rolled) in repose. The pronotum extends ventrally to the vicinity of the lateral coxal articulation, with the reduced pleural sclerites internal. Larvae are variable but have legs with four segments and a single claw (or legs vestigial or absent).

The Polyphaga are divided into numerous superfamilies. Most of these are relatively homogeneous in terms of both morphological and ecological characteristics, the major exceptions being the Cucujoidea and Tenebrionoidea. The following key is modified from that of E. B. Britton (1970); the taxonomic arrangement is essentially that of Crowson (1960). The most important taxonomic characters include the configuration of the antennae (Fig. 39.6), the number of abdominal segments, and tarsal structure. The numbers of tarsal segments on the fore, middle, and hind legs, respectively, are indicated by three-digit formulas in the key (e.g., tarsal segmentation 5-5-4). The procoxal cavities are considered to be **open** if not enclosed posteriorly by the prothorax (Fig. 39.16g); they are **closed** if a sclerotized portion of the notum extends medially behind them to the procoxal process (Fig. 39.16c). Keying members of the Polyphaga requires great care because of the many exceptional genera. For a more detailed treatment of North American Coleoptera adults, Arnett (1960) is the most complete reference, although the higher classification does not reflect many recent changes that have been made.

KEY TO THE SUPERFAMILES OF
THE SUBORDER

Polyphaga

1. Antennae with terminal 3 to 7 segments enlarged and flattened as a discrete, lamellate, one-sided club (Figs. 39.6i, 39.9b)
   Scarabaeoidea (p. 466)
   Antennae variable, but not lamellate, or if rarely lamellate, more than 7 segments involved   2
2(1). Antennae with segment 6 (occasionally segment 4 or 5) cup-shaped, transverse (Fig.

39.7b); maxillary palpi frequently as long as antennae or longer (Fig. 39.7b)
   Hydrophiloidea (p. 464)
   Antennae with segment 6 unmodified; maxillary palpi almost never elongate   3
3(2). Elytra truncate, exposing 2 or more abdominal tergites, which are sclerotized (Fig. 39.8a-d)   4
   Elytra usually covering entire abdomen, or rarely exposing 1 or 2 tergites   10
4(3). Wings at rest extending beyond elytra, without transverse folds   5
   Wings at rest folded transversely beneath elytra or absent   6
5(4). Maxillary palpi flabellate; eyes nearly contiguous dorsally   15
   Maxillary palpi filiform; eyes widely separated   27

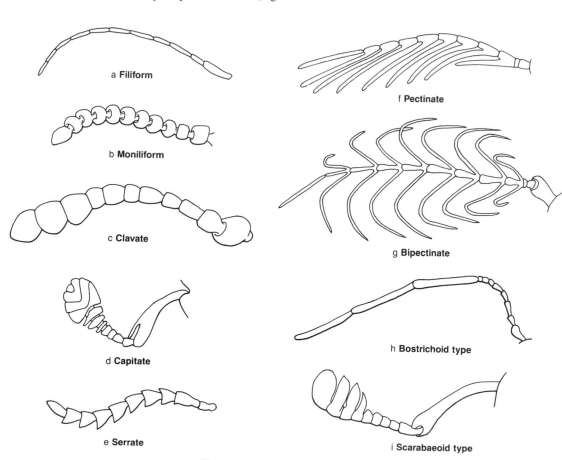

Figure 39.6   Antennae of Coleoptera.

6(4). Claws dentate (Figs. 39.15c, 39.16d) or appendiculate (Fig. 39.16i)   **10**

      Claws unmodified   **7**

7(6). Antennae elbowed with last 3 segments forming capitate club (Fig. 39.6d)

      Histeroidea (p. 465)

      Antennae not elbowed   **8**

8(7). Abdomen with 6 or 7 visible sternites; antennae usually clavate or filiform (Fig. 39.6a,c)   **Staphylinoidea** (p. 465)

      Abdomen with 5 visible sternites; antennae with capitate club (Fig. 39.14c,d)   **9**

9(8). Abdomen with 3 or more tergites exposed

      Staphylinoidea (p. 465)

      Abdomen with 2 or fewer tergites exposed   **10**

10(3,6,9). Abdomen with 6 or 7 visible sternites

      11

      Abdomen with 3 to 5 visible sternites

      16

11(10). Tarsal segmentation 3–3–3

      Cucujoidea (Coccinellidae, p. 476)

      Tarsal segmentation 5–5–5   **12**

12(11). Antennae with distinct, usually, capitate club (Fig. 39.14c,d)   **13**

      Antennae filiform, serrate, or pectinate (Fig. 39.6a,e,f)   **15**

13(12). Tarsi with segment 4 deeply bilobed; claws frequently toothed (Fig. 39.15c)

      Cleroidea (p. 474)

      Tarsi with segment 4 not bilobed; claws simple   **14**

14(13). Antennal club with 3 segments

      Staphylinoidea (p. 465)

      Antennal club with 5 segments

      Hydrophiloidea (p. 464)

15(5,12). Tarsi filiform (Fig. 39.16f), slender, at least as long as tibiae

      Lymexyloidea (p. 475)

      Tarsi stout, not filiform, shorter than tibia

      Cantharoidea (p. 471)

16(10). Metacoxae with posterior face vertical, concavely grooved for reception of femora (Fig. 39.11b)   **17**

      Metacoxae flat or convex, without excavated posterior face (Fig. 39.14a)   **23**

17(16). Procoxae globular round or oval (Fig. 39.16g), not projecting from coxal cavities   **18**

      Procoxae transversely or dorsoventrally elongate (Fig. 39.16c), sometimes projecting from coxal cavities (Fig. 39.8d)   **19**

18(17). Abdomen with first 2 sternites fused, sutures obscured   **Buprestoidea** (p. 470)

      Abdomen with all sternites free, separated by distinct sutures   **Elateroidea** (p. 471)

19(17). Antennae with last 3 to 5 segments differentiated as strong, capitate club (Fig. 39.14d), or greatly elongate (Fig. 39.6h)   **20**

      Antennae filiform, serrate, or clavate, but not with last 3 segments strongly differentiated   **21**

20(19). Antennae with last 3 segments greatly elongate   **Bostrichoidea** (p. 474)

      Antennae with last 3 to 5 segments forming short, capitate club (Fig. 39.14c,d)

      Dermestoidea (p. 473)

21(19). Tarsi with next to last segment deeply bilobed (Fig. 39.10e)   **Dascilloidea** (p. 467)

      Tarsi with next to last segment not bilobed; tarsi usually filiform (Fig. 39.16f)   **22**

22(21). Head with clypeus distinctly separated from frons by fine suture

      Dryopoidea (p. 469)

      Head with clypeus fused with frons

      Byrrhoidea (p. 468)

23(16). Tarsal segmentation 5-5-5   **24**

      Tarsal segmentation 5-5-4, 4-4-4, or 4-4-3

      27

24(23). Prothorax usually hoodlike, produced forward over head (Fig. 39.14a); legs usually with trochanters elongate; antennal insertions usually very close

      Bostrichoidea (p. 474)

      Prothorax not hoodlike; trochanters small; antennal insertions distant   **25**

25(24). Tarsi with large, bisetose empodium (Fig. 39.15f); procoxae conical, projecting from cavities; body usually with numerous, erect bristles   **Cleroidea** (p. 474)

      Tarsi with empodium absent or small, inconspicuous (Fig. 39.15c); procoxae not projecting; body usually without erect bristles   **26**

26(25). Tarsi with fourth segment minute, concealed in groove of bilobed third segment so that tarsi appear 4-segmented (Fig. 39.18b)   **30**

      Tarsi with fourth segment subequal to third; if small, not concealed in deep groove of third segment   **Cucujoidea** (p. 476)

27(5,23). Tarsal segmentation 5–5-4

Tenebrionoidea (p. 477)

Tarsal segmentation 4–4-4 or 3–3-3   **28**

28(27). Tarsal segments not lobed beneath   **34**

Tarsi with 1 to 3 basal segments lobed beneath (Figs. 39.10e, 39.15c)   **29**

29(28). Metasternum without transverse suture

Cleroidea (p. 474)

Metasternum with transverse suture just before metacoxae   **30**

30(26,29). Head prolonged as a beak on which the antennae are inserted (Fig. 39.19a,b)

Curculionoidea (p. 481)

Head not prolonged as a beak; antennae inserted on frons near eyes (Fig. 39.19d)   **31**

31(30). Antennae filiform, serrate, or occasionally clavate   **Chrysomeloidea** (p. 479)

Antennae strongly capitate   **32**

32(31). Antennae elbowed (Fig. 39.6d)

Curculionoidea (p. 481)

Antennae straight   **33**

33(32). Body densely covered by scales or hairs

Curculionoidea (**Anthribidae** p. 481)

Body glabrous or sparsely pubescent

Cucujoidea (**Erotylidae, Languriidae**) (p. 476)

34(28). Abdominal sternites all freely movable; antennal insertions usually exposed

Cucujoidea (p. 476)

Abdominal with basal 3 or 4 sternites connate   **Tenebrionoidea** (p. 477)

Superfamily **Hydrophiloidea** (Fig. 39.7). Predominantly aquatic beetles, characterized by short antennae, with elongate scape and compact club with 3 to 5 segments. In many species the maxillary palpi are elongate, frequently longer than the antennae (Fig. 39.7b).

Hydrophilidae (Fig. 39.7a,b) occupy all aquatic situations. The most familiar species are relatively large, free-swimming forms. Others burrow in bottom debris or hide in aquatic vegetation. Many hydrophilids inhabit the littoral zone, especially along gravelly margins, and do not swim. The numerous members of the subfamily Sphaeridiinae inhabit dung or moist humus in terrestrial habitats, and a few species breed in piles of marine algae stranded on beaches. Aquatic species obtain air by breaking the surface film with an antenna, which communicates with the subelytral air reservoir by means of an air film held on a **plastron** on the ventral surface of the body. Hydrophilids are mostly scavengers as adults, predatory as larvae.

Hydraenidae (Fig. 39.7c) are small beetles (1 to 2 mm) which occur in a variety of littoral habitats, including seeps, wet gravel, or coarse sand banks, splash zones, and marine rock pools. Adults are aquatic, using an antenna to break the surface film to obtain air, while larvae frequent adjacent, moist, terrestrial habitats. **Georyssidae** creep about mud or sandbars along streams. Adults coat the elytra with a camouflaging of mud. Both families are widespread within their restricted habitats.

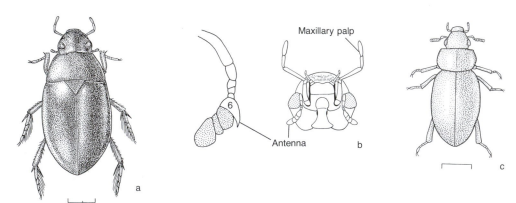

Figure 39.7 Hydrophiloidea: **a**, Hydrophilidae (*Tropisternus ellipticus;* scale equals 2 mm); **b**, ventral aspect of head of *Tropisternus* with cup-shaped sixth antennal segment marked; **c**, Hydraenidae (*Ochthebius rectus;* scale equals 0.5 mm).

KEY TO THE FAMILIES OF

Hydrophiloidea

1. Abdomen with 5 visible sternites; antennae with club 3-segmented **2**

   Abdomen with 6 to 7 visible sternites; antennae with club 5-segmented

   Hydraenidae (including Limnebiidae)

2. Tarsal segmentation 4–4–4   **Georyssidae**

   Tarsal segmentation 5–5–5   **Hydrophilidae**

Superfamily **Histeroidea** (Fig. 39.8c). Adults distinguished by compact body, frequently with retractile appendages; antennae elbowed with abrupt, capitate club. Appendages frequently bear enlarged, spinose setae in characteristic patterns, and the cuticle is typically very hard, black, and polished.

Histeridae are cosmopolitan predators, abundant about carrion and dung, where they are most frequently encountered. Numerous species burrow in sand, many occur beneath bark of dead trees, and others are restricted to mammal burrows or ant or termite nests. Many of the free-living species are predators of fly larvae or pupae. Species frequenting dung or living in sand are usually globular, while those in subcortical habitats are either flattened or cylindrical and elongate. The latter conformation is an adaptation for moving through burrows made by scolytids or other beetles. **Sphaeritidae,** with three known species, are occasionally found in association with decaying plant and animal products. Their biology is unknown. Sphaeritids are distinguished from histerids in having the antennae straight (not elbowed) and the front tibiae simple (not dentate).

Superfamily **Staphylinoidea** (Fig. 39.8a,b,d). Adults with wing venation reduced by loss of the medio-cubital loop and with only 4 Malpighian tubules. Larvae usually with articulated urogomphi and 4-segmented legs with single claws. In external features the Staphylinoidea are extremely variable and difficult to recognize. Most have the elytra abbreviated, exposing 3 or more abdominal tergites, which are sclerotized, but there are many exceptions.

Staphylinoid beetles are found mainly in moist habitats, frequently in association with fungi or in leaf litter. **Ptiliidae, Pselaphidae,** and **Scydmaeni-** dae are largely restricted to fungus/leaf mold environments, but in all these families a few species are known to be commensals in ant or termite nests and others may be encountered beneath bark.

Silphidae, as well as some Leiodidae and many Staphylinidae, are associated with carrion, either as scavengers or as predators, especially of Diptera. *Nicrophorus* (Fig. 39.8b) (Silphidae) is well known for burying corpses of small vertebrates, which are then prepared as food for the grublike, sedentary larvae. Many **Leiodidae** breed in fungi or slime molds, but others feed on carrion or inhabit litter. A few occur in ant or termite nests. **Leptinidae** are all specialized as nest inhabitants or ectoparasites of mammals. Adult leptinids have reduced eyes and antennae, flattened body with leathery cuticle, and mouthparts adapted for feeding on cutaneous debris. The best known species, *Platypsylla castoris,* inhabits the lodges and fur of beavers. **Scaphidiidae** (sometimes considered a subfamily of Staphylinidae) are distinguished by the short, convex body and elytra that largely cover the abdomen and the pointed, conical abdominal apex. Adults and larvae feed on basidiomycete fungi or slime molds.

The **Staphylinidae** are a very large, diverse group, encompassing all the modes of life described above. Perhaps the largest number of species is associated with fungi, including molds and rusts. Many are found near water, and staphylinids are the commonest marine Coleoptera. Very few staphylinoid beetles are phytophagous; although none is presently of economic significance, many Staphylinidae are predaceous, especially on Diptera, and may become important as biological control agents.

KEY TO THE FAMILIES OF

Staphylinoidea

1. Elytra truncate, exposing at least 3 abdominal tergites **2**

   Elytra exposing no more than 2 tergites, frequently covering entire abdomen **3**

2(1). Abdomen swollen, at least twice width of pronotum; tarsi with 1 claw or with claws unequal

   Pselaphidae

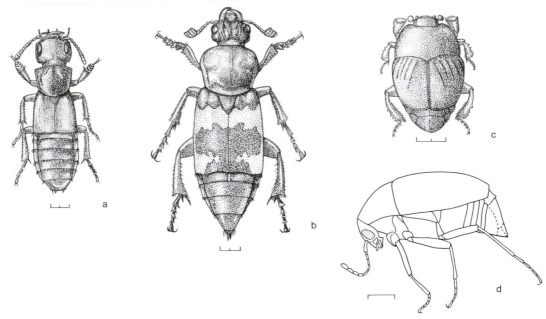

Figure 39.8 Staphylinoidea and Histeroidea: **a**, Staphylinidae *(Platycratus maxillosus);* **b**, Silphidae *(Nicrophorus marginatus);* **c**, Histeridae *(Saprinus lugens);* **d**, Scaphidiidae *(Scaphisoma quadriguttatum)*. Scales equal 2 mm, except that of d equals 0.5 mm.

Abdomen slender, less than 1.5 times width of pronotum; tarsi with claws of equal size
Staphylinidae (Fig. 39.8a)

3(1). Last abdominal tergite conical, as long as 3 preceding segments (Fig. 39.8d); first abdominal sternite as long as 2 to 4 combined
Scaphidiidae

Last abdominal tergite about equal to preceding segment; first sternite about equal to second **4**

4(3). Metacoxae with posterior surface concave, excavated for reception of femora; length less than 1.5 mm   **Ptiliidae** (incl. **Limulodidae**)

Metacoxae with posterior surface flat or convex; length greater than 1.5 mm   **5**

5(4). Procoxae conical, prominent (Fig. 39.8d)   **6**

Procoxae globular, not projecting from cavities
Leptinidae

6(5). Metacoxae contiguous or approximate   **7**

Metacoxae widely separated   **Scydmaenidae**

7(6). Tibial spurs small, indistinct   **8**

Tibial spurs large, conspicuous; body frequently convex, capable of being rolled into a ball
Leiodidae (incl. **Leptodiridae**)

8(7). Antennae clavate with segment 8 smaller than 7 or 9 (39.6c)   **Leiodidae**

Antennae clavate or capitate; segments 7 to 9 subequal
Silphidae (incl. Agyrtidae) (Fig. 39.8b)

Superfamily **Scarabaeoidea** (Fig. 39.9). Distinguished from nearly all other Coleoptera by short antennae with asymmetrical, lamellate club (Figs. 36.6i; 39.9b). Most scarabaeoid beetles are stout-bodied, with the head sunk deeply into the prothorax and anterior tibiae expanded and serrate for digging. Larvae are stout, with large, hypognathous head, well-developed legs, and large, soft abdomen, usually curled ventrally beneath the forebody (Fig. 39.9d).

Lucanidae (stag beetles) and **Passalidae** inhabit decaying wood, where larvae and adults are usually associated in common galleries. The mandibles of most male Lucanidae vary allometrically, being similar to those of the female in small individuals and disproportionately enlarged in large individuals. These organs are used in agonistic encounters between males in some species.

Passalids are flattened and heavily sclerotized, with striate elytra. Larvae, which always occur with adults, stridulate by rubbing the tiny, digitate hind leg over the microscopically ridged mesocoxa. Adults apparently macerate food for the larvae. The only species found in the United States occur in the south and southeast.

Scarabaeidae (chafers, June beetles, etc.) constitute the dominant family. Some scarabaeid larvae feed on rotting wood, but mammal dung, humus, carrion, and roots of living vegetation are consumed by others. Most of the dung feeders excavate a burrow and larval feeding chamber beneath the dung pile, but species of *Canthon* form a mass of dung into a ball and roll it away before excavating the burrow. *Trox* are found about dry animal carcasses, where they feed on the debris (hair, dried skin, etc.). The larvae inhabit burrows beneath the food, coming to the surface at night to feed. Adults of many species, such as the Japanese beetle *(Popillia japonica),* attack living foliage, especially of the Rosaceae. A few species are specialized as predators or commensals in ant or termite nests. Scarabs range in size from burrowing species less than 2 mm long to giants over 10 cm in length, including the bulkiest extant insects.

Males of many species, especially from the Tropics, bear horns on the head or prothorax or have the legs modified in various ways. Most species are brown or black, but many are brightly pastel or metallic, often in striking patterns.

KEY TO THE FAMILIES OF

Scarabaeoidea

1. Lamellae of antennal club thick, rounded, not capable of close apposition (Fig. 39.6i)   **2**
   Lamellae of antennal club flattened, apposable (Fig. 39.9b)   **Scarabaeidae (including Trogidae, Geotrupidae)**

2(1). Antennae elbowed, first segment longer than next 5 segments combined (Fig. 39.6i)   **Lucanidae**
   Antennae straight, first segment shorter than next 4 combined   **Passalidae**

Superfamily **Dascilloidea** (Fig. 39.10). Antennae filiform or serrate, procoxae conical, projecting from cavities; procoxal cavities open; intercoxal process narrow; metacoxae posteriorly concave for reception of femora; tarsal segmentation 5-5-5; 5 visible abdominal sternites.

The Dascilloidea contains a diverse assemblage of beetles from a few small families. Adults of

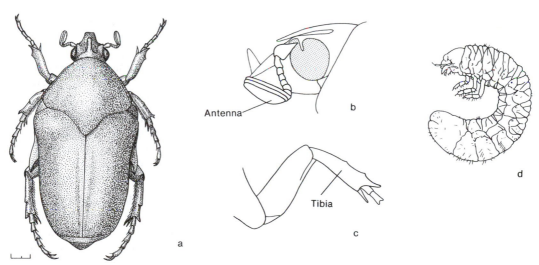

Figure 39.9 Scarabaeoidea, Scarabaeidae: **a**, *Cotinis nitida* (scale equals 2 mm); **b**, antenna and **c**, foreleg of *Cotinis nitida;* **d**, larva of *Popillia japonica*. (d, modified from Böving and Craighead, 1931.)

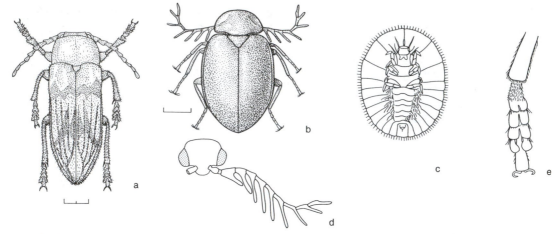

Figure 39.10 Dascilloidea and Dryopoidea: **a**, Dascillidae (*Dascillus davidsoni;* scale equals 2 mm); **b**, Psephenidae, adult (*Eubrianax edwardsi;* scale equals 1 mm); **c**, Psephenidae, larva *(Eubrianax);* **d**, antenna of adult *Eubrianax;* **e**, tarsus of *Dascillus* (Dascillidae).

most families tend to occur in moist situations, frequently around water, but only the larvae of **Scirtidae** (= Helodidae) are aquatic, sometimes living in streams, but usually in tree holes (water-filled cavities in trunks or large branches of trees). Their multiarticulate antennae are unique among immatures of endopterygote insects. Adults of some scirtids have the hind femora enlarged for jumping. Larvae of **Dascillidae** are scarabaeiform and occur in soil feeding on roots *(Dascillus).* Adults are usually seen clinging to the tops of grass stems.

Rhipiceridae adults (Fig. 39.12c) are robust, elongate, brown or black beetles 16 to 24 mm long. Their scarabaeiform larvae are internal parasitoids of cicada nymphs. **Eucinetidae** occur under bark in association with fungi or slime molds. Adults are flattened, ovoid beetles with the head deflexed and the hind femora enlarged for jumping. The **Clambidae,** included in the Dascilloidea because of similarities of the larvae to that of *Eucinetus* (Crowson, 1960), are minute beetles with 4-4-4 tarsal segmentation and clubbed antennae. Adults are capable of rolling into a ball. The life histories and biology of clambids are almost unknown.

KEY TO THE FAMILIES OF

Dascilloidea

1. Length less than 1 mm; antennae clubbed
   Clambidae
   Length greater than 3 mm; antennae filiform, serrate (Fig. 39.6a,e), or flabellate (Fig. 39.12c)
   2

2(1). Antennae filiform or serrate   **3**
      Antennae flabellate (Fig. 39.12c)
      Rhipiceridae

3(2). Metacoxae broad, platelike, oblique
      Eucinetidae
      Metacoxae narrow, transverse   **4**

4(3). Tarsi with segments 2 to 4 lobed ventrally (Fig. 39.10e)   **Dascillidae**
      Tarsi with only segment 4 lobed   **Scirtidae**

Superfamily **Byrrhoidea.** Strongly convex beetles with the head strongly deflexed, the clypeus fused with the frons, antennae filiform; procoxae transverse, intercoxal process broad, procoxal cavities open; tarsal segmentation 5-5-5; 5 visible abdominal sternites.

The single family, **Byrrhidae,** includes about 40 North American species. They are small to moderate in size. Adults are convex with retractile appendages, accounting for the common name, "pill beetles." Adults and larvae are herbivorous, many feeding in moss or liverworts. Larvae are scarabae-iform or onisciform.

Superfamily **Dryopoidea** (Fig. 39.10). Procoxal cavities open posteriorly, prosternal process fitting into concavity in mesosternum; metacoxae concavely excavate posteriorly for reception of femora, tarsal claws very elongate; abdomen with 5 visible sternites. Larvae eruciform, with retractable anal gills, or flattened, onisciform [Psephenidae (Fig. 39.10c)].

Dryopoidea are aquatic or subaquatic in some stage of their lives, with the exception of Callirhipidae, and Artematopidae and Chelonariidae. Adult **Artematopidae** are encountered on foliage, and the larvae live in moss. Adults and larvae of **Callirhipidae** occur under the bark or in the wood of dead logs. The single North American species, *Zenoa picea,* is in Florida. The predominantly tropical **Chelonariidae,** represented in North America by a single species, *Chelonarium lecontei,* are myrmecophilous. The strongly sclerotized, oval adults have the head retracted into the underside of the pronotum and grooves on the venter for receiving the retractile appendages.

Heteroceridae and **Limnichidae** occur along stream margins. Adults and larvae of heterocerids burrow in mud or sand, feeding on algae, diatoms, and detritus. Limnichids are most frequently en-countered on mud or sand banks or on low vegetation near water, feeding on detritus. A few species are intertidal.

Dryopidae are unusual among insects in having terrestrial larvae and aquatic adults. The larvae burrow in damp sand or soil. Newly emerged adults fly to streams, where they cling to the undersides of submerged stones or logs, often in association with elmid adults and psephenid larvae (water pennies). In contrast to the Hydrophiloidea, aquatic dryopoids respire with tracheal gills (larvae) or cuticular plastrons (adults), and may remain submerged indefinitely without replenishing their air supply. Most of these beetles are infrequently collected, but many are common stream inhabitants.

Psephenidae (water pennies) are most commonly encountered as the extremely flattened, disk-shaped larvae (Fig. 39.10c), which occur under stones, submerged logs, and the like in flowing water. Larvae move to crevices or overhangs just above the waterline and pupate pharate in the larval cuticle. Adults are short-lived brownish or blackish beetles usually seen on streamside rocks or vegetation. Males usually have pectinate antennae. Most adults and all larvae of **Elmidae** are aquatic, most frequently inhabiting lotic situations with rocky bottoms where they feed on decaying

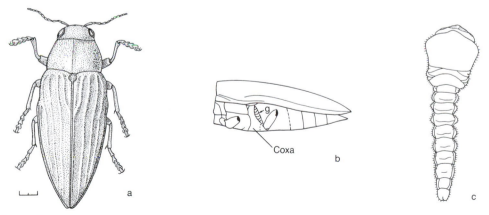

Figure 39.11 Buprestoidea, Buprestidae: **a**, *Buprestis aurulenta* (scale equals 2 mm); **b**, metacoxal region of *Buprestis;* **c**, dorsal aspect of larva of *Chrysobothris.* (c, modified from Böving and Craighead, 1931.) g, groove in coxa for reception of femur.

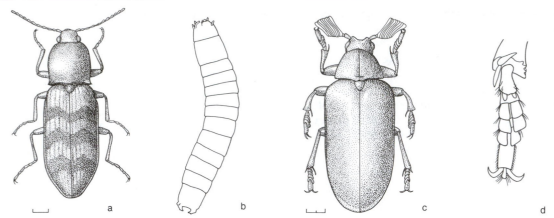

Figure 39.12   Elateroidea and Dascilloidea: **a**, Elateridae (*Ctenicera tigrina;* scale equals 1 mm); **b**, Elateridae, larva *(Cryptohypnus);* **c**, Rhipiceridae (*Sandalus niger,* male; scale equals 2 mm); **d**, tarsus of *Sandalus* (Rhipiceridae). (b, modified from Böving and Craighead, 1931.)

vegetation. Larvae leave the water to pupate. **Ptilodactylidae** are small to medium sized brownish or blackish beetles typically found on vegetation near water. Antennae of females are serrate, those of males usually pectinate. Larvae mostly tunnel through very moist, rotten wood or live in very moist leaf litter, but *Anchytarsus* is aquatic, feeding on submerged wood. Only a few genera and species of ptilodactylids occur in North America, but the family is abundant in moist tropical environments.

KEY TO THE FAMILIES OF

Dryopoidea

  1. Procoxae conical, projecting from cavities   **2**
Procoxae transverse or rounded, not projecting   **6**

2(1). Maxillary palpi with second segment as long as next 2 segments combined; abdomen usually with 6 to 7 visible sternites   **Psephenidae**
Maxillary palpi with second segment subequal to third; abdomen with 5 visible sternites   **3**

3(2). Antennae flabellate (as in Fig. 39.12c), inserted on a strong protuberance on the frons
                                                Callirhipidae
Antennae filiform, clavate, or serrate (Fig. 39.6a,c,e) frequently concealed within prosternal cavity; frons without antennal protuberance   **4**

4(3). Posterior margin of pronotum wrinkled   **5**
Posterior margin of pronotum not wrinkled
                                              Artematopidae

5(4). Ventral surface of body with grooves for retraction of legs and antennae   **Chelonariidae**
Ventral surface of body without grooves for retraction of appendages   **Ptilodactylidae**

6(1). Tarsi with 4 segments   **Heteroceridae**
Tarsi with 5 segments   **7**

7(6). Tarsi with last segment longer than preceding segments combined   **8**
Tarsi with last segment shorter than preceding segments combined   **Limnichidae**

8(7). Antennae filiform, elongate   **Elmidae**
Antennae short, with stout, pectinate club
                                                  Dryopidae

Superfamily **Buprestoidea** (Fig. 39.11). Fore coxae globular, procoxal cavities open posteriorly, prosternal process articulated with mesosternum; metasternum with transverse suture; metacoxae grooved for reception of femora; tarsal segmentation 5-5-5; abdomen with 5 visible sternites, the first pair rigidly fused.

The single family **Buprestidae,** is characterized by the closely articulated, heavily sclerotized body with short serrate antennae and short legs. Buprestids are active, diurnal insects, found either about their host plants or on flowers or foliage, on which most adults feed. Unlike nearly all other Coleoptera the elytra are lifted slightly but remain closed during flight, with the wings protruding below emarginations just behind the humeri. Many adult buprestids are metallic bronze or green; *Acmaeodera,* marked with contrasting black, yellow and red, mimic bees and wasps in flight (Silberglied and Eisner, 1968). Larvae are narrowly adapted for boring in tissues of perennial plants. The head and thorax are flattened (hence the name "flat-headed borers"), legs are absent, and the abdomen is long, slender, and unsclerotized. Most buprestids bore in wood of trees or shrubs, but some mine the pithy stems of rosaceous plants, others mine leaves or form galls, and a few genera have larvae that feed externally on roots.

Superfamily **Elateroidea** (Fig. 39.12). Fore coxae globular or rounded, cavities open posteriorly, prosternal process elongate, articulated with mesosternum; metasternum without transverse suture; metacoxae grooved for reception of femora; tarsal segmentation 5-5-5; 5 visible abdominal sternites, all freely articulated. Larvae mostly elongate, cylindrical or subcylindrical, coriaceous, prognathous with well-developed legs; urogomphi frequently present.

This superfamily includes one large, familiar family, the **Elateridae** (click beetles), and several small, obscure families. Most elateroid beetles are associated with wood, at least as larvae, but the family Elateridae has many soil-dwelling forms. Many of these wireworms were important pests of cereal and root crops before the advent of soil insecticides. Other elaterid larvae are predatory or omnivorous. Adult elaterids are distinguished from all other beetles by the loosely articulated prothorax and the modification of the prosternal process and mesosternum into a jumping mechanism. The beetle first deflexes the prothorax, then suddenly straightens the body, producing an audible click, suggesting the name "click beetles." Small click beetles are able to propel the body 30 cm or more into the air, but in larger species jumps are much shorter and the click may function primarily to startle predators. Some Throscidae and Eucnemidae are also able to click. Most adult elaterids are small to medium, brownish or blackish beetles, but *Alaus,* up to 40 mm long, is marked with a

pair of prominent, eyelike spots on the pronotum, apparently as a protective startle mechanism against vertebrate predators. The tropical genus *Pyrophorus,* with a few species in southern United States, bears a pair of luminescent spots on the pronotum. The larvae, called railroad worms, luminesce through a large transparent spot on the head and a row of smaller spots on each side of the body.

Most **Cebrionidae,** occasionally seen in southwestern United States, develop in soil. *Selenodon and Scaptolenus* are termitophilous and fly only during rain. **Throscidae** are small beetles with the body form of buprestids. Their biology is poorly known. **Eucnemidae** is a diverse but rare family whose larvae live in wood or bark. The larvae have highly modified mouthparts and an enlarged thorax with the legs reduced or absent so that they superficially resemble dipterous larvae.

KEY TO THE FAMILIES OF

Elateroidea[1]

1. Head with labrum freely articulated; antennae inserted close to eyes   **2**
   Head with labrum fused to clypeus, indistinguishable; antennae closer to mandibles than eyes
        Eucnemidae
2(1). Mandibles large, prominent, sickle-shaped
        Cebrionidae
   Mandibles small, not noticeably protruding from mouth.   **3**
3(2). Prothorax loosely joined to mesothorax
        Elateridae
   Prothorax tightly joined to mesothorax; prosternal process fused with mesosternum
        Throscidae

Superfamily **Cantharoidea** (Fig. 39.13). Procoxae prominent, conical, projecting; procoxal cavities open posteriorly, intercoxal process reduced, seldom reaching mesosternum; tarsal segmentation 5-5-5; abdomen almost always with 6 to 7 visible sternites. Larvae eruciform (Fig. 39.13c) or highly modified, with suctorial mandibles.

The cantharoid beetles are characterized by their leathery, flexible integument and loosely articu-

---

[1] Perothopidae and Cerophytidae, each represented by a single rare species in North America, are not included in the key.

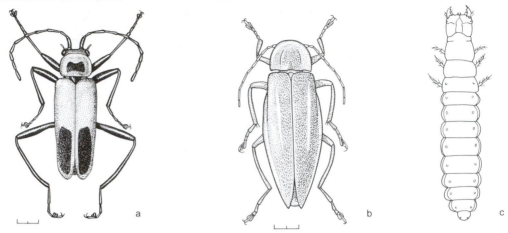

Figure 39.13 Cantharoidea: **a,** Cantharidae *(Chauliognathus pennsylvanicus);* **b,** Lampyridae *(Photurus pennsylvanicus);* **c,** larva of *Chauliognathus.* (c, modified from Böving and Craighead, 1931.) Scales equal 2 mm.

lated, flattened bodies. Antennae are usually serrate, less commonly pectinate or flabellate. The elytra are approximated in repose, but not interlocked as in many other beetles. Larvae are eruciform, with small head and soft abdomen, or have the body segments platelike and usually flattened dorsally. Some of the abdominal segments may be strongly differentiated from the thoracic region, as in Lycidae and Brachypsectridae, resulting in distinctive "trilobite larvae." Most cantharoid beetles are predators, at least as larvae, which have the mandibles medially channeled for uptake of liquid food. **Cantharidae** (Fig. 39.13a) and **Brachypsectridae** are nonspecific predators of small, soft-bodied arthropods. Adult cantharids are common on flowers and foliage; the larvae live in leaf litter. Brachypsectrid larvae are locally abundant in dry areas beneath loose bark; the adults are short-lived and rarely seen. **Lampyridae** (Fig. 39.13b) feed almost exclusively on terrestrial snails, and some **Phengodidae** attack only millipedes, which are immobilized by toxin produced by oral glands. **Lycidae** larvae feed on fluids associated with rotten wood, often in aggregations of dozens of individuals. Adults are usually encountered on flowers or foliage. Biology of Telegeusidae is unknown.

The hemolymph of cantharoid beetles contains various antifeedant compounds, and cantharids also release defensive secretions from segmental glands. The bright, conspicuous coloration of these diurnal insects is aposematic, and many members of Cantharidae, Lycidae, and Lampyridae serve as models for mimicking insects ranging from Hemiptera to Lepidoptera as well as other Coleoptera, especially Cerambycidae.

Many Lampyridae luminesce from abdominal organs and use flash patterns in communication between the sexes, which are sometimes strongly dimorphic, with wingless, larviform females. Where several species occur together at the same time, they differ in color, pattern, or duration of their signals. Females of some species of *Photuris,* however, have more complex flashing repertoires, using one set of signals to attract conspecific males and mimicking the flashing patterns of other fireflies to attract their males, which are then eaten. Female *P. versicolor* prey on at least 11 other species. Males of some *Photuris* have also been observed to mimic the signals of males of other species and may be predators as well. Some larval and adult female phengodids also luminesce, but communication between the sexes is by pheromones.

KEY TO THE FAMILIES OF

Cantharoidea

1. Abdomen with 5 sternites visible
   Brachypsectridae
   Abdomen with 6 to 7 sternites visible   **2**
2(1). Elytra with reticulate sculpture; mesocoxae distant   **Lycidae**
   Elytra striate, punctate, or smooth, mesocoxae usually contiguous   **3**
3(2). Antennae inserted laterally in front of eyes; maxillary palpi greatly elongate   **Telegeusidae**
   Antennae inserted on frons between eyes; maxillary palpi normal   **4**
4(3). Head retracted beneath pronotum, concealed in dorsal aspect (Fig. 39.13b)   **Lampyridae**
   Head fully visible in dorsal aspect   **5**
5(4). Antennae with 12 segments, usually flabellate, pectinate (Fig. 39.6f) or bipectinate (Fig. 39.6g)   **Phengodidae**
   Antennae with 11 segments, usually filiform or serrate   **Cantharidae**

Superfamily **Dermestoidea** (Fig. 39.14c,d). Antennae abruptly capitate, procoxae prominent, projecting or transverse; metacoxae excavated for reception of femora; tarsal segmentation 5-5-5, with simple tarsomeres; abdomen with 5 visible sternites. Larvae eruciform, usually with long spinose setae on the tergites.

These small to moderate-sized beetles are compact, oval, and usually densely covered with setae or appressed scales. Many species bear a single median ocellus. Typically the posterior margin of the pronotum is medially produced, concealing the scutellum. Three families are included: Dermestidae, Derodontidae, and Nosodendridae.

As larvae, **Dermestidae** (Fig. 39.14c,d) are scavengers of both plant and animal material but are especially abundant on drying skin, feathers, fur, and other proteinaceous substances. The smaller species frequently occur in old nests of Hymenoptera, feeding on cast skins, pollen, and dead insects. Several species are pests on stored products, woolen or fur articles and in museum collections of insects and plants. Females of *Thylodrius contractus* are larviform, and a few species are myrmecophilous. Adults of many dermestids are common on flowers.

Adult and larval **Nosodendridae** occupy slime fluxes (flows of fermenting sap) on trees, both

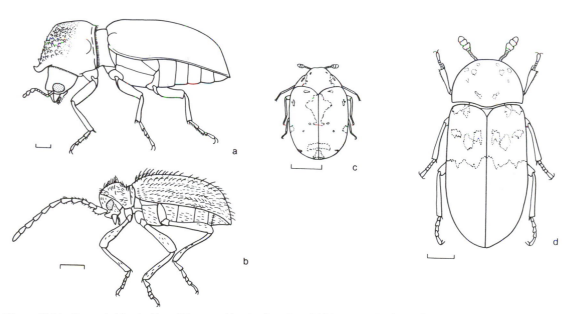

Figure 39.14   Bostrychoidea (a, b) and Dermestoidea (c, d): **a**, Bostrichidae *(Apatides fortis);* **b**, Ptinidae *(Ptinus clavipes);* **c**, Dermestidae *(Anthrenus lepidus);* **d**, Dermestidae *(Dermestes lardarius).* Scales equal 1 mm, except that of b equals 0.5 mm.

conifers and angiosperms, probably feeding on fungi. For example, *Nosodendron californicum* is common in sap flows on white fir, *Abies concolor.*

Derodontidae are rarely collected beetles that mostly inhabit the fruiting bodies of slime molds and wood-rotting fungi, frequently under the bark of dead trees. Members of the genus *Laricobius* feed on Adelgidae (Homoptera).

KEY TO THE FAMILIES OF

Dermestoidea

1. Head with pair of ocelli near inner margins of compound eyes **Derodontidae**
Head with single median ocellus or without ocelli **2**

2(1). Forelegs with tibiae flat, serrate; median ocellus absent **Nosodendridae**
Forelegs with tibiae cylindrical or oval in cross section, with straight margins; median ocellus frequently present **Dermestidae**

Superfamily **Bostrichoidea** (Fig. 39.14a,b). Antennae filiform, frequently with last three segments enlarged or elongate; head usually deflexed, frequently concealed dorsally by hoodlike prothorax, tarsal segmentation 5-5-5; abdomen with 5 visible sternites. Larvae weakly sclerotized, glabrous, with reduced head capsule and legs; urogomphi absent; body stout, C-shaped, weakly sclerotized.

The beetles of this superfamily mostly feed on dead plant material. **Bostrichidae, Lyctidae,** and some **Anobiidae** bore in dead wood of both dicotyledons and monocotyledonous plants such as bamboo. Bostrichids mostly attack recently felled trees, while anobiids and lyctids may enter seasoned timbers, usually affecting primarily the sapwood. *Rhyzopertha dominica* (Bostrichidae) is a cosmopolitan pest of stored grain. Ptinids and some anobiids are more polyphagous, consuming a variety of plant and animal products, including leather, seeds, tobacco, fish meal, spices, certain drugs, and especially stored cereal products. Anobiids also feed in fungi, including puffballs. A significant number of ptinids are commensals in ant nests, and many others frequent bird or mammal nests. Some ptinids and anobiids form cocoons from chitin filaments derived from the peritrophic membrane.

KEY TO THE FAMILIES OF

Bostrichoidea

1. Metacoxae with posterior face excavated, concave, for reception of femora (as in Fig. 39.11b) **Anobiidae**
Metacoxae with posterior face flat or slightly convex **2**

2(1). Antennae filiform, inserted in close proximity between eyes (Fig. 39.14b) **Ptinidae**
Antennae clubbed or with 2 to 3 terminal segments longer than preceding 7 segments (Fig. 39.6h) **3**

3(2). Prothorax hoodlike, concealing head in dorsal aspect (Fig. 39.14a), or, if not hoodlike, first abdominal sternite subequal to second **Bostrichidae**
Prothorax not hoodlike; first abdominal sternite nearly as long as second and third combined **Lyctidae**

Superfamily **Cleroidea** (Fig. 39.15). Head usually prognathous, with antennae clubbed or filiform; procoxae transverse; metacoxae usually extending laterally beyond metasternum, prominent or flat, but not posteriorly concave for reception of femora; tarsal segmentation 5-5-5, with large bisetose empodium frequently present (Fig. 39.15f); abdomen with 5 (rarely 6) visible sternites. Larvae eruciform, with prognathous head, sclerotized prothorax, well-developed legs, urogomphi arising from sclerotized tergal plate.

Cleroids are nearly all predatory, at least as larvae. Exceptions include the Corynetinae (**Cleridae**) and several **Trogossitidae** *(Ostoma, Calitys, Thymalus).* Most Corynetinae feed on carrion, including preserved meat, hence the name "ham beetles," but some species are predators of *Dermestes* larvae. *Ostoma* (Fig. 39.15e), *Calitys,* and *Thymalus* are fungus feeders. Many clerids and trogossitids, such as *Temnochila* (Fig. 39.15d), are broad-spectrum predators, while others specialize on one or a few prey organisms. For example, some species of *Enoclerus* (Cleridae; Fig. 39.15a) locate scolytid (bark beetle) burrows by following sex pheromone gradients produced by the female bark beetles, while *Trichodes* prey on bee larvae in their galleries or on grasshopper eggs. Several small trogossitids have cylindrical bodies that allow them to move through the galleries of bark beetle prey.

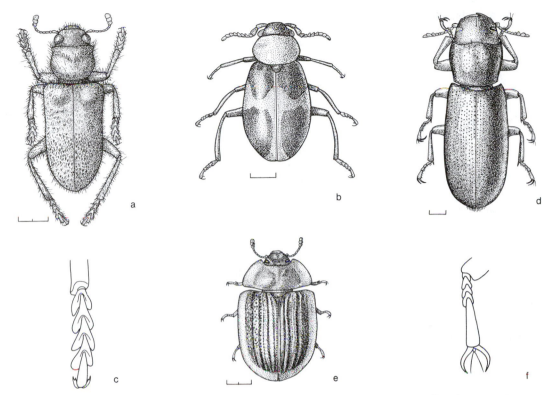

Figure 39.15 Cleroidea: **a**, Cleridae (*Enoclerus moestus;* scale equals 2 mm); **b**, Melyridae (*Collops histrio;* scale equals 1 mm); **c**, tarsus of *Enoclerus* (Cleridae); **d**, Trogossitidae (*Temnochila virescens;* scale equals 1 mm); **e**, Trogossitidae (*Ostoma pippingskoeldi;* scale equals 2 mm); **f**, tarsus of *Ostoma* (Trogossitidae).

Adults of *Eronyxa* (Trogossitidae) are commonly found on flowers, feeding on pollen; the larvae are predators of soft-bodied insects such as coccids. Larvae of **Melyridae** are predators in diverse situations, including beneath bark, in litter or soil, in galls and termite nests, and in the refuse chambers of harvester ants. *Endeodes* inhabits the upper intertidal zone along the Pacific coast. Some **Melyridae** (*Collops;* Fig. 39.15b) are of minor benefit as predators of aphids and other soft-bodied insects on crops; the adults of many species also feed on pollen.

KEY TO THE FAMILIES OF

Cleroidea

1. Procoxae prominent, conical, projecting (as in Fig. 39.8d)   **2**

    Procoxae transversely elongate (as in Fig. 39.16c), not projecting   **Trogossitidae**

2(1). Antennae usually clubbed; tarsi with segments 1 to 4 ventrally lobed (Fig. 39.15c)

        Cleridae

    Antennae usually filiform, rarely with loose 2- to 3-segmented clubs; tarsi simple, filiform, or with fourth segment lobed   **Melyridae**

Superfamily **Lymexyloidea.** Elongate, with 5 to 7 visible abdominal sternites; procoxal cavities confluent, mesocoxal cavities contiguous; antennae very short; tarsi at least as long as tibiae, with 5-5-5 segmentation.

A single family, **Lymexylidae.** The extremely slender, elongate larvae tunnel through hard wood of moribund or dead trees or sawn lumber, apparently feeding on fungus. Adult males are remarkable for the elaborate, flabellate maxillary palpi. The elytra of *Atractocerus* (Neotropical) are extremely short, almost completely exposing the longitudinally folded wings.

Superfamily **Cucujoidea** (Figs. 39.16, 39.17). Antennae filiform, moniliform, or clubbed, very rarely serrate; metacoxae flat or convex, lacking concave posterior face for reception of femora; tarsal segmentation usually 5-5-5, occasionally 4-4-4 or 3-3-3; abdomen almost always with 5 sternites, rarely with 6 or 7. Larvae eruciform or highly modified; head capsule sclerotized, thoracic legs well developed. Abdomen usually with urogomphi on one or more terminal segments.

The Cucujoidea, containing about 25 families, is exceedingly diverse morphologically and ecologically, yet the number of species is relatively low, including less than 5% of all the Coleoptera. Coccinellidae and Nitidulidae are the largest families, with about 4500 and 3000 species, respectively; the other families are mostly small and frequently quite restricted biologically, to narrow, specialized niches.

Habits and habitats of Cucujoidea are extremely variable, but the majority frequent the subcortical region of dead trees and shrubs. This habitat is extremely rich in fungi, molds, and other saprophytic organisms, and many Cucujoidea are fungivorous. For example, **Erotylidae** and **Endomychidae** are strictly dependent on fungi, mostly those associated with rotting wood. Many other families are less obviously associated with fungi but probably feed on molds, yeasts, or other saprophytes, at least as larvae (e.g., **Lathridiidae, Rhizophagidae, Cryptophagidae, Phalacridae,** some **Cucujidae, Nitidulidae**). Both adult and larval **Sphindidae** feed on slime mold spores. The biology of many families, though clearly associated with the subcortical habitat, is almost unknown. Several families, such as **Nitidulidae** and **Cucujidae,** include predatory forms. *Oryzaephilus* and several other cucujids are pests in stored grains, while many nitidulids feed in fruiting structures of higher plants. **Coccinellidae** (Color photo 3d) are familiar as predators of aphids and other soft-bodied arthropods, but also include fungus-feeding

forms (*Psyllobora* and relatives) and plant feeders (*Epilachna* and relatives). Coccinellidae are unusual among Cucujoidea in that the larvae feed exposed on plant surfaces.

Classification of the Cucujoidea is extremely complex, and identification is difficult without a reference collection. Many families are only rarely encountered or are poorly differentiated. The following key includes only the more familiar groups and should be used with care. For identification of the smaller, more obscure families, Arnett (1960) and J. F. Lawrence and Britton (1991, 1994) are the most useful references.

KEY TO THE COMMON FAMILIES OF

Cucujoidea

1. Tarsal segmentation 5-5-5   **8**
   Tarsal segmentation 4-4-4 or less   **2**
2(1). All tarsi with 4 segments   **4**
   Tarsi with 3 segments or fewer   **3**
3(2). Elytra truncate, exposing terminal abdominal segment (Fig. 39.16b)   **4**
   Elytra covering entire abdomen   **5**
4(2,3). Procoxae transversely elongate (Fig. 39.16c)
       Nitidulidae (Fig. 39.16b)
   Procoxae globular, round (Fig. 39.16g)
       Rhizophagidae
5(3). Tarsi with second segment slender   **7**
   Tarsi with second segment dilated (Fig. 39.17c)
       6
6(5). First abdominal sternite with impressed lines parallel to coxal cavities
       Coccinellidae (Fig. 39.17a)
   First abdominal sternite without impressed lines
       Endomychidae
7(5). Antennae filiform; body strongly flattened
       Cucujidae (Fig. 39.17g)
   Antennae capitate, with basal segment (scape) usually large, globular (Fig. 39.17f); minute to small in size   **Lathridiidae**
8(1). Antennae filiform or moniliform (Fig. 39.6a,b); body usually strongly flattened
       Cucujidae (Fig. 39.17g)
   Antennae clubbed (Figs. 39.6c, 39.16b); body usually convex   **9**
9(8). Epipleura continuing to elytral apices   **11**
   Epipleura terminating before elytral apices   **10**

Figure 39.16 Cucujoidea (b,c) and Tenebrionoidea (a,d–i): **a**, Mordellidae (*Mordella albosuturalis;* scale equals 1 mm); **b**, Nitidulidae (*Carpophilus hemipterus;* scale equals 0.5 mm); **c**, ventral aspect of prothorax of *Carpophilus* (Nitidulidae); **d**, tarsus of *Stereopalpus* (Pedilidae); **e**, Meloidae (*Pleuropompha tricostata;* scale equals 2 mm); **f**, tarsus of Tenebrionidae *(Telacis opaca);* **g**, ventral aspect of prothorax of *Synchroa* (Melandryidae); **h**, tarsus of *Ditylus quadricollis* (Oedemeridae); **i**, tarsus of *Pleuropompha tricostata* (Meloidae).

10(9). Antennae with 10 segments (10 and 11 fused)
                                  Rhizophagidae[1]
        Antennae with 11 segments
                       Nitidulidae (Fig. 39.16b)

11(9). Body convex, short, hemispherical; claws appendiculate or toothed (Figs. 39.16i, 39.17c)
                                Phalacridae
        Body elongate, ovoid; claws simple   **12**

12(11). Body with upper surface smooth, without setae   **13**

[1] Males of Rhyzophagidae have 5-5-4 tarsal configuration.

Body with upper surface pubescent
                       Cryptophagidae

13(12). Procoxal cavities closed posteriorly (Fig. 39.16c)   **Erotylidae**
        Procoxal cavities open posteriorly (Fig. 39.16g)
                       Languriidae

Superfamily **Tenebrionoidea** (Figs. 39.16a,d–i; 39.17b,e). Antennae filiform, moniliform, clubbed, or serrate; metacoxae flat or convex, without concave posterior face for reception of femora; tarsal segmentation 5-5-4, reduced to 4-4-4 in a few lineages; abdomen with 5 or 6 sternites. Larvae cruciform or highly modified, especially in parasitoid forms; head capsule sclerotized, thoracic legs almost always well developed, abdomen with or without terminal urogomphi.

Various authorities recognize 25 to 30 families of Tenebrionoidea, but in number of species they constitute only about 7 to 8 percent of the Coleoptera, and Tenebrionidae is the only major family, with about 20,000 species. Most families of Tenebrionoidea have some members that inhabit rotting wood, as larvae and often as adults, and many of

the smaller families are more or less restricted to this habitat. Most of these species are probably feeding on fungi, which may be the primitive food for the entire superfamily. All species of **Ciidae** feed in fruiting bodies of basidiomycete fungi; **Mycetophagidae** and many **Melandryidae** and **Zopheridae** also feed directly on or in fungi. The most successful families by numbers of species and individuals, however, are those that are not primarily restricted to fungal feeding. **Tenebrionidae** includes several mycetophagous lineages and

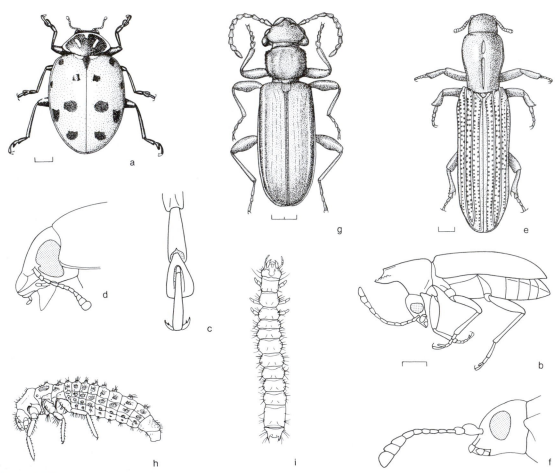

Figure 39.17   Cucujoidea (a,c–d, f–i) and Tenebrionoidea (b,e): **a**, Coccinellidae (*Hippodamia convergens;* scale equals 1 mm); **b**, Anthicidae (*Notoxus monodon;* scale equals 0.5 mm); **c, d**, tarsus and antenna of *Hippodamia* (Coccinellidae); **e**, Colydiidae (*Deretaphrus oregonensis;* scale equals 1 mm); **f**, antenna of *Enicmus* (Lathridiidae); **g**, Cucujidae (*Cucujus clavipes;* scale equals 2 mm); **h**, larva of Coccinellidae *(Coccinella);* **i**, larva of Cucujidae *(Cucujus).* (h, i, modified after Böving and Craighead, 1931.)

others that are associated with rotting wood; but the larger part of the family comprises species that have become specialized for a ground-dwelling existence in arid habitats. Many of these arid-adapted forms are large, conspicuous ambulatory beetles, and a few are important pests in field crops, especially in the Old World. **Meloidae** (Fig. 39.16e) consists entirely of parasitoid forms, the larvae either feeding in the buried egg masses of grasshoppers or on immature bees and wasps in their brood cells. The first instars are vagile **triungulins** that seek the host on which later sessile instars will feed. Species with hymenopteran hosts are usually phoretic on adult bees or wasps as a means of entering their nests. **Anthicidae** adults (Fig. 39.17b) are encountered in various habitats (flowers or foliage; soil surfaces; under bark); larvae have been found in leaf litter. **Colydiidae** (Fig. 39.17e) have diverse feeding habits, including rotting wood, fungi, and lichens; many species inhabit leaf litter, and a few are predators of wood-boring beetles. Larvae of **Oedemeridae** and **Mordellidae** inhabit punky rotten wood.

KEY TO THE COMMON FAMILIES OF

Tenebrionoidea

  1. Procoxal cavities closed posteriorly by a process of the hypomeron (Fig. 39.16c) **9**
    Procoxal cavities open posteriorly (Fig. 39.16g) **2**

2(1). Head abruptly constricted behind eyes, exserted **6**
    Head gradually narrowed behind eyes, usually retracted into prothorax **3**

3(2). Tarsi with fourth segment dilated, ventrally pubescent (Fig. 39.16h) **Oedemeridae**
    Tarsi with no dilated segments **4**

4(3). Mesocoxal cavities closed laterally by mesosternum (as in Fig. 39.14a,b) **Cryptophagidae**
    Mesocoxal cavities closed laterally by mesepimeron (Fig. 39.1b) **5**

5(4). Body very strongly flattened **Cucujidae** (Fig. 39.17g)
    Body rounded in transverse section **Melandryidae**

6(2). Prothorax with sharply defined lateral margins;

last abdominal tergite conical, projecting beyond elytral apex **Mordellidae** (Fig. 39.16a)
    Prothorax rounded, without distinct lateral margins **7**

7(6). Base of pronotum narrower than elytra **8**
    Base of pronotum at least as wide as elytra **Rhipiphoridae**

8(7). Tarsal claws toothed or serrate or bifid (Fig. 39.16i) **Meloidae** (Fig. 39.16e)
    Tarsal claws simple, not toothed or serrate **Anthicidae** and **Pedilidae** (Fig. 39.17b)

9(1). Tarsal formula 5-5-4 **10**
    Tarsal formula 4-4-4 **Colydiidae** (incl. Cerylonidae) (Fig. 39.17e)

10(9). Prosternal process expanded laterally behind coxae, closing cavities **Zopheridae**
    Prosternal process expanded only slightly behind coxae; cavities closed by medial extension of postcoxal process of hypomeron **11**

11(10). Tarsal claws simple **12**
    Tarsal claws pectinate **Tenebrionidae** (Alleculinae)

12(11). Apical antennal segment 2 to 4 times length of preceding segments **Tenebrionidae** (Lagriinae)
    Apical antennal segment about as long as preceding segments **Tenebrionidae** (Fig. 39.1a)

Superfamily **Chrysomeloidea** (Fig. 39.18). Antennae filiform, serrate, or clavate, rarely capitate; metacoxae flat or convex, without posterior concavity for retraction of femora; tarsal segmentation 5-5-5, but with fourth segment almost always extremely small, concealed in bilobed third segment; abdomen with 5 visible sternites. Larvae eruciform, with 3-segmented antennae and well-developed legs (externally feeding forms) or with antennae and legs reduced (borers, internally feeding forms).

Nearly all the 40,000 members of this superfamily are phytophagous. As a group they feed on all parts of higher plants, from the roots on the one hand to the flowers and seeds on the other. Most species of higher plants are probably utilized by at least one of these beetles, and the economic importance of chrysomeloid beetles is almost entirely due to their consumption of plant parts. Chrysomelidae such as *Diabrotica* (cucumber beetles, corn root worms; Fig. 39.18a), *Haltica* and *Epitrix* (flea beetles), and *Leptinotarsa* (Colorado potato beetle) are significant pests on many crops. A few chrysomelids, such as *Diabrotica,* have been implicated in transmission of plant viruses. Cerambycidae (Fig. 39.18e; Color photo 3a) include many minor forest pests. The apple tree borer, *Saperda candida,* infests the living trunks and large branches of several rosaceous trees.

Chrysomelidae, popularly called "leaf beetles," actually attack most parts of plants. The majority of species are external feeders on leaves, flowers, buds, or stems, but many occur in the soil as larvae, feeding on roots or underground stems. Members of the predominantly tropical subfamily Hispinae are leaf miners, and species of Cryptocephalinae specialize on dead leaves. Chrysomelids also feed on aquatic plants *(Donacia),* form galls, and live as commensals in ant nests. Adults

are extraordinarily variable in body form, ranging from disk-shaped (subfamily Cassidinae) to cylindrical types (subfamilies Cryptocephalinae, Chlamisinae). All species of Alticinae (flea beetles) have the hind femora enlarged for jumping. The hind femora are also greatly enlarged in the tropical subfamily Sagrinae, which do not jump.

Cerambycidae as larvae are primarily borers in woody plants, either living or dead. They attack trunks, branches, stems, and roots. Some feed on stems or roots of herbs, a few mine in leaf petioles; there appear to be no true leaf miners. Adults may be either nocturnal or diurnal. The diurnal forms often visit flowers to feed on pollen, and some of these resemble Hymenoptera, particularly in flight. Other cerambycids mimic lycids and lampyrids. *Parandra* and relatives are unusual in having 5 subequal tarsomeres. **Bruchidae** (bean weevils, Fig. 39.18c; often included in Chrysomelidae) are

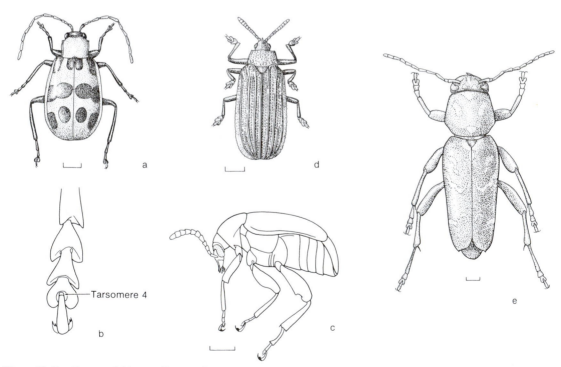

Tarsomere 4

Figure 39.18 Chrysomeloidea: **a**, Chrysomelidae *(Diabrotica undecimpunctata);* **b**, tarsus of *Chrysochus* (Chrysomelidae); **c**, Bruchidae *(Bruchus rufimanus);* **d**, Chrysomelidae *(Odontata dorsalis);* **e**, Cerambycidae *(Xylotrechus nauticus).* Scales equal 1 mm.

borers in seeds, mostly of Leguminosae and Palmaceae, but many other plant families are attacked. The great majority oviposit in immature seed pods or capsules, but *Acanthoscelides obtectus,* the bean weevil, and a few other species attack dried peas and beans and are important pests of stored products.

KEY TO THE FAMILIES OF

Chrysomeloidea

1. Antennae usually inserted on prominent tubercles, capable of being reflexed backward over body, and frequently very elongate
   Cerambycidae
   Antennae not usually inserted on tubercles, usually less than half the length of body   **2**
2(1). Head usually deeply retracted into prothorax; fore coxae usually distinctly separated
   Chrysomelidae
   Head prominent, not deeply set in prothorax; fore coxae usually contiguous   **Bruchidae**

Superfamily **Curculionoidea** (Fig. 39.19). Antennae elbowed and capitate, or occasionally filiform, serrate, or clavate; oral region more or less prolonged as a distinct rostrum; prothorax without distinct lateral margins; metacoxae flat or convex, without posterior concavity for reception of femora; tarsal segmentation 5-5-5, with fourth segment minute, hidden in bilobed third; abdomen with 5 visible sternites. Larvae apodous (with reduced legs in a few primitive genera), with papilliform, 1- or 2-segmented antennae; no urogomphi.

The Curculionoidea, comprising over 65,000 species, is easily the largest superfamily group of Coleoptera, and the most important from an economic standpoint. The evolutionary success of these beetles apparently stems at least in part from the elongation of the stomal region of the head into a rostrum, which is used in preparing oviposition holes as well as in feeding. The rostrum is most specialized in the Curculionidae, where the labrum is absent and maxillae and labium are reduced and partly concealed. The most specialized members of Curculionoidea are the weevils (Curculionidae) and bark beetles (Scolytidae). Both families are characterized by capitate, elbowed antennae. The remaining curculionoid families lack one or both of the antennal specializations.

The **Curculionidae** (weevils; Fig. 39.19a,b), with an estimated 60,000 described species, are probably the most highly evolved family of Coleoptera. Weevils are typically found on foliage or flowers as adults, but some species are ground dwellers, some burrow in sand dunes, and a few are aquatic or marine. Larvae are mostly internal feeders or subterranean, and are rarely encountered without searching, but some, such as *Hypera,* feed externally on foliage.

Weevils include many serious economic pests, such as the alfalfa and clover weevils *(Hypera),* the cotton boll weevil *(Anthonomus grandis),* Fuller's rose weevil *(Pantomorus godmani;* infests many ornamentals), strawberry weevils *(Brachyrhinus;* infests roots of ornamentals and row crops), and the plum curculio *(Conotrachelus nenuphar;* infests fruit of many orchard crops). Species of *Sitophilus* are important pests of stored whole grains.

Brentidae and **Anthribidae** are relatively small families, closely related to Curculionidae. Most Brentidae are elongate, cylindrical beetles that feed in decaying wood as larvae, the adults usually being encountered under bark. Anthribidae feed in fungi, decaying wood, or, occasionally, seeds or fruit. Males of some species have antennae considerably longer than the body. The small, compact, black adults of **Apionidae** are extremely common on flowers.

Nemonychidae infest the fruiting structures of gymnosperms and a few primitive angiosperms. Adults are not uncommon on Pinaceae when the staminate cones are mature. **Attelabidae** (including Rhynchitidae), with about 60 species in North America, mostly feed in rose hips, acorns or other fruiting structures of plants. *Attelabus* females form a leaf roll into which they oviposit and in which the larva feeds. **Ithyceridae** includes a single species, which is associated with oak, hickory, and beech in eastern North America.

The **Scolytidae** (bark beetles; Fig. 39.19c,d) attack woody plants by boring into the cambial tissue, where adults feed and prepare oviposition

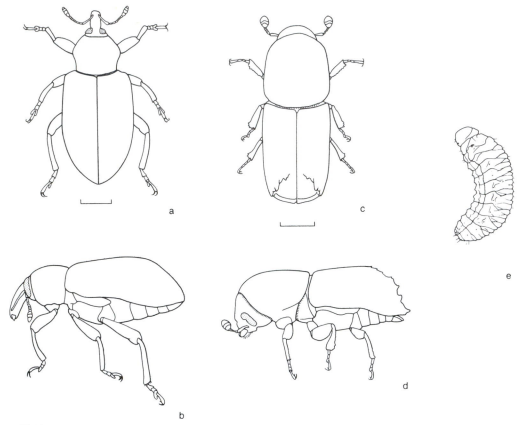

Figure 39.19 Curculionoidea: **a**, **b**, Curculionidae *(Hypera postica);* **c**, **d**, Scolytidae *(Ips);* **a**, **c**, adults, dorsal aspect; **b**, **d**, adults, lateral aspect; **e**, larva of Curculionidae *(Naupactus)*. Scales equal 1 mm. (e, modified from Böving and Craighead, 1931).

galleries. The adult beetles transport yeast spores in special mycangia, inoculating the walls of the galleries. The growths that result are consumed by both adults and larvae. Most species infest necrotic or freshly killed plants, but some invade apparently healthy trees and are the world's major forest pests. In North America species of *Ips* and *Dendroctonus* cause severe damage on conifers. *Scolytus multistriatus* is responsible for transmitting a fungal disease that has largely exterminated American elm trees. Morphologically, bark beetles are virtually indistinguishable from some weevils, and it may be pointed out that Cossoninae (Curculionidae) are biologically very similar to bark beetles. The two groups are treated separately here because

of the massive amount of applied literature keyed to traditional classifications.

KEY TO THE NORTH AMERICAN FAMILIES OF

Curculionoidea[1]

1. Antenna straight; scape about as long as suc-
    ceeding segments   **3**
    Antennae elbowed; scape at least as long as next
    several segments combined (Fig. 39.19a)
                                                   2

2(1). Antennae inserted distant from eyes (Fig. 39.19b);
       rostrum short to very long   **Curculionidae**

[1] Oxycorinidae, with one species in southern Florida, is not included.

Antennae inserted contiguous to eyes (Fig. 39.19d); rostrum short

Scolytidae (incl. Platypodidae)

3(1). Maxillary palp segmented; tarsal claws usually toothed or bifid **7**

Maxillary palp fused into single, rigid article; gular sutures fused or absent **4**

4(3). Antenna with 11 segments **5**

Antenna with 10 segments **Apionidae**

5(4). Antenna with compact club, segments appearing partly fused **Ithyceridae**

Antenna with loose club, segments appearing separate **6**

6(5). Body elongate, cylindrical (more than 3 times as long as wide); rostrum long, narrow

Brentidae

Body about twice as long as wide, narrowed anteriorly; rostrum long, narrow or short, broad **Attelabidae** (incl. Rhynchitidae)

7(3). Gular sutures separated; pronotum without lateral carinae **Nemonychidae**

Gular sutures absent; pronotum with short to complete lateral carinae **Anthribidae**

SELECTED REFERENCES

General

Arnett, 1960 (biology and systematics of North American species); Britton, 1970 (excellent general discussion); *Coleopterists Bulletin,* 1947 to date; Crowson, 1981; M. E. G. Evans, 1975 (biology); Hinton, 1945 (species associated with stored products); J. F. Lawrence and Britton, 1991, 1994.

Morphology

Doyen, 1966 *(Tenebrio);* Caveney, 1986 (eye structure); Duporte, 1960 (cranial structures); M. E. G. Evans, 1961 (musculature, reproductive systems); Kukalová-Peck and Lawrence, 1993 (wing venation) (see also Whiting and Kathirithamby, 1995); Larsen, 1966 (locomotor organs).

Biology and Ecology

Alexander et al., 1963 (stridulation); Balduf, 1935 (bionomics of predaceous species); H. P. Brown, 1987 (biology of riffle beetles); Desender, 1994 (carabid ecology); Dettner, 1987 (chemical defenses); Eberhard, 1979, 1982 (courtship behavior); M. E. G. Evans, 1975 (general); Hanski and Cambefort, 1991 (dung beetle ecology); Hespenheide, 1991 (leaf mining); Hinton, 1955b, 1969a (respiration); Hodek, 1973 (Coccinellidae); Jolivet, 1996 (Chrysomelidae); Kent and Simpson, 1992 (social behavior in Curculionidae); J. F. Lawrence, 1989 (associations with fungi); Linsley, 1959a (mimicry) and 1959b (Cerambycidae); Lloyd, 1971, 1983, 1984 (bioluminescence, communication); Pearson, 1988 (biology of tiger beetles); Pearson and Cassola, 1992 (tiger beetles as indicator taxa); Ritcher, 1958 (Scarabaeidae); Rudinsky, 1962 (Scolytidae); Thayer, 1992 (wing dimorphism); Tschinkel, 1975a,b (chemical defenses).

Systematics and Evolution

Arnett, 1960 (North American genera); Arnol'di and Vandenberg, 1992 (Mesozoic fossils); Boving and Craighead, 1930 (larvae); Bright, 1993 (Curculionidae of Canada); Britton, 1974 (larvae); Caveney, 1986; Coope, 1979 (Cenozoic fossils); Costa et al., 1988 (larvae); Crowson, 1955, 1960 (higher classification); Dillon and Dillon, 1961 (eastern United States species); Gorham, 1991 (stored products pests); Hansen, 1991 (systematics of world genera); Hatch, 1957–1973 (Pacific Northwest species); Hinton, 1945 (species associated with stored products); Kukalova-Peck and Lawrence, 1993 (wing venation) (but see Whiting and Kathirithamby, 1995); J. F. Lawrence, 1982, 1993a (larvae, world), 1993b (larvae, interactive identification and information retrieval); J. F. Lawrence and Britton, 1994 (Australian fauna); J. F. Lawrence and Newton, 1982; Leech and Chandler, 1956 (aquatic families); Leech and Sanderson, 1959 (aquatic families); Merritt and Cummins, 1996 (aquatic families); Pakaluk and Slipinski, 1995; Peterson, 1951 (larvae); Stehr, 1991 (larvae); White, 1983 (field guide); Whiting and Kathirithamby, 1995; Zimmerman, 1991–1994 (series of volumes on weevils of Australia).

40
ORDER STREPSIPTERA

Male *Stylops pacifica* (Stylopidae) mating with female partly inside the host bee, *Andrena* sp. (Andrenidae), which was taking nectar in a buttercup.

Minute to small Endopterygota specialized for endoparasitism. **Males** (Fig. 40.1a) highly distinctive, with large, transverse head; antennae with 4 to 7 pectinate segments; bulging eyes with large facets; ocelli absent. Prothorax and mesothorax very small, mesothorax with wings reduced, elytra-like, without veins, functioning as halteres; metathorax very large, bearing broad, fan-shaped wings with a few thick, radiating veins; legs usually without tarsal claws, first two pairs without trochanters, hind pair without coxae. Abdomen cylindrical, tapering, concealed basally by large postscutellum. **Females** wingless, coccidlike (Mengenillidae) (Fig. 40.2c) or larviform, grublike. **First-instar larva** (Fig.

40.1b,c) (triungulin) spindle-shaped, lacking antennae, mandibles; 3 pairs thoracic legs without trochanters; abdomen with bristlelike caudal styles. Later instars grublike (Fig. 40.1d), without appendages or distinct mouthparts and with alimentary canal ending blindly. **Pupa** exarate, adecticous, pharate in puparium formed by cuticle of last larval instar.

While numerous members of several orders live as parasitoids, Strepsiptera are among the very few insects that internally parasitize other arthropods.

They infest or stylopize a variety of insects, including Thysanura, Blattodea, Mantodea, Orthoptera, Hemiptera, Diptera, and Hymenoptera. Of these, Hemiptera (especially auchenorrhynchous Homoptera) and Hymenoptera are the most common hosts. Host specificity of Strepsiptera varies. Stylopidae, which parasitize Hymenoptera, tend to be narrowly host-specific, whereas Halictophagidae, which parasitize auchenorrhynchous Homoptera, often have host ranges that include many species in several genera. In the great majority of species both sexes use the same host, but in the family Myrmecolacidae males parasitize ants while females parasitize Orthoptera and Mantodea. Unlike parasitoid Hymenoptera and Diptera, which eventually kill their host and usually consume most of its internal organs, Strepsiptera mature within active, living insects. The females of the great majority of species never leave the host's body.

Morphologically Strepsiptera are superficially more similar to Coleoptera than to any other order of insects. The antennae and elytriform forewings resemble the same structures in parasitoid beetles (Rhipiphoridae), and the first-instar larvae (Fig. 40.1b,c) are similar to triungulins of Meloidae and Rhipiphoridae. Furthermore, subsequent larval

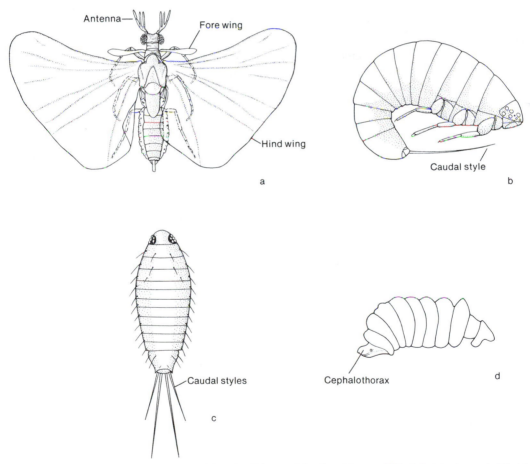

Figure 40.1 Various instars of Strepsiptera: **a**, adult male of *Eoxenos laboulbenei* (mengenillidae); **b**, lateral aspect of first-instar larva or triungulin of *Halictophagus tettigometrae* (Halictophagidae); **c**, dorsal aspect of first instar of *Eoxenos* (mengenillidae); **d**, lateral aspect of third-instar larva of *Halictophagus*. (a, c, modified from Parker and Smith, 1934; b, d, modified from Silvestri, 1940.)

instars of parasitoid beetles are grublike, as in Strepsiptera. However, the reduced prothorax and enlarged postscutellum are not features that occur in Coleoptera. Moreover, purported homologies in wing venation of Strepsiptera and Coleoptera have been disputed (Whiting and Kathirithamby, 1995). Strepsiptera are unique in lacking trochanters in all instars and, unlike Coleoptera, lack a gula. Elenchidae, a derived family, is known from Cretaceous amber from Lebanon, suggesting a possible divergence from Coleoptera or some other endopterygote order as early as the Carboniferous. As a further bit of conflicting evidence, however, the pupal stage of some Strepsiptera is preceded by pharate instars with wing rudiments, suggesting that they may not be true endopterygotes at all (Kristensen, 1991).

The life cycle is strongly modified for endoparasitism. Males spend their life of 1 or 2 days seeking females, which in typical species remain inside the host with only the anterior end or **cephalothorax** exposed. Virgin females secrete a sex pheromone from a gland on the cephalothorax, which provides an efficient means of collecting. Males will be attracted for many days unless copulation is allowed, whereupon pheromone production ceases. Numerous eggs are produced (as many as 75,000 have been recorded) and are retained within the female until hatching. The minute triungulins emerge to seek appropriate hosts, which are usually immature instars of the proper species. Movement is by crawling or by springing, which is effected by curving the abdomen ventrally, then suddenly snapping it backward. If an appropriate host is encountered, the triungulin immediately enters its body, aided by an oral secretion that softens the host cuticle. The triungulin soon molts, producing the legless, grublike second instar; subsequent growth and development occur within the host body cavity. There are three grublike feeding instars (Fig. 40.1d), with the cuticle from each molt remaining as a persistent sheath around the larva. The larval alimentary canal ends blindly, probably to prevent the feces from entering and injuring the host (Hymenoptera and Neuroptera show an analogous adaptation). The fully grown last-instar larva extrudes its anterior end through the intersegmental membrane of the host abdomen (Fig. 40.2d), and pupation occurs in this position. Pupation usually occurs in adult hosts, but sometimes occurs in immatures of exopterygote hosts.

In males the protruding head becomes tanned, forming the **cephalotheca,** which the adult later pushes aside in order to emerge. The neotenic, grublike adult female remains pharate within the saclike cuticle of the last larval instar, through which respiration and food absorption continue to take place. The anterior portion of the body, equivalent to the head, prothorax, and mesothorax, is extruded as the cephalothorax. The **brood passage** (Fig. 40.2b) provides an opening through which copulation occurs and through which the emerging triungulins later escape. The actual genital openings are pores on abdominal segments two to five.

The typical life cycle described above differs in certain species. In the suborder Mengenillidia, with about 20 species in three genera in the family Mengenillidae, the females (Fig. 40.2c), though wingless, have well-developed legs and live free of their hosts, which are Lepismatidae (Thysanura). Last-instar larvae of both sexes leave the host and pupate externally. The Mengenillidia undoubtedly represent the most primitive Strepsiptera, although they occur relatively late (Eocene) in the fossil record. Extant species are known from the Mediterranean region, tropical Africa, Asia, and Australia.

In Stylopidae, which parasitize bees and wasps, the triungulins are deposited on flowers. When adults of suitable host species visit the flowers, the triungulins usually cling to the bee or wasp, or in *Stylops pacifica* (see Chap. 11, Fig. 11.3), the triungulins are ingested with nectar. Once in the host's nest, the triungulins disembark (or are regurgitated) and parasitize the host's eggs or larvae. It may be noted that many meloid and rhipiphorid triungulins are likewise phoretic on adults of their hosts. Stylopidae is the largest family of Strepsiptera.

Stylopized hosts frequently differ from unparasitized members of the same species in mor-

Figure 40.2 Cephalothoraces of mature females of **a**, Halictophagidae *(Halictophagus tettigometrae);* **b**, Corioxenidae *(Triozocera mexicana).* **c**, mature female, Mengenillidae *(Eoxenos laboulbenei);* **d**, puparia of Stylopidae *(Xenos vesparea)* in the abdomen of a vespid wasp *(Polistes nympha).* (b, modified from Silvestri, 1940; c, modified from Parker and Smith, 1940; d, modified from Szekessy, 1960.)

phological details. Especially affected are secondary sexual features, which may be modified to resemble homologous structures of the opposite sex. Internal organs, especially the gonads, are stunted, suggesting the term "parasitic castration." However, at least some male hosts appear to be capable of reproduction. When the strepsipteran pupates in an immature host, the host does not molt to the adult stage. It is known that host life expectancy is sometimes increased by stylopization, but it is not clear whether differential survival has a physiological basis or is mediated by behavioral changes which decrease predation.

About 525 species are known worldwide; several times that many probably remain undescribed.

KEY TO NORTH AMERICAN FAMILIES OF

Strepsiptera

1. Males with tarsi 5-segmented, the terminal segment narrow with minute claws; females with brood passage opening terminally (Fig. 40.2b) **Corioxenidae**

   Males with tarsi 2- to 4-segmented, without claws; females with brood passage opening preterminally (Fig. 40.2a) **2**

2(1). Males with tarsi 4-segmented; parasites of Hymenoptera **Stylopidae**

   Males with tarsi 2- or 3-segmented; parasites of Hemiptera or Orthoptera **3**

3(2). Males with tarsi 2-segmented; parasites of Fulgoroidea **Elenchidae**

   Males with tarsi 3-segmented; parasites of Homoptera, Orthoptera **Halictophagidae**

Corioxenidae. In the United States occurring uncommonly in the southern states. Hosts are Heteroptera.

Stylopidae. Common parasites of bees (Andrenidae, Halictidae, Colletidae) and wasps (Sphecidae, Vespidae). Over 150 species occur in North America.

Halictophagidae. In North America locally common parasites of Homoptera; pygmy grasshoppers

(Tetrigidae) are also infested, and exotic species parasitize Tridactylidae, Blattodea, Heteroptera and Diptera.

Elenchidae. Similar to Halictophagidae. Elenchidae parasitize Fulgoroidea, especially Delphacidae in North America; uncommon.

SELECTED REFERENCES

Afzelius and Dallai, 1994 (spermatozoa); Bohart, 1941 (North American species); Kathirithamby, 1989; Kathirithamby and Carcupino, 1992; Kinzelbach, 1971a,b, 1978 (general), 1990 (phylogenetic relationships); Kinzelbach and Pohl, 1994 (fossils); Kukalová-Peck and Lawrence, 1993 (wing venation) (see also Whiting and Kathirithamby, 1995); Linsley and MacSwain, 1957 (biology of *Stylops*); Manuel and Bohart, 1993 (trichopteran host); Stehr, 1991 (immatures); Ulrich, 1966 (classification); Whiting and Kathirithamby, 1995.

41
ORDER MECOPTERA (SCORPIONFLIES)

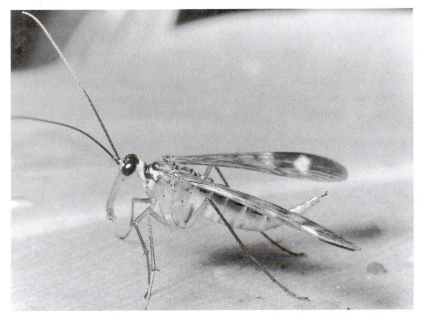

Female of *Neopanorpa tuberosa* (Panorpidae) from Thailand. Adults of all panorpids have a similar appearance. Males curl the abdomen with its large genital bulb forward like a scorpion's tail, hence the name scorpionfly.

Small to medium-sized Endopterygota.

Adult (Fig. 41.1a) with slender elongate body; elongate hypognathous head capsule; mandibles and maxillae slender, elongate, apically serrate; labium elongate, with fleshy 1- to 3-segmented palps. Thoracic segments subequal or prothorax reduced, mesothorax and metathorax fused; wings narrow, elongate, subequal, crossveins numerous. Abdomen cylindrical, elongate, with 1- or 2-segmented cerci.

Larvae (Fig. 41.1c) eruciform, or scarabaeiform with sclerotized head capsule, mandibulate mouthparts, lateral compound eyes; thoracic segments subequal, prothorax with distinct notal sclerite; thoracic legs short with fused tibia and tarsus and single claw; abdomen usually with prolegs on segments 1 to 8, segment 10 modified as a suction disk or with a pair of hooks *(Nannochorista)*.

Pupa decticous, exarate, nonmotile.

In their simple mandibulate mouthparts and membranous wings of similar shape and venation, scorpionflies are similar to the primitive neuropteroid orders Neuroptera and Megaloptera. As in those orders, mecopteran fossils first appear in the Permian Period. Before the end of the Paleozoic Era a diverse fauna, including three suborders and numerous families, had developed. Most of these taxa, including a fourth suborder that appeared in the Mesozoic Era, are now extinct. Mecoptera are presently widespread, especially in humid temperate and subtropical regions, but a high degree of endemism characterizes local regions. The Australian fauna is highly distinct, with many primitive species and one endemic family, Choristidae. Five families occur in North America, including Meropeidae, otherwise known only from Australia. There are about 500 species worldwide, most in the families Bittacidae and Panorpidae.

Despite the superficial similarity of Mecoptera to Neuroptera, the structural details of the mouthparts of adult scorpionflies indicate a relationship with Diptera. The slender mandibles and maxillae and elongate face of bittacids (Fig. 41.1b) are not greatly different from the homologous structures in primitive Diptera (compare with Fig. 42.3a,b), and the labial palps of *Chorista* (Choristidae) are enlarged and fleshy, suggesting a primitive labellum. In addition, the fossils *Permotanyderus* and *Choristotanyderus* are essentially intermediate between Mecoptera and Diptera. In contrast, the larvae of some Mecoptera are similar to those of Lepidoptera and Trichoptera, a relationship that is also indicated by certain fossils.

Most Mecoptera prefer moist conditions, especially in forests, where the adults are to be found on rank vegetation or flying about. In species that inhabit arid or hot regions, such as *Apterobittacus* in California, adult activity is usually restricted to the cool, wet part of the year. Most scorpionflies are winged, but various genera are brachypterous [*Boreus,* Boreidae (Figs. 9.5, 41.2a)] or completely wingless [*Apterobittacus,* Bittacidae (Fig. 41.2b)]. *Boreus* is a winter form, most commonly found about moss, which is used as adult and larval food. *Apterobittacus* frequents low vegetation, clambering nimbly about with the long legs. In *Brachypanorpa* (Panorpodidae) females are brachypterous, males fully winged.

Bittacidae are predatory as adults, but most

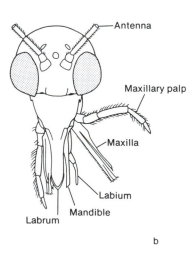

Figure 41.1 Mecoptera: **a**, adult Panorpidae (*Panorpa anomala,* male; scale equals 2 mm); **b**, facial aspect of head and mouthparts of *Bittacus chlorostigma,* antennae partly removed (Bittacidae); **c**, larva of *Panorpa nuptialis* (Panorpidae). (c, modified from G. W. Byers, 1963.)

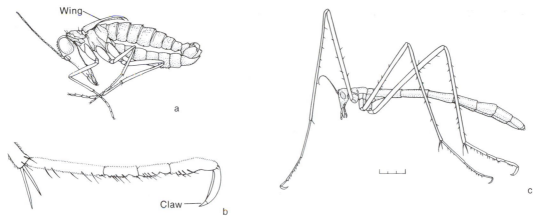

Figure 41.2   Mecoptera: **a**, Boreidae *(Boreus brevicaudus);* **b**, tarsus of *Apterobittacus;* **c**, Bittacidae *(Apterobittacus apterus;* scale equals 3 mm). (a, modified from G. W. Byers, 1961.)

Mecoptera, such as *Boreus,* are apparently herbivores or scavengers. In Bittacidae the single large tarsal claws fold bladelike against the terminal tarsomeres (Fig. 41.2b,c) and are used to snare small, soft-bodied insects. Panorpidae occasionally visit flowers, probably for nectar, but feed chiefly on dead insects and snails, which they locate on the ground or in low vegetation. *Panorpa* also steal prey items from spiderwebs and are able to move about in webs with some alacrity.

Larvae of Panorpidae and Bittacidae are caterpillar-like, with prolegs on the first eight abdominal segments. Larvae of Nannochoristidae (Australia and Chile) are elongate and slender, and those of Boreidae and Panorpodidae are scarabaeiform, all without abdominal legs.

Life histories have been described for only a few species. Copulation takes place on vegetation or the soil. In *Panorpa* and many Bittacidae, mating is accompanied by complex courtship behavior involving a sex attractant secreted by abdominal glands of males. During copulation males of some species present the female with a previously acquired food item, which she partially consumes. Males of *Panorpa* may also produce a salivary mass that hardens and is then consumed by the female during copulation, which may last several hours in some species. In Boreidae classical courtship is lacking; the male uses his shortened wings

to position the female and hold her on his back during copulation. Egg deposition is variable. Panorpidae and Choristidae oviposit small or large batches of delicate, smooth eggs in crevices, while Bittacidae scatter their tough, leathery, polygonal or spherical eggs on the soil. *Nannochorista* oviposits along stream margins. Eggs of most Mecoptera absorb water from the substrate and increase substantially in size subsequent to oviposition. Larvae emerge after a brief period or, as in *Harpobittacus* and *Apterobittacus,* after a diapause of several months. Habits of immatures are poorly known; most larvae apparently occur in moist litter, consuming dead insects or plant material. The larva of *Nannochorista* is aquatic and predatory; *Boreus* larvae feed on moss, and *Brachypanorpa* on herbaceous vegetation. Where known, pupation occurs in a cell in soil, decaying wood, or litter, without a cocoon. Prepupal diapause of several months is recorded in *Harpobittacus* and in *Panorpa.*

Mecoptera are a distinctive group of insects, almost always classified as a single order. It has been suggested, however (Hinton, 1981), that the family Nannochoristidae (Australia, New Zealand, South America) should be considered to be a separate order, Nannomecoptera, based on differences in adult mouthpart and larval leg structure. Nannomecoptera has been proposed as the sister taxon to Diptera or Siphonaptera (Wood and Borkent, 1989).

KEY TO NORTH AMERICAN FAMILIES OF

Mecoptera

1. Tarsi with single large claw about as long as apical tarsomere (Fig. 41.2b,c) **Bittacidae**

   Tarsi with claws paired, usually much shorter than apical tarsomere **2**

2(1). Ocelli absent or very small; wings with reticulate crossveining in costal space, or rudimentary, without veins **3**

   Ocelli present; wings with a few simple crossveins in costal space (wings may be reduced, but venation is preserved) **4**

3(2). Wings large, broad with complex venation; body dorsoventrally flattened **Meropeidae**

   Wings scalelike or bristlelike; body cylindrical or laterally flattened (Fig. 41.2a) **Boreidae**

4(2). Rostrum more than twice as long as width at base; wings fully developed in males and females **Panorpidae**

   Rostrum slightly longer than width at base; wings fully developed in males, reduced in females **Panorpodidae**

Bittacidae (Fig. 41.2b,c). Slender-bodied, long-legged Mecoptera resembling crane flies. Widespread and locally common in woodlands throughout the world, with nine species in North America. Adults commonly hang by their forelegs from vegetation, hence the name "hangflies" or "hanging flies."

Panorpidae (Fig. 41.1a,c). The most abundant and diverse family in the Northern Hemisphere, with about 40 species of *Panorpa* in eastern North America. The common name, "scorpionfly," refers to the enlarged genital bulb of males.

Panorpodidae. A single genus, *Brachypanorpa,* with two species in the Appalachian Mountains and two in the Pacific Northwest. Males are normally winged, females brachypterous.

Meropeidae. A single species, *Merope tuber* in the eastern states, and one species in Australia. Body relatively robust, flattened.

Boreidae (Fig. 41.2a) Restricted to the Northern Hemisphere, with about a dozen North American species, mostly in western mountain ranges. The dark-colored adults, about 4 mm long, sometimes crawl about on the surface of snow, hence the name "snow fleas."

SELECTED REFERENCES

G. W. Byers, 1963 (life history), 1965, 1971, 1989 (higher classification); G. W. Byers and Thornhill, 1983 (biology); Carpenter, 1931 (systematics of North American species); Cooper, 1972 (morphology), 1974 (life history); Ennos and Wootton, 1989 (flight dynamics); Hepburn, 1969 (morphology); Hinton, 1958 (phylogenetic relationships); Melzer et al., 1994 (larval eye); Penny, 1975 (evolution and phylogeny); Russell, 1982 (*Boreus* biology); Setty, 1940 (morphology, biology); Stehr, 1987 (immatures); Thornhill and Sauer, 1991, 1992 (mating behavior); Willman, 1987 (phylogeny).

42

ORDER DIPTERA (FLIES)

This fly, *Rhaphiomidas trochilus* (Apioceridae), exhibits the halteres (white object beneath the base of the fore wings) that characterize the Diptera. Adult apiocerids suck nectar of flowers, but spend much time simply sitting on sandy surfaces.

Adult. Minute to moderately large Endopterygota, usually with functional wings on the mesothorax. Head free, mobile, usually with large compound eyes, 3 ocelli; antennae multiarticulate, filiform or moniliform, or reduced, with terminal segment bristle-shaped; mouthparts diversely modified, usually as a proboscis for imbibing liquid food (Fig. 42.3). Prothorax reduced to a small, collarlike region, usually immovably fused to the mesothorax; mesonotum large, convex, usually divided into prescutum, scutum, and scutellum by transverse sutures (Fig. 42.2); metathorax reduced to a narrow, transverse band or lateral plates bearing halteres and metathoracic spiracles; wings with few or numerous longitudinal veins, few closed cells; legs usually with pulvilli, empodia, or other pretarsal structures.

Larva (Fig. 42.5). Usually maggotlike or vermiform, without legs, but occasionally with unsegmented pseudopods on various segments, especially in Nematocera; head capsule sclerotized as discrete tagma or undifferentiated; eyes and antennae reduced or absent. Thorax usually undifferentiated from abdominal region; abdomen almost always without cerci or urogomphi.

Pupa. Adecticous; obtect (Nematocera, Orthorrhapha) or exarate, enclosed in sclerotized **puparium** (Cyclorrhapha).

The Diptera comprise approximately 85,000 species, considerably fewer than in the Coleoptera, Hymenoptera, or Lepidoptera. Yet the flies must be considered one of the dominant groups of insects because of their extreme abundance over a great variety of ecological situations. Moreover, because of the small size and obscure habits of many flies, it seems likely that the eventual number will exceed 150,000 species.

Diptera include predators, parasites, and parasitoids, but the majority of species are saprophytic, and Diptera are usually the dominant invertebrate consumers of decaying vegetation or decomposing animal products. Flies are common visitors to flowers, which provide the only food for adults of some families (e.g., Bombyliidae, Conopidae, Acroceridae, Apioceridae). However, few Diptera have engaged in the intimate type of symbiotic relationship displayed by the angiosperm plants and Hymenoptera, and flies appear to be of sporadic importance in pollination except in Arctic and high montane regions. Almost no adult Diptera directly damage living plants, except by oviposition.

Larvae occur predominantly in moist or subaquatic situations, less frequently in either dry or strictly aquatic habitats. Many species of arid or subarid regions exist in moist microhabitats, such as animal burrows, rotting succulent vegetation, or feces, and many are specialized for quick exploitation of ephemeral sources of food and water. In such species the larval growth phase may span 1 to 2 weeks or less. Immatures of several families are internal parasites of vertebrates, the only significant group of insects to acquire this habit. A relatively small number of fly larvae attack living plants, as miners or borers in various plant parts, or by causing galls.

The chief economic importance of Diptera stems from their transmission of virulent diseases of humans and domestic animals. Malaria, yellow fever, and encephalitis, all transmitted exclusively by mosquitoes, have been three of most persistent and debilitating human diseases, particularly in tropical climates. House flies (Fig. 42.1), as well as other species, mechanically vector numerous microorganisms, especially those causing dysen-

Figure 42.1 *Musca domestica,* the house fly (Muscidae). (Modified from U.S. Department of Agriculture.)

teries and other enteric ailments. With notable exceptions such as the Hessian fly [*Mayetiola destructor* (Cecidomyiidae)], root maggots (Psilidae), and fruit flies (Tephritidae, Otitidae), Diptera cause relatively minor damage to crops, as would be expected from their predominance as decomposers rather than herbivores. In addition, many phytophagous Diptera, such as fruit flies (Tephritidae), tend to be highly host-specific and susceptible to control by cultural practices such as staggering crops, isolating plantings, and removing crop debris.

MORPHOLOGICAL ADAPTATIONS

The most conspicuous characteristic of the body plan of adult Diptera is the high degree of adaptation for rapid, efficient flight. For most flies, flight is required for location of food or other resources, escape from predators, and successful reproduction. Diptera, along with Hymenoptera and Lepidoptera, are notable for the strong partitioning of resources between immatures and adults. For most species, highly developed flying ability has facilitated such partitioning.

Obvious adaptations to aerial life include foreshortening and streamlining of the body and shortening of the antennae, as in most Cyclorrhapha.

More profound modifications involve structural changes in the thorax. In all Diptera, the wing-bearing mesothorax has become enormously enlarged, while the prothorax and metathorax are reduced to narrow, collarlike regions (Fig. 42.2). The prothorax, marked by the attachments of the forelegs, is largest in Nematocera; in most Brachycera it is represented by narrow pleural sclerites and dorsolateral prominences, or **humeri.** The metathorax is represented in most species by small lateral sclerites supporting the halteres. The metapleura and sterna are closely associated with the mesothorax and difficult to distinguish.

The architecture of the mesothorax is complicated by the development of secondary sutures, as well as by distortion caused by enlargement, especially of the notal region. Both the notum and postnotum are usually distinct (Fig. 42.2a,b). The notum is divided into **prescutum, scutum,** and **scutellum** by transverse sutures, which are represented internally by ridges, providing braces against the stress of the highly developed flight muscles. In Nematocera the postnotum is a relatively large, dorsal sclerite (Fig. 42.2b). In Brachycera the scutellum is prominent, becoming especially enlarged in Cyclorrhapha, where it may

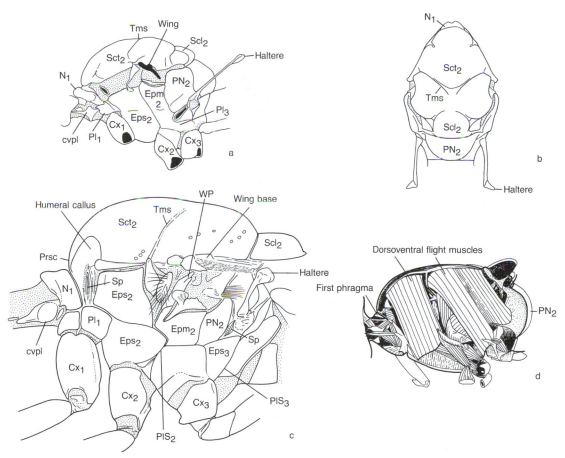

Figure 42.2 Structural features of Diptera: **a**, lateral aspect of pterothorax of Nematocera (Tipulidae, *Holorusia*); **b**, dorsal aspect of same; **c**, lateral aspect of pterothorax of Brachycera (Asilidae, *Mallophora*); **d**, longitudinal section through pterothorax of *Tabanus* (Tabanidae). cvpl, cervical plates; Cx, coxa; Epm, epimeron; Eps, episternum; N, notum; Pl, pleuron; PlS, pleural suture; PN, postnotum; Prsc, prescutum; Scl, scutellum; Sct, scutum; Sp, spiracle; Tms, transmesonotal suture; WP, pleural wing process. (c, modified from Bonhag, 1949; d, modified from Cole, 1969.)

be divided into distinct scutellum and subscutellum (Figs. 42.2d, 42.18b–e). In these flies the postnotum is represented by small, lateral plates. Enlargement of the scutellum probably provides additional area for muscle attachment, which is further increased in most Diptera by the large, laminate **phragmata** that project ventrally into the thoracic cavity (Fig. 42.2c). The **prephragma** is often small or ridge-shaped, but the **postphragma** (postnotum) is almost always a prominent, convex plate, sometimes nearly separating the internal thoracic and abdominal spaces.

The pleural region of the mesothorax is exceedingly variable and complex, frequently with secondary sutures, incorporation of portions of the coxae, and radical shifts in size, shape, or position of certain sclerites. The principal landmarks are the pleural wing suture, straight in Nematocera (Fig. 42.2a) but sinuous in most Brachycera (Fig. 42.2d), and the metathoracic spiracle, located at the anterior dorsal margin of the metapleuron, usually just below the haltere. The homology of certain structures and their functional significance are obscure, but the topographic features illustrated in Fig. 42.2 are important for identification of many taxa.

The thoracic specializations of Diptera are correlated, of course, with two-winged flight. Whereas the wings of most Hymenoptera and Lepidoptera are coupled during flight, functioning as a single airfoil, the hind wings of Diptera are transformed into **halteres** (Fig. 42.2), which are vibrated rapidly during flight, apparently functioning as gyroscopic balancing organs. In some Brachycera (the Calyptratae), the halteres are enclosed in cavities covered by **calypters,** foldings of the trailing edge of the wing base (Fig. 42.18d). Venation is highly variable. Some Nematocera have extensive, complex venation, not greatly different from that in Mecoptera (Figs. 42.6a, 42.7a). In most families, some reduction in number of veins and closed cells is apparent (Figs. 42.10, 42.14), often accompanied by a crowding of veins (usually the costal, subcostal, and first radial) into the leading edge of the wings, a strengthening configuration that occurs repeatedly in insects with

a rapid wing beat. Also correlated with the development of more efficient flight is an increasing consolidation of the nervous system. In Nematocera several abdominal and three thoracic ganglia are usually present. In Brachycera these numbers are variously reduced, and in the most specialized Cyclorrhapha all are incorporated into a single, large ganglionic mass in the thorax.

Diptera all imbibe liquid food, but unlike the Hemiptera, which are narrowly specialized for piercing and sucking, flies have evolved a variety of modes of feeding, and some Syrphidae and Bombyliidae ingest finely particulate material such as pollen along with liquids. In certain Nematocera mouthparts (Fig. 42.3a) are similar to those of Mecoptera, with bladelike mandibles, maxillae with lacinia and multiarticulate palpi, and elongate labium with one- or two-segmented fleshy palpi or **labellar** lobes. This configuration appears to be primitive. In Tabanidae and some bloodsucking Nematocera, the bladelike mandibles, maxillae (laciniae), labrum, and hypopharynx are enclosed in the stout U-shaped labium, which acts as a guide for the cutting mouthparts. The labellum bears numerous minute surface channels, or **pseudotracheae,** for sponging up oozing blood by capillarity. In Culicidae the mandibles and laciniae are fine, styliform blades, and the labrum and hypopharynx are delicate tubes for uptake of food and delivery of saliva (Fig. 42.3b). All are ensheathed by the flexible labium and form a piercing structure analogous to that of Hemiptera (see Fig. 34.2e–g). In predatory Brachycera (Asilidae, Empididae), the labial sheath is rigidly sclerotized, and the laciniae are stout blades used to pierce the integument of prey. Mandibles are absent, and the labellum vestigial. The salivary glands have become specialized in various bloodsucking families to deliver anticoagulants, or in predators to produce paralytic toxins, which are delivered through the hypopharynx.

In many flies, including nearly all Cyclorrhapha, the mouthparts show a highly derived configuration, consisting of the labrum, labium, and hypopharnyx combined to form a thick, fleshy **rostrum,** or proboscis, terminated by a greatly

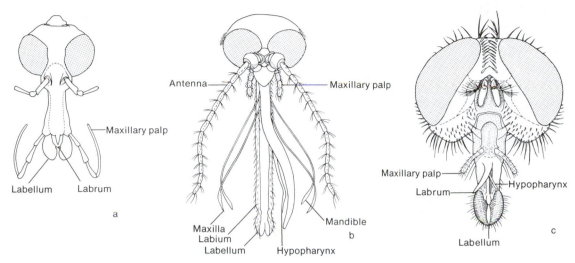

Figure 42.3 Mouthparts of adult Diptera: **a**, primitive nematoceran type (Tipulidae, *Ctenacroscelis*); **b**, piercing-sucking type (Culicidae, *Culex*); **c**, sponging type (Muscidae, *Musca*). (Modified from Crampton, 1925 and Snodgrass, 1932.)

enlarged labellum (Fig. 42.3c). Soluble constituents of solid foods are dissolved in saliva or regurgitated gut contents before being swallowed. Predation and parasitism in these specialized forms, which have lost the laciniae, are accomplished by the secondary development of sharp, sclerotized projections of the pseudotracheae or preoral region of the labellum, which are used to rasp or pierce the host integument. In the highly specialized tsetse fly *(Glossina),* the rostrum has secondarily become slender and elongate, with highly reduced labella. The stout, elongate maxillary palpi are closely applied to the rostrum, providing additional rigidity.

Most Diptera have one or sometimes three diverticula of the foregut, which function as a crop. Liquid meals, large in volume, may be temporarily held in the diverticula, to be gradually released into the intestine.

A peculiarity of the order is the temporary or permanent rotation of the male genitalia 90 to 360° from their original position. Permanent rotation, termed torsion, occurs during pupal development. When rotation proceeds through 180° ("inversion"), the anus comes to lie below the genitalia. If rotation has proceeded through 360° ("circum-

version"), the original dorsal position of the anus is restored, but the sperm duct is twisted around the hindgut. The function of torsion is not understood, but the degree of twisting is of taxonomic significance, as explained below. The male terminalia are also often bent anteriorly or laterally (flexion). In Cyclorrhapha both torsion and flexion are apparent and are associated with marked asymmetry of the external genitalia. In various taxa rotation and flexion may be in part permanent and in part temporary at the time of mating.

Diptera couple with the male and female initially facing the same direction, usually with the male dorsal. After coupling, however, the pair turn tail to tail, and in flies without flexion or rotation, the male is turned onto its back. Rotation and flexion are apparently adaptations that allow the male to remain upright in the tail-to-tail position and have apparently occurred independently many times in Diptera, producing the complexity of male genitalic orientation. Circumversion, however, appears to have occurred a single time and is one of the defining characters of the Cyclorrhapha.

Diptera are remarkable for the profound morphological difference between adults and immatures. Larvae usually have 3 thoracic and 8 or 9 discernible abdominal segments, but these may be fused or secondarily increased by subdivision, obscuring the original segmentation. In Bibionidae 10 pairs of spiracles are present (prothorax, metathorax, 8 abdominal segments) but in other Diptera the number is reduced. Often only the anterior and posterior spiracles are retained in an **amphipneustic** arrangement, or only the anterior or posterior are present (**propneustic** and **metapneustic,** respectively). Metapneustic spiracles may be associated with a breathing siphon, as in Culicidae. In several families with aquatic larvae, such as Blephariceridae and Simuliidae, all spiracles have been lost, producing an **apneustic** arrangement.

In various families the configuration of functional spiracles changes in different instars. All dipterous larvae are without true legs, and the great majority lack trunk appendages entirely. Leglessness among insect larvae is mostly restricted to borers in plant tissues (Coleoptera, Lepidoptera, Hymenoptera), parasitoids (Hymenoptera, Coleoptera), and those inhabiting other protected microenvironments with assured food supply (nest-making Hymenoptera). Reduction of locomotory and cranial structures in fly larvae is probably related to their widespread occurrence in rich nutrient sources, where it is advantageous to feed and grow as rapidly as possible. Under such circumstances extreme divergence between adults and larvae is possible, the larvae being specialized as a feeding stage, the adults as a reproductive and dispersal stage. Divergence not only allows specialization for rapid exploitation of temporary food sources, but also decreases competition between larvae and adults.

Larvae of the major taxonomic groups of Diptera differ in important morphological features. In most Nematocera the **eucephalic** head capsule is sclerotized, with functional mandibles, maxillae, stemmata, and antennae (Fig. 42.4a). Mandibles move in a transverse plane, as in most mandibulate adult insects. In Orthorrhapha and some Nematocera, such as Tipulidae, the head capsule is unsclerotized posteriorly (**hemicephalic)** (Fig. 42.4b) and usually retracted within the thorax, which is tough and leathery. Mandibles move in a more nearly vertical plane. Cyclorrhapha are **acephalic,** without a differentiated head. Mouthparts consist of a pair of protrusible, curved "mouth hooks" and associated internal sclerites (Fig. 42.4c). To a certain extent larval morphology corresponds to habitat. For example, free-swimming nematoceran larvae frequently have pseudopodial appendages, sometimes jointed and bearing terminal bristles or hairs (Fig. 42.5a,b). Internal parasitoids and larvae inhabiting rich food sources, such as rotting flesh or dung, usually lack appendages (Fig. 42.5c).

LIFE HISTORY

Diptera are predominantly bisexual, with most cases of parthenogenesis recorded from the Nematocera. Eggs are usually small and ovoid, with thin, unsculptured chorion, and are typically deposited singly or in small masses on or near the larval food. Eggs usually hatch quickly, and egg diapause is infrequent. Cyclorrhapha include several ovoviviparous (larviparous) forms, as well as others which oviposit eggs that hatch immediately after deposition (ovolarviparous). Considerable larval development may occur in the mother's oviducts, and in Glossinidae (tsetse flies), Hippoboscidae, Streblidae, and Nycteribiidae, the larvae feed to maturity on oviduct secretions, pupating almost immediately after deposition. Larval development in many families is extremely rapid, with many generations per year, but many aquatic species and free-living predators, such as Therevidae and Asilidae, have a single generation each year. Four larval-feeding instars are typical, more being recorded for some species and only three in Cyclorrhapha, where the fourth instar has become specialized to produce the protective puparium that encloses the pupa. Aquatic families include some that use dissolved oxygen for respiration (Chironomidae, Simuliidae, Blephariceridae), but most obtain air either by rising periodically to the surface (Culicidae) or extending a breathing siphon to the surface (Syrphidae). Some mosquito larvae have spiracles modified to pierce the tissues of aquatic plants in order to obtain the contained air.

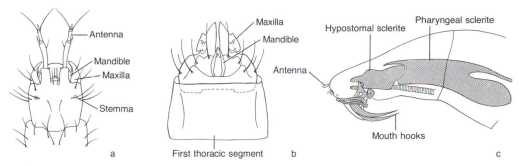

Figure 42.4   Cephalic structure of larval Diptera: **a**, Nematocera (Chironomidae, *Tanytarsus fatigans*); **b**, Brachycera, Orthorrhapha (Asilidae, *Dasyllis*), **c**, Brachycera, Cyclorrhapha (Anthomyiidae, *Spilogona riparia*). (a, modified from Branch, 1923; b, modified from Malloch, 1917; c, modified from Keilin, 1917.)

Many endoparasitic larvae maintain an opening through the host's integument for breathing (Tachinidae, Oestridae), while others use dissolved oxygen.

Pupation occurs either in the larval substrate or a short distance away, where drier conditions prevail. Obtect pupae frequently bear tubercles, horns, or other prominences on various parts of the body. In most cases these are probably used by the pharate adult to work its way through loose substrates before emerging from the pupal skin. In forms with coarctate pupae the puparium is burst along a transverse line of weakness by expansion of an eversible elastic sac, or **ptilinum,** which is extruded through the **ptilinal suture,** just above the antennal bases of the pharate adult. In the mature adult this suture is permanently sealed. Orthorrhapha and Cyclorrhapha pupate naked; many Nematocera spin silk cocoons. Adult emergence in aquatic forms usually occurs at the surface (Culicidae, Chironomidae, Tabanidae. etc.), but in cases in which the puparium is attached to the substrate the adult emerges under water and floats quickly to the surface (Simuliidae, Blephariceridae).

Adult Diptera characteristically have rather short life spans of a few days to a few weeks, and some ephemeral forms do not feed. Courtship and mating commonly take place in the air, at least in part. Aerial swarming behavior, in which males form dense, dancing masses, often near character-istic landmarks or over water, precedes mating in many Nematocera and some Brachycera, notably Empididae. A female entering the swarm pairs with a male; usually the coupled flies then leave the swarm. Elaborate courtship behavior typifies various families. Empidid males frequently present females with corpses of prey. Many drosophilids and tephritids show complex premating behavioral displays involving posturing or visual presentation of the pictured wings by the males. In both families courtship differences are important isolating mechanisms among species.

CLASSIFICATION

There has been continuing proliferation of family names in the Diptera, often as designations for one or a few small genera that did not entirely conform to larger families that were better known and more widely accepted. There have also been proposals for revolutionary reclassifications of all or part of the order (Griffiths, 1972). The classification presented here is conservative on both counts. Most of the family names omitted apply to very uncommonly encountered insects, often with extremely specialized habits. Exhaustive treatments of North American Diptera include those of Curran (1934), Cole (1969), Griffiths (1980–), and McAlpine et al. (1981–89). Larvae are treated in Stehr (1991) and Ferrar (1988).

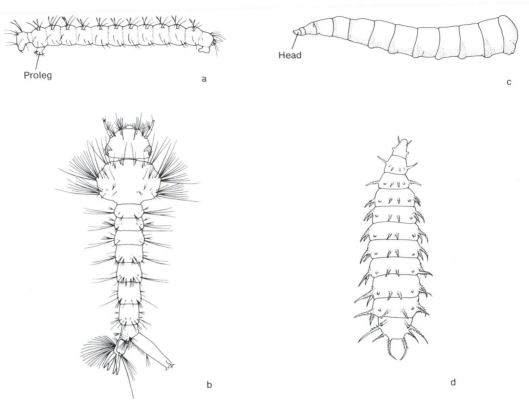

Figure 42.5   Representative Diptera larvae: **a**, Nematocera (Ceratopogonidae, *Forcipomyia specularis*); **b**, Nematocera (Culicidae, *Aëdes stimulans*); **c**, Brachycera, Cyclorrhapha (Muscidae, *Musca domestica*); **d**, Brachycera, Cyclorrhapha (Anthomyiidae, *Fannia canicularis*). (a, modified from Malloch, 1917; b, modified from Matheson, 1945; c, modified from Patton, 1931; d, modified from Detweiler, 1929.)

Although flies are predominantly aerial as adults, about 20 families include members that are brachypterous or apterous. The following keys do not accommodate these flightless forms.

KEY TO THE SUBORDERS OF

Diptera

1. Antennae with at least 6 segments, usually filiform or moniliform and frequently longer than thorax; maxillary palpi with 3 to 5 segments; pleural suture straight (Fig. 42.2a)   **Nematocera** (p. 500)
Antennae with fewer than 6 segments,[1] terminal

[1] The terminal segment is sometimes annulate, giving appearance of more than six segments. The annulations are shallower and less well defined than the true divisions between the basal segments.

segment usually either elongate or with bristlelike style or arista; maxillary palpi with 1 to 2 segments; pleural suture with two right-angle bends (Fig. 42.2d)   **Brachycera** (p. 505)

SUBORDER NEMATOCERA
Adults. Antennae with 6 to 14 or more segments, usually filiform or moniliform, frequently longer than head and thorax, never with style or arista; maxillary palpi usually elongate, drooping, with 3 to 5 segments; mesothorax with pleural suture straight or slightly curved (Fig. 42.2a); wings usually without discal cell, and with veins Cu_2 and 2A never intersecting, usually diverging (Fig. 42.7a).

Larvae. Head capsule sclerotized, with labium, maxillae, and horizontally biting, toothed mandibles.

Pupa. Obtect, free (not enclosed in larval cuticle).

The flies comprising the suborder Nematocera are mostly frail-bodied, gnatlike insects with long, slender legs and thin, delicate cuticle. Body size is commonly about 1 cm or less, with the notable exception of some Tipulidae, which have wingspans greater than 100 mm. The great majority of Nematocera are dull yellowish, brown, or black, usually nocturnal insects.

Structurally, Nematocera include the most primitive Diptera. In adult Blephariceridae, which are predatory, the mandibles are stout, bladelike cutting organs. The maxillary palpi are long, with four segments, and the labium is elongate, not distally enlarged or specially textured as a labellum. Such mouthparts, which superficially are not greatly different from those of Bittacidae (Mecoptera) (see Fig. 41.1b), except in the development of the maxillae, may represent the primitive condition in Diptera. In Simuliidae, Culicidae, Ceratopogonidae, and Psychodidae, the mouthparts are only slightly modified from this condition and are used to puncture the integument, usually of vertebrate hosts. Specialization for ectoparasitism on vertebrates probably represents the initial adaptive radiation of the Diptera. In nonbiting Nematocera, the mouthparts are either vestigial (Deuterophlebiidae, Cecidomyiidae, Chironomidae) or adapted for imbibing nectar or other free liquids (Tipulidae, Scatopsidae, Bibionidae, Anisopodidae), with reduced mandibles, maxillae, and labellum.

Relatively large, articulated prothoracic sclerites are present in Tipulidae, Tanyderidae, Bibionidae, and Scatopsidae; a distinct metanotal sclerite is present in Psychodidae. Wing venation is highly variable. Four radial branches are characteristic of several families, and five radials are present in Tanyderidae, together with four branches of the media (Fig. 42.6a). This is probably the most generalized venation in the order; most families show fusion or loss of certain veins (Figs. 42.6b, 42.11, 42.14), and in Cecidomyiidae and Simuliidae all but two or three anterior veins are extremely weak or absent. Foldings or creases in the wing membrane form a faint network between the normal veins in Blephariceridae and Deuterophlebiidae (Fig. 42.6d). This peculiarity is apparently an adaptation to allow the fully hardened, folded wings to be used immediately after the adults emerge from the torrents and cataracts where pupation occurs.

Adult Nematocera are predominantly nocturnal, or if diurnally active, usually restrict their activity to deep shade or to periods of overcast skies. The major exception is the family Bibionidae, which consists of diurnal, flower-visiting

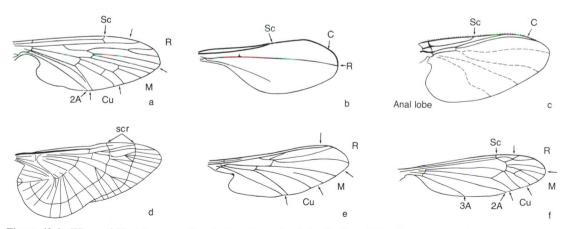

Figure 42.6 Wings of Nematocera: **a**, Tanyderidae *(Protoplasa);* **b**, Cecidomyiidae *(Contarinia sorghicola);* **c**, Simuliidae *(Simulium argus);* **d**, Deuterophlebiidae *(Deuterophlebia);* **e**, Blephariceridae *(Agathon comstocki);* **f**, Tipulidae *(Nephrotoma wulpiana).* (Sc, subcosta; scr, secondary creases; arrows at margin of wings (between veins) delimit branches of major veins. (a, modified from Alexander; d, modified from Cole, 1969.)

members. Food habits of most adults are obscure, but members of families such as Scatopsidae, Anisopodidae, Tipulidae, Sciaridae, and Mycetophilidae probably imbibe liquid decay products from plant or animal material or nectar or honeydew from plants. Most Chironomidae, Deuterophlebiidae, Cecidomyiidae, as well as some members of other families do not feed as adults. Some Blephariceridae prey on soft-bodied flying insects, including other flies and mayflies. Ceratopogonidae, commonly known for their habit of sucking blood from vertebrates, are predominantly parasites of other insects, usually much larger than themselves. Female Simuliidae (black flies) and Culicidae (mosquitoes) feed on vertebrate blood, attacking reptiles, birds, and mammals. In many species a blood meal is required for production of fertile eggs. Members of both families transmit serious diseases of humans (see Chap. 13), and this association constitutes the chief economic importance of nematocerous flies.

Swarming behavior occurs in many families, being especially pronounced in Chironomidae, Bibionidae, and Mycetophilidae. Eggs are deposited on or near the larval substrate, usually singly, occasionally in adhering masses. Egg rafts, formed by one to many females, are characteristic of many mosquitoes. The great majority of Nematocera reproduce bisexually, but paedogenetic parthenogenesis occurs in *Miastor* (Cecidomyiidae), in which larviform females reproduce ovoviviparously. *Miastor* inhabits the fermenting phloem beneath bark of dead trees. Parthenogenesis is apparently a mechanism for rapidly exploiting this rich food source. When the tree begins to dry out, winged, bisexual adults appear.

Larvae of Nematocera include aquatic, subaquatic, and terrestrial forms. Blephariceridae and Deuterophlebiidae occupy torrents and cataracts, clinging to stones with suckers and grazing on diatoms and algae. Simuliidae occur primarily in moving water, including rapids, attaching either to stones or larger aquatic organisms with the single sucker on the abdominal apex. All pupate near the site of larval attachment. Chironomidae and Culicidae occur in a diverse array of aquatic habitats. Chironomid larvae are usually bottom dwellers, frequently forming tubes in mud, silt, or under stones in still or moving water. Both well-oxygenated and stagnant waters are occupied, and many species have oxygen-binding pigments, including hemoglobin in the blood. Most chironomid larvae are detritivores, but the entire subfamily Tanypodinae is predaceous. *Clunio* larvae (Chironomidae) inhabit intertidal cracks and crevices throughout the world. Emergence of the terrestrial adults is usually synchronized with full moons, ensuring favorable low tides. *Pontomyia,* inhabiting coral reefs in the Indo-Pacific region, is marine in all instars. The highly modified males superficially resemble Veliidae (Hemiptera). They skate about on the surface of tide pools, carrying the vermiform females.

Culicid larvae, or mosquito wrigglers, are primarily inhabitants of standing water, but backwaters or eddies in slowly moving streams, rivers, or ditches offer suitable habitats as well. Many species are specialized for utilizing very small, enclosed bodies of water, such as rot holes in trees, potholes in rocks, crab burrows, or the water entrapped by bromeliads, pitcher plants, or certain other plants. Rainwater held by discarded cans and bottles provides the preferred larval habitat for some species, which have, in effect, become domesticated. Mosquito larvae are mostly filter feeders on plankton; a few species, including the entire subfamily Chaoborinae (phantom midges), are predatory.

Most dipterous larvae, aquatic or not, require atmospheric oxygen for respiration. The major exceptions are the Blephariceridae, Deuterophlebiidae, and Chironomidae, which respire either through tracheal gills or, occasionally, with plastrons. Some mosquito larvae can respire slowly under water and may overwinter in this way, but most other "aquatic" Nematocera will drown if long separated from the air. Most of these subaquatic larvae inhabit littoral regions of streams or ponds, semiliquid decaying vegetation or sewage, or other wet situations. A few, such as Ptychopteridae, extend a flexible breathing tube to the surface for air, an adaptation encountered again in some syrphid larvae, but most have only short breathing tubes or none. Many Tipulidae and Ceratopogonidae, as well as the Psychodidae, Anisopodidae, Thaumaleidae, Tanyderidae, and Trichoceridae, can properly be classified as subaquatic. Within their semiliquid habitats these larvae function mostly as decomposers, less frequently as herbivores or predators.

The families Scatopsidae, Bibionidae, Sciaridae, and Mycetophilidae, as well as some Ceratopogonidae and Tipulidae, are best considered terrestrial, though the larval medium is usually moist. Larvae in this category are predominantly decomposers, differing chiefly in the type of material consumed. Mycetophilidae and Sciaridae are commonly associated with fungi, ranging from molds to mushrooms and bracket fungi; Scatopsidae frequent dung, and Bibionidae, moist humus. Bibionid larvae may feed on the roots of plants, occasionally becoming economically important. In all these families exceptional members are predatory. For example, the luminescent "glowworms" of Waitomo Cave in New Zealand are predatory mycetophilid larvae that dangle mucilaginous filaments to entangle their prey. The mycetophilid genus *Planivora* is an internal parasite of terrestrial planarians.

The Cecidomyiidae cannot be accommodated in a simple classification by larval habitat. These minute, fragile flies are best known as the causative agents of galls, which the larvae produce on a wide variety of herbaceous plants. Other members mine leaves or stems, and include agriculturally important species, such as the Hessian fly *(Mayetiola)*. A few are predators or parasitoids of aphids, coccids, and other soft-bodied insects, and some live in forest litter and soil, probably feeding on fungi.

In Simuliidae and Mycetophilidae pupation occurs within silken cocoons, but the obtect pupae of most families are naked. Blephariceridae, Deuterophlebiidae, and Simuliidae, which often pupate in waterfalls or rapids, surface and immediately take flight without an initial period of wing expansion and hardening of the cuticle. Culicids are exceptional in remaining fully vagile throughout the pupal period.

KEY TO THE FAMILIES OF THE SUBORDER

Nematocera

1. Wings about 3 times or less as long as broad, with fringe of marginal setae less than half as long as wing width   **2**
   Wings at least 5 times as long as broad, with fringe of marginal setae more than half as long as wing width   **Nymphomyiidae**

2(1). Wings with secondary network of fine creases or folds in addition to venation (Fig. 42.6d); mesonotum with transverse suture straight
   Deuterophlebiidae and **Blephariceridae**
   Wings without secondary network of creases or folds; mesonotum with transverse suture straight or V-shaped   **3**

3(2). Mesonotum with transverse suture V-shaped, extending to vicinity of scutellum (Fig. 42.2b)   **4**
   Mesonotum with transverse suture straight or weakly curved   **6**

4(3). Head with 2 to 3 ocelli   **Trichoceridae**
   Head without ocelli   **5**

5(4). Wing with 2 anal veins reaching margin (Fig. 42.6f); antennae usually with 12 to 13 segments   **Tipulidae**
   Wing with 1 anal vein reaching margin (Fig. 42.6a); antennae with 15 to 25 segments
   Ptychopteridae and **Tanyderidae**

6(3). Ocelli present   **7**
   Ocelli absent   **13**

7(6). Costal vein ending at or near wing tip (Fig. 42.6c–f)   **8**
   Costal vein weakened behind wing tip but continuing around posterior margin of wing (Fig. 42.6b); wings with 3 to 6 longitudinal veins, usually without crossveins; minute flies with long, many-segmented antennae
   Cecidomyiidae (Fig. 42.8b)

8(7). Wing with discal cell (M_2) present (Fig. 42.7a)   **9**
   Wing without discal cell (Fig. 42.7b)   **10**

9(8). Wing with cell M_3 open externally (Fig. 42.7a–e)   **Anisopodidae** (incl. **Pachyneuridae**)
   Wing with cell M_3 closed (Fig. 42.11a) (aberrant orthorrhaphous Brachycera)
   Xylophagidae

10(8). Tibiae without apical spurs (tibial apex may be pointed, spurlike); wings with anterior veins

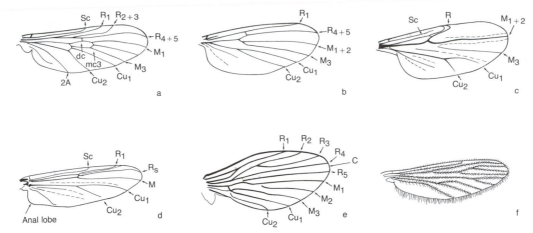

Figure 42.7 Wings of Nematocera: **a**, Anisopodidae *(Sylvicola fenestralis);* **b**, Mycetophilidae *(Mycetophila fungorum);* **c**, Ceratopogonidae *(Culicoides variipennis);* **d**, Chironomidae; **e**, Psychodidae *(Pericoma truncata);* **f**, Culicidae *(Culex tarsalis).* dc, discal cell or cell M_2; mc3, cell M_3.

usually much thicker than posterior veins
      18

Tibia with articulated spurs near apex, wings usually with all veins of about same thickness
      11

11(10). Tarsi with broad, fleshy pads (pulvilli) beneath claws (Fig. 42.9a)   **Bibionidae**

Tarsi without pulvilli   **12**

12(11). Eyes meeting above antennal bases (Fig. 42.9b); coxae moderate in length (Fig. 42.8b); occasionally elongate   **Sciaridae**

Eyes separated dorsally (Fig. 42.9c); coxae nearly always greatly elongate
    Mycetophilidae[1]

13(6). Costal vein ending at or near wing tip (Fig. 42.7c,d)   **14**

Costal vein weakened behind wing tip, but continuing around posterior margin (Fig. 42.7e)   **16**

14(13). Wings broad, oval, with large anal lobe; anterior veins always much stronger than faint posterior veins (Fig. 42.6c)
    Simuliidae (Fig. 42.8c)

Wings usually lanceolate, elongate, with anal lobe small or absent (Fig. 42.7c,d); venation variable   **15**

15(14). Wing with radial veins intersecting costal margin near midpoint (Fig. 42.7c); radial veins usually stronger than posterior veins; mouthparts of piercing type
    Ceratopogonidae (Fig. 42.8e)

Wing with radial veins extending nearly to apex (Fig. 42.7d); radial and posterior veins similar; mouthparts without piercing mandibles
    Chironomidae

16(13). Wings broad, with pointed tip; membrane usually densely hairy; radial vein usually with 5 branches (Fig. 42.7e)
    Psychodidae (Fig. 42.8d)

Wings usually narrow, lanceolate, and bare (veins may bear short hairs); radial vein usually with 3 to 4 or fewer branches   **17**

17(16). Wings with 7 or fewer longitudinal veins reaching margin (Fig. 42.6b); minute, frail flies some   **Cecidomyiidae** and **Thaumaleidae**

Wings with at least 9 longitudinal veins reaching margin (Fig. 42.7f)   **Culicidae**
    (incl. **Chaoboridae, Dixidae** (Fig. 42.8a))

18(10). Wing with costal vein ending much before apex; costal and radial veins darkly pigmented, prominent, others faint
    Scatopsidae (including **Synneuridae**)

Wing with costal vein continuing to apex; anterior and posterior veins of about equal prominence   **Axymyiidae**

---

[1] A few Anisopodidae and Pachyneuridae will key out here.

Figure 42.8 Representative Nematocera: **a**, Culicidae *(Anopheles freeborni);* **b**, Cecidomyiidae *(Mayetiola destructor);* **c**, Simuliidae *(Prosimulium mixtum);* **d**, Psychodidae *(Psychoda);* **e**, Ceratopogonidae *(Culicoides variipennis).* (a, modified from Wilson, 1904; c, modified from Peterson, 1970; b, d, modified from Cole, 1969; e, modified from Hope, 1932.)

SUBORDER BRACHYCERA
Adults. Mostly stouter-bodied than Nematocera, with antennae short, with terminal style or arista; maxillary palpi usually short, with 1 to 2 porrect segments; pleural suture bent around sternopleural sclerite (Fig. 42.2d); wings with discal cell present (Orthorrhapha) or absent (Cyclorrhapha), anal cell narrowed before wing margin (Figs. 42.10, 42.11, 42.14).

    Larvae. Head capsule incomplete or vestigial, usually retractable into thorax; mandibles biting in vertical plane (Fig. 42.4b), or mouthparts represented by specialized "mouthhooks" (Fig. 42.4c).

Brachycera comprise the great majority of flies, including many familiar groups such as horse flies, house flies, and blow flies. The suborder consists of two groups of families, termed **divisions,**[2] which are difficult to separate by adult features but are well differentiated by larval and

[2] The taxonomic divisions of Diptera are groups of families and should not be confused with the divisions Exopterygota and Endopterygota. The taxonomic category **series** is used to designate major subdivisions of the Cyclorrhapha.

Figure 42.9 Taxonomic characters of Nematocera: **a**, tarsus of Bibionidae; **b**, frontal view of head of Sciaridae; **c**, frontal view of head of Mycetophilidae.

pupal characteristics. The division **Orthorrhapha** is characterized by hemicephalic larvae with vertically biting mandibles. The pupa is obtect (exception: exarate, coarctate in Stratiomyidae). In the division **Cyclorrhapha,** larvae are acephalic, without true mandibles. The pupa is coarctate, enclosed in a hardened puparium formed from the cuticle of the last larval instar. Cyclorrhapha are further subdivided as the series **Schizophora,** characterized by the presence of the ptilinal suture just above the antennae. In the series **Aschiza** the ptilinum and ptilinal suture are absent, although pupation is within a puparium. In the following key the Aschiza segregate with the Orthorrhapha, with which they share several adult features.

KEY TO THE MAJOR DIVISIONS OF
THE SUBORDER

Brachycera

1. Head with U-shaped ptilinal suture just above antennae (Fig. 42.15a); antennae usually with 3 segments, terminal segment with arista or style attached dorsally before apex (Fig. 42.15a–h)
        Cyclorrhapha, series **Schizophora** (p. 513, 515)
    Head without U-shaped ptilinal suture; antennae 3- to 5-segmented, terminal segments elongate or with style or arista attached terminally (Fig. 42.13c)   **Orthorrhapha** and **Cyclorrhapha,** series **Aschiza** (p. 506, 510)

DIVISION ORTHORRHAPHA

Differentiated from Cyclorrhapha primarily by the obtect pupae (pupae exarate, enclosed in puparium in

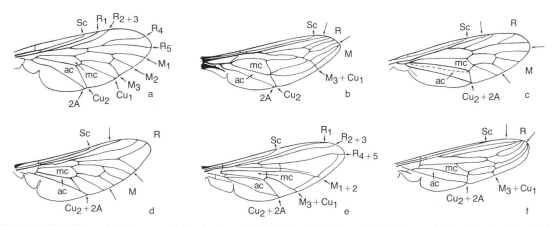

Figure 42.10 Wings of Brachycera, Orthorrhapha: **a**, Rhagionidae *(Symphoromyia);* **b**, Nemestrinidae *(Neorhynchocephalus sackeni);* **c**, Xylophagidae *(Xylophagus decorus);* **d**, Tabanidae *(Tabanus laticeps);* **e**, Syrphidae *(Syrphus opinator);* **f**, Apioceridae *(Rhaphiomidas maehleri).* ac, anal cell; mc, median cell; marginal arrows delimit branches of major veins.

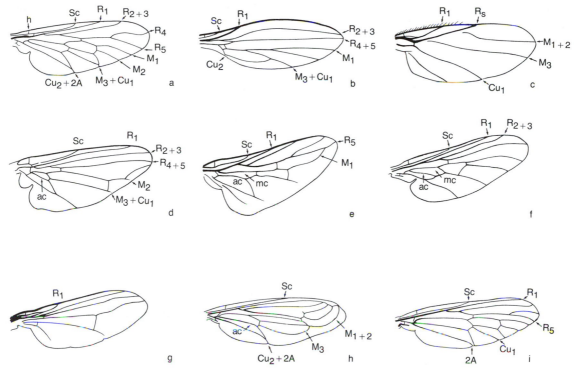

Figure 42.11   Wings of Brachycera, Orthorrhapha: **a**, Therevidae *(Thereva vialis)*; **b**, Lonchopteridae *(Lonchoptera)*; **c**, Phoridae *(Chaetoneurophora variabilis)*; **d**, Platypezidae *(Calotarsa insignis)*; **e**, Scenopinidae *(Metatrichia bulbosa)*; **f**, Empididae *(Rhamphomyia)*; **g**, Dolichopodidae *(Tachytrechus angustipennis)*; **h**, Mydidae *(Nemomydas pantherinus)*; **i**, Asilidae *(Cyrtopogon montanus)*. ac, anal cell; mc, median cell.

Cyclorrhapha). Exception: Stratiomyidae, with pupa obtect, enclosed in unmodified larval cuticle.

The Orthorrhapha comprise an extremely diverse assemblage of flies, including several of the largest families. Body form varies from rather frail, slender types similar to Nematocera (e.g., Rhagionidae) to those with stout, compact shape resembling that of Cyclorrhapha (e.g., Tabanidae and Dolichopodidae). Many Empididae and Dolichopodidae are minute, but body size in Orthorrhapha averages larger than in Nematocera or Cyclorrhapha. Members of the Asilidae (robber flies), Mydidae, and Apioceridae include the largest flies, with body lengths up to 60 mm. Most Orthorrhapha are active diurnal insects, often conspicuous on flowers or foliage (Color photo 7b).

Most of their morphological features suggest

that the Orthorrhapha are intermediate between the Nematocera and Cyclorrhapha, from which some Orthorrhapha cannot be distinguished on the basis of adult characteristics. For example, antennae in some Coenomyiidae have up to 10 segments, and in some Rhagionidae have 8 segments, recalling the multiarticulate antennae of Nematocera. In contrast, Dolichopodidae and many Empididae, as well as some members of several other families, have short antennae with 3 to 4 segments with a terminal arista, as in some Cyclorrhapha. This intermediacy involves all the characteristics used to separate the Cyclorrhapha and Nematocera, including mouthpart structure, thoracic structure and wing venation, genitalic features, and internal organ systems.

The Rhagionidae seem to represent the least specialized Orthorrhapha. Many of these usually soft-bodied flies have piercing-sucking mouthparts with well-developed, bladelike mandibles and maxillae. Similar mouthparts occur in Tabanidae (horse flies). In the predatory Asilidae and Empididae, at least the mandibles are absent, and usually only the hypopharynx is used to pierce the prey. Many Orthorrhapha [Bombyliidae (bee flies), Mydidae, Apioceridae, Acroceridae, Nemestrinidae] have mouthparts adapted for sucking up nectar through an elongate proboscis formed by the labium. In a few Empididae and in Dolichopodidae the labium is produced as a short, flexible proboscis tipped by a fleshy labellum. These flies are predatory, the prey being pierced by cuticular pseudotracheal teeth along the medial groove of the labellum.

The wing venation is very generalized in several families, notably the Rhagionidae (Fig. 42.10a), with the radial vein four-branched and the medial vein four-branched. In a number of families one or more branches of the radial or medial veins are looped forward to intersect the wing margin before its apex (Fig. 42.10b,f) In several of these families (e.g., Bombyliidae, Asilidae, Apioceridae, Mydidae), venation is highly distinctive, and useful in identification. In many Empididae and Dolichopodidae, the venation is essentially the same as in the Cyclorrhapha, with the radius three-branched, the median two-branched, and R_1 extremely short (Fig. 42.11g). Calypters are typically small or absent in Orthorrhapha, but are large and conspicuous in Tabanidae and Acroceridae.

Two features in which Orthorrhapha are clearly intermediate between the more generalized Nematocera and the Cyclorrhapha are (1) torsion of the external genitalia in males and (2) the degree of consolidation of the central nervous system. In most Nematocera rotation is not evident; but torsion of 90 to 180°, frequently accompanied by asymmetry of genitalic structures, is typical of most Orthorrhapha. In Dolichopodidae and Cyclorrhapha, torsion (here termed circumversion) has proceeded through 360°, so that the original dorsoventral relationships are restored. Male terminalia of Cyclorrhapha are also usually folded ventrally or ventrolaterally, and markedly asymmetrical, making the structures difficult to homologize with those in other Diptera.

The consolidation of the central nervous system is probably related to flying ability. Nematocera, which are often weak, periodic fliers, have three thoracic and six to seven abdominal ganglia. In Orthorrhapha a variable amount of condensation has occurred, and in Cyclorrhapha a single large thoracic ganglion or one thoracic and one abdominal ganglion remain.

BIOLOGY

Parasitic blood feeding, of common occurrence in adult Nematocera, reappears in many Tabanidae and in a few Rhagionidae *(Symphoromyia)*. In these flies, as in Nematocera, only females are bloodsuckers, opening a wound with the blade-shaped mandibles and lapping up the oozing blood rather than inserting the entire piercing apparatus as in mosquitoes. Females of some species and males have the mandibles atrophied and feed on nectar. Predation, encountered in such nematocerous families as Ceratopogonidae and Blephariceridae, is universal among Asilidae and widespread among Empididae and Dolichopodidae. Asilidae are aggressive, strong-flying predators, attacking any insects small enough to subdue, including wasps and bees, large Orthoptera, and beetles. The legs are long and robust, frequently bearing strong bristles useful in holding struggling prey. The posterior pairs are successively longer, and together the legs form a basket much as in the Odonata. The strong bristles about the eyes and face probably prevent thrashing prey from injuring the fly's head. Prey items are clasped tightly against the robber fly's body, punctured by the stout, sharp oral stylets and injected with rapidly acting toxic saliva. Many collectors have learned that the pain from a robber fly's bite is as instantaneous and severe as that of stinging Hymenoptera. A number of Asilidae strikingly resemble bumble bees and wasps, doubtless gaining protection from potential predators, but possibly using the resemblance to facilitate predation on flower visiting insects. One group of asilids (Leptogastrinae) superficially resemble ichneumonid wasps, with long slender body and legs. They pick small insects off surfaces and some species pluck spiders from their webs. Empididae and Dolichopodidae never achieve the large size of Asilidae, and tend to inhabit shadier or moister situations, frequently about seeps or running or standing water. Most members of both families are nimble runners, and many run down their prey, rather than pursuing it in the air. Empididae commonly use prey items in courtship, as discussed below.

Adults of Acroceridae, Bombyliidae, Mydidae, Nemestrinidae, Apioceridae, and occasional species of several other families of Orthorrhapha feed on nectar, and some Bombyliidae eat pollen. Some of these nectar-feeding species are equipped with elongate proboscises and hover above flowers while feeding. Other species have only a short proboscis or labellum, and alight on the flower. In several families, notably Therevidae, Scenopinidae, Coenomyiidae, and Xylophagidae, adult food is unknown and some of these may not feed as adults. Saprophagy, extremely widespread among Cyclorrhapha, is uncommon among adult Orthorrhapha.

Most Orthorrhapha are solitary, but male swarming is highly developed in Empididae and occurs in some Rhagionidae. Individual females enter the swarms of males before pairing takes place. As in Nematocera, the swarms are often located above characteristic objects such as shrubs or sunlit spots in forests, or over water. During courtship, males of many empidids present females with prey, among which smaller nematocerous Diptera are especially favored. In more specialized species courtship behavior has become ritualized to the extent that males use the following strategies: (1) present the female with the prey plastered to the side of a silken or frothy balloon; (2) present the balloon with a nonprey item (small stick, sand grain); (3) present the balloon alone or present a nonprey item alone. Both sexes of empidids also take nectar, but in many species the only proteinaceous food is that received from males during courtship. Courtship in Dolichopodidae often in-

volves signaling by the males, using the brushes and flanges on the tibiae and tarsi.

Most Orthorrhapha oviposit moderate numbers of eggs on or near the larval substrate. Bombyliidae, Acroceridae, and Nemestrinidae, whose larvae are internal parasitoids of other insects, oviposit extremely large numbers of eggs (up to 900 recorded in Acroceridae and over 4000 in Nemestrinidae) in habitats where the active first-instar larvae are likely to encounter hosts. Most Orthorrhapha apparently deposit the eggs singly or in small batches, but horse flies form tiered egg masses on emergent aquatic vegetation or other objects near aquatic habitats. *Atherix* (Rhagionidae) forms communal egg masses on twigs, small branches, culverts, or other objects overhanging water. The flies die in place after oviposition, their bodies combining with the eggs to form encrustations that were formerly used as food by some Native Americans.

Larvae of Orthorrhapha occupy a broad array of habitats, ranging from aquatic to terrestrial. Aquatic larvae (some Rhagionidae, Stratiomyidae, Tabanidae, Empididae, and Dolichopodidae) require atmospheric oxygen, hence are restricted to shallow water. Most are inhabitants of seeps, marshes, or the littoral regions of lakes or streams, rather than of open water situations. Empididae, Dolichopodidae, Asilidae, and Xylophagidae usually inhabit moist soil, leaf litter, or rotten wood. Therevidae, Apioceridae, and Mydidae frequent rotten wood, soil, or sand, frequently in quite arid circumstances. Scenopenid larvae are most frequently encountered in carpets, where they feed on moth larvae and other insects.

Larvae of Orthorrhapha are commonly predaceous. For example, therevid larvae, which mostly inhabit loose sandy soils, are general predators of other arthropods, especially the larvae of Elateridae, Scarabaeidae, and Tenebrionidae, and mydids also feed on immature Coleoptera, as far as known. Biologies of Rhagionidae, Apioceridae, and Xylophagidae are very poorly known, but most are believed to be predators as immatures. The larvae of *Vermileo* (Rhagionidae) construct steep-walled pits in dry, friable soil and trap small arthropods, much in the manner of antlions (Neuroptera, Myrmeleontidae). Several families include both predatory and herbivorous species (Stratiomyidae, Tabanidae), or scavengers (Empididae, Dolichopodidae, Xylomyidae). Tabanid larvae mostly inhabit aquatic or very moist soils, feeding on diverse arthropod prey. Large larvae of *Tabanus* are known to kill newly emerged toads as well. Larvae of Dolichopodidae and Empididae are believed to be generalized predators. Dolichopodids inhabit moist soil, rotting wood, and similar situations. Empididae include both aquatic species and others that live in moist soil.

Larvae of Acroceridae, Nemestrinidae, and Bombyliidae are all internal parasitoids of other arthropods. Acroceridae utilize spiders exclusively; nemestrinid larvae feed on grasshoppers and larvae of scarab beetles. Bombyliidae feed on grasshopper egg masses and are parasitoids of the larvae or pupae of a variety of immature endopterygote insect hosts, including Hymenoptera, Lepidoptera, Diptera, Coleoptera, and Neuroptera. First instars of these flies are vagile, planidium-like larvae that locate the host on which the later maggotlike instars feed.

Pupation occurs in or near the larval substrate. The conspicuous, posteriorly directed spines of many orthorrhaphous pupae aid the pharate adult in working its way to the surface of loose sand or soil, from which the cast pupal skins may frequently be seen protruding. Stratiomyidae are unusual in having the exarate pupa enclosed in the cuticle of the last larval instar, which is not modified as a puparium as in the Cyclorrhapha. This arrangement probably functions in floating the pupa in the stagnant waters where many Stratiomyidae develop and is apparently not homologous to the cyclorrhaphous puparium.

The families of the Cyclorrhapha Aschiza display diverse biological adaptations. Lonchopteridae are apparently nectar or honeydew feeders as adults, while the larvae are saprophytic in soil, decaying leaves, and similar situations. Platypezid adults imbibe honeydew or various exudates from leaves in forest understory. The males aggregate in mating swarms in forest openings. The larvae feed in the fruiting bodies of various soft or hard fungi and pupate in the soil. Phoridae are scavengers in many situations, including feces, decomposing vegetation, fungi, and the nests of bees, wasps, termites, and rodents. Other species feed in corpses (including those of humans) or parasitize insects, spiders, and millipedes.

Syrphidae are familiar as adults, which often mimic stinging Hymenoptera and frequently hover motionless above vegetation or flowers. Most of the adults imbibe nectar and eat pollen; the larvae have diverse habits. Most familiar are the taxa that feed on aphids and other Homoptera. Related forms are predators of Thysanoptera and immature Coleoptera and Lepidoptera. Other groups live in decaying succulent plants, especially cacti, or infest the bulbs of monocotyledonous plants, where they are of some economic importance. Still other species inhabit moist feces or water with high organic content, as that of tree holes, cesspools, or sewage. These larvae are known as rat-tailed maggots after their elongate breathing siphons.

Pipunculidae, which superficially resemble syrphids with enormous heads, as larvae are endoparasites of Homoptera, especially Cicadellidae and Delphacidae. The adult females capture immature hosts and oviposit into their hemocoels. The mature larva leaves the host to pupate in the soil.

KEY TO THE FAMILIES OF THE DIVISIONS

Orthorrhapha and Cyclorrhapha
(Series Aschiza)

   1. Tarsi with 3 subequal pads below claws (Fig. 42.9a)   **2**

      Tarsi with 2 or fewer pads below claws (Fig. 42.13a)   **8**

2(1). Head very small, thorax and abdomen inflated; calypters very large   **Acroceridae**

      Body not so shaped; calypters small or vestigial   **3**

3(2). Antennae with third segment annulate (Fig. 42.13b)   **5**

      Antennae with third segment not annulate; usually elongate or with terminal bristle (Fig. 42.13c)   **4**

Figure 42.12 Representative Brachycera, Orthorrhapha: **a**, Tabanidae *(Chrysops proclivus);* **b**, Asilidae *(Metapogon pictus);* **c**, Bombyliidae *(Bombylius lancifer);* **d**, Dolichopodidae (male *Dolichopus).* (From Cole, 1969.)

4(3). Wings with branches of median vein parallel to posterior margin (Fig. 42.10b)

                  Nemestrinidae

    Wings with branches of median vein intersecting posterior margin (Fig. 42.10a)

                  some **Rhagionidae**[1]

5(3). Calypters large, conspicuous   **6**

    Calypters small or vestigial   **7**

6(5). Anal cell open (as in Fig. 42.10a); eyes densely hairy   some **Rhagionidae**[1]

    Anal cell closed at or before wing margin (Fig. 42.10c,d); eyes not usually hairy

              Tabanidae (Fig. 42.12a)

[1]Including Vermilionidae, Athericidae, Pelecorhynchidae.

7(5). Tibial spurs present at least on middle legs

                         20

    Tibial spurs absent   **Stratiomyidae**

8(1). Head hemispherical, larger than thorax, with antennae usually inserted below midline of long, narrow face   **Pipunculidae**

    Head much smaller than thorax, usually with broad face; antennae inserted above midline of head   **9**

9(8). Cu_2 long, reaching wing margin or joining 2A near margin; anal cell much longer than median cell (Fig. 42.10e)   **10**

    Cu_2 absent (Fig. 42.11c), or, if present, joining 2A more than a quarter of its length in from margin (Fig. 42.11d,e,f); anal cell subequal to median cell (Fig. 42.11d,e,f)   **15**

10(9). Medial veins looped forward with M_{1+2} joining R_{4+5} before wing margin (Fig. 42.10e); antennae usually with dorsal bristle
Syrphidae (Fig. 42.16a)

Medial veins not joining R_{4+5} (Fig. 42.10f) (but radial veins may unite before wing margin); antennae usually with terminal style   **11**

11(10). Vertex concave between eyes (Fig. 42.13d)   **12**

Vertex flat or convex between eyes (Fig. 42.15a–c)   **13**

12(11). Head with 3 ocelli; antennae usually with 3 segments   **Asilidae** (Fig. 42.12b)

Head with 1 or no ocelli; antennae with 4 segments   **Mydidae**

13(11). Veins R_5 and M_1 looped forward, terminating anterior to wing apex (Fig. 42.10f)
Apioceridae

Veins R_5 and M_1 intersecting wing margin posterior to apex (Fig. 42.11a)   **14**

14(13). Wings with 3 median and cubital cells (Fig. 42.12c); usually scaly or hairy flies with long proboscis
Bombyliidae (including **Hilarimorphidae**) (Fig. 42.12c)

Wings with 4 median and cubital cells (Fig. 42.11a); body bare or pubescent; proboscis short   **Therevidae**

15(9). Wings with pointed apex; crossveins restricted to extreme basal portion of wing (Fig. 42.11b)   **Lonchopteridae**

Wings with rounded apex; distal crossveins present or absent   **16**

16(15). Radial veins much thicker than medial and cubital veins, which are without crossveins (Fig. 42.11c); antennae appearing 1-segmented   **Phoridae**

Veins all of approximately equal thickness; mediocubital region of wing with at least 1 crossvein; antennae with 2 or more obvious segments   **17**

17(16). Hind tibiae and tarsi dilated, flattened; wings with anal cell pointed distally (Fig. 42.11d)   **Platypezidae**

Hind tibiae and tarsi cylindrical or rounded; if dilated, anal cell not pointed   **18**

18(17). Vein M_1 curved anteriorly, intersecting wing margin before apex (Fig. 42.11e); antennae without arista or style   **Scenopinidae**

Vein M_1 straight or curved posteriorly, intersecting wing margin behind apex (Fig. 42.11f); arista or style frequently present   **19**

19(18). Wings with r-m crossvein very close to base; R_1 usually joining C in basal half of wing (Fig. 42.11g); calypters usually conspicuous, fringed; proboscis usually terminating as a fleshy labellum.
Dolichopodidae (Fig. 42.12d)

Wings with r-m crossvein located more than one-fourth distance from base to apex; R_1 usually joining C in distal half of wing (Fig. 42.11f); calypters small; proboscis usually a pointed beak   **Empididae**

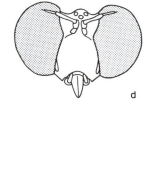

Figure 42.13 Taxonomic characters of Brachycera, Orthorrhapha: **a**, tarsus of Syrphidae; **b**, antenna of Tabanidae; **c**, antenna of Rhagionidae; **d**, head of Asilidae. (b, modified from Brennan, 1935.)

20(7). Costa ending near apex of wing   **21**

Costa continuing around posterior margin of wing   **22**

21(20). Fore tibia with ventral apical spur

some **Xylophagidae**[2]

Fore tibia without ventral apical spur

Xylomyidae

22(20). Clypeus convex, exposed

some **Rhagionidae**[1]

Clypeus recessed in deep facial groove and flattened   some **Xylophagidae**[2]

DIVISION CYCLORRHAPHA

Differing from Orthorrhapha chiefly in having the pupae coarctate and enclosed in a puparium formed by the modified cuticle of the last-instar larva.

The Cyclorrhapha comprise a vast diversity of mostly small, compactly built flies. The great majority of species emerge by pushing off the end of the puparium by inflating the ptilinum. The ptilinal opening is represented by a faint crescentic scar just above the antennal sockets in the fully hardened flies, hence the series name Schizophora. A few families, comprising the series Aschiza, considered in the preceding section, lack the ptilinum.

The series Schizophora may be further divided into sections (groups of families). The section **Calyptratae** includes flies with relatively large calypters (muscids, tachinids, etc.; Fig. 42.18d). The section **Acalyptratae** contains flies with the calypters small or absent. Included here are numerous families of small flies that are usually difficult to identify.

Mouthparts are much reduced in all Cyclorrhapha, with mandibles absent and maxillae represented by palpi. The labium is modified as a labellum, which bears sharp pseudotracheal or prestomal teeth in predatory or parasitic forms. The head typically bears bristles in characteristic positions which are of great taxonomic utility.

Wing venation is reduced, leaving no more than three radial and three medial branches. The costa is often broken at one or more places; in several families one or more radial or medial veins

are looped anteriorly, sometimes joining the next anterior vein. Such anomalies produce highly characteristic venational patterns, which are very useful in identifying many families.

The abdomen of Cyclorrhapha is characteristically divided into an anterior **preabdomen** and the **postabdomen,** which is flexed anteroventrally beneath the preabdomen in males. Male genitalia are rotated 360°. This extreme degree of torsion, termed circumversion, differentiates Cyclorrhapha from most Orthorrhapha.

BIOLOGY

The most common feeding mode of cyclorrhaphous larvae is decomposition; these larvae are the primary insect consumers of most decaying plant and animal material. High vagility of adults allows prompt location and exploitation of temporary food sources. High mobility may also be used to exclude competitors. An extreme example is the horn fly *(Haematobia irritans),* which resides on cattle, immediately ovipositing in freshly dropped dung before other insects are attracted.

The variety of feeding substrates of saprophagous Cyclorrhapha is very great. Some species are quite catholic, but most restrict themselves to either plant or animal material, and many are highly specific. For example, Neriidae (in North America) are limited to rotting cacti, Coelopidae to decaying seaweed, and many species of other families have narrow substrate ranges, including such items as fermenting tree sap, frass in burrows of wood-boring beetles, or the debris that collects in bird nests or animal burrows. Many of these saprophagous species probably feed largely on fungi or bacteria, and some will mature on agar bacterial cultures. It may be noted that *Drosophila melanogaster,* an extremely important experimental animal for research in genetics and nutrition, is a decomposer of decaying fruits in nature, accounting for its easy maintenance in confined spaces.

The saprophytic Cyclorrhapha are of considerable economic significance because of their frequent close associations with humans or domesti-

---

[1]Including Vermilionidae, Athericidae, Pelechorhynchidae.
[2]Including Coenomyiidae.

cated animals. Muscidae and Anthomyiidae (as well as several less significant families) commonly breed in animal dung, often reaching epidemic numbers in stock-raising areas. Many of these species are mechanical vectors of communicable diseases, ranging from dysenteries to poliomyelitis. A few species such as the horn fly *(Haematobia irritans),* the stable fly *(Stomoxys calcitrans),* and the tsetse flies *(Glossina* spp.) have secondarily acquired bloodsucking habits. *Glossina* are intermediate hosts for several trypanosomiases, including sleeping sickness of humans.

The Calliphoridae and Sarcophagidae are known as blow flies and flesh flies, from their habit of ovipositing on carrion. Some species, especially Calliphoridae, may oviposit in wounds of living animals, including humans. Some of these, such as *Cochliomyia hominovorax* and *Lucilia cuprina,* cause serious livestock losses, and the former has been the subject of intense control efforts using sterile-male techniques (see Chap. 14).

Oestridae (including the subfamilies Gasterophilinae, Hypodermatinae, and Cuterebrinae), the bot flies and warbles, are almost unique among insects in parasitizing mammals internally. *Gasterophilus* (Gasterophilinae) larvae attach to the stomach lining of horses, donkeys, rhinoceroses, and elephants. *Hypoderma* (Hypodermatinae) larvae live in cutaneous cysts or warbles in cattle, breathing through a small hole maintained in the skin. Most larvae of the subfamilies Oestrinae (primarily Old World) and Cuterebrinae (New World) inhabit the nasal or respiratory passages of rodents, lagomorphs, monkeys, and ungulates. The tropical *Dermatobia hominis* (Cuterebrinae), attacking humans as well as livestock, is unusual in that the adult females attach their eggs to mosquitoes or other flies, which they capture. When the mosquito later feeds on a warm-blooded animal, the host's body heat stimulates the egg to hatch. The larva then drops to the host and penetrates the skin. Other species oviposit directly on the host's body, usually cementing the eggs to hairs. It seems likely that these parasitic species evolved from flesh flies, and some authorities classify *Dermatobia* as a calliphorid.

Adults of three specialized families of Cyclorrhapha are ectoparasitic. Hippoboscidae are widespread on birds, with a lesser number infesting mammals, especially ungulates, but also dogs. Nycteribiidae and Streblidae are restricted to bats and almost never encountered away from their hosts. All these flies are superficially similar to lice, with flattened bodies and tough, leathery cuticle; many species dehisce their wings after reaching a host.

Braulidae live in the hives of honey bees. Adults are apterous and are usually found on the queen or worker bees. The larvae tunnel in the wax combs.

Parasitoid Cyclorrhapha attack a variety of insect orders, as well as spiders; Sciomyzidae are unusual in attacking freshwater and terrestrial snails. Parasitoid species oviposit directly on or in the host (Chloropidae, Pyrgotidae, Tachinidae) or in locations where the first-instar larvae will encounter the host (Tachinidae, Sarcophagidae). Internal parasitoids frequently tap the host tracheal system, emphasizing the dependence of most flies on atmospheric respiration.

Relatively few Cyclorrhapha are predatory. Carnivorous species include Chamaemyiidae and many Syrphidae (discussed in the preceding section; larvae predatory on Homoptera, Color photo 6c), some Sciomyzidae (larvae predatory on snails), as well as occasional members of many other families. Predatory adults (some Anthomyiidae) usually attack organisms much smaller than themselves.

Eggs of most Cyclorrhapha are deposited directly into appropriate food sources, hatching after a brief time. Most Sarcophagidae, many Calliphoridae, as well as occasional other species, are viviparous or ovoviviparous, depositing larvae onto decomposing organic material. Larvae of *Glossina* (Glossinidae), Hippoboscidae, Nycteribiidae, and Streblidae are retained internally by the female until fully mature. Larval nourishment is by means of a placenta-like organ in the common oviduct. The newly deposited larvae pupate immediately.

Cyclorrhapha pass through three functional

instars, the fourth instar being suppressed in conjunction with the development of the puparium. Many species pupate in the larval food, protected by the hard, rigid puparial shell, which deters cannibalism, as well as offering protection against predation, parasitism, and physical trauma. In contrast to most cocoons and other pupation shelters, the puparium requires no external support or material in its construction; it is probably responsible in large part for the success of the Cyclorrhapha.

Biology of the commoner families of acalyptrate Cyclorrhapha is summarized in Table 42.1.

KEY TO THE COMMON FAMILIES OF THE DIVISION

Cyclorrhapha (Series Schizophora)

1. Body flattened, louselike, with hind coxae widely separated or mesonotum reduced, similar to abdominal segments

Hippoboscidae, Braulidae, Nycteribiidae, Streblidae

Body not louselike; middle and hind coxae closely approximated **2**

2(1). Mouthparts vestigial or lacking; mouth cavity small (Fig. 42.18a) pubescent flies at least 15 mm long

Oestridae (including Cuterebridae, Hypodermatidae, Gasterophilidae)

Mouthparts functional, usually with large labellum, and labial palps protruding from broad mouth cavity (Fig. 42.3c); usually less than 15 mm long **3**

3(2). Mesothorax with complete transverse suture (sometimes faint) anterior to wing attachments (Figs. 42.18b,c; 42.17d); lower calypter usually large; greater ampulla present below wing base (Fig. 42.18d,e) (section **Calyptratae**) **29**

Mesothorax with transverse suture present at lateral margins or absent (Fig. 42.17a); lower calypter small or vestigial; greater ampulla usually absent (section **Acalyptratae**) **4**

4(3). Vein Sc reaching costal margin of wing (Fig. 42.14a–d), separate from R_1 except at extreme base; anal cell present (Fig. 42.14a–c,e) **5**

Vein Sc not reaching costal margin, frequently fused with R_1 for most of its length (Fig.

42.14f,g); anal cell present (Fig. 42.14a-c,e) or absent (Fig. 42.14f) **21**

5(4). Mesonotum strongly flattened; head closely appressed to thorax, legs bristly

Coelopidae

Mesonotum convex, if flattened, head loosely attached to thorax, legs not conspicuously bristly **6**

6(5). Head with at least a pair of stout bristles (oral vibrissae) just in front of mouth cavity (Fig. 42.15a,b) **7**

Head without oral vibrissae (as in Fig. 42.15g) **11**

7(6). Palpi vestigial; metathoracic spiracle with at least 1 peripheral bristle; abdomen usually narrowed at base

Sepsidae (Fig. 42.17a)

Palpi functional, carried lateral to rostrum **8**

8(7). Head with postvertical bristles parallel or divergent (sometimes absent) (Fig. 42.15b–d) **9**

Head with postvertical bristles converging distally; ocellar bristles between median and lateral ocelli (Fig. 42.15a); tibiae with dorsal spines near apex; costal vein spinose

Heleomyzidae

9(8). Antennae with subapical arista on third segment; eyes round; head with 1 to 4 pairs of fronto-orbital bristles (Fig. 42.15b)

Clusiidae

Antennae with arista attached near base of third segment; eyes elongate, oval; head with 0 to 2 pairs of fronto-orbital bristles **10**

10(9). Antennae with third segment at least 2 to 3 times as long as broad (Fig. 42.15c); vein 2A slightly sinuous, reaching wing margin; tibiae usually without preapical bristles; usually stout black or blue-black flies less than 5 mm long **Lonchaeidae**

Antennae with third segment no more than twice as long as broad (as in Fig. 42.15b); vein 2A straight or curved, not reaching wing margin; costal vein never spinose

Piophilidae

11(6). Wings with vein M_{1+2} curving anteriorly, approaching or joining R_{4+5} (Fig. 42.14a); legs often long, stiltlike **12**

Wings with vein M_{1+2} not closely approaching R_{4+5} (Fig. 42.14b), or legs not long stiltlike **13**

TABLE 42.1.
Biological Characteristics of Common Families of Cyclorrhapha Acalyptratae

Family	Larval habitat/habits	Larval food habits	Adult habitat	Adult food habits
Agromyzidae	Miners in fruits, seeds, twigs, stems, roots and especially leaves; mostly monophagous or oligophagous	Plant tissues	Near host plants	Probably nectar, honeydew, and various plant or animal exudates
Anthomyzidae	Borers in grass stems	Phytophagous or saprophagous	Moist, grassy areas	Unknown
Chamaemyiidae	Predators of aphids, adelgids, coccids, and scale insects	On plants near prey	Near larval hosts	Probably honeydew and nectar
Chloropidae	Mostly borers in grass stems or decaying plants; some in insect frass, in galls, in fungi or predators on insect larvae or eggs. One species is a cutaneous parasite of frogs	Mostly phytophagous or saprophagous	Common on low plants; some attracted to animal exudates	Mostly nectar; some feed on animal exudates
Clusiidae	Rotting wood and termite galleries	Probably saprophagous	On decaying tree trunks	Nectar, sap, rotting plants
Coelopidae	In rotting seaweed	Rotting seaweed	Usually along seacoasts	Rotting seaweed
Conopidae	Internal parasitoids of aculeate wasps, cockroaches, and calyptrate flies	Host tissues	On flowers or near larval hosts	Nectar
Dryomyzidae (including Helcomyzidae)	Carrion, dung, decaying fungi and seaweed; predators of barnacles	Saprophagous or zoophagous	Abundant in many habitats	Unknown
Ephydridae	Aquatic or semi-aquatic, especially in shallow water near shore lines; a few saprophagous in excrement, etc. or leafminers, parasitoids or predators	Mostly microphagous filter feeders on bacteria, algae, yeasts	Abundant near shorelines	Mostly microphagous on algae, bacteria, etc.; some scavengers on dead insects; a few predators

TABLE 42.1 (*continued*)

Family	Larval habitat/ habits	Larval food habits	Adult habitat	Adult food habits
Heleomyzidae	Many types of decaying plant or animal material, including dung, carcasses, fungi	General saprophages	Near larval food and in moist, shady areas	Unknown
Lauxaniidae	Decaying leaf litter or other plant material; dung	Microphagous on fungi associated with decay	Low vegetation in moist, shady areas	Plant exudates; occasionally nectar
Lonchaeidae	Decaying flowers, fruits, tubers, succulent plants, or subcortical region of decaying trees	General saprophages	Near larval habitats	Unknown
Micropezidae	Decaying plant matter; dung; tree holes	Saprophagous	Marshes, moist woodlands	Unknown
Milichiidae	Decaying plants; dung; fungi in leaf-cutting ant nests	Saprophagous, coprophagous	Many on flowers; some commensal with large predatory insects (Asilidae, Reduviidae)	Nectar; commensals feed on juices from host prey
Neriidae	Rotting succulent plants or fermenting sap	Saprophagous	On rotting succulent plants	Exudates from plant decay?
Otitidae	Dung, decaying bulbs, tubers, roots, stems; subcortical region of decaying trees; occasionally in live plants	Mostly saprophagous	Near larval habitats; on flowers	Nectar; probably decay exudates
Piophilidae	Protein rich decaying plants or animals, including meat, bones, hides, fungi, cheese; one species ectoparasitic on nesting birds.	Saprophagous or blood feeding	Near larval habitats	Probably decay exudates
Psilidae	Roots and stems of plants	Phytophagous	Moist or wooded areas	Unknown
Pyrgotidae	Internal parasitoids of adult Scarabaeidae	Entomophagous	Nocturnal near hosts; oviposit into host while in flight	Unknown

TABLE 42.1. (Continued)

Family	Larval habitat/ habits	Larval food habits	Adult habitat	Adult food habits
Sciomyzidae	Associated with freshwater or terrestrial mollusks as predators, parasitoids, or scavengers	Zoophagous	Moist areas near larval hosts or prey	Unknown?
Sepsidae	Dung; sometimes in carcasses, sewage sludge	Coprophagous or saprophagous on protein-rich material	Usually near larval food	Decay exudates
Sphaeroceridae	Dung, carcasses, decaying plants, fungi, seaweed, leaf litter; dung balls of Scarabaeidae	Saprophagous	In moist to wet habitats near larval food	Probably decay exudates
Tanypezidae	Unknown; probably decaying plants	Unknown	Deciduous woods	Unknown
Tephritidae	Flowers, fruits, seeds, stems, leaves, and roots, especially of dicots; some make galls	Phytophagous	Usually near host plants	Nectar or plant exudates
Tethinidae	Mostly unknown; probably in soil or algal masses; some probably in bird dung	Unknown	Mostly marine shorelines and saline areas	Mostly unknown; some feed on nectar

12(11). Proboscis long, slender, jointed, and bent anteriorly (Fig. 42.17b); body stout or slender
        Conopidae (Fig. 42.17b)
  Proboscis short, thick, not jointed as above; body slender; legs long, slender
      Neriidae, Micropezidae, Tanypezidae

13(11). Some or all tibiae with dorsal bristles near apex; ovipositor short, membranous, retractile
      14
  Tibia usually without preapical bristles; if present, ovipositor long, sclerotized   **16**

14(13). Head with postvertical bristles parallel or diverging (rarely absent) (Fig. 42.15d); anten-

nae without dorsal bristle on second segment   **15**
  Head with postvertical bristles converging (Fig. 42.15a); antennae with dorsal bristle on second segment   **Lauxaniidae**

15(14). Femora without bristles; vein R_1 intersecting costa beyond middle of wing
      Dryomyzidae (including **Helcomyzidae**)
  Femora with fine setae and stout bristles, usually with bristle near middle of anterior surface of midfemur; vein R_1 intersecting costa near middle of wing (Fig. 42.14b)
      Sciomyzidae

16(13). Costa with break near intersection of Sc (Fig. 42.14e) **19**

Costa not broken near intersection of Sc (Fig. 42.14c,d) **17**

17(16). Vein Cu$_2$ straight, meeting 2A at about a right angle (Fig. 42.14c); vein R$_1$ bare; ovipositor short, membranous; small gray flies

Chamaemyiidae

Vein Cu$_2$ angulate, meeting 2A at an acute angle (Fig. 42.14d); vein R$_1$ usually bearing setae; ovipositor sclerotized, usually projecting **18**

18(17). Ocelli absent; base of ovipositor conical

Pyrgotidae

Ocelli present; ovipositor flattened; wings usually banded or spotted

Otitidae (Fig. 42.17c)

19(16). Vein Sc bent apically toward costa at nearly a right angle, usually not reaching costa (Fig.

42.14e); wings frequently patterned

Tephritidae

Vein Sc bent toward costa at much less than a right angle, usually meeting costa at about 45° angle (Fig. 42.14d) **20**

20(19). Femora thickened, spiny; abdomen with bristles on second segment

Otitidae (Fig. 42.17c)

Femora slender, not spiny; abdomen without bristles on second segment

Lonchaeidae

21(4). Vein Sc bent apically toward costa at nearly a right angle, usually not reaching costa (Fig. 42.14e); Cu$_2$ angulate, meeting 2A at an acute angle **Tephritidae**

Vein Sc bent toward costa at much less than a right angle, usually meeting costa at about 45° (as in Fig. 42.14c) or ending before costa

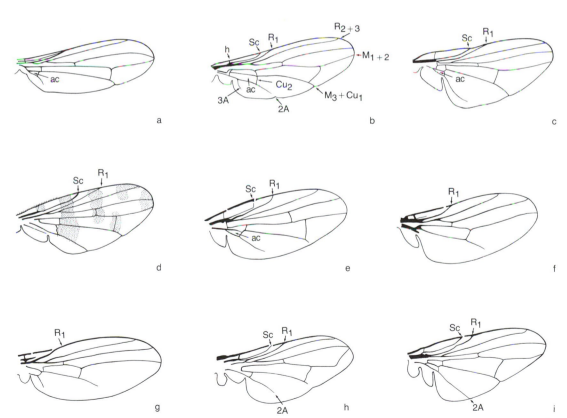

Figure 42.14 Wings of Brachycera, Cyclorrhapha: **a**, Micropezidae (*Compsobata mima*); **b**, Sciomyzidae (*Sepedon pacifica*); **c**, Chamaemyiidae (*Leucopis bivitta*); **d**, Otitidae (*Ceroxys latiusculus*); **e**, Tephritidae (*Trupanea nigricornis*); **f**, Chloropidae (*Thaumatomyia pulla*); **g**, Agromyzidae (*Cerodontha dorsalis*); **h**, Muscidae (*Musca domestica*); **i**, Anthomyiidae. ac, anal cell.

Figure 42.15 Heads of Brachycera, Cyclorrhapha, showing bristle patterns and lateral views of antennae; **a**, Heleomyzidae *(Suillia limbata);* **b**, Clusiidae *(Clusia occidentalis);* **c**, Lonchaeidae *(Earomyia brevistylata);* **d**, Sciomyzidae *(Sepedon pacifica);* **e**, Agromyzidae *(Cerodontha dorsalis);* **f**, Psilidae *(Psila microcera);* **g**, Ephydridae *(Paracoenia bisetosa);* **h**, Drosophilidae *(Drosophila simulans).* fob, fronto-orbital bristles; ivb, inner vertical bristles; ob, ocellar bristles; ov, oral vibrissae; ovb, outer vertical bristles; ps, ptilinal suture; pvb, postvertical bristles; scb, subcranial bristles.

(Fig. 42.14f); Cu_2 straight, meeting 2A at about a right angle **22**

22(21). Hind legs with 2 basal tarsal segments thickened, larger than succeeding segments; first segment shorter than second

Sphaeroceridae

Hind legs with basal tarsal segments not markedly thicker than succeeding segments; first segment usually longer than second **23**

23(22). Costal vein with single break near Sc or R_1 (Fig. 42.14f) **24**

Costal vein with breaks near Sc and near humeral crossvein (Fig. 42.14e) **27**

24(23). Wings with anal cell closed or nearly closed (Fig. 42.14c,e); antenna with arista covered with short hairs **25**

Anal cell absent (Fig. 42.14f); arista variable; ocellar triangle large, conspicuous

Chloropidae

25(24). Head with postvertical bristles converging apically (Fig. 42.15a)

Tethinidae, Anthomyzidae

Head with postvertical bristles diverging (Fig. 42.15c)or absent **26**

26(25). Head with pair of stout bristles (oral vibrissae) just in front of mouth cavity (Fig. 42.15e);

postvertical bristles diverging; small gray or black flies with ovipositor sclerotized, not retractable   **Agromyzidae**

Head without oral vibrissae (Fig. 42.15f); postvertical bristles frequently absent; small brownish or yellowish flies with very strong break at basal third of costa   **Psilidae**

27(23). Head with postvertical bristles diverging (Fig. 42.15g); pair of stout bristles (oral vibrissae) usually absent in front of mouth cavity (Fig. 42.15g); anal cell absent   **Ephydridae**

Head with postvertical bristles parallel or converging (Fig. 42.15h) (occasionally absent); oral vibrissae usually present; anal cell present or absent   **28**

28(27). At least 1 pair of fronto-orbital bristles converging toward midline of head (Fig. 42.15b,e); antennae with arista usually bare   **Milichiidae**

Fronto-orbital bristles not directed medially (Fig. 42.15g,h); arista plumose   **Drosophilidae** (Fig. 42.16b) (incl. **Camillidae, Curtonotidae, Diastatidae**)

29(3). Hypopleuron with row or cluster of large hypopleural bristles below metathoracic spiracle (Fig. 42.18e); wing with vein M_1 most always abruptly curving forward (Figs. 42.14h, 42.17d)   **30**

Hypopleuron usually without bristles; if bristled, the pteropleuron lacks bristles, and vein M_1 is straight or gently curved (Fig. 42.14i)   **32**

30(29). Thorax with postscutellum large, convex, transverse (Fig. 42.18d,e); abdominal terga usually bearing strong bristles medially   **Tachinidae** (Fig. 42.17d)

Thorax with postscutellum small; abdominal terga rarely with medial bristles   **31**

31(30). Body usually metallic green or blue; arista almost always plumose along entire length; mesothorax with 2 or 3 notopleural bristles (Fig. 42.18b)   **Calliphoridae**

Body not metallic, usually black or gray with dark longitudinal stripes on thorax; arista usually plumose in basal half; mesothorax with 4 notopleural bristles (Fig. 42.18c)   **Sarcophagidae**

32(29). Vein $Cu_2 + 2A$ not reaching wing margin (Fig. 42.14h); lower calypter longer than upper calypter   **Muscidae**

Vein $Cu_2 + 2A$ reaching wing margin (Fig. 42.14i); lower calypter subequal to or smaller than upper calypter   **Anthomyiidae**

a

b

Figure 42.16 Representative Brachycera, Cyclorrhapha; **a**, Syrphidae *(Syrphus opinator,* scale equals 2 mm); **b**, Drosophilidae *(Drosophila busckii).* (b, modified from Cole, 1969.)

SELECTED REFERENCES

General

Askew, 1971 (parasitic families); Colless and McAlpine, 1991; Felt 1940 (gall-forming families); Foote et al., 1993 (Tephritidae); Gillett, 1971, 1972 (mosquitoes); McAlpine et al., 1981–88; Oldroyd, 1964 (comprehensive discussions of biology and ecology); D. B. Roberts, 1986 *(Drosophila);* Seguy, 1950, 1951.

Figure 42.17 Representative Brachycera, Cyclorrhapha: **a**, Sepsidae *(Sepsis violacea);* **b**, Conopidae *(Physocephala texana);* **c**, Otidae *(Tritoxa flexa);* **d**, Tachinidae *(Trichopoda plumipes).* (Modified from Cole, 1969.)

Morphology

Bonhag, 1949 (horse fly); Crampton et al., 1942; Seguy, 1951; M. J. Roberts, 1971 (structure and function of larval mouthparts); Snodgrass, 1944 (mouthparts).

Biology and Ecology

Bateman, 1972 (fruit flies); Demerec, 1950 *(Drosophila);* Downes, 1969 (swarming behavior), 1971 (bloodsucking); Gillett, 1972 (mosquitoes); Hering, 1951 (leafminers); Hespenheide, 1991 (leafminers); Horsfall et al., 1973 (bionomics of *Aedes*); James, 1947 (myiasis); McAlpine et al., 1981–88; Norris, 1965 (blow flies); Oldroyd, 1964 (general); Oliver, 1971 (Chironomidae); Sacca, 1964 (house flies); Seguy, 1950, 1951 (general); Spencer, 1990 (Agromyzidae); West, 1951 (house fly).

Systematics and Evolution

Cole, 1969 (western United States species); Curran, 1934 (North American genera); Desalle and Grimaldi, 1991 (molecular systematics of *Drosophila*); *Diptera of Connecticut,* 1942–1964; Ferrar, 1988 (immature Cyclorrhapha); Gagne, 1989 (Cecidomyiidae); Griffiths, 1972 (higher classification), 1980–present (fauna of North America), 1994 (higher classification of Brachycera); Hall, 1948 (Calliphoridae of North America); Hennig, 1948–1952 (larvae); Hull, 1962 (Asilidae); McAlpine et al., 1981–88 (North America); Merritt and Cummins, 1996 (aquatic families); Peterson, 1951 (larvae); Skarlato, 1985; Spencer and Steyskal, 1986 (Agromyzidae); Steffan, 1966 (Sciaridae of North America); Stehr, 1991 (immatures); Wilcox et al., 1989 (Mydidae); Wirth and Grogan, 1988 (Ceratopogonidae); Wirth and Stone, 1956 (aquatic families).

Figure 42.18 Taxonomic characters of Brachycera, Cyclorrhapha: **a**, head of Oestridae *(Cuterebra);* **b**, dorsal view of pterothorax of Calliphoridae; **c**, dorsal view of pterothorax of Sarcophagidae; **d, e**, lateral view of thorax of Calliphoridae *(Phormia regina).* acr, acrostichal bristles; Cx, coxa; dc, dorsocentral bristles; Epm, epimeron; hb, humeral bristles; Hc, humeral callus; hyb, hypopleural bristles; Hypl, hypopleuron (also called the meron); iab, intra-alar bristles; mpb, mesopleural bristles; nb, notopleural bristles; Npl, notopleuron; N, notum; pab, postalar bristles; Pc, postalar callus; pb, posthumeral bristles; Pl, pleuron; psb, prescutal bristles; Pscl, postscutellum; ppb, propleural bristle; ptb, pteropleural bristles; Pt, pteropleuron; scb, scutellar bristles; Scl, scutellum; Sp, spiracle; stpl, sternopleural bristles; Stpl, sternopleuron; spb, supra-alar bristles; Tms, transmesonotal suture.

43
ORDER SIPHONAPTERA (FLEAS)

A female oriental rat flea, *Xenopsylla cheopis* (Pulicidae), as seen on a microscope slide. This species is the vector of bubonic plague.

Minute to small, wingless Endopterygota with laterally flattened body.

Adults (Fig. 43.1a) with hypognathous head capsule, 3-segmented antennae, frequently with small lateral ocelli (no compound eyes or dorsal ocelli); labrum and laciniae slender, bladelike (Fig. 43.1b); maxillary and labial palps usually 4- to 5-segmented, mandibles absent. Thoracic segments subequal or metanotum slightly enlarged; legs stout with large coxae; hind legs enlarged, modified for jumping. Abdomen 10-segmented, tenth tergite bearing a complex sensilium and associated guard hairs (Fig. 43.1c).

524

Larva apodous, vermiform, usually with distinct head capsule, 1-segmented antennae, mandibulate mouthparts (Fig. 43.1d).

Pupa adecticous, exarate, in wispy cocoon.

About 2400 species of flea occur throughout the world, parasitizing mostly mammals (94 percent); about 130 species (6 percent) infest various birds, including domestic fowls. The species that infest birds appear to be specialized derivatives of lineages that otherwise occur on mammals. Unlike lice, which depend on the host for shelter, warmth, and oviposition sites as well as food, fleas are intermittently parasitic. Larvae are mostly nest inhabitants rather than parasites, but larvae of a few species are ectoparasites, and one is an endoparasite. Adults frequently leave the host for various periods of time, and those associated with migratory birds survive in their nests from season to season. Aquatic as well as terrestrial mammals may serve as hosts, but fleas are not known from seals and other marine mammals that do not occupy nests. The chigoe flea, *Tunga penetrans,* of tropical regions, is exceptional in that the females permanently attach to the skin, usually of the feet, of various mammals including man. Fertilized females burrow into the flesh and distend to the size of a small pea.

Host specificity varies greatly among fleas, with up to 35 hosts known for some species. The number of flea species supported by a given host is

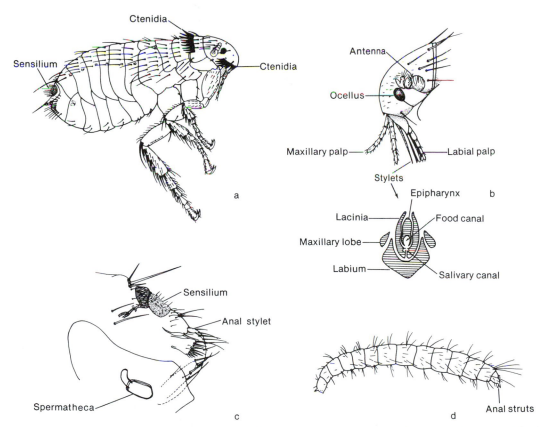

Figure 43.1 Siphonaptera: **a**, adult of *Ctenocephalides felis,* the cat flea; **b**, mouthparts of *Xenopsylla cheopis;* **c**, abdominal sensilium of *Megabothris;* **d**, larva of *Xenopsylla cheopis.* (a, courtesy of D. P. Furman; b, c, modified from Furman and Catts, 1961; d, modified from Patton, 1931.)

likewise highly variable, with 22 species recorded from the rat, *Rattus fuscipes*. Species with broad host ranges readily change host species, and many fleas with restricted host ranges will temporarily infest unsuitable hosts, particularly if hungry. Transfers among individuals of the same species may be frequent, especially among colonial or gregarious hosts. Dead animals are quickly deserted, and in general, high vagility and infidelity to a single host are responsible for the importance of fleas in transmitting bubonic plague from rodents to humans (see Chap. 13). Fleas are also important vectors of murine typhus *(Rickettsia)* and tularemia *(Francisella)* and may serve as intermediate hosts of tapeworms.

Adult fleas are so highly modified that their relationship to other orders of insects is uncertain. However, the absence of mandibles and presence of suctorial mouthparts suggest a possible origin from Diptera or Mecoptera, and flea larvae strongly resemble immatures of the dipterous suborder Nematocera.

Adult morphology is strongly correlated with the ectoparasitic life. The laterally compressed body and short, retractable antennae allow easy passage through the host's pelage or plumage, and the wedge-shaped coxae appear to be adapted for pushing through hair. Various parts of the body, especially the pronotum, bear combs of stout, backwardly directed, modified setae, or **ctenidia,** which further aid forward progress. Smaller bristles on various parts of the body serve the same function. In general, fleas avoid the preening or scratching of the host by rapidly running over the host's body. As in most other insect ectoparasites of vertebrates, the cuticle is tough and leathery, protecting the internal organs by a hydrostatic cushion provided by the enclosed hemolymph. The jumping hind legs are used to escape or to mount a potential host. The jumping mechanism (Figs. 43.2a–c) functions after the manner of the release mechanism of a camera shutter. Prior to jumping the hind leg is rotated dorsally and anteriorly at the coxotrochanteral joint by contraction of the trochanter levator muscle. Rotation continues until the tendon of the large notum-trochanter muscle

passes backward across the pivot between the trochanter and the coxa (Fig. 43.2b). This position causes the force of the notum-trochanter and coxal remotor muscles, which are now contracted, to be transmitted through the sclerotized ridges on the coxa and pleuron to the resilin block, which is situated between the pleuron and notum. With the leg in this cocked position the flea is ready to leap. The jump is initiated by contraction of the trochanter depressor muscle, which pulls the tendon of the notum-trochanter muscle anteriorly across the trochanteral-coxal pivot. The combined force of the contracted muscles and the compressed resilin block causes the hind leg to rotate rapidly downward, striking the substrate and causing the flea to be propelled into the air.

Adult life spans as long as a year have been recorded, and *Pulex irritans,* the human flea, has survived over 4 months without feeding. Mating and oviposition take place on the host or within its nest or burrow. In species associated with solitary animals, reproduction may be synchronized with hormonal changes in the host, so that a fresh generation of fleas is available to infest the young animals leaving the nest. Blood meals are probably necessary for egg maturation in all species.

Pulex irritans may produce over 400 eggs per individual, and cat fleas up to 1000 eggs, but eggs are laid a few at a time. Eclosion ensues after a few days or weeks, and larval development is complete after about 1 to 3 weeks under favorable conditions such as the warm, humid environs of a mammal nest. Under harsher conditions, especially low temperatures, larval growth may require up to 6 months. Larvae are scavengers and predators, feeding on dried flea excrement, small arthropods, and other debris that collects in animal nests. In houses the cracks and crevices in floors covered by carpets are favorable developmental sites. The third-instar larva spins a thin cocoon, to which dust and debris adhere. Pupation requires a few weeks in most species. Newly emerged adults frequently remain in the cocoon for some time, especially if no host is available. In some species massive emergence of such quiescent adults is apparently stimulated by mechanical disturbances

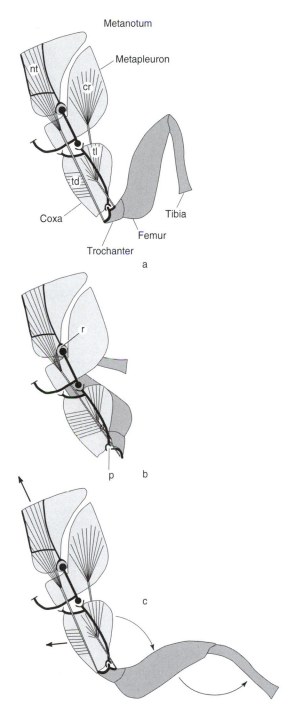

Metanotum

Metapleuron

nt

cr

tl

td

Coxa

Tibia

Femur

Trochanter

a

r

p   b

c

Figure 43.2 Diagrams of the jumping mechanism of a flea, showing part of the hind leg (excluding the tarsus) and musculature as seen from inside the metathorax and with the head to the left; especially note the position of the trochanter relative to the trochanteral-coxal pivot (p): **a**, leg in relaxed position; **b**, mechanism with leg being drawn up into the cocked position (coxal remotor muscle not shown); **c**, mechanism during the jump, with leg extended by muscular contractions and the rebound of the compressed resilin block. Thick arrows show directions of forces from contracting muscles; thin arrows show direction of movement of the leg; thick lines on sclerites are strong ridges; nt, notum-trochanter muscle; td, trochanter depressor muscle; tl, trochanter levator muscle; cr, coxal remotor muscle; p, trochanteral-coxal pivot; r, resilin block between pleuron and notum. (Modified from Bennet-Clark and Lucey, 1967.)

such as movements of a potential host—for example, swallows returning to their nest in spring or vacationers to an empty house.

Positive identification of fleas is difficult, requiring slides of cleared specimens. The North American fauna includes 9 of 17 families. Nearly all the familiar species that attack humans and domestic animals belong to a single family **Pulicidae.** Small rodents are the richest source of additional taxa, providing hosts to about 200 species in seven families.

SELECTED REFERENCES

Askew, 1971 (biology); Cheetham, 1988 (genitalia and phylogeny); Ewing and Fox 1943 (North American species); G. P. Holland, 1949 (Canadian species), 1964 (evolution and classification); Hubbard, 1947 (western North American species); R. E. Lewis et al., 1988 (Pacific Northwest species); Mead-Briggs, 1977 (interaction with hosts); Rothschild, 1965 (hormonal interaction with hosts), 1975 (general); Snodgrass, 1946 (morphology); Traub and Starcke, 1980 (general).

44

ORDER LEPIDOPTERA (BUTTERFLIES AND MOTHS)

This anise swallowtail butterfly, *Papilio zelicaon* (Papilionidae), belongs to a group of related species occuring across the northern Hemisphere.

Adult (Figs. 44.1–44.3). Minute to large Endopterygota with body and wings covered with scales or hairs, and mouthparts almost always modified as a proboscis; head hypognathous with large lateral eyes, frequently with ocelli and/or **chaetosemata,** antennae with numerous segments; mouthparts (Fig. 44.2c) consisting of elongate galeae, maxillary palpi (frequently very small), and large labial palpi. Prothorax usually small, collarlike, with **patagia** (Fig. 44.1) dorsolaterally; mesothorax large with clearly defined scutum and scutellum, and with large, lateral tegulae covering wing bases; metathorax usually much reduced, with lateral scutal

sclerites. Wings covered with scales (flattened macrotrichia); hind wing coupled to forewing by **frenulum, jugum** (Fig. 44.3), or basal overlap; venation mostly of longitudinal veins with few crossveins, usually with large discal cell in basal half of wing (Figs. 44.7e, 44.10a). Legs adapted for walking, usually with 5 tarsomeres; prothoracic legs occasionally much reduced, almost always with **epiphysis** on tibia. Abdomen with 10 segments, first segment reduced, with vestigial sternum; segments 9 and 10 strongly modified in relation to genitalia.

　Larva (Fig. 44.5a–d). Usually eruciform with scle-

529

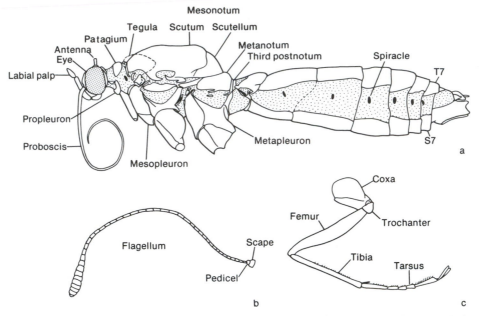

Figure 44.1   Anatomy of Lepidoptera (*Danaus plexippus,* Nymphalidae, modified from Ehrlich, 1958): **a**, lateral view of body; **b**, antenna; **c**, mesothoracic leg; T, tergum; S, sternum.

rotized head capsule; hypognathous, mandibulate mouthparts, short, 3-segmented antennae and usually 6 lateral stemmata; thorax with short, 5-segmented legs with single tarsal claws; abdomen 10-segmented, with short fleshy **prolegs** on some or all of segments 3 to 6 and 10. Larval morphology often highly modified in leaf mining forms.

   Pupa (Fig. 44.5e). Rarely decticous, exarate; typically adecticous, obtect, with body parts fused, except for 2 or more free abdominal segments, which remain movable; pupa usually enclosed in silken cocoon or shelter.

––––––––––––––––

Members of the order Lepidoptera have received more attention from naturalists and entomologists than any other group of insects. Most of this effort has been directed toward the butterflies (Color photos 1a–f, 3b, 3c, 5a) and showier moths (Color photos 3f, 7f) and knowledge of the smaller, more obscure species, which constitute the great majority of the order, is largely lacking.

   Biologically, as well as morphologically, the Lepidoptera are perhaps the most uniform of the large holometabolous orders. Adults, with few ex-

ceptions, feed on nectar, honeydew, fermenting sap, or other similar products, or have the mouthparts atrophied. Larvae of nearly all species feed on foliage or fruiting structures of angiosperms and gymnosperms; it is larval feeding that is almost entirely responsible for the great economic importance of the order. It seems likely that nearly all higher plants are host to at least one lepidopteran species. In general, Lepidoptera are the dominant insect herbivores in most habitats, having extensively invaded all available niches except that of sucking plant sap. In North America larvae of about 175 species feed on fungi or detritus.

   Certain Lepidoptera have been important in the development of ideas in evolutionary biology and ecology. The butterfly *Papilio dardanus,* whose females mimic danaine nymphalids of several species, is one of a few organisms for which the genetic basis of polymorphism is understood. Complex mimetic relationships have been intensively investigated in the monarch (*Danaus plexippus,* Color photo 3b), queen (*Danaus gilippus),* and viceroy (*Limenitis archippus)* (see Chap. 12).

The monarch is also famous for its migratory behavior, documented by Urquhart (1987). Sex-limited genes were first described in the genus *Colias*. The evolution of concealing coloration has been intensively studied by Kettlewell and others in the geometrid moth *Biston betularia* (see Chap. 12), and studies of Ehrlich and others (Ehrlich et al., 1975) on the population biology and adaptive ecology of *Euphydryas editha* (Nymphalidae) are important in explanations of how populations differentiate. These are just a few examples. In general, butterflies have been favorite subjects for much recent experimental and analytical work, especially regarding plant herbivore coevolution, geographic variation, and population structure.

A distinctive morphological feature of most Lepidoptera is the body covering of scalelike mac-

rotrichia. In most species these are of dull brown or gray hues in mottled patterns that help to conceal the resting insects. In butterflies the scales produce bright contrasting patterns, which may function aposematically or may provide characteristic patterns allowing easy recognition between the sexes. Scales are easily detached and probably enable these insects to slip from the grasp of many potential predators. In some of the larger moths, scales may provide an insulation that allows the internal body to more quickly reach the elevated temperatures at which flight occurs. Body temperatures 10 to 18C° above ambient have been recorded in large Hepialidae and Sphingidae.

The majority of Lepidoptera have the mouthparts adapted as a long coiled proboscis through which liquid food is ingested (Fig. 44.2c). The proboscis consists of the greatly elongate maxillary galeae, structured as a series of sclerotized rings connected by membranes and two sets of muscles, which provide the flexibility required for coiling and uncoiling. The two galeae are tightly interlocked by hooks and spines, forming a double-walled tube. Small maxillary palpi are usually present at the base of the proboscis, and large three- to four-segmented labial palpi are present on all but some nonfeeding forms. Other oral structures are absent in typical Lepidoptera except as vestiges, but functional mandibles, accompanied by maxillae with both laciniae and unmodified galeae, occur in Micropterygidae and a few other families (Fig. 44.2a,b). Reduced, toothless mandibles also occur in Eriocraniidae, in combination with a short proboscis whose galeae are held together only while feeding.

As in most four-winged insects adapted to an aerial existence, Lepidoptera have evolved mechanisms to synchronize their wing beats. The simplest type of wing coupling consists of the jugum, a lobe projecting from the posterior base of the forewing and overlying the hind wing (Fig. 44.3b). In most species coupling is achieved by one or more stout bristles (the frenulum) that arise from the base of the hind wing and interlock with a cuticular flap or row of setae (the retinaculum) on the underside of the forewing (Fig. 44.3a). In

Figure 44.2 Mouthparts of Lepidoptera: **a**, mandibles of Zeugloptera (Micropterygidae, *Palaeomicroides*); **b**, maxilla of same; **c**, head of moth, *Synanthedon*. (a, b, modified from Issiki, 1931; c, modified from Snodgrass, 1935.)

Figure 44.3 Wing coupling mechanisms of Lepidoptera: **a**, frenulum (Geometridae, *Neoalcis californica*); **b**, jugum (Hepialidae, *Hepialus*).

Papilionoidea and a few other Ditrysia, coupling is amplexiform. In these species the forewing, especially in the basal region, broadly overlaps the hind wing, which may bear special humeral veins that stiffen the region of overlap (Fig. 44.10e).

Female Lepidoptera are remarkable for the complexity of the internal reproductive system. Distinct organizational plans typify the suborders Monotrysia and Ditrysia. In all Lepidoptera, sperm are transmitted as a spermatophore, which is stored in the bursa copulatrix. In Monotrysia (Fig. 44.4a) the bursa and common oviduct are confluent and may join the alimentary canal in a cloaca. Eggs are fertilized from the bursa copulatrix as they pass down the common oviduct. The small suborders Zeugloptera and Dacnonypha also have the monotrysian arrangement. In Ditrysia (Fig. 44.4b) the bursa, common oviduct, and rectum open independently. The bursal opening is situated medially on

sternite 8, which may be highly modified, along with segments 9 and 10. As in the Monotrysia, spermatophores are stored in the bursa, but the spermatozoa migrate via the ductus seminalis to the spermatheca, whence the eggs are fertilized. In the suborder Exoporia, the openings of the bursa and common oviduct are situated close together on segments 9 and 10, being connected by a groove along which the sperm migrate to the spermatheca. In Trichoptera, the sister taxon to Lepidoptera, the female reproductive tract is of the same general plan as in other insects, with the ovaries and bursa copulatrix branching from the common oviduct and the single genital opening functioning in both copulation and oviposition. Ontogeny of the ditrysian structures during metamorphosis suggests that a single opening on the eighth sternite represents the primitive configuration in Lepidoptera, the common oviduct secondarily extending posteriad to the region of the anus.

The body plan shared by nearly all lepidopterous larvae, or caterpillars, comprises an elongate trunk with large, sclerotized head capsule, three pairs of thoracic legs, and up to five pairs of unjointed abdominal appendages, or prolegs (Fig. 44.5). The prolegs bear terminal patches of short hooked spines, or **crochets** (Fig. 16.2b), which assist the larvae in clinging to the exposed surfaces of vegetation or to silk strands attached to the substrate. Analogous tarsal modifications occur in a variety of other insects that frequent smooth surfaces of leaves. Abdominal segments 3 to 6 and 10 typically bear prolegs, but in many families, such as Geometridae, the number is reduced. Crochets may be arranged in single or multiple circles or in rows or bands. They may be of uniform size, may alternate large and small, or may be larger at the ends of the rows. They are lacking only in a few burrowing forms and in highly specialized leaf-mining families such as Nepticulidae, Heliozelidae, and Eriocraniidae (Fig. 44.5c,d), and are consequently of great taxonomic utility. Besides reduction or loss of abdominal prolegs leaf-mining caterpillars often lack thoracic legs and have highly modified, prognathous mouthparts and flattened bodies. Prolegs and sometimes thoracic legs

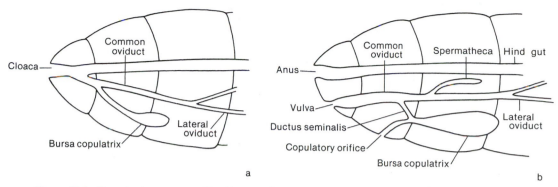

Figure 44.4   Reproductive systems of Lepidoptera (diagrammatic); **a**, monotrysian system; **b**, ditrysian system.

are occasionally lost or reduced in caterpillars that bore in woody substrates.

Caterpillars characteristically bear setae on the head and trunk. Nearly all larvae emerge with a complement of primary setae that are relatively constant in number and position across families. The primary setae are usually long, tactile structures. Internally feeding forms may have only these primary setate, but in later instars additional setae usually appear, either in addition to or replacing the primary setae. Secondary setae are frequently grouped on elevations, processes, or flat or sclerotized plates and may occur on almost any part of the body, including the mandibles. Most often they are simple, but plumose, spatulate, or knobbed secondary setae occur in many taxa. Often the stiff, bristly secondary setae serve a mechanical protective function, as in many Arctiidae, but the dehiscent setae of caterpillars of several families (Limacodidae, Megalopygidae, some Saturniidae and Lymantriidae) are strongly urticating.

Silk production is universal among lepidopterous larvae, being utilized in construction of a wide variety of larval shelters or galleries, as well as in formation of the pupal shelter or cocoon. Many caterpillars also attach silk strands to the substrate to assist locomotion, and some lower themselves on silk lines to escape predators. Silk is produced

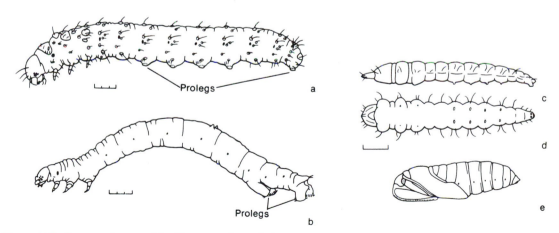

Figure 44.5   Larvae and pupae of Lepidoptera: **a**, larva of Hepialidae (*Hepialus;* scale equals 3 mm); **b**, larva of Geometridae (*Synaxis pallulata;* scale equals 3 mm); **c**, **d**, lateral and ventral views, respectively, of a leaf-mining larva (Tischeriidae, *Tischeria;* scale equals 1 mm); **e**, pupa of Sphingidae *(Manduca).*

by large paired glands, usually extending beside the gut posteriorly into the abdomen and opening through a median spinneret on the labium. The glands are extremely variable in size and configuration, but the evolutionary significance of this variation is not known.

Pupae are exarate in many primitive forms, including Zeugloptera and Dacnonypha. Obtect pupae (Fig. 44.5e; Color photo 2d), with the appendages variably fused to the body, are characteristic of the more derived families, where pupal appendages may be received in grooves in the body wall, and the body regions may be ill defined because of loss of external sutures. Rudiments of mandibles, maxillae, or other mouthparts are present in many families; functional mandibles, occurring in the suborders Zeugloptera and Dacnonypha, are used to open the cocoon just before adult emergence. Members of other suborders escape through a weakened area left by the larva at the time of cocoon formation, or the newly emerged adult regurgitates a liquid which dissolves the silk. At least one pupal abdominal joint remains flexible, and wriggling of the abdomen is the only means of motility in most pupae. Segment 10 may be modified as a cremaster, a series of hooked setae used to anchor pupae of some specialized families to the cocoon or to silk strands attached to the substrate.

BIOLOGY

Monarch butterflies migrate in autumn to overwintering sites, then disperse and reproduce the following spring, and long adult diapause is known in some moths. However, most adult Lepidoptera are rather short-lived. Many species with atrophied mouthparts do not feed. Others imbibe nectar, honeydew, or exudates from fermenting fruit or sap, while a very few are known to suck blood or pierce fruit, using the proboscis to penetrate the skin. Both sexes of nearly all species possess functional wings, but many female Psychidae and Lymantriidae, as well as occasional species of other families, are flightless, especially those that

are active in winter or at high elevations. Brachyptery in males is limited to a very few species, mostly on oceanic islands. Activity is predominantly nocturnal or at dusk and dawn, but nearly all members of the Papilionoidea (butterflies) are sunshine-loving, and certain families or genera of moths are diurnal. In general, nocturnal species tend toward brown or gray and white mottled color patterns, which are important in concealing the insects on their daytime perches. Tree trunks or branches, lichen, and other irregular surfaces are favorite spots for daytime sheltering, and the scale or hair tufts on many moths enhance their camouflage. At least some moths are able to select backgrounds matching their own ground color (Kettlewell, 1973). The brightly colored hind wings of some nocturnal species are covered by the forewings at rest, being suddenly revealed if the moth is disturbed. Such hind-wing patterns sometimes strikingly resemble eyes of vertebrates and are apparently used to startle birds or other potential predators (Fig. 12.4; Color photo 7f). Aposematism and mimicry, perhaps best understood in the monarch, queen, and viceroy, occur very widely among other butterflies, especially in the families Papilionidae and Nymphalidae, particularly in the Tropics (see Papilionoidea, below). Diurnal moths often display aposematic color patterns, sometimes mimicking bees, wasps, or other distasteful Lepidoptera, but the chemicals involved and other details have been little studied.

Lepidoptera are almost exclusively bisexual. Female sex pheromones serving to attract males have been isolated from a great number of species, especially moths, and are probably present in most species. Pheromones may also be produced by males, as in the family Hepialidae, where males may form leks to which the females are attracted. Pheromones may be effective over relatively short distances, as in many of the smaller moths, or may attract males from distances of hundreds of meters, as in Saturniidae. Many butterflies apparently rely on visual cues for initial recognition, frequently followed by behavioral sequences or release of short-range pheromones by males. Copulation takes place on the ground or other perches.

Eggs are characteristically deposited on or near the host plant, either singly or in masses. In Eriocraniidae, Incurvariidae, and a few other families in which ovipositors are present, eggs are inserted into plant tissue, and in some Hepialidae they are scattered during flight. The number of eggs per individual varies from about a dozen to more than 18,000 in some Hepialidae. Eggs are extremely variable, ranging from cylindrical or barrel-shaped to flat wafers. The chorion may be smooth and featureless or regularly or irregularly sculptured with ribs, tubercles, or pits. Eggs may be laid singly or in small groups, or deposited in large masses, usually on or near the larval food. Egg diapause is a common feature of lepidopteran life cycles. A few European Psychidae, including one species *(Apterona helix)* introduced into North America, are parthenogenetic.

Larvae are predominantly herbivores, but scavengers (many Tineidae, Oecophoridae) and a few predators (a few Lycaenidae, Cosmopterygidae, Noctuidae, Geometridae) are known, feeding on eggs of spiders or other Lepidoptera or on scale insects. Predatory Lycaenidae are symbiotic with ants, feeding on their larvae. Hawaiian species of *Eupithecia* (Geometridae) have become specialized to feed on small flies, especially Drosophilidae. The caterpillars poise on twigs, which they grasp with the terminal prolegs, holding the forebody outstretched, then make sudden striking motions to grasp flies with their enlarged forelegs. Larvae of Epipyropidae are unique in living externally on the bodies of leafhoppers, especially Fulgoroidea, feeding on hemolymph exuding through punctures made by the caterpillar's slender, sharply pointed mandibles. The caterpillars maintain their purchase on the host by clinging with the crochets to lines of silk they attach to the host's body.

Roots, stalks and stems, bark and wood, leaves, buds and flowers, fruits, and seeds are consumed by phytophagous forms. In general, larvae of specialized families of the suborder Ditrysia live in exposed situations and are frequently colored or patterned to resemble their host plants or substrates on which they rest. Larvae feeding exposed often show additional defensive mecha-

nisms. For example, papilionid caterpillars evert bright colored, aromatic osmeteria (Color photo 7d). Sphingid caterpillars (Color photo 2c) often rear up and strike at disturbances with the forepart of the body, at the same time regurgitating copious amounts of gut contents. Other defensive mechanisms include dense body coverings of stiff, spiny setae (Color photos 1d,e), sometimes urticarious, and contrastingly colored spots concealed in folds in the body wall and revealed by arching the back. Many caterpillars, such as those of the monarch butterfly, sequester toxic compounds from their food and are aposematically patterned to warn predators of their inedibility. Many caterpillars attempt to escape predators or parasites by feigning death, dropping off their perches or rapidly lowering themselves by silk threads.

Larvae of primitive families of Ditrysia as well as those of the suborders Zeugloptera, Dacnonypha, and Monotrysia frequently feed in leaf mines (Color photo 5f), tunnels, galls, self-constructed shelters, or other concealed situations. Fossil leaves with mines show that the leaf-mining habit was established by the Cretaceous Period. Shelters commonly consist of tied or rolled leaves or twigs, parchmentlike shields or covers, or silken webs. Most species are solitary, but Lasiocampidae (tent caterpillars), among others, are highly gregarious. Many externally feeding caterpillars show variable color patterns, depending on population density, usually becoming darker or contrastingly striped in denser populations. Host specificity varies from acceptance of a single part of a single plant species, as in many leaf miners, to extreme polyphagy, as in many Noctuidae and Geometridae. Lepidoptera are heavily parasitized by Diptera and Hymenoptera, probably because of the exposed position of many of the larvae.

Pupation frequently occurs on the host plant, usually in a cocoon formed by the prepupal larva. Leafminers often pupate within the larval mines, and borers in the larval tunnels, with or without a silken shelter. Noctuidae, Sphingidae (Color photo 2d), and some others pupate naked in soil or litter, and most Papilionoidea (butterflies) and occasional members of other groups hang exposed as a chrysalis on plants (Color photo 1f) or other erect objects.

CLASSIFICATION

A confusing variety of classifications is currently available. Older arrangements recognize the Frenatae and Jugatae as suborders, based on the type of wing coupling. The terms "macrolepidoptera" and "microlepidoptera" refer primarily to body size, which is without phylogenetic significance, and the names of the suborders Rhopalocera and Heterocera refer to antennal shape (clubbed or not clubbed). The classification presented here, which

follows that of Hodges et al. (1983), emphasizes the important structural features of pupal and adult mouthparts and of the female reproductive system. The number of families recognized is conservative, many of the small, obscure families being combined with more generally accepted taxa.

KEY TO THE NORTH AMERICAN SUBORDERS OF

Lepidoptera

1. Forewings and hind wings ovate, similar in venation, size, and shape (Fig. 44.6a); hind wings with at least 10 veins reaching margin   **2**
   Forewings and hind wings conspicuously different

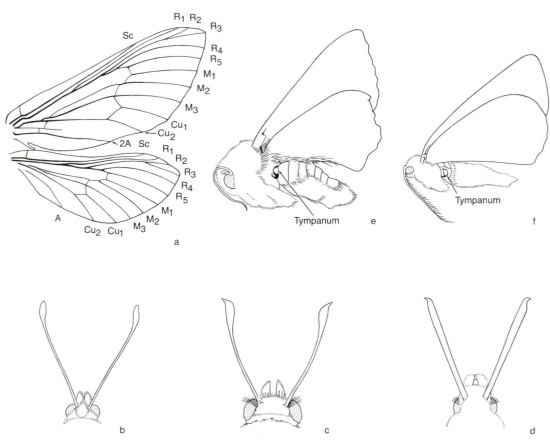

Figure 44.6 Taxonomic characters of Lepidoptera: **a**, wings of Hepialidae *(Hepialus sequoiolus);* **b**, antenna of Papilionidae *(Papilio);* **c**, antenna of Hesperiidae *(Hesperia);* **d**, antenna of Sphingidae *(Celerio);* **e**, tympanal organ of Noctuidae *(Orthosia pacifica);* **f**, tympanal organ of Geometridae *(Drepanulatrix).*

in venation (Figs. 44.7 to 44.11); usually different in shape and size, or narrow, lanceolate (Fig. 44.8f); hind wings usually with 9 or fewer veins reaching margin   **4**

2(1). Large moths at least 10 mm long; maxillary palpi minute or vestigial

Monotrysia, in part (p. 538)

Small moths less than 6 mm long; maxillary palpi 5-segmented, conspicuous (Fig. 44.2b); iridescent, diurnal moths   **3**

3(2). Middle tibiae without spurs (apical tufts of hairs may be present); proboscis absent

Zeugloptera (p. 537)

Middle tibiae with single spur; proboscis short, noncoiled   **Dacnonypha** (p. 537)

4(1). Wings narrow, lanceolate (Fig. 44.8f); hind wings with fringe of setae broader than wing membrane; small moths less than 6 mm long   **5**

Wings broader, ovate or truncate; hind wings with setal fringe narrower than membrane; small to very large lepidopterans   **6**

5(4). Proboscis absent or vestigial; maxillary palpi 5-segmented, conspicuous; female with sclerotized, large external ovipositor

Monotrysia, in part (p. 538)

Proboscis present; maxillary palpi usually very small, 1- to 4-segmented; ovipositor membranous, inconspicuous or absent

Ditrysia, in part (p. 539)

6(4). Maxillary palpi large, folded, with 5 segments (as in Fig. 44.7c); if palpi inconspicuous, antennae much longer than body; ovipositor large, sclerotized, conspicuous; small moths 5 to 10 mm long   **Exoporia** (p. 538) and **Monotrysia,**

in part (p. 538)

Maxillary palpi usually inconspicuous, with 1 to 4 segments (Figs. 44.7f, 44.8b); if palpi 5-segmented, then antennae shorter than body and ovipositor membranous, inconspicuous, or absent; small to very large lepidopterans 3 to 50 mm long   **Ditrysia,** in part (p. 539)

SUBORDER ZEUGLOPTERA

Small moths without proboscis but with functional mandibles in pupa and adult (Fig. 44.2a); adult with ocelli, 5-segmented maxillary palp, 4-segmented labial palp; subcosta branching near middle, jugum with humeral vein. Larva slug-like with prolegs on first 8 abdominal segments, large suckers on ninth and tenth segments. Pupa exarate.

The family **Micropterygidae** contains about 100 species in nine genera of diurnal, frequently iridescent moths that occur sporadically throughout the world. Adults feed on pollen of various flowers; larvae feed on moss, liverworts, and possibly detritus and angiosperms. Pupation is in a tough, parchment-like cocoon, which is cut open with the pupal mandibles. The North American fauna includes three or four species of *Epimartyria,* which are mostly encountered in damp woodlands and bogs and have a 2-year life cycle. *Agathiphaga* and *Heterobathmia,* each with a single genus in Australia and South America, respectively, have functional mandibles as adults, but differ as larvae and are placed in the suborders Aglossata and Heterobathmiina in some classifications. *Agathiphaga* mines the seeds of *Agathis* (Araucariaceae); *Heterobathmia* mines the leaves of *Nothophagus* (southern beech). Both hosts have typical Gondwanan distributions, emphasizing the early divergence of these primitive moths.

SUBORDER DACNONYPHA

Small moths with or without proboscis, mandibles vestigial; maxillary palpi 5-segmented, labial palpi 3- to 4-segmented; forewing with fingerlike jugum. Larva apodous. Pupa exarate with large functional mandibles.

Dacnonypha are small or very small, mostly diurnal moths, often with iridescent vestiture. The suborder contains an assemblage of primitive moths with many generalized features. *Eriocraniella* and *Dyseriocrania* (family **Eriocraniidae**), with species throughout North America, have vestigial adult mandibles and maxillae lacking the laciniae. The galea is modified as a short proboscis. Larvae initiate a linear mine in leaves of oak, chestnut, or other woody plants, later expanding the mine into a large blotch, then leaving the mine to pupate in the soil. Adults appear in late winter or early spring. *Acanthopteroctetes* (family **Acanthopteroctetidae**), with four species in western North America, forms blotch mines on *Ceanothus* (Rhamnaceae). The two families of Dacnonypha may be distinguished by the presence (Eriocraniidae) or absence (Acanthopteroctetidae) of ocelli. Dacnonypha are restricted to the Northern Hemisphere.

SUBORDER EXOPORIA

Exoporia are very largely restricted to the southern continents. The superfamily Hepialoidea, with the single family **Hepialidae,** is represented by about 20 species in North America. Hepialids are moderate or large moths with hairy, elongate bodies and long narrow wings (Fig. 44.9a). The proboscis is vestigial or absent, and adults do not feed. Venation is very similar in the forewings and hind wings (Fig. 44.6a); wing coupling is with a jugum (Fig. 44.3b). The family is cosmopolitan but especially well represented in the Indo-Australian region, where larval feeding habits include external foliage feeding as well as boring in stems or trunks of woody plants. Some adults are showy, diurnal moths. North American species are borers in the stems or root crowns of various shrubs and trees or feed externally on their roots. Pupation is in a cocoon in the larval tunnel, which is also lined with silk. Adults are somber colored and most are crepuscular. Hepialids are unusual in broadcasting the numerous eggs in the general vicinity of the larval food, rather than ovipositing on the host plant.

SUBORDER MONOTRYSIA

Very small to large moths with mandibles and laciniae absent, galeae as a short proboscis or absent; maxillary palpi 1- to 5-segmented, frequently minute; labial palpi with 2 to 3 segments. Larva caterpillar-like (Fig. 44.5a) or flattened, apodous (leafminers: Fig. 44.5c,d); pupa adecticous, obtect.

The Monotrysia includes extremely diverse taxa and is probably polyphyletic.

Nepticulidae and **Opostegidae** are small to exceedingly small moths with linear wings with reduced venation, characteristics common to various unrelated small lepidopterous leafminers. Adults have rudimentary mouthparts, and many probably do not feed. The ovipositor is reduced and the eggs deposited on the surface of the feeding substrate. Larvae of a few species cause twig or petiole galls or mine in bark, but the great majority excavate serpentine mines in the leaves of a broad array of woody plants. Larvae leave the mine to pupate in a tough cocoon in soil or leaf litter.

Larvae of **Tischeriidae** mine the leaves of shrubs of several families of angiosperms and also occur on annual Compositae. Eggs are deposited on the leaf surface, the newly emerged larvae boring into the leaf, eventually forming trumpet-shaped mines. The larval mine is lined with silk and may include a silken retreat used when the larva is threatened. Pupation is in the mine.

Females of the superfamily Incurvarioidea have sclerotized ovipositors that are used to insert the eggs into plant tissues. **Heliozelidae** and some **Incurvariidae** are leafminers as early-instar larvae, later leaving the mine and feeding externally, usually cutting out oval shelters from the mine. Larvae of *Adela* (Incurvariidae) (Fig. 44.9b) first feed in developing ovules, later inhabiting cases constructed from leaf or flower parts, and feeding externally, often on detritus. Other Incurvariidae *(Tegeticula* and *Prodoxus)* bore the seedpods or stems of *Yucca,* the adult moths ovipositing in the immature flower ovaries or stalks. *Tegeticula* females have specialized, elongate maxillary palpi, which are used to collect the glutinous pollen of yucca, which is pollinated only by these moths (see Fig. 10.4). The suborder Monotrysia includes about 1000 species worldwide.

KEY TO THE NORTH AMERICAN FAMILIES OF THE SUBORDERS

Monotrysia and Exoporia

1. Mouthparts atrophied or absent; wings with fingerlike jugum (Fig. 44.3b); head small, attached anteroventrally   **Hepialidae**

   Mouthparts with haustellum and maxillary and labial palps; wings with frenulum; head large, attached anteriorly.   **2**

2(1). Antenna with scape expanded into broad eye cap (Fig. 44.7b); forewing lacking discal cell (superfamily **Nepticuloidea**)   **3**

   Antennae with scape not expanded; forewing with discal cell (Fig. 44.7e) (superfamilies **Incurvarioidea** and **Tischerioidea**)   **4**

3(2). Forewing without branched veins

                                                              Opostegidae

   Forewing with some veins branched

                                                              Nepticulidae

4(2). Cranium with erect, irregular hairs
Incurvariidae and **Tischeriidae**
Cranium with smooth, flattened scales
Heliozelidae

SUBORDER DITRYSIA

An exceedingly variable assemblage characterized by the ditrysian organization of the female reproductive organs (Fig. 44.4b). Ditrysia have the galea produced as a proboscis (occasionally atrophied). The wings are scaled (rarely with a few aculeae), and wing coupling is either frenulate or amplexiform (never jugate). Wing venation is usually strikingly different in the forewings and hind wings (heteroneurous), in contrast to the similar, homoneurous venation of the more primitive suborders.

The Ditrysia comprise the great majority of Lepidoptera, including nearly all the familiar or economically important species. Venation, an important differentiating feature for many families, is frequently invisible without descaling the wings. The following key, based as far as possible on characteristics other than venation, is modified in part from Common (1970).

KEY TO THE COMMON NORTH AMERICAN SUPERFAMILIES OF THE SUBORDER

Ditrysia

1. Wings absent or vestigial   **24**
Wings large, suitable for flying   **2**
2(1). Antennae threadlike, with abruptly swollen knob at tip, or with tip abruptly hooked (Fig. 44.6b,c); frenulum and ocelli absent   **3**
Antennae filiform, feathery, or gradually enlarged, very rarely knobbed; if thickened apically, frenulum present; ocelli present or absent   **4**
3(2). Distance between bases of antennae greater than diameter of eye (Fig. 44.6c); hind tibia usually with intermediate spur; all veins arising individually from discal cell in both wings (Fig. 44.10a)   **Hesperioidea** (p. 548)
Distance between bases of antennae less than diameter of eye; hind tibiae without intermediate spur; forewings with some peripheral veins branched (Fig. 44.10b, at arrow)
Papilionoidea (p. 550)

4(2). Antennae spindle-shaped, thickest near middle or in apical third (Fig. 44.6d); body long, stout, pointed at both ends; ocelli absent; heavy-bodied moths more than 20 mm long
Sphingoidea (p. 554)
Antennae threadlike, featherlike, or comblike, never spindle-shaped; body rarely spindle-shaped; ocelli present or absent   **5**
5(4). Wings deeply cleft (Fig. 44.9k) with hind wing divided into 3 or more lobes   **6**
Wings not deeply cleft; hind wings consisting of single undivided lobe   **7**
6(5). Hind wing divided into 3 lobes (Fig. 44.9k)
Pterophoroidea (p. 548)
Hind wing divided into 6 to 7 lobes
Copromorphoidea (Alucitidae, p. 545)
7(5). Hind wing narrow or linear (Fig. 44.8f), with marginal fringe of hairs or scales subequal to or broader than wing membrane; venation frequently reduced; small to minute moths with wingspan almost always less than 15 mm   **8**
Hind wing oval or triangular, with marginal fringe of scales much shorter than wing breadth; venation usually complete; wingspan usually greater than 10 mm   **12**
8(7). Hind wing with vein $Sc+R_1$ fused with R_S beyond discal cell, then diverging (Fig. 44.8d); abdomen with paired tympana on first sternite (Fig. 44.6f)   **Pyraloidea** (p. 548)
Hind wing with veins $Sc + R_1$ and R_S separate beyond discal cell (Fig. 44.8e,g) or Sc occasionally joined to discal cell by short R_1 (Fig. 44.8h, at arrow); abdomen without tympana   **9**
9(8). Proboscis not covered by scales; maxillary palpi large or small   **10**
Proboscis densely covered by overlapping scales, at least basally (Fig. 44.8b,c); maxillary palpi short, 4-segmented, folded about base of proboscis (Fig. 44.8c)
Gelechioidea, in part (p. 546)
10(9). Antenna with scape enlarged, cup-shaped on underside, forming cover over eye (Fig. 44.7b)   **Tineoidea**, in part (p. 543)
Antennae with scape not expanded, not cup-shaped   **11**
11(10). Maxillary palpi usually long, 5-segmented, folded in repose (Fig. 44.7c); hind tibiae with long, stiff bristles only on dorsal surface, or

smooth-scaled

Tineoidea, in part (p. 543)

Maxillary palpi with 3 to 4 segments, not folded in repose (Fig. 44.8a); hind tibiae frequently with whorls of long stiff bristles

Yponomeutoidea and **Copromorphoidea**, in part (p. 545)

12(7). Tympanal organs present in thorax or abdomen (Fig. 44.6e,f) **13**

Tympanal organs absent **15**

13(12). Tympanal organs in thorax (Fig. 44.6e)

Noctuoidea (p. 554)

Tympanal organs in abdomen (Fig. 44.6f) **14**

14(13). Maxillary palpi minute or vestigial; proboscis bare, not covered with scales

Geometroidea and **Drepanoidea** (p. 552)

Maxillary palpi with 3 or 4 segments, usually

about as long as head and held horizontally in front of head; proboscis densely covered with scales at base (Fig.44.8b)

Pyraloidea (p. 548)

15(12). Vein M present in discal cell of one or both wings (Fig. 44.7d, at arrows) **16**

Vein M absent from discal cell of both wings (Fig. 44.7e) **20**

16(15). Vein M branched in discal cell of both wings; R with 5 branches in forewing (Fig. 44.7d, at arrows); proboscis vestigial or absent; large, heavy-bodied moths at least 20 mm long

Cossoidea (p. 541)

Vein M not branched in discal cell of both wings; R usually with 4 branches in forewing (Fig. 44.7a); mostly small, frail moths less than 15 mm long **17**

17(16). Hind wing with marginal hairs or scales much

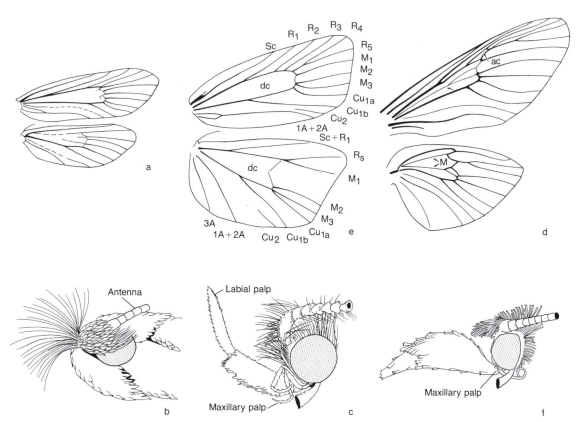

Figure 44.7 Taxonomic characters of Lepidoptera: **a**, wings of Tineidae *(Scardia gracilis);* **b**, head of Lyonetiidae *(Bucculatrix althaeae);* **c**, head of Tineidae *(Scardia gracilis);* **d**, wings of Cossidae *(Prionoxystus robiniae;* arrows indicate vein M in discal cells); **e**, wings of Tortricidae *(Sparganothis);* **f**, head of Tortricidae *(Sparganothis senecionana).* Left labial palp (near side) removed on heads. ac, accessory cell; dc, discal cell.

longer near wing base than at apex   **18**

Hind wing with marginal hairs or scales near wing base subequal in length to those at apex   **19**

18(17). Labial palpi held horizontally in front of head (Fig. 44.7f); frequently longer than head

Tortricoidea, in part (p. 541)

Labial palpi drooping or upright, not held horizontally (Fig. 44.7c); usually short

Tineoidea, in part (p. 541)

19(17). Forewing with vein Cu_2 intersecting 1A before margin   **Tineoidea**, in part (p. 543)

Forewing with vein Cu_2 not intersecting 1A, separate to margin (as in Fig. 44.8g–i, vein 1A at arrow)   **Zygaenoidea** (p. 547)

20(15). Proboscis bare or with few, sparse scales near base (Fig. 44.8a); if scaled, ocelli are large and prominent and maxillary palpi are not folded about base of proboscis (Fig. 44.8b)   **21**

Proboscis densely covered by overlapping scales, at least basally (Fig. 44.8b,c); ocelli always small or absent; maxillary palpi short, folded around base of proboscis (Fig. 44.8c)   **Gelechioidea** (p. 546)

21(20). Hind wing with marginal hairs or scales much longer near wing base than at apex   **22**

Hind wing with marginal hairs or scales near wing base subequal in length to those at apex   **23**

22(21). Labial palpi held horizontally in front of head, frequently as long as head or longer (Fig. 44.7f)   **Tortricoidea**[1] (p. 541)

Labial palpi drooping or upright, not held horizontally (Fig. 44.8a), usually shorter than head

Yponomeutoidea and **Copromorphoidea**, in part (p. 545)

23(21). Forewing at least 4 times as long as broad; wings held together by recurved bristles along margins and by frenulum (as in Fig. 44.3a)   **Sesioidea** (Sesiidae) (p. 545)

Forewing no more than twice as long as broad; wings lacking both marginal bristles and frenulum, or with vestigial frenulum

Bombycoidea (p. 553)

24(1). Legs greatly reduced, nonfunctional; moth de-

[1] Carposcinidae (Copromorphoidea), with seven North American species, will also key here.

veloping in case constructed by larva

Tineoidea (p. 543)

Legs well developed, used in walking; moth not developing in larval case   **25**

25(24). Proboscis vestigial or absent

Noctuoidea (p. 554)

Proboscis present, longer than head

Geometroidea (p. 552)

Superfamily **Cossoidea**. Small to large moths, North American species moderate or large (Fig. 44.9c). Body stout, long, usually exceeding hind wings; head with erect hairs or scales, ocelli absent, proboscis very short or absent, antennae bipectinate in male; forewing with large accessory cell, discal cell divided by forked M vein (Fig. 44.7d, at arrows). Larva stout, cylindrical, with prognathous mandibles; thorax with large, sclerotized dorsal shield.

The family **Cossidae** occurs in North America. These moths show many primitive features, particularly the extensive wing venation with poorly defined discal cell, and the chaetotaxy of the larva. At the same time, they are specialized in the reduction of the mouthparts and loss of ocelli. Adult cossids are short-lived. They produce large numbers of eggs (up to 18,000 have been recorded from a single individual). Larvae bore in heartwood or beneath the bark of various trees or shrubs, or invade the crowns of woody herbs, requiring up to several years to mature. Frass and debris are pushed outside the tunnel, and pupation is within the larval gallery. The family is particularly diversified in Australia. About 40 species are known in North America.

Superfamily **Tortricoidea**. Small moths, mostly with broad, truncate, gray, brown, or tan mottled wings (Fig. 44.9e). Head with rough scales on vertex, ocelli and chaetosemata usually present, proboscis short or vestigial (Fig. 44.7f); maxillary palpi minute, labial palpi 3-segmented, porrect. Forewings with accessory cell often present (as in Fig. 44.7d), discal cell sometimes divided by unbranched M.

The single family, **Tortricidae,** comprises a large cosmopolitan group of considerable economic importance. The **Tortricinae** and **Olethreutinae** (sometimes recognized as families) contain many species whose larvae bore into fruit, nuts, or seeds,

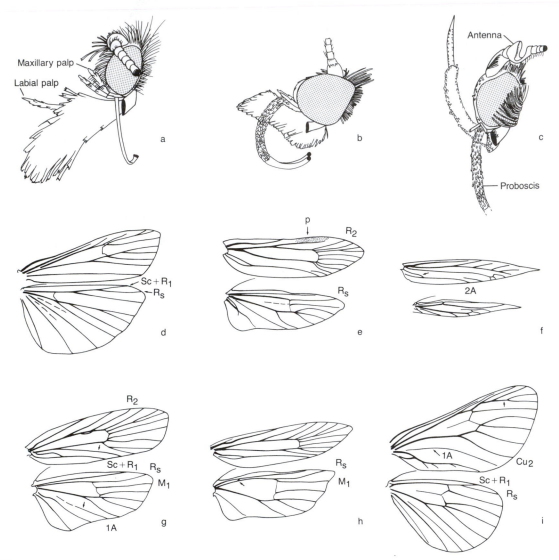

Figure 44.8 Taxonomic characters of Lepidoptera: **a**, head of Yponomeutidae *(Abebaea cervella);* **b**, head of Pyralidae *(Nomophila nearctica);* **c**, head of Blastobasidae *(Holcocera gigantella);* **d**, wings of Pyralidae *(Dioryctria);* **e**, wings of Blastobasidae *(Holcocera);* **f**, wings of Coleophoridae *(Coleophora;* arrow indicates basal fork of 2A); **g**, wings of Oecophoridae *(Agonopteryx fusciterminella;* arrows indicate veins 1A in forewings and hind wings); **h**, wings of Gelechiidae *(Filatima demissae;* arrow indicates short R_1); **i**, wings of Megalopygidae *(Norape tener;* arrow indicates branched radial vein). Left labial palp (near side) removed on heads. p, pterostigma.

including the codling moth, *Cydia pomonella*, and the oriental fruit moth, *Grapholita molesta*, among the more destructive forms. The movements of Mexican jumping beans are caused by the sudden jerks of the larvae of *Cydia deshaisiana*, which lives in the seeds. Many other tortricid larvae feed externally, usually sheltering in rolled or tied leaves; some species feed internally in early instars and then externally in later instars. The external feeders also include many economically important species, such as the spruce budworm, *Choristoneura fumiferana*, and the fruit tree leaf roller, *Archips argyrospilus*. Pupation usually occurs within the larval food or shelter. In the **Phaloniinae** the forewing is usually buckled and bent downward at the end of the discal cell (wings usually flat in Tortricinae and Olethreutinae). The larvae hollow seeds, root crowns, or foliage terminals of a variety of herbs, trees and shrubs.

Superfamily **Tineoidea.** Very small to medium-sized moths, usually somber gray or brown with elongate rounded or pointed wings (Fig. 44.9d). Head rough-scaled (most families, Fig. 44.7c) or smooth-scaled (Gracilariidae); ocelli present or absent, chaetosemata absent; proboscis usually present, without covering of scales; maxillary palpi large, usually 5-segmented, labial palpi short, drooping or ascending (not porrect) (Fig. 44.7c). Forewings with extensive venation (Fig. 44.7a), frequently including branched or unbranched M vein in discal cell; accessory cell absent.

Both generalized and highly specialized taxa are included in this superfamily. As in most other primitive groups of Lepidoptera, the larvae tend to occur in protected situations. For example, **Tineidae** includes species that tunnel in fungi or fungus-infested wood, as well as others that scavenge beneath loose bark, in bat caves and animal burrows, or in human habitations; several species are major pests of stored grains. Many members of this family are capable of digesting keratin, and they consume material of animal origin, especially fur and feathers, including articles of apparel; hence their common name, "clothes moths." Many tineid larvae construct tubes over their food or inhabit a portable protective case, in which pupation later occurs. In the **Psychidae**, which are plant feeders, this case-making behavior is universal. The cases, sometimes of bizarre configuration, may be constructed entirely of silk or may incorporate fragments of plants or substrate material. A wide variety of angiosperms serves as hosts, and a few species feed on gymnosperms, lichens, and mosses. Females of all North American species are wingless, remaining in the case, where the eggs are deposited. However, in primitive forms, including many Australian species, the females are

winged. Psychidae do not feed as adults, and the proboscis is vestigial or absent.

Larvae of **Lyonetiidae** and **Gracilariidae** are phytophagous. Most species mine leaves; they may be highly modified, with legs reduced or absent. Some species form serpentine mines, while others later enlarge the mine into a blotch. *Bucculatrix*, probably the most familiar genus, initiates a serpentine mine, then abandons the mine to tie adjacent leaves together and skeletonize them. Adult lyonetiids, like many other very small moths, have narrow, lanceolate wings with reduced venation. The wing tip is suddenly attenuate in many species. Many gracilariid larvae show striking hypermetamorphosis. The early leaf-mining instars, which feed on sap, often lack both prolegs and thoracic legs, and have reduced head capsule and mouthparts. The full-grown larva has normal head and mouthparts, thoracic legs with claws, and prolegs on abdominal segments 3 to 5 and 10.

Ochsenheimeriidae (not included in following key) is represented in eastern North America by a single introduced species, *Ochsenheimeria vacculella*, whose larvae mine the stems of a variety of grasses. Other members of this small family are Palearctic.

KEY TO THE NORTH AMERICAN FAMILIES OF

Tineoidea

1. Wings about as long as body or longer   2
   Wings absent or shorter than body
                              Psychidae (females only)

2(1). Antenna with scape enlarged, ventrally concave, forming eyecap (Fig. 44.7b); very small moths (<4 mm long) with lanceolate wings
                              Lyonetiidae
   Antenna with scape slender, cylindrical or slightly flattened, not forming cover for eye; size variable   3

3(2). Head with erect, bushy scales covering vertex and face (Fig. 44.7c)   4
   Head with smooth, appressed scales at least on face (as in Fig. 44.7b); vertex smooth-scaled or with posterior fringe of erect scales; very small moths (<4 mm long) with lanceolate wings
                              Gracilariidae

Figure 44.9 Representative Lepidoptera: **a**, Hepialidae *(Hepialus sequoiolus);* **b**, Incurvariidae *(Adela trigrapha);* **c**, Cossidae *(Prionoxystus robiniae);* **d**, Tineidae *(Tinea pellionella);* **e**, Tortricidae *(Archips argyrospilus);* **f**, Yponomeutidae *(Atteva punctella);* **g**, Gelechiidae *(Phthorimaea operculella);* **h**, Sesiidae *(Ramosia resplendens);* **i**, Pyralidae *(Udea profundalis);* **j**, Pyralidae *(Desmia funeralis);* **k**, Pterophoridae *(Platyptilia carduidactyla);* **l**, Ethmiidae *(Ethmia arctostaphylella).*

4(3). Proboscis and maxillary palps vestigial; antennal scape without pecten   **Psychidae** (males)

Proboscis and maxillary palps almost always large, functional (Fig. 44.7c); if vestigial, pecten present on antennal scape   **Tineidae**

Superfamilies **Yponomeutoidea and Copromorphoidea** (Fig. 44.9f). Small or occasionally moderate in size; head with smooth scales; ocelli and chaetosemata present or absent; proboscis moderate, without covering of scales; labial palpi short or prominent, either drooping (Fig. 44.8a) or ascending. Forewing usually elongate with rounded or angulate apex; discal cell rarely divided by M vein.

Yponomeutoidea and Copromorphoidea include moths that are superficially similar, and several families have been included in both superfamilies in different classifications. The arrangement followed here is conservative.

The moths of these superfamilies are mostly uncommon or obscure, with some notable exceptions mentioned below. Many species, including most **Glyphipterygidae** and **Heliodinidae**, are brightly colored, diurnal insects. In contrast to the morphologically similar Tineoidea, which contains many scavengers, nearly all Yponomeutoidea and Copromorphoidea are phytophagous.

Douglasiidae (Yponomeutoidea) mine leaves or stems as larvae. Heliodinidae and Epermeniidae include leafminers, borers in fruit terminals, and leafskeletonizers, and **Yponomeutidae** show a comparable diversity in larval habits, including external foliage feeding. The external feeders usually spin a more or less extensive web over the food, sometimes tying leaves together. Although many are widely distributed, such as the cosmopolitan diamondback moth (*Plutella xylostella*) (Yponomeutidae), which feeds on most Cruciferae, few are of economic significance.

Copromorphoidea are represented by a few small families of rarely encountered moths in North America. **Alucitidae** are readily distinguished from all other Lepidoptera in having the hind wing divided into at least 6 plumes. Larval habits are largely unknown. **Carposinidae** and **Copromorphidae** (not included in following key) are superficially similar to Tortricidae, but lack

vein M in the hind wing. Known larvae of both families are internal feeders in fruits, buds, galls, or necrotic bark. Fewer than 10 species are known from North America. Larvae of **Glyphipterigidae** are mostly seed and stem borers, especially in grasses and sedges.

KEY TO THE COMMONER NORTH AMERICAN FAMILIES OF

Yponomeutoidea and Copromorphoidea

1. Hind wings entire   2

   Hind wings divided into at least 6 plumes   **Alucitidae**

2(1). Tibia and tarsus of hind legs bearing stiff, erect bristles   **Heliodinidae** and **Epermeniidae**

   Tibia and tarsus of hind legs with smooth scales   3

3(2). Ocelli large, conspicuous; maxillary palpi minute (as in Fig. 44.7f)   4

   Ocelli absent; if ocelli present, maxillary palpi large (Fig.44.8a)

   Yponomeutidae (including Argyresthiidae, Acrolepidae, Plutellidae)

4(3). Body length less than 3 mm; hind wings narrow, lanceolate, with fringe of setae broader than membrane   **Douglasiidae**

   Body length greater than 3 mm; hind wings with membrane broader than fringe of setae

   Glyphipterygidae (incl. Choreutidae)

Superfamily **Sesioidea** (Fig. 44.9h). Small to moderate-sized moths; ocelli usually present, chaetosemata present or absent; proboscis naked or scaled at base; maxillary palpi small, labial palpi recurved, ascending; forewing with discal cell divided by M vein or not.

Sesiidae (Fig. 44.9h) includes several species destructive to agricultural crops. The moths of this family are mostly diurnal, fast-flying mimics of stinging Hymenoptera; usually the wing membrane is free of scales except at the margins. Most North American species occur in the arid west. Larvae bore in trunks, bark, stems or roots of herbs, shrubs, or trees, causing significant damage to peach, currant, gooseberry, and blackberry, and to squash and other cucurbits. Pupation is in the larval gallery. Adults usually visit flowers.

Superfamily **Gelechioidea** (Fig.44.9g,l). Small or very small moths; ocelli present or absent, chaetosemata absent. Proboscis clothed with scales, at least basally, and embraced by small, 3- to 4-segmented maxillary palpi; labial palpi recurved over head, usually extending to vertex (Fig. 44.8c). Forewing with discal cell undivided by M vein.

The Gelechioidea represents the largest superfamily of "microlepidoptera," the family **Gelechiidae** alone containing at least 4000 species. North American forms are predominantly small, somber-colored, rather obscure moths, but some Old World species reach 75 mm in wingspan, with bright, conspicuous markings. In terms of larval feeding the Gelechioidea are perhaps more diverse than any other superfamily. Included are leafminers, tiers, and rollers; borers in stems, seeds, nuts, tubers, and fruits; and gall makers, scavengers, and predators of scale insects. In addition, many species feed exposed or protected by a silken web or case. Many of the larvae wriggle vigorously when exposed, often bouncing erratically as a result. Most gelechioid families, especially those with large numbers of species, are variable both in feeding mode and in host plant preferences. Several of the smaller, more specialized families are more restricted. For example, **Coleophoridae** are leafminers in the first instar, but in later instars they construct a case from silk, frass, and plant material. **Elachistidae** are predominantly leafminers, but some bore into stems. The more primitive species feed in dicotyledonous plants, but most form blotch mines on the leaves of grasses, sedges, and rushes. Pupation occurs in the larval shelter in many species, or exposed in others, sometimes with a silk girdle attaching the pupa to the substrate. **Blastobasidae** include a few species that feed in flowers or fruits of *Yucca* and *Agave*, but most appear to be opportunistic scavengers that inhabit diverse situations such as the webs of foliage-feeding caterpillars and the galleries of other insects, or even feeding on scale insects or other small prey. Most **Scythridae** spin frail webs from which they feed on buds and leaves of a variety of plants. *Areniscythris brachypteris* of central coastal California is notable for the greatly abbreviated wings of both sexes. The larvae inhabit silken tubes, which are attached to sand dune vegetation and covered by shifting sand.

Several families, notably Gelechiidae, and to a less extent **Oecophoridae** and **Cosmopterygidae**, show a marked tendency to attack the reproductive structures of their hosts (buds, flowers, tubers, etc.), but other species are leafminers, leaf tiers or rollers, stem borers, or gall formers. Some species feed communally beneath webs on foliage, and some are scavengers. Most economically important species feed in reproductive structures. These include the pink bollworm, *Pectinophora gossypiella* (Gelechiidae), possibly the most important insect pest on cotton. Gelechiids of lesser importance include the potato tuber worm, *Phthorimaea operculella* (Fig. 44.9g), which mines leaves as well as boring into the tubers, and the Angoumois grain moth, *Sitotroga cerealella*, which infests various whole grains, either fresh or dried. A few Oecophoridae and Cosmopterygidae occasionally cause minor damage on a few crops or ornamentals. *Euclemensia* and *Antequera* (Cosmopterygidae) are unusual in being internal parasitoids of *Kermes* (Coccidae) on oaks.

KEY TO THE COMMON NORTH AMERICAN FAMILIES OF

Gelechioidea

1. Hind wings narrow, linear (Fig. 44.8e,f), with posterior fringe of setae much wider than membrane   **2**

   Hind wing broader, oval, with setal fringe narrower than membrane   **7**

2(1). Forewing with thickening (pterostigma) on costal margin near apex (Fig. 44.8e); vein R_2 arising from near apex of discal cell; antennae with scape usually expanded, sometimes concave beneath   **Blastobasidae**

   Forewing without pterostigma; antenna with scape slender, never concave beneath   **3**

3(2). Maxillary palpi with two segments, exceedingly small, apparently absent   **4**

Maxillary palp with 3 to 4 segments; folded around base of proboscis (as in Fig. 44.8c)   **5**

4(3). Forewings with vein 2A forked at base (Fig. 44.8f, at arrow)   **Coleophoridae**
Forewings with vein 2A undivided at base   **Elachistidae**

5(3). Forewing with vein R_2 arising well before apex of discal cell (Fig. 44.8g)   **6**
Forewing with vein R_2 arising very near apex of discal cell (as in Fig. 44.8e); mostly small, dark moths; hind wing with closed discal cell   **Scythridae**

6(5). Hind wing with veins R_S and M_1 arising separately on discal cell (Fig. 44.8g); abdomen often with dorsal spines on segments 1 to 7   **Oecophoridae**
Hind wing with veins R_S and M_1 arising conjointly from discal cell (as in Fig. 44.8h); abdomen without dorsal spines   **Cosmopterygidae**
(incl. Momphidae, Walshiidae)

7(1). Both wings with vein 1A present, at least in distal half of wing; hind wing not terminating as a narrow appendage (Fig. 44.8g, at arrows)   **Oecophoridae** (incl. Ethmiidae, Stenomidae)
Both wings with vein 1A absent; hind wing often narrowed, apically appendiculate (Fig. 44.8h)   **Gelechiidae**

Superfamily **Zygaenoidea**. Small to medium-sized moths with bipectinate antennae, at least in males; ocelli, chaetosemata present or absent; proboscis large (Zygaenidae) or absent; maxillary palpi small or absent, labial palpi small or minute. Forewings and hind wings with discal cells divided by M vein (Fig. 44.8i).

Adults of this small, predominantly tropical superfamily are mostly densely hairy, heavy-bodied moths, frequently brownish or tan, sometimes with pectinate antennae. **Zygaenidae**, however, include brightly colored, diurnal species that visit flowers for nectar. Larvae are highly modified, often sluglike in appearance, with semiconcealed head and sticky cuticle or tufts or bands of hairs that may produce severe skin irritation on contact. Most zygaenid caterpillars are leafskeletonizers, especially on plants of the grape family. Caterpillars of *Harrisina americana*, the grape leafskeletonizer, which often causes economic damage in North

America, feed communally, lining up side by side and moving backward as a leaf is consumed. The larvae of **Epipyropidae** are ectoparasites of Fulgoroidea, primarily in the Old World tropics. Large numbers of eggs (up to 3000) are laid on foliage frequented by the hosts, which are located by the active first instar. Later instars are grublike, the caterpillar residing beneath one wing of the host and lapping up hemolymph that oozes through punctures made by the styliform mandibles. A single species, *Fulgoraecia exigua*, occurs in North America. The larvae of **Megalopygidae** (puss caterpillars) are covered with tufts of dense, silky setae that often form a posterior tail-like extension. Beneath the silky tufts are strongly urticating setae. Puss caterpillars feed exposed on a variety of trees and shrubs. **Limacodidae** and **Dalceridae** (slug caterpillars) feed exposed on woody plants or occasionally on herbs. Limacodidae include naked, sluglike species and others with spiny lobes that are sometimes dehiscent and urticating. Dalcerid caterpillars (one species in extreme southern Arizona, not included in following key) are covered with soft, detachable tubercles of jellylike consistency, presumable defensive. About 100 species of Zygaenoidea occur in North America, mostly in the families **Zygaenidae** and **Limacodidae**.

KEY TO THE NORTH AMERICAN FAMILIES OF

Zygaenoidea

1. Proboscis and maxillary palpi vestigial or absent; vertex without chaetosemata   **2**
Proboscis and maxillary palpi large, functional; chaetosemata large   **Zygaenidae**

2(1). Hind wing with veins Sc + R_1 and R_S fused along most of the discal cell (Fig. 44.8i)   **Megalopygidae**
Hind wing with veins Sc + R_1 and R_S separate or fused for a very short distance near base (as in Fig. 44.8g)   **3**

3(2). Tibial spurs absent; all mouthparts absent or vestigial; forewing with all veins arising separately from discal cell   **Epipyropidae**
Tibial spurs almost always present on middle and hind legs; maxillary and labial palpi usually apparent; forewing with at least one radial vein

branched after leaving discal cell (as in Fig. 44.8i, at upper arrow)   **Limacodidae**

Superfamily **Pyraloidea**. Small to large moths with ocelli and chaetosemata present or absent; maxillary palpi usually moderate, 4-segmented; labial palpi usually beaklike, held porrect in front of head (Fig. 44.8b); legs usually long, slender; wings without M vein in discal cell (Fig. 44.8d).

Three families occur in North America. **Thyrididae,** including small, dusky moths, usually with transparent patches in the wings, are distinguished from Pyralidae in lacking vein CuP in the hind wing. Fewer than 10 species occur in North America. Thyridid larvae bore into stems or roll leaves of various plants. **Hyblaeidae,** with a single introduced species in southeastern North America, are superficially similar to Noctuidae but have two anal veins in the forewings. In contrast, the **Pyralidae** constitute an enormous grouping, with over 1400 species recorded north of Mexico. Adults are mostly small, dull-colored insects with elongate, triangular wings (Fig. 44.9i,j) but large bright-colored or metallic forms occur, especially in the Tropics. Pyralidae are unique among "microlepidoptera" in possessing tympanal auditory organs on the abdomen. Analogous structures occur on the metathorax of Geometroidea and Noctuoidea.

Larval biology is highly variable, including aquatic as well as terrestrial forms, external feeders on webbed foliage, borers in stems, seeds, or fruit, gall-inhabiting forms, scavengers, and a few predators. Many species fold leaves or construct silk shelters or tubes, usually incorporating frass or food particles. Most Pyralidae are solitary, but dense populations of larvae are common in forms that infest concentrated food sources such as stored products. A few foliage-feeding species construct communal webs. Pupation is in a silk cocoon, either on the host or in or on surrounding litter or soil. Aquatic species occupy habitats ranging from lakes to hot springs to swiftly flowing streams, mostly feeding on submerged vegetation, which they web or tunnel, but females of *Petrophila* crawl down into the water to oviposit beneath stones,

where the larvae spin silk webs and feed on algae and diatoms. Pupation is in a silk cocoon. The newly emerged adult swims to the surface with the posterior pairs of legs.

Numerous economically important species attack stored products, in addition to a wide array of agricultural crops and ornamental plants. More destructive species in North America include *Ostrinia nubilalis,* the European corn borer, and *Desmia funeralis,* the grape leaffolder, as well as several cosmopolitan granary pests *(Pyralis farinalis, Ephestia kuehniella,* and *Plodia interpunctella). Galleria mellonella* and *Achroia grisella* infest beehives and are capable of digesting wax with the aid of intestinal bacteria. A few species are beneficial, including *Cactoblastis cactorum,* which has been utilized to control *Opuntia* cactus in Australia.

Superfamily **Pterophoroidea.** Small to moderate-sized moths, ocelli and chaetosemata absent; maxillary palpi minute, labial palpi long, held horizontally in front of head; fore and hind wings each divided into 2 to 4 lobes.

The only family **Pterophoridae,** includes slender-bodied moths with characteristic, deeply cleft wings, held at right angles to the body at rest (Fig. 44.9k). The wings of pterophorids are split into two or three plumes, while those of alucitids (Copromorphoidea) are divided into at least six plumes. Pterophorid larvae are sometimes leafminers initially, later folding or rolling leaves, boring in stems, or occasionally feeding exposed. In North America fewer than 100 species utilize a variety of hosts in about 15 families of plants.

Superfamily **Hesperioidea** (skippers). Medium-sized, with ocelli absent, chaetosemata present; antennae scaled, gradually clubbed, usually with hooked, bare tip (Fig. 44.6c); maxillary palpi absent, labial palpi large, erect; forewing with 5 radial veins arising separately from discal cell (Fig. 44.10a).

A single family, **Hesperiidae** (Fig. 44.11a,b). Over 3000 species of these stout-bodied butterflies are known, with about 200 in the United States. North American species are diurnal, but some tropical

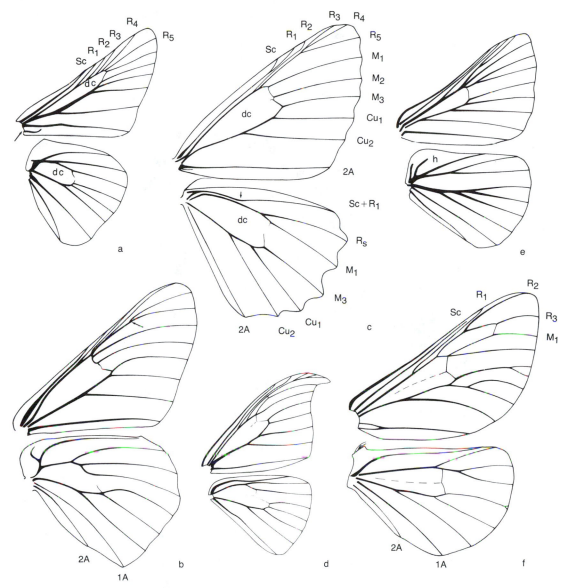

Figure 44.10 Wings of Lepidoptera: **a**, Hesperiidae (*Hesperia harpalus;* arrow indicates vein 3A); **b**, Nymphalidae (*Euphydryas chalcedona;* arrow indicates branched vein); **c**, Geometridae (*Neoalcis californica;* arrow indicates Sc + R₁ recurved toward R$_S$); **d**, Drepanidae *(Drepana);* **e**, Lasiocampidae *(Malacosoma californica);* **f**, Saturniidae *(Hemileuca nevadensis).* dc, discal cell; h, humeral veins.

species fly at dusk or night. The wing beat is rapid and flight direct, rather than fluttering, as in most butterflies (Papilionoidea). Adults sip nectar from flowers; the larvae mostly feed from webbed foliage. The more primitive species utilize a variety of dicotyledonous plants, but the more specialized ones, comprising about 60 percent of the superfamily, eat monocotyledonous plants, especially grasses. Larvae of the subfamily Megathyminae (sometimes treated as a family) bore in stalks and roots of *Yucca* and *Agave*. Pupation is in a cocoon of webbed foliage or in the larval gallery.

Superfamily **Papilionoidea** (butterflies) (Fig. 44.11c–i). Medium to large, occasionally small lepidop-

terans; ocelli absent, chaetosemata present; antennae gradually clubbed (Fig. 44.6b), rarely with hooked tip; maxillary palpi small, 1-segmented or absent; labial palpi short to much longer than head; forewing with 2 or more radial veins arising conjointly from discal cell (Fig. 44.10b).

Figure 44.11 Representative Lepidoptera, Papilionoidea and Hesperioidea: **a**, Hesperiidae *(Hesperia harpalus);* **b**, Hesperiidae *(Pyrgus scriptura);* **c**, Papilionidae *(Papilio rutulus);* **d**, Nymphalidae *(Cynthia virginiensis);* **e**, Nymphalidae *(Euphydryas editha);* **f**, Pieridae *(Colias philodice);* **g**, Lycaenidae *(Strymon melinus);* **h**, Lycaenidae *(Apodemia mormo);* **i**, Nymphalidae (Satyrinae) *(Cercyonis peglerae).*

The butterflies, because of the aesthetic appeal of their brilliant colors or striking patterns, have received more attention than any other group of insects. Accurate species lists exist for most temperate regions, and comprehensive biological information is available for many species. Recent treatments of the North American species include Ferris and Brown (1981), Howe (1975), Opler and Krizek (1984), Opler and Malikul (1992), and Pyle (l981). The Papilionoidea are among the most specialized Lepidoptera in terms of behavior and ecology. Perhaps more than any other insects, butterflies display aposematic or warning coloration and engage in mimicry relationships. In North America the most familiar examples are (1) the monarch (*Danaus plexippus,* Color photo 3b), the queen *(Danaus gilippus),* and the viceroy *(Limenitis archippus)* (Nymphalidae), and (2) *Battus philenor* (Papilionidae) and *Limenitis astyanax* (Nymphalidae). Mimicry is especially common in tropical regions. At single localities in Central and South America up to 20 mimetic species of Ithomiinae and Heliconiinae (Nymphalidae) are commonly encountered, along with occasional members of other families, including moths of families such as Ctenuchidae. Such mimetic complexes include both Müllerian relationships, in which several distasteful species are mutually protected by displaying a common, easily recognized pattern, and Batesian relationships, in which edible species resemble distasteful ones. Batesian mimics may be polymorphic, with the different color forms resembling distasteful species with strikingly different patterns, as in the African *Papilio dardanus* (Papilionidae), in which males have the appearance of typical swallowtails while females lack tails on the wings and mimic several species of distasteful Nymphalidae. Seasonal polymorphism, regulated by external factors such as temperature and photoperiod, and usually comprising paler and darker generations, also occurs in a number of butterflies, especially Pieridae.

Adults of several species are migratory. Most notable is the monarch, which has spread over the globe, following introductions of its host plants. In North America two-way migrations between Mexico and southern Canada annually cover at least 2000 km. In North America *Cynthia, Libytheana,* and *Colias* migrate over shorter distances. Local movements include aggregation on hilltops and patrolling behavior in males, both involved with mate location. Males of many species have a home range in which they occupy centralized perches from which exploratory flights are made to locate females.

Larvae of butterflies are predominantly external foliage feeders. Most are cryptically colored, but species feeding on distasteful plants may be brightly ringed or banded. Papilionidae evert an odoriferous osmeterium when disturbed, while larvae of many Nymphalidae are protected by hairs or spines. Certain lycaenid larvae are myrmecophilous, producing special glandular secretions and sometimes stridulatory signals to attract ants, which protect them from predators and parasites. Most of these myrmecophilous caterpillars are foliage feeders, but some of them spend daylight hours in the host galleries, migrating nightly to feed. A few species reside permanently in the host nest, feeding on ant larvae or pupae. A few other lycaenids are predatory on aphids or coccids. Naked pupae or chrysalises, attached to the substrate by the cremaster, are typical of Papilionoidea. Some chrysalises are also supported by a silken girdle about the midregion.

The Papilionidae, or swallowtails, are among the most familiar butterflies but constitute a relatively small family, with only about 30 species north of Mexico. Pieridae, the whites and sulphurs, number only about 60 North American species, but include the economically important *Pieris rapae,* the cabbage butterfly (Color photo 3c), whose caterpillars feed widely on garden crucifers. Libytheidae, the snout butterflies, are primarily tropical, with a single North American species, *Libytheana bachmanii.* The largest families of butterflies are the Lycaenidae and Nymphalidae, each with over 200 species in the United States.

The butterflies were formerly classified as the suborder Rhopalocera, a special recognition that is contradicted by their overall similarity to certain families of moths, notably Hedylidae, a small neotropical family superficially similar to Geometridae (Scoble, 1986). Some authorities recognize considerably more families than are treated here, but recent comparative studies of skeletal anatomy show that most of the obvious variation involves superficial characteristics.

KEY TO THE FAMILIES OF

Papilionoidea

1. Hind wing with 2 anal veins (Fig. 44.10b); fore tibiae without epiphysis (articulated spur)

   2

   Hind wing with 1 anal vein (as in Fig. 44.10c); fore tibiae with epiphysis   **Papilionidae**

2(1). Tarsal claws simple, not divided   **3**
   Tarsal claws bifid   **Pieridae**

3(2). Labial palpi shorter than thorax; usually erect

   4

   Labial palpi longer than thorax, held horizontally in front of head   **Libytheidae**

4(3). Antennal base and eye contiguous; eye usually notched around antennal base

   Lycaenidae (including Riodinidae)

   Antennal base and eye not contiguous; eye not notched around antenna   **(Nymphalidae) 5**

5(4). Hind wing with humeral vein recurved toward wing base   **Heliconiinae**

   Hind wing with humeral vein straight or curved toward wing apex   **6**

6(5). Forewing with vein 3A absent (Fig. 44.10b); antennae with scales dorsally   **7**

   Forewing with vein 3A appearing as a short basal fork of A_2 (as in Figs. 44.10a,c, at arrow in a)

   Danainae

7(6). Forewing with vein Sc strongly swollen at base (Fig. 44.11i)   **Satyrinae**

   Forewing with veins at most slightly swollen at base   **Nymphalinae**

Superfamilies **Geometroidea** and **Drepanoidea.** Mostly small to medium-sized moths with broad wings and slender body; ocelli usually absent; chaetosemata present in Geometroidea, absent in Drepanoidea, except in Drepanidae; maxillary palpi minute or absent; labial palpi usually short, ascending; abdomen with tympanal auditory organs located ventrolaterally in second segment (Fig. 44.6f).

In North America these moths are represented predominantly by **Geometridae** (Fig. 44.13d,e), one of the largest families of moths, with about 12,000 species, 1400 recorded north of Mexico. Accumulations of moths at lights normally contain a high proportion of Geometridae. Many species assume characteristic postures, with the wings pressed against the substrate or held at various angles to the long axis of the body, and spend the daylight hours resting on bark, lichen, twigs, or other surfaces where posturing apparently disrupts the body configuration, helping to camouflage the moths. The abdominal tympanal organs serve the same function as those of the Noctuidae—detection of the high-pitched sounds produced by bats in the echolocation of their prey. Females of several species are brachypterous, living only long enough to mate and deposit their eggs near the pupation site.

Geometrid larvae are commonly known as inchworms or measuring worms, in reference to their looping gait, which is necessitated by the lack of abdominal prolegs in the middle of the body (Fig. 44.5b). Many bear a strong resemblance to twigs, which they can enhance by adopting a rigid posture at rest. Caterpillars of *Eupithecia* in the Hawaiian Islands have utilized this cryptic behavior to become predators of Drosophilidae and other small flies, which are seized with sudden strikes when they alight on the caterpillar's hindquarters. Several species, such as the omnivorous looper, cause minor feeding damage on ornamentals. The cankerworms *Paleacrita vernata* and *Alsophila pometaria* become serious pests, but most Geometridae do not cause appreciable economic loss. Pupation is usually in a flimsy cocoon.

Whereas most members of the superfamily have cryptically patterned brown, gray, or green wings and bodies, the tropical, diurnal Uraniinae, frequently with brilliantly metallic wings, often resemble papilionid butterflies. Larvae of many species feed on Euphorbiaceae, which contain alkaloids toxic to vertebrates. Presumably these toxins are retained by the adults. Larvae of Thyatiridae, Drepanidae, and Epiplemidae differ from those of Geometridae in having four pairs of abdominal prolegs (one or two pairs or one or two pairs plus two or three reduced pairs in Geometridae). They feed on foliage of various shrubs and trees.

KEY TO THE COMMON NORTH AMERICAN FAMILIES OF

Geometroidea and Drepanoidea[1]

1. Wings about as long as body   **2**
   Wings less than half as long as body, not functional in flying   **Geometridae,** in part

2(1). Hind wings with veins $Sc + R_1$ and R_S diverging or parallel throughout length; frenulum always present   **Epiplemidae**
   Hind wings with veins $Sc + R_1$ recurved toward vein R_S at or beyond discal cell (Fig. 44.10c,d, at arrow in c); frenulum present or absent   **3**

3(2). Hind wings with veins $Sc + R_1$ and R_S parallel or fusing beyond middle of discal cell (Fig. 44.10d); proboscis usually absent; forewings usually with hooked apex   **Drepanidae**
   Hind wings with veins $Sc + R_1$ and R_S usually contiguous or fused before or near the middle of discal cell (Fig. 44.10c, at arrow); proboscis almost always present; forewings not usually hooked
        Geometridae, in part (including Uraniidae)

Superfamily **Bombycoidea.** Medium-sized to large, stout-bodied moths with broad wings and without ocelli or chaetosemata (Fig. 44.13g,h); proboscis and maxillary palpi rudimentary or absent; antennae bipectinate, featherlike; frenulum reduced or absent; tympanal organs absent.

Adult Bombycoidea are specialized for reproduction, with atrophied mouthparts. The frenulum is almost always absent, wing coupling being accomplished by overlap, as in butterflies. Nevertheless, many of these moths are strong fliers, the males traversing distances of several miles to reach females. Males fly upwind and locate nearby females by following gradients of pheromones released by the moths. Despite their inability to feed, the moths may remain active for many days, subsisting on stored fats. Females emerge with the eggs at an advanced stage of development and usually remain near the pupation site until mating occurs.

Larvae feed predominantly on foliage of shrubs and trees. Most species are solitary, but some Saturniidae feed gregariously, and Lasiocampidae, the tent caterpillars, construct large, communal webs or tents that envelop entire branches and frequently cause severe defoliation. They are the only significantly destructive members of the superfamily. In North America several species of *Malacosoma* (Fig. 44.13h) are important tent caterpillars.

The pupal period is passed in a tough cocoon, and *Bombyx mori,* the silkworm (Bombycidae) is propagated commercially for its silk.

In North America **Saturniidae** is the commonest family. Many of the species are elegant moths, subtly shaded in muted colors with contrasting eyespots in one or both wings (Fig. 44.13g). **Mimallonidae** and **Apatelodidae** are primarily tropical families with only a few genera and species in North America. Mimallonidae are notable for the portable cases, open at both ends, constructed by their larvae. When resting, the larva blocks the anterior opening with its head capsule and the posterior opening with the circular, thickened posterior abdominal segment. Adults of both families are similar to small Saturniidae.

KEY TO THE NORTH AMERICAN FAMILIES OF

Bombycoidea

1. Hind wing with costal margin greatly expanded basally and supported by 1 to 3 short, stout humeral veins (Fig. 44.10e, marked "h")
        Lasiocampidae
   Hind wing with costal margin not broadened, without humeral veins (Fig. 44.10f)   **2**

2(1). Forewing with vein R with 3 branches (Fig. 44.10f); hind wing usually with 1 anal vein
        Saturniidae (incl. Citheroniidae)
   Forewing with vein R with 4 to 5 branches; hind wing with 2 anal veins   **3**

3(2). Forewing with vein R_{2+3} and R_{4+5} with separate stalks; hind wing without crossvein between veins Sc and R_S **Mimallonidae**

---

[1] Sematuridae, lacking abdominal tympanal organs, are classified as Geometroidea. A single species occurs in southern Arizona. Thyatiridae, represented by about 15 uncommon species in North America, superficially resemble noctuids but have the cubitus three-branched in the forewings and four-branched in the hind wings.

Forewing with veins R_2, R_3, R_4, and R_5 arising from a common stalk (as in Fig. 44.12c)

Bombycidae[1] and **Apatelodidae**

Superfamily Sphingoidea. Medium-sized to large, heavy-bodied moths with long, narrow forewings; ocelli and chaetosemata absent; proboscis usually very long (Fig. 10.1a); antennae usually thickened in middle (Fig. 44.6d); maxillary palpi 1-segmented, labial palpi stout, 2- or 3-segmented.

The single family, **Sphingidae** is most diverse in tropical regions, but the large, fast-flying moths are familiar insects throughout the world (Fig. 44.13f). The larger species exceed small hummingbirds in size and share the avian habit of extracting nectar while hovering above flowers. Certain plants, including some Onagraceae in temperate regions, are adapted for pollination by sphingid moths, with deep-throated, usually pale-colored flowers that open at dusk as the moths become active. Other sphingids are diurnal, some closely resembling bumblebees or other stinging Hymenoptera.

The stout-bodied larvae often bear a tapering caudal projection; hence the common name "hornworms." They consume prodigious quantities of foliage, and *Manduca quinquemaculata* and *M. sexta* are pests on solanaceous plants such as tobacco, tomato, and potato. Pupation occurs free in a cell, in soil, or in a wispy cocoon.

Superfamily Noctuoidea. Small to large, mostly stout-bodied moths without chaetosemata, ocelli absent; proboscis long or reduced or absent, maxillary palpi minute or absent; tympanal organs present in metathorax (Fig. 44.6e).

The Noctuoidea constitute the largest group of Lepidoptera, with more than 20,000 species recognized. Included are many extremely familiar insects, such as the gypsy moth, *Lymantria dispar* (Lymantriidae, Fig. 44.13i), and various species of cutworms (Noctuidae, Fig. 44.13a,b). In terms of economic damage to crops, the Noctuoidea are easily the most important group of Lepidoptera in North America.

[1]*Bombyx mori*, the silkworm (Bombycidae), is sporadically reared in North America.

Adults are typically strong-flying, nocturnal insects that almost always constitute a major proportion of the insects attracted to lights. Certain species of **Noctuidae** annually migrate distances of many hundreds of miles north from mild winter regions in southeastern United States. Similar migrations have been documented in other regions. Tympanal organs (Fig. 44.6e) are present in all but a few exceptional species. In Noctuidae they have been shown to function in detecting the high-pitched sounds emitted by bats to locate potential prey. The moths respond by evasive flight patterns, including sudden diving or erratic looping. Aposematic Arctiidae produce ultrasonic signals to warn bats of their inedibility. Females of some species, including many Lymantriidae (tussock moths), are brachypterous or have such heavy bodies that they fly short distances at most. Early instar larvae of such species are dispersed by wind.

Most Noctuoidea are mottled brown or black, very often with prominent tufts or mounds of hair or scales on the thorax or wings. As in other moths that rest on irregular surfaces, such patterning is important in camouflaging the body and disrupting its outline and contours. Many **Arctiidae** (Fig. 44.13c) and most **Ctenuchidae** are aposematically patterned in white, yellow, or red and black. Many species, especially of Ctenuchidae, are diurnal, sometimes with clear wings, providing a close resemblance to stinging Hymenoptera. Some arctiid adults feign death when disturbed, distending the abdomen to reveal bright colored patches of cuticle. The chemical basis for such apparently aposematic behavior in these families is largely unknown.

Larvae of Noctuoidea are herbivores on a great variety of plants, including herbaceous and woody angiosperms and gymnosperms. Many species are highly polyphagous, and this characteristic greatly augments their destructive potential. For example, the noctuid *Helicoverpa zea* is known as the corn earworm, the tomato fruitworm, and the cotton bollworm in different parts of the country or by different growers. Several species of **Lymantriidae,** most notably the introduced gypsy moth,

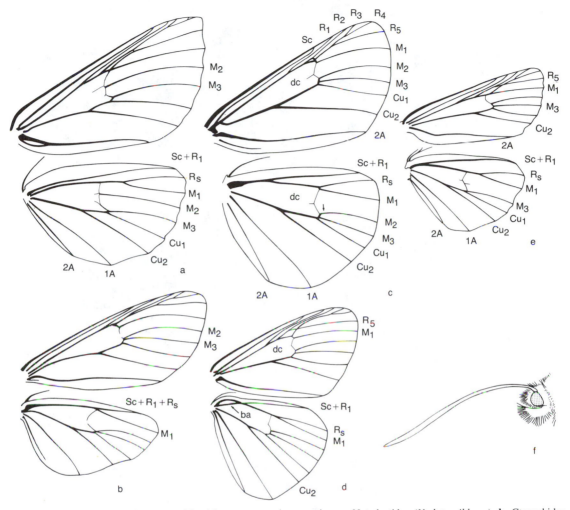

Figure 44.12 Taxonomic characters of Lepidoptera: **a–e**, wing venation; **a**, Notodontidae *(Nadata gibbosa);* **b**, Ctenuchidae *(Ctenucha rubroscapus);* **c**, Arctiidae *(Apantesis ornata;* arrow indicates basal arch of M_2); **d**, Lymantriidae *(Orgyia vetusta);* **e**, Noctuidae *(Lacinipolia quadrilineata);* **f**, antenna of Agaristidae *(Alypia ridingsi).* ba, basal areole; dc, discal cell.

are important defoliators of shade trees in eastern North America. Noctuid, notodontid, and agaristid larvae are typically smooth-skinned, whereas arctiid, ctenuchid, and lymantriid caterpillars are closely covered with bristly or silky hairs, which are brightly colored and urticating in some species. Noctuids frequently hide under debris or litter, emerging at night to feed on or near the surface, where they often chew through stems—hence the name "cutworms." Species of several families,

including Noctuidae, Nolidae, and Arctiidae, construct feeding webs on foliage; some Arctiidae feed communally in webs.

Pupation occurs naked in a cell in the ground or bark (Agaristidae, some Noctuidae), or in a cocoon. Larval hair is usually incorporated into the cocoon, which may then contain little or no silk. The urticating hairs of some Lymantriidae are transferred by the adult female from the cocoon to her freshly deposited egg mass.

Figure 44.13 Representative Lepidoptera: **a**, Noctuidae *(Trichoplusia brassicae);* **b**, Noctuidae *(Catocala andromache);* **c**, Arctiidae *(Apantesis ornata);* **d**, Geometridae *(Sabulodes caberata);* **e**, Geometridae *(Dichorda iridaria);* **f**, Sphingidae *(Celerio lineata);* **g**, Saturniidae *(Hemileuca electra);* **h**, Lasiocampidae *(Malacosoma californica);* **i**, Lymantriidae *(Lymantria dispar).*

KEY TO COMMON NORTH AMERICAN FAMILIES OF

Noctuoidea

1. Wings absent or much less than half the length of body   **Lymantriidae** (females only)
   Wings about as long as body or longer   **2**

2(1). Forewing with veins M_2 and M_3 parallel (Fig. 44.12a); tympanal organ directed ventrally
        Notodontidae[1]

[1] Dioptidae key here. The single North American species, *Phryganidia californica,* occurs in California. The slender-bodied, diurnal adults are common about the larval food plants (various species of oaks).

Forewing with veins M_2 and M_3 converging near discal cell, frequently adjacent (Fig. 44.12b–e); tympanal organ directed posteriorly   **3**

3(2). Hind wing with veins Sc + R_1 and R_S confluent (Fig. 44.12b); brightly colored, frequently metallic, or wasplike diurnal species

Ctenuchidae

Hind wing with veins Sc + R_1 and R_S separate beyond base (Fig. 44.12c–e); usually gray or brown, rarely metallic or wasplike   **4**

4(3). Hind wing with vein M_2 arising closer to M_3 than M_1; M_2 usually arched basally (Fig. 44.12c–e, arch at arrow in c)   **5**

Hind wing with vein M_2 arising midway between M_1 and M_3, or closer to M_1; M_2 not usually arched (as in Fig. 44.12a)

Arctiidae, in part

5(4). Antennae gradually enlarged apically (Fig. 44.12f); veins Sc + R_1 and R_S confluent in basal fourth of discal cell in hind wing; black moths with large white or yellow wing spots   **Agaristidae**

Antennae threadlike or feathery, not gradually enlarged; veins Sc + R_1 and R_S variable   **6**

6(5). Ocelli present   **7**

Ocelli absent   **9**

7(6). Hind wing with vein Sc + R_1 usually swollen, bulbous at base (Fig. 44.12c), usually fused with R_S to about middle of discal cell

Arctiidae, in part

Hind wing with vein Sc + R_1 not noticeably swollen at base; separating before middle of discal cell   **8**

8(7). Black moths with several large white dots on wings; tympanal hoods very large, prominent, dorsolateral on first abdominal segment

Pericopidae

Dark brown or gray moths with mottled color pattern or contrastingly banded or patterned, but not with large, white dots; tympanal hoods lateral   **Noctuidae**

9(6). Forewings with smooth scaling   **10**

Forewings with tufts or lines of raised scales

Nolidae

10(9). Proboscis vestigial or absent; hind wings with Sc + R_1 and R_S forming basal areole (Fig. 44.12d)   **Lymantriidae**

Proboscis large, functional, coiled; veins Sc + R_1 and R_S not separated at base (Fig. 44.12e)   **Noctuidae**

SELECTED REFERENCES

General

Common, 1970; Hering, 1951 (leafminers); *Journal of the Lepidopterists Society,* 1947 to date; *Journal of Research on the Lepidoptera,* 1962 to date; Nielsen and Common, 1991; Scoble, 1992 (form, function, and diversity); Urquhart, 1960 (monarch butterfly).

Morphology

Dugdale, 1974 (female genitalia); Eaton, 1988 (Sphingidae); Ehrlich, 1958 (butterflies); Hinton, 1955a (structure and distribution of larval legs); Kiriakoff, 1963 (tympana); Kristensen, 1968 (mouthparts); Miskimen et al., 1983 (female genitalia); Mosher, 1916 (pupae); Oseto and Helms, 1976 (Noctuidae); Sorensen, 1980 (butterflies).

Biology and Ecology

Ackery, 1984 (milkweed butterflies); Birch and Hefetz, 1987, Birch et al., 1990 (pheromone production in male moths); Brower et al., 1971 (mimicry); Brower and Brower, 1964 (mimicry); J. R. Byers and Hinks, 1973 (adaptive significance of surface sculpturing in larvae); Cook and Scoble, 1992 (structure and function of tympanal organs); Dennis, 1992 (ecology of British butterflies); DeVries, 1990, 1992 (symbioses between ants and larvae of Lycaenidae); Ehrlich and Raven, 1967 (coevolution of butterflies and plants); Fenton and Roeder, 1974 (defense against bats); Ford, 1971 (summary of studies of mimicry in *Papilio dardanus);* Gilbert and Singer, 1975 (butterfly ecology); Hering, 1951 (leafminers); Hespenheide, 1991 (leaf mining); Kettlewell, 1959, 1961 (industrial melanism); Malcolm and Zalucki, 1993 (monarch butterfly biology and conservation); Montgomery, 1983 (biology of predatory Geometridae); New, 1991a (butterfly conservation); Powell, 1964 (life histories of many Tortricidae), 1980 (evolution of larval food preferences); Powell and Mackie, 1966 (biology of yucca moths); Roeder, 1965, 1966 (bat evasion); Rienks, 1985 (seasonal polymorphism); Rothschild, 1985 (aposematism and mimicry in butterflies); Scoble, 1992 (form, function, diversity); Sheppard, 1962 (mimicry and polymorphism); Spangler, 1988 (bat evasion by moths); Stamp and Casey, 1993 (caterpillar ecology); Tietz, 1973 (bibliography); Urquhart, 1987 (monarch butterfly).

Systematics and Evolution

Ackery, 1984 (milkweed butterflies); Common, 1975 (higher classification); D. R. Davis, 1986 (phylogeny of Monotrysia); Dugdale, 1974 (higher classification) 1961 (ecology), 1970 (coevolution); Ehrlich, 1958 (higher classification), 1961 (ecology), 1970 (Coevolution); Ehrlich and Ehrlich, 1961 (butterflies); Ferris and Brown, 1981 (butterflies of Rocky Mountains); Forbes, 1923–1960 (outstanding survey of comparative external morphology of order); Fracker, 1930 (larvae); Grote, 1971 (Noctuidae of North America); Hinton, 1946, 1955a, 1958 (phylogenetic relationships); Hodges et al., 1983; Howe, 1975 (butterflies of North America); Holland, 1968 (large moths of North America); Klots, 1951 (butterflies); Kristensen, 1968, 1971, 1984 (higher classification); Labandeira et al., 1994 (evolution of leaf mining); Merritt and Cummins, 1996 (aquatic species); Mosher, 1916 (pupae); *Moths of America North of Mexico,* 1972 to date; Opler and Krizek, 1984 (butterflies of eastern North America; includes much biological information); Opler and Malikul, 1992 (butterflies of eastern North America); Peterson, 1948 (larvae); Pyle, 1981 (butterflies of North America); Scoble, 1986 (phylogeny of Papilionoidea); Stehr, 1987 (larvae).

45
ORDER TRICHOPTERA (CADDISFLIES)

A caddisfly, *Psychoglypha* sp. (Limnephilidae), on streamside vegetation. Caddisflies resemble moths, but have hairy rather than scaly wings.

Adult. Small to moderate-sized Endopterygota with 2 pairs of wings. Head free, mobile, with large compound eyes, 2 to 3 ocelli present or absent, antennae filiform, elongate, with numerous segments; mandibles vestigial; maxillae with single small lobe or **mala,** maxillary palpi 3- to 5-segmented; labium with large mentum and 3-segmented palps. Thorax with first segment reduced, collarlike; mesothorax largest, metathorax large to small (Fig. 45.1a); wings with macrotrichia or hairs densely covering both veins and membrane; venation mostly of longitudinal veins with very few crossveins (Fig. 45.1b); legs long, slender, with long coxae (Fig. 45.1a); tibiae with both apical and preapical spurs; tarsi

with 5 segments; abdomen with 9 segments in males, 10 segments in females.

Larvae (Fig. 45.2). Eruciform or campodeiform, frequently occurring in protective case or covering; head capsule sclerotized, with strongly developed mandibles, maxillae, and labium, very short antennae and lateral ocelli. Three distinct, short thoracic segments of variable sclerotization; legs long, slender, usually with some segments subdivided; abdomen soft, weakly sclerotized, with hook-shaped prolegs on terminal segment.

Pupa. Decticous, exarate, occurring in larval case or in specially constructed pupal case or cocoon.

With an estimated 7000 species worldwide, over 1200 in North America, the Trichoptera are one of the smaller orders of Endopterygota. The adults are slender, mostly somber-colored insects that somewhat resemble moths; the maxillae are never modified to form a proboscis, however, and hair covers the body and wings, rather than scales. The basal portions of the radial and median veins are not obliterated to form a discal cell, as in most Lepidoptera, and in Trichoptera veins 2A and 3A in the forewing are usually looped forward to vein 1A (Fig. 45.1b). A useful diagnostic feature is the length of the antennae, which are longer than the forewing in most Trichoptera, whereas they are shorter in most Lepidoptera.

By day adult caddisflies rest concealed on vegetation, under bridges, or in culverts, and are seldom seen without searching. In contrast, the larvae, which often inhabit portable cases, are conspicuous in many aquatic habitats, where they are of considerable importance as food for fish and other predators. The great majority of species inhabit freshwater, but smaller numbers occur in brackish coastal waters, and *Philanisus* lives in intertidal rock pools in Australia and New Zealand.

Adult Trichoptera are relatively homogeneous in external morphological features, seemingly adapted for a relatively short life. Although the mouthparts appear to be adapted for lapping up liquids, few Trichoptera have been observed to

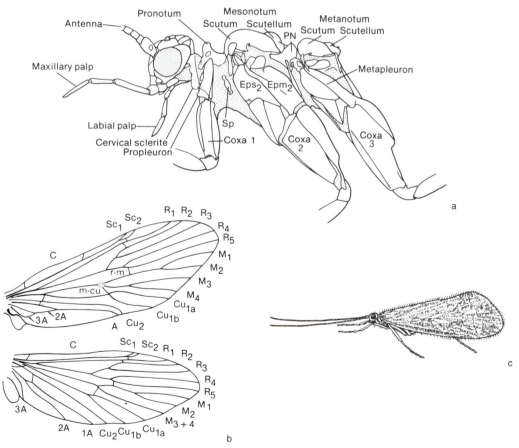

Figure 45.1   Trichoptera: **a**, head and thorax of *Rhyacophila torrentium* (Rhyacophilidae); **b**, wings of *Rhyacophila torrentium;* **c**, *Frenesia missa* (Limnephilidae). (a, b, modified from Schmid, 1970; c, modified from Ross, 1944). Eps, episternum; Epm, epimeron; PN, postnotum; Sp, spiracle.

feed, and most species probably reproduce without taking any sustenance. Most caddisflies are weak, fluttery fliers, but others may fly great distances. In general the weaker fliers have relatively broad forewings that broadly overlap the hind wings, or have the jugal lobe enlarged, as in primitive Lepidoptera. In stronger flying species the macrotrichia of the hind-wing base are developed to form a frenulum, which interlocks with the jugal lobe of the forewing. In several families a row of enlarged, upcurved macrotrichia along the costal margin of the hind wing hooks over a ventral, longitudinal ridge on the anal region of the forewing. In Leptoceridae, which include many active, strong-flying diurnal species, and in Helicopsychidae, the modified hind-wing macrotrichia are developed as hamuli (Fig. 45.6c), much as in Hymenoptera, and the antennae are as much as three times the length of the body.

Larvae are of two general types. Free-living and net-spinning larvae are campodeiform, while those that inhabit cases tend to be eruciform. However, many intermediate forms are known. In Hydropsychidae, which appear to be primitive in a number of features, all the thoracic terga, as well as the prosternum, are sclerotized. The legs are robust and widely spaced, suited for walking and the anal prolegs are stout with simple claws (Fig. 45.2a). Many of these non-case-making larvae are predatory and have slender elongate mandibles, superficially similar to those of Carabidae and other predatory Coleoptera. In some Rhyacophilidae the fore femora and tibiae are strongly raptorial, for grasping prey.

Several modifications are apparent in larvae that construct cases. The terga of the mesothorax and metathorax and the prosternum are membranous. The thoracic segments are short, so that the legs are attached close together near the anterior end of the body and can be easily protruded from the case (Fig. 45.5a). The middle and especially the hind legs are relatively long, further facilitating movement from within the case. In most caddisfly larvae the forelegs are relatively short and are used more for manipulation of food than for walking.

In case-making forms the abdomen is always extremely soft and vulnerable. Gills are more frequently present than in free-living species, and in some case makers a lateral fringe of abdominal hairs is used to circulate water through the case. Three tubercles, which position the larva within the case, are frequently present on the first abdominal segment (Fig. 45.2b). The terminal prolegs are fused basally, with the appendage like distal portions ending in complexly pointed claws adapted for holding the larva inside the case.

Case construction and configuration are quite variable and apparently have arisen independently several times. Hydrobiosidae and Rhyacophilidae are free-living, while the closely related Glossosomatidae construct saddle-shaped cases of silk and sand grains, open at both ends (Fig. 45.3b), the protruding abdominal apex being used in locomotion. Hydroptilidae are free-living in early instars and morphologically similar to the Rhyacophilidae. In later instars they spin flattened cases from silk, sometimes incorporating fine sand or other material, and the abdomen enlarges. Hydroptilid cases are purse-shaped in general configuration, with the head and legs projecting from a slit in the anterior end and the abdominal prolegs projecting from a slit in the posterior margin (Fig. 45.4a).

Hydropsychidae, Psychomyiidae, and Philopotamidae construct fixed shelters or retreats and feed by filtering small particulate matter from the water. These larvae are similar to the free-living Rhyacophilidae. Details of shelter construction vary enormously, from masonry tubes or burrows strengthened with silk to a variety of silken webs, funnels, or tubes (Fig. 45.3c–e). The constructions are arranged so that water moves through a net, either placed at right angles to the stream flow or spun across a burrow.

Cases of other families are generally in the form of elongate, portable tubes completely enclosing their inhabitants (Fig. 45.4c–e), Construction always involves binding particles from the substrate with silk produced from labial glands and extruded through a spinneret. Larvae of these typical case makers have soft, membranous abdo-

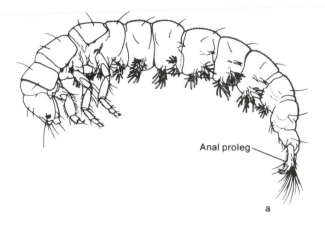

Anal proleg

a

Figure 45.2   Trichoptera larvae: **a**, Rhyacophilidae *(Smicridea fusciatella);* **b**, Limnephilidae *(Philarctus quaeris).* (a, modified from Flint, 1974; b, modified from Wiggins, 1963.)

b

mens and hypognathous heads, as well as a number of other specialized features mentioned above; apparently they represent a monophyletic lineage. Configuration of the cases varies from stout, straight tubes (many Limnephilidae; Color photo 2e) to long, slender, curved tubes (Leptoceridae). Cases are usually circular in cross section but may be polygonal [Brachycentridae, Leptoceridae (Fig. 45.4e)]. Members of the Helicopsychidae use sand grains or small pebbles to build spiral tubes (Fig. 45.4b) that resemble snail shells, and Calamoceratidae build flat shelters from two pieces of leaf. In general, species inhabiting riffles and rapids build the most solidly constructed cases, those occupying lakes or ponds the loosest ones. Since choice of building materials depends in part on local availability, significant variation exists within species. Yet most Trichoptera choose particles within a certain size range and with a certain composition, and cases are often somewhat characteristic at the species level.

It is unclear whether the free-living mode of life or the case-making habit is ancestral in Trichoptera, and the evolutionary relationships among the various types of case makers are disputed.

Caddisfly adults are encountered on vegetation and on stone overhangs and similar places near water. Most are nocturnal or fly at dusk or dawn, and many are attracted to lights. Copulation takes place in the air or on substrates such as low vegetation, and swarming behavior occurs in some families, including Leptoceridae. A few forms, such as *Dolophilodes* (Philopotamidae), are brachypterous or apterous. About 100 to over 1000 eggs, depending on the species, are laid in compact masses or long strings, directly into the water by most species, and females may oviposit while submerged. The eggs are enclosed in a polysaccharide matrix that swells to form a protective covering in the water. Limnephilidae may deposit egg masses on objects above the water, even on twigs high in trees. Apparently rain dissolves the matrix holding the eggs, which then drop into the water below. The Australian genus *Philanisus,* whose larvae inhabit intertidal rock pools, oviposits into the coelomic cavity of starfish. The larvae are free-living, feeding on coralline algae.

Larvae pass through five to seven instars. Non-case-making forms may shelter beneath submerged stones or in debris (Rhyacophilidae) or may spin nets used for filtering food from moving water, as described above (Psychomyiidae, Hydro-

psychidae). In numbers of individuals these filter feeders are the dominant Trichoptera in larger streams in North America, and the larvae or their exuviae may become abundant, clogging intakes in generating plants or municipal water-works. Case-bearing species may be found exposed on stony, sandy, or silty bottoms, sheltered beneath stones or logs, or clambering over vegetation, depending on the species. Some Leptoceridae are capable of active swimming by using the fringed hind legs. Free-living and net-spinning forms tend

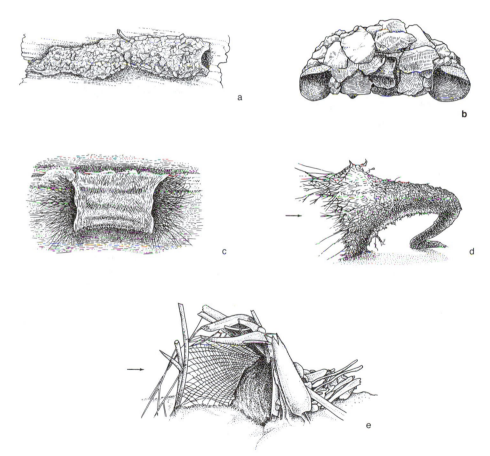

Figure 45.3 Larval cases and retreats of Trichoptera: **a**, retreat of *Lype* sp. (Psychomyiidae) is on a piece of wood, open at both ends, with the roof of silk and bits of detritus; **b**, case of *Culoptila* sp. (Glossosomatidae) is made of small stones; **c**, retreat of *Nyctiophylax* sp. (Psychomyiidae, Polycentropinae) has a silk roof, open at both ends, over a depression in a rock or piece of wood, from which the larva attacks prey that touch the threads at the entrances; **d**, retreat of *Neureclipsis* sp. (Psychomyiidae, Polycentropinae) is a funnel-shaped silk tube that serves to filter prey from water entering the large opening; **e**, capture net or retreat of *Hydropsyche* sp. (Hydropsychidae) is made in flowing water and catches small prey. The direction of the current is indicated by arrows. (Redrawn, with permission, from *Larvae of the North American Caddisfly Genera (Trichoptera),* 2nd ed., by Glenn B. Wiggins, University of Toronto Press, 1996.)

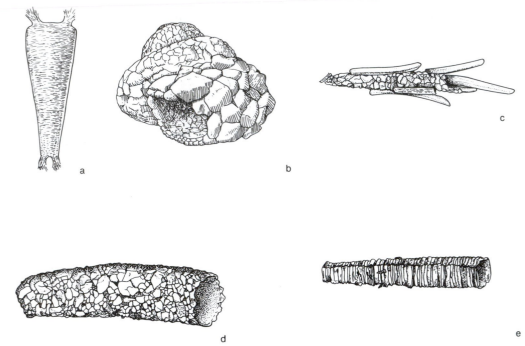

Figure 45.4   Cases of Trichoptera larvae: **a**, Hydroptilidae *(Oxyethira serrata);* **b**, Helicopsychidae *(Helicopsyche borealis);* **c**, Leptoceridae *(Mystacides sepulchralis);* **d**, Limnephilidae *(Pseudostenophylax edwardsi);* **e**, Brachycentridae *(Brachycentrus).* (a, b, modified from Ross, 1944; c, modified from Yamamoto and Wiggins, 1964; d, modified from Wiggins and Anderson, 1968; e, modified from Wiggins, 1965.)

to be predatory, while most case bearers are probably scavengers. Algae and diatoms are commonly consumed by many species, as well as detritus of plant and animal origin. Larval instars are similar in most species, but in the Hydroptilidae, the first four instars are free-living, with short abdomen and long, slender abdominal prolegs. The fifth and sixth instars, which have a bloated, soft abdomen with short prolegs, inhabit cases (Fig. 45.5). Early instars of some Lepidostomatidae build cases of sand grains, then switch to cases of leaf fragments as they grow.

Distribution of larval caddisflies is strongly influenced by the qualities of the substrate, as well as by current, presence and type of vegetation, and water temperature. If closely adjacent bodies of water, or even different portions of the same stream, differ in one or more of these characteristics, the environments so distinguished may sup-

port extremely different faunas. Because of this specificity, caddisflies are important indicators of water quality.

Pupation occurs in water within a shelter—either the larval case or retreat, which may be shortened and closed, or a specially created pupal case. The saddle case making larvae (Glossosomatidae) remove the bottom of the case, attach the top to a rock and spin a cocoon inside. Trichoptera use the pupal mandibles to open the case just before adult emergence. The pharate adult (Color photo 2e) swims to the surface, using setal fringes on the middle legs, and immediately emerges and takes flight, or first climbs onto emergent vegetation or other objects.

Identification of most Trichoptera depends on structures that shrivel in dried specimens; both larvae and adults should be collected into liquid preservatives. Pupae can frequently be identified

Figure 45.5 Larva and case of *Ithytrichia* sp. (Hydroptilidae): **a**, larva, lateral view, larva is naturally flattened from side to side; **b**, silk case, lateral view, opening is at left, silken attachment to substrate is at right; **c**, silken case, ventral view. (Redrawn, with permission, from *Larvae of the North American Caddisfly Genera (Trichoptera)*, 2nd ed., by Glenn B. Wiggins, University of Toronto Press, 1996.)

by the features of the pharate adult. They are a valuable means of associating adults and larvae, since the cast skin of the last larval instar frequently remains in the case. The following key is based on those of H. H. Ross (1944, 1959), Wiggins (1987, 1996), and Morse and Holzenthal (1996).

KEY TO THE COMMON NORTH AMERICAN FAMILIES OF

Trichoptera

ADULTS[1]

   1. Wings without clubbed hairs; antennae about as long as wings or much longer; body length 5 to 40 mm   **2**

      Wings with numerous erect clubbed hairs; antennae shorter than wings; densely hairy caddisflies less than 6 mm long   **Hydroptilidae**

2(1). Ocelli present   **3**

      Ocelli absent   **9**

3(2). Maxillary palpi with 3 to 4 segments   **4**

      Maxillary palpi with 5 segments   **5**

4(3). Maxillary palpi with 3 segments
              Limnephilidae (males; incl. Goeridae)

      Maxillary palpi with 4 segments
                   Phryganeidae (males)

5(3). Maxillary palpi with fifth (apical) segment subequal to fourth segment   **6**

      Maxillary palpi with fifth segment 2 to 3 times longer than fourth segment
                   Philopotamidae

6(5). Maxillary palpi with first and second segments short, subquadrate, approximately equal in length   **7**

      Maxillary palpi with second segment about twice length of first segment, distinctly longer than wide   **8**

7(6). Fore tibia with preapical spur   **21**

      Fore tibia without preapical spur
                  Glossosomatidae

8(6). Fore tibiae with at least 2 spurs; middle tibiae with 4 spurs   **Phryganeidae** (females)

      Fore tibiae with 1 spur or without spurs; middle tibiae with 2 to 3 spurs
         Limnephilidae (females; incl. Goeridae)

9(2). Maxillary palpi with at least 5 segments
                     10

      Maxillary palpi with 3 to 4 segments   **12**

10(9). Maxillary palpi with fifth (apical) segment at least twice as long as fourth segment; fifth

[1] The families Hydrobiosidae, Uenoidae, and Xiphocentridae, each with a few species in North America are omitted.

TABLE 45.1
Characteristics of Larval Trichoptera of North America

Family	Case structure	Case material	Current	Temperature	Substrate
Philopotamidae	Long tunnels or nets	Silk—no extraneous materials	Rapids, riffles	Cold or cool	Stony
Psychomyiidae	Tunnels or nets	Silk	Streams, ponds, lakes	Variable	Sandy bottoms, stones, aquatic vegetation
Hydropsychidae	Nets	Silk	Rivers, streams, moderate current	Variable	Stones, logs, debris
Rhyacophilidae	Free-living	No case	Riffles, moving water	Cold	Irregular bottoms, stones, debris
Glossosomatidae	Tortoise-shell-shaped; open both ends	Small stones, pebbles	Moving water	Cold	Stony, irregular bottoms, stones, debris
Hydrobiosidae	Free living	No case	Moving water	Variable	Irregular bottoms, stones, debris
Hydroptilidae	Barrel-shaped or purse-shaped, open both ends	Variable	Variable	Variable	Variable
Phryganeidae	Stout, tubular case	Grass stem segments, spirally arranged	Marshes, stream eddies	Cold	Plant debris, silt
Brachycentridae	Cylindrical or square cases	Mostly silk (cylindrical cases) or rocks or woody material (square cases)	Rapids, riffles in streams, rivers	Cold	Variable
Limnephilidae	Tubular; variable in proportions	Extremely variable	Variable	Variable	Variable
Lepidostomatidae	Four-sided cases or slender tubes, or irregular cases	Sticks, twigs, sand, or varied materials	Springs, streams, rivers, lake shores	Cold	Variable
Calamoceratidae	Tubular, often triangular in cross section	Sticks, leaf fragments	Springs, slow streams	Cool or cold	Variable
Helicopsychidae	Helical, snail-shell-shaped cases	Stone	Springs, rapid streams, lake margins	Variable	Variable
Odontoceridae	Elongate tubular cases	Minute stones, compactly webbed together	Rapid streams	Cold	Sand, gravel, stones

TABLE 45.1 (*continued*)

Family	Case structure	Case material	Current	Temperature	Substrate
Sericostomatidae	Stout, tubular cases	Minute stones	Rapid streams, springs	Cold	Sand, gravel, stones
Leptoceridae	Variable	Variable	Lakes, streams	Variable, often warm	Variable
Molannidae	Flattened, oblong cases	Minute stones, sand	Lakes, streams	Cool to cold	Stony
Beraeidae	Curved, tapering cases	Sand grains	Springs, seeps, small streams	Cool	Plant debris

segment annulate, appearing multisegmented   **11**
Maxillary palpi with fifth segment subequal
to fourth segment, not annulate   **12**

11(10). Hind wings at least as broad as forewings,
with anal region occupying one-third to
one-half of wing; mesoscutum without
raised, wartlike areas
                                Hydropsychidae
Hind wings narrower than or occasionally
subequal to forewings, with anal region
usually occupying less than one-fourth of
wing area; mesoscutum with a pair of
raised, wartlike tubercles (as in Fig.
45.6b)
                Psychomyiidae (incl. **Polycentropodidae**)

12(9,10). Middle tibiae without preapical spurs   **13**
Middle tibiae with preapical spurs   **16**

13(12). Antennae 1.5 to 3 times length of forewing;
mesoscutum with irregular raised spots
bearing setae (Fig. 45.6a)
                                Leptoceridae
Antennae about as long as forewing; mesoscutum with small paired tubercles (Fig. 45.6b)
or without tubercles   **14**

14(13). Hind wings with anterior margin straight or
evenly curved   **15**
Hind wings with anterior margin straight basally, with row of stout curved setae, then
abruptly curved (Fig. 45.6c)
                                Helicopsychidae

15(14). Middle tibiae with apical spurs about half the
length of basal tarsomere; distal segments

of middle and hindtarsi with spines in apical crown on each segment   **Beraeidae**
Middle tibiae with apical spurs about one-
fourth the length of basal tarsomere; distal
segments of middle and hind tarsi with
scattered spines
            Sericostomatidae and **Brachycentridae**

16(12). Middle femora with a row of 6 to 10 dark
spines on apex of anterior face
                                Molannidae
Middle femora with 2 or fewer dark spines
on anterior face   **17**

17(16). Mesoscutellum small, arcuate (Fig. 45.6d);
mesoscutum with a pair of longitudinal
rows of setate spots   **Calamoceratidae**
Mesoscutellum large, triangular or acutely
rounded, pointed apically; mesoscutum
with tubercles round or oval (Fig. 45.6e,f)
                                          18

18(17). Mesoscutellum with a single large, raised area
medially (Fig. 45.6e,f)   **19**
Mesoscutellum with paired, lateral raised areas   **20**

19(18). Tibial spurs hairy; mesoscutellum triangular,
with medial raised area (Fig. 45.6e)
                                some **Limnephilidae**
Tibial spurs without hairs; mesoscutellum
acutely rounded with raised wart occupying
most of area (Fig. 45.6f)
                                Odontoceridae

20(18). Middle tibiae with irregular row of spines;
middle tarsi with double row of spines
                                Brachycentridae

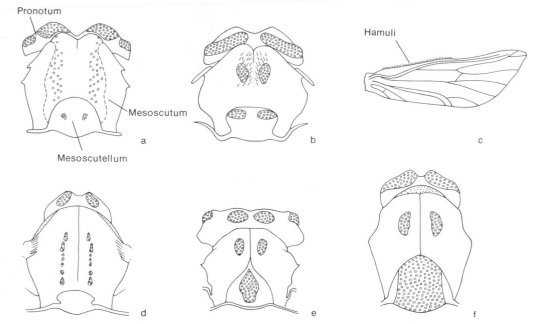

Figure 45.6 Taxonomic characters of Trichoptera, pronotum and mesonotum, and wings: **a**, Leptoceridae *(Athripsodes tarsipunctatus);* **b**, Helicopsychidae *(Helicopsyche borealis);* **c**, Helicopsychidae *(Helicopsyche):* **d**, Calamoceratidae *(Ganonema americanum);* **e**, Limnephilidae *(Goera calcarata);* **f**, Odontoceridae *(Psilotreta frontalis).* (Modified from Ross, 1944.)

Middle tibiae without spines; middle tarsi with sparse, scattered spines
Lepidostomatidae

21(7). Maxillary palpi with second segment rounded and globose   **Rhyacophilidae**
Maxillary palpi with second segment cylindrical, similar to first   **Hydrobiosidae**

LARVAE

1. Metathorax with single tergal plate   2
Metathorax entirely membranous or with dorsum divided into at least 2 sclerites   3

2(1). Abdomen without gills; abdominal leg with 2 to 3 bristles at base of claw
Hydroptilidae
Abdomen with numerous branched gills ventrally; abdominal leg with broad fan of setae at base of claw   **Hydropsychidae**

3(1). Abdomen with sclerotized plate on ninth segment   5
Abdomen entirely membranous   4

4(3). Labrum membranous, narrowed basally
Philopotamidae

Labrum sclerotized, broadest basally
Psychomyiidae (incl. **Polycentropodidae**)

5(3). Abdominal prolegs fused basally, appearing as tenth abdominal segment; claws directed laterally, much shorter than basal portion of proleg   17
Abdominal prolegs free; claws directed ventrally, about as long as basal portion of proleg   6

6(5). Claws on all legs similar in structure   7
Claws of hind legs much smaller than those of front and middle legs, modified as short, setose clubs   **Molannidae**

7(6). Antennae 1 to 4 times as long as wide   8
Antennae at least 8 times as long as wide
Leptoceridae

8(7). Mesonotum with conspicuous sclerotized plates   9
Mesonotum membranous or with minute sclerites   **Phryganeidae**

9(8). Mesonotum with a single pair of narrow, arcuate sclerites oriented longitudinally near midline (Fig. 45.7a)   **Leptoceridae**

Figure 45.7   Taxonomic characters of Trichoptera larvae: **a**, thorax of Leptoceridae *(Athripsodes);* **b**, labrum of Calamoceratidae *(Ganonema);* **c**, labrum of Limnephilidae *(Limnephilus).* (Modified from Ross, 1944.)

Mesonotum with 1 to several subquadrate, transverse, or oval sclerites   **10**

10(9). Labrum with about 16 long, stout setae in a closely set, transverse row across middle (Fig. 45.7b)   **Calamoceratidae**
Labrum with 6 to 8 large setae arranged in a loose, transverse arc or irregularly scattered (Fig. 45.7c)   **11**

11(10). Larva inhabiting a spirally twisted case (Fig. 45.4b); claws of abdominal legs with a comb of subequal teeth   **Heliocopsychidae**
Larval case not spirally twisted; claws of abdominal legs with teeth not arranged as a comb   **12**

12(11). Metanotum with broad anterior sclerite, longitudinal lateral sclerites, and narrow, transverse posterior sclerite   **Odontoceridae**
Metanotum usually with 1 to several small, rounded, poorly defined sclerites   **13**

13(12). Pronotal sclerite bearing strong, transverse furrow   **14**
Pronotal sclerite without transverse furrow
15

14(13). Hind leg with tarsal claw longer than tibia; claw without basal tooth   **Beraeidae**
Hind leg with tarsal claw much shorter, usually about half the length of tibia; claw with prominent basal tooth   **Brachycentridae**

15(13). Antennae inserted adjacent to anterior margin of eye   **Lepidostomatidae**
Antennae inserted midway between eyes and mandibles or closer to mandibles   **16**

16(15). Antennae inserted midway between eyes and mandibles; prosternum with distinct, medial hornlike projection

Limnephilidae (including **Goeridae**)
Antennae inserted very close to base of mandibles; prosternum without hornlike projection   **Sericostomatidae**

17(5). Anal prolegs broadly joined with abdominal segment 9; anal claw with one or more dorsal accessory hooks   **Glossosomatidae**
Anal prolegs free from segment 9; anal claw without dorsal accessory hooks   **18**

18(17). Free part of abdominal prolegs much longer than fused portion; claw without small dorsal accessory hook   **Rhyacophilidae**
Free part of abdominal proleg shorter than fused portion; claw with small dorsal accessory hook   **Hydrobiosidae**

SELECTED REFERENCES

Bowles and Allen, 1992 (life histories); Denning, 1956 (California species); B. J. Downes and Jordan, 1993 (ecology of net-building species); Hildrew and Wagner, 1992 (colonial behavior); Lloyd, 1921 (biology of larvae); Mackay and Wiggins, 1979 (ecological diversity); Merritt and Cummins, 1996 (North America); Morse, 1984; H. H. Ross, 1944 (Illinois species), 1956 (montane species and general evolution), 1959 (North American species), 1964 (nets and cases of larvae), 1967 (biogeography); St Clair, 1993 (life histories); Stehr, 1987 (larvae); Wallace and Merrit, 1980 (filter feeding); Weaver and Morse, 1986 (evolution of feeding and case making); Wiggins, 1996 (larvae); Wiggins and Wichard, 1989 (phylogeny of pupation).

46
ORDER HYMENOPTERA (BEES, WASPS, ANTS, ETC.)

A thread-waisted wasp, *Ammophila* sp. (Sphecidae), using a pebble to close its nest burrow. Underground, the wasp has stored paralyzed caterpillars as food for its larva.

Adult. Minute to large Endopterygota, usually with 2 pairs of wings. Head free, mobile, with mandibles adapted for chewing; galeae or labium frequently elongate, suited for imbibing liquids; antennae multiarticulate, usually filiform or moniliform; compound eyes large, lateral; 3 ocelli usually present. Mesothorax enlarged, fused with small, collarlike pronotum and small metathorax (Fig. 46.1); wings membranous, stiff, hyaline, with relatively few veins, which usually delimit a few large, closed cells; forewings about 1.5 to 2 times as long as hind wings; wing coupling effected by **hamuli** (hooklike setae on leading edge of hind wing). Abdomen with first segment inflexibly joined with metathorax, usually incorporated into thorax as propodeum, second abdominal segment usually constricted as narrow petiole, followed by swollen gaster, female with ovipositor modified as slicing or piercing organ for inserting eggs into tissue, or adapted as sting.

Larva. Eruciform, with distinct head capsule, chewing mandibles, 3 pairs thoracic legs and 6 to 8 pairs abdominal prolegs; or apodous, grublike, often with head capsule and mouthparts reduced or vestigial.

Pupa. Adecticous, usually exarate; frequently enclosed in silken cocoon secreted from labial glands.

Certain Hymenoptera, such as honey bees, ants, and some of the larger wasps, are among the most familiar insects. In contrast to most other orders, many Hymenoptera are beneficial to agriculture, through either pollination of crops or destruction of phytophagous insects. Parasitoid species have proved extremely useful in biological control, especially of introduced pests, and have been transported extensively about the earth. *Apis mellifera*, the honey bee, among the most thoroughly domesticated insects, is extremely important as a pollinator, and its products, honey and wax, are still commercially important, despite competition from alternative materials. Bee and wasp stings, annoyances from ants, and occasional damage by sawflies, gall wasps, or other phytophagous species are of only nuisance significance compared with the positive attributes of this order. Hymenoptera comprise about 108,000 species, but this number is certain to increase to about 150,000 as the numerous small, parasitoid species are discovered.

The most important adaptive features in Hymenoptera involve the mouthparts, thoracic and wing structure, and abdomen, including the ovipositor, all these structures apparently being modified in relation to predation. All Hymenoptera possess well-developed mandibles, which in many species are used for tasks other than feeding and may be modified for specific functions. For example, some megachilid bees and some ants cut foliage with the sharp scissorslike mandibles, while in the soldier castes of many ants the mandibles are designed for offense or defense. In most species, however, multiple functions are served. For instance, a worker ant may dig, transport food or soil particles, subdue or manipulate prey, defend the colony, or tend immatures—using the mandibles as a primary tool. Many Hymenoptera, especially the more advanced families of wasps and all bees, have evolved the ability to ingest liquid food, primarily nectar, through a proboscis formed by the glossae (part of the labium) and the maxillary palpi and galeae (Fig. 46.2b). In anthophorid and apid bees the proboscis may be longer than the body.

As in other aerial insects, the hymenopterous thorax is strongly modified for efficient flight. The mesothorax is much the largest segment, and in most species is solidly fused to the reduced pronotum and metathorax (Fig. 46.1). The head and forelegs are attached to the propleura, which are usually flexibly articulated to the pronotum and mesopleura. In general, the prothorax is best developed in ambulatory species such as ants, or in those which use the forelegs to dig (Pompilidae, Scoliidae). In most strong fliers the pronotum is much smaller than the mesothorax, and in bees and many wasps it may appear as a narrow collar

Figure 46.1 Lateral aspect of pterothorax of *Apis mellifera*, the honey bee. (Modified from Snodgrass, 1942.)

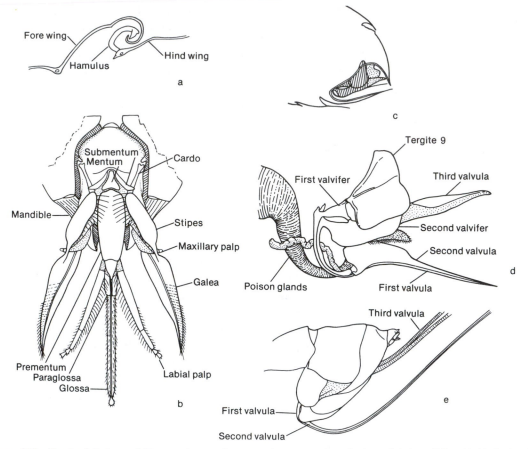

Figure 46.2 Structural features of Hymenoptera: **a**, diagrammatic cross section of wings of *Apis mellifera* (Apidae), showing wing coupling; **b**, mouthparts of *Apis mellifera;* **c**, medial section through abdomen of *Apis mellifera,* showing sting in repose; **d**, dissected sting of *Apis mellifera;* **e**, end of abdomen and ovipositor of *Megarhyssa atrata* (Ichneumonidae). (Modified from Snodgrass, 1933, 1942.)

in dorsal aspect. The metanotum is represented by a narrow sclerite situated between the mesopostnotum and the first abdominal tergite, which is inflexibly incorporated into the thorax as the **propodeum.** The propodeum may be recognized by the presence of the first abdominal spiracle (Fig. 46.1). In Apocrita the metapostnotum or second phragma is an internal cowling situated just beneath the propodaeum. The dorsal longitudinal flight muscles attach to the large second phragma, which is held close to the propodeum by a very short muscle. The functional significance of this complex arrangement is obscure. The pleural sclerites are usually large, distinct plates, bearing the meso-

coxal and metacoxal attachments. In general the extensive variation in size and shape of the various thoracic sclerites is distinctive of families and superfamilies and is of some use in classification.

The hind wings are always coupled to the much larger forewings by a series of hooked setae (hamuli), which extend forward from the costal margin of the hind wing and catch beneath a ventral fold of the anal margin of the forewing (Fig. 46.7a,c,d). The hamuli number from two or three in small species such as Chalcidoidea, to many in the larger wasps where wing coupling may be very tenacious. Wing folding is rare in Hymenoptera but occurs in several families (Vespidae, some Gasteruptiidae) in which wing coupling is so strong that the wings do not separate when held over the back. Simplification of wing venation by reduction is typical of Hymenoptera, being broadly correlated with body size. Very small species, such as some Chalcidoidea, may retain only a single longitudinal vein in the forewing, with the hind wing veinless (Fig. 46.12e). In general, venation is highly variable and characteristic of various families or superfamilies, but aerodynamic significance of the different configurations is not obvious.

The morphological specializations most responsible for the success of the Hymenoptera involve the abdomen. In the relatively primitive suborder Symphyta, the abdomen is broadly and relatively inflexibly joined to the thorax. The ovipositor is bladelike or sawlike in most symphytan families, adapted for slicing plant tissues. In the suborder Apocrita the abdomen is strongly constricted at the second segment, greatly increasing its mobility (Fig. 46.1). The ovipositor is always developed as a piercing organ, which can be inserted deeply into host (usually animal) tissues. Although highly specialized for slicing or piercing tissues, the hymenopteran ovipositor shows obvious homologies to the ovipositors of relatively primitive insects such as Orthoptera and even Thysanura (compare Figs. 46.2e and 2.16a,c). Morphologically the ovipositor consists of the modified appendages of abdominal segments eight and nine. Each appendage consists of a short triangular or trapezoidal **gonocoxite** or valvifer, which rocks about its articulation with the abdominal sternite (Fig. 46.2c–e). Each gonocoxite bears a bladelike **gonapophysis** or valve. Rapid, repeated rocking of the gonocoxites is translated to linear movement of the gonapophyses, which generally bear barbs that assist their penetration. Gonocoxite 9 bears an additional apophysis, the **gonostyle** or ovipositor sheath. The gonostyles, which are frequently fused, do not enter the oviposition substrate, but provide a tubular support for the more delicate gonapophyses. The lumen of the ovipositor is very narrow, and eggs of parasitoid Hymenoptera are capable of great deformation, regaining their normal elongate oval shape once inside the host. The gonapophyses vary from short, relatively stout blades to slender, flexible stylets several times the body length, enabling the wasps to reach hosts concealed in cocoons or in galleries in fruiting structures or wood. Insertion of such elongate ovipositors may require bending into a loop, steadied in some species by the hind coxae (Fig. 46.3a). Many parasitoid species lap up hemolymph exuding from oviposition wounds. If the host is secreted in a cocoon or gallery, the female may secrete material that hardens around the ovipositor, forming a tube that functions as a drinking straw when the ovipositor is withdrawn (Fig. 46.3c). In stinging wasps and

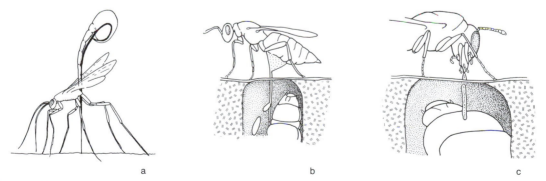

| a | b | c |

Figure 46.3 Postures of female wasps during oviposition. **a**, Ichneumonidae, the ovipositor plus sheath is supported by the hind coxae as the ovipositor is driven into the substrate; **b**, Chalcididae, wasp ovipositing on host larva inside a kernel of grain; **c**, Chalcididae, wasp drinking host's hemolymph through a feeding tube made by her ovipositor. (a, c, modified from Askew, 1971; b, modified from Fulton, 1933.)

bees the first and second gonapophyses are further specialized as a stinger (Fig. 46.2d), discharging only venom, the eggs issuing at the base. In most wasps the stinger is used to immobilize prey, but in Vespoidea (yellowjackets, hornets) and Apoidea (bees), its primary function is defense against vertebrates, and the venoms produce intense, instantaneous pain. Their stings provide efficient protection against many predators, and bees and wasps are probably the most familiar examples of warning coloration. Stinging Hymenoptera are widely mimicked by insects of many other orders and occasionally by spiders (see Chap. 12).

In general, larvae of Hymenoptera conform to two body plans. Larvae of many Symphyta, such as sawflies, are highly similar to lepidopterous caterpillars and, like most caterpillars, are external foliage feeders (Fig. 46.4a). The large, chewing mandibles are set in a strongly sclerotized head capsule. The thick, cylindrical trunk bears appendages on most abdominal segments, as well as on the thoracic segments. The abdominal legs, unlike those of lepidopterous larvae, lack grasping spines or crochets. Other useful differentiating characters include the presence of a single stemma (six stemmata usually present in lepidopterous larvae) and the reduction of the antennae to papillae (an-

tennae usually three-segmented in lepidopterous larvae).

The symphytan families Siricidae and Cephidae, as well as certain Tenthredinidae, are internal plant feeders that lack abdominal legs and have the thoracic legs reduced. All larvae of the suborder Apocrita are apodous, with soft, maggotlike bodies; sometimes the head capsule is unsclerotized or retracted into the thorax (Fig. 46.4b). These larvae all inhabit protected environments (usually the body cavity of a host insect, or a cell prepared by the mother); hence the structural simplification. The midgut is not connected with the hindgut in most apocritan larvae. This is evidently an adaptation for avoiding fecal contamination of the host body or larval cell. The alimentary canal becomes complete at the last larval molt, and the accumulated fecal material, or **meconium,** is voided just before pupation.

Various parasitoid Hymenoptera are hypermetamorphic, the first instar, or occasionally more than one instar, being differentiated from the later maggotlike stages. For example, in some chalcidoid wasps that do not deposit their eggs in a host, the first instar is a **planidium** (Fig. 46.5a). Planidia resemble the first-instar triungulin larvae of Coleoptera (Meloidae, Rhipiphoridae) and Strepsiptera and perform the same function—host location. Locomotion is accomplished with the movable ventral extensions of the segmental sclerites, as legs are absent. **Caudate larvae,** found in some Ichneumonidae and Braconidae, and **nauplioid larvae** (Fig. 46.5b,c), in some Proctotrupoidea, hatch internally in the host. They may function in eliminating supernumerary parasites, as suggested by the large, sickle-shaped mandibles of the nauplioid type. A variety of other first-instar forms, sometimes bizarrely spined or shaped and of uncertain function, occur in different families of parasitoids.

Adult Hymenoptera, with the notable exception of some nocturnal ichneumonids, bees, and a few others, are largely sunshine-loving creatures that frequent flowers, sunlit vegetation, or open soil. However, as parasitoids and predators, they have followed other insects into almost every niche

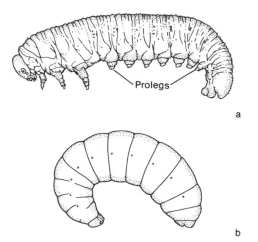

Figure 46.4 Larvae of Hymenoptera: **a,** eruciform larva of Symphyta (Diprionidae, *Neodiprion lecontei*); **b,** apodous larva of Apocrita (Apidae, *Xylocopa virginica*). (a, modified from Middleton, 1921; b, modified from W. P. Stephen et al., 1969.)

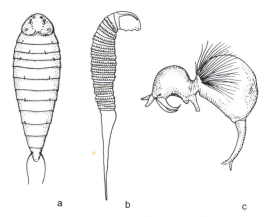

Figure 46.5 First-instar larvae of parasitoid Hymenoptera: **a,** planidium larva of Chalcidoidea (Perilampidae, *Perilampus hyalinus*); **b,** caudate larva of Ichneumonoidea (Ichneumonidae, *Cremastus flavoorbitalis*); **c,** eucoiliform larva of Proctotrupoidea (Scelionidae, *Scelio fulgidus*). (a, modified from Smith, 1912; b, modified from Bradley, 1934; c, modified from Noble, 1938.)

imaginable. As adults, many Hymenoptera consume nectar and pollen, but many sawflies (Tenthredinidae) are predators of small, soft-bodied insects, and some parasitoid forms imbibe hemolymph that oozes from wounds made during oviposition.

Mating may occur in the air (as in honey bees, ants) or on some substrate. Males seek out females about their emergence sites or in areas the females will visit, such as stands of blooming plants. In some species males establish territories including emergence sites and defend them against other males. In some Tiphiidae and Mutillidae the male is much larger than the wingless female and carries her about during copulation. In Hymenoptera, males (if present) are produced parthenogenetically from unfertilized eggs and are haploid, while the diploid females are produced from fertilized eggs. This mode of reproduction is called **arrhenotoky.** Sperm are stored in a spermatheca, where they may remain viable for long periods, their release being controlled by the female. Female control of fertilization allows unusual sex ratios, which may be extreme in social species, where relatively small numbers of males are produced only once or a few times a year. Various species

of Hymenoptera are entirely parthenogenetic **(thelytokous),** being known only from diploid females. Cynipidae (gall wasps) are **cyclically parthenogenetic,** with generations of bisexual males and females alternating with generations of asexual females.

Arrhenotoky allows recombination in each generation of females, promoting heterozygosity, while causing recessive lethal alleles to be eliminated from each generation of males. Arrhenotoky also provides a mechanism for the formation of female societies in bees and wasps, because of the asymmetrical degrees of relatedness, with sisters sharing more genes than do daughters and mothers (see Chap. 7). Thelytoky is of particular advantage in populations that are highly dispersed, as in colonizing species, since male/female encounters may be rare. Thelytoky would also seem to be advantageous in hyperparasitoid species, where populations would normally be sparse.

Several distinct types of life history occur within the order (Table. 46.1). As mentioned above, most or all members of the suborder Symphyta are strictly herbivorous as larvae, and this mode of feeding is apparently primitive for the order. The vast majority of Apocrita are zoophagous as larvae. The most primitive species are parasitoids, living internally or sometimes externally on a host that is almost always selected by the mother. The host is always at least as large as the parasitoid, so that a single victim is sufficient to allow a single parasitoid to develop. It is uncertain how parasitoidism arose from phytophagous ancestors, but **inquilinism** could have been an intermediate stage. Inquilines inhabit galls or nests of other species, consuming the food intended for the original occupant, which is incidentally killed and eaten as well, usually as an egg or early-instar larva. Several primitive families, principally in the superfamilies Proctotrupoidea and Chalcidoidea, include inquiline species.

Parasitoidism is practiced by about half the families of Hymenoptera, including some eminently successful taxa such as Ichneumonidae, Braconidae, and most families of Chalcidoidea. Most orders of insects as well as spiders and

TABLE 46.1
Evolution of Food Habits of Hymenoptera

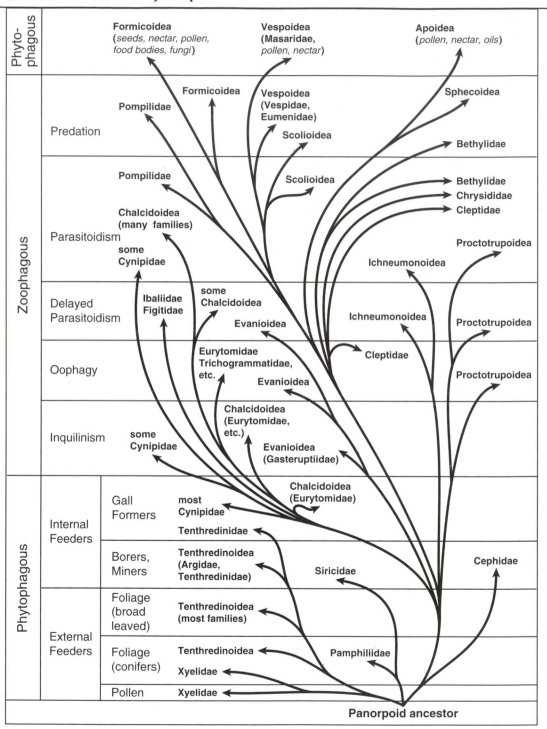

ticks are attacked. All instars, including eggs, are utilized as hosts. In typical life histories, larval development of the parasite begins soon after oviposition, and in some species eggs hatch during oviposition. Eggs of some species are deposited into immature hosts (eggs or early-instar larvae), where they may eclose, but larval development is postponed, sometimes as much as a year, until the host is nearly full-grown. Such **delayed parasitoidism** occurs in Ichneumonoidea and especially in the Proctotrupoidea.

Many chalcidoids, as well as occasional members of other superfamilies, are **hyperparasitoids.** That is, the host is another parasitoid rather than a primary victim. Secondary and even tertiary levels of hyperparasitoidism are known. Most hyperparasitoid species oviposit into insects that already have the required primary parasitoid. A few oviposit speculatively into unparasitized individuals, their eggs developing only if the host is subsequently attacked by the proper primary parasitoid. Some species can complete development as either primary or secondary parasitoids. Most parasitoids develop alone in a host (solitary parasitoids), and many species chemically mark hosts in which they have oviposited. When additional eggs are laid, only a single parasitoid larva survives to maturity.

Other parasitoid species, especially in Ichneumonoidea and Chalcidoidea, live gregariously, many individuals in or on a single host. The gregarious condition may result from multiple oviposition or from polyembryonic development of a single egg. Typically eggs are deposited in or on hosts or very close to them. Some species of Chalcidoidea and Ichneumonoidea oviposit on vegetation. The motile first-instar larvae (planidia; Fig. 46.5a) search for a suitable host to which they attach.

Most of the stinging wasps are predators that immobilize or kill one or more prey, which are installed in a specially constructed cell in which the larva matures. The Pompilidae (spider wasps; Color photo 7a), Bethylidae, Scoliidae, and Tiphiidae clearly bridge the difference between parasitoidism and typical predation. For example, some pompilids oviposit on a temporarily paralyzed host, which may regain partial motility before being consumed. Some species attack spiders in their burrows, where larval development occurs. The basic behavioral sequence of hunting for prey, stinging, and oviposition (Table 46.2) is repeated by an individual wasp many times, just as in parasitoids. Other pompilids prepare their own burrows after subduing the prey. The immobilized prey must be concealed during gallery preparation to prevent it from being stolen by other wasps or scavengers such as ants. All Pompilidae and many Sphecidae utilize prey is that too large to be carried in flight (Color photo 6f). In the more typical mode of predation, most highly developed in the superfamilies Vespoidea and Sphecoidea, prey size is reduced, permitting numerous prey items to be quickly flown to the larval gallery, which is prepared before provisioning begins. Many of these more specialized hunters temporarily close the gallery between provisioning trips and permanently close it when provisioning is complete, excluding parasitoids and scavengers. Many wasps have evolved specialized methods of rapid transport of prey: some remove nonnutritious body parts such as antennae and legs; others hold the prey on the barbed stinger so that the legs are immediately available to open the gallery.

Behavior in hunting wasps has undergone two other specializations. First, each burrow leads to several cells, each containing a developing larva. Each cell is individually closed after provisioning has been completed, and the nest is closed after all cells have been provisioned. Second, many Sphecidae and Vespidae **progressively provision** their larvae, adding prey items as required. This practice contrasts with **mass provisioning,** in which the required amount of prey is provided before an egg is laid. The progressive provisioning of sphecids and vespids, where a female tends many cells with larvae at different developmental stages, is probably the most complex behavior of all nonsocial insects. Progressive provisioning has the advantage of allowing the mother to periodically inspect the cell for parasitoids or scavengers or to remove spoiled provisions and, through the increased contact between mother and offspring,

TABLE 46.2

Evolution of Predatory Behavior and Nest Construction in Aculeates

						Behavior sequence
Larval development in cells prepared by adult female	**Numerous host individuals used**	**Progressive provisioning**		Vespidae, Eumenidae Masaridae	Sphecidae (Bembicinae, Philanthinae)	→ Gallery preparation → Cell preparation Oviposition → Hunting, stinging Transport [Temporary closure] Cell closure [Final nest closure]
		Mass provisioning	Many cells per nest	Eumenidae	Sphecidae (many subfamilies)	→ Gallery preparation → Cell preparation → Hunting, stinging Transport [Temporary closure] Oviposition Cell closure [Final nest closure]
			Single cell per nest		Sphecidae (many subfamilies)	→ Gallery preparation → Hunting, stinging Transport [Temporary closure] Oviposition [Final nest closure]
	Single host individual	**Nest preparation before locating host**		Pompilidae	Sphecidae (Sphecinae, Ampulicinae) — Tiphiidae Scoliidae	→ Gallery preparation Hunting, stinging Transport Oviposition [Nest closure]
		Nest preparation after locating host		Pompilidae	Sphecidae (Sphecinae, Ampulicinae) — Tiphiidae Scoliidae — Bethylidae	→ Hunting, stinging [Transport] Gallery preparation Oviposition [Nest closure]
Ovipositor for stinging, no special behavior				Pompilidae	Tiphiidae Scoliidae Mutillidae — Bethylidae Dryinidae	→ Hunting, stinging Oviposition
Ovipositor for egg laying					Cleptidae, Chrysididae	

was probably prerequisite to the development of sociality.

Bees (superfamily Apoidea) are secondarily phytophagous. Like the morphologically similar Sphecoidea, bees construct larval galleries, but they provision their larvae with pollen and nectar, rather than with paralyzed insects. By thus moving to a lower level on the food chain they have been able to achieve much larger populations than the zoophagous wasps. Nevertheless, the families Megachilidae, Apidae, and Anthophoridae include some **cleptoparasitic** species, which appropriate the nests and provisions of phytophagous species. Bees are notable, along with the ants and vespoid wasps, in having evolved complex societies. In all these taxa, sociality involves repression of reproduction in most female individuals, which function in obtaining food and maintaining and defending the colony. Small numbers of males are produced periodically, functioning only in reproduction. Sociality is discussed in greater detail in Chapter 7.

KEY TO THE SUBORDERS OF

Hymenoptera

1. Abdomen with base broadly attached to thorax (Fig. 46.8); thoracic nota almost always with **cenchri** (Fig. 46.8b); hind wings almost always with 3 basal cells (Fig. 46.7b–d)
   Suborder Symphyta (p. 579)
   Abdomen with base narrowly constricted at attachment to thorax (Fig. 46.10c,d); thoracic nota without cenchri; hind wings with 2 basal cells or fewer (Fig. 46.10e)   Suborder **Apocrita** (p. 583)

SUBORDER SYMPHYTA

Abdomen with base broadly attached to thorax, but with first abdominal tergite distinct, not incorporated as part of thorax; metanotum bearing raised, roughened, knob-like cenchri (Fig. 46.8b) (absent in Cephidae); hind wings usually with 3 basal cells (Fig. 46.7c–d) (except in Orussoidea, Fig. 46.9). Larvae eruciform, caterpillar-like, with large, round head capsule and 6 to 8 pairs of abdominal legs (external feeders, Fig. 46.4a), or abdominal legs absent (some internal feeders).

Symphyta include the most generalized Hymenoptera, both in terms of morphological characteristics and biological adaptations. Larvae are herbivorous,

with the possible exception of the Orussoidea. In number of species, only the superfamily Tenthredinoidea, whose members are largely adapted as external foliage feeders, is dominant today.

KEY TO THE NORTH AMERICAN SUPERFAMILIES OF

Symphyta

1. Antennae inserted beneath a transverse ridge below the eyes; wings with veins strong, thick in basal half, faint apically (Fig. 46.9)
   Orussoidea (p. 581)
   Antennae inserted between the eyes; veins equally developed throughout wing   **2**
2(1). Anterior tibia with 1 apical spur   **3**
   Anterior tibia with 2 apical spurs   **4**
3(2). Pronotum much wider than long (Fig. 46.6a,b), often with posterior margin strongly incurved; abdomen cylindrical   **Siricoidea** (p. 580)
   Pronotum longer than wide or subquadrate (Fig. 46.8c); abdomen slightly compressed in lateral plane   **Cephoidea** (p. 581)
4(2). Costal cell divided by intercostal vein, or if fused with subcostal vein, forewing with 3 marginal cells (Fig. 46.7d)   **Xyeloidea** (p. 579)
   Costal cell without intercostal vein (Fig. 46.7c); forewing with 1 or 2 marginal cells
   Tenthredinoidea (p. 582)

Superfamily **Xyeloidea.** Small to moderate sawflies with the wing venation very generalized (Fig. 46.7d) and the ovipositor short or rudimentary.

The earliest known fossil hymenopteran, *Liadoxyela praecox,* from the Lower Triassic Period, is very

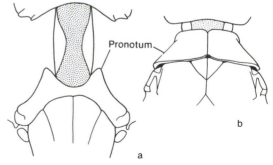

Figure 46.6 Taxonomic characters of Hymenoptera, Symphyta: **a**, **b**, dorsal aspects of pronota of Xiphydriidae and Anaxyelidae, respectively.

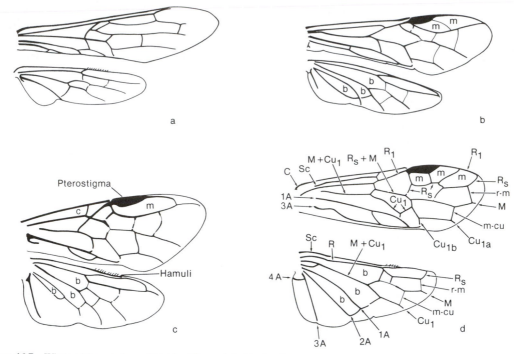

Figure 46.7 Wings of Symphyta: **a**, Siricidae *(Tremex);* **b**, Tenthredinidae *(Tenthredo);* **c**, Argidae *(Arge dulciana);* **d**, Xyelidae *(Macroxylea ferruginea).* b, basal cell; c, costal cell; m, marginal cell; Sc, intercostal vein (subcosta).

similar to existing xyelids, which have the most generalized wing venation in the order. The superfamily contains two families in North America.

Xyelidae are distinguished by the elongate third antennal segment, longer than the remaining segments combined. The larvae, which have prolegs on all abdominal segments, feed on staminate (male) cones of pines *(Xyela),* needle buds of firs *(Macroxyela),* or foliage of other trees such as walnut, pecan, and elm. Mature larvae drop from the trees to pupate in earthen cells, where they remain until the following year. Adults appear very early in the spring and may visit early flowers, especially those with massed blooms, such as chokecherry.

Pamphiliidae have at least 13 antennal segments of approximately equal length. Larvae, which lack abdominal legs and have the thoracic legs greatly reduced, move by a wriggling, peristaltic action. They tie or roll leaves of various trees (both angiosperms and gymnosperms), some species feeding gregariously beneath self-constructed webs, eventually forming large, unsightly tangles of silk, partly consumed vegetation, frass, and cast skins. Pamphiliid larvae depend entirely on the silk scaffolding they produce for locomotion and must extend the webbing to reach new foliage. Pupation is in cells in the soil.

Superfamily **Siricoidea** (wood wasps, horntails). Moderate or large wasps with robust, cylindrical body; anterior tibia with single apical spur.

The Siricoidea are associated with wood, where the short-legged larvae excavate tunnels. Females possess a cylindrical ovipositor, which is used to bore oviposition holes into living or felled trees. The most familiar insects in this superfamily are the **Siricidae** (Fig. 46.8a), whose larvae bore into the heartwood of various trees. The larvae actually feed on symbiotic fungi, which are transmitted by the mother at the time of egg laying and do not digest wood. Siricids normally attack damaged or dead trees but occasionally cause minor damage. Pupation in Siricoidea is inside the larval gallery. **Anaxyelidae** includes a single species, *Syntexis libocedrii,* whose larvae tunnel incense cedar *(Calocedrus),* western juniper, and western red cedar *(Thuja)* in the Pacific Coast states. Females oviposit in trunks still warm and smoking from forest fires. Larvae of **Xiphydriidae** bore in angiosperm trees, including birch, maple, oak, elm, and poplar. A symbiotic fungus, on which they presumably feed, is associated with their galleries. The wasps are fairly common in the eastern states.

KEY TO THE FAMILIES OF

Siricoidea

1. Abdomen with last segment modified as short, upturned prong or spine (Fig. 46.8a); mesonotum bearing 2 diagonal grooves converging toward scutellum **Siricidae**

   Abdomen with last segment not modified as a prong or spine; mesonotum without grooves **2**

2(1). Pronotum with posterior margin strongly incurved (Fig. 46.6a) **Xiphydriidae**

   Pronotum with posterior margin nearly straight (Fig. 46.6b) **Anaxyelidae**

Superfamily **Orussoidea** (Fig. 46.9). A single family, **Orussidae,** of uncommon wasps whose biology and evolutionary relationships are unclear.

Adults are similar to other Symphyta in having a broad thoracic-abdominal articulation. Unlike Symphyta, orussids have the wing venation reduced, and the very long ovipositor is coiled like a watchspring inside the abdomen. The insertion of the antennae beneath a ridge below the eyes is unique among Symphyta. The apodous larvae occur in galleries of buprestid beetles, siricid wood wasps, or other wood-boring insects. At least one Australian species is parasitic, but it is uncertain whether the North American representatives are parasitoids or subsist on the borings and fungus filling old burrows.

Superfamily **Cephoidea** (stem sawflies). Very slender, elongate insects with the abdomen slightly compressed in the lateral plane.

The single family, **Cephidae,** includes delicate wasps, either black or patterned in black and yellow (Fig. 46.8c). Adults are encountered on vegetation or flowers. Larvae of most species bore the stems of grasses, including small grains, and *Cephus cinctus* is a significant pest on wheat. Rose, blackberry, gooseberry, and other shrubs

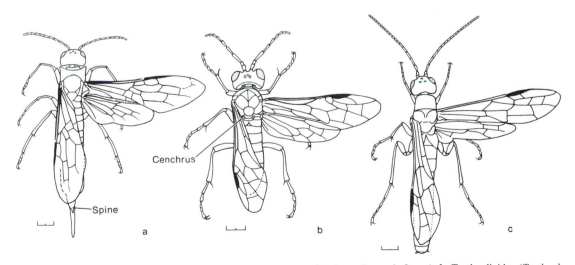

Figure 46.8 Representative Symphyta: **a,** Siricidae (*Sirex juvencus,* female; scale equals 2 mm); **b,** Tenthredinidae (*Tenthredo;* scale equals 2 mm); **c,** Cephidae (*Cephus clavatus;* scale equals 1 mm).

Figure 46.9 Orussidae (*Orussus;* scale equals 2 mm).

with pithy stems are attacked by a few species. Pupation occurs within the larval gallery. The lack of cenchri and the relatively great constriction of the abdomen of adults suggest that the cephoid line of evolution may have given rise to the Apocrita. This relationship is reinforced by the legless, internally feeding larvae.

Superfamily **Tenthredinoidea.** Small to moderate, mostly stout-bodied sawflies; the ovipositor blades have sawtooth serrations for slitting vegetation.

Tenthredinoidea is the dominant superfamily of symphytan Hymenoptera. In north temperate regions the family **Tenthredinidae** (Fig. 46.8b), with over 4000 described species, many abundant and widespread, includes nearly all commonly collected sawflies. Sawfly larvae are characteristically external foliage feeders. Some consume herbs or occasionally ferns, but trees and shrubs are especially favored. Externally feeding larvae are usually caterpillar-like in general appearance, with six to eight pairs of abdominal legs, which lack crochets (Fig. 46.4a). Sawfly larvae are often camouflaged in shades of green, sometimes with contrasting tubercles or spines in red, white, or other colors. Many species evert from the abdominal sternites glandular invaginations that resemble the osmeteria of Lepidoptera. Some tenthredinid and

cimbicid larvae are covered with a whitish powder, and some tenthredinids are sluglike and covered with slimy secretions. Internally feeding larvae, which include leafminers (Argidae, Tenthredinidae) and gall formers (Tenthredinidae), may lack abdominal legs. Sawflies typically have a single generation per year. Eggs are deposited in slits made in leaves or needles, and mature larvae usually leave the host plant to pupate in a cocoon or a cell in soil or leaf litter. Adults are commonly found on flowers and may imbibe nectar or eat pollen, but many Tenthredinidae are predatory.

The most important families are Tenthredinidae and Diprionidae, each with a number of economically significant species. Among the more familiar tenthredinid pests are *Caliroa cerasi,* the pear and cherry slug, *Endelymyia aethiops,* the rose slug, and *Nematus ribesii,* which consumes currant leaves. Many other genera may defoliate or cause cosmetic damage on a wide variety of trees and shrubs. Tenthredinidae feed on a great variety of other plants, including ferns, sedges, grasses, and horsetails *(Equisetum),* but they seldom cause serious damage. More important, especially in North America, are forest pests of conifers, which include *Pristiphora erichsonii* (Tenthredinidae) on larch and various species of *Diprion* and *Neodiprion* (Diprionidae) on spruce and pine. Larvae of **Diprionidae** feed exclusively on conifers (Pinaceae and Cupressaceae), almost always as external foliage feeders. They are usually gregarious, at least in early instars, and often aposematically colored. When disturbed they synchronously rear up with a jerky motion and may regurgitate resinous substances.

Argidae is the second largest family of Symphyta with over 800 species, the great majority tropical. About 60 species occur in North America, the larvae feeding on a variety of shrubs, trees, and less frequently on herbs. The 12 species of North American **Cimbicidae** feed on foliage of trees, especially of the families Rosaceae, Salicaceae (willows), and Betulaceae (birches). The family **Pergidae** is largely restricted to South America and Australia, where the other families of Tenthredinoidea are relatively uncommon. The few North

American species are leafskeletonizers on oak, walnut, and a few other hardwoods.

KEY TO THE NORTH AMERICAN FAMILIES OF

Tenthredinoidea

1.	Antennae with 6 or more segments	**2**
	Antennae with 3 segments; apical segment elongate, sometimes bifurcate	**Argidae**
2(1).	Antennae with apical club; abdomen with separate pleural sclerites	**Cimbicidae**
	Antennae filiform or pectinate; abdomen without pleural sclerites	**3**
3(2).	Antennae with 6 segments	**Pergidae**
	Antennae with 7 or more segments	**4**
4(3).	Antennae with 7 to 10 segments	**Tenthredinidae**
	Antennae with 13 or more segments	**Diprionidae**

Suborder Apocrita

Abdomen with first segment incorporated into thorax as propodeum, second segment narrowly constricted as petiole, bearing the enlarged gaster (Fig. 46.10c,d); thorax without cenchri; hind wings with no more than two basal cells. Larvae lacking legs, usually with head capsule unsclerotized; maxillae and antennae reduced to papillae (Fig. 46.4b).

Apocrita comprise all the diversified groups of dominant Hymenoptera, with the exception of the symphytan superfamily Tenthredinoidea. Parasitoidism, a specialized form of predation, is the primitive mode of larval feeding. In the most advanced members (superfamilies Vespoidea, Sphecoidea), typical predation is the dominant mode of feeding. Several groups of Apocrita, most notably the bees (superfamily Apoidea) and gall wasps (Cynipoidea), have reverted to phytophagy. The families of Apocrita that have the ovipositor developed as a stinger collectively constitute the Aculeata, a monophyletic group including the bees and most of the familiar wasps.

KEY TO THE NORTH AMERICAN SUPERFAMILIES OF COMMON

Apocrita

1.	Abdomen with basal 1 or 2 segments differentiated as a distinct node or nodes (Fig. 46.15b); minute to moderate-sized insects, usually with antennae elbowed and wings absent	**Formicoidea** (p. 594)
	Abdomen with basal segments not modified as a node; antennae elbowed or straight; wings usually present	**2**
2(1).	Hind leg with trochanter 1-segmented (**Aculeata,** except Formicoidea)	**3**
	Hind leg with trochanter 2-segmented	**12**
3(2).	Hind tibia with inner spur hooked, toothed, or with comb of hairs on inner surface (Fig. 46.18d) (or one or both spurs absent)	**4**
	Hind tibia with spur or spurs simple, without teeth or hairs (Fig. 46.10b)	**9**
4(3).	Hind wings without closed cells	**5**
	Hind wings with at least 1 closed cell	**6**
5(4).	Pronotum short, not extending posteriorly as far as tegulae (small sclerites at wing bases) (Fig. 46.13b); abdomen concave below	**Chrysidoidea** (p. 592)
	Pronotum longer, dorsal lobes extending posteriorly to tegulae (Fig. 46.13a); abdomen flat or convex below	**Bethyloidea** (p. 591)
6(4).	Pronotum with dorsal lobes extending posteriorly to tegulae (small sclerites at wing bases) (Fig. 46.15a,c); ventral lobes short, not covering anterior thoracic spiracle	**8**
	Pronotum with dorsal lobes short, not reaching tegulae (Figs. 46.1, 46.16a); rounded ventral lobes covering anterior thoracic spiracles	**7**
7(6).	Body with some hairs branched; hind tarsi with basal segment wider than following segments (Fig. 46.17b)	**Apoidea** (p. 597)
	Body without branched hairs; hind tarsi without basal segments broadened (Fig. 46.16a,b)	**Sphecoidea** (p. 597)
8(6).	Eyes deeply notched, or if round, then antennae clavate	**Vespoidea** (p. 595)
	Eyes round or oval; antennae filiform	**Pompiloidea** (p. 593)
9(3).	Pronotum large, extending posteriorly as far as tegulae (small sclerites at wing bases) (Figs. 46.13a, 46.14a)	**10**
	Pronotum small, not extending posteriorly as far as tegulae (Fig. 46.13b)	**Chrysidoidea** (p. 592)
10(9).	Hind wings without closed cells; venation almost always much reduced (Fig. 46.10c,d)	**11**

Hind wings with at least 1 closed basal cell; forewings usually with extensive venation (Fig. 46.14c)   **Scolioidea** (p. 592)

11(10). Abdomen attached high on propodeum, remote from hind coxae (Fig. 46.10d)
          Evanioidea (p. 586)

Abdomen attached near base of propodeum, contiguous with or very close to hind coxae (Fig. 46.11a)   **Proctotrupoidea** (p. 587)

12(2). Head with deep groove below base of antennae and circle of teeth on vertex (Fig. 46.10a)
          Stephanoidea (p. 584)

Head without groove below antennal base or circle of teeth on vertex   **13**

13(12). Forewing with distinct pterostigma; costal vein present between wing base and pterostigma (Fig. 46.10d)   **16**

Forewing without pterostigma; costal vein absent (Fig. 46.11c)   **14**

14(13). Antennae elbowed, with first segment as long as succeeding 2 to 3 segments; abdomen cylindrical or subcylindrical   **15**

Antennae filiform, with first segment subequal to second, or with second segment small and first and third segments subequal; abdomen laterally compressed
          Cynipoidea (p. 588)

15(14). Pronotum extending posteriorly as far as tegulae (Fig. 46.11a)   **Proctotrupoidea** (p. 587)

Pronotum short, not approaching tegulae (Fig. 46.12a–d)   **Chalcidoidea** (p. 589)

16(13). Abdomen attached high on thorax, remote from coxae (Fig. 46.10d)
          Evanioidea (p. 586)

Abdomen attached low on thorax, close to coxae (Fig. 46.10a)   **17**

17(16). Hind wings without closed cells (Fig. 46.11a)
          Proctotrupoidea (p. 587)

Hind wings with at least 1 closed cell   **18**

18(17). Forewing with costal cell reduced or absent (Fig. 46.10b,e,f); posterior lobe of pronotum not edged by hairs
          Ichneumonoidea (p. 585)

Forewing with distinct costal cell; posterior lobe of pronotum edged by short, close-set hairs
          Trigonaloidea (p. 584)

Superfamily **Stephanoidea.** Small to moderate parasitoids with trochanters 2-segmented; forewing with 3 to 6 closed cells, pterostigma and costal cell present, hind wing without closed cells; ovipositor long, external.

The single family, **Stephanidae,** is represented by six uncommon species in North America. Superficially stephanids (Fig. 46.10a) resemble certain ichneumonids, but they are distinguished by the crown of teeth on the head, by the swollen hind femora, and by the brush of preening hairs on the hind tibial apex. The subantennal grooves, the crown of teeth on the cranium, and the reduced wing venation suggest relationships to the Orussidae. Stephanids are usually encountered on trunks or large branches of dead, frequently charred trees, searching for their hosts, which are the larvae of various wood-boring beetles and siricids. The long ovipositor is used to parasitize the victims in their galleries. Pupation is within the host gallery without a cocoon.

Superfamily **Trigonaloidea.** Stout-bodied, moderate-sized, wasplike parasitoids with long antennae with at least 14 segments; forewing with at least 6 closed cells, costal cell, and pterostigma present; hind wing with closed basal cells; trochanters 2-segmented; ovipositor short, internal.

The single family, **Trigonalidae,** includes rare species that deposit the eggs in slits cut in developing foliage by the sharp, pointed apical sternite. Eggs hatch when consumed by caterpillars of Lepidoptera or sawflies, and the latter are hosts for a few species. Most species are hyperparasites. They hatch in the guts of caterpillars and enter the hemocoels but do not develop further unless the caterpillar is secondarily parasitized by certain ichneumonid wasps or tachinid flies, which become the trigonalid's host. Some species parasitize larvae of vespoid wasps, being consumed with caterpillars provided as food by the adult wasps. The first three instars feed internally, the last two externally.

In body form and wing venation trigonalids are similar to the higher wasps, especially the Sphecoidea, but they share with the parasitoid families the long, multiarticulate antennae, two-

Figure 46.10   Representative Apocrita: **a**, Megalyroidea, Stephanidae (*Schlettererius cinctipes;* scale equals 3 mm); **b**, Ichneumonoidea, Ichneumonidae (*Netelia leo;* scale equals 2 mm); **c**, Evanioidea, Gasteruptiidae (*Rhydinofoenus occidentalis;* scale equals 2 mm); **d**, Evanioidea, Evaniidae (*Evania appendigaster;* scale equals 2 mm); **e**, wings of Ichneumonidae (*Catadelphus atrax*); **f**, wings of Braconidae *(Cremnops haematodes).* rv, recurrent veins, c, costal cell; b, basal cell.

segmented trochanters, and unmodified hind tibiae. Because of this combination of characteristics, trigonalids have usually been considered an extremely primitive family, ancestral to most apocritan Hymenoptera.

Superfamily **Ichneumonoidea.** Minute to large parasitoids with 2-segmented trochanters, antennae with at least 13 segments (usually > 16); forewing with at least 3 closed cells, costal cell absent, pterostigma present or absent; hind wing with 2 closed cells, or occasionally without closed cells. Ovipositor short, internal or long, external.

Ichneumonoidea constitute one of the large, diversified superfamilies of parasitoid Hymenoptera, the other being the Chalcidoidea. The family Ichneumonidae contains perhaps 60,000 species (mostly undescribed), a number exceeded only in the largest families of Coleoptera. Most Ichneumonoidea have the abdominal cuticle soft and flexible, at least ventrally, unlike most other Apocrita. The flexible abdomen apparently assists in oviposition. Most Ichneumonoidea apparently parasitize a variety of hosts, and competition among species is reduced by directing the search for hosts to restricted habitats, rather than by selecting specific hosts. Probably because of the relative nonspecificity of hosts, Ichneumonoidea have proven less successful than Chalcidoidea as biological control agents, with some notable exceptions, especially in Braconidae. The larvae may develop either internally (endoparasitoids) or externally (ectoparasitoids), but pupation is nearly always outside the host in a silken cocoon. Most species are primary parasitoids, but many hyperparasitoids are recorded, especially in the Ichneumonidae.

The superfamily contains two families. **Ichneumonidae** include mostly moderate-sized to large species having two recurrent veins in the forewings (Fig 46.10b,e). They are parasitoids of immature endopterygote insects, especially Lepidoptera and symphytan Hymenoptera, and also attack spiders, spider egg sacs, and pseudoscorpions. The eggs may be laid externally on the host or placed inside the host, which may be paralyzed by injected venom. A few species leave the eggs apart from the larval host, which is then located by the motile first-instar larva. Larvae are mostly solitary parasitoids, but some are gregarious. Ichneumonidae are especially abundant in moist situations with rank vegetation, where caterpillars are abundant. However, they occur in a wide spectrum of habitats, including arid situations. Some diurnal species are strikingly similar in color or pattern to stinging wasps, though very few ichneumonids are able to sting humans. *Rhyssa* and related genera are remarkable for the extraordinarily long ovipositor (Fig. 46.3a), which is used to reach wood-boring Coleoptera or Hymenoptera deep in their galleries. Ovipositors of some species may be up to 15 cm long, about three times body length.

Braconidae, with a single recurrent vein in the forewing (Fig. 46.10f), are generally smaller than Ichneumonidae. Braconids are parasitoids of immature and occasionally adult or pupal endopterygote insects, as well as various Homoptera. The great majority are primary parasitoids, but a few hyperparasitize other ichneumonoids and tachinid flies. Some braconids attack hosts only slightly larger than themselves. For example, *Praon* and *Aphidius* utilize aphids. Other species may parasitize hosts hundreds of times their own bulk: *Microctonus* (2 to 3 mm) oviposits in adult tenebrionid beetles as long as 50 mm. Polyembryony is not uncommon among species with very large hosts. *Aphidius* and a few other genera pupate in a thin cocoon within the mummified remains of the host, but most spin a cocoon externally, often attached to the host's body. *Praon* spins a cocoon beneath the husks of consumed aphids, while *Apanteles* cocoons attached to caterpillars are a familiar sight in eastern North America.

Superfamily **Evanioidea.** Small to moderate-sized parasitoids with the abdomen attached high on the propodeum, remote from the hind coxae (Fig. 46.10c,d); hind trochanters 1- or 2-segmented; forewing with pterostigma, costal cell, and several other closed cells; hind wing with only narrow costal cell closed; ovipositor short, internal or long, external.

The insects placed in this superfamily are widespread but seldom common. All are parasitoids with restricted host ranges. For example, **Evaniidae** (Fig. 46.10d) oviposit in cockroach oothecae. The larva consumes the cockroach eggs, then pupates inside the ootheca without a cocoon. **Aulacidae** are internal parasitoids of xiphydriid wood wasps or of Cerambycidae or Buprestidae (Coleoptera). The larvae do not complete their development until the host has reached the last instar (i.e., about a year). Pupation is in the host gallery. **Gasteruptiidae** (Fig. 46.10c) oviposit in the cells of solitary bees or of sphecid or vespid wasps. Their larvae consume the host egg or young larva, but the host's provisions form the bulk of their food. Pupation is in a cocoon in the host cell.

KEY TO THE FAMILIES OF

Evanioidea

1. Abdomen with gaster subquadrate, laterally compressed, attached to a slender petiole (Fig. 46.10d)   **Evaniidae**
   Abdomen with gaster elongate, gradually enlarged, round in cross section, petiole not distinct from gaster (Fig. 46.10c)   **2**
2(1). Abdomen clavate; forewing with 2 recurrent veins (as in Fig. 46.10e)   **Aulacidae**
   Abdomen cylindrical, slender to apex (Fig. 46.10c); forewing with 1 recurrent vein or none (as in Fig. 46.10f)   **Gasteruptiidae**

Superfamily **Proctotrupoidea.** Small to moderate sized parasitoids of extremely variable body configuration; trochanters 1- or 2-segmented; forewings usually without closed cells or with narrow costal cell, or with 2 to 5 closed cells in a few primitive species, pterostigma present or absent; hind wing almost always without closed cells.

The Proctotrupoidea constitute a large, diverse assemblage of parasitoids whose biologies and relationships are poorly known. They are also one of the more ancient lineages of Hymenoptera with two families known from Jurassic fossils. **Heloridae,** probably the most primitive family, resemble braconids in the general body shape and wing venation. Similar venation is found in Vanhorniidae and Roproniidae, but in the other families venation is reduced, frequently to a single vein. Body size is extremely variable, reaching 60 mm in **Pelecinidae.** The great majority of proctotrupoids are small or minute, however, and many superficially resemble Chalcidoidea, except in the large prothorax (cf. Figs. 46.11a and 46.12a). The Ceraphronidae, Megaspilidae, Scelionidae, Platygastridae, Diapriidae, and Proctotrupidae are diverse families of several hundred to several thousand species worldwide, while the Heloridae, Pelecinidae, Roproniidae, and Vanhorniidae are rare, relictual taxa of one to a few species each.

Because of their small size and specialized habits, most Proctotrupoidea are uncommonly encountered by the general investigator. Numerous individuals of many species may frequently be found, however, by using flight intercept traps or soil extraction methods (see collecting methods, Chap. 47).

Proctotrupoidea are internal parasitoids of a variety of insects, including larval Chrysopidae (Heloridae), coleopterous larvae (Heloridae, **Proctotrupidae, Vanhorniidae,** Pelecinidae), sawflies (**Roproniidae**), and dipterous larvae (many **Diapriidae**). **Scelionidae** are parasitoids of the eggs of numerous orders of insects and of spiders. Females of the genus *Mantibaria* have the unusual habit of riding about on adult mantids, then ovipositing in freshly laid egg masses before the oothecal covering has hardened. Some **Platygastridae** parasitize the eggs of Coleoptera, Homoptera, and Diptera (especially Cecidomyiidae) and probably other insects, while others are parasitoids of immatures of Coccoidea and Aleyrodoidea (Homoptera). **Ceraphronidae** and **Megaspilidae** are internal parasitoids of a variety of insects (Coccoidea, Aphidoidea, Neuroptera, Thysanoptera, Mecoptera, and Diptera). In addition, many species are hyperparasitoids of other Hymenoptera.

Many adult Proctotrupoidea are short-winged or wingless, antlike inhabitants of leaf litter. Winged species are commonly encountered on foliage. More than 500 North American species have been recognized, and it is certain that many remain undescribed.

KEY TO THE NORTH AMERICAN FAMILIES OF

Proctotrupoidea[1]

1. Antennae elbowed, with first segment usually longer than succeeding 3 to 5 segments   **6**
   Antennae not elbowed, first segment about as long as third   **2**
2(1). Body length less than 20 mm; hind leg with basal tarsomere longer than succeeding segments   **3**

[1] Ceraphronidae and Megaspilidae are sometimes placed in a separate superfamily, Ceraphronoidea.

Body length 20 to 70 mm; hind leg with basal tarsomere shorter than succeeding segments   **Pelecinidae**

3(2). Antennae with 14 or 16 segments   **4**
Antennae with 13 segments   **5**

4(3). Mandibles with tips pointing out, not meeting when closed   **Vanhorniidae**
Mandibles with tips pointing inward, meeting or overlapping when closed   **Proctotrupidae**

5(3). Antennae with 14 segments; abdomen strongly flattened in lateral plane   **Roproniidae**
Antennae with 13 segments; abdomen slightly wider than high   **Heloridae**

6(1). Antennae inserted on a prominence between the eyes, remote from clypeus; midtibiae with 2 spurs   **Diapriidae**
Antennae inserted below the eyes, near the clypeus, without a prominence; midtibiae with 1 or 2 spurs   **7**

7(6). Abdomen flattened, with lateral margins keeled; midtibiae with single apical spur   **8**
Abdomen rounded laterally; midtibiae with 2 spurs   **Ceraphronidae** and **Megaspilidae**

8(7). Antennae with 10 or fewer segments, forewing venation consisting of short submarginal vein   **Platygastridae**
Antennae with 11 to 12 segments (10 segments rarely present); forewing venation consisting of

marginal vein and short vein extending beyond pterostigma   **Scelionidae**

Superfamily **Cynipoidea.** Small to moderate-sized insects with the abdomen usually compressed in the lateral plane and the tergites extending ventrally to conceal the sternites (Fig. 46.11b); hind trochanters 1- or 2-segmented; forewing with marginal vein and pterostigma lacking, usually without closed cells; hind wing almost always with a single vein; ovipositor almost always short, internal.

The great majority of these wasps may be recognized by the reduced forewing venation with a characteristic marginal cell in the distal third (Fig. 46.11c). They differ from Chalcidoidea and most Proctotrupoidea in having filiform rather than elbowed antennae.

Ibaliidae, the most primitive members of Cynipoidea, are endoparasitoids of siricid and anaxyelid wood wasps. The larva initially feeds internally, then externally in later instars. **Figitidae** are internal parasitoids of dipterous larvae and puparia and of hemerobiid lacewings. **Eucoilidae,** the largest family of Cynipoidea, are internal parasitoids of calyptrate Diptera, the adults emerging from the puparia. The family is mainly tropical,

Figure 46.11 Proctotrupoidea and Cynipoidea: **a**, Proctotrupoidea, Proctotrupidae (*Proctotrupes;* scale equals 0.5 mm); **b**, Cynipoidea, Cynipidae (*Andricus californicus;* scale equals 1 mm); **c**, wings of Ibaliidae (Cynipoidea). c, costal cell; m, marginal cell (radial cell).

but adults may often be found around fly-infested dung pats. **Liopteridae** are uncommon wasps whose biologies are almost unknown. A relatively small number of **Cynipidae** are parasitoids of Diptera or hyperparasitoids of Hymenoptera (Braconidae, Aphelinidae) attacking aphids. The large subfamily Cynipinae, comprising about 600 North American species, all inhabit galls, either as the primary causative agent or as inquilines.

Galls vary enormously in size, shape, color, and texture, as well as in internal structure, and are frequently diagnostic of the species of wasp that formed them. Typically an external epidermis, which may be extremely hard or tough, encloses a thick layer of spongy parenchymal tissue. At the center of the gall is usually a hard shell lined with nutritive tissue, within which the larva feeds. A single gall may contain one to many larval cells and may occur on any part of the host plant. About 75 to 85 percent of all cynipid species attack oaks (*Quercus*) and closely related trees and shrubs; another 7 percent attack the genus *Rosa* (roses); and the remainder utilize a variety of other plants, especially Compositae (sunflower family). Inquilines oviposit into galls formed by other species, frequently causing modifications of the original structure. The original occupant is either killed outright or starved by appropriation of food. Cynipoidea pupate inside the galls or hosts, usually without a cocoon.

In most Cynipidae, sexual generations, occurring in the warm part of the year, alternate with asexual (parthenogenetic) generations, which appear during winter. The individuals representing the sexual and asexual generations are often strikingly different in appearance and usually produce dissimilar galls on different parts of the host. Consequently, alternate generations have often been described as different species, which may be associated only as the biological relationships are discovered.

KEY TO THE FAMILIES OF

Cynipoidea

1. Abdomen with fourth, fifth, or sixth segment of gaster larger than other segments; body length, 5 to 15 mm   2

Abdomen with second or third segment of gaster larger than other segments; body length seldom greater than 6 mm   3

2(1). Gaster with segments strongly compressed in lateral plane, with sharp dorsal edge   **Ibaliidae**

Gaster not strongly compressed; dorsal edge rounded   **Liopteridae**

3(1). Scutellum raised into round or teardrop-shaped plate; head and mesoscutum smooth   **Eucoilidae**

Scutellum without raised plate; head and mesonotum usually roughly textured   4

4(3). First segment of gaster shorter than second   **Figitidae**

First segment of gaster much longer than second   5

5(4). First segment of gaster at least half as long as entire abdomen excluding petiole   **Cynipidae** (incl. Charipidae)

First segment of gaster less than half as long as entire abdomen   **Figitidae**

Superfamily **Chalcidoidea.** Minute to small or occasionally moderate-sized wasps with the antennae elbowed (exception, Eucharitidae), trochanters 1- or 2-segmented; wings with venation reduced to a single vein, without closed cells; pronotum with posterior lobe usually rounded, not contiguous with tegulae; ovipositor short, internal or exserted, elongate.

The Chalcidoidea, which probably comprise the largest superfamily of Hymenoptera, are remarkable for their great variety of body configurations (Fig. 46.12a–d), which are diagnostic for some families (e.g., **Leucospidae, Chalcididae, Agaonidae**) and may verge on the fantastic or capricious. In most species the cuticle is strongly sclerotized and coarsely sculptured; many are iridescent blue or green. Chalcidoidea range to about 30 mm in length, but the great majority are less than 3 mm and many less than 1 mm long. Nearly all Chalcidoidea may be recognized by the combination of reduced wing venation (Fig. 46.12e) and short pronotum (Fig. 46.12a).

Some **Eurytomidae, Pteromalidae, Torymidae, Perilampidae,** and **Eulophidae** are phytophagous, either tunneling in seeds or forming galls. A few species, such as *Bruchophagus platyptera*,

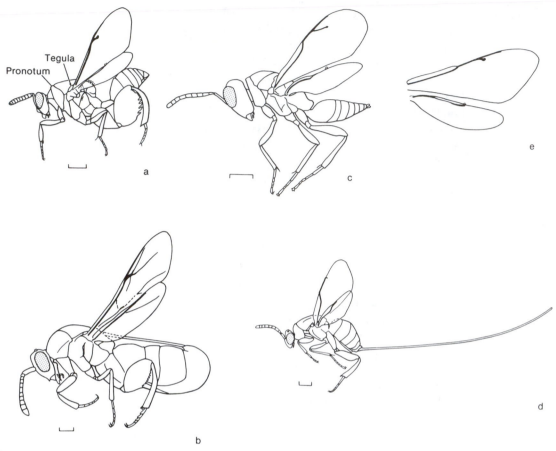

Figure 46.12 Chalcidoidea: **a**, Chalcididae (*Chalcis;* scale equals 1 mm); **b**, Leucospidae (*Leucospis birkmani;* scale equals 1 mm); **c**, Pteromalidae (*Pteromalus vanessa;* scale equals 0.5 mm); **d**, Torymidae (*Torymus californicus;* scale equals 1 mm); **e**, wings of Chalcididae *(Chalcis divisa).*

which infests clover seeds, are of minor economic importance. Members of the family Agaonidae engage in an obligatory symbiotic relationship with species of figs *(Ficus),* whose flowers they both gall and pollinate. Certain commercial varieties of figs, such as Smyrna, are pollinated only by the wasps, which have been transported about the world for this purpose (see Fig. 10.3). The great majority of Chalcidoidea, however, are internal or external parasitoids of insects of all orders, but especially Homoptera and immature stages of Endopterygota. Coleoptera, Diptera, and other orders are included in the host range, but Lepidoptera are especially favored among the Endopterygota.

Eggs, larvae, and pupae are utilized by various species, and some families attack only a single developmental stage (e.g., **Trichogrammatidae, Mymaridae** on eggs). Numerous species are hyperparasitoids of primary parasitoids, and care must be exercised in deducing host relationships.

Many chalcidoids are hypermetamorphic, with the first one or two larval stages specialized for host location or elimination of competitors within the host (Fig. 46.5a). Later instars are always maggotlike. Parthenogenesis occurs in various families, and polyembryony has been described in many species. Both features are common components of the life cycles of hymenopterous parasitoids. Pupation occurs within the host remains or externally in soil or litter or on plants. Very few chalcidoids spin a cocoon.

Different authorities recognize from 9 to 22 families of Chalcidoidea. A few are easy to identify, but some families are difficult to separate with existing keys. An additional difficulty is the small or minute size of most species, necessitating excellent optical equipment to see the taxonomic characters. Interested students are referred to Peck et al. (1964), Grissell and Schauff (1990), Gauld and Bolton (1988), and Goulet and Huber (1993) for identification.

Superfamily **Bethyloidea.** Small to moderate-sized (1 to 10 mm) wasps with hind trochanters 1-segmented; forewings with at least 2 to 4 closed cells (veins may be faint), including costal cell (Fig. 46.13c); hind wings without closed cells; ovipositor internal.

Bethyloidea are infrequently encountered, although **Bethylidae** are common inhabitants of leaf litter, tree trunks, and fungi, where they are parasi-toids of larvae of Lepidoptera and Coleoptera. While Bethyloidea do not prepare special nests for larval development, some species of Bethylidae move the immobilized prey to a sheltered spot before oviposition. Unlike the fossorial wasps (Scolioidea, Pompiloidea, Sphecoidea), they deposit more than one egg on a single host. Except for their reduced wing venation, Bethylidae are morphologically similar to Tiphiidae and in general are intermediate between typical parasitoids such as the Proctotrupoidea and the primitive stinging wasps.

Dryinidae are internal parasitoids of Homoptera; females are apterous, with the protarsus modified as a pincer for clasping the host. The larvae develop internally, but later instars form a characteristic sac bulging from the host abdomen. **Sclerogibbidae** are external parasitoids of Embiidina, and **Cleptidae** attack sawfly prepupae and eggs of Phasmatodea. Bethyloidea include many wingless species and others in which only the females are wingless. Sexual dimorphism in other

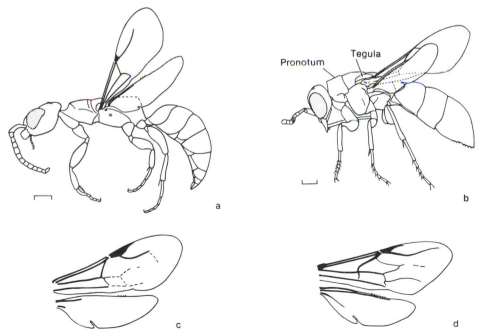

Figure 46.13 Bethyloidea and Chrysidoidea: **a**, Bethylidae (*Anisepyris williamsi*; scale equals 0.5 mm); **b**, Chrysididae (*Parnopes edwardsi*; scale equals 1 mm); **c**, wings of Bethylidae (*Pristocera armi*); **d**, wings of Chrysididae (*Chrysis*).

characteristics may be extreme, and sexes may be difficult to associate. Winglessness in females is common in Tiphiidae and Mutillidae (both in Scolioidea), emphasizing the close relationship to Bethyloidea.

KEY TO COMMON NORTH AMERICAN
FAMILIES OF

Bethyloidea

1. Antennae with 11 to 13 segments; fore tarsi unmodified   **2**

     Antennae with 10 segments; females with fore tarsi claw-shaped   **Dryinidae**

2(1). Abdomen with 7 to 8 segments in gaster; pronotum extending posteriorly to tegulae (Fig. 46.13a)   **Bethylidae**

     Abdomen with 4 to 6 segments in gaster; pronotum not reaching tegulae   **Cleptidae**

Superfamily **Chrysidoidea.** Small to moderate-sized parasitoids with antennae elbowed, hind trochanters 1-segmented; forewings with 3 or more closed cells, including costal cell; hind wings without closed cells; ovipositor an elongate, telescoping tube, concealed internally.

The single family, **Chrysididae** (Fig. 46.13b), is readily recognized by the metallic green and blue colors and the concave ventral surface of the abdomen. Chrysidids roll the heavily sclerotized body into a ball when molested, thus concealing the head and thoracic venter in the abdominal concavity. Morphologically Chrysididae resemble some Cleptidae and are sometimes classified in Bethyloidea, but they differ in leg and thoracic structure.

Chrysididae attack larvae of bees and wasps in their cells, pupating within a cocoon in the host cell. The most familiar species attack mud daubers (*Sceliphron:* Sphecidae) and vespids, which reuse mud dauber nests. Adult chrysidids are most commonly encountered on the surface of soil or tree trunks or occasionally on flowers.

Superfamily **Scolioidea.** Small to large parasitoids with stout body; hind trochanters 1-segmented; forewings with at least 7 closed cells, including costal cell; hind wings with at least 1 closed cell or wingless; ovipositor developed as internal sting.

The scolioid wasps are all parasitoids, biologically similar to Bethylidae. Morphologically the superfamily is diverse, showing similarities to Bethyloidea, Formicoidea (ants), Pompiloidea (spider wasps), and Vespoidea. Tiphiidae are usually considered to be the ancestral group from which all other stinging Hymenoptera differentiated.

Most scolioid wasps are highly fossorial, with strong forelegs and large prothorax extending posteriorly to the region of the wing bases (Fig. 46.14a). Sexual dimorphism is frequently marked, especially by winglessness in females, which may be very difficult to associate with the winged males. In contrast to many other Hymenoptera, the females are often smaller than the males, probably an adaptation that allows increased dispersal of the flightless females, which are carried about by the males during a nuptial flight.

Tiphiidae and **Scoliidae** mostly attack ground-dwelling beetle larvae. Scarabaeidae are the commonest hosts, but Tenebrionidae and Cicindelidae (tiger beetles), as well as Orthoptera and bees and wasps, are utilized by Tiphiidae. **Mutillidae** (velvet ants) whose powerfully stinging females are always wingless, are mostly parasitoids of bees and wasps. Most diurnal species of velvet ants are brilliantly patterned in red or orange and black pile, rendering them highly conspicuous as they rapidly run about the ground searching for host nests. Nocturnal species are tan, brown, or black, resembling tiphiids. **Sapygidae** are parasitoids of bees and wasps.

KEY TO THE FAMILIES OF

Scolioidea

1. Wings with apical membrane bearing numerous fine, longitudinal creases; sterna of mesothorax and metathorax forming a broad plate overlapping middle and hind coxae   **Scoliidae**

     Wings without creases in apical membranes, or wings absent; sterna of mesothorax and metathorax not overlapping coxae   **2**

2(1). Abdomen with lateral felt line on second gastral segment; females wingless, males with wings lacking anal lobes (Fig. 46.14d)   **Mutillidae**

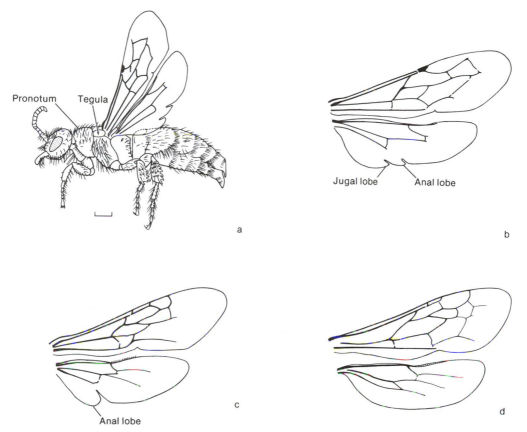

Figure 46.14 Scolioidea: **a**, Tiphiidae (*Paratiphia verna;* scale equals 1 mm); **b**, wings of Tiphiidae *(Tiphia);* **c**, wings of Scoliidae *(Scolia nobilitata);* **d**, wings of Mutillidae *(Timulla vagans).*

Abdomen without lateral felt line; females winged or wingless; wings of both sexes with anal lobe separated from rest of wing by notch (Fig. 46.14b)   **3**

3(2). Abdomen with gaster usually constricted between first and second segments; eyes usually round   **Tiphiidae**

Abdomen with gaster not constricted between first and second segments; eyes usually deeply emarginate   **Sapygidae**

Superfamily **Pompiloidea.** Small to very large wasps with large, mobile prothorax, hind trochanters 1-segmented; forewings with 7 to 10 closed cells, including costal cell, hind wing with at least one closed cell; ovipositor developed as internal sting.

This superfamily comprises two families of greatly unequal size and abundance. **Rhopalosomatidae,** represented by only two rare species in North America, parasitize Gryllidae. The larvae cause a hernia-like pouch in the abdominal wall of the host, much in the manner of Dryinidae. **Pompilidae** (spider wasps, Fig. 46.15a; Color photo 7a), characterized by the presence of a horizontal furrow on the mesopleuron (absent in Rhopalosomatidae), all utilize spiders as larval food. Pompilids provide each larva with a single prey item, which is necessarily larger than the wasp. A wide variety of spiders is attacked, including tarantulas, which serve as prey for some *Pepsis.*

Because of the large body size of their prey,

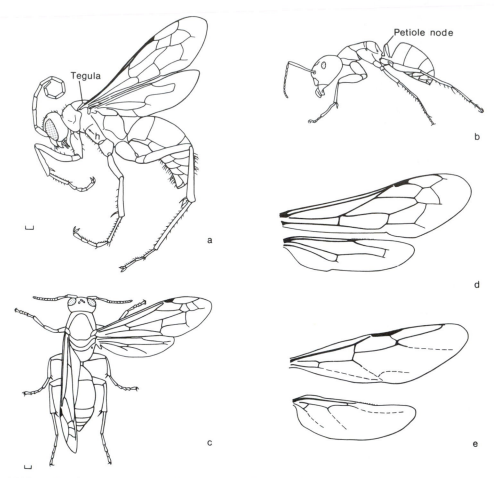

Figure 46.15   **a**, Pompiloidea, Pompilidae *(Priocnemoides unifasciatus);* **b**, Formicoidea, Formicidae *(Camponotus);* **c**, Vespoidea, Vespidae *(Polistes metricus);* **d**, Vespidae, wings *(Pseudomasaris vespoides);* **e**, Formicidae, wings *(Camponotus morosus)*. Scales equal 1 mm. h, horizontal furrow on mesopleuron.

pompilids usually construct a burrow near the site of attack and immobilization. The paralyzed spider is frequently concealed to avoid attack by parasites and scavengers, then quickly moved a short distance into the completed larval gallery. Other species simply conceal the prey in an appropriate crevice or in the host's own tunnel, and some only temporarily paralyze the spider, which regains activity before being killed by the maturing wasp larva. Such species are essentially parasitoid, showing no important biological differences from Bethylidae or Tiphiidae. More specialized pompilids resemble primitive sphecids in preparing a larval gallery before prey acquisition, while a number of species oviposit on prey of other pompilids. The family is cosmopolitan but is especially diverse in tropical and subtropical regions.

Superfamily **Formicoidea.** Minute to moderate-sized Hymenoptera with basal one or two segments of abdominal gaster modified as a nodelike petiole (Fig. 46.15b); prothorax large, mobile; antennae almost always strongly elbowed.

Ants are among the most ubiquitous and familiar insects (Color photos 4c, 6e). Currently about 8800 species and 297 genera are recognized; possibly twice as many species will be known ultimately. The greatest variety exists in the Tropics, but ants are found north to the Arctic tree line, south to the southern tips of the continents, and on all but the highest mountains. Most oceanic islands within these latitudinal extremes are also inhabited by ants. Wilson (1971) estimates that 1 percent of all insects are ants (i.e., 10^{15} individuals). Several species have become closely associated with humans, occupying their dwellings and other structures throughout the world, including metropolitan areas. Despite great variation in size and habits, practically all ants are recognizable by the petiolate abdomen and elbowed antennae, and are classified as a single family, **Formicidae.**

Without exception, ants are social insects. Colonies consist of one or a few sexual females, or **queens,** specialized for reproduction, and a variable number of apterous, neuter females that function as **workers** or **soldiers.** Males are normally produced once or a few times a year, surviving only long enough to mate. Males and females are usually larger than nonsexual castes and have the thorax enlarged with the wing muscles. Female sexuals dehisce their wings after a nuptial flight; males are permanently winged but die shortly after mating. Worker and soldier castes are always wingless. Soldiers, which function in colony defense, are larger than workers, often with the head disproportionately enlarged. Colony organization and structure, which are extremely complex, are discussed in detail in Chapter 7. We mention here that colony formation, which is carried out by single females, is critical in the perpetuation of most species.

Ant colonies, which may persist for many years, vary from assemblages of a single queen and a few dozen individuals to multitudes of hundreds of thousands of individuals with numerous queens. Typically, colonies are established in fixed locations. Subterranean galleries are most common, but many species tunnel dead wood or twigs, and *Oecophylla* species use silk produced by their own larvae to construct nests of living leaves of trees and shrubs.

A few tropical species have evolved highly specialized symbiotic relationships with plants. Especially interesting is the relationship between ants of the genus *Pseudomyrmex* and certain species of acacia tree. The large swollen thorns of bull's horn acacia and other species are inhabited by various species of *Pseudomyrmex.* Extrafloral nectaries and Beltian bodies on the leaves supply the ants with sugar and protein. The ants aggressively protect the trees from vertebrate and invertebrate predators (Janzen, 1966, 1967a,b). Army ants do not occupy a fixed colony but lead a nomadic existence, bivouacking in litter or dense vegetation.

A number of species of ants, especially in the genus *Formica,* enslave other species. For example, *F. sanguinea* workers attack the nests of other species of *Formica,* abducting the pupae and carrying them back to their own nest. Still other species live as inquilines in the nests of other ants, a relationship sometimes termed social parasitism.

Ants include herbivores, scavengers, and predators. Most ants are occasionally opportunistic predators, but army ants are exclusively predatory, including a wide variety of small invertebrates in their diet. A few ants are specific predators of spider eggs, termites, or other arthropods. Many species tend aphids or other Homoptera, from which they obtain honeydew, and several, such as the argentine ant (*Linepithema humile*), are pests in gardens for this reason. Harvester ants specialize on seeds, and leaf-cutting ants (tribe Attini) cut out pieces of leaves and flowers on which are cultured fungi. The ants eat and replace the gardens in a manner analogous to the fungal gardening of some Termitidae (Isoptera).

Ants are sufficiently abundant that certain birds, mammals, and reptiles (horned lizards, anteaters) are specialized as predators on them. Ant colonies are widely exploited by **myrmecophilous** insects, especially beetles, which may act as predators of the ant brood or as nest scavengers. The more specialized of these myrmecophilous insects may be morphologically modified to provide a tactile resemblance to their hosts.

Superfamily **Vespoidea.** Small to large wasps with the prothorax with large, triangular lobes extending posteriorly to the tegulae (Fig. 46.15c); hind trochanters

1-segmented; forewings with at least 7 closed cells, including costal cell (Fig. 46.15d); hind wing with at least 2 closed cells; ovipositor developed as internal sting.

The single family, **Vespidae,** includes diverse wasps that are mostly predatory, usually on caterpillars, occasionally on larvae of sawflies, phytophagous beetles, or other soft-bodied insects. Species of *Vespula* (yellowjackets and hornets) are also aggressive scavengers of all sorts of proteinaceous material, often becoming pests about refuse sites, picnic areas, or pet feeding locations. Members of the subfamily Masarinae, which provision their nests with pollen and nectar, are locally common in the western states. Most vespids may be recognized by the emarginate eyes that partially surround the antennal bases.

Vespids include many solitary forms such as potter wasps (subfamily Eumeninae), but the familiar species, including yellowjackets, hornets, and paper wasps, form annual colonies. Vespid nests may be suspended from branches, rafters, or other aerial structures (Color photo 4d); may be concealed beneath stones or in hollow logs, or they may be constructed in cavities the wasps excavate in the ground. A few solitary species use mud as a building material, but nests of social species are commonly constructed of masticated wood or paper. Colonies range in size from a few cells arranged in a single, open comb to multicombed structures enclosed in tough, external jackets of paper. Different species display all stages of intermediacy between solitary and social organization.

Vespids are commonly colored in contrasting patterns of black and yellow, in advertisement of their severe stings. The potter wasps (subfamily Eumeninae) paralyze their caterpillar prey by stinging, in the manner of the sphecoid wasps. Most social Vespidae, however, kill the prey with the mandibles. Stinging is largely used to deter mammalian predators, which are very sensitive to the pain-producing substances in vespid venoms. The large number of insects that mimic vespids

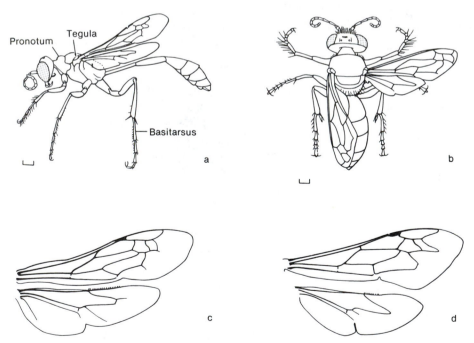

Figure 46.16 Sphecoidea, Sphecidae: **a,** *Ammophila cleopatra;* **b,** *Bembex americanus;* **c,** wings of *Ammophila;* **d,** wings of *Tachytes.* Scales equal 1 mm.

attests to the efficiency of their stings (see Chap. 12).

Superfamily **Sphecoidea.** Small to large wasps with the prothorax collarlike with rounded posterior lobes that do not reach the tegulae (Fig. 46.16a); hind trochanters 1-segmented; forewings with at least 7 closed cells, including costal cell; hind wings with at least 2 closed cells (Fig. 46.16c,d); ovipositor developed as internal sting.

The Sphecoidea are commonly called "solitary wasps" or "digger wasps," although some nest in twigs, old beetle burrows, or self-constructed mud cells. The single large family, **Sphecidae,** includes an astonishing diversity of body form and size, as well as behavior, but taxonomic subdivision is impractical because of the many intermediates.

A few sphecids (subfamily Nyssoninae) are inquilines in the nests of bees or other sphecids, but predation on other insects is much the dominant mode of life (Color photo 6f). The adult female wasps construct nests containing one or more cells in which the larvae will mature while feeding on prey provided by the mother. A great variety of insects is utilized as prey, and host specificity varies greatly among different species. The most primitive sphecids use a single prey item for each larva. Because of the large prey size, such wasps either drag the prey over the ground or transport it with a series of short flights initiated from trees, shrubs, or other prominences. Except in the construction of a special larval gallery, these primitive sphecids are not greatly different from bethylids, tiphiids, or other parasitoid wasps.

In the great majority of sphecids, prey size is much reduced, allowing rapid transport by air, so that such abundant insects as aphids, leafhoppers, and thrips may be utilized. Provisioning follows two modes. **Mass-provisioning** species accumulate the total amount of prey required before oviposition. Normally only a single larval cell is tended at one time. **Progressive-provisioning** species oviposit on one of the first prey items acquired, or in an unprovisioned cell, later supplying additional food as needed by the growing larva. The most advanced species tend several larvae simultane-

ously. The resultant close contact between the adult females and their offspring is probably requisite to true sociality (eusociality, occurring in Vespoidea and Apoidea: see Chap. 7). Communal nesting occurs in several subfamilies of Sphecidae, and division of labor among communal members has been recorded in the subfamilies Pemphredoninae, Philanthinae, and Crabroninae. Eusociality, with reproductive division of labor, has been well documented in *Microstigmus comes* (Pemphredoninae) and probably occurs in other sphecids as well.

Superfamily **Apoidea** (bees). Small to large Hymenoptera with branched, plumose hairs on at least part of the body; prothorax with rounded posterior lobes that do not reach the tegulae; hind trochanters 1-segmented; forewings with at least 7 closed cells; hind wings with at least 2 closed cells; ovipositor developed as internal sting.

In general morphological characters, Apoidea are very similar to sphecoid wasps. The most reliable differentiating features are the presence of branched hairs on the body and the specialization of the hind basitarsus as a part of the pollen-collecting apparatus (Figs. 46.17b, 46.18c,d). Some primitive genera, such as *Hylaeus* (Colletidae), carry pollen in the crop. In these and in cleptoparasitic (using the galleries and provisions of other bees) genera of other families, the body is relatively hairless and the tarsi are unmodified, rendering separation from some sphecids difficult.

In terms of nest architecture and behavior, bees (except the social species) are not greatly different from solitary wasps, with the major difference that most bees provision their larval cells with nectar and pollen rather than animal material. Their lower position in food chains (as primary rather than secondary or tertiary consumers) is probably responsible for the great abundance of bees in most habitats. As pollinators, bees are more important than any other group of insects. The maxillae and labium are modified as a proboscis for sucking nectar, and may be very elongate in some Apidae (Fig. 10.1b) and Anthophoridae. **Scopae,** or pollen-collecting structures, may occur

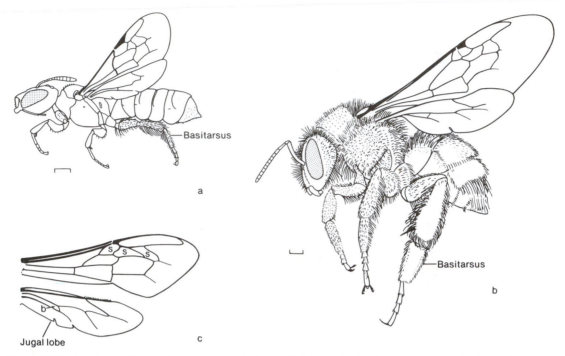

Figure 46.17   Apoidea, Apidae: **a**, *Ceratina strenua* (scale equals 0.5 mm); **b**, *Bombus americanorum* (scale equals 1 mm); **c**, *Apis mellifera,* wings. b, Basal cell in hind wings; s, submarginal cells.

on the hind tarsi and tibiae [Anthophoridae (Fig. 46.18d); Apidae (Figs. 46.17b, 46.18c), Andrenidae, Halictidae, Melittidae] or on the venter of the abdomen (Megachilidae) of females. Scopae are lacking only in some Colletidae *(Hylaeus)* and in cleptoparasitic or inquiline species.

The honey bee *(Apis mellifera)* is the best-known pollinator (Color photos 4a, 4b, 5b). Commercial hives are seasonally rotated among many plantings, including fruit trees, alfalfa, and certain row crops. Honey bees visit a wide variety of non-cultivated plants as well. Wild bees are more efficient pollinators than honey bees for some crops. For example, alkali bees *(Nomia)* and leafcutter bees *(Megachile)* are utilized for alfalfa pollination. Pollination of many uncultivated plants is accomplished largely by wild bees, which may be highly specialized for one plant species or a group of closely related species **(oligolectic).** For example, certain species of *Andrena* (Andrenidae) collect pollen only from night-blooming species of *Oe-*

nothera (Onagraceae), which they visit at night or very early in the morning, before other pollinators have had an opportunity. **Polylectic** species of bees collect pollen from many plant species, but frequently only a single species is visited on sequential trips for pollen. For example, in the highly polylectic honey bee, any individual bee normally visits only one type of flower until that source of pollen and nectar is exhausted; then an alternative flower type is selected (see Chap. 10).

Eusociality has apparently evolved independently several times in bees (see Chap. 7). The largest and most complex colonies are maintained by honey bees *(Apis)* and tropical stingless bees (Apidae: Meliponinae), but many other apids, such as bumble bees, as well as some halictids, form small annual colonies. As in other social Hymenoptera, these are female societies, the male bees, or drones, appearing only once a year in relatively small numbers. Besides these eusocial forms, many bees, especially Apidae and Halictidae, are presocial, showing various degrees of altruistic or cooperative behavior. The biology of social Hymenoptera is extensively discussed by Wilson (1971), Michener (1974), and K. G. Ross and Matthews (1991).

KEY TO THE NORTH AMERICAN FAMILIES OF

Apoidea

1. Face with 2 vertical sutures between each antennal socket and clypeus (Fig. 46.18a)
   Andrenidae
   Face with a single suture between each antennal socket and clypeus   2

2(1). Hind wing with jugal lobe as long as basal cell or longer (Fig. 46.18b)   3
      Hind wing with jugal lobe shorter than basal cell (Fig. 46.17c) or lacking (Fig. 46.17b)   4

3(2). Forewing with the basal vein (basal sector of M vein) strongly arched (Fig. 46.18b)
      Halictidae
      Forewing with the basal vein straight or feebly curved (Fig. 46.17c)   **Colletidae**

4(2). Labial palpi with all segments cylindrical, subequal in length   **Melittidae**
      Labial palpi with 2 basal segments flattened, at least twice as long as distal segments (Fig. 46.2b)   5

5(4). Forewing usually with 3 submarginal cells (Fig.

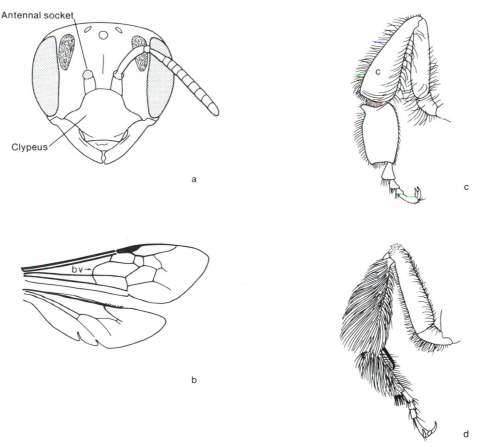

Figure 46.18   Apoidea: **a**, frontal view of head of Andrenidae *(Andrena gardineri)*; **b**, wings of Halictidae, bv, basal vein *(Halictus ligatus)*; **c**, hind tarsus of Apidae, showing corbicula (c) *(Apis mellifera)*; **d**, hind tarsus of Anthophoridae, showing scopa *(Anthophora curta)*.

46.17c); labrum usually broader than long; subantennal suture extending from center of antennal socket; females with scopa on hind legs
(Fig. 46.18c,d)   **6**

   Forewing with 2 submarginal cells; labrum longer
than broad; subantennal suture extending from
lateral margins of antennal sockets; females
with scopa on venter of abdomen
                                            Megachilidae

6(5). Hind tibiae without spurs (exception, *Bombus,*
bumblebees, Fig. 46.17b); hind tibiae of females
with scopa forming a corbicula, or basket (Fig.
46.18c)   **Apidae**

   Hind tibiae with apical spurs; hind tibiae of females with scopa forming a dense brush of
setae (Fig. 46.18d)   **Anthophoridae**

SELECTED REFERENCES

General

Askew, 1971 (parasitoid species); Clausen, 1940 (predaceous and parasitoid species); Dadant and Sons,
1975 (honey bee); Debach, 1974 (biological control);
Entomophaga, 1956 to date (many articles on Hymenoptera); H. E. Evans and Eberhard, 1970
(wasps); Felt, 1940 (gall-making families); Gauld
and Bolton, 1988; Goulet and Huber, 1993; Huffaker,
1971 (biological control); O'Toole and Raw, 1991
(bees); Spradberry, 1973 (wasps); Stehr, 1987
(larvae).

Morphology

Michener, 1944 (bees); Daly 1963, 1964 (muscles);
Field and Austin, 1994 (ovipositor); Quicke et al.,
1992 (sperm ultrastructure); Snodgrass, 1956 (honey
bee); Stephen et al., 1969 (bees).

Biology and Ecology

Beattie, 1985 (ant-plant mutualism); Billen, 1992 (biology of social insects); Carroll and Janzen, 1973 (ant
foraging); Cook, 1993 (sex determination); Dadant
and Sons, 1975; Danforth, 1989 (evolution of wing
morphology); Dressler, 1982 (biology of orchid
bees); Eickwort and Ginsberg, 1980 (foraging and
mating in Apoidea); H. E. Evans and Eberhard, 1970
(wasps); Fletcher and Ross, 1985 (reproduction in
social species); Frisch, 1967, 1971 (honey bee orientation and communication); Godfray, 1994 (parasitoids); Goodman and Fisher, 1991 (behavior and
physiology of bees); Holldobler and Wilson, 1990
(ants); Huxley and Cutler, 1991 (ant-plant interactions); Ito, 1993 (wasp behavior); Iwata, 1976 (behavior); Janzen, 1966, 1967a,b (coevolution of ants
and acacias); Jeanne, 1980 (social behavior in Vespoidea); Krombein 1967 (cavity-nesting bees,
wasps); McIver and Stonedahl, 1993 (mimicry of
ants); Malyshev, 1968 (extensive life history information); Matthews, 1974 (Braconidae); Michener,
1969, 1974 (social behavior of bees), 1979 (biogeography of bees); Michener and Sakagami, 1990 (classification of bees); Morgan, 1968 (Siricidae); K. G.
Ross and Matthews, 1991 (social biology of wasps);
Roubik, 1988 (tropical bee biology); Spradberry,
1973 (wasps); Sudd and Franks, 1987 (behavioral
ecology of ants); Traniello, 1989 (foraging in ants);
Urbani, 1989 (phylogeny and behavioral evolution),
1993 (recruitment behavior); Wagner and Raffa,
1993 (sawfly biology); Way, 1963 (ant-Homoptera
mutualism); Wenzel, 1990 (Cretaceous wasp nest);
Wiebes, 1979 (fig pollination); Wilson, 1971 (social
behavior), 1975 (slavery in ants); Winston, 1987
(honey bee biology).

Systematics and Evolution

Bohart and Menke, 1976 (keys, sphecids of world);
Bolton, 1995 (keys to ant genera); Creighton, 1950
(ants of North America); Boucek, 1988 (higher classification of Chalcidoidea); Brothers, 1975 (phylogeny of Aculeata); Grissell and Schauff, 1990 (keys
to Nearctic Chalcidoidea); Hinton, 1946, 1971 (phylogenetic relationships); Holldobler and Wilson,
1990 (ants); Malyshev, 1968 (higher classification);
Michener, 1944 (bees), 1953 (bee larvae), 1974 (social behavior of bees); Michener et al., 1994 (genera
of bees of North America); Mitchell, 1960–1962
(bees of Eastern states); Peck et al., 1964 (family
keys to Chalcidoidea); Peterson, 1948 (larvae);
O. W. Richards, 1956 (keys to families); H. H. Ross,
1937a (Symphyta of North America); Stephen et al.,
1969 (bees of Pacific Northwest); Townes, 1969
(Ichneumonidae).

47
COLLECTING AND PRESERVATION

Insects are often best captured by carefully stalking them with net bag held extended so that the bag is open when swinging the net.

Because of the extraordinarily large number of species, the study of insects usually entails some collecting of specimens. Collecting is an integral part of systematics, but it is also usually necessary in ecological studies, including those of an applied nature. Species identification in many groups is extremely difficult, even for specialists, and it is not uncommon for specimens to be misidentified or for classifications to change because new species are discovered. Even in physiological or morphological work carried out entirely in the laboratory, it is important to preserve voucher specimens, to verify the identity of the species at a later date. Collecting is per

haps the most effective way to associate taxonomic names with the insects themselves and to gain a knowledge of insect natural history. Living animals are always more interesting than preserved specimens, and characteristic behaviors such as the clicking of elaterid beetles are usually much easier to recall than morphological details found in keys. Finally, the pleasure and satisfaction that come from pursuits centered on natural history are considerable. Insect collecting or photography is a life-long avocation for many people who are not professional entomologists, and has the advantage that it can be enjoyed almost anywhere, so ubiquitous are the subjects.

COLLECTING EQUIPMENT

Basic Equipment

The most basic item for most collectors is the **aerial net,** which is used to snare flying insects and also to capture specimens perched on flowers or branch tips, or on the ground. Nets are conveniently purchased from supply houses or easily constructed from locally available materials (Fig. 47.1a,b). The exact dimensions are not critical, but handles about 1 m long and rims about 35 cms in diameter are preferred by most collectors. The bags of aerial nets are sewn from an open mesh netting that allows the captured insects to be seen. When an insect is caught, the net should be swung vigorously to force the specimen to the bottom. The net is then turned so the bag folds across the rim, preventing the insect from flying out the opening. The tip of the net and the enclosed insect may then be placed into a killing jar, or one may slip the jar into the net and maneuver the insect into it.

The basic net may be modified in various ways. Longer handles may be used to advantage to catch insects that often perch out of reach, such as dragonflies, or those that visit the flowers of trees, such as butterflies or bees. Nets with bags of muslin or very light canvas are used to sweep insects from coarse vegetation, which would damage the fabric of an aerial net. Such **sweep nets** often have a larger diameter rim and shorter handle than a standard aerial net. Sweeping usually accumulates vegetation and flowers as well as insects; normally the entire contents is placed into a large killing jar until the insects have been stunned and can be sorted from the debris. Aerial nets may also be used in water, but a separate **aquatic net** is convenient. Aquatic nets may be constructed in the manner of aerial nets, but using a heavier netting material. Alternatively the aquatic net bag may be a cylinder of muslin with a flat bottom of netting (Fig. 47.1c).

Many insects hide in leaf axils, flower clusters, or otherwise concealed spots on plants, and many drop to the ground when the plant is disturbed, where they are difficult to find without extensive searching. The sweep net described above is used on low, soft vegetation, but stiff shrubs, brambles, or the lower branches of trees are most conveniently sampled with the **beating sheet,** which is a rectangle of muslin or light canvas suspended from crossed sticks (Fig. 47.2). The beating sheet is held under the canopy of a shrub or beneath a branch, which is then tapped with a stick or the net handle. The insects fall onto the sheet, where

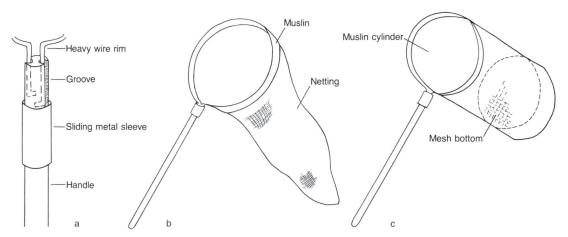

Figure 47.1 Insect nets: **a,** detail of attachment of rim of aerial net to handle. **b,** aerial net. **c,** aquatic net, with muslin cylinder extending below the net bottom.

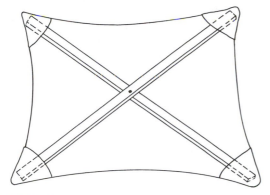

Figure 47.2 Beating sheet in inverted position to show arrangement of crossed sticks held in reinforced pockets of sheet.

are mercury vapor and ultraviolet. Mercury vapor lights require high current, necessitating a generator if an electrical outlet is unavailable. Inexpensive devices to convert direct current to alternating allow ultraviolet lights to be run from a car battery. Most insects usually alight on the sheet near the light, but many beetles sit on the ground or shrubbery several meters away and must be located by flashlight.

A stout trowel, small pry bar, or similar tool will be very useful for removing the bark from dead trees, breaking apart rotten wood, doing light digging, or extracting insects from many of the other places where they hide.

they are readily seen. Beating sheets work especially well during cool, overcast weather or early in the morning when active, flying insects are sluggish. Beating sheets may also serve to separate insects from tangles of vines and from vegetation overhanging streams or road cuts, providing, as well, a useful surface on which to break apart rotten wood, fungus, or other materials containing hidden insects.

An **aspirator** (Fig. 47.3) or pooter is an extremely useful device for removing small insects from a beating sheet. The aspirator is also useful for gathering insects from light sheets or from irregular surfaces such as bark or fungus, or from cracks and crevices in rotten wood. Sold by supply houses, aspirators are also easily constructed from plastic tubing, corks or rubber stoppers, and pill bottles (Fig. 47.3). The tube should be long enough to allow the collector's arm to be extended, and the corks should fit tightly against the bottle, leaving no crack where small insects could wedge themselves and be crushed when the aspirator is opened. A number of insects may be aspirated into the collecting vial before it is emptied into a killing jar.

The collector of moths or other nocturnal insects will probably wish to employ a **light sheet.** This is simply a muslin bedsheet or other large cloth hung vertically with one or more lights suspended at or near the top. The most effective lights

Traps

Many insects are most effectively captured with special trapping devices. Even experienced entomologists are often surprised by the number and diversity of the catches of various traps, and some taxa are seldom encountered without trapping. While these devices are more elaborate than the simple equipment described above, most may be constructed from commonly available materials.

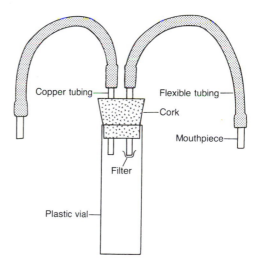

Figure 47.3 Aspirator: the lengths of the tubes may be customized to fit individual preferences. The neck of the plastic receiving vial should be reinforced with tape.

The **Malaise trap,** consisting of vertical panels of fine netting intersecting at right angles with a catcher at the top (Fig. 47.4), is extremely useful for capturing a variety of flying insects. The trap is pegged to the ground, and insects fly against the panels, then move upward, eventually entering the catcher, which contains a killing agent.

Malaise traps are highly effective for Diptera, Hymenoptera, some Lepidoptera and other aerial insects. For Coleoptera and other insects that drop to the ground when their flight is interrupted, the **windowpane trap** is more useful. Windowpane traps consist of vertical transparent panels (usually plastic) with a trough at the bottom to catch the falling insects (Fig. 47.5). If the trap is to be emptied daily, the trough may be filled with water containing a little soap solution (to relieve tension at the surface film). Windowpane traps may be left for long periods if the trough is filled with a nonvolatile preservative such as ethylene glycol (antifreeze). If rain may be expected, the trap should be shielded to prevent the trough from filling with water.

Insects that inhabit leaf litter or soil are most conveniently extracted with **Berlese funnels,** also known as separators. This device consists of a cone or cylinder, usually made of sheet metal, with a coarse mesh wire platform for holding samples of litter or soil (Fig. 47.6). A lightbulb or heating element placed above the sample drives the insects and other arthropods downward, where they are collected in a catcher containing alcohol. Berlese funnels or similar devices provide practically the only means of collecting Protura, Diplura, and many families of beetles and other insects that inhabit litter. Rodent or bird nests, some types of fungus, and other porous substrates may also be processed using Berlese funnels.

Drift nets are fine mesh traps left in streams to capture insects carried by currents. Old nylon stockings are effective for this purpose. **Pitfalls** are simply containers that are buried to the rim and left to trap ground-dwelling arthropods. Pitfalls may be baited with dung, rotten meat, or other substances to lure specific types of insects, but often they are used without attractants. The

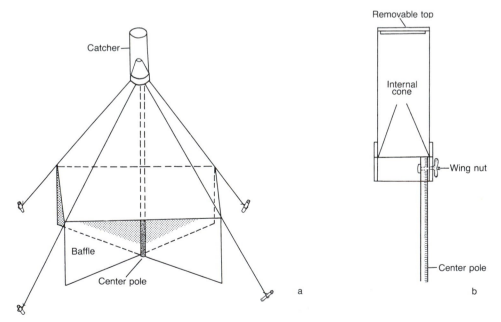

Figure 47.4 Diagram of Malaise trap (**a**): the baffles should reach the soil surface and may be pegged down; the plastic catcher (**b**) may be attached to the center pole by a wing nut and bolt (height of trap is usually about 6 ft).

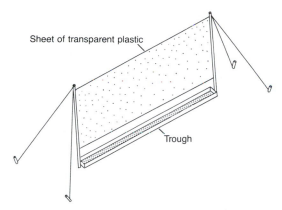

Sheet of transparent plastic

Trough

Figure 47.5 Diagram of a windowpane trap (vertical dimension usually about 2 to 3 ft). If left for more than a few days, or depending on weather, the trap should be covered with a rain deflector.

containers may be shielded with covers supported by small stones or blocks to deter insectivorous mammals and to deflect rain. Unless they are to be tended frequently, pitfalls should be partially filled with a nonvolatile preservative such as ethylene glycol; otherwise many of the trapped insects will be damaged. Pitfalls are especially effective in catching beetles and work especially well in sand dunes or very sandy soils.

Light traps enhance the catch of nocturnal insects. The typical light trap entails a mercury vapor or ultraviolet light suspended above a funnel with a catcher at the lower end (Fig. 47.7). The trap is covered to keep out rain and may have a pair of vertical vanes to obstruct flying insects and direct them downward. A hardware cloth screen placed in the funnel will prevent large beetles from entering the catcher and damaging more delicate insects.

The **Lundgren trap** (Fig. 47.8) consists of a series of conical sleeves, open at the bottom and arranged in a column with spacers between each cone. At the bottom of the column is a catcher containing killing agent. The entire apparatus is suspended by cords or wire and hung from a convenient support such as a tree limb. The Lundgren trap is especially proficient at catching forest Coleoptera, which apparently mistake the colum-

nar shape for a tree trunk and attempt to alight, falling through the series of cones to the bottom container. Lundgren traps may be left for long periods if a nonvolatile preservative such as ethylene glycol is used in the catcher.

KILLING METHODS

For purposes of general collecting all harder bodied insects are best killed dry in **killing jars.** Several different chemicals may be used to quickly kill captured insects. All killing jars should be labeled "POISON" and stored in a secure place. Typically the toxic agent is held in the bottom of the jar by a layer of cardboard or plaster (Fig. 47.9). The insects are placed in the space above the plaster. The most effective killing substance is potassium or calcium cyanide, which has the disadvantage of extreme toxicity to humans. Cyanide in any form is extremely dangerous. Do not

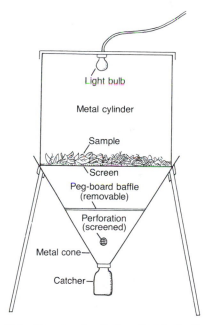

Light bulb

Metal cylinder

Sample

Screen

Peg-board baffle (removable)

Perforation (screened)

Metal cone

Catcher

Figure 47.6 Diagram of a Berlese funnel: the Peg-Board baffle prevents debris from entering the catcher. The metal cone should have several perforations to ensure good air circulation through the sample material, thus greatly speeding operation

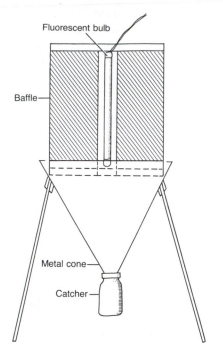

Figure 47.7 Diagram of a light trap: there are four baffles, radiating at right angles from the fluorescent tube; a rain deflector may be added.

inhale the fumes or touch the chemicals. Killing jars containing cyanide must have the bottoms securely covered with tape to prevent breakage and must have the cyanide held beneath plaster. Cyanide jars will last for weeks or months before needing replacement and do not sweat like jars charged with ethyl acetate or other liquids. Most insects may be left in cyanide jars for several hours without physical deterioration, but the yellow pigment in Hymenoptera irreversibly turns red with prolonged exposure to cyanide. Cyanide also causes most specimens to become rigid, and relaxing them before further processing may be desirable.

Ethyl acetate kills insects more slowly than cyanide and causes the jars to sweat, but it is relatively nontoxic. Specimens killed in ethyl acetate are relaxed and may be immediately processed.

Several killing jars are needed to accommo-

date insects of a variety of types and sizes. Small delicate flies or wasps should be kept separate from beetles, grasshoppers, and other large, robust insects. Butterflies and moths should be placed individually in small killing vials; once stunned, they can be transferred to a larger holding jar. All killing jars should contain a loose, shock-absorbing wad of tissue paper to keep specimens apart and take up excess moisture. Jars should never be overloaded with specimens, or the more delicate ones will surely be damaged.

Larvae and soft-bodied adults (Table 47.1) should be killed in 95 percent ethyl alcohol or in special solutions such as Kahle's solution or Bouin's fluid. Large specimens need to be slit open or punctured to allow penetration of the preservative and to prevent deterioration due to enzymatic action. Excellent preservation of color and body

Figure 47.8 Diagram of the Lundgren trap: the number of cones may be varied, but at least six are normally used.

Cork

Plaster of Paris

Sawdust
Cyanide
Sawdust

Figure 47.9 Diagram of a killing vial. The chemical agent is held below a cap of plaster or tightly fitting cardboard. The rim of the vial should be reinforced with clear tape and the bottom protected with strong paper or cloth tape. The vial should be labeled "POISON."

configuration is usually obtained if larger specimens are dipped momentarily into boiling water, then transferred to 80 percent alcohol. After a day or so the old alcohol should be replaced with fresh to avoid dilution by water contained in the specimens.

PRESERVING INSECTS

All insects may be preserved in liquids such as alcohol, which must be used for specimens in which the tissues are to be examined or soft part anatomy observed. For general use, 80 percent ethyl alcohol is excellent, but for cytological purposes or DNA extraction mixtures such as Kahle's solution and Bouin's fluid are preferable. These fixatives, which contain formalin and acetic acid, result in better tissue preservation but cause excessive hardening of muscles and other tissues over time. Cytology manuals provide additional information. Larvae and soft bodied adults (Table 47.1) are permanently preserved in liquids, usually 80 percent alcohol. To ensure the integrity of alcohol-

preserved specimens, it is only necessary to maintain the level of alcohol in the containers.

The great majority of adult insects are mounted on pins (Fig. 47.10). Dried, pinned insects retain much of their natural appearance and may be conveniently arranged in boxes or trays. The pin provides a handle that eliminates the need to touch the specimen when it is moved or examined. Nonrusting pins are specially manufactured in various sizes for mounting insects and are longer than sewing pins, which are unsuitable. Insect pin sizes 2 and 3 are used for practically all routine pinning. Specimens should be mounted so that about 1 cm of the pin extends above the dorsal surface. A convenient surface for mounting is a block of high-density foam, but many persons prefer to hold the specimen between the thumb and fingers while inserting the pin with the other hand. Most specimens are pinned in horizontal position with the pin just to the right of the midline of the mesothorax. Grasshoppers, roaches, and beetles are usually pinned through the right wing or elytra, and Heteroptera through the right side of the scutellum. Dragonflies are best pinned through the sides with the wings folded over the back. Lepidoptera are traditionally pinned with the wings spread horizontally, exposing the venation and color pattern.

Special **spreading boards** (Fig. 47.11) are used to keep the wings at a slight dihedral angle, and paper strips held in place with pins retain them in a symmetrical position with the trailing edges of the forewings at approximately right angles to the body. Spread specimens should be dried for about 3 to 7 days, depending on size, before removal from the boards. Insects other than Lepidoptera are not usually arranged in any particular manner, but for purposes of handling and storage, the appendages and wings should be drawn close to the body and the abdomen should not sag. Appendages may be retained with pins until the specimen dries; strips of stiff paper pinned beneath the specimen will hold the abdomen in horizontal position.

Standard pinning techniques work well with most insects larger than about 5 mm but damage

TABLE 47.1.
Killing and preserving of insects for study

Order	Recommended method

Protura, Diplura, Collembola
  Archeognatha, Thysanura
  Ephemeroptera
	Kill in 95% ethanol & store in 80% ethanol

Odonata	Larvae as above; use killing jar for adults & pin (option: spread wings) or fold wings and store in envelopes
Blattodea	Killing jar & pin (option: eviscerate large specimens); or kill in 95% ethanol & store in 80% ethanol
Mantodea	Killing jar & pin (option: eviscerate large specimens)
Isoptera	Kill in 95% ethanol & store in 80% ethanol
Grylloblattodea	As above
Dermaptera	Killing jar & pin; or collect in 95% ethanol & store in 80% ethanol
Plecoptera	Kill in 95% ethanol & store in 80% ethanol
Embiidina	As above
Orthoptera	Killing jar & pin (options: eviscerate large specimens)
Phasmatodea	As above
Zoraptera	Kill in 95% ethanol & store in 80% ethanol
Psocoptera	As above
Phthiraptera	As above
Hemiptera, Heteroptera	
terrestrial species:	Killing jar & pin; small species mounted on paper points
aquatic species:	Killing jar & pin; or collect in 95% ethanol & store in 80% ethanol
parasites of vertebrates:	Kill in 95% ethanol & store in 80% ethanol
Hemiptera, Homoptera	
larger, hard-bodied species:	Killing jar & pin; small species mounted on paper points
aphids:	Kill in 95% ethanol & store in 80% ethanol (option: collect on piece of food plant)
scales, whiteflies, meally bugs:	Place in dry box or as above (option: collect on piece of food plant)
winged males of scales, etc.:	Kill in 95% ethanol & store in 80% ethanol
Thysanoptera	Kill in 95% ethanol & store in 80% ethanol
Megaloptera, Raphidioptera Neuroptera	Larvae as above; adults in killing jar & pin or store large specimens in envelopes
Coleoptera	Larvae in 95% ethanol & store in 80% ethanol; adults in killing jar & pin; small species glued to paper points or put in 80% ethanol
Strepsiptera	Kill in 95% ethanol & store in 80% ethanol
Mecoptera, Diptera	Larvae in 95% ethanol & store in 80% ethanol; adults in killing jar & pin; small species mounted on points or minuten pins
Siphonaptera	Kill in 95% ethanol & store in 80% ethanol
Lepidoptera	Larvae in 95% ethanol & store in 80% ethanol; adults in killing jar & pin (option: spread both wings) or fold wings & store in envelopes; small species mounted on minuten pins
Trichoptera	Kill in 95% ethanol & store in 80% ethanol
Hymenoptera	Larvae in 95% ethanol & store in 80% ethanol; adults in killing jar & pin; small species mounted on paper points or minuten pins

through the broad end and the specimen glued to the pointed end, which is bent with fine forceps to conform to the side of the specimen (Fig. 47.12b). Fingernail polish is a convenient source of glue, or special glues that remain flexible may be purchased from supply houses. Small Diptera and Lepidoptera are usually mounted through the venter or the sides on tiny pins called **minuten,** that are first thrust through a small block of pith that is mounted on a standard insect pin (Fig. 47.12a).

Insects should be pinned as soon as possible after death. If pinning is not possible within about 12 hours, specimens should be held in a relaxing chamber or placed in a deep freeze. Relaxing chambers may be made of any tight-fitting ceramic or plastic container in which the specimens may be held apart from the water that saturates the atmosphere. If water alone is present, specimens may be held for only about a day. For this reason, a saturated solution of phenol is normally used and will prevent decay for about a week. Frozen specimens may be held indefinitely but will gradually dry and require relaxing.

LABELING AND STORAGE

All pinned specimens and all alcohol vials should receive an individual label bearing the collection

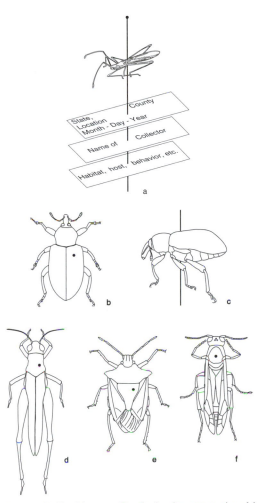

Figure 47.10 Diagrams for pinning insects: **a,** pinned insect with labels (long appendages should be tucked under the body to prevent breakage); **b, c,** dorsal and lateral views of a beetle, showing position of the pin in the right elytron and the position of the body axis at right angles to the pin; **d,** position of pin through the right side of the prothorax of a grasshopper; **e,** position of the pin through the right side of the scutellum of Heteroptera; **f,** position of the pin through the middle of the mesothorax in a wasp.

or destroy many smaller specimens, which are best prepared with double mounts. Double mounts are of two types. **Paper points,** small triangles cut or punched from light card stock, are used with small Coleoptera, Hemiptera, and Hymenoptera, and optionally for most other orders. The point is pinned

Figure 47.11 Diagram of a spreading block for Lepidoptera: several blocks of differing dimensions are needed to spread specimens of different size.

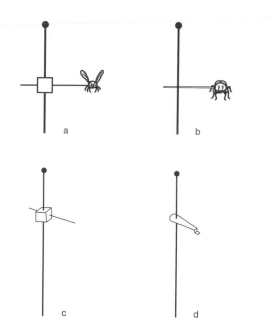

Figure 47.12 Diagrams of special mounts for small insects: **a**, **c**, mounts on a polyporous block and minuten pin, used especially for small Diptera and Lepidoptera; **b**, **d**, mounts on a paper point used for small Coleoptera, Hymenoptera, and Hemiptera; the tip of the point is bent to roughly fit the specimen.

be used to house a small insect collection. Bottoms suitable for pinning may be fashioned from polyethylene foam, balsa wood or other materials. Most such containers, however, do not exclude dermestids and other museum pests that will destroy unprotected insects. High-quality boxes or trays purchased from supply houses give better protection, but are expensive. Regardless of the type of container, periodic inspection is necessary to detect and eliminate dermestids and other pests. The safest way of killing such invaders is to place each container overnight in a freezer. Alternatively, containers may be fumigated with ethylene dichloride, carbon disulfide, or Vapona-type insecticide strips.

HOW TO FIND INSECTS

Despite the frequently written advice that insects are everywhere, much experience is required to efficiently sample an area. The continual discovery of new species, even in relatively well-studied regions such as North America, testifies to the difficulty of finding many insects.

Probably the greatest diversity of insects of all types is encountered on plants, especially those in flower or in fruit. Some orders, such as Hemiptera, occur predominantly on plants, and many are quite host-specific. Thus as many plants as possible should be sampled, either by searching or by using the sweep net or beating sheet. Flowering plants are a productive place to search for Hymenoptera, especially Apoidea, which, once again, are often highly host-specific. Other insects will be found in or on other plant parts, such as bark, leaf axils, rolled or tied leaves, in mines in the leaves or stems, or on the roots.

Two other especially productive terrestrial habitats are the subcortical region of dying or dead trees and leaf litter and the upper layers of soil. The subcortical habitat is most effectively sampled by hand searching. Rotting wood of all ages should be investigated, since insects appear in succession as decay proceeds. Insects inhabiting the subcortical region are predominantly Coleoptera, but many

locality and date. The locality should include the country, state or province, county or district (if appropriate), and a specific locality with orientation to a place name on standard maps. Dates should include year, month, and day. This locality label is usually placed on the pin just below the specimen (Fig. 47.10a). The collector's name usually appears on a smaller separate label, but may be included on the locality label if space permits. Additional information, such as a host plant name, time of day, or other specific ecological information appears on one or more additional labels. For large lots of insects, labels are printed, while individual specimens or very small lots are labeled by hand. In either case it is important to keep the labels as small as possible (usually no more than about 3 cm long or 1.5 cm broad). Labels should be printed on acid-free, light card stock using indelible black ink.

Practically any tight-fitting, lidded box may

adult and immature Apterygota, Orthoptera, Dermaptera, and Hemiptera may be found, as well as larvae of many other orders. Litter and soil are best sampled with Berlese funnels and produce a multitude of insects and other arthropods. Especially predominant are Myriapoda, Parainsecta, Entognatha, Apterygota, larval Diptera, and Coleoptera, but some members of practically every order are occasionally encountered. Even a few species of normally aquatic groups such as Trichoptera and Odonata (immatures) inhabit saturated litter in high-rainfall areas such as Hawaii. Some orders, such as Diplura and Protura, are found almost exclusively in these habitats.

Several orders and some members of most other orders are aquatic and are usually specific to certain types of habitats, such as vegetation in ponds, sandy or stony bottoms in either still or running water, on or under submerged stones, or in the interstices of gravelly bottoms. Tree holes, vernal pools, and the marine intertidal also support a variety of aquatic or subaquatic insects. A habitat that should not be overlooked is the littoral zone along streams, ponds, and seacoasts, where numerous insects will be found on the soil surface as well as burrowing into sandy or muddy edges.

Rotting or decaying plant or animal material is inhabited by many scavenging insects as well as their predators and parasites. A backyard compost bin or carcass beside a highway will often produce many insects rarely encountered anywhere else. Deteriorating mushrooms support a broad selection of insects, especially Diptera, and the bracket fungi are host to numerous others, especially Coleoptera.

Many insects are uncommonly encountered as adults but are readily reared from larvae, which are easily found. Perhaps the best example is gall formers, which are usually small, are often relatively short-lived as adults, and may be active in winter. The galls, however, are relatively conspicuous, and gall-bearing twigs placed in water will often produce adults of the gall makers as well as parasitoids and predators. Parasitoid forms in general are often easier to obtain by rearing from hosts than by general collecting, and this method provides valuable biological information, as well. Lepidopterous caterpillars are especially productive of parasitoids, but practically every insect that has been intensively studied has proven to have at least one parasitoid.

Perhaps the most rewarding method of collecting for many people is searching. Hiding places of all sorts should be investigated, including animal burrows or nests, undersides of logs, stones, boards, pieces of cardboard, old buildings, animal dung, old wasp or bee nests, ant nests, inside crevices in rocks, and in plant parts such as cones, galls or hollow stems or twigs, in pads of moss or lichens, around the bases of plants, in the middens left by rodents, or in practically any other place offering food or shelter. Insects even occur in the briny evaporating ponds used to produce salt and in pools of raw petroleum.

LEGAL REQUIREMENTS

Until recently insects could be freely collected in most parts of the world and moved from one nation to another as long as they had been killed and properly prepared. Shrinking natural habitats and consequent declines in populations of many organisms, including some insects, have prompted passage of guidelines or rules regulating collecting and transportation of practically all natural history objects, including preserved insects. The intent of these laws is to protect endangered or threatened species or populations. While protection may seem unnecessary for abundant organisms such as insects, many species occur only in localized, narrow habitats. Many insects have been listed as **threatened** or **endangered** by the U.S. Fish and Wildlife Service, based on detailed studies of distribution and life history, and at least one species has become extinct.

Laws vary greatly from one country to another, and regulations may vary among states or districts in the United States. Furthermore, many of the regulations, particularly those governing transportation of specimens, are under review and subject to change. In general, however, a few practices will satisfy the legal requirements.

- No collecting should be undertaken on publicly owned land without written consent of the agency responsible. Parks (national, state, and local), wildlife refuges and reserves, and national forests are all public lands.

- Collecting on private land is unregulated except for endangered species. Permission is needed from the landowner.

- Collecting in nearly all foreign nations requires permission from the responsible government agency and usually entails a fee, sometimes a considerable one. Collecting without the necessary permits exposes the collector to the risks of prosecution by local wardens or police, as well as by federal (U.S.) officials if the specimens are brought into this country.

GLOSSARY

Absolute sample: Collection or counting of every individual belonging to a designated group within a volume of habitat.

Accessory cell: In Lepidoptera, a small cell between two branches of the radial vein usually R_{2+3} and R_{4+5}; areole; see Fig. 44.7d.

Acephalic: Without a distinct sclerotized head capsule, as in many immature Diptera.

Acrotergite: The anterior precostal part of the tergal plate of a secondary segment.

Aculea: A minute, needlelike hair on the wings of Lepidoptera.

Adecticous: Without the ability to move the mouthparts, as in many holometabolous pupae.

Adjuvants: Chemicals combined with insecticides to improve their storage, safety, or effectiveness.

Aedeagus: The penis or male intromittent organ.

Alinotum: The wing-bearing sclerite of a pterothoracic notum.

Allomone: A substance transmitted in chemical communication between individuals of different species that benefits the sender or both the sender and receiver.

Ambulatory: Adapted for walking.

Ametabolous: Without noticeable metamorphosis, as in Protura, Collembola, etc.

Amphipneustic: Pertaining to a larval respiratory system in which only the anterior and posterior spiracles are functional, as in certain Diptera.

Amplexiform: In Lepidoptera, the mode of wing coupling in which the humeral region of the hind wing projects beneath the forewing.

Anal lobe: The posterior portion of the wing, occupied by the anal veins; see Fig. 2.14b.

Anamorphosis: Postembryonic development in which abdominal segments are added at the time of molting.

Angulate: Forming an angle, as opposed to rounded.

Anlage: A group of embryonic cells that will form an organ in the adult; also known as imaginal buds or disks.

Annular: Ring-shaped or ringlike.

Annulate: Ringed, bearing ringlike divisions.

Anteclypeus: Ventral part of frontoclypeus of sucking insects.

Antecosta: The internal ridge marking the original intersegmental boundary and on which the longitudinal muscles of secondary segments insert.

Antecostal suture: The external suture tracing the internal antecosta.

Antennifer: An articulatory process in the antennal socket.

Antibiosis: A characteristic that reduces longevity or fecundity.

Antixenosis: Deterring feeding or oviposition.

Apodeme: An ingrowth of the integument to which muscles attach.

Apodous: Without legs.

Apophysis: An ingrowth of the integument to which muscles attach.

Aposematic: Warning, as in conspicuous aposematic coloration or behavior of poisonous or dangerous animals

Appendiculate: Bearing an appendage or appendages.

Apposed: Having surfaces adjacent to or against one another.

Aptery: The condition of lacking wings; see Brachyptery (apterous).

Arbovirus: Arthropod-borne virus.

Arista: A large bristle, usually dorsal, on the apical antennal segment of some Diptera; see Fig. 2.7m.

Arolium: A cushionlike pad between the tarsal claws, as in Orthoptera, or paired pads, as in Hemiptera; see Empodium; Pulvillus.

Arrhenotoky: Parthenogenetic reproduction in which the offspring are all males.

Asexual: Not having separate sexes; reproducing by parthenogenesis.

Atrophied: Undeveloped, rudimentary, wasted away through lack of growth (atrophy).

Attenuate: Tapering apically.

Augmentation: Reinstating or supplementing populations of natural enemies.

Autotomize: To cast off a limb, usually at a predetermined zone of weakness (autotomy).

Axillary sclerites: Four sclerites at the wing base; see Fig. 2.14a.

Bacteriocytes: Host cells that harbor bacterial symbionts.

Basal cell: In Hymenoptera, a cell near the base of the wing, bordered partly by the longitudinal veins; see Fig. 46.7b–d.

Basal vein: In Hymenoptera, a vein connecting the subcostal and cubital veins; see Fig. 46.18b.

Basalare: The episternal epipleurite, on which insert the anterior pleural wing muscles.

Basitarsus: The basal tarsomere.

Bifid, bifurcate: Cleft; forked; divided into two parts.

Biological amplification: The increase in concentration of chemicals in organisms at higher trophic levels.

Biological control: Use of predators, parasites, and pathogens for pest control.

Biotype: Population of pest species able to overcome resistance genes in plants or animals; also equivalent to a strain or distinctive population of a species.

Bipectinate: With comblike processes or extensions on both sides, as in bipectinate antennae; see Fig. 39.6g.

Bisexual: Having two distinct sexes.

Bivoltine: Having 2 generations in a year.

Blood gill: Hollow process of the body wall through which blood circulates but which lacks tracheae; present in endopterygote larvae, probably functioning in maintaining ionic balance; see Tracheal gills.

Blotch mines: Broad, irregular excavations produced in leaves especially by larvae of Lepidoptera; see Serpentine mines.

Brachyptery: The condition of having short, atrophied wings; see Aptery (brachypterous).

Brood passage: In Strepsiptera, the space between the venter of the female and the puparium through which the triungulins emerge.

Bursa copulatrix: A copulatory chamber, as in Lepidoptera.

Calypter: A small basal fold or lobe in the hind margin of the wing of Diptera, covering the haltere; see Squama.

Campodeiform: Having the general body shape of Campodeidae (Diplura); applied to larvae.

Capitate: Club-shaped, with an abrupt, apical enlargement; see Figs. 2.7e, 39.6d.

Caudal lamellae: The caudal gills of damselflies (Odonata, Zygoptera).

Caudate: Bearing a tail-like extension or process.

Caudate larva: In certain Hymenoptera, first-instar larvae in which the terminal abdominal segment forms a tail-like extension.

Cenchrus: (pl., cenchri) A pale, membranous lobe or area on each side of the metanotum of symphytan Hymenoptera.

Centrolecithal egg: An egg with a large amount of yolk in the center.

Cephalotheca: The head covering in the pupal stage, especially in reference to Strepsiptera.

Cephalothorax: The fused head and thorax of Arachnida and Crustacea; that portion of an obtect pupa covering the head and thorax, as in Strepsiptera.

Cercus: (pl., cerci) An appendage of the eleventh abdominal segment.

Cervical sclerites: Small sclerites in the membrane between the head and thorax.

Cervix: The neck.

Chaeta: An articulated hair or spine.

Chaetosema: (pl., chaetosemata) In Lepidoptera, sense organs located on the head between the eyes and ocelli.

Chaetotaxy: The arrangement and nomenclature of bristles or chaetae on the exoskeleton.

Characteristic abundance: The population density within upper and lower limits typical of a species within a particular type of habitat. Also called equilibrium density.

Cheliform: Shaped like a pincer or chela (chelate).

Chitin synthesis inhibitors: Insecticides that interfere with cuticle development and molting.

Chorion: The outer shell of an insect egg.

Chrysalis: The naked pupal stage of a butterfly.

Cibarium: The preoral cavity between the hypopharynx and epipharynx.

Circulative transmission: Transmission by vectors that continues for days to weeks without reacquiring a pathogen.

Circumversion: In male Diptera, the 360° rotation of the genitalia about their longitudinal axis; see Torsion.

Classical biological control: The establishment of exotic natural enemies or diseases to increase natural control of a targeted species.

Clavate: Gradually thickening toward the apex to form a club; see Figs. 2.7d, 39.6c.

Clavus: In Hemiptera, the oval or triangular anal portion of the front wing; see Fig. 34.5.

Cleavage: See Holoblastic, Meroblastic.

Cleptoparasite: A species that uses the nest and provisions of another species, as in cleptoparasitic bees.

Cloaca: A common chamber into which the anus and gonopore open.

Clypeogenal suture: A groove separating the clypeus and gena.

Clypeus: The anterior region of the head to which the labrum is attached.

Coarctate pupa: A pupa that is enclosed in the hardened cuticle of the last larval instar, as in cyclorrhaphan Diptera.

Cocoon: A case formed partly or wholly of silk, in which pupation occurs.

Collophore: In Collembola, a large, ventral projection from the first abdominal sternite; see Fig. 17.2.

Commensal: One of the organisms participating in a commensalism.

Commensalism: The association of two species without harm to either, and with benefit to at least one; see Symbiosis.

Common name: vernacular, as opposed to scientific, name.

Common oviduct: The unpaired, median duct through which the eggs pass.

Competitive exclusion principle: No two species with the same [ecological] niche can occupy the same resource without adverse effect on one or both. A result is the complete exclusion of one species by another with the same niche requirements.

Confluent: Running together, merging.

Conservation: A method of biological control that increases the effectiveness of existing natural or biological control.

Contact poisons: Poisons that kill after direct contact with the exterior of a target organism.

Corbiculum: In bees, a basket-shaped fringe of hairs on the hind tibia for carrying pollen; see Fig. 46.18c.

Coriaceous: Thick, tough, and leathery.

Corium: In Heteroptera, the elongate, basal part of the forewing, usually thickened, coriaceous; see Fig. 34.5.

Cornified: Tough, horny.

Corporotentorium: A broad, platelike tentorium.

Cosmetic damage: Injury that damages the appearance but not other economic values of a crop.

Cosmopolitan: Occurring throughout most of the world.

Coxa: The basal segment of the leg.

Cranium: The head capsule, excluding the mouthparts.

Cremaster: In Lepidoptera, the terminal hooks by which the pupa suspends itself.

Crepuscular: Active at dusk.

Crochet: In Lepidoptera, the curved spines or hooks on the abdominal prolegs of larvae.

Crop rotation: Change from one season to next of the type of crop grown in a field.

Cross-resistance: Resistance to one insecticide that confers resistance to another insecticide.

Crumena: In certain Hemiptera with feeding stylets much longer than the rostrum, a pouch in the back part of the head, sometimes extending into the thorax, as in Coccoidea.

Crypsis: Concealment, camouflage (cryptic).

Cryptic coloration: Concealing coloration, camouflage.

Cryptonephry: The close association of the Malpighian tubules with the hindgut, an adaptation to increase water retention.

Ctenidium: (pl., ctenidia) In Siphonaptera and certain other ectoparasites, a comb of short, flat spines (ctenidia).

Cucujiform: Having the body extremely flat, with the legs directed laterally; applied to larvae.

Cultural control: Changes in normal agricultural practices for pest control.

Cuneus: In Heteroptera, a small, triangular region of the corium delimited by the cuneal fracture; See Fig. 34.5.

Cursorial: Adapted for running.

Cuticle: The noncellular outer layer of the integument.

Cyclical parthenogenesis: Parthenogenetic reproduction that recurs every other generation, as in Cynipidae.

Cyclodevelopmental transmission: Developmental changes of parasites without multiplication within the vector.

Cyclomorphosis: Seasonal change of body form, as in Aphididae, Cynipidae, and other insects.

Cyclopropagative transmission: Processes in which the parasites change in form as they multiply within and invade organs of the vector.

Decomposer: An organism that consumes dead organic material.

Decticous: Having the ability to move the mandibles; applied to the pupal stage.

Degree-days: Units of physiological time.

Delayed parasitoidism: Parasitoidism in which the parasitoid egg remains dormant in an early stage of development until the host is near maturity, then develops rapidly.

Density-dependent: Changing in intensity proportional to population density.

Density-independent: Not changing in intensity with population density.

Denticle: A small tooth or tooth-shaped structure.

Deutocerebrum: Part of the brain that innervates the antennae.

Development rate: The reciprocal of development time ($1/H$).

Diapause: A period of arrested development.

Dicondylic: Having two articulations.

Digitate: Fingerlike.

Dimorphic: Occurring in two distinct forms.

Diploid: Double; having the full complement of maternal and paternal chromosomes.

Direct flight muscles: Dorsoventral muscles attaching directly to sclerites at the wing bases.

Direct pests: Pests that cause immediate injury or disturbance

Discal cell: A cell, frequently enlarged, in the central part or the wing; see Figs. 42.7a, 44.7e.

Disruptive coloration: Color patterning that tends to interrupt the outlines of the body.

Ditrysian: In Lepidoptera, referring to the presence of separate genital openings for copulation and oviposition; see Fig. 44.4b.

Diurnal: Active during the day.

Ductus seminalis: In female Lepidoptera, the tube connecting the bursa copulatrix and the common oviduct; see Fig. 44.4b.

Ecdysial line: A preformed line of weakness in the cuticle where the cuticle splits during ecdysis.

Ecdysis: The actual shedding of the old cuticle during molting.

Eclosion: The act of emerging from the egg; act of emerging from pupal cuticle; or act of emerging from any ecdysis.

Economic injury level (EIL): The amount of pest damage equal to the cost of preventing damage.

Economic threshold (ET): The level of pest density where action must be taken to prevent reaching the economic injury level

Ectoparasite: A parasite that lives externally on its host.

Ectoparasitoid: A parasitoid that lives externally on its host.

Ectosymbionts: Extracellular symbionts; symbionts that live only outside the hemocoel.

Elateriform: In larvae, having a long, cylindrical body with firm cuticle, as in Elateridae.

Elytriform: Having the form of elytra.

Elytron: A leathery or rigid forewing, especially of Coleoptera.

Emarginate: Notched or indented.

Empodium: A spinelike or cushionlike structure between the tarsal claws, especially in Diptera; see Fig. 42.13a.

Endemic: Peculiar to a given geographic region.

Endite: A mesal lobe of a limb segment.

Endocuticle: The inner unsclerotized layer of the procuticle.

Endoparasite: A parasite that lives inside its host.

Endoparasitoid: A parasitoid that lives inside its host.

Endopterygota: Those insects in which the wings develop internally.

Endopterygote: Having the wings develop from internal anlagen; member of the Endopterygota; see holometabolous.

Endoskeleton: An internal part of the skeleton.

Endosymbionts (Endosymbiotes): Symbionts that live intracellularly or within hemocoel.

Entognathous: Having the mandibles and maxillae retracted into pouches in the head, as in Protura, Collembola, Diplura.

Epicuticle: The outermost layer of cuticle.

Epidermis: The single layer of cells beneath the cuticle of the integument.

Epimeron: The posterior division of a thoracic pleuron.

Epipharynx: The lobelike inner surface of the labrum.

Epiphysis: In some Lepidoptera, an articulated process on the anterior tibia.

Epipleurites: The basalare and subalare sclerites of the pleuron.

Epiproct: The dorsal plate of the eleventh abdominal segment.

Episternum: The anterior division of a thoracic pleuron.

Eruciform: Caterpillar-like.

Eucephalic: In Diptera, referring to the presence of a distinct head capsule.

Eusocial behavior: The most advanced kind of social behavior: members cooperate in caring for young; at least two generations overlap; and the labor of reproduction is divided between reproductive and nonreproductive individuals.

Exarate pupa: Pupa in which the appendages are free of the body, as in Coleoptera.

Exite: An outer lobe of a limb segment.

Exocuticle: The outer sclerotized layer of the procuticle.

Exopterygota: Those insects in which the wings develop externally.

Exopterygote: Having the wings develop externally; member of the Exopterygota; see Hemimetabolous.

Exponential growth: Growth of a population that increases proportional to time and other variables as an exponent.

Exuviae: The cast skin of nymphs or larvae at the time of metamorphosis.

Facultative: Having the ability to live under more than one set of conditions.

Facultative symbiont: Association not necessary for survival of host.

Filariasis: Disease of vertebrates caused by nematodes.

Filiform: Threadlike; slender and of nearly constant diameter; see Figs. 2.7a, 39.6a.

Filter chamber: In Homoptera, a complex organ in which the two ends of the midgut and the beginning of the intestine are associated.

Filter feeding: Feeding by straining small particles from water, as in culicid larvae or Corixidae.

Flabellate: Fanlike, with thin, platelike processes lying flat against one another, as in antennae of Scarabaeidae; see Figs. 2.7i, 39.9a,b.

Flagellum: The part of the antenna beyond the pedicel.

Foliate: Leaflike.

Fontanelle: In termites, the shallow depression in which the frontal pore opens.

Foramen magnum: The posterior opening of the cranium.

Forcipate: Bearing forcepslike structures.

Forcipiform: Forceps-shaped.

Fossorial: Adapted for digging.

Frass: Solid excrement of insects, often mixed with plant or wood fragments.

Frenate: Bearing a frenulum.

Frenulum: In Lepidoptera, the spinelike process arising from the base of the hind wing and projecting beneath the forewing, serving to couple the wings during flight; see Fig. 44.3a.

Frons: The anterior part of the cranium below the vertex and above the clypeus.

Frontoclypeal suture: A transverse groove separating the frons and the clypeus; also known as the epistomal suture.

Frontogenal suture: A groove separating the frons and gena.

Fumigants: Volatile poisons that are lethal in the gas phase.

Functional response: An increase in prey consumption rates as a result of increased prey density.

Fungivorous: Consuming fungi.

Furca: A forked endosternal process (see Sternal apophysis); in Collembola, the appendage on abdominal segment 6, used for leaping.

Furcula: A forked process.

Fusiform: Spindle-shaped, tapering at both ends.

Galea: The outer maxillary lobe.

Gall: An abnormal growth of plant tissue, caused by an external stimulus, frequently from insects.

Gaster: In Hymenoptera, the distal portion of the abdomen, usually globular (see Petiole).

Gena: The cheek or posterior part of the head behind and below the eye to the gular or occipital sutures.

Gene for gene theory: Parasites have a gene that corresponds to every resistance gene in a host.

Genetic engineering: Artificial transformation of a genome by molecular methods.

Geniculate: Abruptly bent; see Fig. 2.7g.

Genital segments: Abdominal segments 8 and 9.

Germ anlage: In an insect embryo, the part of the blastoderm that will develop into the germ band.

Glossa: One of the paired inner lobes of the labium (see Paraglossa).

Gonapophysis: In female insects the median process of abdominal segment 8 (first gonapophysis) or 9 (second gonapophysis), forming the ovipositor in Hymenoptera and many other orders; in males the processes forming the parameres of the aedeagus or intromittent organ.

Gondwanan: Pertaining to Gondwanaland, the Mesozoic supercontinental land-mass consisting of South America, Africa, Madagascar, India, Australia, and Antarctica.

Gonocoxite: The proximal segment of a paramere.

Gonopore: The external opening of the ejaculatory duct (males) or oviduct (females).

Gonostylus: The distal segment of a paramere.

Gradate veins: A series of transverse crossveins, each before or beyond the next; see Fig. 38.3b,d.

Gula: The midventral sclerotized area on the posterior part of the cranium of some insects.

Gular sutures: The paired sutures forming the lateral boundaries of the gula.

Halteres: The reduced, knob-shaped hind wings of Diptera; see Fig. 42.2a–c.

Hamate: Wing coupling by hamuli.

Hamuli: (hamulus) In Hymenoptera, the hook-shaped setae on the leading edge of the hind wing, which catch beneath a posterior fold of the forewing; see Fig. 46.2a.

Haploid: Having only a single complement of chromosomes.

Hemelytron: In Hemiptera, the anterior, coriaceous part of the forewing; see Fig. 34.5.

Hemicephalic: In Diptera, the condition in which the reduced head capsule is intermediate between eucephalic and acephalic.

Hemimetabolous: Developing by incomplete or gradual metamorphosis, as in Orthoptera; see Exopterygote.

Hemocoel: The body cavity of insects and other arthropods.

Hemolymph: The fluid filling the hemocoel.

Herbivorous: Feeding on living plants; phytophagous.

Heteroneurous: Having the forewings and hind wings different in venation.

Hexapoda: Those arthropods with three pairs of walking legs, namely: Insecta, Collembola, Diplura, and Protura.

Hexapodous: Having three pairs of walking legs.

Holoblastic cleavage: The type of embryonic cell division in which the entire egg is divided into cells; see Meroblastic.

Holometabolous: Developing by complete metamorphosis; see Endopterygote.

Homology: A likeness of structure, physiology, behavior, or other characteristic due to a common evolutionary origin.

Homoneurous: Having the forewings and hind wings similar in venation.

Horizontal resistance: Resistance to pests or pathogens that is not overcome by distinct populations (biotypes) of pests or pathogens.

Hormone: A substance produced in one part of the body that has a profound effect on another part of the body.

Humeral plate: A plate in the humeral region of the wing.

Humeral region: The anterior base of the wing.

Humeral suture: In termites, the line of weakness along which the wings break after the nuptial flight.

Humeral veins: In Lepidoptera, veins that strengthen the basal portion of the leading edge of the hind wing; see Fig. 44.10e.

Humerus: (pl., humeri) The shoulder; in Diptera, the anterior angles of the mesothorax; in Coleoptera, the outer angles of the elytra.

Hyaline: Transparent, glassy.

Hydrofuge: Repelling water, unwettable.

Hypermetamorphosis: The type of development in which an insect passes through more than a single distinct larval stage.

Hyperparasitoidism: The condition in which one parasitoid feeds on another parasitoid; hyperparasitism.

Hyphae: Filamentous extensions of fungi.

Hypognathous: Having the mouthparts directed ventrally; see Prognathous; Opisthognathous.

Hypomeron: In Coleoptera, the portion of the pronotum that is inflected below the lateral margin.

Hypopharynx: The tonguelike structure in the preoral cavity between the mouthparts.

Hypostoma: The part of the subgena delimited by the hypostomal suture.

Hypostomal bridge: The median fusion of the hypostomal areas behind the mouthparts.

Hypostomal suture: Part of the subgenal suture posterior to the mandibles.

Immature: An immature insect; customarily refers to larval instars, not sexually immature adult insects.

Indirect fertilization: In Collembola, Diplura, Archeognatha, and Thysanura, the transfer of spermatozoa by means of a spermatophore that is deposited onto the substrate and picked up by the female.

Indirect flight muscles: The tergosternal and longitudinal tergal muscles of the pterothorax.

Indirect pests: Pests that create conditions leading to crop damage by other factors.

Inquilinism: The habit of living as a guest in the nest or abode of another organism.

Insect growth regulators: Chemical analogues of insect juvenile hormones that perpetuate immature stages of development

Instar: The developmental stage of the insect between molts; see Stadium.

Integrated pest management (IPM): A widescope approach to managing pest populations by combining compatible pest control methods to optimize pest control with minimal negative consequences to the environment.

Integument: The cuticular and cellular bodywall of arthropods.

Intercoxal process: In Coleoptera, the process of the basal abdominal segment that extends between the metacoxae.

Inundation: Release of mass-reared natural enemies to control a target pest.

Isolating mechanism: Differences in geographical distribution, temporal occurrence, behavior, or genetic composition that prevent interbreeding of populations.

Johnston's organ: A sense organ in the pedicel of the antenna.

Jugal fold: The posterior basal fold between the jugal and anal regions of the wing; see Fig. 2.13.

Jugal lobe: The posterior basal portion of the wing delimited by the jugal fold; see Fig. 2.14.

Jugate: In Lepidoptera and Trichoptera, having the wings coupled by a jugum.

Jugum: In certain Lepidoptera and Trichoptera, a basal lobe of the forewing that overlaps the hind wing, coupling the wings during flight; see Fig. 44.3b.

K-**selected:** Species that are adapted to maintaining stable population densities near the carrying capacity (*K*). Also called *K*-strategists.

Kairomone: A substance transmitted in chemical communication between individuals of different species that benefits the receiver of the substance.

Keel: A sharply elevated ridge.

Key factors: The mortality or environmental factors that best explain major population changes for a species.

Key pests: Pest species whose control is crucial for profitable crop production.

Labellum: In Diptera, the fleshy apex of the labium; see Fig. 42.3.

Labium: The fused second maxillae, forming the lower lip of the mouth in Hexapoda.

Labroclypeal suture: A line of articulation between the labrum and clypeus.

Labrum: The upper lip, attached to the clypeus.

Lacinia: The medial lobe of the maxilla; see Galea.

Lamellate: Sheetlike or leaflike, or composed of thin sheets; see Fig. 2.7h.

Lanceolate: Spear-shaped, tapering to a sharp point.

Larva: The immature stage of any insect; may be restricted to immature stage of a holometabolous insect, before the pupal stage; see Naiad; Nymph.

Larviform: Resembling a larva.

Laterotergite: A lateral sclerite on the dorsum, distinct from the principal median tergite.

LC$_{50}$: Concentration of a substance that kills half of a subject population.

LD$_{50}$: The estimated dosage that kills half of a subject population.

Lek: A site where males gather to display and compete for females, as in Hawaiian *Drosoph-*

ila, euglossine bees (Apidae), and hepialid moths.

Lerp: In certain nymphal Psyllidae a delicate, complex shelter produced from dried carbohydrate secretions from the anus.

Life tables: Accounting of mortality rates and reproduction throughout the life of a defined even-age population in a particular habitat.

Logistic growth: Exponential population growth that asymptotically approaches a constant limit (*K*).

Macrolecithal: Having a large yolk mass in the egg, as in most Hexapoda other than Collembola.

Macrotrichia: In Diptera, the larger hairs on the wings.

Mala: The single maxillary lobe of some endopterygote larvae.

Malpighian tubules: Insect urinary organs: long slender, blindly ending ducts, opening into the hindgut.

Mandibulate: Having the mandibles adapted for biting or chewing.

Marginal cell: In Diptera and Hymenoptera, a cell bordering the distal anterior wing margin, beyond the pterostigma.

Marginal vein: In Hymenoptera, the vein delimiting the marginal cell posteriorly; any vein near the wing margin.

Mark-release-recapture: A technique of estimating population size by marking and releasing captured individuals and counting their proportional abundance in later samples.

Mass provisioning: In solitary bees and wasps, the storage of sufficient food for larval development before oviposition.

Mating disruption: Use of pheromones to interfere with mating for pest control.

Meconium: The accumulated excretory wastes from the larval stage, which are voided shortly after adult emergence in certain insects.

Median caudal filament: A long process of the epiproct.

Median plates: Two plates at the wing base near the media and cubitus veins.

Membrane: Any flexible, nonsclerotized integument; the part of the wing between the veins.

Mentum: Distal sclerite of postmentum.

Meroblastic cleavage: The superficial type of embryonic cell division in which only the nucleus and nuclear cytoplasm of the egg are divided.

Meron: In Diptera, the lateral, basal portion of the coxa, which is incorporated into the thorax (also called the hypopleuron); the lateral, basal area of the coxa of any insect.

Meropleuron: The combined meron and ventral portion of the epipleuron.

Merozoites: Form of malarial parasite that invades liver cells or red blood cells.

Mes(o)-: Prefix indicating the middle or middle member of a series (e.g., mesothorax).

Mesothorax: The second or middle segment of the thorax.

Meta-: Prefix indicating the posterior part of a structure or posterior member of a series (e.g., metathorax).

Metamorphosis: The change of body form through which insects or other organisms pass in developing to the adult; see Holometabolous; Hemimetabolous.

Metapneustic: Pertaining to a larval respiratory system in which only the terminal pair of spiracles are functional, as in Culicidae.

Metathorax: The last segment of the thorax.

Microfilariae: First development stage of filarial nematodes taken up by mosquito vectors.

Microlecithal: Having little yolk in the egg, as in Collembola.

Microtrichia: Minute hairs on the wings of Diptera and certain other insects.

Mimicry: The resemblance of an organism to another, distantly related organism.

Model: In Batesian mimicry, the distasteful organism that is mimicked.

Molar lobe, mola: The basal portion of the mandible, usually modified as a grinding surface.

Molt: The entire process of shedding the old cuticle.

Moniliform: Composed of beadlike segments; beaded; see Figs. 2.7c, 39.6b.

Monocondylic: Having a single articulation; see Dicondylic.

Monophyletic: Evolved from a single ancestral form; a taxonomic group including the ancestor and all descendants.

Monotrysian: In Lepidoptera, referring to the condition in which the genital system opens through a single pore into a cloaca; see Fig. 44.4a.

Mouth hooks: In cyclorrhaphan Diptera, the secondary, mandible-like structures located laterally in the oral cavity; see Fig. 42.4c.

Müllerian mimicry: Mimicry in which several distasteful organisms share points of similarity.

Multiple resistance: Resistance to insecticides in more than one chemical class.

Mutualism: Association that benefits both partners.

Mycangium: Specialized cavity for transporting fungi or spores.

Mycetocytes: Host cells specialized to house endosymbiontic microorganisms.

Mycetome: Aggregation or organ containing mycetocytes.

Myrmecophile: An organism that lives with ants, as either predator or commensal (myrmecophilous).

Naiad: In hemimetabolous insects, any aquatic nymph.

Nasute: In termites, a type of soldier in which the frontal gland opens upon a median horn.

Natural control: Regulation of a species' population by natural factors.

Nauplioid larva: In parasitic Hymenoptera, a first-instar larva that resembles the nauplius larva of Crustacea.

Necrosis: Decay.

Nectariferous: Feeding on nectar.

Neoptera: Those insects possessing a neopterous flight mechanism.

Neopterous: Having a flight mechanism involving indirect musculature and the ability to flex the wings over the dorsum at rest.

Neoteny: Reaching sexual maturity during a larval stage (neotenic).

Niche: (ecological niche) The combination of all environmental requirements for a species existence. Also the ecological position of an organism, such as "aquatic predator in small streams."

Nocturnal: Active at night.

Node, nodus: In Hymenoptera, the knoblike segments at the base of the abdomen; in Odonata, a crossvein near the costal margin; see Fig. 21.2.

Nonpersistent transmission: Transmitting to only one or a few plants in succession or for no longer than minutes to hours.

Notal wing process: Anterior and posterior processes on the alinotum where the wings articulate.

Notum: Tergum; dorsal part of a thoracic segment.

Numerical response: An increase in predator populations as a reaction to increased prey density.

Nuptial flight: In Isoptera and Hymenoptera, the dispersal flight of sexual forms; a mating flight.

Nygma: (pl., nygmata) In Neuroptera, Trichoptera, and some Hymenoptera, a small sensory organ, sometimes present in the radial or median portion of the wing membrane.

Nymph: An immature hemimetabolous insect.

Obligate symbiont: Association necessary for host.

Oblongum: In Coleoptera, a closed cell in the median part of the wing.

Obtect: In pupae, having the appendages rigidly adherent to the body wall.

Occipital suture: An annular suture on the posterior part of the insect cranium, ending near the insertions of the mandibles.

Occiput: The posterior part of the head capsule, defined by an anterior occipital suture and a posterior postoccipital suture.

Ocellar triangle: In Diptera, the region defined by the three ocelli.

Ocellus: (pl., ocelli) The simple eye of an adult insect.

Oligolectic: In Hymenoptera, especially bees, visiting a restricted range of plants for pollen or nectar; see Polylectic.

Ommatidium: (pl., ommatidia) An individual unit of the compound eye.

Omnivorous: Having variable feeding preferences, including both animal and vegetable material; see Predaceous; Herbivorous.

Oncogenic: Causing cancer.

Onisciform: Having a body shaped like a sowbug, as in larvae of Byrrhidae.

Oocyst: An outpocketing of mosquito gut lamina containing a malarial parasite form that produces haploid sporozoites.

Ookinete: The motile gamete formed by fusion of micro- and macrogametes of malarial parasite that invades vector's gut.

Ootheca: A protective covering for an egg mass.

Operculum: A lid or cover, usually circular.

Opisthognathous: Having the mouthparts directed posteroventrally; see Prognathous; Hypognathous.

Oral vibrissae: In Diptera, a pair of stout bristles arising laterally near the oral region; see Fig. 42.15.

Osmeterium: (pl., osmeteria) Fleshy glandular structure that may be everted from the bodies of certain caterpillars to release an odoriferous defensive secretion.

Oviduct: One of the (usually) paired tubes that pass the eggs from the ovaries to the vagina.

Ovolarviparous: Producing eggs that hatch immediately after deposition, as in certain parasitic Diptera.

Ovoviviparous: Producing live offspring by the hatching of eggs retained in the mother's body.

Paleoptera: Those insects possessing a paleopterous flight mechanism.

Paleopterous: Having a flight mechanism involving direct musculature and lacking the ability to flex the wings over the back at rest.

Palpifer: Prominence on which maxillary palps attach.

Palpiger: Prominence on which labial palps attach.

Papilla: A small, fleshy, nipple-shaped projection.

Papilliform: Papilla-shaped, nipple-shaped.

Paraglossa: (pl., paraglossae) A paired lateral lobe of the labium, corresponding to the galea; see Fig. 27.2a,d.

Paramere: A paired copulatory appendage of the male.

Paraphyletic: A taxonomic group that does not include all the descendants of a common ancestor.

Paraproct: A lateral plate of the eleventh abdominal segment.

Parasitic castration: The induction by internal parasites of morphological traits of the opposite sex; stylopization.

Parthenogenesis: Reproduction without fertilization; see Arrhenotoky; Thelytoky.

Patagium: (pl., patagia) In Lepidoptera, one of the small lobes that overlie the forewings; see Fig. 44.1a.

Pathogen: Microorganism or virus that causes disease.

Pecten: A comb; in Hymenoptera, the rigid setae on the basal parts of the maxillae and labium; in Lepidoptera, a row of setae on the antennal scape; pectinate antenna, see Figs. 2.7k, 39.6f.

Pedicel: The second segment of the antenna; in Hymenoptera, the basal one or two segments of the abdomen.

Penis: The male intromittent organ.

Penultimate: Next to the last.

Periproct: The unsegmented part of the body surrounding the anus.

Pest resurgence: The rapid increase of pest populations to even higher than normal levels following control actions.

Petiole: In Hymenoptera, the slender, stalklike basal abdominal segment (petiolate, stalked).

Pharate: Referring to a developmental stage enclosed in the integument of the preceding instar, especially adults that are pharate in the pupal integument.

Pheromone: A substance that is secreted externally, functioning in communication among individuals of the same species; see Allomone; Kairomone.

Phoresy: An interrelationship in which one organism is transported by another.

Phragmata: (s., phragma) Transverse, platelike invaginations of the skeleton, usually increasing the area available for muscle attachment; homologous to the antecosta of a secondary segment.

Physiological time: The combination of heat or temperature needed for development.

Physogastry: Having a swollen abdomen (physogastric).

Phytophagous: Feeding on plants.

Phytotoxicity: Causing a toxic reaction in plants.

Planidium: In Hymenoptera, a type of primary larva with sclerotized, fusiform body and spinelike locomotory organs; see Fig. 46.5a.

Plastron: A respiratory organ in which special hydrofuge cuticle holds a layer of air through which gas exchange occurs.

Pleural apophysis: An invagination of the pleural ridge.

Pleural coxal process: A ventral process of the pleuron on which the coxa articulates.

Pleural ridge: An internal ridge that strengthens the pleuron.

Pleural suture: The suture running between the coxa and the pleural wing process and separating the episternum and epimeron.

Pleural wing process: A dorsal process of the pleuron on which the wing articulates.

Pleuron: The platelike lateral part of a segment, including the episternum and epimeron.

Pleurosternal suture: A suture between the pleuron and sternum.

Pleurum: Pleuron.

Plumose: Feathery, like a plume; see Fig. 2.7l.

Polyembryony: The production of more than one embryo from an egg.

Polyethism: In social insects, different behaviors among individuals of a colony according to caste or age, resulting in a division of labor.

Polygamous: Mating more than once; strictly, referring to males that mate with more than one female.

Polylectic: In Hymenoptera, especially bees, visiting a broad range of plants for pollen and nectar; see Oligolectic.

Polymorphism: Having more than one body form.

Polyphagous: Having broad feeding tolerances; in phytophagous insects, accepting many host plants.

Polyphyletic: Arising from more than one evolutionary lineage.

Polyploid: Having a multiple, greater than 2, of the haploid number of chromosomes.

Population dynamics: Changes in population size over time.

Populations: The set of individual organisms in a defined space, as a population of an insect species within a particular field.

Porrect: Extending forward horizontally.

Postabdomen: In Diptera, the modified, posterior abdominal segments, including the genitalia; see Preabdomen.

Postclypeus: Enlarged frontoclypeal region of sucking insects.

Postgena: The lower portion of the occiput.

Postgenal bridge: The median fusion of postgenal areas behind the mouthparts.

Postgenital segments: Abdominal segments beyond the ninth.

Postmentum: Basal part of the labium, proximal to the labial suture.

Postnotum: An intersegmental dorsal sclerite associated with the tergum of the preceding segment; see Acrotergite.

Postoccipital suture: A groove bordering the foramen magnum, running dorsally between the posterior tentorial pits.

Postocciput: The cranium posterior to the postoccipital suture.

Postphragma: The posterior phragma of a segment.

Postscutellum: The posterior-most dorsal sclerite of a segment, located just posterad of the scutellum.

Postvertical bristles: In Diptera, a pair of bristles arising behind the lateral ocelli; see Fig. 42.15.

Preabdomen: In Diptera, the unmodified, anterior portion of the abdomen; see Postabdomen.

Predaceous: Preying on other organisms, usually animals.

Pregenital segments: Abdominal segments 1 to 7.

Prehensile: Adapted for grasping.

Prementum: Sclerite of labium distal to labial suture and to which the palps attach.

Prephragma: The anterior phragma of a segment.

Prepupa: A quiescent, nonfeeding stage, passed during the last larval instar.

Prescutum: The anterior portion of the scutum, usually delimited by a suture.

Prestomal teeth: In Diptera, sclerotized, abrasive processes on the labellum around the oral opening.

Pretarsus: The terminal leg segment, including tarsal claws and medial structures such as arolium, pulvillus, or empodium.

Primary reproductives: In termites, the original pair that founded a colony.

Pro-: Before, anterior; prefix indicating the first part of a structure or the first member of a series (e.g., prothorax).

Proboscis: An extended, tubular mouth structure, as in many Diptera, Hymenoptera, Lepidoptera, Hemiptera; see Rostrum.

Procuticle: The cuticle minus the epicuticle.

Prognathous: Having the mouthparts directed anteriorly.

Progressive provisioning: In solitary bees and wasps, the practice of providing the larval food over a period of time.

Proleg: The fleshy, unsegmented legs, usually on the abdominal segments, of holometabolous larvae.

Propagative: Pathogens that multiply within their vectors.

Propneustic: Pertaining to a larval respiratory system in which only the anterior (mesothoracic) spiracles are functional.

Propodeum: In Hymenoptera, the anterior-most abdominal segment, fused with the metathorax; see Fig. 46.1.

Prostomium: The unsegmented part of the body in front of the mouth.

Prothorax: The first segment of the thorax.

Protocephalon: Part of the primitive head, including the prostomium and one to three true segments.

Protocerebrum: Part of the brain that innervates the compound eyes and ocelli.

Protozoa: Subphylum of single-celled eucaryotes in the kingdom Protista.

Proventriculus: The posterior part of the foregut.

Pseudoculus: In Protura, cephalic sensory organs; see Fig. 17.1b.

Pseudopod: (pseudopodia) Fleshy, unsegmented appendages on holometabolous larvae; prolegs.

Pseudotracheae: In Diptera, the minute grooves on the labellum through which liquid food is ingested.

Pterostigma: A thickening on the costal margin of the wing of many insects.

Pterothorax: The wing-bearing portion of the thorax, usually the mesothorax and metathorax.

Pterygota: Those insects that primitively bear wings (sometimes lost secondarily).

Ptilinal suture: In Diptera, a crescentic groove just above the antennal insertions; see Fig. 42.15.

Ptilinum: In Diptera, a membranous sac that is everted through the ptilinal suture of the teneral adult to force open the puparium.

Pubescent: Covered with short, densely set hairs.

Pulvillus: Paired, padlike structures between or beneath the tarsal claws.

Pupa: In holometabolous insects, the (usually) nonmotile instar during which transformation from larva to adult occurs.

Puparium: In Diptera, the pupal case formed from the strongly sclerotized cuticle of the last larval instar.

Quadrate: Square or nearly square.

Queen: In social insects, the primary female reproductive.

r-selected: Species that optimize exponential growth with minimal adaptation to a stable maximum population near carrying capacity of the environment. Also called r-strategists.

Raptorial: Adapted for grasping prey, as in the legs of Mantodea.

Rectum: The posterior part of the hindgut or proctodaeum.

Relative sample: Collection or set of counted individuals belonging to a designated group (e.g., a species) to determine population densities or other characteristics by using methods that show proportional changes in population density.

Relictual: Occupying a geographic or ecological range much more restricted than formerly.

Remigium: The stiff, anterior area of the wing.

Reproductive incompatibility: Inability to produce offspring; may be caused by microbial infection.

Reticulate: Covered with a meshwork of lines or ridges; netlike.

Retinaculum: In Collembola, the unsegmented appendage of the third abdominal segment; in Lepidoptera, the loop into which the frenulum fits; see Fig. 44.3a.

Rickettsiae: Intracellular, gram-negative bacterial parasites of animals.

Rostrum: An elongate, tubular projection of the head; cf. Hemiptera, Coleoptera.

Rudimentary: Undeveloped.

Salivarium: A cavity between the hypopharynx and labium into which the salivary duct opens.

Saltatorial: Adapted for jumping.

Sampling: The collection of a small number of individuals within a population to make inferences about the entire population.

Sanitation: As pest control method, the removal of weeds, crop residues, trash, or other sources of food and shelter to prevent pest infestation or breeding.

Saprophagous: Feeding on decaying organic matter (saprophytic).

Saprophytic: Saprophagous.

Scape: The basal segment of the antenna.

Scarabaeiform: Having the general form of a scarabaeid beetle larva; see Fig. 39.9d.

Schizont: Form of malarial parasite in liver cells that produces merozoites.

Sclerite: A hardened or sclerotized plate of the cuticle, or an area of hardened cuticle surrounded by sutures.

Sclerotization: The polymerization and cross-bonding of protein and chitin to produce the hardened insect exoskeleton.

Scopa: In Hymenoptera, the pollen-collecting apparatus on the hind legs, consisting of a brush of stiff setae.

Scutellum: The posterior division of the notum; in Hemiptera and Coleoptera, the triangular or shield-shaped posterior portion of the mesothorax.

Seasonal polymorphism: Change of body form in different seasons, as in Aphididae and certain other Homoptera.

Secondary pests: Species that reach damaging levels only after control has been used against a pest species.

Semidicondylic: Referring to mandibles of Archeognatha and Ephemeroptera, where the weakly developed dorsal articulation fits into a very shallow notch in the oral rim of the head capsule.

Semisocial behavior: Adults of same generation cooperate in brood care and some become workerlike.

Sensilium: In fleas, a sensory plate on the apex of the abdomen.

Sensillum: (pl., sensilla) A simple sense organ; one of the units of a compound sense organ.

Serpentine mines: In Lepidoptera and Diptera, winding larval feeding tunnels in leaves or stems of plants.

Serrate: Sawlike, with toothed margins, as in serrate antennae; see Figs. 2.7f, 39.6e.

Seta: (pl., setae) A hair or bristle.

Setaceous: Hairlike or bristlelike; see Fig. 2.7b.

Setose: Covered with setae.

Sex ratio organism: A microbe that affects the sex ratio of host offspring.

Sister group: The taxonomic group most closely related to another taxonomic group.

Sociogenesis: In social insects, development of a colony throughout its life.

Solitary: Occurring singly or in pairs, not socially; applied especially to wasps and bees.

Sperm precedence: Having sperm that fertilize the eggs of a female, regardless of multiple matings (see Odonata).

Spermatheca: In female insects, the saclike organ that receives and stores the sperm from the male.

Spermatophore: The capsule or covering that males of some insects secrete around the spermatozoa.

Spine: A rigid multicellular outgrowth of the integument.

Spinneret: An organ with which silk is produced.

Spiracles: The openings through which air enters an insect's tracheae.

Spittle insects: Immature Cercopidae, so-called because of the frothy secretion in which these insects live.

Spur: An articulated, multicellular outgrowth of the integument.

Squama: In some Diptera, a small, scalelike folding in the posterior basal wing margin.

Stadium: The interval between molts; see Instar.

Stellate: Star-shaped.

Sterile insect method: Pest control that uses the intentional sterilization of insects to prevent reproduction.

Sternal apophysis: An invagination of the thoracic sternum to which muscles attach.

Sternite: Any sclerite in the sternal region of a segment; a subdivision of the sternum.

Sternum: The ventral portion of a body segment, delimited from the pleura by sutures (sternal).

Stigma: (pl. stigmata) A thickened, usually pigmented region on the costal margin of the wings of many insects.

Sting: In some Hymenoptera, the modified ovipositor, which is adapted for injecting venom; see Fig. 46.2c,d.

Stipes: One of the basal sclerites of the maxilla.

Stomach poisons: Poisons that kill after being ingested.

Stomodaeum: The anterior portion of the alimentary canal; foregut.

Stria: In Coleoptera, the longitudinal, impressed lines usually visible on the elytra; any fine longitudinal line or furrow.

Striate: Sculptured with fine, parallel lines or furrows.

Stridulation: The production of sounds by rubbing one surface against another (stridulate, stridulatory).

Style: Any small, fine, needle-shaped or sharply pointed organ; in Diptera, the fine, dorsal appendage on the third antennal segment (Fig. 2.7j); in entognathous arthropods and Apterygota, the rudimentary abdominal appendages (stylus, styli; Figs. 18.1a, 19.1a).

Stylet: A small style; in Hemiptera, the flexible filaments forming the piercing mouthparts.

Styliform: Having the shape of a style or stylet.

Stylopization: Parasitization by Strepsiptera.

Sub-: Slightly less than, or about equal to, or just below.

Subalare: The posterior epipleurite.

Subcortical: The region between the bark and wood of trees and shrubs.

Subfamily: A taxonomic subdivision of a family.

Subgena: The cranium below the subgenal suture.

Subgenal suture: A groove above the bases of the mouthparts, running between the anterior and posterior tentorial pits.

Subgenital plate: A ventral sclerite covering the gonopore, as in Blattodea and some other insects; usually the eighth sternum in females and the ninth in males.

Subimago: In Ephemeroptera, the winged instar immediately preceding the reproductively mature adult.

Submarginal cells: In Hymenoptera, the cells just behind the marginal cell; see Fig. 46.18c.

Submarginal vein: Especially in Hymenoptera, a vein just behind the costal wing margin.

Submentum: Proximal sclerite of postmentum.

Suborder: A taxonomic category intermediate to order and family.

Subquadrate: Nearly square.

Subscutellum: In some Diptera, the infolded portion of the postscutellum; see Fig. 42.18.

Subsocial behavior: Parents care for young offspring for some period of time.

Superfamily: A taxonomic category intermediate to family and order.

Superorganism: A social colony of many individuals that exhibits features analogous to those of a single organism (reproduction, circulation of nutrients, excretion, defense, etc.).

Suture: A line, seam, or furrow in the cuticle; may indicate an articulation or internal ridge.

Symbiont (symbiote): One partner in a symbiosis.

Symbiosis: Close association or living together of two dissimilar organisms.

Synanthropic: Living in close association with humans.

Synergist: With insecticides, a chemical added to a toxicant to increase its toxicity greater than the additive toxicities of the synergist and toxicant separately.

Systemic: Taken up and distributed in the vascular system (of plants or animals).

Tagma: (pl., tagmata) A distinct region of the body; in insects the head, thorax, or abdomen.

Tagmosis: The evolutionary differentiation of the body into distinct regions, or tagmata.

Tarsomeres: Subsegments of the tarsus.

Tegmen: (pl., tegmina) The leathery forewing of Orthoptera, Blattodea, and other orthopteroid insects.

Tegula: A small articulated sclerite at the costal base of the wing, especially in Hymenoptera; see Fig. 46.1.

Teneral: The condition of an insect immediately after molting and before the cuticle has hardened.

Tentorial bridge: The median fusion of the tentorial arms inside the cranium.

Tentorial pit: The point of invagination of a tentorial arm.

Tentorium: The endoskeleton of the insect head.

Teratogenic: Causing defects in developing embryos.

Tergite: Any sclerite in the tergal region of a segment; a subdivision of the tergum.

Tergosternal: Referring to the tergum and the sternum, as in tergosternal muscles.

Tergum: The dorsal portion of a body segment (tergal).

Termitarium: The structure or nest inhabited by a termite colony.

Termitophile: An organism that lives with termites, as either predator or commensal (termitophilous).

Thelytoky: Parthenogenetic reproduction in which all the offspring are females.

Thigmotactic: Contact-loving, as in insects that inhabit crevices, cracks, or other tight quarters.

Thorax: The intermediate tagma of the insect body, bearing the legs and wings.

Tillage: Physical disturbance of the soil (e.g., plowing, disking).

Tolerance: Applied to plant resistance, minimizes loss of yield despite pest damage.

Tormogen: An epidermal cell that develops into the socket for a seta or hair.

Torsion: A twisting; in many male Diptera, the rotation of the genitalia 90 to 360° from their primitive position; see Circumversion.

Toxicity: Ability of a substance to sicken or kill.

Tracheae: The internal, cuticular air tubes of insects.

Tracheal gill: Gills supplied with tracheae; more loosely, any gill involved with respiration.

Tracheal system: The tracheae and tracheoles.

Tracheole: An extremely fine trachea.

Transmission efficiency (vector competence): Measure of how frequently an infectious vector successfully transmits a pathogen to a host.

Transovarial: Passing via the ovary (Transovarian).

Transverse: Intersecting the longitudinal axis at right angles.

Traumatic insemination: In Cimicoidea and Nabidae, the deposition of the sperm in the hemocoel through the abdominal body wall of the female.

Trichogen: An epidermal cell that develops into a seta or hair.

Tritocerebrum: Part of the brain that innervates the labrum; originally a postoral ventral ganglion.

Triungulin: The campodeiform first-instar larva of Strepsiptera and some parasitoid beetles; see Fig. 40.1b.

Trochanter: The leg segment between the coxa and femur.

Trochantin: A small, movable pleurite articulating anteriorly with the coxa.

Trophallaxis: In social insects the mutual exchange of food between colony members.

Trophic: Pertaining to eating, food, or nutrition.

Trophont: Free-living stage of a ciliated protozoan.

Trophozoites: Form of malarial parasite in the red blood cells of humans that produce merozoites.

Truncate: Having a square margin.

Trunk: The thorax and abdomen as a whole.

Trypanosomiasis: Disease of vertebrates caused by trypanosomes.

Tubercle: A small nipple-shaped protrusion (tuberculate).

Tymbal: In Homoptera, the membrane of the sound-producing organ.

Tympanal organ: An ear or auditory organ.

Tympanum: Any tightly stretched membrane, especially the membranes of the auditory organs.

Univoltine: Having a single generation per year.

Urogomphus: Any fixed or movable terminal abdominal process of insect larvae; applied especially to Coleoptera.

Urticate: To sting or nettle (urticating).

Valvifer: The base of an ovipositor appendage.

Valvula: One blade of the shaft of an ovipositor appendage.

Vasiform orifice: In Aleyrodidae, a dorsal glandular opening on the last abdominal segment; see Fig. 34.14d.

Vector: An animal or agency that transports and transmits something else, usually a microorganism; also applied to insects that carry pollen between flowering plants.

Vector specificity: Degree of exclusivity of vectors in transmitting a particular pathogen.

Vermiform: Worm-shaped.

Vertex: The top of the head, between the eyes and above the frons.

Vertical resistance: Resistance to pests or pathogens that is overcome by distinct populations (biotypes) of pests or pathogens.

Vesicle: A small sac or bladder.

Vestigial: Rudimentary, small; the remnants of a formerly functional organ.

Vibrissae: A group of bristles inserted around the oral fossa in many Diptera; see Fig. 42.15.

Viviparous: Bearing live young.

Zoonosis: Disease of man acquired from other animals via arthropod vectors.

Zoophagous: Feeding on material of animal origin.

REFERENCES CITED

Achtelig, M., and N. P. Kristensen. 1973. A re-examination of the relationships of the Raphidioptera (Insecta). Z. Zool. Syst. Evol. Forsch. **11**:268–274.

Ackery, P. R. 1984. *Milkweed Butterflies.* Br. Mus. Nat. Hist., London. 425 pp.

Adams, J. (ed.) 1992. *Insect Potpourri: Adventures in Entomology.* Sandhill Crane Press Gainesville, FL. 336 pp.

Afzelius, B. A., and R. Dallai. 1988. Spermatozoa of Megaloptera and Raphidioptera (Insecta, Neuropteroidea). J. Ultrastruct. Mol. Struct. Res. **101**:185–191.

Afzelius, B. A., and R. Dallai. 1994. Characteristics of the flagellar axoneme in Neuroptera, Coleoptera, and Strepsiptera. J. Morphol. **219**:15–20.

Alba-Tercedor, J., and A. Sanchez-Ortega. 1991. *Overview and Strategies of Ephemeroptera and Plecoptera.* Sandhill Crane Press, Gainesville, FL. 588 pp.

Albrecht, F. O. 1953. *The Anatomy of the Migratory Locust.* Athlone Press, London. 118 pp.

Alcock, J. 1994. Postinsemination associations between males and females in insects: the mate-guarding hypothesis. Annu. Rev. Entomol. **39**:1–21.

Aldrich, J. R. 1988. Chemical ecology of the Heteroptera. Annu. Rev. Entomol. **33**:211–238.

Alexander, R. D. 1957. Sound production and associated behavior in insects. Ohio J. Sci. **57**:101–113.

Alexander, R. D. 1962. Evolutionary change in cricket acoustical communication. Evolution **16**:443–467.

Alexander, R. D., and W. L. Brown. 1963. Mating behavior and the origin of insect wings. Occas. Pap. Mus. Zool. Univ. Michigan No. 628:1–19.

Alexander, R. D., and T. E. Moore. 1962. The evolutionary relationships of 17-year and 13-year cicadas, and three new species (Homoptera, Cicadidae, *Magicicada*). Misc. Publ. Mus. Zool. Univ. Michigan **121**:1–59.

Alexander, R. D., T. E. Moore, and R. E. Woodruff. 1963. The evolutionary differentiation of stridulatory signals in beetles (Insecta: Coleoptera). Anim. Behav. **11**:111–115.

Alford, D. V. 1975. *Bumblebees.* Davis-Poynter, London. 352 pp.

Allan, J.D. 1995. *Stream Ecology. Structure and Function of Running Waters.* Chapman & Hall, London. 388 pp.

Alonso-Mejia, A., and L. P. Brower. 1994. From model to mimic: age-dependent unpalatability in monarch butterflies. Experientia (Basel) **50**:176–181.

Amos, W. H. 1967. *The Life of the Pond.* McGraw-Hill, New York. 232 pp.

Ananthakrishnan, T. N. 1984. *Bioecology of Thrips.* Indira Publishing House, Oak Park, MI. 233 pp.

Ananthakrishnan, T. N. 1990. *Reproductive Biology of Thrips.* Indira Publishing House, Oak Park, MI. 158 pp.

Ananthakrishnan, T. N. 1993. Bionomics of thrips. Annu. Rev. Entomol. **38**:71–92.

Ananthakrishnan, T. N., and R. Gopichandran. 1993. *Chemical Ecology in Thrips—Host Plant Interactions.* International Science Publishers, New York. 125 pp.

Ananthakrishnan, T. N., and A. Raman. 1989. *Thrips and Gall Dynamics.* E.J. Brill, Leiden. 120 pp.

Ananthakrishnan, T. N., and A. Raman. 1993. *Chemical Ecology of Phytophagous Insects.* International Science Publishers, New York. 332 pp.

Andersen, N. M., and T. A. Weir. 1994. *Austrobates rivularis*, gen. et sp. nov., a freshwater relative of *Halobates* Eschscholtz (Hemiptera: Gerridae), with a new perspective on the evolution of sea skaters. Invertebr. Taxon. **8**:1–15.

Anderson, D. T. 1973. *Embryology and Phylogeny in Annelids and Arthropods.* Pergamon Press, Oxford. 495 pp.

Anderson, N. H., and J. R. Sedell. 1979. Detritus processing by macroinvertebrates in stream ecosystems. Annu. Rev. Entomol. **24**:351–377.

Andersson, M. 1984. The evolution of eusociality. Annu. Rev. Entomol. **15**:165–189.

Ando, H. 1962. *The Comparative Embryology of Odonata, with Special Reference to a Relic Dragonfly, Epiphlebia superstes* Selys. Society for the Promotion of Science, Tokyo, Japan. 205 pp.

Ando, H. 1982. *Biology of the Notoptera.* Kashiyo Insatsu, Nagano, Japan. 194 pp.

Arnett, R. H., Jr. 1960. *The Beetles of the United States (A Manual for Identification).* Catholic University Press, Washington, DC. 1112 pp. (Reprinted, 1968, American Entomological Institute, Ann Arbor, MI.)

Arnett, R. H., Jr. 1985. *American Insects: A Handbook of the Insects of America North of Mexico.* Van Nostrand Reinhold, New York. 850 pp.

Arnol'di, L. V., and N. J. Vandenberg. 1992. *Mesozoic Coleoptera.* Smithsonian Institution Libraries, Washington, DC. 284 pp.

Askew, R. R. 1971. *Parasitic Insects.* American Elsevier, New York. 316 pp.

Askew, R. R. 1984. The biology of gall wasps. In *Biology of Gall Insects,* T. N. Ananthakrishnan (ed.). Edward Arnold, London. pp. 223–271.

Aspock, H. 1986. The Raphidioptera of the world: a review of present knowledge. In *Recent Research in Neuropterology. Proc. 2nd Int. Symp. Neuropterology, Hamburg,* J. Gepp (ed.). Privately published, Graz. pp. 15–29.

Aspock, H., and U. Aspock. 1975. The present state of knowledge on the Raphidioptera of America (Insecta: Neuropteroidea). Pol. Pismo Entomol. **45**:537–546.

Atkins, M. D. 1978. *Insects in Perspective.* Macmillan, New York. 513 pp.

Atkins, M. D. 1980. *Introduction to Insect Behavior.* Macmillan, New York. 237 pp.

Atkinson, T. H., P. G. Koehler, and R. S. Patterson. 1991. *Catalog and Atlas of the Cockroaches (Dictyoptera) of North America North of Mexico.* Misc. Publ. Entomol. Soc. Am. no. 78. Entomological Society of America, Lanham, MD. 85 pp.

Baccetti, B. 1987. *Evolutionary Biology of Orthopteroid Insects.* Ellis Horwood series in entomology and acarology. Halsted Press, Chichester, UK. 612 pp.

Bailey, S. F. 1957. The thrips of California, Part 1: Suborder Terebrantia. Bull. California Insect Survey **4**:143–220.

Bailey, W. J., and D. C. Rentz. 1990. *The Tettigoniidae: Biology, Systematics, and Evolution.* Crawford House Press, Bathurst, UK. 395 pp.

Balduf, W. V. 1935. *The Bionomics of Entomophagous Coleoptera.* John S. Swift Co., St. Louis, MO. 220 pp.

Balduf, W. V. 1939. *The Bionomics of Entomophagous Insects, part II.* John S. Swift Co., St. Louis, MO. 384 pp.

Ball, E. D., E. R. Tinkham, R. Flock, and C. T. Vorheis. 1942. The grasshoppers and other Orthoptera of Arizona. Arizona Agric. Exp. Stn. Tech. Bull. **93**:257–373.

Banks, N. 1927. Revision of nearctic Myrmeleontidae. Bull. Mus. Comp. Zool., Harvard Univ. **68**:1–84.

Bao, N., and W. R. Robinson. 1990. Morphology and mating configuration of genitalia of the oriental cockroach *Blatta orientalis* (Blattodea; Blattidae). Proc. Entomol. Soc. Washington **92**:416–421.

Barbosa, P., and D. K. Letourneau. 1988. *Novel Aspects of Insect-Plant Interactions.* John Wiley & Sons, New York. 362 pp.

Barker, S. C. 1994. Phylogeny and classification, origins, and evolution of host associations of lice. Int. J. Parasitol. **24**:1285–1291.

Barlet, J. 1989. Study of the musculature and nervous system of the japygid (Insecta, Apterygotes, Diplura). Bull. Inst. R. Sci. Nat. Belg. Entomol. **59**:23–42.

Barnhart, C. S. 1961. The internal anatomy of the silverfish *Ctenolepisma campbelli* and *Lepisma saccharinum* (Thysanura: Lepismatidae). Ann. Entomol. Soc. Am. **54**:177–196.

Baron, S. 1972. *The Desert Locust.* Eyre Methuen, London. 228 pp.

Barr, T. C., Jr. 1967. Observations on the ecology of caves. Am. Nat. **101**:475–491.

Barr, T. C., Jr. 1974. Cave insects. Insect World Digest (September/October):11–21.

Barth, R. 1954. Untersuchungen an den Tarsaldrusen von *Embolyntha batesi* MacLachlan, 1877 (Embioidea). Zool. Jahrb. **74**:172–188.

Bateman, M. A. 1972. The ecology of fruit flies. Annu. Rev. Entomol. **17**:493–518.

Bates, H. W. 1862. Contributions to an insect fauna of the Amazon Valley. Lepidoptera: Heliconidae. Trans. Linn. Soc. Zool. **23**:495–566.

Baumann, P., L. Baumann, C.-Y. Lai, D. Rouhbakhsh, N. A. Moran, and M. A. Clark. 1995. Genetics, physiology, and evolutionary relationships of the genus *Buchnera:* intracellular symbionts of aphids. Annu. Rev. Microbiol. **49**:55–94.

Baumann, R. W. 1987. Order Plecoptera. In *Immature Insects,* vol. 1, F. W. Stehr (ed.). Kendall/Hunt, Dubuque IA. pp. 186–195.

Baust, J. G. 1972a. Influence of low temperature acclimation in cold hardiness in *Pterostichus brevicornis.* J. Insect Physiol. **18**:1935–1947.

Baust, J. G. 1972b. Mechanisms of insect freezing protection: *Pterostichus brevicornis.* Nature **236**:219–221.

Baust, J. G. 1976. Temperature buffering in an Arctic microhabitat. Annu. Entomol. Soc. Am. **69**:117–119.

Baust, J. G., and L. K. Miller. 1970. Seasonal variations in glycerol content and its influence in cold hardiness in the Alaskan carabid beetle, *Pterostichus brevicornis.* J. Insect Physiol. **16**:979–990.

Bay, E. C. 1974. Predator-prey relationships among aquatic insects. Annu. Rev. Entomol. **19**:441–454.

Baz, A. 1991. Observations on the biology and ecology of Psocoptera found in different kinds of leaf litter in East-Central Spain. Pedobiologia **35**:89–100.

Baz, A. 1992. Phenology and crypsis as possible determinants of habitat selection in populations of *Hemineura bigoti* from central Spain (Psocoptera: Elipsocidae). Entomol. Gen. **17**:293–298.

Beardsley, J. W., Jr., and R. H. Gonzalez. 1975. The biology and ecology of armored scales. Annu. Rev. Entomol. **20**:47–73.

Beattie, A. J. 1985. *The Evolutionary Ecology of Ant-Plant Mutualisms.* Cambridge Studies in Ecology. Cambridge University Press, Cambridge. 182 pp.

Beck, S. D. 1965. Resistance of plants to insects. Annu. Rev. Entomol. **10**:207–232.

Beckage, N. E., S. N. Thompson, and B. A. Federici.

1993. *Parasites and Pathogens of Insects, vol. 2. Pathogens.* Academic Press, New York. 294 pp.

Bedford, G. O. 1978. Biology and ecology of the Phasmatodea. Annu. Rev. Entomol. **23**:125–130.

Begon, M., M. Mortimer, and D. J. Thompson. 1996. *Population Ecology: A Unified Study of Animals and Plants.* 3rd ed. Blackwell Scientific, Oxford. 247 pp.

Beier, M. 1955. Ordnung: Saltatoptera m. (Saltatoria Latrielle 1817). Bronn's Kl. Ordn. Tierreichs (5)(3)**6**:34–304.

Ben-Dov, Y. 1993. *A Systematic Catalogue of the Soft Scale Insects of the World (Homoptera:Coccoidea:Coccidae): With Data on Geographical Distribution, Host Plants, Biology, and Economic Importance.* Flora & Fauna Handbook no. 9. Sandhill Crane Press, Gainesville, FL. 536 pp.

Benedetti, R. 1973. Notes on the biology of *Neomachilis halophila* on a California sandy beach (Thysanura: Machilidae). Pan-Pac. Entomol. **49**:246–249.

Berenbaum, M. 1995. Putting on airs. Am. Entomol. **41**:199–200.

Bernard, E. C., and S. L. Tuxen. 1987. Class and Order Protura. In *Immature Insects,* vol. 1, F. W. Stehr (ed.). Kendall/Hunt, Dubuque, IA. pp. 47–54.

Bernays, E. A. (ed.). 1989. *Insect-Plant Interactions.* CRC Press, Boca Raton, FL. vol. 1, 164 pp.; vol. 2, 199 pp.

Bernays, E. A., and R. F. Chapman. 1994. *Host-Plant Selection by Phytophagous Insects.* Contemporary Topics in Entomology, vol. 2. Chapman & Hall, New York. 312 pp.

Berner, L. 1950. *The Mayflies of Florida.* University of Florida Press, Gainesville. 267 pp.

Berner, L. 1959. A tabular summary of the biology of North American mayfly nymphs (Ephemeroptera). Bull. Florida State Mus. Biol. Ser. **4**:1–58.

Berner, L., and M. L. Pescador. 1988. *The Mayflies of Florida.* University Presses of Florida, Gainesville. 415 pp.

Berry, R. J. 1990. Industrial melanism and peppered moths *Biston-betularia* L. Biol. J. Linn. Soc. **39**:301–322.

Billen, J. 1992. *Biology and Evolution of Social Insects.* Leuven University Press, Leuven, Belgium. 390 pp.

Billingsley, P. F. 1990. The midgut ultrastructure of hematophagous insects. Annu. Rev. Entomol. **35**:219–248.

Binnington, K. and A. Retnakaran. 1991. *Physiology of the Insect Epidermis.* Commonwealth Scientific and Industrial Research Organization, Canberra, Australia. 344 pp.

Birch, C., G. M. Poppy, and T. C. Baker. 1990. Scents and eversible scent structure of male moths. Annu. Rev. Entomol. **35**:25 58.

Birch, M. C., and A. Hefetz. 1987. Extrusible organs in male moths and their role in courtship behavior. Bull. Entomol. Soc. Am. **33**:222–229.

Birch, M. C., and D. L. Wood. 1975. Mutual inhibition of the attractant pheromone response by two species of *Ips* (Coleoptera: Scolytidae). J. Chem. Ecol. **1**:101–113.

Birket-Smith, S. J. P. 1974. On the abdominal morphology of Thysanura (Archeognatha and Thysanura s. str.). Entomol. Scand. Suppl. **6**:1–67.

Bishop, J. A., and L. M. Cook. 1975. Moths, melanism and clean air. Sci. Am. **232**(1):90–99.

Blackman, R. 1974. *Aphids.* Ginn and Company, London. 175 pp.

Blackman, R. L., and V. F. Eastop. 1994. *Aphids on the World's Trees: An Identification and Information Guide.* CAB International, Wallingford, UK. 1024 pp.

Blatchley, W. D. 1920. *Orthoptera of Northeastern America.* Nature Publishing Company, Indianapolis, IN. 784 pp.

Blatchley, W. D. 1926. *Heteroptera or True Bugs of Eastern North America, with Special Refrence to the Faunas of Indiana and Florida.* Nature Publishing Company, Indianapolis, IN. 1116 pp.

Blest, A. D. 1957. The function of eyespot patterns in the Lepidoptera. Behaviour **11**:209–256.

Blum, M. S. 1987. Biosynthesis of arthropod exocrine compounds. Annu. Rev. Entomol. **32**:381–413.

Bohart, R. M. 1941. A revision of the Strepsiptera with special reference to the species of North America. Univ. California Publ. Entomol. **7**:91–160.

Bohart, R. M., and A. S. Menke. 1976. *Sphecid Wasps of the World.* University of California Press, Berkeley. 695 pp.

Bolton, B. 1995. *Identification Guide to the Ant Genera of the World.* Harvard University Press, Cambridge, MA. 224 pp.

Bonhag, P. F. 1949. The thoracic mechanism of the adult horsefly. Cornell Univ. Agric. Exp. Stn. Mem. **285**:1–39.

Borden, J. H. 1982. Aggregation pheromones. In *Bark Beetles in North American Conifers. A System for the Study of Evolutionary Biology,* J. B. Mitton and K. B. Sturgeon (eds.). University of Texas Press, Austin. pp. 74–139.

Borden, J. H. 1984. Semiochemical-mediated aggregation and dispersion in the Coleoptera. In *Insect Communication,* T. Lewis (ed.). Academic Press, London. pp. 123–149.

Borror, D.J., C. A. Triplehorn, and N. F. Johnson. 1989. *An Introduction to the Study of Insects.* 6th ed. Saunders College Publishing, Philadephia. 875 pp.

Boucek, Z. 1988. An overview of the higher classification of the Chalcidoidea (parasitic Hymenoptera). In

Advances in Parasitic Hymenoptera Research, V.-K. Gupta (ed.). E. J. Brill, Leiden. pp. 11–23.

Böving, A. G., and F. C. Craighead. 1930. An illustrated synopsis of the principal larval forms of the order Coleoptera. Entomol. Am. **11:**1–351.

Bowles, D. E., and R. T. Allen. 1992. Life histories of six species of caddisflies (Trichoptera) in an Ozark stream, USA. J. Kansas Entomol. Soc. **65:**174–184.

Brackenbury, J. 1992. *Insects in Flight.* Blanford, London. 192 pp.

Bradley, T. J. 1987. Physiology of osmoregulation in mosquitoes. Annu. Rev. Entomol. **32:**439–462.

Brakefield, P. M. 1987. Industrial melanism: do we have the answers? Trends Ecol. and Evol. **2:**117–122.

Brakefield, P. M. 1990. A decline of melanism in the peppered moth *Biston betularia* in The Netherlands. Biol. J. Linn. Soc. **39:**327–334.

Brakefield, P. M., and D. R. Lees. 1987. Melanism in *Adalia* ladybirds and declining air pollution in Birmingham, England [UK]. Heredity **59:**273–278.

Breed, M. D., and R. E. Page. 1989. *The Genetics of Social Evolution.* Westview Studies in Insect Biology. Westview Press, Boulder, CO. 213 pp.

Breed, M. D., C. D. Michener, and H. E. Evans (eds.). 1982. *The Biology of Social Insects.* Westview Press, Boulder, CO. 419 pp.

Breznak, J. A., and A. Brune. 1994. Role of microorganisms in the digestion of lignocellulose by termites. Annu. Rev. Entomol. **39:**453–487.

Briceno, R. D., and W. G. Eberhard. 1994. The functional morphology of male cerci and associated characters in 13 species of tropical earwigs (Dermaptera: Forficulidae, Labiidae, Carcinophoridae, Pygidicranidae). Smithsonian Contrib. Zool. **555:**1–63.

Bright, D. E. 1993. *The Weevils of Canada and Alaska.* The Insects and Arachnids of Canada, part 21. Research Branch, Agriculture Canada, Ottawa.

Brinck, P. 1957. Reproductive systems and mating in Ephemeroptera. Opusc. Entomol. **22:**1–37.

Brindle, A. 1987. Dermaptera. In *Immature Insects,* vol. 1, F. W. Stehr (ed.). Kendall/Hunt, Dubuque, IA. pp. 171–178.

Britt, N. W. 1962. Biology of two species of Lake Erie mayflies, *Ephoron album* (Say) and *Ephemera simulans* Walker. Bull. Ohio Biol. Survey **1:**1–70.

Brittain, J. E. 1982. Biology of mayflies. Annu. Rev. Entomol. **27:**119–148.

Britton, E. B. 1970. Coleoptera (beetles). In *The Insects of Australia,* Commonwealth Scientific and Industrial Research Organization (ed.). Melbourne University Press, Carlton. pp. 495–621.

Britton, E. B. 1974. Coleoptera (beetles). In *The Insects of Australia, Supplement 1974,* CSIRO (ed.). Melbourne University Press, Carlton. pp. 62–89.

Britton, W. E., et al. 1923. The Hemiptera or sucking insects of Connecticut. Guide to the insects of Connecticut. Connecticut State Geol. Nat. Hist. Survey **4:**1–807.

Brodsky, A. K. 1994. *The Evolution of Insect Flight.* Oxford University Press, Oxford. 229 pp.

Brooks, A. R. 1958. Acridoidea of southern Alberta, Saskatchewan, and Manitoba (Orthoptera). Can. Entomol. Suppl. **9:**1–92.

Brooks, A. R., and L. A. Kelton. 1967. Aquatic and semiaquatic Heteroptera of Alberta, Saskatchewan and Manitoba (Hemiptera). Mem. Entomol. Soc. Can. **51:**1–92.

Brothers, D. J. 1975. Phylogeny and classification of the aculeate Hymenoptera with special reference to Mutillidae. Univ. Kansas Sci. Bull. **50:**483–648.

Brower, L. P., and J. V. Z. Brower. 1964. Birds, butterflies, and plant poisons: a study in ecological chemistry. Zoologica **49:**137–159.

Brower, L. P., W. N. Ryerson, L. L. Coinger, and S. C. Glazier. 1968. Ecological chemistry and the palatability spectrum. Science **161:**1349–1351.

Brower, L. P., J. Alcock, and J. V. Z. Brower. 1971. Avian feeding behavior and the selective advantage of incipient mimicry. In *Ecological Genetics and Evolution,* R. Creed (ed.). Appleton-Century-Crofts, New York. pp. 261–274.

Brown, H. P. 1987. Biology of riffle beetles. Annu. Rev. Entomol. **32:**253–273.

Brown, K. S., Jr. 1981. The biology of *Heliconius* and related genera. Annu. Rev. Entomol. **26:**427–456.

Brown, W. L., Jr., T. Eisner, and R. H. Whittaker. 1970. Allomones and kairomones: transspecific chemical messengers. BioScience **20:**21.

Brues, C. T., A. L. Melander, and F. M. Carpenter. 1954. Classification of insects. Bull. Mus. Comp. Zool. Harvard Univ. **73:**1–917.

Brushwein, J. R. 1987. Bionomics of *Lomamayia hamata* (Neuroptera: Berothidae). Ann. Entomol. Soc. Am. **80:**671–679.

Buchmann, S. L. 1987. The ecology of oil flowers and their bees. Annu. Rev. Ecol. Syst. **18:**343–369.

Buchner, P. 1965. *Endosymbiosis of Animals with Plant Microorganisms,* rev. ed. Interscience Publishers, New York. 909 pp.

Buckley, R. C. 1987. Interactions involving plants, Homoptera, and ants. Annu. Rev. Ecol. Syst. **18:**111–136.

Burks, B. D. 1953. The mayflies, or Ephemeroptera, of Illinois. Illinois Nat. Hist. Survey Bull. **26:**1–216.

Burn, A. J., T. H. Coaker, and P. C. Jepson (eds.). 1987. *Integrated Pest Management.* Academic Press, London. 474 pp.

Busvine, J. R. 1948. The "head" and "body" races of *Pediculus humanus* L. Parasitology **39:**1–16.

Butcher, J. W., R. Snider, and R. J. Snider. 1971. Bio-ecology of edaphic Collembola and Acarina. Annu. Rev. Entomol. **16**:249–288.

Buxton, P. A. 1947. *The Louse. An Account of the Lice Which Infest Man, Their Medical Importance and Control.* 2nd ed. Edward Arnold, London. 164 pp.

Byers, G. W. 1963. The life history of *Panorpa nuptialis* (Mecoptera: Panorpidae). Ann. Entomol. Soc. Am. **56**:142–149.

Byers, G. W. 1965. Families and genera of Mecoptera. Proc. l2th Int. Congr. Entomol. **1964**:123.

Byers, G. W. 1971. Ecological distribution and structural adaptation in the classification of Mecoptera. Proc. l3th Int. Congr. Entomol. 1968, **1**:486.

Byers, G. W. 1989. Homologies in wing venation of primitive Diptera and Mecoptera. Proc. Entomol. Soc. Washington **91**:497–501.

Byers, G. W., and R. Thornhill. 1983. Biology of the Mecoptera. Annu. Rev. Entomol. **28**:203–228.

Byers, J. A. 1991. Pheromones and chemical ecology of locusts. Biol. Rev. Cambridge Philos. Soc. **66**:347–378.

Byers, J. R., and C. F. Hinks. 1973. The surface sculpturing of the integument of lepidopterous larvae and its adaptive significance. Can. J. Zool. **51**:1171–1179.

Byrne, D. N., and T. S. Bellows. 1991. Whitefly biology. Annu. Rev. Entomol. **36**:431–457.

Caltagirone, L. E. 1981. Landmark examples in classical biological control. Annu. Rev. Entomol. **26**:213–232.

Caltagirone, L. E., and R. L. Doutt. 1989. The history of the vedalia beetle importation to California and its impact on the development of biological control. Annu. Rev. Entomol. **34**:1–16.

Cameron, E. 1961. *The Cockroach (Periplaneta americana, L.), An Introduction to Entomology for Students of Science and Medicine.* William Heinemann, London. 111 pp.

Campbell, B. C. 1989. On the role of microbial symbiotes in herbivorous insects. In *Insect-Plant Interactions,* vol. 1, E. A. Bernays (ed.). CRC Press, Boca Raton, FL. pp. 1–44.

Campbell, B. C., J. D. Steffen-Campbell, and R. J. Gill. 1994. Evolutionary origin of whiteflies (Hemiptera: Sternorrhyncha: Aleyrodidae) inferred from 18s rDNA sequences. Insect Mol. Biol. **3**:73–88.

Campbell, B. C., J. D. Steffen-Campbell, J. T. Sorensen, and R. J. Gill. 1995. Paraphyly of Homoptera and Auchenorrhyncha inferred from 18s rDNA nucleotide sequences. Syst. Entomol. **20**:175–194.

Campos-Ortega, J. A., and V. Hartenstein. 1985. *The Embryonic Development of Drosophila melanogaster.* Springer-Verlag, Berlin. 227 pp.

Cantwell, G. E. (ed.). 1974. *Insect Diseases,* vols. 1 and 2. Marcel Dekker, New York. 595 pp.

Cappuccino, N. and P. W. Price (eds.). 1995. *Population Dynamics: New Approaches and Synthesis.* Academic Press, New York. 429 pp.

Carayon, J. 1966. Traumatic insemination and the paragenital system. In *Monograph of Cimicidae,* R. L. Usinger (ed.). Entomological Society of America, Baltimore. pp. 81–166.

Carey, J. R. 1993. *Applied Demography for Biologists.* Oxford University Press, Oxford. 206 pp.

Carlberg, U. 1987. Chemical defense in Phasmida vs. Mantodea (Insecta). Zool. Anz. **218**:369–373.

Carpenter, F. M. 1931. Revision of nearctic Mecoptera. Bull. Mus. Comp. Zool. Harvard Univ. **72**:205–277.

Carpenter, F. M. 1936. Revision of the nearctic Raphidioidea (recent and fossil). Proc. Am. Acad. Arts Sci. **71**:89–157.

Carpenter, F. M. 1940. Revision of the nearctic Hemerobiidae, Berothidae, Sisyridae, Polystoechotidae, and Dilaridae (Neuroptera). Proc. Am. Acad. Arts Sci. **74**:193–280.

Carpenter, F. M. 1953. The evolution of insects. Am. Sci. **41**:256–270.

Carpenter, F. M. 1970. Adaptations among paleozoic insects. Proc. North Am. Paleontol. Conv., 1969 **2**:1236–1251.

Carpenter, F. M. 1977. Geological history and evolution of the insects. Proc. 15th Int. Congr. Entomol. Washington (1976):63–70.

Carpenter, F. M. 1992. Part R. Arthropoda 4. Vols. 3 and 4: Superclass Hexapoda. In *Treatise on Invertebrate Paleontology.,* R. L. Kaesler (ed.). Geological Society of America, Boulder, CO. and Lawrence, KS. pp. 1–655.

Carroll, C. R., and D. H. Janzen. 1973. Ecology of foraging by ants. Annu. Rev. Ecol. Syst. **4**:231–257.

Carson, R. 1962. *Silent Spring.* Houghton Mifflin, Boston. 368 pp.

Carter, W. 1973. *Insects in Relation to Plant Disease.* 2nd ed. John Wiley & Sons, New York. 759 pp.

Carver, M., G. F. Ross, and T. E. Woodward. 1991. Hemiptera. In *The Insects of Australia, vol. 1,* Commonwealth Scientific and Industrial Research Organization (ed.). Melbourne University Press, Carlton. pp. 429–509.

Cassagnau, P. 1990. Four hundred million years old Hexapoda, the Collembola: 2. Biogeography and ecology. Annee Biol. **29**:39–70.

Catts, E. P., and M. L. Goff. 1992. Forensic entomology in criminal investigations. Annu. Rev. Entomol. **37**:253–272.

Caudell, A. N. 1920. Zoraptera not an apterous order. Proc. Entomol. Soc. Washington **22**:84–97.

Caveney, S. 1986. The phylogenetic significance of ommatidium structure in the compound eyes of polyphagan beetles. Can. J. Zool. **64**:1787–1819.

Chandler, H. P. 1956. Megaloptera. In *Aquatic Insects*

of California, R. L. Usinger (ed.). University of California Press, Berkeley. pp. 229–233.

Chapco, W., R. A. Kelln, and D. A. Mcfadyen. 1994. Mitochondrial DNA variation in North American melanopline grasshoppers. Heredity **72:**1–9.

Chapela, I. H., S. A. Rehner, T. R. Schultz, and U. G. Mueller. 1994. Evolutionary history of the symbiosis between fungus-growing ants and their fungi. Science **266:**691–694.

Chapman, R. F. 1982. *The Insects. Structure and Function.* 3rd ed. Harvard University Press, Cambridge, MA. 919 pp.

Chapman, R. F. 1991. General anatomy and function. In *The Insects of Australia,* vol. 1, Commonwealth Scientific and Industrial Research Organization (ed.). Melbourne University Press, Melbourne. pp. 33–67.

Chapman, R. F., and G. de Boer (eds.) 1995. *Regulatory Mechanisms in Insect Feeding.* Chapman & Hall, New York. 398 pp.

Chapman, R. F., and A. Joern. 1990. *Biology of Grasshoppers.* John Wiley & Sons, New York. 563 pp.

Cheetham, T. B. 1988. *Male Genitalia and Phylogeny of the Pulicoidea (Siphonaptera).* Theses zoologicae, vol. 8. Koeltz Scientific Books, Koenigstein, Germany. 224 pp.

Cheng, L. (ed.). 1976. *Marine Insects.* North-Holland Publishing Company, Amsterdam. 581 pp.

Cheng, L. 1985. Biology of *Halobates* (Heteroptera: Gerridae). Annu. Rev. Entomol. **30:**111–135.

Cherry, R. H. 1987. History of sericulture. Bull. Entomol. Soc. Am. **33:**83–84.

Choe, J. C. 1989. *Zorotypus gurneyi,* new species, from Panama and redescription of *Z. barberi* Gurney (Zoraptera: Zorotypidae). Ann. Entomol. Soc. Am. **82:**149–155.

Choe, J. C. 1994a. Sexual selection and mating system in *Zorotypus gurneyi* Choe (Insecta: Zoraptera): I. Dominance hierarchy and mating success. Behav. Ecol. Sociobiol. **34:**87–93.

Choe, J. C. 1994b. Sexual selection and mating system in *Zorotypus gurneyi* Choe (Insecta: Zoraptera): II. Determinants and dynamics of dominance. Behav. Ecol. Sociobiol. **34:**233–237.

Choe, J. C. 1994c. Communal nesting and subsociality in a webspinner, *Anisembia texana* (Insecta: Embiidina: Anisembiidae). Anim. Behav. **47:**971–973.

Chopard, L. 1938. La biologie des Orthopteres. Encycl. Entomol. (A) **20:**1–541.

Chopard, L. 1949a. Ordre des Dictyopteres Leach, 1818 (= Blattaeformia Werner, 1906; = Oothecaria Karny, 1915). Traite Zool. **9:**355–407.

Chopard, L. 1949b. Ordre de Cheleutopteres. Traite Zool. **9:**594–616.

Chopard, L. 1949c. Ordre de Orthopteres. Traite Zool. **9:**617–722.

Chopard, L. 1949d. Ordre de Dermapteres. Traite Zool. **9:**745–770.

Christiansen, K. 1964. Bionomics of Collembola. Annu. Rev. Entomol. **9:**147–178.

Christiansen, K. 1992. Springtails. Kansas School Nat. **39:**1–16.

Christiansen, K., and P. Bellinger. 1988. Marine littoral Collembola of North and Central America. Bull. Mar. Sci. **42:**215–245.

Christiansen, K. A., and P. F. Bellinger. 1992. *Collembola.* Insects of Hawaii, vol. 15. University of Hawaii Press, Honolulu. 445 pp.

Claassen, P. W. 1931. Plecoptera nymphs of America (North of Mexico). Thomas Say Found. Publ. **3:**1–199.

Claridge, M. F. 1985a. Acoustic signals in the Homoptera: behavior, taxonomy, and evolution. Annu. Rev. Entomol. **30:**297–317.

Claridge, M. F. 1985b. Acoustic behavior of leafhoppers and planthoppers: species problems and speciation. In *The Leafhoppers and Planthoppers,* L. R. Nault and J. G. Rodriquez (eds.). John Wiley & Sons, New York. pp. 103–125.

Claridge, M. F., and J. Den Hollander. 1982. Virulence to rice cultivars and selection for virulence in populations of the brown planthopper, *Nilaparvata lugens.* Entomol. Exp. Appl. **32:**213–221.

Clark, E. J. 1948. Studies in the ecology of British grasshoppers. Trans. R. Entomol. Soc. London **99:**173–222.

Clarke, C. A., F. M. M. Clarke, and H. C. Dawkins. 1990. *Biston betularia* (the peppered moth) in West Kirby, Wirral, 1959–1989: updating the decline in f. *carbonaria.* Biol. J. Linn. Soc. **39:**323–326.

Clausen, C. P. 1940. *Entomophagous Insects.* McGraw-Hill, New York. 688 pp.

Clausen, C. P. 1976. Phoresy among entomophagous insects. Annu. Rev. Entomol. **21:**343–368.

Clay, T. 1970. The Amblycera (Phthiraptera: Insecta). Bull. Br. Mus. Nat. Hist. Entomol. **25:**73–98.

Cleveland, L. R., S. R. Hall, E. P. Saunders, and J. Collier. 1934. The wood-feeding roach, *Cryptocercus,* its protozoa, and the symbiosis between protozoa and roach. Mem. Am. Acad. Sci. **17:**185–342.

Cloudsley-Thompson, J. L. 1975. Adaptations of arthropoda to arid environments. Annu. Rev. Entomol. **20:**261–283.

Cobben, R. H. 1968. *Evolutionary Trends in Heteroptera. Part I: Eggs, Architecture of the Shell, Gross Embryology, and Eclosion.* Centre for Agricultural Publishing and Documentation, Wageningen, Netherlands. 475 pp.

Cobben, R. H. 1978. Evolutionary trends in Heteroptera. Part II. Mouthpart structures and feeding strategies. Meded. Landb.-Hoogesch. Wageningen **78:**1–407.

Cochran, D. G. 1985. Nitrogen excretion in cockroaches. Annu. Rev. Entomol. **30**:29–49.

Cohen, A. C. 1995. Extra-oral digestion in predaceous terrestrial Arthropoda. Annu. Rev. Entomol. **40**:85–103.

Cole, F. R. 1969. *The Flies of Western North America.* University of California Press, Berkeley. 693 pp.

Colless, D. H., and D. K. McAlpine. 1991. Diptera (flies). In *The Insects of Australia,* vol. 2, Commonwealth Scientific and Industrial Research Organization (ed.). Melbourne University Press, Carlton. pp. 717–786.

Common, I. F. B. 1970. Lepidoptera (moths and butterflies). In *The Insects of Australia,* Commonwealth Scientific and Industrial Research Organization (ed.). Melbourne University Press, Carlton. pp. 765–866.

Common, I. F. B. 1975. Evolution and classification of the Lepidoptera. Annu. Rev. Entomol. **20**:183–203.

Comstock, J. H. 1918. *The Wings of Insects.* Comstock Publishing, Ithaca, NY. 430 pp.

Cook, J. M. 1993. Sex determination in the Hymenoptera: a review of models and evidence. Heredity **71**:421–435.

Cook, M. A., and M. J. Scoble. 1992. Tympanal organs of geometrid moths: a review of their morphology, function, and systematic importance. Syst. Entomol. **17**:219–232.

Coope, G. R. 1979. Late Cenozoic fossil Coleoptera: evolution, biogeography, and ecology. Annu. Rev. Entomol. **10**:247–268.

Cooper, K. W. 1972. A southern California *Boreus, B. notoperates* n. sp. I. Comparative morphology and systematics (Mecoptera: Boreidae). Psyche **79**:269–283.

Cooper, K. W. 1974. Sexual biology, chromosomes, development, life histories and parasites of *Boreus,* especially *B. notoperates,* a southern California *Boreus.* II. (Mecoptera: Boreidae). Psyche **81**:84–120.

Cope, O. B. 1940. The morphology of *Psocus confraternus* Banks. Microentomology **5**:91–115.

Corbet, P. S. 1962. *A Biology of Dragonflies.* Quadrangle Books, Chicago. 247 pp.

Corbet, P. S. 1980. Biology of Odonata. Annu. Rev. Entomol. **25**:189–217.

Corbet, P. S., C. Longfield, and N. W. Moore. 1960. *Dragonflies.* Collins, London. 260 pp.

Corbet, P. S., C. Longfield, and N. W. Moore. 1985. *Dragonflies.* Collins, London. 260 pp.

Cordero, A., and P. L. Miller. 1992. Sperm transfer, displacement and precedence in *Ischnura graellsii* (Odonata: Coenagrionidae). Behav. Ecol. Sociobiol. **30**:261–267.

Cornwell, P. B. 1968. *The Cockroach.* Vol. I. *A Laboratory Insect and an Industrial Pest.* Hutchinson Publishing Group, London. 391 pp.

Costa, C., S. A. Vanin, and S. A. Casari-Chen. 1988. *Larvas de Coleoptera do Brasil.* Universidad de Sao Paulo, Sao Paulo. 282 pp.

Cox, J. M. 1987. *Pseudococcidae (Insecta: Hemiptera).* Fauna of New Zealand, no. 11. Science Information Publishing Centre, DSIR, Wellington, N.Z. 230 pp.

Craighead, F. C. 1950. Insect Enemies of Eastern Forests. U.S. Dept. Agric. Misc. Publ. **657**:1–679.

Crampton, G. C., et al. 1942. Guide to the insects of Connecticut. Part VI. The Diptera or true flies of Connecticut. First fascicle. Bull. Connecticut Geol. Nat. Hist. Survey **64**:1–509.

Crane, E. 1990. *Bees and Beekeeping. Science, Practice, and World Resources.* Comstock Publishing Associates, Ithaca, NY. 614 pp.

Crane, J. 1952. A comparative study of innate defensive behavior in Trinidad mantids (Orthoptera, Mantodea). Zoologica **37**:239–293.

Creighton, W. S. 1950. The ants of North America. Bull. Mus. Comp. Zool., Harvard Univ. **104**:1–585.

Crespi, B. 1987. Fighting behavior in male tubuliferan thrips. In *Population Structure, Genetics and Taxonomy of Aphids and Thysanoptera; International Symposia, Smolenice, Czechoslovakia, September 9–14, 1985.* J. Holman (ed.). Spb Academic Publishing, The Hague. pp. 424–425.

Crespi, B. J. 1992. Behavioural ecology of Australian gall thrips (Insecta, Thysanoptera). J. Nat. Hist. **26**:769–809.

Croft, B. A. 1990. *Arthropod Biological Control Agents and Pesticides. Environmental Science and Technology.* John Wiley & Sons, New York. 723 pp.

Cromartie, R. I. T. 1959. Insect pigments. Annu. Rev. Entomol. **4**:59–76.

Crowson, R. A. 1955. *The Natural Classification of the Families of Coleoptera.* Nathaniel Lloyd, London. 187 pp.

Crowson, R. A. 1960. The phylogeny of the Coleoptera. Annu. Rev. Entomol. **5**:111–134.

Crowson, R. A. 1970. *Classification and Biology.* Heinemann Educational Books, London. 350 pp.

Crowson, R. A. 1981. *The Biology of the Coleoptera.* Academic Press, London. 802 pp.

Cruz, Y. P. 1981. A sterile defender morph in a polyembryonic hymenopterous parasite. Nature **294**:446–447.

Commonwealth Scientific and Industrial Research Organization (CSIRO). 1991. *The Insects of Australia,* 2 vols. 2nd ed. Melbourne University Press, Carlton. 1137 pp.

Cummins, K. W. 1973. Trophic relations of aquatic insects. Annu. Rev. Entomol. **18**:183–206.

Curran, C. H. 1934. *The Families and Genera of North American Diptera.* Privately published, New York. 512 pp.

Dadant and Sons (eds.). 1975. *The Hive and the Honey Bee.* Dadant and Sons,, Hamilton, IL. 740 pp.

Dafni, A. 1984. Mimicry and deception in pollination. Annu. Rev. Ecol. Syst. **15**:259–278.

Daly, H. V. 1963. Close-packed and fibrillar muscles of the Hymenoptera. Ann. Entomol. Soc. Am. **56**:295–306.

Daly, H. V. 1964. Skeleto-muscular morphogenesis of the thorax and wings of the honey bee, *Apis mellifera* (Hymenoptera: Apidae). Univ. Calif. Publ. Entomol. **39**:1–77.

Danforth, B. D. 1989. The evolution of hymenopteran wings: the importance of size. J. Zool. (London) **218**:247–276.

David, D. T., and M. C. Birch. 1989. Pheromones and insect behavior. In *Insect Pheromones in Plant Protection,* A. R. Jutsum and R. F. S. Gordon (eds.). John Wiley & Sons Ltd., London. pp. 17–35.

Davis, C. 1940. Family classification of the order Embioptera. Ann. Entomol. Soc. Am. **33**:677–682.

Davis, D. R. 1967. A revision of the moths of the subfamily Prodoxinae (Lepidoptera: Incurvariidae). U.S. Natl. Mus., Bull. **255**:1–170.

Davis, D. R. 1986. A new family of monotrysian moths from austral South America (Lepidoptera: Palaephatidae), with a phylogenetic review of the monotrysia. Smithsonian Contrib. Zool. **282**:1–39.

Day, W. C. 1956. Ephemeroptera. In *Aquatic Insects of California,* R. L. Usinger (ed.). University of California Press, Berkeley. pp. 79–105.

DeBach, P. (ed.). 1964. *Biological Control of Insect Pests and Weeds.* Chapman & Hall, London. 844 pp.

DeBach, P. 1974. *Biological Control by Natural Enemies.* Cambridge University Press, New York. 320 pp.

DeBach, P., and D. Rosen. 1991. *Biological Control by Natural Enemies.* 2nd ed. Cambridge University Press, Cambridge. 440 pp.

DeBach, P., and R. A. Sundby. 1963. Competitive displacement between ecological homologues. Hilgardia **34**:105–166.

DeCoursey, R. M. 1971. Keys to the families and subfamilies of the nymphs of North American Hemiptera-Heteroptera. Proc. Entomol. Soc. Washington **73**:413–428.

Decu, V., M. Gruia, S. L. Keffer, and S. M. Sarbu. 1994. Stygobiotic waterscorpion, *Nepa anophthalma,* n. sp. (Heteroptera: Nepidae), from a sulfurous cave in Romania. Ann. Entomol. Soc. Am. **87**:755–761.

DeFoliart, G. R. 1989. The human use of insects as food and as animal feed. Bull. Entomol. Soc. Am. **35**(1):22–35.

Delamare-Deboutteville, C. 1948. Observations sur l'écologie et l'éthologie des Zoraptères. La question de leur vie social et de leurs prétendus rapports avec les termites. Rev. Entomol. **19**:347–352.

Delany, M. J. 1957. Life histories in the Thysanura. Acta Zool. Cracov. **2**:61–90.

Delany, M. J. 1959. The life histories and ecology of two species of *Petrobius* Leach, *P. brevistylis* and *P. maritimus.* Trans. R. Soc. Edinburgh **63**:501–533.

Delcomyn, F. 1985. Factors regulating insect walking. Annu. Rev. Entomol. **30**:239–256.

DeLong, D. M. 1971. The bionomics of leafhoppers. Annu. Rev. Entomol. **16**:179–210.

Demerec, M. (ed.). 1950. *Biology of Drosophila.* John Wiley & Sons, New York. 632 pp.

Denholm, I., and M. W. Rowland. 1992. Tactics for managing pesticide resistance in arthropods: theory and practice. Annu. Rev. Entomol. **37**:91–112.

Denning, D. G. 1956. Trichoptera. In *Aquatic Insects of California,* R. L. Usinger (ed.). University of California Press, Berkeley. pp. 237–270.

Dennis, R. L. H. (ed.). 1992. *The Ecology of Butterflies in Britain.* Oxford University Press, Oxford. 354 pp.

Dent, D. 1991. *Insect Pest Management.* CAB International, Oxon, UK. 604 pp.

Desalle, R., and D. A. Grimaldi. 1991. Morphological and molecular systematics of the Drosophilidae. Annu. Rev. Ecol. Syst. **22**:447–476.

Desalle, R., J. Gatesy, W. Wheeler, and D. Grimaldi. 1992. DNA sequences from a fossil termite in Oligo-Miocene amber and their phylogenetic implications. Science **257**:1933–1936.

Desender, K. 1994. *Carabid Beetles: Ecology and Evolution.* Series Entomologica 51. Kluwer Academic Publishers, Dordrecht, Netherlands. 474 pp.

Dethier, V. G. 1947. The response of hymenopterous parasites to chemical stimulation of the ovipositor. J. Exp. Zool. **105**:199–208.

Dethier, V. G. 1954. Evolution of feeding preferences in phytophagous insects. Evolution **8**:33–54.

Dethier, V. G. 1992. *Crickets and Katydids, Concerts and Solos.* Harvard University Press, Cambridge, MA. 140 pp.

Dethier, V. G., L. Barton-Browne, and C. N. Smith. 1960. The designation of chemicals in terms of the responses they elicit from insects. J. Econ. Entomol. **53**:134–136.

Dettner, K. 1987. Chemosystematics and evolution of beetle chemical defenses. Annu. Rev. Entomol. **32**:17–48.

Dettner, K., and C. Liepert. 1994. Chemical mimicry and camouflage. Annu. Rev. Entomol. **39**:129–154.

Devonshire, A. L., and L. M. Field. 1991. Gene amplification and insecticidal resistance. Annu. Rev. Entomol. **36**:1–23.

DeVries, P. J. 1990. Enhancement of symbioses between

butterfly caterpillars and ants by vibrational communication. Science **248:**1104–1106.

DeVries, P. J. 1992. Singing caterpillars, ants and symbioses. Sci. Am. **249:**76–82.

Dewalt, R. E., and K. W. Stewart. 1995. Life histories of stoneflies (Plecoptera) in the Rio Conejos of southern Colorado. Great Basin Nat. **55:**1–18.

Dillon, E. S., and L. S. Dillon. 1961. *A Manual of Common Beetles of Eastern North America.* Row, Peterson & Company, Evanston, IL. 884 pp.

Dirsh, V. M. 1975. *Classification of the Acridomorphoid Insects.* E. W. Classey, Faringdon, Oxon., UK. 171 pp.

Dixon, A. F. G. 1973, *Biology of Aphids,* Edward Arnold Publishers, London. 58 pp.

Dixon, A. F. G. 1985. *Aphid Ecology.* Blackie, Glasgow. 157 pp.

Dolling, W. R. 1991. *The Hemiptera.* Natural History Museum Publications, Oxford University Press, London. 274 pp.

Douglas, A. E. 1989. Mycetocyte symbiosis in insects. Biol. Rev. Cambridge Philos. Soc. **64:**409–434.

Douglas, A. E. 1994. *Symbiotic Interactions.* Oxford University Press, New York. 148 pp.

Downes, B. J., and J. Jordan. 1993. Effects of stone topography on abundance of net-building caddisfly larvae and arthropod diversity in an upland stream. Hydrobiologia **252:**163–174.

Downes, J. A. 1969. The swarming and mating flight of Diptera. Annu. Rev. Entomol. **14:**271–298.

Downes, J. A. 1971. The ecology of blood-sucking Diptera: An evolutionary perspective. In *Ecology and Physiology of Parasites,* A. M. Fallis (ed.). University of Toronto Press, Toronto. pp. 232–258.

Doyen, J. T. 1966. The skeletal anatomy of *Tenebrio molitor* (Coleoptera: Tenebrionidae). Misc. Publ. Entomol. Soc. Am. **5:**103–150.

Dressler, R. L. 1982. Biology of orchid bees. Annu. Rev. Ecol. Syst. **13:**373–394.

Dugdale, J. S. 1974. Female genital configuration in the classification of Lepidoptera. N. Z. J. Zool. **1:**127–146.

Dunn, A. M. 1969. *Veterinary Helminthology.* Lea & Febiger, Philadelphia. 302 pp.

Duporte, E. M. 1960. Evolution of cranial structures in adult Coleoptera. Can. J. Zool. **38:**655–675.

Durden, L. A., and G. G. Musser. 1994. The sucking lice (Insecta, Anoplura) of the world: a taxonomic checklist with records of mammalian hosts and geographical distributions. Bull. Am. Mus. Nat. Hist. **218:**1–90.

Eaton, J. L. 1988. *Lepidopteran anatomy.* John Wiley & Sons, New York. 257 pp.

Ebeling, W. 1968. Termites: Identification, biology, and control of termites attacking buildings. Calif. Agric. Exp. Stn. Ext. Serv. Man. **38:**1–68.

Eberhard, M. J. W. 1975. The evolution of social behavior by kin selection. Q. Rev. Biol. **50:**1–34.

Eberhard, W. G. 1979. The function of horns in *Podischnus agenor* (Dynastinae) and other beetles. In *Sexual Selection and Reproductive Competition in Insects,* M. S. Blum and N. A. Blum (eds.). Academic Press, New York. pp. 231–258.

Eberhard, W. G. 1982. Beetle horn dimorphism: making the best of a bad lot. Am. Nat. **119:**420–426.

Eberle, M. W., and D. L. McLean. 1983. Observation of symbiote migration in human body lice with scanning and transmission microscopy. Can. J. Microbiol. **29:**755–762.

Edgerly, J. S. 1988. Maternal behavior of a webspinner (order Embiidina): mother-nymph associations. Ecol. Entomol. **13:**263–272.

Edmunds, G. F., Jr. 1959. Ephemeroptera. In *Freshwater Biology,* W. T. Edmondson (ed.). John Wiley & Sons, New York. pp. 908–916.

Edmunds, G. F. 1972. Biogeography and evolution of Ephemeroptera. Annu. Rev. Entomol. **17:**21–42.

Edmunds, G. F., Jr., and R. K. Allen. 1987. Order Ephemeroptera. In *Immature Insects,* vol. 1, F. W. Stehr (ed.). Kendall/Hunt, Dubuque, IA. pp. 75–94.

Edmunds, G. F., Jr. , and W. P. McCafferty. 1988. The mayfly subimago. Annu. Rev. Entomol. **33:**509–529.

Edmunds, G. F., Jr. , and J. R. Traver. 1954. The flight mechanics and evolution of the wings of Ephemeroptera, with notes on the archetype insect wing. J. Washington Acad. Sci. **44:**390–400.

Edmunds, G. F., Jr., K. Allen, and W. L. Peters. 1963. An annotated key to the nymphs of the families and subfamilies of mayflies (Ephemeroptera). Univ. Utah Biol. Ser. **13:**1–55.

Edmunds, G. F., Jr., S. L. Jensen, and L. Berner. 1976. *The Mayflies of North and Central America.* University of Minnesota Press, Minneapolis. 330 pp.

Edmunds, M., and D. Dudgeon. 1991. Cryptic behavior in the oriental leaf mantis *Sinomantis denticulata* Beier (Dictyoptera, Mantodea). Entomol. Mon. Mag. **127:**45–48.

Edney, E. B. 1974. Desert Arthropods. In *Desert Biology, vol. 2,* G. W. Brown, Jr. (ed.). Academic Press, New York. pp. 311–384.

Edwards, J. S. 1987. Arthropods of alpine aeolian ecosystems. Annu. Rev. Entomol. **32:**163–179.

Edwards, J. S. 1992. Adhesive function of coxal vesicles during ecdysis in *Petrobius brevistylis* Carpenter (Archaeognatha, Machilidae). Int. J. Insect Morphol. Embryol. **21:**369–371.

Ehrlich, P. R. 1958. The comparative morphology, phylogeny and higher classification of the butterflies

(Lepidoptera: Papilionoidea). Univ. Kansas Sci. Bull. **39**:305–370.

Ehrlich, P. R. 1961. Intrinsic barriers to dispersal in checkerspot butterfly. Science **134**:108–109.

Ehrlich, P. R. 1970. Coevolution and the biology of communities. In *Biochemical Coevolution,* K. L. Chambers (ed.). Oregon State University Press, Corvallis. pp. 1–11.

Ehrlich, P. R., and A. H. Ehrlich. 1961. *How to Know the Butterflies.* Wm. C. Brown, Dubuque, IA. 262 pp.

Ehrlich, P. R., and P. Raven. 1967. Butterflies and plants. Sci. Am. **216**:104–113.

Ehrlich, P. R., R. R. White, M. C. Singer, S. W. McKechnie, and L. E. Gilbert. 1975. Checkerspot butterflies: a historical perspective. Science **188**:221–228.

Eickwort, G. C., and H. J. Ginsberg. 1980. Foraging and mating behavior in Apoidea. Annu. Rev. Entomol. **25**:421–446.

Eisner, T. 1960. Defense mechanisms of arthropods. II. The chemical and mechanical weapons of an earwig. Psyche **67**:62–70.

Ekblom, T. 1926. Morphological and biological studies of the Swedish families of Hemiptera Heteroptera. Parts I and II. Zool. Bidr. **10**:31–180, **12**:113–150.

Elias, S. A. 1994. *Quaternary Insects and Their Environments.* Smithsonian Institution Press, Washington, DC. 284 pp.

Elliott, J. M. 1968. The daily activity patterns of mayfly nymphs (Ephemeroptera). J. Zool. **155**:201–221.

Elzinga, R. J. 1987. *Fundamentals of Entomology.* 3rd ed. Prentice-Hall, Englewood Cliffs, NJ. 456 pp.

Emerson, A. E. 1952. The biogeography of termites. Bull. Am. Mus. Nat. Hist. **99**:217–225.

Emerson, A. E. 1955. Geographical origins and dispersions of termite genera. Fieldiana, Zool. **37**:465–521.

Emerson, A. E. 1967. Cretaceous insects from Labrador. 3. A new genus and species of termite (Isoptera: Hodotermitidae). Psyche **74**:276–289.

Emerson, K. C. 1956. Mallophaga (chewing lice) occurring on the domestic chicken. J. Kansas Entomol. Soc. **35**:63–79.

Engels, W., and A. Buschinger. 1990. *Social Insects: An Evolutionary Approach to Castes and Reproduction.* Springer-Verlag, Berlin. 264 pp.

Ennos, A. R., and R. J. Wootton. 1989. Functional wing morphology and aerodynamics of *Panorpa germanica* (Insecta: Mecoptera). J. Exp. Biol. **143**:267–284.

Eriksen, C. H. 1966. Ecological significance of respiration and substrate for burrowing Ephemeroptera. Can. J. Zool. **46**:93–103.

Essig, E. O. 1930. *A History of Entomology.* Macmillan, New York. 1029 pp.

Evans, D. L., and J. O. Schmidt (eds.) 1990. *Insect Defenses. Adaptive Mechanisms and Strategies of Prey and Predators.* State University of New York Press, Albany. 482 pp.

Evans, H. E. 1984. *Insect Biology: A Textbook of Entomology.* Addison-Wesley, Reading, MA. 436 pp.

Evans, H. E. 1985. *The Pleasures of Entomology.* Smithsonian Institution Press, Washington, DC. 238 pp.

Evans, H. E., and M. J. W. Eberhard. 1970. *The Wasps.* University of Michigan Press, Ann Arbor. 265 pp.

Evans, J. W. 1963. The phylogeny of the Homoptera. Annu. Rev. Entomol. **8**:77–94.

Evans, J. W. 1982. A review of the present knowledge of the family Peloridiidae and new genera and species from New Zealand and New Caledonia (Hemiptera: Insecta). Rec. Aust. Mus. **34**:381–406.

Evans, M. E. G. 1961. On the muscular and reproductive systems of *Atomaria ruficornis* (Marshall) (Coleoptera, Cryptophagidae). Trans. R. Soc. Edinburgh **64**:297–399.

Evans, M. E. G. 1975. *The Life of Beetles.* George Allen & Unwin, Ltd., London. 232 pp.

Ewing, H. E. 1940. The Protura of North America. Annu. Entomol. Soc. Am. **33**:495–551.

Ewing, H. E., and I. Fox. 1943. The fleas of North America. U. S. Dept. Agric. Misc. Publ. **500**:1–128.

Farrow, R. A. 1990. Flight and migration in acridoids. In *Biology of Grasshoppers,* R. F. Chapman and A. Joern (eds.). John Wiley & Sons, Somerset, NJ. pp. 227–314.

Felt, E. P. 1940. *Plant Galls and Gall Makers.* Comstock Publishing Associates, Ithaca, NY. 364 pp.

Fenton, M. B., and K. D. Roeder. 1974. The microtymbals of some Arctiidae. J. Lepid. Soc. **28**:205–211.

Ferrar, P. 1988. A guide to the breeding habits and immature stages of the Diptera Cyclorrhapha. Entomograph **8**:1–907.

Ferris, C. D., and F. M. Brown. 1981. *Butterflies of the Rocky Mountain States.* University of Oklahoma Press, Norman. 461 pp.

Ferris, G. F. 1931. The louse of elephants, *Haematomyzus elephantis.* Parasitology **23**:112–127.

Ferris, G. F. 1937–1955. *Atlas of the Scale Insects of North America.* Stanford University Press, Palo Alto, CA.

Ferris, G. F. 1940. The morphology of *Plega signata* (Hagen) (Neuroptera, Mantispidae). Microentomology **5**:33–56.

Ferris, G. F. 1951. The sucking lice. Mem. Pac. Coast Entomol. Soc. **1**:1–320.

Ferris, G. F., and P. Pennebaker. 1939. The morphology of *Agulla adnixa* (Hagen) (Neuroptera, Raphidiidae). Microentomology **5**:33–56.

Field, S. A., and A. D. Austin. 1994. Anatomy and mechanics of the telescopic ovipositor system of

Scelio Latreille (Hymenoptera: Scelionidae) and related genera. Int. J. Insect Morphol. Embryol. **23**:135–158.

Fishelson, L. 1985. *Orthoptera, Acridoidea.* Fauna Palaestina. Insecta, 3. Israel Academy of Sciences and Humanities, Jerusalem. 229 pp.

Fisk, F. W. 1987. Order Blattodea. In *Immature Insects,* vol. 1, F. W. Stehr (ed.). Kendall/Hunt, Dubuque, IA. pp. 120–131.

Fletcher, D. J. C., and K. G. Ross. 1985. Regulation of reproduction in eusocial Hymenoptera. Annu. Rev. Entomol. **30**:319–344.

Flint, M. L. (ed.). 1981. *Integrated Pest Management for Alfalfa Hay.* University of California, Agricultural Sciences Publications, Richmond. 96 pp.

Flower, J. W. 1964. On the origin of flight in insects. J. Insect Physiol. **10**:81–88.

Foote, R. H., F. L. Blanc, and A. L. Norrbom. 1993. *Handbook of the Fruit Flies (Diptera: Tephritidae) of America North of Mexico.* Comstock Publishing Associates, Ithaca, NY. 571 pp.

Forbes, W. T. M. 1923–1960. Lepidoptera of New York and neighboring states. Mem. Cornell Univ. Agric. Exp. Stn. **68**:1–729; **274**:1–263; **329**:1–433; **371**:1–188.

Ford, E. B. 1971. *Ecological Genetics.* 3rd ed. Chapman & Hall, London.

Ford, N. 1926. On the behavior of *Grylloblatta.* Can. Entomol. **58**:66–70.

Fowler, H. G., and J. C. A. Pinto. 1989. Life history of the "Northern" mole cricket, *Neocurtilla hexadactyla* (Perty) (Orthoptera, Gryllotalpidae), in southeastern Brazil. Rev. Bras. Entomol. **33**:143–148.

Fox, H. M., and G. Vevers. 1960. *The Nature of Animal Colours.* Macmillan, New York. 246 pp.

Fracker, S. B. 1930. The classification of lepidopterous larvae. Contrib. Entomol. Lab., Univ. Illinois. **43**:1–161.

François, J., R. Dallai, and W. Y. Yin. 1992. Cephalic anatomy of *Sinentomon erythranum* Yin (Protura: Sinentomidae). Int. J. Insect Morphol. Embryol. **21**:199–213.

Fraser, F. C. 1957. *A Reclassification of the Order Odonata.* Royal Zoological Society of New South Wales, Sydney. 133 pp.

Free, J. B. 1987. *Pheromones of Social Bees.* Comstock Publishing Associates, Ithaca, NY. 218 pp.

Free, J. B. 1993. *Insect Pollination of Crops.* 2nd ed. Academic Press, London. 768 pp.

Freilich, J. E. 1991. Movement patterns and ecology of *Pteronarcys* nymphs (Plecoptera): Observations of marked individuals in a Rocky Mountain stream. Freshwater Biol. **25**:379–394.

Fremling, C. R. 1960. Biology of a large mayfly, *Hexa-genia bilineata* (Say), of the upper Mississippi River. Iowa State Univ. Agric. Home Econ. Exp. Stn. Res. Bull. **482**:841–852.

Frisbie, R. E., K. El-Zik, and L. T. Wilson (eds.). 1989. *Integrated Pest Management Systems and Cotton Production.* John Wiley & Sons, New York. 437 pp.

Frisch, K. von. 1967. *The Dance Language and Orientation of Bees.* Harvard University Press, Cambridge, MA. 566 pp.

Frisch, K. von. 1971. *Bees, Their Vision, Chemical Senses, and Language.* Cornell University Press, Ithaca, NY. 157 pp.

Frisch, K. von. 1974. Decoding the language of the bee. Science **185**:663–668.

Frison, T. H. 1935. The stoneflies, or Plecoptera, of Illinois. Illinois Nat. Hist. Survey Bull. **20**:281–471.

Frison, T. H. 1942. Studies on North American Plecoptera with special reference to the fauna of Illinois. Illinois Nat. Hist. Survey Bull. **22**:235–355.

Froeschner, R. C. 1954. The grasshoppers and other Orthoptera of Iowa. Iowa State Coll. J. Sci. **29**:163–354.

Froeschner, R. C. 1960. Cydnidae of the Western Hemisphere. Proc. U.S. Natl. Mus. **111**:337–680.

Fulton, B. B. 1924. Some habits of earwigs. Ann. Entomol. Soc. Am. **17**:357–367.

Furman, D. P., and E. P. Catts. 1961. *Manual of Medical Entomology.* National Press Books, Palo Alto, CA. 162 pp.

Fuzeau-Braesch, S. 1972. Pigments and color changes. Annu. Rev. Entomol. **17**:403–424.

Gagne, R. J. 1989. *The Plant-Feeding Gall Midges of North America.* Comstock Publishing Associates, Ithaca, NY. 356 pp.

Galil, J., and G. Neeman. 1977. Pollen transfer and pollination in the common fig (*Ficus carica* L.). New Phytol. **79**:163–171.

Garman, P. 1927. The Odonata or dragonflies of Connecticut. Guide to the Insects of Connecticut. Connecticut Geol. Nat. Hist. Survey **5**:1–331.

Gaufin, A. R., A. V. Nebeker, and J. Sessions. 1966. The stoneflies (Plecoptera) of Utah. Univ. Utah Biol. Ser. **14**:1–93.

Gaugler, R, and H. K. Kaya (eds.) 1990. *Entomopathogenic Nematodes in Biological Control.* CRC Press, Boca Raton, FL. 365 pp.

Gauld, I., and B. Bolton. 1988. *The Hymenoptera.* Oxford University Press, Oxford. 332 pp.

Georghiou, G. P., and A. Lagunes-Tejeda. 1991. *The Occurrence of Resistance to Pesticides in Arthropods.* Food and Agriculture Organization of the United Nations, Rome. 318 pp.

Giberson, D. J., and R. J. Mackay. 1991. Life history and distribution of mayflies (Ephemeroptera) in some acid streams in south central Ontario, Canada. Can. J. Zool. **69:**899–910.

Gilbert, C. 1994. Form and function of stemmata in larvae of holometabolous insects. Annu. Rev. Entomol. **39:**323–349.

Gilbert, L. E., and M. C. Singer. 1975. Butterfly ecology. Annu. Rev. Ecol. Syst. **6:**365–397.

Giles, E. T. 1963. The comparative external morphology and affinities of the Dermaptera. Trans. R. Entomol. Soc. London **115:**95–164.

Giles, E. T. 1974. The relationship between Hemimerina and the other Dermaptera: a case for reinstating Hemimerina within the Dermaptera, based on numerical procedure. Trans. R. Entomol. Soc. London **126:**189–206.

Gillett, J. D. 1971. *Mosquitoes.* Weidenfeld and Nicolson, London. 274 pp.

Gillett, J. D. 1972. *The Mosquito: Its Life, Activities and Impact on Human Affairs.* Doubleday, Garden City, NY. 358 pp.

Gillott, C. 1995. *Entomology.* 2nd ed. Plenum Publishing, New York. 744 pp.

Glick, P. A. 1939. The distribution of insects, spiders, and mites in the air. U.S. Dept. Agric., Tech. Bull. **673:**1–150.

Gloyd, L. K., and M. Wright. 1959. Odonata. In *Freshwater Biology,* W. T. Edmondson (ed.). John Wiley & Sons, New York. pp. 917–940.

Godfray, H. C. J. 1994. *Parasitoids, Behavioral and Evolutionary Ecology.* Monographs in Behavior and Ecology. Princeton University Press, Princeton, NJ. 473 pp.

Goldsmith, M. R., and A. S. Wilkins (eds.). 1995. *Molecular Model Systems in the Lepidoptera.* Cambridge University Press, Cambridge. 542 pp.

Goodman, L. J., and R. C. Fisher. 1991. *The Behaviour and Physiology of Bees.* CAB International, Wallingford, Oxon, UK. 362 pp.

Gordon, H. T. 1961. Nutritional factors in insect resistance to chemicals. Annu. Rev. Entomol. **6:**27–54.

Gorham, J. R. (ed.). 1991. *Insect and Mite Pests in Food. An Illustrated Key.* U. S. Departments of Agriculture and Health and Human Services, Agriculture Handbook No. 655, 2 vols. Washington, DC. pp. 1–310, 311–767.

Gould, J. L. 1975. Honey bee recruitment: the dance-language controversy. Science **189:**685–693.

Goulet, H., and J. T. Huber. 1993. *Hymenoptera of the World: An Identification Guide to Families.* Research Branch, Agriculture Canada, Publication 1894/E. 608 pp.

Graham, S. A., and F. B. Knight. 1965. *Principles of Forest Entomology,* 4th ed. McGraw-Hill, New York. 417 pp.

Grandocolas, P., and P. Deleport. 1992. La position

systématique de *Cryptocercus* Scudder au sein de Blattes et ses implications évolutives. C. R. Acad. Sc., Paris **315:**317–322.

Grbic, M., P. J. Ode, and M. R. Strand. 1992. Sibling rivalry and brood sex ratios in polyembryonic wasps. Nature (London) **360** (6401):254–256.

Greathead, D. J., and A. H. Greathead. 1992. Biological control of insect pests by insect parasitoids and predators: the BIOCAT database. Biocontrol News Inf. **13** (4):61N–68N.

Greathead, D. J., and J. K. Waage. 1986. *Insect Parasitoids: 13th Symposium of the Royal Entomological Society of London, 18–19 September 1985 at the Department of Physics Lecture Theatre, Imperial College, London.* Academic Press, London. 389 pp.

Greenslade, P. J. 1991. Collembola. In *The Insects of Australia,* vol. 1, Commonwealth Scientific and Industrial Research Organization (ed.). Melbourne University Press, Carlton. pp. 252–264.

Greenwood, S. R. 1988. Habitat stability and wing length in two species of arboreal Psocoptera. Oikos **52:**235–238.

Griffiths, G. C. D. 1972. The phylogenetic classification of the Diptera Cyclorrhapha with special reference to the structure of the male postabdomen. Ser. Entomol. Hague **8:**1–340.

Griffiths, G. C. D. 1980–. *Flies of the Nearctic Region.* E. Schweizerbartsche Verlag, Stuttgart.

Griffiths, G. C. D. 1994. Relationships among the major subgroups of Brachycera (Diptera): a critical review. Can. Entomol. **126:**861–880.

Grissell, E. E., and M. E. Schauff. 1990. *A Handbook of the Familes of Nearctic Chalcidoidea (Hymenoptera).* Entomological Society of Washington, Washington, DC. Handbook No. 1. 85 pp.

Grote, A. R. 1971. *Noctuidae of North America.* E. W. Classey, London. 85 pp.

Guenther, K. K., and K. Herter. 1974. Ordnung Dermaptera (Ohrwurmer). Handb. Zool. **4** (2):1–158.

Gullen, P. J., and P. S. Cranston. 1994. *The Insects: An Outline of Entomology.* Chapman & Hall, London. 491 pp.

Guppy, R. 1950. Biology of *Anisolabis maritima* (Gene) the seaside earwig, on Vancouver Island (Dermaptera, Labiidae). Proc. Entomol. Soc. Br. C. **46:**14–18.

Gupta, A. P. 1985. Cellular elements in the hemolymph. In *Comprehensive Insect Physiology, Biochemistry and Pharmacology,* vol. 3, G. A. Kerkut and L. I. Gilbert (eds.). Pergamon Press, New York. pp. 401–451.

Gupta, A. P. (ed.) 1990. *Morphogenetic Hormones of Arthropods.* Rutgers University Press, New Brunswick, NJ. 635 pp.

Gurney, A. B. 1938. A synopsis of the order Zoraptera, with notes on the biology of *Zorotypus hubbardi*

Caudell. Proc. Entomol. Soc. Washington **40**:57–87.

Gurney, A. B. 1948a. Praying mantids of the United States. Smithsonian Inst. Rep. **1950**:339–362.

Gurney, A. B. 1948b. The taxonomy and distribution of the Grylloblattidae. Proc. Entomol. Soc. Washington **50**:86–102.

Gurney, A. B. 1950. Corrodentia. In *Pest Control Technology.* Entomological Section. National Pest Control Association, New York. pp. 129–163.

Gurney, A. B., and F. W. Fisk. 1986. Chap. 2. Cockroaches. In *Insects and Mites in Foods.,* J. R. Gorham (ed.). U.S., Department of Agriculture. Washington, DC.

Guthrie, D. M., and A. R. Tindall. 1968. *The Biology of the Cockroach.* St. Martins Press, New York. 408 pp.

Gwynne, D. T., and G. K. Morris (eds.). 1983. *Orthoptera Mating Systems. Sexual Competition in a Diverse Group of Insects.* Westview Press, Boulder, CO. 376 pp.

Hadley, N. F. 1994. *Water Relations of Terrestrial Arthropods.* Academic Press, San Diego, CA. 356 pp.

Hagan, H. R. 1951. *Embryology in the Viviparous Insects.* Ronald Press, New York. 472 pp.

Hagen, K. 1987. Nutritional ecology of terrestrial insect predators. In *Nutritional Ecology of Insects, Mites, Spiders, and Related Invertebrates,* F. J. Slansky and J. G. Rodriguez (eds.). John Wiley & Sons, New York. pp. 533–577.

Hagen, K. S., T. L. Tassan, and E. F. Sawall, Jr. 1970. Some ecophysiological relationships between certain *Chrysopa,* honeydews and yeasts. Boll. Lab. Entomol. Agric., Portici. **28**:113–134.

Hagen, K. S., S. Bombosch, and J. A. McMurtry. 1976. The biology and impact of predators. In *Theory and Practice of Biological Control,* C. B. Huffaker and P. S. Messenger (eds.). Academic Press, New York. pp. 93–142.

Hale, W. G. 1965. Observations on the breeding biology of Collembola. Pedobiologia **5**:146–152.

Hall, D. G. 1948. The blow flies of North America. Thomas Say Found. Publ. **4**:1–477.

Hamilton, K. G. A. 1971. The insect wing. I. Origin and development of wings from notal lobes. J. Kansas Entomol. Soc. **44**:421–433.

Hamilton, K. G. A. 1972. The insect wing. Part IV. Venational trends and the phylogeny of the winged orders. J. Kansas Entomol. Soc. **45**:295–308.

Hamilton, K. G. A. 1981. Morphology and evolution of the rhynchotan head (Insecta: Hemiptera, Homoptera). Can. Entomol. **113**:953–974.

Hamilton, K. G. A. 1983. Classification, morphology and phylogeny of the family Cicadellidae (Rhynchota: Homoptera). In *Proceedings of the 1st International Workshop on Biotaxonomy, Classification and Biology of Leafhoppers and Planthoppers (Auche-*

norrhyncha) of Economic Importance, W. J. Knight (ed.). Commonwealth Institute of Entomology, London. pp. 15–37.

Handlirsch, A. 1908. *Die fossilen Insekten und die Phylogenie der rezenten Formen.* W. Englemann, Leipzig. 1430 pp.

Handlirsch, A. 1926. Vierter Unterstamm des Stammes der Arthropoda. Insecta = Insekten. Handbuch Zool. **4**:403–592.

Hanover, J. W. 1975. Physiology of tree resistance to insects. Annu. Rev. Entomol. **20**:75–95.

Hansen, M. 1991. The hydrophiloid beetles: Phylogeny, classification and a revision of the genera (Coleoptera, Hydrophiloidea). K. Dan. Vidensk. Selsk. Biol. Skr. **40**:1–367.

Hanski, I., and Y. Cambefort. 1991. *Dung Beetle Ecology.* Princeton University Press, Princeton, NJ. 481 pp.

Happ, G. M. 1992. Maturation of the male reproductive system and its endocrine regulation. Annu. Rev. Entomol. **37**:303–320.

Harden, P. H., and C. E. Mickel. 1952. The stoneflies of Minnesota (Plecoptera). Univ. Minn. Agric. Exp. Stn. Tech. Bull. **201**:1–84.

Harland, W. B., R. L. Armstrong, A. V. Cox, L. E. Craig, A. G. Smith, and D. G. Smith. 1990. *A Geologic Time Scale 1989.* Cambridge University Press, Cambridge. 263 pp.

Harris, K. F. (ed.). 1986–. *Advances in Vector Research.* Springer-Verlag, New York.

Harris, W. V. 1961. *Termites. Their Recognition and Control.* Longmans Group, London. 186 pp.

Hassell, K. A. 1990. *The Biochemistry and Uses of Insecticides.* 2nd ed. VCH, Weinheim. 536 pp.

Hatch, M. H. 1957–1973. *The Beetles of the Pacific Northwest.* 5 vols. University of Washington Publications Biology and University of Washington Press, Seattle.

Haub, F. 1980. Concerning "phylogenetic" relationships of parasitic Psocodea and taxonomic position of the Anoplura by K. C. Kim and H. W. Ludwig. Ann. Entomol. Soc. Am. **73**:3–6.

Haunerland, N. H., and P. D. Shirk. 1995. Regional and functional differentiation in the insect fat body. Annu. Rev. Entomol. **40**:121–145.

Hayashi, F. 1988. Prey selection by the dobsonfly larva, *Protohermes grandis* (Megaloptera: Corydalidae). Freshwater Biol. **20**:19–30.

Hayashi, F. 1989. Microhabitat selection by the fishfly larva, *Parachauliodes japonicus,* in relation to its mode of respiration. Freshwater Biol. **21**:489–496.

Hayashi, F. 1992. Large spermatophore production and consumption in dobsonflies *Protohermes* (Megaloptera, Corydalidae). Jpn. J. Entomol. **60**:59–66.

Hayashi, F., and M. Nakane. 1989. Radio tracking and

activity monitoring of the dobsonfly larva, *Protohermes grandis* (Megaloptera: Corydalidae). Oecologia (Berlin) **78**:468–472.

Hebard, M. 1934. The Dermaptera and Orthoptera of Illinois. Illinois Nat. Hist. Survey Bull. **20**:125–279.

Heinrich, B. 1973. The energetics of the bumblebee. Sci. Am. **228**(4):96–102.

Heinrich, B. 1974. Thermoregulation in endothermic insects. Science **185**:747–756.

Heinrich, B. 1993. *The Hot-Blooded Insects: Strategies and Mechanisms of Thermoregulation.* Harvard University Press, Cambridge, MA 601 pp.

Heinrich, B., and G. A. Bartholomew. 1972. Temperature control in flying moths. Sci. Am. **226**(6):69–77.

Heise, B. A., and J. F. Flannagan. 1987. Life histories of *Hexagenia limbata* and *Ephemera simulans* (Ephemeroptera) in Dauphin Lake, Manitoba [Canada]. J. North Am. Benthol. Soc. **6**:230–240.

Helfer, J. R. 1987. *How to Know the Grasshoppers, Crickets, Cockroaches, and Their Allies.* Dover Publications, New York. 363 pp.

Heliövaara, K., and R. Vaisanen. 1993. *Insects and Pollution.* CRC Press, Boca Raton, FL. 393 pp.

Hellenthal, R. A., and R. D. Price. 1991. Biosystematics of the chewing lice of pocket gophers. Annu. Rev. Entomol. **36**:185–203.

Hendry, L. B., B. Piatek, L. E. Browne, D. L. Wood, J. A. Byers, R. H. Fish, and R. A. Hicks. 1980. *In vivo* conversion of a labelled host plant chemical to pheromones of the bark beetle *Ips paraconfusus.* Nature **284**:485.

Hennig, W. 1948–1952. *Die Larvenformen der Dipteren.* Part 1. Akademie-Verlag, Berlin.

Hennig, W. 1953. Kritische Bemerkungen zum phylogenetischen System der Insekten. Beitr. Entomol. **3**:1–85.

Henry, C. S. 1989. The unique purring song of *Chrysoperla comanche* (Banks), a green lacewing of the Rufilabris spp. group (Neuroptera: Chrysopidae). Proc. Entomol. Soc. Washington **91**:133–142.

Henry, T. J., and R. C. Froeschner (eds.). 1988. *Catalog of the Heteroptera, or True Bugs, of Canada and the Continental United States.* St. Lucie Press, Delray Beach, FL. 958 pp.

Hepburn, H. R. 1969. The skeleto-muscular system of Mecoptera: The head. Kansas Univ. Sci. Bull. **48**:721–765.

Hering, M. 1951. *Biology of the Leaf Miners.* The Hague, Netherlands. 420 pp.

Hermann, H. R. (ed.). 1979–82. *Social Insects,* vols. 1–4. Academic Press, New York.

Herring, J. L., and P. D. Ashlock. 1971. A key to the nymphs of the families of Hemiptera (Heteroptera) of America north of Mexico. Florida Entomol. **54**:207–213.

Hespenheide, H. H. 1991. Bionomics of leaf-mining insects. Annu. Rev. Entomol. **36**:535–560.

Hildrew, A. G., and R. Wagner. 1992. The briefly colonial life of hatchlings of the net-spinning caddisfly *Plectrocnemia conspersa.* J. North Am. Benthol. Soc. **11**:60–68.

Hille Ris Lambers, D. 1966. Polymorphism in Aphididae. Annu. Rev. Entomol. **11**:47–78.

Hinton, H. E. 1945. *A Monograph of the Beetles Associated with Stored Products.* British Museum of Natural History, London. 443 pp.

Hinton, H. E. 1946. On the homology and nomenclature of the setae of the lepidopterous larvae, with some notes on the phylogeny of the Lepidoptera. Trans. R. Entomol. Soc. London **97**:1–37.

Hinton, H. E. 1955a. On the structure and distribution of the prolegs of the Panorpoidea with a criticism of the Berlese-Imms theory. Trans. R. Entomol. Soc. London **106**:455–540.

Hinton, H. E. 1955b. On the respiratory adaptions, biology and taxonomy of the Psephenidae, with notes on some related families (Coleoptera). Proc. Zool. Soc. London **125**:543–568.

Hinton, H. E. 1958. The phylogeny of the panorpoid orders. Annu. Rev. Entomol. **3**:181–206.

Hinton, H. E. 1963a. The origin and function of the pupal stage. Proc. R. Entomol. Soc. London (A) **38**:77–85.

Hinton, H. E. 1963b. The origin of flight in insects. Proc. R. Entomol. Soc. London (C) **28**:23–32.

Hinton, H. E. 1969a. Plaston respiration in adult beetles of the suborder Myxophaga. J. Zool. Soc. London **159**:131–137.

Hinton, H. E. 1969b. Respiratory systems of insect egg shells. Annu. Rev. Entomol. **14**:343–368.

Hinton, H. E. 1971. Some neglected phases in metamorphosis. Proc. R. Entomol. Soc. London (C) **35**:55–64.

Hinton, H. E. 1977. Enabling mechanisms. XV Int. Congr. Entomol., 71–83.

Hinton, H. E. 1981. *Biology of Insect Eggs.* Pergamon Press, Oxford. 3 vols., 1125 pp.

Hodek, I. 1973. *Biology of Coccinellidae.* Academia, Prague, and Junk's, The Hague. 260 pp.

Hodges, R. W. et al. 1983. *Checklist of the Lepidoptera of America North of Mexico.* Wedge Entomological Research Foundation, National Museum of Natural History, Washington, DC. 284 pp.

Hodgson, C. J. 1994. *The Scale Insect Family Coccidae: An Identification Manual to Genera.* CAB International, Wallingford, Oxon, UK. 639 pp.

Hoeppli, R. 1969. *Parasitic Diseases in Africa and the Western Hemisphere.* Verlag für Recht und Gesellshaft, Basel. 240 pp.

Holdsworth, R. P. 1941. The life history and growth of

Pteronarcys proteus Newman (Pteronarcidae: Plecoptera). Ann. Entomol. Soc. Am. **34:**394–502.

Holland, G. P. 1949. The Siphonaptera of Canada. Can. Dept. Agric. Publ. 817, Tech. Bull. **70:**1–306.

Holland, G. P. 1964. Evolution, classification and host relationships of Siphonaptera. Annu. Rev. Entomol. **9:**123–146.

Holland, W. J. 1968. *The Moth Book.* Dover Publications, New York. 479 pp.

Holldobler, B., and E. O. Wilson. 1990. *The Ants.* Belknap Press, Cambridge, MA. 747 pp.

Holman, G. M., R. J. Nachman, and M. S. Wright. 1990. Insect neuropeptides. Annu. Rev. Entomol. **35:**201–217.

Holsinger, J. R. 1988. Troglobites: The evolution of cave-dwelling organisms. Am. Sci. **76:**147–153.

Homberg, U., T. A. Christensen, and J. G. Hildebrand. 1989. Structure and function of the deutocerebrum in insects. Annu. Rev. Entomol. **34:**477–501.

Hopkins, G. H. E. 1949. The host associations of the lice of mammals. Proc. Zool. Soc. London **119:**387–604.

Horn, D. K. 1988. *Ecological Approach to Pest Management.* Guilford Press, New York. 285 pp.

Horsfall, W. R., Jr., H. W. Fowler, L. J. Moretti, and J. R. Larsen. 1973. *Bionomics and Embryology of the Inland Floodwater Mosquito Aedes vexans.* University of Illinois Press, Urbana. 211 pp.

Houk, E. J., and G.W. Griffiths. 1980. Intracellular symbiotes of the Homoptera. Annu. Rev. Entomol. **25:**161–187.

Howarth, F. G. 1983. Ecology of cave arthropods. Annu. Rev. Entomol. **28:**365–389.

Howarth, F. G. 1991. Environmental impacts of classical biological control. Annu. Rev. Entomol. **36:**485–509.

Howe, W. H. 1975. *The Butterflies of North America.* Doubleday, Garden City, NY. 633 pp.

Howse, P. E. 1970. *Termites: A Study in Social Behavior.* Hutchinson University Library, London. 150 pp.

Hoy, M. A. 1994. *Insect Molecular Genetics: An Introduction to Principles and Applications.* Academic Press, San Diego, CA. 546 pp.

Hoy, M. A., and Herzog, D. C. (eds.). 1985. *Biological Control in Agricultural IPM Systems.* Academic Press, New York. 589 pp.

Hsu, C.-Y., and C.-W. Li. 1994. Magnetoreception in honeybees. Science **265:**95–97.

Hubbard, C. A. 1947. *Fleas of Western North America.* Iowa State College Press, Ames. 533 pp.

Hubbard, M. D. 1990. *Mayflies of the World: A Catalog of the Family and Genus Group Taxa (Insecta: Ephemeroptera).* Flora & Fauna Handbook No. 8. Sandhill Crane Press, Gainesville, FL. 119 pp.

Hubbell, S. 1993. *Broadsides from the Other Orders. A Book of Bugs.* Random House, New York. 276 pp.

Huber, I., E. P. Masler, and B. R. Rao. 1990a. *Cockroaches as models for neurobiology: applications in biomedical research.* CRC Press., Boca Raton, FL. vol. 1, 266 pp.; vol 2, 326 pp.

Huber, F., T. E. Moore, and W. Loher. 1990b. *Cricket Behavior and Neurobiology.* Cornell University Press, Ithaca, NY. 536 pp.

Huffaker, C. B. (ed.). 1971. *Biological Control.* Plenum Publishing New York. 511 pp.

Huffaker, C. B., and P. S. Messenger (eds.). 1976. *Theory and Practice of Biological Control.* Academic Press, New York. 788 pp.

Hull, F. M. 1962. Robberflies of the world: the genera of the family Asilidae. U.S. Natl. Mus. Bull. **224:**1–907 (2 vols.).

Hungerford, H. B. 1959. Hemiptera. In *Freshwater Biology,* W. T. Edmondson (ed.). John Wiley & Sons, New York. pp. 958–972.

Hunt, B. P. 1950. The life history and economic importance of the burrowing mayfly, *Hexagenia limbata,* in southern Michigan lakes. Inst. Fisheries Res. Bull. Michigan Dept. Conserv. **4:**1–151.

Hunt, J. H., and C. A. Nalepa (eds.) 1994. *Nourishment and Evolution in Insect Societies. Studies in Insect Biology.* Westview Press, Boulder. CO. 449 pp.

Huxley, C. R., and D. F. Cutler. 1991. *Ant-Plant Interactions.* Oxford University Press, Oxford. 601 pp.

Hynes, H. B. N. 1970. The ecology of stream insects. Annu. Rev. Entomol., **15:**25–42.

Hynes, H. B. N. 1976. Biology of Plecoptera. Annu. Rev. Entomol. **21:**135–154.

Ide, F. P. 1935. The effect of temperature on the distribution of the mayfly fauna of a stream. Univ. Toronto Studies, Biol. Ser. No. 39, Publ. Ontario Fisheries Res. Lab. **50:**3–76.

Illies, J. 1965. Phylogeny and zoogeography of the Plecoptera. Annu. Rev. Entomol. **10:**117–140.

Imms, A. D. 1936. The ancestry of insects. Trans. Soc. Br. Entomol. **3:**1–32.

Iseley, F. B. 1944. Correlation between mandibular morphology and food specificity in grasshoppers. Ann. Entomol. Soc. Am. **37:**47–67.

Istock, C. A. 1966. The evolution of complex life cycle phenomena: an ecological perspective. Evolution **21:**592–605.

Ito, Y. 1993. *Behaviour and Social Evolution of Wasps: the Communal Aggregation Hypothesis.* Oxford Series in Ecology and Evolution. Oxford University Press, Oxford. 59 pp.

Iwata, K. 1976. *Evolution of Instinct: Comparative Ethology of Hymenoptera.* Amerind Publishing, New Delhi. 535 pp.

Jacobi, D. I., and A. C. Benke. 1991. Life histories and abundance patterns of snag-dwelling mayflies in a blackwater Coastal Plain river. J. North Am. Benthol. Soc. **10**:372–387.

James, M. T. 1947. The flies that cause myiasis in man. U.S. Dept. Agric. Misc. Publ. **631**:1–175.

James, M. T., and R. F. Harwood. 1969. *Herm's Medical Entomology.* 6th ed. Macmillan, London. 484 pp.

Janzen, D. H. 1966. Coevolution of mutualism between ants and acacias in Central America. Evolution **20**:249–275.

Janzen, D. H. 1967a. Fire, vegetation structure, and the ant-acacia interaction in Central America. Ecology **48**:26–35.

Janzen, D. H. 1967b. Interaction of the bull's-horn acacia (*Acacia cornigera* L.) with an ant inhabitant (*Pseudomyrmex ferruginea* F. Smith) in eastern Mexico. Kansas Univ. Sci. Bull. **47** (6):15–558.

Jeanne, R. L. 1980. Evolution of social behavior in the Vespidae. Annu. Rev. Entomol. **25**:371–396.

Jewett, S. G. 1956. Plecoptera. In *Aquatic Insects of California,* R. L. Usinger (ed.). University of California Press, Berkeley. pp. 155–181.

Jewett, S. G. 1959. The stoneflies (Plecoptera) of the Pacific Northwest. Oregon State Mongr. Stud. Entomol. **3**:1–95.

Jewett, S. G., Jr. 1963. A stonefly aquatic in the adult stage. Science **139**:484–485.

Johannsen, O. A., and F. H. Butt. 1941. *Embryology of Insects and Myriapods.* McGraw-Hill, New York. 462 pp.

Johnson, J. B., and K. S. Hagen. 1981. A neuropteran larva uses an allomone to attack termites. Nature **289**:506–507.

Jolivet, P. 1996. *Biologie de Coleoptères Chrysomélides.* Boubée, Paris. 248 pp.

Jones, T. 1954. The external morphology of *Chirothrips hematus* (Trybom) (Thysanoptera). Trans. R. Entomol. Soc. London **105**:163–187.

Joosse, E. N. G. 1976. Littoral apterygotes (Collembola and Thysanura). In *Marine Insects,* L. Cheng (ed.). North-Holland Publishing Company, Amsterdam. pp. 151–186.

Joosse, E. N. G. 1983. New developments in the ecology of Apterygota. Pedobiologia **25**:217–234.

Joosse, E. N. G., and H. A. Verhoef. 1987. Developments in ecophysiological research on soil invertebrates. Adv. ecol. Res. **16**:175–248.

Judd, W. W. 1948. A comparative study of the proventriculus of orthopteroid insects with reference to its use in taxonomy. Can. J. Res. **26**:93–161.

Juniper, B., and R. Southwood (eds.). 1986. *Insects and the Plant Surface.* Edward Arnold, London. 360 pp.

Kaiser, W., and J. Steiner-Kaiser. 1987. Sleep research on honeybees: neurophysiology and behavior. In *Neurobiology and Behavior of Honeybees.,* R. Men-

zel and A. Mercer (eds.). Springer-Verlag, Berlin. pp. 112–120.

Kamp, J. W. 1963. Descriptions of two new species of Grylloblattidae and of the adult of *Grylloblatta barberi,* with an interpretation of their geographic distribution. Ann. Entomol. Soc. Am. **56**:53–68.

Kamp, J. W. 1973. Numerical classification of the orthopteroids, with special reference to the Grylloblattodea. Can. Entomol. **105**:1235–1249.

Kathirithamby, J. 1989. Review of the order Strepsiptera. Syst. Entomol. **14**:41–92.

Kathirithamby, J., and M. Carcupino. 1992. Comparative spermology of four species of Strepsiptera and comparison with a species of primitive Coleoptera Rhipiphoridae. Int. J. Insect Morph. Embryol. **22**:459–470.

Kauffman, W. C., and J. E. Nechols (eds.). 1992. *Selection Criteria and Ecological Consequences of Importing Natural Enemies.* Proceedings, Thomas Say Publications in Entomology, Entomological Society of America, Lanham, MD. 117 pp.

Kaufmann, T. 1971. Hibernation in the arctic beetle, *Pterostichus brevicornis,* in Alaska. J. Kansas Entomol. Soc. **44**:81–92.

Keister, A. R., and E. Strates. 1984. Social behavior in a thrips from Panama. J. Nat. Hist. **18**:304–314.

Keler, S. von. 1969. 17. Ordnung Mallophaga (Federlinge und Haarlinge). Handb. Zool. **4**(2):1–72.

Kelsey, L. P. 1954. The skeleto-motor mechanism of the Dobson fly, *Corydalus cornutus.* Part I. Head and prothorax. Mem. Agric. Exp. Stn. Cornell Univ. **334**:1–51.

Kelsey, L. P. 1957. The skeleto-motor mechanism of the Dobson fly, *Corydalus cornutus.* Part II. Pterothorax. Mem. Agric. Exp. Stn. Cornell Univ. **346**:1–31.

Kennedy, J. S., and H. L. G. Stroyan. 1959. Biology of aphids. Annu. Rev. Entomol. **4**:139–160.

Kent, D. K., and J. A. Simpson. 1992. Eusociality in the beetle *Austroplatypus incompertus* (Coleoptera, Curculionidae). Naturwissenschaften **79**:86–87.

Kerkut, G. A., and L. I. Gilbert (eds.). 1985. *Comprehensive Insect Physiology, Biochemistry, and Pharmacology.* Pergamon Press, Oxford.

Kettlewell, H. B. D. 1959. Darwin's missing evidence. Sci. Am. **200**:48–53.

Kettlewell, H. B. D. 1961. The phenomenon of industrial melanism in Lepidoptera. Annu. Rev. Entomol. **6**:245–262.

Kettlewell, H. B. D. 1973. *The Evolution of Melanism.* Clarendon Press, Oxford. 423 pp.

Kevan, D. K. McE. 1962. *Soil Animals.* Philosophical Library, New York. 237 pp.

Kevan, D. K. McE. 1977. The higher classification of the orthopteroid insects. Mem. Lyman Entomol. Mus. Res. Lab. **4**:1–52.

Kevan, D. K. McE. 1982. Orthoptera (pp. 352–379), Phasmatoptera (pp. 379–383). In *Synopsis and Classification of Living Organisms,* S. P. Parker (ed.). McGraw-Hill, New York.

Kevan, P. G., and H. G. Baker. 1983. Insects as flower visitors and pollinators. Annu. Rev. Entomol. **28:**407–453.

Kim, K. C., and H. W. Ludwig. 1982. Parallel evolution, cladistics, and classification of parasitic Psocodea. Ann. Entomol. Soc. Am. **75:**537–548.

Kim, K. C., H. D. Pratt, and C. J. Stojanovich. 1986. *The Sucking Lice of North America: An Illustrated Manual for Identification.* Pennsylvania State University Press, University Park. 241 pp.

King, P. E., and K. S. Ahmed. 1989. Sperm structure in the Psocoptera. Acta Zool. (Stockholm) **70:**57–62.

Kingsolver, J. G., and M. A. R. Koehl. 1985. Aerodynamics, thermoregulation, and the evolution of insect wings: differential scaling and evolutionary change. Evolution. **39:**488–504.

Kingsolver, J. G., and M. A. R. Koehl. 1994. Selective factors in the evolution of insect wings. Annu. Rev. Entomol. **39:**425–451.

Kinzelbach, R. 1990. The systematic position of Strepsiptera (Insecta). Am. Entomol. **36:**292–303.

Kinzelbach, R., and H. Pohl. 1994. The fossil Strepsiptera (Insecta: Strepsiptera). Ann. Entomol. Soc. Am. **87:**59–70.

Kinzelbach, R. K. 1971a. Strepsiptera (Facherflugler). Handb. Zool. **42**(24):1–68.

Kinzelbach, R. K. 1971b. Morphologische Befunde an Facherfluglern und ihre phylogenetische Bedeutung (Insecta: Strepsiptera). Zoologica **41**(119):1–256.

Kinzelbach, R. K. 1978. Insecta: Facherflugler (Strepsiptera). Tierwelt Dtschl. **65:**1–166.

Kirchner, W. H. 1993. Acoustical communication in honeybees. Apidologie **24:**297–307.

Kirchner, W. H., and W. F. Towne. 1994. The sensory basis of the honeybee's dance language. Sci. Am. **270:**74–80.

Kirchner, W. H., M. Lindauer, and A. Michelsen. 1988. Honeybee dance communication. Acoustical indication of direction in round dances. Naturwissenschaften **75:**629–630.

Kiriakoff, S. G. 1963. The tympanic structures of Lepidoptera and the taxonomy of the order. J. Lepid. Soc. **17:**1–6.

Kistner, D. H. 1979. Social and evolutionary significance of social insect symbionts. In *Social Insects,* vol. 1, H. R. Herman (ed.). Academic Press, New York. pp. 339–415.

Kistner, D. H. 1982. The social insects bestiary. In *Social Insects,* vol. 2, H. R. Herman (ed.). Academic Press, New York. pp. 1–244.

Klimaszewski, J., and D. K. McE. Kevan. 1985. The brown lacewing flies of Canada and Alaska (Neuroptera: Hemerobiidae). Memoir, Lyman Entomological Museum and Research Laboratory, no. 15.

Klots, A. B. 1951. *A Field Guiide to the Butterflies.* Houghton Mifflin, Boston. 349 pp.

Knight, K.L. (ed.). 1989. Entomology. Serving Society 1889–1989. Bull. Entomol. Soc. Am. **35** (3):1–215.

Kofoid, C. A. (ed.). 1934. *Termites and Termite Control.* University of California Press, Berkeley. 734 pp.

Kormondy, E. J. 1961. Territoriality and dispersal in dragonflies (Odonata). J. New York. Entomol. Soc. **69:**42–52.

Koss, R. W. 1968. Morphology and taxonomic use of Ephemeroptera eggs. Ann. Entomol. Soc. Am. **61:**696–721.

Kosztarab, M. 1987. Everything unique or unusual about scale insects (Homoptera: Coccoidea). Bull. Entomol. Soc. Am. (winter):215–220.

Kosztarab, M. 1996. *Scale Insects of Northeastern North America: Identification, Biology, and Distribution.* Virginia Museum of Natural History, Martinsville. 650 pp.

Kosztarab, M., and C. W. Schaefer (eds.). 1990. *Systematics of the North American Insects and Arachnids: Status and Needs.* Virginia Agric. Exp. Stn. Inf. Ser. 90-1. Blacksburg. 247 pp.

Kosztarab, M., and M. P. Kosztarab. 1988. *A Selected Bibliography of the Coccoidea (Homoptera): Third Supplement (1970–1985).* Studies on the Morphology and Systematics of Scale Insects no. 14. Virginia Polytechnic Institute and State University, Blacksburg. 252 pp.

Kosztarab, M., Y. Ben-Dov, M. P. Kosztarab, and H. Morrison. 1986. *An Annotated List of Generic Names of the Scale Insects (Homoptera, Coccoidea): Third Supplement.* Studies on the Morphology and Systematics of Scale Insects no. 13. Virginia Polytechnic Institute and State University, Blacksburg. 38 pp.

Kramer, S. 1950. Morphology and phylogeny of the auchenorrhynchous Homoptera (Insecta). Illinois Biol. Monogr. **20:**1–78.

Krishna, K. 1990. Isoptera. Bull. Am. Mus. Nat. Hist. **195:**76–81.

Krishna, K., and F. M. Weesner. 1969a. *Biology of Termites,* vol. 1. Academic Press. New York. 598 pp.

Krishna, K., and F. M. Weesner. 1969b. *Biology of Termites,* vol. 2. Academic Press, New York. 643 pp.

Kristensen, N. P. 1968. The morphological and functional evolution of the mouthparts in adult Lepidoptera. Opusc. Entomol. **33:**69–72.

Kristensen, N. P. 1971. The systematic position of the Zeugloptera in the light of recent anatomical investigations. Proc. 13th Int. Congr. Entomol. (1968) **1:**261.

Kristensen, N. P. 1975. The phylogeny of hexapod

"orders": a critical review of recent accounts. Z. Zool. Syst. Evol. Forsch. **13**:1–44.

Kristensen, N. P. 1981. Phylogeny of insect orders. Annu. Rev. Entomol. **26**:135–158.

Kristensen, N. P. 1984. Studies on the morphology and systematics of primitive Lepidoptera (Insecta). Steenstrupia **10**:141–191.

Kristensen, N. P. 1991. Phylogeny of extant hexapods. In *The Insects of Australia,* vol. 1, Commonwealth Scientific and Industrial Research Organization (ed.). Melbourne University Press, Carlton. pp. 125–140.

Krombein, K. V. 1967. *Trap-Nesting Wasps and Bees. Life Histories, Nests, and Associates.* Smithsonian Institution, Washington, DC. 570 pp.

Kuhnelt, W. 1961. *Soil Biology, with Special Reference to the Animal Kingdom.* Faber & Faber, London. 397 pp.

Kukalová-Peck, J. 1978. Origin and evolution of insect wings and their relation to metamorphosis, as documented by fossil record. J. Morphol. **156**:53–126.

Kukalová-Peck, J. 1983. Origin of the insect wing and wing articulation from the arthropodan leg. Can. J. Zool. **61**:1618–1669.

Kukalová-Peck, J. 1985. Ephemeroid wing venation based on new gigantic Carboniferous mayflies and basic morphology, phylogeny, and metamorphosis of pterygote insects (Insecta, Ephemerida). Can. J. Zool. **63**:933–955.

Kukalová-Peck, J. 1987. New Carboniferous Diplura, Monura and Thysanura, the hexapod ground plan, and the role of thoracic side lobes in the origin of wings (Insecta). Can. J. Zool. **65**:2327–2345.

Kukalová-Peck, J. 1991. Fossil history and the evolution of hexapod structures. In *The Insects of Australia,* vol. 1, Commonwealth Scientific and Industrial Research Organization (ed.). Melbourne University Press, Carlton. pp. 141–179.

Kukalová-Peck, J. 1992. The "Uniramia" do not exist: the ground plan of the Pterygota as revealed by Permian Diaphanopterodea from Russia (Insecta: Paleodictyopteroidea). Can. J. Zool. **70**:236–255.

Kukalová-Peck, J., and C. Brauckmann. 1992. Most Paleozoic Protorthoptera are ancestral hemipteroids: major wing braces as clues to a new phylogeny of Neoptera (Insecta). Can. J. Zool. **70**:2452–2473.

Kukalová-Peck, J., and J. F. Lawrence. 1993. Evolution of the hind wing in Coleoptera. Can. Entomol. **125**:181–258.

Kukalová-Peck, J., and S. Peck. 1993. Zoraptera wing structure: evidence for new genera and relationship with the blattoid orders (Insecta: Blattoneoptera). Syst. Entomol. **18**:333–350.

Labandeira, C. C., and J. J. Sepkoski. 1993. Insect diversity in the fossil record. Science **261**:310–315.

Labandeira, C., B. Beall, and P. Hueber. 1988. Early insect diversification: evidence from a lower Devonian bristletail from Quebec. Science **242**:913–916.

Labandeira, C. C., D. L. Dilcher, D. R. Davis, and D. L. Wagner. 1994. Ninety-seven million years of angiosperm-insect association: paleobiological insights into the meaning of coevolution. Proc. Natl. Acad. Sci. U.S.A. **91**:12278–12282.

Lakshminarayana, K. V. 1985. Glossary of taxonomic characters for the study of chewing-lice (Phthiraptera: Insecta). Rec. Zool. Survey India., Misc. Publ., Occas. Pa. **82**:1–59.

Landa, V., and T. Soldan. 1985. *Phylogeny and Higher Classification of the Order Ephemeroptera: A Discussion from the Comparative Anatomical Point of View.* Academia, Pragues. 121 pp.

Langston, R. L., and J. A. Powell. 1975. The earwigs of California (order Dermaptera). Bull. California Insect Survey **20**:1–25.

Larsen, O. 1966. On the morphology and function of the locomotor organs of the Gyrinidae and other Coleoptera. Opusc. Entomol. Suppl. **30**:1–242.

Lavialle, M., and B. Dumortier. 1975. Mise en evidence de deux caractères inhabituels dans les rhythmes circadiens: existence d'un photorécepteur extraoculaire et particularités du libre cours dans le rhythme de ponte du phasme: *Carausius morosus.* C. R. Acad. Soc. Paris **281**:1489–1492.

Lavine, M. D., and N. E Beckage. 1995. Polydnaviruses: potent mediators of host immune dysfunction. Parasitol. Today **11**:368–378.

Lawrence, J. F. 1993a. *Beetle Larvae of the World.* Commonwealth Scientific and Industrial Research Organization, Division of Entomology, Canberra, Australia. 48 pp.

Lawrence, J. F. 1993b. *Beetle Larvae of the World: Interactive Identification and Information Retrieval for Families and Sub-families.* Commonwealth Scientific and Industrial Research Organization, Division of Entomology, Canberra, Australia.

Lawrence, J. F. 1982. Coleoptera. In *Synopsis and Classification of Living Organisms,* vol. 2, S. P. Parker (ed.). McGraw-Hill, New York. pp. 482–553.

Lawrence, J. F. 1989. Mycophagy in the Coleoptera: feeding strategies and morphological adaptations. In *Insect-Fungus Interactions,* N. Wilding (ed.). Royal Entomological Society of London no. 14. pp. 1–23.

Lawrence, J. F., and E. B. Britton. 1991. Coleoptera (beetles). In *The Insects of Australia,* vol. 2, Commonwealth Scientific and Industrial Research Organization (ed.). Melbourne University Press, Carlton. pp. 543–683.

Lawrence, J. F., and E. B. Britton. 1994. *Australian Beetles.* Melbourne University Press, Carlton. 192 pp.

Lawrence, J. F., and A. F. Newton, Jr. 1982. Evolution and classification of beetles. Annu. Rev. Ecol. Syst. **13:**261–290.

Lawrence, P. A. 1992. *The Making of a Fly: The Genetics of Animal Design.* Blackwell Scientific, Oxford. 228 pp.

Lawrence, R. F. 1953. *The Biology of the Cryptic Fauna of Forests.* A. A. Balkema, Cape Town, South Africa. 408 pp.

Layne, J. N. 1971. Fleas (Siphonaptera) of Florida. Florida Entomol. **54:**35–51.

Lee, K. E., and T. G. Wood. 1971. *Termites and Soils.* Academic Press, London. 251 pp.

Leech, H. B., and H. P. Chandler. 1956. Aquatic Coleoptera. In *Aquatic Insects of California,* R. L. Usinger (ed.). University of California Press, Berkeley. pp. 293–371.

Leech, H. B. , and M. W. Sanderson. 1959. Coleoptera. In *Freshwater Biology,* W. T. Edmondson (ed.). John Wiley & Sons, New York. pp. 981–1023.

Lehane, M. J. 1991. *Biology of Blood-Sucking Insects.* HarperCollins Academic, London. 288 pp.

Leonard, D. E. 1974. Recent developments in ecology and control of the gypsy moth. Annu. Rev. Entomol. **19:**197–230.

Leonard, J. W., and F. A. Leonard. 1962. *Mayflies of Michigan Trout Streams.* Cranbrook Institute of Science, Bloomfield Hills. 139 pp.

Levin, M. D. 1983. Value of bee pollination to U.S. agriculture. Bull. Entomol. Soc. Am. **29**(4):50–51.

Lewis, R. E., J. H. Lewis, and C. Maser. 1988. *The Fleas of the Pacific Northwest.* Oregon State University Press, Corvallis, Or. 296 pp.

Lewis, T. 1973. *Thrips, Their Biology, Ecology and Economic Importance.* Academic Press, London. 366 pp.

Lewis, T. (ed.). 1984. *Insect Communication.* Academic Press, London. 414 pp.

Liebhold, A., V. Mastro, and P. W. Schaefer. 1989. Learning from the legacy of Leopold Trouvelot. Bull. Entomol. Soc. Am. **35**(2):20–22.

Linit, M. J. 1988. Nematode-vector relationships in the pine wilt disease system. J. Nematol. **20:**227–235.

Linsley, E. G. 1959a. Mimetic form and coloration in the Cerambycidae (Coleoptera). Ann. Entomol. Soc. Am. **52:**125–131.

Linsley, E. G. 1959b. Ecology of Cerambycidae. Annu. Rev. Entomol. **4:**99–138.

Linsley, E. G., and J. W. MacSwain. 1957. Observations on the habits of *Stylops pacifica* Bohart. Univ. Calif. Publ. Entomol. **11:**395–430.

Littig, K. S. 1942. External anatomy of the Florida walking stick *Anisomorpha buprestoides* Stoll. Florida Entomol. 25:33–41.

Lloyd, J. E. 1971. Bioluminescent communication in insects. Annu. Rev. Entomol. **16:**97–122.

Lloyd, J. E. 1983. Bioluminescence and communication in insects. Annu. Rev. Entomol. **28:**131–160.

Lloyd, J. E. 1984. Occurrence of aggressive mimicry in fireflies. Florida Entomol. **67:**368–376.

Lloyd, J. L. 1921. The biology of North American caddis fly larvae. Lloyd Libr. Bot., Pharm. Mater. Med. Bull. **21:**1–124.

Lockwood, A., L. D. Debrey, C. D. Thompson, C. M. Love, R. A. Nunamaker, S. R. Shaw, S. P. Schell, and C. R. Bomar. 1994. Preserved insect fauna of glaciers of Fremont County in Wyoming: insights into the ecology of the extinct rocky mountain locust. Environ. Entomol. **23:**220–235.

Loher, W. 1960. The chemical acceleration of the maturation process and its hormonal control in the male of the desert locust. Proc. R. Entomol. Soc. London, Ser. B **153:**380–397.

Loher, W. 1972. Circadian control of stridulation in the cricket *Teleogryllus commodus* Walker. J. Comp. Physiol. **79:**173–190.

Loher, W., I. Ganjian, I. Kubo, D. Stanley-Samuelson, and S. S. Tobe. 1981. Prostaglandins: their role in egg-laying of the cricket *Teleogryllus commodus.* Proc. Natl. Acad. Sci. U.S.A. **78:**7835–7838.

Lomer, C. J., and C. Prior. 1992. *Biological Control of Locusts and Grasshoppers: Proceedings of a Workshop Held at the International Institute of Tropical Agriculture, Cotonou, Republic of Benin, 29 April– 1 May 1991.* CAB. International, Wallingford, Oxon, UK. 394 pp.

Louw, G. N., and M. K. Seely. 1982. *Ecology of Desert Organisms.* Longmans Groups, London. 194 pp.

Luscher, M. 1961. Social control of polymorphism in termites. Symp. R. Entomol. Soc. London **1:**57–67.

Luttrell, R. G. 1994. Cotton pest management Part II. A U. S. perspective. Annu. Rev. Entomol. **39:**527–542.

Luttrell, R. G., G. P. Fitt, F. S. Ramlaho, and E. S. Sugonyaev. 1994. Cotton pest management. Part I. A worldwide perspective. Annu. Rev. Entomol. **39:**515–526.

Lyal, C. H. C. 1985. Phylogeny and classification of the Psocodea with particular reference to the lice (Psocodea: Phthiraptera). Syst. Entomol. **10:**145–165.

Lyman, F. E. 1955. Seasonal distribution and life cycles of Ephemeroptera. Ann. Entomol. Soc. Am. **48:**380–391.

Mackay, R. J., and G. B. Wiggins. 1979. Ecological diversity in Trichoptera. Annu. Rev. Entomol. **24:**185–208.

MacKerras, I. M. 1970. Evolution and classification of the insects. In *The Insects of Australia,* Commonwealth Scientific and Industrial Research Organization (ed.). Melbourne University Press, Carlton. pp. 152–167.

MacLeod, E. G., and P. A. Adams. 1967. A review of the taxonomy and morphology of the Berothidae, with a description of a new subfamily from Chile (Neuroptera). Psyche **74**:237–265.

Malcolm, S., and P. Zalucki. 1993. *Biology and Conservation of the Monarch Butterfly.* Natural History Museum of Los Angeles, Los Angeles. 419 pp.

Mallis, A. 1971. *American Entomologists.* Rutgers University Press, New Brunswick, NJ. 549 pp.

Malyshev, S. I. 1968. *Genesis of the Hymenoptera and Phases of Their Evolution.* Methuen, London. 319 pp.

Mangan, B. P. 1994. Pupation ecology of the dobsonfly *Corydalus cornutus* (Corydalidae: Megaloptera) along a large river. J. Freshwater Ecol. **9**:57–62.

Mani, M. S. 1962. *Introduction to High Altitude Entomology.* Methuen, London. 302 pp.

Mani, M. S. 1968. *Ecology and Biogeography of High Altitude Insects.* Series Entomologica. Junk, The Hague. 527 pp.

Mani, M. S., and L. E. Giddings. 1980. *Ecology of Highlands.* Monographiae Biologicae. Junk, The Hague. 249 pp.

Manton, S. M. 1964. Mandibular mechanisms and the evolution of arthropods. Philos. Trans. R. Soc. London, Ser. B **247**:1–183.

Manton, S. M. 1972. The evolution of arthropod locomotory mechanisms. Part 10. Locomotory habits, morphology and evolution of the hexapod classes. Zool. J. Linn. Soc. **51**:203–400.

Manton, S. M. 1973. Arthropod phylogeny—a modern synthesis. J. Zool. **171**:111–130.

Manuel, K. L., and R. M. Bohart. 1993. First report of a twisted-wing insect (Strepsiptera) larva in a caddisfly (Trichoptera). Entomol. News **104**:139.

Marden, J. H., and M. G. Kramer. 1994. Surface-skimming stoneflies: a possible intermediate stage in insect flight evolution. Science **266**:427–429.

Margulis, L, and R. Fester (eds.). 1991. *Symbiosis as a Source of Evolutionary Innovation: Speciation and Morphogenesis.* MIT Press, Cambridge, MA. 454 pp.

Markin, G. P. 1970. The seasonal life cycle of the Argentine ant, *Iridomyrmex humilis* (Hymenoptera: Formicidae), in southern California. Ann. Entomol. Soc. Am. **63**:1238–1242.

Marshall, A. G. 1981. *The Ecology of Ectoparasitic Insects.* Academic Press, London. 459 pp.

Marshall, A. G. 1987. Nutritional ecology of ectoparasitic insects. In *Nutritional Ecology of Insects, Mites, Spiders, and Related Invertebrates,* F. Slansky, Jr.

and J. G. Rodriguez (eds.). John Wiley & Sons, New York. pp. 721–739.

Marshall, A. T. 1966. Histochemical studies on a muco-complex in the Malpighian tubules of cercopid larvae. J. Insect Physiol. **12**:925–932.

Martin, M. 1934. Life history and habits of the pigeon louse [*Columbicola columbae* (Linn.)]. Can. Entomol. **66**:1–16.

Martin, M. M. 1987. *Invertebrate-Microbial Interactions.* Explorations in Chemical Ecology. Cornell University Press, Ithaca, NY. 148 pp.

Martynov, A. V. 1925. Über zwei Grundtypen des Flugel bei den Insekten und ihre Evolution. Zh. Morphol. Oekol. Tiere **4**:465–501.

Martynova, O. 1961. Palaeoentomology. Annu. Rev. Entomol. **6**:285–294.

Matsuda, R. 1965. Morphology and evolution of the insect head. Mem. Am. Entomol. Inst. **1**:1–334.

Matsuda, R. 1970. Morphology and evolution of the insect thorax. Mem. Entomol. Soc. Can. **76**:1–431.

Matthews, G. A., and J. P. Tunstall (eds.). 1994. *Insect Pests of Cotton.* CAB International, Oxon, UK. 593 pp.

Matthews, R. W. 1974. Biology of Braconidae. Annu. Rev. Entomol. **19**:15–32.

Matthews, R. W., and J. R. Matthews. 1978. *Insect Behavior.* John Wiley & Sons, New York. 507 pp.

Matthysse, J. G. 1946. Cattle lice, their biology and control. Cornell Univ. Agric. Exp. Stn. Bull. **823**:1–67.

Maynard, E. A. 1951. *A Monograph of the Collembola, or Springtails, of New York State.* Comstock Publishing Associates, Ithaca, NY. 339 pp.

Mazzini, M., and V. Scali. 1987. *Stick Insects: Phylogeny and Reproduction.* University of Siena, Bologna. 222 pp.

McAlpine, J. F., B. V. Peterson, G. E. Shewell, H. J. Teskey, J. R. Vockeroth, and D. M. Wood (eds.). 1981–89. *Manual of Nearctic Diptera.* Canadian Govt. Publications, Agriculture Canada, Quebec. Monographs 27, 28, 32; vol. 1, 674 pp.; vol. 2, 1332 pp.; vol. 3, 1581 pp.

McGavin, G., and K. Preston-Mafham. 1993. *Bugs of the World.* Facts on File, New York. 192 pp.

McIver, J. D., and G. Stonedahl. 1993. Myrmecomorphy: morphological and behavioral mimicry of ants. Annu. Rev. Entomol. **38**:351–379.

McKenzie, H. L. 1967. *The Mealybugs of California.* University of California Press, Berkeley. 525 pp.

McKeown, K. C., and V. H. Mincham. 1948. The biology of an Australian mantispid (*Mantisspa vittata* Guerin). Aust. Zool. **11**:207–224.

McKittrick, F. A. 1964. Evolutionary studies of cockroaches. Cornell Univ. Agric. Exp. Stn., Mem. **389**:1–197.

Mead-Briggs, A. R. 1977. The European rabbit, the European rabbit flea and myxomatosis. Appl. Biol. **2**:183–261.

Meinander, M. 1972. A revision of the family Coniopterygidae (Planipennia). Acta Zool. Fenn. **136**:1–357.

Melzer, R. R., H. F. Paulus, and N. P. Kristensen. 1994. The larval eye of nannochoristid scorpionflies (Insecta, Mecoptera). Acta Zool. (Copenhagen) **75**:201–208.

Membiela, P. 1990. The mating calls of *Perla madritensis* Rambur, 1842 (Plecoptera, Perlidae). Aquat. Insects **12**:223–226.

Menzel, R., and A. Mercer (eds.). 1987. *Neurobiology and Behavior of Honeybees.* Springer-Verlag, Berlin. 334 pp.

Merritt, R. W. , and K. W. Cummins (eds.). 1996. *An Introduction to the Aquatic Insects of North America.* 3rd ed. Kendall/Hunt, Dubuque, IA. 862 pp.

Metcalf, R. L., and W. H. Luckmann (eds.). 1982. *Introduction to Insect Pest Management.* 2nd ed. John Wiley & Sons, New York. 577 pp.

Metcalf, R. L., and R. A. Metcalf. 1993. *Destructive and Useful Insects: Their Habits and Control.* 5th ed. McGraw-Hill, New York. 971 pp.

Michener, C. D. 1944. Comparative external morphology, phylogeny, and a classification of the bees (Hymenoptera). Bull. Am. Mus. Nat. Hist. **82**:151–326.

Michener, C. D. 1953. Comparative morphological and systematic studies of bee larvae with a key to the families of hymenopterous larvae. Univ. Kansas Sci. Bull. **35**:987–1102.

Michener, C. D. 1969. Comparative social behavior of bees. Annu. Rev. Entomol. **14**:299–342.

Michener, C. D. 1974. *The Social Behavior of the Bees.* Harvard University Press, Cambridge, MA. 404 pp.

Michener, C. D. 1979. Biogeography of the bees. Annu. Missouri Bot. Garden **66**:277–347.

Michener, C. D., and D. A. Grimaldi. 1988. The oldest fossil bee: apoid history, evolutionary stasis, and antiquity of social behavior. Proc. Natl. Acad. Sci. U.S.A. **85**:6424–6426.

Michener, C. D., and S. F. Sakagami. 1990. Classification of the Apidae (Hymenoptera). Univ. Kansas Sci. Bull. **54**(4):76–164.

Michener, C. D., R. J. McGinley, and B. N. Danforth. 1994. *The Bee Genera of North and Central America (Hymenoptera: Apoidea).* Smithsonian Institution Press, Washington, DC. 209 pp.

Mickoleit, G. 1973. Über den Ovipositor des Neuropteroidea und Coleoptera und seine phylogenetische Bedeutung (Insecta, Holometabola). Z. Morphol. Tiere **74**:37–64.

Miller, D. R., and M. Kosztarab. 1979. Recent advances in the study of scale insects. Annu. Rev. Entomol. **24**:1–27.

Miller, L. K. 1969. Freezing tolerance in an adult insect. Science **166**:105–106.

Miller, N. C. E. 1956. *Biology of the Heteroptera.* Leonard Hill, London. 162 pp.

Mills, H. B. 1932. The life history and thoracic development of *Oligotoma texana* (Mel.). Ann. Entomol. Soc. Am. **35**:648–652.

Mills, H. B. 1934. *A Monograph of the Collembola of Iowa.* Iowa State College Press, Ames. 143 pp.

Minks, A., and P. Harrewijn. 1987. *Aphids: Their Biology, Natural Enemies and Control.* Elsevier Science Publishers, Amsterdam. 450 pp.

Miskimen, G. W., N. L. Rodriguez, and M. L. Nazario. 1983. Reproductive morphology and sperm transport facilitation and regulation in the female sugarcane borer, *Diatrea saccháralis* (F.) (Lepidoptera: Crambidae). Ann. Entomol. Soc. Am. **76**:248–252.

Mitchell, R. W. 1969. A comparison of temperate and tropical cave communities. Southwest. Nat. **14**:73–88.

Mitchell, T. B. 1960–1962. Bees of eastern United States. 2 vols. North Carolina. Agric. Exp. Stn. Tech. Bull. **141**:1–538.

Mitton, J. B., and K. B. Sturgeon (eds.). 1982. *Bark Beetles in North American Conifers. A System for the Study of Evolutionary Biology.* University of Texas Press, Austin. 527 pp.

Mizuta, K. 1988a. Adult ecology of *Ceriagrion melanurum* Selys and *Ceriagrion nipponicum* Asahina (Zygoptera: Coenagrionidae): 1. Diurnal variations in reproductive behavior. Odonatologica (Utrecht) **17**:195–204.

Mizuta, K. 1988b. Adult ecology of *Ceriagrion melanurum* Selys and *Ceriagrion nipponicum* Asahina (Zygoptera: Coenagrionidae): 2. Movements and distribution. Odonatologica (Utrecht) **17**:357–364.

Mockford, E. L. 1951. The Psocoptera of Indiana. Proc. Indiana Acad. Sci. **60**:192–204.

Mockford, E. L. 1957. Life history studies on some Florida insects of the genus *Archipsocus* (Psocoptera). Bull. Florida State Mus. **1**:253–274.

Mockford, E. L. 1993. *North American Psocoptera (Insecta).* Flora & Fauna Handbook no. 10. Sandhill Crane Press, Gainesville, FL. 455 pp.

Mockford, E. L., and A. B. Gurney. 1956. A review of the psocids or book-lice and bark-lice of Texas (Psocoptera). J. Washington Acad. Sci. **46**:353–368.

Montgomery, S. L. 1983. Carnivorous caterpillars: the behavior, biogeography and conservation of *Eupithecia* in the Hawaiian Islands. Geojour. **7**.6:549–556.

Moore, J. C., and D. E. Walter. 1988. Arthropod regulation of micro- and mesobiota in below-ground detrital food webs. Annu. Rev. Entomol. **33**:419–439.

Moore, J. M., and M. D. Picker. 1991. Heuweltjies

(earth mounds) in the Clanwilliam district, Cape Province, South Africa: 4000-year-old termite nests. Oecologia (Heidelburg) **86:**424–432.

Moore, T. E., W. Loher, and F. Huber. 1989. *Cricket Behavior and Neurobiology.* Comstock Publishing Associates, Ithaca, NY. 565 pp.

Moran, N. A. 1992. The evolution of aphid life cycles. Annu. Rev. Entomol. **37:**321–348.

Morgan, F. D. 1968. Bionomics of Siricidae. Annu. Rev. Entomol. **13:**239–256.

Moritz, R. F. A., and E. E. Southwick. 1992. *Bees as Superorganisms: An Evolutionary Reality.* Springer-Verlag, Berlin. 395 pp.

Morse, J. C. (ed.). 1984. *Proceedings, IV International Symposium on Trichoptera, Clemson University, South Carolina.* Junk, The Hague. 486 pp.

Morse, J. C., and R. W. Holzenthal. 1996. Trichoptera genera. In *An Introduction to the Aquatic Insects of North America,* R. W. Merritt and K. W. Cummins (eds.). Kendall/Hunt, Dubuque, IA. pp. 350–386.

Mosher, E. 1916. A classification of the Lepidoptera based on characters of the pupa. Bull. Illinois Nat. Hist. Survey **12:**15–159.

Mound, L. A., B. S. Heming, and J. M. Palmer. 1980. Phylogenetic relationships between the families of recent Thysanoptera (Insecta). Zool. J. Linn. Soc. **69:**111–141.

Mukerji, S. 1927. On the morphology and bionomics of *Embia minor* sp. n., with special reference to its spinning organs. Rec. Indian Mus., Calcutta **29:**253–282.

Mukhopadhyay, A. 1993. Chemical ecology of some seed-feeding Hemiptera. In *Chemical Ecology of Phytophagous Insect.,* T. N. Ananthakrishnan and A. A. Raman (eds.). International Science Publisher, New York. pp. 179–195.

Mulkern, G. B. 1967. Food selection by grasshoppers. Annu. Rev. Entomol. **12:**59–78.

Müller, F. 1879. *Ituna* and *Thyridia;* a remarkable case of mimicry in butterflies. Proc. Entomol. Soc. London:xx–xxix.

Murray, M. D. 1960. The ecology of lice on sheep. I-II. Aust. J. Zool. **5:**13–29, 173–187.

Murray, M. D. 1987. Arthropods—the pelage of mammals as an environment. Int. J. Parasitol. **17:**191–195.

Myers, J. G. 1929. *Insect Singers. A Natural History of the Cicadas.* George Routledge and Sons, London. 304 pp.

Narang, S. K., A. C. Bartlett, and R. M. Faust (eds.). 1993. *Applications of Genetics to Arthropods of Biological Control Significance.* CRC Press, Boca Raton, FL. 199 pp.

Nault, L. R., and J. G. Rodriguez (eds.). 1985. *The Leafhoppers and Planthoppers.* Wiley, New York. 500 pp.

Needham, G. R., R. E. Page, Jr., M. Delfinado-Baker, and C. E. Bowman (eds.). 1988. *Africanized Honey Bees and Bee Mites.* Ellis Harwood Ltd., Chichester, UK. 572 pp.

Needham, J. G., and P. W. Claassen. 1925. A monograph of the Plecoptera or stoneflies of America north of Mexico. Thomas Say Found. Publ. **2:**1–397.

Needham, J. G., and M. J. Westfall, Jr. 1955. *A Manual of Dragonflies of North America (Anisoptera).* University of California Press, Berkeley. 615 pp.

Needham, J. G., J. R. Traver, and Y.-C. Hsu. 1935. *The BIology of Mayflies.* Comstock Publishing Associates, Ithaca, NY. 759 pp.

Neumann, D. 1976. Adaptations of chironomids to intertidal environments. Annu. Rev. Entomol. **21:**387–414.

New, T. R. 1975. The biology of Chrysopidae and Hemerobiidae (Neuroptera), with reference to their usage as biocontrol agents: a review. Trans. R. Entomol. Soc. London **127:**115–140.

New, T. R. 1987. Biology of the Psocoptera. Orient. Insects **21:**1–109.

New, T. R. 1991a. *Butterfly Conservation.* Oxford University Press, Oxford. 224 pp.

New, T. R. 1991b. *Insects as predators.* Australian Studies in Biology Series 5. New South Wales University Press, Kensington, Australia. 178 pp.

New, T. R., and G. Theischinger. 1993. *Megaloptera (Alderflies, Dobsonflies).* Handbuch der Zoologie. Band IV, Arthropoda: Insecta, T.33. de Gruyter, Berlin. 102 pp.

Newcomer, E. G. 1918. Some stoneflies injurious to vegetation. J. Agric. Res. **13:**37–41.

Nichols, S. W. 1989. *The Torre-Bueno Glossary of Entomology.* The New York Entomological Society in cooperation with the American Museum of Natural History, New York. 840 pp.

Nicolas, G., and D. Sillans. 1989. Immediate and latent effects of carbon dioxide on insects. Annu. Rev. Entomol. **34:**97–116.

Nieh, J. C. 1993. The stop signal of honey bees: reconsidering its message. Behav. Ecol. Sociobiol. **33:**51–56.

Nielsen, E. S., and I. F. B. Common. 1991. Lepidoptera (moths and butterflies). In *The Insects of Australia,* vol. 2, Commonwealth Scientific and Industrial Research Organization (ed.). Melbourne University Press, Carlton. pp. 817–915.

Niering, W. A. 1966. *The Life of the Marsh.* McGraw-Hill, New York. 232 pp.

Nijhout, H. F. 1994. *Insect Hormones.* Princeton University Press, Princeton, NJ. 280 pp.

Noble-Nesbitt, J. 1989. Spiracular closing mechanisms

in the firebrat, *Thermobia domestica* (Packard) (Thysanura). Tissue Cell **21**:93–100.

Norris, D. M. 1986. Anti-feeding compounds. In *Chemistry of Plant Protection,* vol. 1. *Sterol Biosynthesis Inhibitors and Anti-feeding Compounds.* G. Haug and H. Hoffmann (eds.). Springer-Verlag, Berlin. pp. 97–146.

Norris, K. R. 1965. The bionomics of blow flies. Annu. Rev. Entomol. **10**:47–68.

Nur, U. 1980. Evolution of unusual chromosome systems in scale insects (Coccoidea: Homoptera). In *Insect Cytogenetics,* R. L. Blackman (ed.). Blackwell, Oxford. pp. 97–117.

O'Neill, S. L., H. A. Rose, and D. Rugg. 1987. Social behavior and its relationship to field distribution in *Panesthia cribrata* Saussure (Blattodea: Blaberidae). J. Aust. Entomol. Soc. **26**:313–321.

O'Toole, C., and A. Raw. 1991. *Bees of the World.* Blandford, London. 192 pp.

Ohnishi, E., and H. Ishizaki (eds.). 1990. *Molting and Metamorphosis.* Springer-Verlag, Berlin. 270 pp.

Oken, L. 1831. *Lehrbuch der Naturphilosophie. (2nd. ed.).* Jena.

Oldroyd, H. 1964. *The Natural History of Flies.* Weidenfeld and Nicolson, London. 324 pp.

Oliver, D. R. 1971. Life history of the Chironomidae. Annu. Rev. Entomol. **16**:211–230.

Opler, P. A., and G. O. Krizek. 1984. *Butterflies East of the Great Plains.* Johns Hopkins University Press, Baltimore. 360 pp.

Opler, P. A., and V. Malikul. 1992. *Eastern Butterflies.* Peterson Field Guides, Houghton Mifflin, New York. 396 pp.

Oseto, C. Y., and T. J. Helms. 1976. Anatomy of the adults of *Loxagrotis albicosta.* Univ. Nebraska Stud. **52**:1–127.

Ossiannilsson, F. 1949. Insect drummers: a study of the morphology and function of the sound-producing organs of Swedish Homoptera, Auchenorrhyncha. Opusc. Entomol. Suppl. **10**:1–146.

Oswald, J. D. 1993a. Revision and cladistic analysis of the world genera of the family Hemerobiidae (Insecta: Neuroptera). J. New York Entomol. Soc. **101**:143–299.

Oswald, J. D. 1993b. Phylogeny, taxonomy, and biogeography of extant silky lacewings (Insecta: Neuroptera: Psychopsidae). Mem. Am. Entomol. Soc. **40**:1–65.

Otte, D. 1981a. *The North American Grasshoppers.* vol. 1. *Acrididae: Gomphocerinae and Acridinae.* Harvard University Press, Cambridge, MA. 368 pp.

Otte, D. 1981b. *The North American Grasshoppers.* vol. 2. *Acrididae:* Oedipodinae. Harvard University Press, Cambridge, MA. 352 pp.

Pages, J. 1989. Abdominal sclerites and appendages of Diplura (Insecta, Apterygota). Arch. Sci. (Geneva) **42**:509–552.

Painter, R. H. 1951. *Insect Resistance to Crop Plants.* University of Kansas Press, Lawrence. 520 pp.

Painter, R. H. 1958. Resistance of plants to insects. Annu. Rev. Entomol. **3**:267–290.

Pakaluk, J., and S. A. Slipinski. 1995. *Biology, Phylogeny, and Classification of Coleoptera: Papers Celebrating the 80th Birthday of Roy A. Crowson.* Muzeum i Instytut Zoologii PAN, Warsaw. 2 vols. 1092 pp.

Palmer, J. M., L. A. Mound, G. J. Du Heaume, and C. R. Betts. 1989. *Thysanoptera.* CIE Guides to Insects of Importance to Man, vol. 2. CAB International, Wallingford, UK. 73 pp.

Pannabecker, T. 1995. Physiology of the Malpighian tubule. Annu. Rev. Entomol. **40**:493–510.

Papaj, D. R., and A. C. Lewis (eds.). 1993. *Insect Learning. Ecological and Evolutionary Perspectives.* Chapman & Hall, New York. 398 pp.

Parfin, S. 1952. The Megaloptera and Neuroptera of Minnesota. Am. Midl. Nat. **47**:421–434.

Pasteels, J. M., J.-C. Gregoire, and M. Rowell-Rahier. 1983. The chemical ecology of defense in arthropods. Annu. Rev. Entomol. **28**:263–289.

Pasteur, G. 1982. A classificatory review of mimicry systems. Annu. Rev. Entomol. **13**:169–199.

Paulson, D. R. 1974. Reproductive isolation in damselflies. Syst. Zool. **23**:40–49.

Payne, T. L., M. C., Birch, and C. E. J. Kennedy. 1989. *Mechanisms of Insect Olfaction.* Clarendon Press, Oxford. 364 pp.

Pearman, J. V. 1928. On sound-production in the Psocoptera and on a presumed stridulatory organ. Entomol. Mon. Mag. **64**: 179–186.

Pearson, D. L. 1988. Biology of tiger beetles. Annu. Rev. Entomol. **33**:123–147.

Pearson, D. L., and F. Cassola. 1992. World-wide species richness patterns of tiger beetles (Coleoptera: Cicindelidae): indicator taxon for biodiversity and conservation studies. Conserv. Biol. **6**:376–391.

Peck, O., Z. Boucek, and A. Hoffer. 1964. Keys to the Chalcidoidea of Czechoslovakia. Mem. Entomol. Soc. Can. **34**:1–120.

Pedigo, L. P. 1989. *Entomology and Pest Management.* Macmillan, Inc., New York. 646 pp.

Pellmyr, O., J. Leebens-Mack, and C. J. Huth. 1996. Non-mutualistic yucca moths and their evolutionary consequences. Nature **380**:155–156.

Penny, N. D. 1975. Evolution of the extant Mecoptera. J. Kansas Entomol. Soc. **48**:331–350.

Peters, W. L., and P. G. Peters (eds.). 1973. *Proceedings of the First International Conference on Ephemeroptera.* E. J. Brill, Leiden. 312 pp.

Petersen, H. 1994. A review of collembolan ecology in ecosystem context. Acta Zool. Fenn. **195**:111–118.

Peterson, A. 1948. *Larvae of Insects. Part I. Lepidoptera and Plant Infesting Hymenoptera.* Edwards Brothers, Ann Arbor, MI. 315 pp.

Peterson, A. 1951. *Larvae of Insects.* Part II. *Coleoptera, Diptera, Neuroptera, Siphonaptera, Mecoptera, Trichoptera.* Edwards Brothers, Ann Arbor, MI. 416 pp.

Pfadt, R. E. (ed.). 1985. *Fundamentals of Applied Entomology.* Macmillan, New York. 742 pp.

Pickett, J. A., L. J. Wadhams, C. M. Woodcock, and J. Hardie. 1992. The chemical ecology of aphids. Annu. Rev. Entomol. **37**:67–90.

Pinder, L. C. V. 1986. Biology of freshwater Chironomidae. Annu. Rev. Entomol. **31**:1–23.

Platt, A. P. 1983. Evolution of North American admiral butterflies. Bull. Entomol. Soc. Am. **29**(3):11–22.

Plumstead, E. P. 1963. The influence of plants and environment on the developing animal life of Karroo times. S. Afr. J. Sci. **59**:147–152.

Poinar, G. O., Jr. 1979. *Nematodes for Biological Control of Insects.* CRC Press, Boca Raton, FL. 277 pp.

Poinar, G. O., Jr. 1983. *The Natural History of Nematodes.* Prentice-Hall, Englewood Cliffs, NJ. 323 pp.

Poinar, G. O. Jr. 1988. *Zorotypus palaeus,* new species, a fossil Zoraptera (Insecta) in Dominican amber. J. New York Entomol. Soc. **96**:253–259.

Poinar, G. O., Jr. 1992. *Life in Amber.* Stanford University Press, Palo Alto, CA. 350 pp.

Poinar, G. O., Jr. 1993. Insects in amber. Annu. Rev. Entomol. **39**:145–159.

Poisson, R., and P. Pesson. 1951. Super-ordre des Hemipteroides (Hemiptera Linné, 1758, Rhynchota Bermeister, 1835). Traite Zool. **10**:1385–1803.

Polcyn, D. M. 1994. Thermoregulation during summer activity in Mojave desert dragonflies (Odonata: Anisoptera). Funct. Ecol. **8**:441–449.

Polhemus, J. T. 1996. Aquatic and semiaquatic Hemiptera. In *An Introduction to the Aquatic Insects of North America,* R. W. Merritt and K. W. Cummins (eds.). Kendall/Hunt, Dubuque, IA. pp. 267–297.

Popham, E. J. 1965. A key to the Dermaptera subfamilies. Entomologist **98**:126–136.

Poulson, D. F. 1950. Histogenesis, organogenesis and differentiation in the embryo of *Drosophila melanogaster* (Meigen). In *Biology of Drosophila,* M. Demerec (ed.). John Wiley & Sons, New York. pp. 168–274.

Powell, J. A. 1964. Biological and taxonomic studies on tortricine moths, with reference to the species in California. Univ. California Publ. Entomol. **32**:1–317.

Powell, J. A. 1980. Evolution of larval food preferences in microlepidoptera. Annu. Rev. Entomol. **25**:133–160.

Powell, J. A. 1984. Biological interrelationships of moths and *Yucca schottii.* Univ. California Publ. Entomol. **100**:1–93.

Powell, J. A. 1992. Interrelationships of yuccas and yucca moths. Trends Ecol. Evol. **7**:10–15.

Powell, J. A., and R. A. Mackie. 1966. Biological interrelationships of moths and *Yucca whipplei.* Univ. California Publ. Entomol. **42**:1–59.

Preston-Mafham, K., and R. Preston-Mafham. 1990. *Grasshoppers and Mantids of the World.* Facts on File, New York. 192 pp.

Prestwich, G. D. 1983. Chemical systematics of termite exocrine secretions. Annu. Rev. Ecol. Syst. **14**:287–312.

Prestwich, G. D. 1984. Defense mechanisms of termites. Annu. Rev. Entomol. **29**:201–232.

Prezler, R. W., and P. W. Price. 1988. Host quality and sawfly populations: a new approach to life table analysis. Ecology **69**:2012–2020.

Price, P. W. (ed.). 1974. *Evolutionary Strategies of Parasitic Insects and Mites.* Plenum Press, New York. 224 pp.

Price, P. W. 1984. *Insect Ecology.* 2nd ed. John Wiley & Sons, New York. 607 pp.

Price, P. W., W. J. Mattson, and Y. N. Baranchikov. 1994. *The Ecology and Evolution of Gall-Forming Insects.* General technical report NC, 174. U.S. Dept. of Agriculture, Forest Service, North Central Forest Experiment Station, St. Paul, MN. 222 pp.

Pringle, J. A. 1938. A contribution to the knowledge of *Micromalthus debiliis* Lec. (Coleoptera). Trans. R. Entomol. Soc. London. **87**:271–286.

Pritchard, G. 1983. Biology of Tipulidae. Annu. Rev. Entomol. **28**:1–22.

Proctor, M., and P. Yeo. 1973. *The Pollination of Flowers.* William Collins Sons, London. 418 pp.

Prokopy, R. J. 1991. Epideictic pheromones that influence spacing patterns of phytophagous insects. In *Semiochemicals: Their Role in Pest Control,* D. A. Nordlund, R. L. Jones, and W. J. Lewis (eds.). John Wiley & Sons, New York. pp. 181–197.

Purcell, A. H. 1989. Homopteran transmission of xylem-inhabiting bacteria. In *Advances in Disease Vector Research,* vol. 6, K. F. Harris (ed.). Springer-Verlag, New York. pp. 243–266.

Pyle, R. M. 1981. *The Audubon Society Field Guide to North American Butterflies.* Alfred A. Knopf, New York. 916 pp.

Quicke, D. L. J., S. N. Ingram, H. S. Baillie, and P. V. Gaitens. 1992. Sperm structure and ultrastructure in the Hymenoptera (Insecta). Zool. Scripta **21**:381–402.

Raffa, K. F., T. W. Phillips, and S. M. Salom. 1993. Strategies and mechanisms of host colonization by bark beetles. In *Beetle-Pathogen Interactions in Co-*

nifer Forests, T. D. Schowalter and G. M. Filip (eds.). Academic Press, London. pp. 103–128.

Raikhel, A. S., and T. S. Dhadialla. 1992. Accumulation of yolk proteins in insect oocytes. Annu. Rev. Entomol. **37:**217–251.

Rainey, R. C. 1965. The origin of insect flight: some implications of recent findings from palaeoclimatology and locust migration. Proc. 12th Int. Congr. Entomol. **1964:**134.

Ramsay, G. W. 1990. *Mantodea (Insecta): With a Review of Aspects of Functional Morphology and Biology.* Fauna of New Zealand no. 19. DSIR Pub., Wellington, N.Z. 96 pp.

Rathman, R. J., R. D. Akre, and J. F. Brunner. 1988. External morphology of a species of *Metajapyx* (Diplura: Japygidae) from Washington [USA]. Pan-Pac. Entomol. **64:**185–192.

Redborg, Kurt E., and Ellis G. MacLeod. 1985. The developmental ecology of *Mantispa uhleri* Banks (Neuroptera:Mantispidae). Illinois Biolo. Monog. **53:**1–130.

Rehn, J. A. G., and H. J. Grant, Jr. 1961. A monograph of the Orthoptera of North America (North of Mexico), vol. I. Monogr. Acad. Natl. Sci., Philadelphia **12:**1–255.

Rehn, J. W. H. 1939. Studies in North American Mantispidae (Neuroptera). Trans. Am. Entomol. Soc. **65:**237–263.

Rehn, J. W. H. 1950. A key to the genera of North American *Blattaria,* including established adventives. Entomol. News **61:**64–67.

Rehn, J. W. H. 1951. Classification of the *Blattaria* as indicated by their wings (Orthoptera). Mem. Am. Entomol. Soc. **14:**1–134.

Reitze, M., and W. Nentwig. 1991. Comparative investigations into the feeding ecology of six Mantodea species. Oecologia (Heidelburg) **86:**568–574.

Rembold, H. 1991. Roles of morphogenetic hormones in caste polymorphism in sting bees. In *Morphogenetic Hormones of Arthropods,* vol. 3, A. P. Gupta (ed.). Rutgers University Press, New Brunswick, NJ. pp. 325–345.

Remington, C. L. 1954. The suprageneric classification of the order Thysanura (Insecta). Ann. Entomol. Soc. Am. **47:**277–286.

Remington, C. L. 1956. The "Apterygota." In *A Century of Progress in the Natural Sciences, 1853–1953,* E. L. Kessel (ed.). California Academy of Science, San Francisco. pp. 495–505.

Resh, V. R., and D. M. Rosenberg (eds.). 1984. *The Ecology of Aquatic Insects.* Praeger Publishers, New York. 625 pp.

Retnakaran, A., and K. Binnington. 1991. *Physiology of the Insect Epidermis.* Commonwealth Scientific and Industrial Research Organization, East Melbourne, Victoria. 344 pp.

Ribeiro, J. M. C. 1987. Role of saliva in blood-feeding by arthropods. Annu. Rev. Entomol. **32:**463–478.

Richards, O. W. 1956. *Hymenoptera. Introduction and Key to Families.* Royal Entomological Society of London, Handbook for Identifying British Insects, no. 6, London. 94 pp.

Richards, O. W., and R. G. Davies. 1977. *Imms's General Textbook of Entomology.* 10th ed. Chapman & Hall, London. 1354 pp.

Richards, W. R. 1968. Generic classification, evolution, and biogeography of the Sminthuridae of the world (Collembola). Mem. Entomol. Soc. Can. **53:**1–54.

Richardson, J. W., and A. R. Gaufin. 1971. Food habits of some western stonefly nymphs. Trans. Am. Entomol. Soc. **97:**91–121.

Richman, D. B. 1993. *A Manual of the Grasshoppers of New Mexico (Orthoptera: Acrididae and Romaleidae).* New Mexico State University Cooperative Extension Service Handbook, New Mexico State University, Las Cruces. 112 pp.

Ricker, W. E. 1952. Systematic studies in Plecoptera. Indiana Univ. Stud. Sci. Ser. **18:**1–200.

Ricker, W. E. 1959. Plecoptera. In *Freshwater Biology,* W. T. Edmondson (ed.). John Wiley & Sons, New York. pp. 941–957.

Riddiford, L. M. 1995. Hormonal regulation of gene expression during lepidopteran development. In *Molecular Model Systems in the Lepidoptera,* M. R. Goldsmith and A. S. Wilkins (eds.). Cambridge University Press, Cambridge. pp. 293–322.

Riegel, G. T. 1963. The distribution of *Zorotypus hubbardi* (Zoraptera). Ann. Entomol. Soc. Am. **56:**744–747.

Riek, E. F., and J. Kukalová-Peck. 1984. A new interpretation of dragonfly wing venation, based upon early Upper Carboniferous fossils from Argentina (Insecta: Odonatoidea) and basic character states in pterygote wings. Can. J. Zool. **62:**1150–1166.

Rienks, J. H. 1985. Phenotypic response to photoperiod and temperature in a tropical pierid butterfly. Aust. J. Zool. **33:**837–847.

Riley, C. V. 1892. The yucca moth and yucca pollination. Rept. Missouri Bot. Gdn. **3:**99–158.

Ritcher, P. O. 1958. Biology of Scarabaeidae. Annu. Rev. Entomol. **3:**311–334.

Ritchie, M., and D. Pedgley. 1989. Desert locusts cross the Atlantic. Antenna **13:**10–12.

Ritland, D. B. 1994. Variation in palatability of queen butterflies *(Danaus gilippus)* and implications regarding mimicry. Ecology **75:**732–746.

Ritland, D. B. 1995. Comparative unpalatability of mimetic viceroy butterflies *(Limenitis archippus)* from four south-eastern united states populations. Oecologia **103:**327–336.

Ritland, D. B., and L. P. Brower. 1991. The viceroy butterfly is not a Batesian mimic. Nature **350:**497.

Robert, A. 1962. *Les Libellules du Québec.* Ministère du Tourisme, de la Chasse et de la Pêche, Province de Québec. 224 pp.

Roberts, D. B. 1986. *Drosophila: A Practical Approach.* The Practical Approach Series. IRL Press, Oxford [Oxfordshire]. 295 pp.

Roberts, M. J. 1971. The structure of the mouthparts of some calyptrate dipteran larvae in relation to their feeding habits. Acta Zool. **52:**171–188.

Robinson, G. E. 1992. Regulation of division of labor in insect societies. Annu. Rev. Entomol. **37:**637–665.

Robinson, G. E. 1996. Chemical communication in honeybees. Science **271:**1824–1825.

Roeder, K. D. 1935. An experimental analysis of the sexual behavior of the praying mantis (*Mantis religiosa,* L.). Biol. Bull. **69:**203–220.

Roeder, K. D. 1965. Moths and ultrasound. Sci. Am. **212:**94–102.

Roeder, K. D. 1966. Auditory system of noctuid moths. Science **154:**1515–1521.

Rohdendorf, B. B. 1944. A new family of Coleoptera from the Permian of the Urals. C. R. Acad. Sci. URSS **44:**252–262.

Rohdendorf, B. B. 1969. Phylogenie. Handb. Zool. **4:**1/4.

Romoser, W. S., and J. G. Stoffolano, Jr. 1994. *The Science of Entomology.* 3rd ed. Wm. C. Brown Publishers, Dubuque, IA. 532 pp.

Roonwal, M. L. 1962. Recent developments in termite systematics (1949–60). In *Proceedings of the New Delhi Symposium, 1960, Termites in the Humid Tropics.* UNESCO, Paris. pp. 31–50.

Rosen, D. 1990. *Armored Scale Insects: Their Biology, Natural Enemies, and Control.* World Crop Pests, 4. Elsevier, Amsterdam. vol. A, 400 pp.; vol. B, 700 pp.

Rosenthal, G. A. 1986. The chemical defenses of higher plants. Sci. Am. **254**(1):94–99.

Ross, E. S. 1940. A revision of the Embioptera of North America. Ann. Entomol. Soc. Am. **33:**629–676.

Ross, E. S. 1944. A revision of the Embioptera, or webspinners, of the new world. Proc. U.S. Natl. Mus. **94:**401–504.

Ross, E. S. 1957. The Embioptera of California. Bull. California Insect Survey **6:**51–57.

Ross, E. S. 1970. Biosystematics of the Embioptera. Annu. Rev. Entomol. **15:**157–172.

Ross, E. S. 1984a. A synopsis of the Embiidina of the United States. Proc. Entomol. Soc. Washington **86:**82–93.

Ross, E. S. 1984b. A classification of the Embiidina of

Mexico with descriptions of new taxa. Occas. Pap. California Acad. Sci. **140:**1–54.

Ross, E. S. 1991. Embioptera, Embiidina (Embiids, web-spinners, foot-spinners). In *The Insects of Australia,* vol. 1, Commonwealth Scientific and Industrial Research Organization (ed.). Melbourne University Press, Carlton. pp. 405–409.

Ross, H. H. 1937a. A generic classification of Nearctic sawflies (Hymenoptera, Symphyta). Illinois Biol. Monogr. **15:**1–173.

Ross, H. H. 1937b. Studies of Nearctic aquatic insects. I. Nearctic alder flies of the genus *Sialis* (Megaloptera, Sialidae). Illinois Nat. Hist. Survey Bull. **21:**57–78.

Ross, H. H. 1944. The caddisflies, or Trichoptera, of Illinois. Bull. Illinois Nat. Hist. Survey **23:**1–326.

Ross, H. H. 1956. *Evolution and Classification of the Mountain Caddisflies.* University of Illinois Press, Urbana. 213 pp.

Ross, H. H. 1959. Trichoptera. *In Freshwater Biology, W. T. Edmondson (ed.).* John Wiley & Sons, New York. 1024–1049 pp.

Ross, H. H. 1964. Evolution of caddisworm cases and nets. Am. Zool. **4:**209–220.

Ross, H. H. 1967. The evolution and past dispersal of Trichoptera. Annu. Rev. Entomol. **12:**169–206.

Ross, K. G., and R. W. Matthews. 1991. *The Social Biology of Wasps.* Comstock Publishing Associates, Ithaca, NY. 678 pp.

Roth, L. M. 1970. Evolution and taxonomic significance of reproduction in *Blattaria.* Annu. Rev. Entomol. **15:**75–96.

Roth, L. M., and H. B. Hartman. 1967. Sound production and its evolutionary significance in the *Blattaria.* Ann. Entomol. Soc. Am. **60:**740–752.

Roth, L. M., and E. K. Willis. 1960. The biotic associations of cockroaches. Smithsonian Misc. Collect. **141:**1–470.

Rothschild, M. 1965. The rabbit flea and hormones. Endeavor **24:**162–168.

Rothschild, M. 1973. Secondary plant substances and warning colouration in insects. Symp. R. Entomol. Soc. London **6:**59–83.

Rothschild, M. 1975. Recent advances in our knowledge of the order Siphonaptera. Annu. Rev. Entomol. **20:**241–259.

Rothschild, M. 1985. Mimicry, butterflies and plants. Symb. Bot. Uppsala **22:**82–99.

Rothschild, M., and T. Clay. 1952. *Fleas, Flukes and Cuckoos: A Study of Bird Parasites.* William Collins Sons, London. 304 pp.

Roubik, D. W. 1988. *Ecology and Natural History of Tropical Bees.* Cambridge University Press, Cambridge. 514 pp.

Rousch, D. K., and J. A. McKenzie. 1987. Ecological

genetics of insecticide and acaricide resistance. Annu. Rev. Entomol. **32**:361–380.

Rowell, C. H. F. 1971. The variable coloration of the acridoid grasshoppers. Adv. Insect Physiol. **8**:146–199.

Rudd, R. L. 1964. *Pesticides in the Living Landscape.* University of Wisconsin Press, Madison. 320 pp.

Rudinsky, J. A. 1962. Ecology of the Scolytidae. Annu. Rev. Entomol. **7**:327–348.

Rudolph, D. 1982. Occurrence, properties and biological implications of the active uptake of water vapor from the atmosphere in Psocoptera. J. Insect Physiol. **28**:111–121.

Rudolph, D. 1983. The water uptake system of the Phthiraptera. J. Insect Physiol. **29**:15–25.

Rugg, D., and H. A. Rose. 1991. Biology of *Macropanesthia rhinoceros* Saussure (Dictyoptera: Blaberidae). Ann. Entomol. Soc. Am. **84**:575–582.

Russell, L. K. 1982. The life history of *Caurinus dectes* Russell, with a description of the immature stages (Mecoptera: Boreidae). Entomol. Scand. **13**:225–235.

Sacca, G. 1964. Comparative bionomics in the genus *Musca.* Annu. Rev. Entomol. **9**:341–358.

Sailer, R. I. 1950. A thermophobic insect. Science **112**:743.

Salmon, J. T. 1951. Keys and bibliography to the Collembola. Victoria Univ. College Publ. Zool., Wellington, NZ No. **8**:1–82.

Salmon, J. T. 1956. Keys and bibliography to the Collembola. First supplement. Victoria Univ. College. Publ. Zool., Wellington, NZ. No. **20**:1–35.

Salt, G. 1970. *The Cellular Defense Reactions of Insects.* Cambridge University Press, New York. 118 pp.

Schaefer, C. W. (ed.) 1996. *Studies on Hemipteran Phylogeny.* Thomas Say Found. Publ. Proc., Entomol. Soc. Amer., Lanham, MD. 244 pp.

Schal, C., J.-Y. Gautier, and W. J. Bell. 1984. Behavioral ecology of cockroaches. Biol. Rev. **59**:209–254.

Schaller, F. 1971. Indirect sperm transfer by soil arthropods. Annu. Rev. Entomol. **16**:407–446.

Scharrer, B. 1951. The woodroach. Sci. Am. **185**:58–62.

Schneider, D. 1992. 100 years of pheromone research: An essay on Lepidoptera. Naturwissenschaften **79**:241–250.

Schoonhoven, L. M. 1972. Secondary plant substances and insects. In *Structural and Functional Aspects of Phytochemistry,* V. C. Runeckles and T. C. Tso (eds.). Academic Press, New York. pp. 197–224.

Schowalter, T. D., and G. M. Filip (eds.). 1993. *Beetle-Pathogen Interactions in Conifer Forests.* Academic Press, London. 252 pp.

Schuh, R. T., and J. H. Slater. 1995. *True Bugs of the World (Hemiptera: Heteroptera).* Cornell University Press, Ithaca, NY. 416 pp.

Scoble, M. J. 1986. The structure and affinities of the Hedyloidea: a new concept of the butterflies. Bull. Br. Mus. Nat. Hist. (Entomol.) **53**:251–286.

Scoble, M. J. 1992. *The Lepidoptera: Form, Function, and Diversity.* Natural History Museum Publications. Oxford University Press, Oxford. 404 pp.

Scott, H. G. 1961. Collembola: Pictorial keys to the Nearctic genera. Ann. Entomol. Soc. Am. **54**:104–113.

Scriber, J. M., and F. Slansky, Jr. 1981. The nutritional ecology of immature insects. Annu. Rev. Entomol. **26**:183–211.

Scudder, G. G. E. 1971. Comparative morphology of insect genitalia. Annu. Rev. Entomol. **16**:379–406.

Seastedt, T. R. 1984. The role of microarthropods in decomposition and mineralization processes. Annu. Rev. Entomol. **29**:25–46.

Seeley, T. D. 1992. The tremble dance of the honey bee: message and meanings. Behav. Ecol. Sociobiol. **31**:375–383.

Seguy, E. 1950. La biologie de Diptères. Encycl. Entomol. **26**:1–609.

Seguy, E. 1951. Ordre des Diptères (Diptera Linne, 1758). Traite Zool. **10**:449–744.

Setty, L. R. 1940. Biology and morphology of some North American Bittacidae. Am. Midl. Nat. **23**:257–353.

Severin, H. H. P. 1911. The life-history of the walkingstick, *Diaphemeromera femorata* Say. J. Econ. Entomol. **4**:307–320.

Seybold, S. J., D. R. Quilici, J. A. Tillman, D. Vanderwel, D. L. Wood, and G. J. Blomquist. 1995. *De novo* biosynthesis of the aggregation pheromone components ipsenol and ipsdienol by the pine bark beetles *Ips paraconfusus* Lanier and *Ips pini* (Say) (Coleoptera: Scolytidae). Proc. Natl. Acad. Sci. U.S.A. **92**:8393–8397.

Shapiro, J. P., J. H. Law, and M. A. Wells. 1988. Lipid transport in insects. Annu. Rev. Entomol. **33**:297–318.

Sharov, A. G. 1966. *Basic Arthropodan Stock with Special Refrence to Insects.* Pergamon Press, Oxford. 271 pp.

Sharov, A. G. 1971. *Phylogeny of the Orthopteroidea.* Israel Program for Scientific Translations, Jerusalem. 251 pp.

Sharov, A. G. 1975. The phylogenetic relations of the order Thysanoptera. Entomol. Rev. **51**:506–508.

Shattuck, S. O. 1992. Review of the dolichoderine ant genus *Iridomyrmex* Mayr with descriptions of three new genera (Hymenoptera: Formicidae). J. Aust. Entomol. Soc. **31**:13–18.

Shear, W. A., and Kukalová-Peck. 1990. The ecology of paleozoic terrestrial arthropods: the fossil evidence. Can. J. Zool. **68**:1807–1834.

Shear, W. A., P. M. Bonamo, J. D. Grierson, W. D. I. Rolfe, E. L. Smith, and R. H. Norton. 1984. Early land animals in North America: evidence from Devonian age arthropods from Gilboa, New York. Science **224**:492–494.

Sheppard, P. M. 1962. Some aspects of the geography, genetics and taxonomy of a butterfly. In *Taxonomy and Geography,* D. Nichols (ed.). Systematics Association, London. Publ. 4. pp. 135–152.

Shetlar, D. J. 1978. Biological observations on *Zorotypus hubbardi* Caudell (Zoraptera). Entomol. News **89**:217–223.

Silberglied, R. E., and T. Eisner. 1968. Mimicry of Hymenoptera by beetles with unconventional flight. Science **163**:486–488.

Skaife, S. H. 1961. *Dwellers in Darkness.* Doubleday, Garden City, NY. 180 pp.

Skarlato, Orest Aleksandrovich. 1985. *Systematics of Diptera (Insecta): Ecological and Morphological Principles.* Oxonian Press, New Delhi. 185 pp.

Slabaugh, R. E. 1940. A new thysanuran, and a key to the domestic species of Lepismatidae (Thysanura) found in the United States. Entomol. News **51**:95–98.

Slater, J. A., and R. M. Baranowski. 1978. *How to Know the True Bugs (Hemiptera-Heteroptera).* Wm. C. Brown, Dubuque, IA. 256 pp.

Smart, J., and N. F. Hughes. 1973. The insect and the plant: progressive palaeoecological integration. In *Insect/Plant Relationships,* H. F. van Emden (ed.). John Wiley & Sons, New York. pp. 143–155.

Smart, J., R. J. Wootton, and R. A. Crowson. 1967. Insecta. In *The Fossil Record. A Symposium with Documentation,* W. B. Hailand (ed.). Geological Society, London. pp. 508–534.

Smith, C. F., R. W. Eckel, and E. Lampert. 1992. A key to many of the common alate aphids of North Carolina (Aphididae: Homoptera). North Carolina Agric. Res. Serv. Tech. Bull. **299**:1–92.

Smith, C. M. 1989. *Plant Resistance to Insects: A Fundamental Approach.* John Wiley & Sons., New York. 286 pp.

Smith, E. L. 1969. Evolutionary morphology of external insect genitalia. I. Origin and relationships to other appendages. Ann. Entomol. Soc. Am. **62**:1051–1079.

Smith, E. L. 1970. Biology and structure of some California bristletails and silverfish (Apterygota: Microcoryphia, Thysanura). Pan-Pac. Entomol. **46**:212–225.

Smith, K. G. V. 1986. *A Manual of Forensic Entomology.* British Museum (Natural History), London. 205 pp.

Smith, R. C. 1922. The biology of the Chrysopidae. Cornell Univ. Agric. Exp. Stn. Mem. **58**:1291–1372.

Smith, R. F., T. E. Mittler, and C. N. Smith (eds.). 1973. *History of Entomology.* Annual Reviews, Palo Alto. CA. 517 pp.

Smith, R. I., and J. T. Carlton. 1975. *Light's Manual: Intertidal Invertebrates of the Central California Coast.* 3rd ed. University of California Press, Berkeley. 716 pp.

Smithers, C. N. 1990. Keys to the families and genera of Psocoptera (Arthropoda, Insecta). Tech. Rep. Aust. Mus. **2**:1–82.

Smithers, C. N., and C. Lienhard. 1992. A revised bibliography of the Psocoptera (Arthropoda: Insecta). Tech. Rep. Austr. Mus. **6**:1–86.

Snodgrass, R. E. 1935. *Principles of Insect Morphology.* McGraw-Hill, New York. 667 pp.

Snodgrass, R. E. 1944. The feeding apparatus of biting and sucking insects affecting man and animals. Smithsonian Misc. Collect. **104**:1–113.

Snodgrass, R. E. 1946. The skeletal anatomy of fleas (Siphonaptera). Smithsonian Misc. Collect. **104**:1–89.

Snodgrass, R. E. 1950. Comparative studies on the jaws of mandibulate arthropods. Smithsonian Misc. Collect. **116**:1–185.

Snodgrass, R. E. 1952. *A Textbook of Arthropod Anatomy.* Comstock Publishing Associates, Ithaca, NY. 363 pp.

Snodgrass, R. E. 1954. The dragonfly larva. Smithsonian Misc. Collect. **123**:1–38.

Snodgrass, R. E. 1956. *Anatomy of the Honey Bee.* Cornell University Press, Ithaca, NY. 334 pp.

Snodgrass, R. E. 1958. Evolution of arthropod mechanisms. Smithsonian Misc. Collect. **138**:1–77.

Snodgrass, R. E. 1963. A contribution toward an encyclopedia of insect anatomy. Smithsonian Misc. Collect. **146**:1–48.

Snow, K. R. 1974. *Insects and Disease.* John Wiley & Sons, New York. 208 pp.

Snyder, T. E. 1954. *Order Isoptera—The Termites of the United States and Canada.* National Pest Control Association, New York. 64 pp.

Snyder, T. E. 1956. Annotated, subject-heading bibliography of termites 1350 B.C. to A.D 1954. Smithsonian Misc. Collect. **130**:1–305.

Snyder, T. E. 1968. Second supplement to the annotated, subject-heading bibliography of termites 1961–1965. Smithsonian Misc. Collect. **152**:1–188.

Solbrig, O. T., and D. J. Solbrig. 1979. *Population Biology and Evolution.* Addison-Wesley Publishing, Reading, MA. 468 pp.

Sommerman, K. M. 1943. Bionomics of *Ectopsocus pumilis* (Banks) (Corrodentia, Caeciliidae). Psyche **50**:53–63.

Sorensen, J. T. 1980. An integumental anatomy for the butterfly *Glaucopsyche lygdamus* (Lepidoptera:

Lycaenidae): a morphological terminology and homology. Zool. J. Linn. Soc. **70**:55–101.

Sorensen, J. T., B. C. Campbell, R. J. Gill, and J. D. Steffen-Campbell. 1995. Non-monophyly of Auchenorrhyncha ("Homoptera"), based upon 18S rDNA phylogeny: eco-evolutionary and cladistic implications within pre-Heteropterodea Hemiptera (S. L.) and a proposal for new monophyletic suborders. Pan-Pac. Entomol. **71**:31–60.

Sosa, O., Jr. 1989. Carlos J. Finlay and yellow fever: a discovery. Bull. Entomol. Soc. Am. **35**(2):23–25.

Southwood, T. R. E. (ed.). 1979. *Ecological Methods with Particular Reference to the Study of Insect Populations.* 2nd ed. Methuen, London. 524 pp.

Southwood, T. R. E., and D. Leston. 1959. *Land and Water Bugs of the British Isles.* Frederick Warne and Co. London. 436 pp.

Spangler, H. G. 1988. Moth hearing, defense, and communication. Annu. Rev. Entomol. **33**:59–82.

Spence, J. R., and N. M. Andersen. 19   Biology of water striders: interactions between systematics and ecology. Annu. Rev. Entomol. **39**:101–128.

Spencer, K. A. 1990. *Host Specialization in the world Agromyzidae (Diptera).* Series Entomologica, vol. 45. Kluwer Academic Publishers, Dordrecht, Netherlands. 444 pp.

Spencer, K. A., and G. C. Steyskal. 1986. *Manual of the Agromyzidae (Diptera) of the United States.* Agriculture Handbook no. 638. U.S. Department of Agriculture, Agricultural Research Service, Washington, DC. 478 pp.

Spivak, M., D. J. C. Fletcher, and M. D. Breed (eds.). 1991. *The "African" Honey Bee.* Westview Press, Boulder, CO. 435 pp.

Spooner, C. S. 1938. The phylogeny of the Hemiptera based on a study of the head capsule. Illinois Biol. Monogr. **16**:1–102.

Spradbery, J. P. 1973. *Wasps. An Account of the Biology and Natural History of Social and Solitary Wasps.* University of Washington Press, Seattle. 416 pp.

St Clair, R. M. 1993. Life histories of six species of Leptoceridae (Insecta: Trichoptera) in Victoria. Aust. J. Mar. Freshwater Res. **44**:363–379.

Staddon, B. W. 1986. Biology of scent glands in the Hemiptera-Heteroptera. Ann. Soc. Entomol. France **22**:183–190.

Stamp, N. E., and T. M. Casey (eds.) 1993. *Caterpillars. Ecology and Evolutionary Constraints on Foraging.* Chapman & Hall, NY 595 pp.

Stannard, L. J., Jr. 1956. The relationship of the hempteroid insects. Syst. Zool. **5**:94–95.

Stannard, L. J., Jr. 1957. The phylogeny and classification of the North American genera of the suborder Tubulifera (Thysanoptera). Illinois Biol. Monogr. **25**:1–200.

Stannard, L. J., Jr. 1968. The thrips, or Thysanoptera, of Illinois. Illinois Nat. Hist. Surv. Bull. **29**:215–552.

Starks, K. J., R. Muniappan, and R. D. Eikenbary. 1972. Interaction between plant resistance and parasitism against greenbug on barley and sorghum. Ann. Entomol. Soc. Am. **65**:650–655.

Steffan, W. A. 1966. A generic revision of the family Sciaridae (Diptera) of America north of Mexico. Univ. California Publ. Entomol. **44**:1–77.

Stehr, F. W. (ed.). 1987. *Immature Insects.* vol. 1. Kendall/Hunt, Dubuque, IA. 754 pp.

Stehr, F. W. (ed.). 1991. *Immature Insects.* vol. 2. Kendall/Hunt, Dubuque, IA. 975 pp.

Steinhaus, E. A. 1946. *Insect Microbiology.* Comstock Publishing Associates, Ithaca, NY. 763 pp.

Steinhaus, E. A. 1949. *Principles of Insect Pathology.* McGraw-Hill, New York. 757 pp.

Stephen, F. M., C. W. Berisford, D. L. Dahlsten, P. Fenn, and J. C. Moser. 1993. Invertebrate and microbial associates. In *Beetle-Pathogen Interactions in Conifer Forests,* T. D. Schowalter and G. M. Filip (eds.). Academic Press, London. pp. 129–153.

Stephen, W. P., G. E. Bohart, and P. F. Torchio. 1969. *The Biology and External Morphology of Bees with a Synopsis of the Genera of Northwestern America.* Oregon State University, Agricultural Experiment Station, Corvallis. 140 pp.

Stern, D. L., and W. A. Foster. 1996. The evolution of soldiers in aphids. Biol. Rev. Cambridge Philos. Soc. **71**:27–79.

Stewart, K. W., B. P. Stark, and J. A. Stanger. 1988. *Nymphs of North American Stonefly Genera (Pleoptera).* Thomas Say Foundation (Series), vol. 12, Entomological Society of America, College Park, MD. (also reprinted, 1993, University of North Texas Press). 460 pp.

Stokes, D. W. 1983. *A Guide to Observing Insect Lives.* Little, Brown, Boston. 371 pp.

Stouthamer, R., R. F. Luck, and W. D. Hamilton. 1990. Antibiotics cause parthenogenetic *Trichogramma* to revert to sex. Proc. Natl. Acad. Sci. U.S.A. **87**:2424–2427.

Stowe, M. K., T. C. J. Turlings, J. H. Lougrin, W. J. Lewis, and J. H. Tumlinson. 1995. The chemistry of eavesdropping, alarm, and deceit. Proc. Nat. Acad. Sci. U.S.A. **92**:23–28.

Strohecker, H. F., W. W. Middlekauff, and D. C. Rentz. 1968. The grasshoppers of California. Bull. California Insect Survey **10**:1–171.

Strong, D. R., J. H. Lawton, and R. Southwood. 1984. *Insects on Plants. Community Patterns and Mechanisms.* Harvard University Press, Cambridge, MA. 313 pp.

Sturm, H. 1987. Das Paarungsverhalten von *Thermobia*

domestica (Packard) (Lepismatidae, Zygentoma, Insecta). Braunschweig. Naturk. Schr. **2**:673–711.

Sturm, H. 1992. Mating behavior and sexual dimorphism in *Promesomachilis hispanica* Silvestri, 1923 (Machilidae, Archeognatha, Insecta). Zool. Anz. **228**:60–73.

Sturm, H., and C. Bach De Roca. 1993. On the systematics of the Archaeognatha (Insecta). Entomol. Gen. **18**:55–90.

Sudd, J. H., and N. R. Franks. 1987. *The Behavioural Ecology of Ants. Tertiary Level Biology*. Blackie, Glasgow. 206 pp.

Sweeney, B. W., D. H. Funk, and L. J. Standley. 1993. Use of the stream mayfly *Cloeon triangulifer* as a bioassay organism: life history response and body burden following exposure to technical chlordane. Environ. Toxicol. Chem. **12**:115–125.

Symmons, S. 1952. Comparative anatomy of the mallophagan head. Trans. Zool. Soc. London **27**:349–436.

Tabashnik, B. E. 1994. Evolution of resistance to *Bacillus thuringiensis*. Annu. Rev. Entomol. **39**:47–79.

Tallamy, D. W., and T. K. Wood. 1986. Convergence patterns in subsocial insects. Annu. Rev. Entomol. **31**:369–390.

Tanada, Y., and H. K. Kaya. 1993. *Insect Pathology*. Academic Press, New York. 666 pp.

Telfer, W. H., and J. G. Kunkel. 1991. The function and evolution of insect storage hexamers. Annu. Rev. Entomol. **36**:205–228.

Terra, P. S. 1992. Maternal care in *Cardioptera brachyptera* (Mantodea, Vatidae, Photininae). Rev. Bras. Entomol. **36**:493–503.

Terra, W. R. 1990. Evolution of digestive systems of insects. Annu. Rev. Entomol. **35**:181–200.

Thayer, M. K. 1992. Discovery of sexual wing dimorphism in Staphylinidae (Coleoptera): *"Omalium" flavidum,* and a discussion of wing dimorphism in insects. J. New York Entomol. Soc. **100**:540–573.

Thorne, B. L. 1990. A case for ancestral transfer of symbionts between cockroaches and termites. Proc. R. Soc. London Ser. B, Biol. Sci. **241**:37–41.

Thorne, B. L. 1991. Ancestral transfer of symbionts between cockroaches and termites: an alternative hypothesis. Proc. R. Soc. London Ser. B Biol. Sci. **246**:191–196.

Thorne, B. L., and J. M. Carpenter. 1992. Phylogeny of the dictyoptera. Syst. Entomol. **17**:253–268.

Thornhill, R., and K. P. Sauer. 1991. The notal organ of the scorpionfly *(Panorpa vulgaris):* an adaptation to coerce mating duration. Behav. Ecol. **2**:156–164.

Thornhill, R., and P. Sauer. 1992. Genetic sire effects on the fighting ability of sons and daughters and mating success of sons in a scorpionfly. Anim. Behav. **43**:255–264.

Thornton, I. W. B. 1985. The geographical and ecological distribution of arboreal Psocoptera. Annu. Rev. Entomol. **30**:175–196.

Tichy, H. 1988. A kinematic study of front legs' movement in walking Protura (Insecta). J. Exp. Zool. **245**:130–136.

Tiegs, O. W., and S. M. Manton. 1958. The evolution of the Arthropoda. Biol. Rev. **33**:255–337.

Tietz, H. M. 1973. *An Index to the Described Life Histories, Early Stages, and Hosts of the Macrolepidoptera of the Continental United States and Canada.* E. W. Classey, Hampton, England. 1042 pp.

Tol, J. van, and M. J. Verdonk. 1988. *The Protection of Dragonflies (Odonata) and Their Biotopes.* Nature and Environment Series, no. 38. European Committee for the Conservation of Nature and Natural Resources, Strasbourg. 181 pp.

Townes, H. 1969. The genera of Ichneumonidae. Mem. Am. Entomol. Inst. **11**:1–300, **12**:10–537,**13**:1–307.

Traniello, J. F. A. 1989. Foraging strategies of ants. Annu. Rev. Entomol. **34**:191–210.

Traub, R., and H. Starcke (eds.). 1980. *Proceedings of the International Conference on Fleas, Ashton Wold, Peterborough, UK, 21–25 June 1977.* Balkema, Rotterdam. 420 pp.

Trivedi, M. C., B. S. Rawat, and A. K. Saxena. 1991. The distribution of lice (Phthiraptera) on poultry *(Gallus domesticus).* Int. J. Parasitol. **21**:247–250.

Troester, G. 1990. The head of *Hybophthirus notophallus* (Neumann) (Phthiraptera: Anoplura): an analysis of functional morphology and phylogeny. Stuttgart Beitr. Naturkd. Ser A (Biol.) **442**:1–89.

Truman, J. W. 1990. Neuroendocrine control of ecdysis. In *Molting and Metamorphosis,* E. Ohnishi and H. Ishizaki (eds.). Springer-Verlag, Berlin. pp. 67–82.

Truman, J.W. 1995. Lepidoptera as model systems for studies of hormone action on the central nervous system. In *Molecular Model Systems in the Lepidoptera,* M. R. Goldsmith and A. S. Wilkins (eds.). Cambridge University Press, Cambridge. pp. 323–339.

Truman, J. W. 1996. Ecdysis control sheds another layer. Science **271**:40–41.

Tschinkel, W. R. 1975a. A comparative study of the chemical defensive system of the tenebrionid beetles. I. Chemistry of the secretions. J. Insect Physiol. **21**:753–783.

Tschinkel, W. R. 1975b. A comparative study of the chemical defensive system of tenebrionid beetles. II. Defensive behavior and ancillary structures. Ann. Entomol. Soc. Am. **68**:439–453.

Turner, J. R. G. 1981. Adaptation and evolution in *Heliconius*: a defense of NeoDarwinism. Annu. Rev. Ecol. Syst. **12**:99–121.

Tuxen, S. L. 1959. The phylogenetic significance of

entognathy in entognathous apterygotes. Smithsonian Misc. Collect. **137**:379–416.

Tuxen, S. L. 1964. *The Protura. A Revision of the Species of the World with Keys for Determination.* Hermann Press, Paris. 360 pp.

Tuxen, S. L. (ed.). 1970a. *Taxonomist's Glossary of Genitalia in Insects,* 2nd ed. rev. Ejnar Munksgaard, Copenhagen. 359 pp.

Tuxen, S. L. 1970b. The systematic position of entognathous Apterygotes. An. Esc. Nat. Cienc. Biol., Mex. **17**:65–69.

Tuxen, S. L. 1986. *Protura (Insecta).* Fauna of New Zealand, no. 9. Science Information Publishing Centre, DSIR, Wellington, N.Z. 50 pp.

Ulrich, W. 1966. Evolution and classification of the Strepsiptera. Proc. lst Int. Congr. Parasitol. **1**:609–611.

Urbani, C. B. 1989. Phylogeny and behavioral evolution in ants with a discussion of the role of behavior in evolutionary processes. Ethol. Ecol. Evol. **1**:137–168.

Urbani, C. B. 1993. The diversity and evolution of recruitment behaviour in ants, with a discussion of the usefulness of parsimony criteria in the reconstruction of evolutionary histories. Insectes Soc. **40**:233–260.

Urquhart, F. A. 1987. *The Monarch Butterfly: International Traveler.* Nelson-Hall, Chicago. 232 pp.

Usinger, R. L. (ed.). 1956a. *Aquatic Insects of California.* University of California Press, Berkeley. 508 pp.

Usinger, R. L. 1956b. Aquatic Hemiptera. In *Aquatic Insects of California,* R. L. Usinger (ed.). University of California Press, Berkeley. pp. 182–228.

Usinger, R. L. 1967. *The Life of Rivers and Streams.* McGraw-Hill, New York. 232 pp.

Uvarov, B. P. 1966. *Grasshoppers and Locusts. A Handbook of General Acridology.* vol 1. *Anatomy, Physiology and Development, Phase Polymorphism, Introduction to Taxonomy.* Cambridge University Press, Cambridge, MA. 481 pp.

van der Plank, J. E. 1963. *Plant Disease Epidemics and Control.* Academic Press, New York. 309 pp.

van Dreische, R. G., and T. S. Bellows (eds.). 1996. *Biological Control.* Chapman & Hall, New York. 539 pp.

van Driesche, R. G., and T. S. Bellows. 1993. *Steps in Classical Arthropod Biological Control.* Thomas Say Publications in Entomology. Proceedings. Entomological Society of America, Lanham, MD. 88 pp.

van Emden, H. F. (ed.). 1972. *Aphid Technology.* Academic Press, Inc, New York. 344 pp.

Vandel, A. 1965. *Biospeleology, the Biology of Cavernicolous Animals.* Pergamon Press, Oxford. 524 pp.

Vickery, V. R., D. K. McE. Kevan, and C. D. Dondale.

1985. *The Grasshoppers, Crickets, and Related Insects of Canada and Adjacent Regions: Ulonata, Dermaptera, Cheleutoptera, Notoptera, Dictuoptera, Grylloptera, and Orthoptera.* The Insects and Arachnids of Canada, pt. 14. Biosystematics Research Institute, Research Branch, Agriculture Canada, Ottawa. 918 pp.

Waage, J. K. 1979. Dual function of the damselfly penis: sperm removal and transfer. Science **203**:916–918.

Waage, J. K. 1984. Sperm competition and the evolution of odonate mating systems. In *Sperm Competition and the Evolution of Animal Mating Systems,* R. L. Smith (ed.). Academic Press, New York. pp. 251–290.

Wagner, M., and K. F. Raffa. 1993. *Sawfly Life History Adaptations to Woody Plants.* Academic Press, San Diego, CA. 564 pp.

Wajnberg, E., and S. A. Hassan. 1994. *Biological Control with Egg Parasitoids.* CAB International, Wallingford, Oxon., UK. 286 pp.

Waldorf, E. S. 1974. Sex pheromone in the springtail, *Sinella curviseta.* Environ. Entomol. **3**:916–918.

Walker, E. M. 1949. On the anatomy of *Grylloblatta campodeiformis* Walker. V. The organs of digestion. Can. J. Res. **27**:309–344.

Walker, E. M. 1953. *The Odonata of Canada and Alaska,* vol. 1, *General, the Zygoptera-Damselflies.* University of Toronto Press, Toronto, Canada. 292 pp.

Walker, E. M. 1958. *The Odonata of Canada and Alaska,* vol. 2, part III: *The Anisoptera—Four Families.* University of Toronto Press, Toronto, Canada. 318 pp.

Walker, E. M., and P. S. Corbet. 1975. *The Odonata of Canada and Alaska,* vol. 3, part III: *The Anisoptera— Three Families.* University of Toronto Press, Toronto, Canada. 307 pp.

Walker, T. J. 1963. The taxonomy and calling songs of the United States tree crickets (Orthoptera: Gryllidae: Oecanthinae). II. The *nigricornis* group of the genus *Oecanthus.* Ann. Entomol. Soc. Am. **56**:772–789.

Wallace, J. B., and R. W. Merritt. 1980. Filter-feeding ecology of aquatic insects. Annu. Rev. Entomol. **25**:103–132.

Wallwork, J. A. 1970. *Ecology of Soil Animals.* McGraw-Hill, New York. 283 pp.

Ward, J. V. 1992. *Aquatic Insect Ecology.* vol. 1. *Biology and Habitat.* John Wiley & Sons, New York. 438 pp.

Ward, J. V., and J. A. Stanford. 1982. Thermal responses in the evolutionary ecology of aquatic insects. Annu. Rev. Entomol. **27**:97–117.

Ware, G. W. 1989. *The Pesticide Book.* Thomson Publications, Fresno, CA. 386 pp.

Washburn, J. O., M. E. Gross, D. R Mercer, and J. R.

Anderson. 1988. Predator-induced trophic shift of a free-living ciliate: parasitism of mosquito larvae by their prey. Science **240**:1193–1195.

Watson, J. A. L. 1965. The endocrine system of the lepismatid Thysanura and its phylogenetic implications. Proc. 12th Int. Congr. Entomol. **1964**:144.

Way, M. J. 1963. Mutualism between ants and honeydew-producing Homoptera. Annu. Rev. Entomol. **8**:307–344.

Way, M. J. 1968. Intra-specific mechanisms with special reference to aphid populations. In *Insect Abundance,* T. R. E. Southwood (ed.). Blackwell, Oxford. pp. 18–36.

Weaver, J. S., and J. C. Morse. 1986. Evolution of feeding and case-making behavior in Trichoptera. J. North Am. Benthol. Soc. **5**:150–158.

Wehner, R. 1984. Astronavigation in insects. Annu. Rev. Entomol. **29**:277–298.

Wenner, A. M. 1971. *The Bee Language Controversy, An Experience in Science.* Education Programs Improvement Corporation, Boulder, CO. 109 pp.

Wenner, A. M., and P. H. Wells. 1990. *Anatomy of a Controversy: The Question of a "Language" Among Bees.* Columbia University Press, New York. 399 pp.

Wenzel, J. W. 1990. A social wasp's nest from the Cretaceous period, Utah, USA, and its biogeographical significance. Psyche **97**:21–30.

West, L. S. 1951. *The Housefly.* Comstock Publishing Associates, Ithaca, NY. 584 pp.

Westbrook, A. L., and W. E. Bollenbacher. 1990. The prothoracicotropic hormone neuroendocrine axis in *Manduca sexta*: development and function. In *Molting and Metamorphosis,* E. Ohnishi and H. Ishizaki (eds.). Springer-Verlag, Berlin. pp. 3–16.

Wheeler, W. C., R. T. Schuh, and R. Bang. 1993. Cladistic relationships among higher groups of Heteroptera: congruence between morphological and molecular data sets. Entomol. Scand. **24**:121–137.

Wheeler, W. M. 1930. *Demons of the Dust.* W. W. Norton, New York. 378 pp.

Whitcomb, R.F. 1981. The biology of spiroplasmas. Annu. Rev. Microbiol. **26**:397–425.

White, R. E. 1983. *A Field Guide to the Beetles.* Peterson Field Guide Series, Houghton Mifflin, Boston. 368 pp.

Whitfield, J. B. 1990. Parasitoids, polydnaviruses and endosymbiosis. Parasitology Today **6**:381–384.

Whiting, M. F. 1991. A distributional study of *Sialis* (Megaloptera: Sialidae) in North America. Entomol. News **102**:50–56.

Whiting, M. F. 1994. Cladistic analysis of the alderflies of America North of Mexico (Megaloptera: Sialidae). Syst. Entomol. **19**:77–91.

Whiting, M. F., and J. Kathirithamby. 1995. Strepsiptera do not share hind wing venation synapomorphies with Coleoptera: a reply to Kukalová-Peck and Lawrence. J. New York Entomol. Soc. **103**:1–14.

Whittaker, R. H., and P. P. Feeny. 1971. Allelochemics: chemical interactions between species. Science **171**:757–770.

Whitten, W. M., A. M. Young, and N. H. Williams. 1989. Function of glandular secretions in fragrance collection by male euglossine bees (Apidae: Euglossini). J. Chem. Ecol. **15**:1285–1296.

Whitten, W. M., A. M. Young, and D. L. Stern. 1993. Nonfloral sources of chemicals that attract male euglossine bees (Apidae: Euglossini). J. Chem. Ecol. **19**:3017–3027.

Wiebes, J. T. 1979. Co-evolution of figs and their insect pollinators. Annu. Rev. Ecol. Syst. **10**:1–12.

Wiggins, G. B. 1996. *Larvae of the North American Caddisfly Genera (Trichoptera).* University of Toronto Press, Toronto. 457 pp.

Wiggins, G. B. 1987. Order Trichoptera. In *Immature Insects,* vol. 1, F. Stehr (ed.). Kendall/Hunt, Dubuque, IA. pp. 253–287.

Wiggins, G. B., and W. Wichard. 1989. Phylogeny of pupation in Trichoptera, with proposals on the origin and higher classification of the order. J. North Am. Benthol. Soc. **8**:260–276.

Wigglesworth, V. B. 1972. *The Principles of Insect Physiology,* 7th ed. Chapman & Hall, London. 827 pp.

Wigglesworth, V. B. 1976. The evolution of insect flight. In *Insect Flight,* R. C. Rainey (ed.). John Wiley & Sons, New York. pp. 255–269.

Wilcox, J., N. Papavero, and T. Pimentel. 1989. *Studies of Mydidae (Diptera). IVb, Mydas and allies in the Americas (Mydinae, Mydini).* Colecao Emilie Snethlage. SCT/PR, CNPq, Museu Paraense Emilio Goeldi, Belem-Para, Brazil. 139 pp.

Wilding, N., N. M Collins, P. M. Hammond, and J. F. Webber (eds.). 1989. *Insect-Fungus Interactions.* Academic Press, London. 344 pp.

Wilkins, A. S. 1993. *Genetic Analysis of Animal Development.* 2nd ed. Wiley-Liss, New York. 546 pp.

Will, K. W. 1995. Plecopteran surface-skimming and insect flight evolution. Science **270**:1684–1685.

Wille, A. 1960. The phylogeny and relationships between the insect orders. Rev. Biol. Trop. **8**:93–123.

Willemstein, S. C. 1987. *An Evolutionary Basis for Pollination Ecology.* Leiden Botanical Series, vol. 10. E.J. Brill, Leiden. 425 pp.

Williams, D. D. 1987. *The Ecology of Temporary Waters.* Croom Helm Ltd., London. 205 pp.

Williams, D. D., and B. W. Feltmate. 1992. *Aquatic Insects.* CAB Int. Wallingford, Oxon., UK. 358 pp.

Williams, K. S., and C. Simon. 1995. The ecology, behavior, and evolution of periodical cicadas. Annu. Rev. Entomol. **40**:269–295.

Williams, M. A. J. 1994. *Plant Galls: Organisms, Interactions, Populations.* Oxford Science Publications. Oxford University Press, Oxford. 488 pp.

Williams, M. L., and M. Kosztarab. 1972. Morphology and systematics of the Coccidae of Virginia, with notes on their biology (Homoptera: Coccoidea). Virginia Polytech. Inst. State Univ. Res. Div. Bull. **74:**1–215.

Willman, R. 1987. The phylogenetic system of the Mecoptera. Syst. Entomol. **12:**519–524.

Wilson, E. O. 1971. *The Insect Societies.* Harvard University Press, Cambridge, MA. 548 pp.

Wilson, E. O. 1975. Slavery in ants. Sci. Am. **232:**232–236.

Wilson, E. O. 1985. The sociogenesis of insect colonies. Science **228:**1489–1495.

Wilson, E. O. 1990. *Success and Dominance in Ecosystems: The Case of the Social Insects.* Excellence in Ecology, 2. Ecology Institute, Oldendorf/Luhe, Germany. 104 pp.

Wilson, E. O., F. M. Carpenter, and W. L. Brown. 1967. The first Mesozoic ants. Science **157:**1038–1040.

Wingstrand, K. G. 1973. The spermatozoa of the Thysanuran insects *Petrobius brevistylis* Carp. and *Lepisma saccharina* L. Acta Zool. **54:**31–52.

Winston, M. L. 1987. *The Biology of the Honey Bee.* Harvard University Press, Cambridge, MA. 281 pp.

Wirth, W. W., and W. L. Grogan. 1988. *The Predaceous Midges of the World.* Flora & Fauna Handbook no. 4. E. J. Brill, Leiden. 160 pp.

Wirth, W. W., and A. Stone. 1956. Aquatic Diptera. In *Aquatic Insects of California,* R. L. Usinger (ed.). University of California Press, Berkeley. pp. 372–482.

Wissinger, S. A. 1989a. Comparative population ecology of the dragonflies *Libellula lydia* and *Libellula luctuosa* (Odonata: Libellulidae). Can. J. Zool. **67:**931–936.

Wissinger, S. A. 1989b. Seasonal variation in the intensity of competition and predation among dragonfly larvae. Ecology **70:**1017–1027.

Withycombe, C. L. 1925. Some aspects of the biology and morphology of the Neuroptera with special reference to the immature stages and their possible phylogenetic significance. Trans. Entomol. Soc. London **1924:**303–411.

Witteveen, J., H. A. Verhoef, and T. E. A. M. Huipen. 1988. Life history strategy and egg diapause in the intertidal collembolan *Anurida maritima.* Ecol. Entomol. **13:**443–452.

Woglum, R. S., and E. A. McGregor. 1958. Observations on the life history and morphology of *Agulla bractea* Carpenter. Ann. Entomol. Soc. Am. **51:**129–141.

Woglum, R. S., and E. A. McGregor. 1959. Observations on the life history and morphology of *Agulla astuta* (Banks) (Neuroptera: Raphidioidea: Raphidiidae). Ann. Entomol. Soc. Am. **52:**489–502.

Wood, D. L. 1982. The role of pheromones, kairomones, and allomones in the host selection behavior of bark beetles. Annu. Rev. Entomol. **27:**411–446.

Wood, D. M. and A. Borkent. 1989. Phylogeny and classification of the Nematocera. In *Manual of Nearctic Diptera,* J. F. McAlpine et al. (eds.). Canadian Govt. Publications, Quebec. vol. 3, pp. 1133–1170.

Wootton, R. J. 1976. The fossil record and insect flight. In *Insect Flight,* R. C. Rainey (ed.). John Wiley & Sons, New York. pp. 235–254.

Wootton, R. J. 1981. Paleozoic insects. Annu. Rev. Entomol. **26:**135–158.

Wootton, R. J. 1992. Functional morphology of insect wings. Annu. Rev. Entomol. **37:**113–140.

Wright, M., and A. Petersen. 1944. A key to the genera of anisopterous dragonfly nymphs of the United States and Canada (Odonata, suborder Anisoptera). Ohio J. Sci. **44:**151–166.

Wu, C. F. 1923. Morphology, anatomy and ethology of *Nemoura.* Bull. Lloyd Library, Cincinnati **23:**1–81.

Wygodzinsky, P. 1961. On a surviving representative of the Lepidotrichidae (Thysanura). Ann. Entomol. Soc. Am. **54:**621–627.

Wygodzinsky, P. 1972. A revision of the silverfish (Lepismatidae, Thysanura) of the United States and the Caribbean area. Am. Mus. Novitates **2481:**10–26.

Yager, D. D. 1990. Sexual dimorphism of auditory function and structure in praying mantises (Mantodea: Dictyoptera). J. Zool. (London) **221:**517–538.

Zacharuk, R. Y., and V. D. Shields. 1991. Sensilla of immature insects. Annu. Rev. Entomol. **36:**331–354.

Zimmerman, E. C. 1991–4. *Australian Weevils, v. 1–3, 5–6.* Commonwealth Scientific and Industrial Research Organization, Melbourne.

Zitnan, D., T. G. Kingan, J. L. Hermesman, and M. E. Adams. 1996. Identification of ecdysis-triggering hormone from an epitracheal endocrine system. Science **271:**88–91.

Zwick, P. 1990. Emergence, maturation and upstream oviposition flights of Plecoptera from the Breitenbach [Hesse, Germany], with notes on the adult phase as a possible control of stream insect populations. Hydrobiologia **194:**207–224.

TAXONOMIC INDEX

This index includes all scientific names of insects and other hexapodous arthropods. Entries in *italic* indicate genera and species; entries in **bold face** indicate living orders and families. Page numbers in *italic* refer to pages with illustrations and may include text discussion; page numbers in **bold face** have the main discussions and may include illustrations.

SUBJECT INDEX

This index includes all scientific names of organisms other than insects. Page numbers in *italic* indicate illustrations and may include text discussion; page numbers in **bold face** indicate main discussions and may include illustrations.